SVETLANA MASGUTOVA EDUCATIONAL INSTITUTE®
for Neuro-Sensory-Motor & Reflex Integration

Reflexes:
Portal to Neurodevelopment and Learning

Svetlana Masgutova Educational Institute®
for Neuro-Sensory-Motor and Reflex Integration

Reflexes: Portal to Neurodevelopment and Learning. (2015). A Collective Work. Florida, USA: SMEI. 563 pages.

Main Editors: Mary Rentschler, Sally Averkamp
Scientific Editors: Svetlana Masgutova Ph.D, Nelli Akhmatova Ph.D, Patty Shackleford Ph.D, Vicky Poston Roy Ph.D
Graphic layout: Sally Averkamp
Translators: Michal Bujak, Magda Gorecka

Svetlana Masgutova Educational Institute®
6275 Hazeltine National Drive
Orlando, Florida 32822

International Dr. Svetlana Masgutova Institute
Warsaw, Poland

Cover photo by Alexi Curlee-Cashen of her family: Henry, Catherine in the Landau position, and Paulette.

This book is written by scientists, professionals, parents, and individuals about their results using the MNRI® program on different aspects of neurodevelopment for those with challenges. These individuals familiar with the work of MNRI® tell of their experience using this method.

The Masgutova Method® is approved for Continuing Education Credits and hours by the following professional organizations in the USA:

- ASHA – The American Speech-Language-Hearing Association (2010)
- AOTA – The American Occupational Therapy Association (2009)
- PTBC – Physical Therapy Board of California (2009)
- FPTA - The Florida Physical Therapy Association (2010)
- TSI NY – Therapeutic Services Inc, NY (2013)
- NCBTMB – National Certification Board for Therapeutic Massage and Bodywork (2010)
- PTSD ISTSS - Membership in International Society for Traumatic Stress Study

More information is available at: www.MasgutovaMethod.com

20% of the income from book sales from ***Reflexes: Portal to Neurodevelopment and Learning*** will be used for helping survivors of specified tragedies and natural disasters.

DISCLAIMER: This book discusses the MNRI® methods of the Svetlana Masgutova Educational Institute®, presents only general information on MNRI®. This book is not intended to replace the services of qualified practitioners in the field. Any application of the information set forth in the following pages is at the reader's discretion and sole risk. For those who want to have their children treated according to the MNRI® system, we recommend working with an experienced MNRI® Core Specialist who is certified through the Svetlana Masgutova Educational Institute®. Practitioners are available at www.MasgutovaMethod.com.

ISBN: 978-1-61966-009-0

Printed in the USA.

Reflexes:

Portal to Neurodevelopment and Learning

Introduction

1. Theory and History of MNRI® Reflex Integration

2. Scientific Research behind MNRI®

3. MNRI® in Application

4. Case Studies – Professional's Expertise in MNRI®

5. Stories of Winners: Parent's and Professional's Experience with MNRI®

5.1. Winners Coming Back to the 'Here and Now'

5.2. Winners in Sensory Processing, Emotional Stability, Behavior, and Learning

5.3. Winners Overcoming the Limits of Health

5.4. Winners in Speech & Communication – Now I Can Express My Thoughts in Words!

5.5. Winners Overcoming Limits in Motor Development

5.6. Winners Moving Forward – Excellence in Sports

5.7 Winners Overcoming the Limits of Nature

5.8. Winners Beyond Post-Traumatic Stress Survival

Acknowledgements

Many people helped this book become a reality. We wish to thank:

Scientists, whose magnificent research work and discoveries in neurophysiology, psychology, and education moved forward the development of practical disciplines and programs to enrich the lives of children and adults – I. Pavlov, I. Sechenov, N. Bernstein, A. Luria, L. Vygotsky, J. Piaget, and many others.

Our greatest teachers – the Children, for taking part in the work on reflex integration and skills formation, and for being open for positive changes.

Parents of our Children, their Helpers and Sponsors – for their dynamic and positive approach, support and trust of our children's resources for neurodevelopment using the MNRI® Reflex Integration Program.

Authors of this book evidencing the ways for creating a portal to neurodevelopment through brain reflexes: Charles Beadling MD, Ph.D., Scarlett Lewis, Lori Burgess, Pamela Curlee, Patty Shackleford Ph.D., Paul Curlee MD, Elvin Akhmatov, Prof. Witold Pilecki, Joanna Kowalewska, Nelli Akhmatova MD, Ph.D., Ekaterina Sorokina MD, Ph.D., Olga Lebedinskaya MD, Ph.D., Irina Shubina MD, Ph.D., Prof. Michael Kiselevskij ,T. Szawrowicz-Pelka Ph.D., Malgorzata Sobieszczanska Ph.D., Anna Pilecka-Kalamarz Ph.D., Prof. Dariusz Kalka MD, Lech Kipinski Ph.D., Prof. Ludwika Sadowska, Henryk Filipowski Ph.D., Anna Maria Choinska Ph.D., Malgorzata Poreba Ph.D., Piotr Koleda Ph.D., Dr. Leah Light, Lisa Ortego, Emily Pelican, Laurie Callaba, Tina Marks, Sally Averkamp, Eldon Akhmatov, Dr. Dasaratha Rama, Renata Chodyko, Denis Masgutov, Eva Vendelin, Jakub Buraczewski, Jo Anne Tierney, Diane Whiteside, Susan Whitehead, Ewa Kulig, Aleksandra Szalaj, Malgorzata Koczyk, Kim Willkom, Suzanne Amanor-Wilks, Connie Jordan, Teresa-Ewa Busz, Beata Oginska-Dutkiewicz, Karolina Niezgodzka, Stephen Best, Kyle Van Dyke MD, & Annette Van Dyke, Peg Johnson, Mary T. Wirth, Hela Bali, Shaaron Freeburg, Candy Heinrich, Elzbieta Cieplinska, Agata Palasz, Paulina Turzynska, Michalina Lelek, Laura Carlson, Ha Nguyen, Elizabeth Rosalina, Diane Hunter, Nancy Morris, Sarah Davis, Liong Jam Hoa, Priscilla Roberts, Mary Rentschler, Abi Terrell, Rachel Shimer-Lal, Brian Esty, Laura Paquette, Diana Lilja, Heather Stewart, Diane Colonnese, Terri Zartman, Heather Taylor, Tricia Borsch, Patricia Kelly, Theresa Wojtalewicz, Tamlyn Hedemann, Kim Masters, Sophia & George Jaszczyk, Jocelyn M. Pare, Meg & Patric Wallin, Priscilla Roberts, Susan Terris, Asa & Janne Pettersson, Lou Bosch-Wilson, Dawn Powers, Kevin & Samantha Oleniks, Shannon Desilets, Sara Christensen, Donna Sanchez, Magdalena Gorecka, Cindy Sprenger, Stephanie Keene, Brian Kretsch, Melissa Brown, Katarzyna and Tadeusz Kubicki, Barb Gregory, Joan McKenzi, Aneta Michalska, Tania Lemus, Isabelle Renard-

Fontaine, Rok Flander, Halina Muzaj, Kimberly Otero, Anna Nameny, Elizabeth Blake Kelly, Tawni Lawrence, Anthony Grecia, Anna Krasowska, and Kristine Ann Lopez.

Colleagues and Friends, Organizers-Sponsors and MNRI® Core Specialists and Students in the USA, Poland, Russia, Canada, Germany, Sweden, France, Belgium, Denmark, Switzerland, Singapore, Australia, Netherlands, Hong-Kong, Indonesia, Argentina, Brazil, El Salvador, Slovenia, Croatia, Africa, and others countries – who initially inspired us to research the development and integration of reflex patterns.

Our dearest Colleagues, Editors and Creators of this Collective Book – Sally Averkamp, Mary Rentschler, Pamela Curlee, and Patty Shackleford Ph.D; proof readers – Vicky Poston Roy Ph.D, Susan Whitehead, Michael Spalding, and Susan Terris.

– Svetlana Masgutova, Ph.D. and MNRI® Team

Foreword

Charles Beadling, MD, FAAFP, IDHA, DMCC
Director, Center for Disaster and Humanitarian Assistance Medicine,
Uniformed Services University of the Health Sciences, Washington, DC, USA

Charles Beadling, MD

I was deeply honored when Dr. Svetlana Masgutova invited me to write the Foreword for their most recent book, *Reflexes: Portal to Neurodevelopment and Learning*. It was also very humbling, as I am certainly a novice regarding Masgutova Neurosensorimotor Reflex Integration (MNRI®). My clinical background is primary care, military medicine, and, for the past 15 years, disaster medicine. That led to my introduction to MNRI® as I was organizing an international conference on disaster risk reduction. An experienced communication science and MNRI® Specialist, Marie-France Renaud (Canada), provided input to our Psychosocial/Mental Health and Building Community Resilience work group.

I had the opportunity to observe Dr. Masgutova and her team of Core Specialists in action during a Conference for children with a variety of neurodevelopmental challenges. Hearing the first-hand experiences of Shannon Desilets, Core MNRI® Specialist (USA), that provided therapy for the families victimized by the Sandy Hook school shooting in Newtown, Connecticut, was incredibly transformative. Shannon exuded a level of compassion often missing in traditional western medicine. It was such a privilege to meet and learn from Shannon and also Pamela Curlee.

Watching children break through walls of isolation and engage with their therapist during the MNRI® Family Conference was remarkable. More remarkable was watching the parents exuberantly express the joy and relief they felt when they witnessed these changes in their children. These parents traveled cross country to get the only therapy that helped their children. I live in a world of evidence-based practice and welcome this book as a step in providing that for MNRI®.

The book begins with the theory of reflex integration. Dr. Masgutova developed this program over decades, beginning with helping traumatized children following the Ufa train catastrophe and then the nuclear disaster in Chernobyl. The understanding of normal reflex integration, and reflex integration disorder, evolved by her treating victims of a series of other disasters. The first section ends with a discussion of the neurophysiological basis of MNRI®.

The second section captures scientific research, or the evidence base, for MNRI® therapy. The effectiveness of MNRI® with functional disabilities like cerebral palsy (CP), and genetic disorders such as Down syndrome, is discussed. Studies that examined a variety of diagnoses, from dyslexia to immunological problems provide depth to this evidence base. The third section examines the systemic application of MNRI® and the exploitation of technology in a computerized assessment.

The diverse application of MNRI® is demonstrated in the fourth section, a compilation of case studies by

Core Therapists and Core Specialists in Training. Common and uncommon problems, from autism to selective mutism, are covered. Even the use of MNRI® following stroke is discussed. This section brings out the unique perspective of each therapist and their individual application of the theories of MNRI®.

Medicine is often described as 'Art and Science.' The first sections address the science showing clear evidence that MNRI® is effective for a variety of issues from Down syndrome to autism to CP. Section five is the heart and soul, or 'Art' of MNRI®. These last sections provide the perspective of the "Winners:" children, parents, and specialists. The difficulty with this perspective is it is 'anecdotal.' One success is not predictive of success in other patients. However, when coupled with evidence from the scientific research given, we see from these stories how this therapy impacts real people. As a primary care physician, I see the challenge before us is to get enough trained specialists mainstreamed to work in hospitals, schools, prisons, the military, and homes where these people need the help. The stories of these parents and MNRI® specialists are impressive and it is the science behind it that shows that it is clearly effective. We need to spread the word of the restoration available through the MNRI® program and continue to capture scientific data to prove its effectiveness.

Reflexes: Essential Component for Optimal Health and Healing

Grayce Marie Stratton, Ph.D., Clinical Psychologist, CA, USA

Grayce Marie Stratton, Ph.D

As a cognitive scientist, EEG biofeedback provider, and clinical neuropsychologist practicing on the front-line of mental health, I was entirely unaware of Masgutova Neurosensorimotor Reflex Integration (MNRI®) until 2011. This was the year when parents of four different child patients of mine nearly simultaneously introduced me to MNRI® as a pivotal treatment in their child's progression toward vastly improved health. These particular patients, all on the autistic spectrum, had made tremendous gains in speech, socialization, cognitive processes, and physical agility and stability that far exceeded the gains achieved by peers who were not involved in Masgutova Method® programs.

Intrigued by the ostensible power of an intervention outside the fields of medicine and psychology, but which unmistakably seemed to have profound positive effects on the targets of both, I investigated further: I met with MNRI® trained specialists and observed their work, spoke with more families about their experiences, and consulted with colleagues in other parts of the country and across the globe, only to become more impressed by what I saw and learned. Eventually I took a 4-day training course myself, solely to understand the basis for this effective work and to puzzle together just how the process as presented could mechanistically and scientifically bring about such remarkable outcomes as I was witnessing.

Four years later, I am now a devotee of the work of Dr. Svetlana Masgutova. Indeed, an MNRI® referral is often a foundational recommendation in my treatment plans for many conditions and for individuals of all ages. This is because primary reflexes trigger development that has either been arrested or impeded or which has not been prompted at the outset of life to progress naturally toward maturation of the senses and systems of the body. Without maturation (termed 'integration' in Masgutovian terms), these resources meant for wellness are compromised, with corresponding decrements in learning, motor control, physical and mental resilience, and the capacity for joy that all humans deserve. With integration, however, the body-mind complex is liberated to develop – and in cases where there has been injury – to fully heal.

I am pleased to endorse Masgutova Neurosensorimotor Reflex Integration® as an essential component of both optimal health and optimal life performance. I believe every child deserves to be evaluated for reflex integration and that difficulties in this area should be considered as a component of any state of disease or disorder of any significance no matter what the age of the patient.

A Letter from Newtown, Connecticut

Scarlett Lewis, Parent, Author of Nurturing, Healing, Love. A Mother's Journey of Hope and Forgiveness *(2013).*

I lost my son, Jesse Lewis, in the Sandy Hook tragedy on December, 14, 2012. As a single mother of two boys I felt immediately afterwards that I might die as well from the shock, horror, immeasurable loss, and sadness. My remaining son and I were both severely traumatized by the nature of the loss of our beloved Jesse and also from our experience while waiting at the firehouse, surrounded by first responders, military, the anger and confusion around us, and the horrific images that plagued us afterwards of the day as well as the murder scene even though we had not witnessed this personally.

Scarlett Lewis & Jesse

People meet in different ways. I met the MNRI® team on their 'expedition to Newtown' when they arrived to help children and adults proposing their reflex integration treatment protocol…

We had several MNRI® pattern integration treatments by Dr. Svetlana, the creator of the MNRI® program, and her team. They explained the physical reactions to the trauma that were happening in my body, some that I noticed, such as sleeplessness, agitation, depression, memory issues, and inability to focus, and some that I did not, but could identify, once the team patiently explained how my brain was reacting to the trauma. Everything they said made complete sense and resonated within me as truth. They demonstrated the reflexes and responses my body was exhibiting from the trauma and were able to explain both in theory and physically how they were working with my brain to correct these issues.

The teams that Dr. Svetlana brought to Newtown to work with survivors were incredible, compassionate specialists – responsive, knowledgeable, and adept at their process. Their calm and professional demeanor immediately put me but even more so, my young son who had trepidation of more strangers in our presence, at ease.

MNRI® has allowed my son and I to move past the physical, mental, and emotional affects of trauma and loss now and has enabled us to re-enter life as productive and well adjusted people. Both my son and I have been able to step outside our own pain and be in service to others which has expanded our healing... that I credit MNRI®, Dr. Svetlana, and her team for cultivating and then helping to flourish. There is no doubt that MNRI® and its ability to connect with the brain and to counteract the negative affects of trauma has a tremendous impact on our healing and ability to move forward with our life.

I consider the MNRI® team my friends as their sacrifice to bring this healing method to the world speaks to their personal integrity and dedication to heal the world from the terrible trauma that results from disasters that occur every day. I highly recommend MNRI® to anyone fortunate enough to experience this incredible healing tool.

Program of Monumental and Pioneering Impact

Paul G. Curlee, MD, MPH, Farmington, NM, USA

Dr. Svetlana Masgutova's work is of monumental and pioneering impact in the understanding and use of motor and sensory reflexes to balance and integrate physiology and psychology.

Paul G. Curlee, MD

Historically, her awareness in applying neuro-sensory integration techniques came to the aid of many children in 1989, victims of a terrible train wreck in Russia.

Since then, I have learned a great deal about her work, attended her seminars with other health care providers, and witnessed the effect on many children and adults. Her work is the profound understanding of physiological reflex patterns. We all know the sensory stimulation-motor reflex phenomenon. We know it on a rather gross level such as an ankle or knee jerk in response to the percussion hammer. This reflex loop is limited to the peripheral sensory nerve, spinal cord efferent nerve, and reactive skeletal muscles. Dr. Masgutova has taken this phenomenon much more deeply into a variety of domains of the nervous system we do not yet fully understand. She has correlated reflex patterns, some of which we call primitive, with higher levels of function such as speech, behavior, coordinated movement and so forth. Her knowledge and research, more importantly, has generated new understanding of the meaning within reflexes we might just call reactions.

I share two examples to give you easy insight. In the case of autism, toe walking and equinovarus postures of the feet are commonly present and may contribute to the constellation of challenges facing those afflicted. Dr. Masgutova's correlation of this reflex pattern, the Achilles' tendon under extreme tension with the gastrocnemius muscle, is termed the 'Tendon Guard Reflex' (TGR). She has deduced, from thousands of empirical observations, that this reflex is often associated with fear. In the case of encountering an unexpected danger, this reflex is appropriately activated for life-saving reactions. When this TGR response is retained without a personal threat, it becomes abnormal. When the TGR is found to be active in an autistic child it commonly represents latent fear-type reactions. Dr. Masgutova has further developed simple physical exercises which integrate this reflex, creating a more balanced physiology and psychological functioning. When this abnormal reflex is balanced, toe walking improves. Also an autistic individual, perhaps fearful to even simple touch, may commonly become receptive to parents and therapists. These balancing techniques occur rapidly and markedly quicker than years of customary therapy that may not have been successful. Dr. Masgutova, and her trained associates, shares this profound balancing of reflex patterns with enormous skill and empathy to parents and clients. The positive, sometimes unexpected, outcomes are striking.

Another example is the Babinski Reflex, which we do call a primitive reflex in that it usually disappears in children between the age of one and two years. When present in adults, it is considered most abnormal, signifying severe infection, tumors, or intracranial hemorrhage; all life-threatening conditions. Contrarily, Dr. Masgutova has found this reflex present in other conditions, such as speech and developmental delays. Her correlations of the Babinski Reflex with other states and the ability to balance this abnormal reflex pattern with simple movement integration techniques are truly monumental. Children with developmental delays such as cerebral palsy, and others, can benefit enormously with Dr Masgutova's balancing of reflex patterns.

Dr. Masgutova has helped innumerable individuals and families confronting traumatic brain injury, Attention Deficit Disorders, developmental delays, and many other conditions. I have encouraged her to take on the enormous problems of PTSD, especially in this time of warfare and injured soldiers. Behavioral and cognitive therapies go only so far. These new techniques of balancing deep, core, psycho physiological reflex patterns thereby integrating harmony into the whole physiology, will become phenomenal adjunctive modalities to usual care.

Many physicians, physical therapists, occupational, speech therapists, and psychologists have attended Dr. Masgutova's seminars. Her associates are trained to her exacting standards of professionalism and profound knowledge. Seeing this growth in knowledge and skillful learning gives me new hope for the healing arts.

I recommend Dr. Masgutova, her associates, and the Svetlana Masgutova Educational Institute with utmost confidence and trust. Practitioners and lay persons learning this new knowledge will enhance their own lives and those of their patients.

The MNRI® Approach is the 'Missing Link' to True Therapy, Development, and Growth

Lori Burgess, MST, CCC, SLP, Minneapolis, MN, USA

I am a master's level speech pathologist, interpreter for the deaf, and hold many other certifications and awards. I have worked in this field for over 20 years with neurologically impaired and neurotypical individuals with differing impairments including cognitive, speech and language, sensory (auditory, proprioceptive, visual, etc), articulation, cochlear implants, trauma, pragmatic, stuttering, swallowing, feeding – the list can go on and on. I am always striving to find the best techniques and the most well rounded instructors to assist me in my endeavors to make the children and adults I work with as independent and communicative as possible.

Lori Burgess

I count myself very lucky to have been introduced to Dr. Svetlana Masgutova and the Masgutova Method® for Neuro Sensory Motor Reflex integration. I truly feel that her method is the 'missing link' to true speech pathology and other types of therapy and developmental growth practitioners.

Prior to Dr. Masgutova's techniques, I worked as a speech pathologist at a very cognitive level with my caseload, expecting that the right input and the right number of repetitions would provide a lasting result. Much to my dismay, there were patients that I could not fully reach and improve their communication abilities to the level I felt they were capable of. I did not understand that in order for the cortex to function and learn appropriately, the brainstem must be free of protection and automatisms. If that protection is too strong, learning and cognition can not take place utilizing a strong foundation. Instead, the foundation is unstable, resulting in speech and language that is not available in a stressful situation. What would seem perfect in my safe speech clinic at times would not be seen in the classroom or out on the playground or at a work place. I knew something was missing in my technique repertoire, but I did not know what it could be. Then I met Dr. Masgutova. Due to her instruction, I have completely changed my practice to begin my work on the brainstem level of functioning, and not jump to the cortex level of functioning. I have learned how to build that strong foundation so learning is effortless and enjoyable for my patients. My patients have faith in their abilities to perform outside of my clinic, and their families feel empowered by the techniques in their Home Programs. Dr. Masgutova's methods are not taught in the undergraduate and graduate schools of America. I am hoping to help change this by bringing the Masgutova Method® to as many families and professionals as I can.

Dr. Svetlana Masgutova holds her Doctorate in Developmental and Educational Psychology from Russia. She has worked as a lecturer at various Universities and as a scientific researcher at the Russian Educational Academy. She is currently a lecturer of the Early Intervention Faculty of the Wroclaw Medical Academy (Poland), Director of the International Dr. Svetlana Masgutova Institute for Movement Development and Reflex Integration, and Director of the Svetlana Masgutova Educational Institute for Neuro-Sensory-Motor and Reflex Integration (USA). Her work focuses on the concepts of Developmental Movement and Reflex Integration processes to facilitate sensory processing, motor-physical and sensory-motor rehabilitation, emotional recovery, as well as learning and developmental enrichment.

The innate primary movements and reflex patterns that are genetically inherent to every human being are key elements of human development. Dr. Masgutova and her MNRI® team of researchers postulate that many of these reflexes are essential to survival, especially during periods of stress and danger. They are also the fundamental neurological building blocks for all learned movements and skills, from first lifting your head, to crawling, to writing to talking. They crucially influence the development of the brain and many cognitive and intellectual processes as we mature. If the brain is 'stuck' at the brainstem level, the brain is constantly in survival mode – only wanting to protect itself. Therefore, higher levels of learning can not occur. For example, if I learned as an infant that swallowing causes me to choke, I learn that feeding is life threatening, so my brainstem sends a signal to my swallowing reflex to not allow food in so the body is not endangered. The result is that the child comes to a speech pathologist for feeding difficulties and the child may end up with a feeding tube. Through the use of Dr. Masgutova's methods, I now understand where the difficulty more than likely lies due to a thorough assessment of the primary reflexes, including Oral-Facial Reflexes. I now work at a level that is functionally 'safe' and natural, which allows the child to be free to learn and understand how to swallow and enjoy feeding. Dr. Masgutova's methods allow us to train the brain to experience safety, and then build on that experience to grow and develop more efficient pathways that lead to natural development and maturation.

I feel one of the most important aspects of Dr. Masgutova's work, that is missing in most other therapies, is the concept of precise timing and use of a gentle touch that is based on the genetically prescribed reflexes that we all inherently own in our bodies – we just need to learn how to use them appropriately, and then develop them to their highest potential. The patients in my practice enjoy the gentle learning process that does not involve drill, drill, and more drill. In fact, it is just the opposite. I do not drill a specific speech and language concept. Instead, I work on the body at the brainstem level, focusing on the reflexes that are assessed to be pathological or dysfunctional, and through the use of specific techniques, work to bring that reflex into integration which leads to maturation of the limbic system and cortex. Dr. Masgutova's work involves making clear connections between a stimulus, the correct response, and then developing the best pathway to the brain. I use these techniques for all my patients – it doesn't matter their age, type of disability, or behaviors – I work at the brainstem level to bring them out of protection and survival and into growth and development. One of the most significant parts of the Masgutova Method is that the techniques – while not hard to perform are very specific to the reflex. Therefore, family members and care givers can learn to empower their loved ones by working with them at home, so they too feel that they are a part of their family member's improvement.

Dr. Masgutova has developed classes for both therapists and parents, and has Family Conferences for intensive therapy sessions. She empowers you with information on how to specifically assist your child in his or her development. We have highly trained professionals who are MNRI® Core Specialists who continue to learn from Dr. Masgutova, and we have the 'Mom down the street' who is learning the method for her child. Everyone benefits from Dr. Masgutova's expertise. Another important aspect – especially in these economic times – is the cost and time saving factor for the families and the providers. Due to the empowering nature of the MNRI® method, families can be trained to provide a lot of the work at home, under the supervision of a Core Specialist. This leads to less time away from home, less therapy time in a center or clinic, less cost to the family in benefits, insurance costs, and gas for traveling to and from the clinic, and more quality family time at home, working to bring about the true potential of a loved one. This approach is gentle, noninvasive and natural – anyone can be trained, and everyone can benefit.

In conclusion, I feel very strongly that the Masgutova Method® is an exciting, natural and proven developmental approach to the enrichment of human potential. We all have the same innate potential – but this is the method that truly defines the purpose and gives us the tools to bring each person to their highest abilities, where ever that may lie. The possibilities are endless.

Introducing MNRI®: A Simple AND Profound Gift to the World

Pamela Curlee, Co-Founder, Svetlana Masgutova Educational Institute®, MNRI® Instructor and Trainer, Denver, CO, USA

Pamela Curlee

Background: MNRI® Masgutova Neurosensorimotor and Reflex Integration

The Masgutova Method® is simple; and simple is truly profound. The techniques are easily reproducible and are repeatedly effective. The work is aimed at integrating primary infant reflexes by providing a precise stimulus (sensory or proprioceptive) to activate the reflex, followed by an in-depth motor experience of the reflex pattern, thereby activating, awakening, or creating a more efficient neurological pathway.

This book has been written with simplicity in mind. The scientific articles presented here are available in language that is easier to understand than the technical language used in scientific journals. Trustfully, parents and professionals will enjoy this form of presentation of the science behind MNRI®.

This inspirational body of knowledge created by Dr. Svetlana Masgutova is enthusiastically embraced and used around the world. In the 1980's Dr. Masgutova developed a strong interest in reflex maturation and development during her university education in Russia. This fascination became the motivation for her graduate thesis, *Unconditioned Reflexes, Unconscious Processes and Personalities*, completed in 1980. Dr. Masgutova's thesis drove her to investigate all available historical information about reflexes, significant portions of which remained available, until the late 1900's, only in Russian. Her investigation led her to study the work of reflex and development pioneers, whom she now attributes as providing the theoretical basis of MNRI®. These researchers include I. Sechenov, C. Sherrington, L. Vygotsky, A. Luria, A. Leontyev, P. Anokhin, N. Bernstein, and others.

Having completed her Doctorate degree in General and Developmental Psychology in Moscow in 1988, Dr. Masgutova was just beginning her career as a post-graduate professor and researcher when a pivotal train wreck tragedy happened in Ufa, Russia. She traveled there to volunteer her professional services, and when faced with the overwhelming task of helping physically and emotionally traumatized children, Dr. Masgutova quickly concluded that a cognitive approach was insufficient to reach these traumatized children because they were stuck in protection and survival, a function of the brainstem. Her typical psychological therapies of time, space, and the spoken language were functions of the higher brain. The activity of these children's primary reflex functions remained in the past traumatic event which blocked their higher level cognitive, emotional, and physical skills. Dr. Masgutova witnessed again that the only 'language' of the brainstem was actually touch and movement. At Ufa, she was able to develop real-time techniques that included touch and movement to drop the protective guard present in each child's body. Her approach paid off as the children began to reconnect with the world.

Dr. Masgutova remained with the children for months (working 16 hour days) as each child proceeded through their difficult recovery. Their progress was so noticeable that other people in the hospital became interested in her work. Dr. Masgutova's emerging burgeoning integration techniques were thereafter used with adult survivors as well. (See more information about this in the article, *You Are a Winner* in the PTSD section of this book.)

From this initial work with PTSD victims from various places (that included Chernobyl, the Baku War, the Armenian earthquake, and the Chechen War), she began to wonder if the same processes would work with children/adults who had experienced birth trauma, accidents, and illnesses. She worked successfully with children with cerebral palsy and later also expanded her work to include individuals with a variety of 'Reflex Integration Disorders' such as behavioral, sensory, and motor disorders, thereby impacting cognitive, emotional, and mental skills for individuals as well as physical development.

Genius and compassion are a rare combination which have enabled Dr. Masgutova to take the work of these original "neurological giants" mentioned above and connect the dots with USA neuroscience while bridging this scientific knowledge with an unprecedented volume of clinical observations and experiences to serve as a foundation for the development of the MNRI® program.

This work is a new pathway that now offers heretofore undiscovered levels of enrichment to people worldwide by bringing these primary reflexes into or back to a state of integration. She has created techniques to uniquely impact the sensorimotor integration for the growth of nerve nets, myelination, activation of genetic blueprints, and the awakening and revival of motor memory.

Dr. Masgutova is delivering a gift of new wisdom to the world that has the unique ability to enrich lives of children and adults from the range of developmental challenges up to enrichment of the highest athletic levels and gifted individuals. This broad spectrum of influence is due to the common denominator of working with the complete circuit of the most basic primary reflex level of physiology. Many people have worked with the enrichment of the nervous system. And some have worked exclusively with the development of the sensory system. Others have focused on function and motor development. But the uniqueness of the Masgutova Method® is that Dr. Masgutova has brilliantly brought all three pieces together in a simple, reproducible scientific program.

This MNRI® program offers in-depth training for professionals as well as education for parents who have the desire to become part of our dynamically successful team. We are all working with one goal: to awaken innate intelligence to open new doors of opportunity. As of the beginning of 2015, this program has been used by professionals and parents in over 25 countries based in all five continents.

It all began with one woman who simply wanted to find a way to help people to move beyond survival and to uncover or recover their potential. As you will see from these stories, these discoveries and this program has provided us all with the 'missing link.' We whole heartedly invite you to come join us and discover for yourself first hand the profound benefits of working with the Masgutova Neurosensorimotor Reflex Integration® program. You'll be glad you came!

MNRI® – The Method that works for EveryOne.

Why We Need a Reflex Integration Therapy Modality

Jane Smith, MNRI® Parent

Infant reflexes are protective motor responses by the brainstem to internal or external stimuli. Scientific investigation continues to reveal the complexities of their development and how hugely important they are to our ability to exist.

One or a combination of four basic conditions can cause primary infant motor reflex dysfunction or pathology: congenital disorders; disease; birth, physical or emotional trauma; and prolonged, intermittent or chronic stress. Each of these conditions can cause the central nervous system, sensory system, or motor system to become compromised.

When one of the four conditions is present in utero, at birth, or during infancy prior to the their maturation, the neurosensorimotor circuits necessary for primary infant motor reflexes to emerge and function may be blocked or damaged. Depending upon the magnitude of impairment, a reflex may: 1) emerge, mature and integrate with little or no problem, 2) emerge, fail to mature, and remain dysfunctional, 3) emerge pathologically or 4) fail to emerge.

If one of the four conditions occurs after integration of the primary reflex system, the autonomic nervous system can still trigger reflexes to re-surface as an adaptive strategy in service of survival. When a primary infant motor reflex re-surfaces, it often remains present in a dysfunctional way.

The autonomic nervous system generates two different bodily states based on its two subsystems: the sympathetic and parasympathetic systems. During non-alarm states, the parasympathetic system normalizes bodily function (supporting rest, digestion, and healing) to ensure long-term growth and development. When the body is threatened, the sympathetic system mobilizes resources (fight, flight, or freeze) to ensure short-term protection or survival.

These two autonomic subsystems function in a symbiotic fashion alternating control depending on the body's internal alarm state. In cases of chronic low-level trauma, the sympathetic nervous system sometimes dominates without allowing the parasympathetic nervous system to restore normal function. Then, not only can the body sustain internal physiological damage, but its overall ability to function effectively and maintain emotional and behavioral stability weakens. Sympathetic dominance during the first two years of life can disrupt the process of emergence, maturation, and integration in the developing reflex system. If the sympathetic system becomes dominant after primary infant motor reflexes have integrated, then low-level trauma, normally managed effectively, can result in hyper-arousal and trigger reflex patterns to re-surface. In other words, prolonged intermittent or chronic stress can reactivate primary infant reflexes or compromise their maturation and integration.

The importance of fully matured integrated reflexes for optimal motor, cognitive, and social development goes way beyond their role in infants and toddlers. Maturation and integration of the reflex system is especially relevant for building the control, motivation, abstract thinking, creativity, and skillful intentional behavior necessary for academic achievement.

If one of the four basic conditions is present, the question becomes how have the underlying neurosensorimotor systems necessary for primary infant motor reflex integration been impacted? An individual's sensory system may be hyper-reactive, hypo-reactive, or non-reactive. A person with a hypersensitive system may find moderate amounts of stimulation irritating or intolerable. A scratch can feel like a deep cut, a light touch like an aggressive push, an everyday sound like nails on a chalkboard, or simple eye contact like a glaring stare.

In contrast, a hypo-reactive sensory system may fail to respond to a small or even moderate amount of stimulation. A deep cut can feel like a simple scratch, an aggressive push like a light touch, nails on a chalkboard like any mild sound, or a glaring stare like simple eye contact. In either case, the sensory system is misinterpreting the environment and sending unreliable signals to the central nervous system.

The central nervous system (CNS) regulates and directs internal and external responses based on the input it receives. To the outside world, an individual with a challenged sensory system can appear emotionally and behaviorally disregulated, when in fact the evoked emotional or behavioral responses may exactly reflect the level of stimulation registered by the challenged sensory system. Hyperactive senses cause the sympathetic nervous system to over-engage its 'alarm state.' Individuals with such an over-reactive sensory system may tend to withdraw protectively and be too fearful to take the normal risks necessary for new learning.

In contrast, a hypoactive sensory system requires a higher level of stimulation for arousal. While the long-term purpose of the parasympathetic 'non-alarm state' is to support restoration and growth, if a body fails to register dangerous or life threatening sensory stimuli it will not engage necessary protective strategies. People with an under-reactive sensory system can be at risk for harm during dangerous and life-threatening situations and are more likely to engage in high-risk activities.

A motor response from a compromised system can range from a very low level of development to dysfunctional or pathological. A dysfunctional motor response can be hyperactive, causing too much muscle tension, high muscle tone, and rigid muscles restricting the functional range of movement; or hypoactive, causing too little muscle tension, low muscle tone, and muscles too soft to control or support the body. A pathological (more severe than dysfunctional) motor response may be: 1) reversed – the opposite of what is expected, 2) incorrect – a response expected for some other stimulus, or 3) a-reflexive – generating no response at all. Dysfunctional and pathological motor responses will not be appropriately integrated and will not support growth and development.

Under normal conditions, the CNS actively generates appropriate and balanced reactions to internal and external sensory stimulation to ensure that basic bodily functions and activities remain well regulated. Due to congenital defects, disease, trauma, or long-term intermittent or chronic stress, CNS impairment can cause incoming sensory information to be blocked or misinterpreted and outgoing motor responses to be dysfunctional or pathological. For this reason the presence of developmentally inappropriate primary infant motor reflexes has long been viewed by medical professionals as symptomatic of neurological issues.

The widely accepted traditional thinking on infant reflexes is that they exist up until about age three, when they normally disappear, having become 'inhibited' or 'extinguished' through normal maturation. Masgutova Neurosensorimotor Reflex Integration originates from a different way of thinking. Svetlana Masgutova, Ph. D., has based her method on research done by Russian physiologists, neurophysiologists, and psychologists from the 1920's to 1950's (Anokhin, 1973; Bernstein, 1997; Pavlov, 1927/1960; Sechenov, 1863/1995; Vygotsky, L. [no date]; Ukhtomsky, 1950-1952) . These scientists placed reflexes in the frame of both higher and lower nervous system activity. They viewed infant reflexes not only a protective or survival response to stress or danger, but also the neuro-physiological foundation for physical, emotional and cognitive development.

Innate motor reflex programs are part of our genetic human inheritance regardless of what challenges may be present. Alexander Luria demonstrated that blocked and damaged neural pathways could be reactivated and restored through functionally related circuits in the brain's reflex system. This is the essence of MNRI®.

Our goal is healthy development in which the reflex system is ready when needed for protection and survival and, at the same time, provides a solid foundation for physical, motor, emotional, and cognitive development. Welcome to Masgutova Method!

References

Anokhin, P. K. (1973). Biology and neurophysiology of the conditioned reflex and its role in adaptive behavior. *International Series of Monographs in Cerebrovisceral and Behavioral Physiology and Conditioned Reflexes.* USA.

Bernstein, N. (1997). *Bio-mechanics and physiology of the movement.* Moscow, Russia: Moscow-Voronezh.

Luria, A. R. (1966). *The human brain and cognitive processes.* University of Moscow. Transl. New York and London: Harper & Row Publishers.

Pavlov, I. (1960). *Conditioned reflexes: An investigation of the physiological activity of the cerebral cortex.* (Anrep, G. V., D. Sc. Trans., 1960). New York, NY, USA: Dover Publications Inc. (First published in 1927 by Oxford University Press)

Sechenov, I. M. (1965). *Reflexes of the brain.* (Russ. tr. S. Belsky). Cambridge, MA: The M.I.T. Press. (First published in 1863.)

Sechenov, I. M. (1995). *Physiology of behavior.* Scientific Works/Ed.: Moscow, Russia: M.G. Jaroshevsky. 319 p. 26-130. (First published in 1863.)

Vygotsky, L. (1986). *The child psychology. The problems of child development.* In 6 Books. Book – 4. Moscow, Russia: Pedagogika.

Ukhtomsky, A. A. (1950-1952). *The study of dominance.* Collected works in 6 books. Book 6. Leningrad, Russia.

A Historical Approach to Reflex Integration

Patty Shackleford, Ph.D., MNRI® Instructor, Melrose, FL, USA

Introduction

Patty Shackleford, Ph.D

Over the last two decades many new ideas and methods have been researched and put into action to assist children and adults with neurodevelopmental and learning challenges, such as: dyslexia, hyperactivity (ADHD/ADD), multiple physical, mental or emotional delays, Autism Spectrum Disorder (ASD), Down syndrome, CP, PTSD, anxiety, behavioral, and many other recognized developmental disorders and neurodeficits. When discussing these neurodeficits with professionals who work with children and adults exhibiting these delays, the conversation comes back to how the number of individuals diagnosed is growing. The conversation often moves next to the concern that treatment options currently used do not make significant changes with this population. These professionals are looking for more research studies and tools that offer treatment options to provide non-invasive, stable, and positive change. Much of the current research is based on traditional studies oriented on the symptoms and characteristics of each neurodeficit. This book offers a shift from the traditional paradigm of treatment options based on symptom orientation and provides information from therapists, parents, and other helping professionals who have implemented the MNRI® integrating techniques into their daily therapy sessions. These MNRI® integrating techniques are oriented on problem solving concepts and based on the fact that each reflex has its own developmental dynamics and role. Delayed maturation and integration of the reflex can disrupt/inhibit the next level of sensory-motor development and cognitive function.

The research completed by Dr. Masgutova and MNRI® specialists over the last two decades has found that primary sensory-motor and motor development offers a genetically based cause for neurodeficits. The development of these primary sensory-motor schemes are a portal to physical, intellectual, emotional, and personality development. This research and direct observation by family members and professionals discovered the underlying cause of many of the above mentioned physical, cognitive, emotional and social challenges falls under the category of Reflex Integration Disorder (S. Masgutova, 2011). The underlying features of Reflex Integration Disorder (RID) have been established through work with over 30,000 children and adults with challenges by Dr. Masgutova and MNRI® Specialists in Poland, Russia, Canada, USA, and other countries. This research has been ongoing by Dr. Masgutova and others since 1989 and builds on the concept of reflex integration: the work of sensory (afferent) nerve pathways, processing part (Central Nervous System), and motor (efferent) pathways though activation of sensory, motor, vestibular and proprioceptive systems within the patterns (schemes) of a specific reflex that are genetically given. Dr. Masgutova and the MNRI® Specialists used the following definition of a reflex when conducting their research: A reflex is a genetic pattern of physiological function created by nature as a code and structure insuring neurosensorimotor development. In working with reflexes, an underlying net is touched– a net that serves like the infrastructure of the nerve system. A reflex is the informational unit for positive survival and gradual physical, emotional, and cognitive development.

This research is based on the following features of a reflex. A reflex is a feedback response and is not an independent response. It is a result of feedback to a stimulus and on the excitation of the nervous system. Motor neurons transmit nerve impulses from the CNS to muscles and glands all over the body. A reflex responds to changes in the environment.

The repatterning or integration procedures and exercises used in the MNRI® programs offer sensory receptors and motor response pattern activation associated with that reflex in order to create a more efficient neurological pathway for every day successful functioning. Dr. Masgutova's background as a professor in developmental psychology; her extensive investigation into Russian literature; and, her experiences working with hundreds of children whom had experienced significant traumatic events formed her understanding of this area. The following events were the springboard for her life work based on the full spectrum of natural neurosensorimortor development as the door to the Masgutova Method® concept of Reflex Integration and its MNRI® programs.

The Development of the Masgutova Method®

Dr. Masgutova's professional work as a psychologist was based not only on student lectures and seminars, but included real life situational and historical events that shaped the background knowledge she developed in her Master's and Ph.D. programs. Dr. Masgutova worked with victims of different disasters which strongly influence her personal and professional values, life rules, and has contributed to her experience in working with the survivor's escape and survival mechanisms.

One more notable event that has molded Dr. Masgutova's understanding is her work with victims of the Chernobyl disaster which occurred on April 26, 1986 while Dr. Masgutova was working on her doctorate. She received an invitation from the Post-Graduate department of the University of Tula, to work with evacuated children from a radiation damaged area, and also in 1990-1994 she continued to work with families evacuated to Orechovo-Zouevo.

The Chernobyl disaster had two main elements: the atomic explosion and the fire which released a huge amount of radiation into the atmosphere. The radioactive cloud spread over western Russia and several European countries. This was a nuclear disaster, probably the worst of the world after Hiroshima and Nagasaki. The mortality was very high. In the disaster 64 people were killed, mainly the nuclear station workers. Two hundred thousand people died from results of the radiation, 985 of them died from the early neoplastic disease. Dr. Masgutova's contribution was lecturing to professional psychologists, teachers, and specialists at boarding schools on how to support the children in their post-traumatic stress and survival. She taught the professionals how to work with these children within the framework of reflex integration as a program base which had been created from 1977-1980 when she was a student of the University. In 1989 another new program, Educational Kinesiology, was added to her expertise. Most of her work was with children who displayed severe phobias and exhibited highly challenging cases of diseases of their vascular, lymphatic, and respiratory systems which constantly increased anxiety and fear. The work with this tragedy moved Dr. Masgutova and her team to new professional discoveries with survival and unconditioned humanism with emphasis on the importance of reflex integration in supporting survival mechanisms when damaged by post-traumatic stress. Also this tragedy gave evidence of how survival influences information processing and learning in children. This experience led her to a bigger understanding of the potential of recovery through specifically organized motion and kinesthetic anchoring, creating new possibilities for

Chernobyl nuclear disaster in 1986.

'changing the outcome of grief and tragedy' into 'resourceful and unfolded possibilities'. The work with these children was continued in the following years and in 1992 Dr. Carla Hannaford, on suggestion of Dr. Paul Dennison, the Founder of Educational Kinesiology (USA), was invited by Dr. S. Masgutova to Tula to work with professionals dealing with these children.

June 1989 – Historic Train Wreck in the Urals of Russia

This tragedy was one of the most tragic train accidents in the history of the world. Dr. Masgutova heard of this incident and decided to pack up her five-year old son who stayed with relatives as she worked as a volunteer for 3½ months at a hospital turned into clinic for burned children who were victims of this fiery crash. It was this experience that cemented in Dr. Masgutova the concept that each of us have the strength of a WINNER. During the first few days of working with these gravely injured children, every sixth child died in the hands of the medical personnel. Because the children were in such deep shock and phobias, Dr. Masgutova realized that none of the methods she had learned from her Ph.D. training would be of any assistance. She had been trained in what to do *after* children came out of shock. She instinctively picked up a little orange book simply called *Brain Gym®* by Dr. Paul and Gail Dennision (1992) and opened it. (It had been given to her by her scientific leader N. Tolstykh before she left.) Instinct and a deep understanding of the relationship of developmental reflexes with corresponding emotions triggered Dr. Masgutova to use the information in this little orange book. She used words and movements simultaneously to help the children to step into a new reality and know that they were now safe. She found that the use of the Brain Gym® movements with her statements on safety in the current situation pulled many children out of their stuck place of shock in two to four days. She also found that children were demanding tactile touch. In this process, Tactile and Art Therapy were added as each child was capable and ready to move forward. Dr. Masgutova realized that developmental movements reminded the body of its primary strategies that are all based on security and survival. It became very clear that the integration of the infant and life long reflexes were crucial for the growth and development of the emotional and physical body. Working with this group of children, Dr. Masgutova moved forward with a new idea of combining psychology and kinesiology. Once Dr. Masgutova returned to Moscow, she shared the new findings from this experience and inspired many research projects to find the deeper meaning of what she experienced and learned from these children. Her further research has shown that the work with reflex patterns was a missing link in facilitation of survival. This is why she again focused on her Reflex Integration idea, that she started developing as a university student.

Train wreck at Ufa.

Work with Children – Victims of Baku Conflict

Dr. Masgutova's work with children and adult victims of the conflict in Baku of 1990-1991 was emotionally very difficult in her professional experiences. This horrific seven day war caused huge moral and material damages. Over 300 people were killed in this war. Dr. Masgutova worked with several children of families that managed to escape to the Moscow district. No words could express the tragedy and the fear these people, especially the children, suffered: war, darkness, night, fire, being molested or killed, fear of death of parent, and others. The hair of two children aged 10 and 11 turned gray due to the cruel reality of the war and the tension accompanying them every day. They developed neurotic symptoms, phobias, and speech impediments such as stuttering. The basic techniques Dr. Masgutova used were psychomotor including reflex integration and several Educational Kinesiology exercises. The results were splendid and highly beneficial. The therapy for the two children described above turned to be very productive and their problems of stuttering was solved in two weeks! These children received elements of psychotherapy in combination with techniques known today as MNRI® Tactile Therapy, MNRI®Neuro-Structural Reflex Integration, and MNRI®Reflex Repatterning.

Dr. Masgutova continued her work with victims of wars, other tragedies, and social crises. These experiences were instrumental in giving Dr. Masgutova the information she needed to compile the Masgutova Method® and MNRI® program that are now used at MNRI® Family Conferences, Clinics, and taught to parents

Sandy Hook School Tragedy in Newtown, CT

and professionals.

In most recent history, Dr. Masgutova and certified MNRI® Core Specialists had the honor of working with over two hundred children and families of the Sandy Hook tragedy in 2013. All the research and learning that Dr. Masgutova developed during the earlier tragedies was updated and put into action in the development of the *MNRI® Reflex Integration and PTSD Recovery: Advancement for Children and Adults Experiencing Post-Traumatic Stress* manual (2013).

The MNRI® Team also was involved in the recovery efforts of the Philippines which suffered from a major earthquake and Typhoon Haiyan in 2013. They assisted through both direct instruction to the professionals living on the island and hands-on work at many of the islands.

MNRI® Becomes Further Defined

Based on her hands-on experiences and from the research of many Russian psychologists, current MNRI® research and work is devoted to unconditioned sensory motor reflexes present in the earliest stages of human development. Research and observation have shown that the use of primary movements and reflex integration can awaken self-regulation of the sensory, motor, and proprioceptive systems that influence health, learning, and development. Dr. Masgutova's theory on reflexes and protection finds that our reflexes provide each of us with protection and survival; the dual purpose for protection is to activate the sympathetic nervous system and offer a signal for help to avoid harm. The positive protection provided by our reflexes indicate a matured neurologic system, a brain that can recognize stimuli and organize protective motor responses and a natural flow in the development of reasoning skills and overall development. If the reflexes are in negative protection then reflexes fail to mature, the sensory system functions poorly, and the brain is confused by incoming stimuli. Negative protection compromises our overall development and reasoning abilities.

Damage from Typoon Haiyan in the Philippines

In individual sessions with trained professionals and in residential family educational conferences, MNRI® programs have promoted neurosensorimotor development and improved motor, emotional, and cognitive function in children and adults with: dyslexia, LD, ADD, ADHD, PDD, PDD-NOS, OCD, apraxia, Asperger syndrome, autism, Down syndrome, Fetal Alcohol Syndrome, ODD, emotional disturbance, Sensory Processing Disorder, Post-Traumatic Stress Disorder, cerebral palsy, genetic disorders, after-effects of stroke, and many others.

There are many reasons for the failure of reflexes to emerge or integrate at the natural, appropriate time. These include: heredity, difficult or premature birth, disease, physical or psychological trauma, toxicity, and sensory-motor deprivation. The consequences of reflex abnormalities include hyper or hypotonic muscles, vestibular dysfunction, aberrant motor development, difficulties with auditory and visual processing, poor sensory-motor integration, delayed language/cognitive development, and poor social/emotional development.

Every neonate has unconditioned innate reflexes that facilitate adaptation to the external environment and organize protection strategy, and become the foundation of cognitive activity. The Masgutova Method® is a set of programs focused on the restoration and maturation of primary movements, reflexes, and coordination systems for the fulfillment of all potentials and learning skills. The modality for implementing the MNRI® programs is the use of neurosensorimotor reflex integration as the unit of correction for adequate and appropriate nerve system functioning.

The Masgutova Method® system is based on research of the following physiologists:

• Ivan Sechenov (1829-1905) was considered the 'Father' of Russian physiology and whose early works investigated the reflexes of the brain. Sechenov's findings indicated that brain activity is linked to neuro electric currents (electrophysiology), that cerebral inhibition involved spinal reflexes, and chemical factors in the environment of the neuron factored into the way a reflex responded to a given stimulus.

• Ivan Pavlov (1849-1936) was awarded the Nobel Prize in 1904; contributed to many areas of physiological science and specifically to the topic of unconditioned and voluntary reflex actions. Studies of reflex actions involuntary to stress and pain led to the discovery of transmarginal inhibition (too much stress and pain – strength of nerve system) and adaptation of abilities for survival.

Ivan Sechenov

• Charles Sherrington (1857-1952) was awarded the Nobel Prize in 1932; added that reflexes do not occur as isolated activity within a reflex arc, they require integrated reciprocal innervation of muscles.

• Lev Vygotsky (1896-1934) was a psychologist who proposed the theories of natural and cultural development in a child and the development of higher cognitive functions in children with the emergence of reasoning through practical activity in a social environment. Vygotsky also was known for his zone of proximal development and the importance of a 'natural crisis' within age development of children and developmental transitions. His theory added to the natural development of inner control and that the maturation of automatism at the reflex level allows a child to move to language and abstract thinking.

Ivan Pavlov

• Nikolai Bernstein (1896–1966) was a pioneer in the field of motor programing, control and motor learning. He was considered the father of biomechanics and physiology of movements. His studies included information on the role of the central nervous system (CNS) in postural control and movement and on forming stereotypical patterns of kinematics and muscle activation.

These researchers along with A. Luria (1902-1977) and P. Anokhin (1898-1974) helped Dr. Masgutova create and implement the Masgutova Method® which was first used in cases of trauma and then moved into the realm of children and adults with neurodeficits and delays. MNRI® was developed on the above base information and is an interdisciplinary program based on the concepts that the primary motor system interacts with different levels and aspects of motor development; that reflexes become matured and integrated and thereby support the development of controlled motor skills and cognitive functions; neuro-integration is used to correct dysfunctional patterns and to integrate reflexes which are not mature; integration of reflexes and reflex patterns is key to the development of natural protective mechanisms which affects brain functioning, the immune system, stress management, and cognitive skills.

Charles Sherrington

It is interesting to note that studies of reflex action (involuntary to stress and pain) led to the discovery of transmarginal inhibition (too much stress and pain) and the different types of strengths of the nerve system (temperament) as well as the body's ability to make adaptations necessary for survival.

Conceptual Approach to Development

Lev Vygotsky

Problems in the natural course of human development have long attracted the attention of scientific investigators. Extensive research has focused on theories of learning (Piaget, 1976; Montessori, 1949/1995; Clark, 1986; Canfield, 1976; Galperin, 1976) and personality development (Allport, 1967; Bruner, 1960; Cattell, 1971, 1987; Goldberg, 2001; Lewis 2014; Greenwald, 1998; Maslow, 1943; Vygotsky, 1986; Bozovich, 1968). However, in the educational and psychological literature from Russia and other countries, only some research has been conducted examining sensory integration (Ayers, 1971, 1975), and very little research has been devoted to neuro-sensory-motor integration and natural motor development which is based on the development of the reflex pattern. Dr. Masgutova suggests these areas of development are significant factors contributing to delays and dysfunctions in children and adults who have experienced traumatic stress, PTSD, and other neuro-deficits. MNRI® programs propose an approach of correction of developmental and neuro-sensory-motor deficits based on the neural basis and natural strength of integration of the reflex patterns. Although motor pattern development is based on genetic motor programs typical for all humans, it is a unique process for each individual. During individual development, reflexes, as natural genetic programs, expand, integrate, and merge with learned motor experiences. Thus, the reflex's most important task is to support the maturation of the nerve system (neurogenesis, synaptogenesis, and myelination) to allow automatic responses to protect the body and provide freedom for the development of higher cognitive level hands-on tasks. Consciously controlled motor systems develop at the next level, and offer the possibility for motor programming and control. Finally, with spontaneously repeated training, these learned motor programs become the foundation for the habituated individual motor program, and internal control for the emotional and cognitive spheres. (Bobath, 1972, 1984; Piaget, 1973; Sherrington, 1947; Pavlov, 1927/1995; Sechenov, 1961, 1963/1995; Luria, 1969; Vojta, 1989; Ayers, 1971, 1975; Silberg, 1996; Cohen, 1993; Semionova, 1999; Sadowska, 2001). Dr. Masgutova used analysis from the above authors to help her develop the basic concepts which is currently known as the Masgutova Method® using a system approach oriented on creating a plan to solve developmental problems based on functions of the neuro-sensory-motor circuit. It is based on recognized stages of natural child development, sensorimotor, emotional, and cognitive growth, including high level coordination, socialization, and the acquisition of academic skills. Research and direct experience with children and adults allowed development of the concept of Masgutova Neurosensorimotor Reflex Integration (MNRI®) as one formula to assist those who demonstrate the debilitating effects of Reflex Integration Disorder (RID). If controlled movements are in tune with natural basic movements, then controlled behavior and communication occur. Developmental challenges arise when the transition from innate primary sensory-motor patterns to subsequent controlled gross and fine movements has not been achieved. Uncontrolled movements in children are associated with emotional instability and frequent outbursts, and the movements of children who cannot control their emotions tend to be over protective. Such children are trapped in a vicious cycle. Poor sensory motor control can cause a wide spectrum of other challenges as well, such as impulsivity, hyperactivity, low motivation, worry, fear, dyslexia, tendency for aggression, and delays in speech and intellectual development.

The challenge for educators and other professionals is to facilitate a child's shift from innate primary motor patterns and reflexes to subsequent controlled movements. This is the main purpose of the MNRI® programs. The correction of a dysfunctional motor pattern can profoundly and positively change a child's behavior and improve their ability to communicate and learn in a socially appropriate and enjoyable manner. Inappropriate reflexes must be corrected noninvasively by integrating them with the whole kinesthetic-motor system.

Matured and controlled movements coordinate the center of the body and the limbs, and all body planes. Every movement is based on reflex patterns and is connected with other movements, and these connections develop brain function and myelination of its axons. This way movement becomes goal-oriented, easy for voluntary control, effective, strong, and harmonious.

An individual who demonstrates basic integrated sensory-motor reflexes and primary movements is ready to learn more advanced motor skills, to respond to multiple and diverse stimuli, and to attain higher levels of cognitive development. Appropriate and age-specific motor development can lead to the joy of exploring the self and the world through all our senses and movements. The role of reflexes is to give infants learning experiences that build the foundation for motor development and sensory integration. If these reflexes are underdeveloped, hyperactive, or non-integrated then they can interfere with cortical processing and impede development. Educators and parents who understand this developmental mechanism can shape the future of a child. The amazing fact about the role of reflexes is that at ANY age the missed stages of development can be replicated using the MNRI® processes and return to the natural sensory-motor patterns to reconnect or build new neural pathways. This is why maturation of a reflex pattern means freedom from rigid automatic responses and impulsivity, and improved ability to naturally and easily learn on the cortical level.

The mission of the MNRI® program is to support children and adults with theoretical knowledge of neurosensorimotor development and practical applications for the use of natural, genetic neurosensory motor resources for successful motor, emotional, and cognitive development and joyful learning.

References

Allport, G. W. (1967) *Studies in expressive movement.* New York, NY, USA: Hafner Pub. Co.

Anokhin, P. K. (1973). Biology and neurophysiology of the conditioned reflex and its role in adaptive behavior. *International Series of Monographs in Cerebrovisceral and Behavioral Physiology and Conditioned Reflexes.* USA.

Ayres, J. (1971). Characteristics of types of sensory integrative functions. *The American Journal of Occupational Therapy,* 25(27). 329-334.

Ayres, J. (1975). *Sensory integration and the child.* Los Angeles, CA, USA: Western Psychological Services

Bernstein, N. (1997). *Bio-mechanics and physiology of the movement.* Moscow-Voroniez. Russia.

Bobath, K. (1972). The normal postural reflex mechanism and this deviation in children with cerebral palsy. *Physiotherapy,* 57, p. 515.

Bobath, K., Bobath B. (1984). The neurodevelopmental treatment. Scrutton D., editor. *Management of the Motor Disorders of Children with Cerebral Palsy.* Oxford: Blackwell Scientific Publications Ltd. p. 6–18.

Bozovich, L. I. (1968). *Basis of personality formation. Selected psychological works.* Moscow. Russia.

Bruner, J. S. (1975/76). From communication to language: a psychological perspective. *Cognition,* 3, p. 255-287.

Bruner, J. (1960). *The process of education.* Cambridge, MA, USA: Harvard University Press.

Center for Disease Control and Prevention *CDC: Trends in the prevalence of developmental disabilities in U.S. children, 1997–2008.*

Canfield, J. (1976). *One hundred ways to enhance self-concept in the classroom: A handbook for teachers and parents.* Englewood Cliffs, NJ., USA, Prentice-Hall.

Cranfield, J. and Wells, H. (1976). 100 Ways to enhance self-concept in the classroom. ... Influence of teacher expec.. on children's academic performance: a cross-lagged panel analysis, *Journal of Educational Psychology,* 70, 39-49.

Cattell, R. B. (1971). *Abilities: Their structure, growth, and action.* Boston, USA: Houghton Mifflin.

Cattell, R. B. (1987). *Intelligence: Its structure, growth, and action.* New York, USA: Elsevier.

Clark, B. (1986). *Optimizing learning.* Columbus, OH, USA: Merrill Publishing Co.

Cohen, B. B. (1993). *Sensing, feeling, and action. The experiential anatomy of body-mind centering.* Northampton, MA., USA (Contact Edition)

Delacato, C. H. (1959). *The treatment and prevention of reading problems.* Springfield, IL, USA: Charles C. Thomas.

Delacato, C. H. (1974). *The diagnosis and treatment of speech and reading problems.* Garden City, NY, USA: Doubleday.

Dennison, P., Dennison G. (2007). *Brain Gym 101 handbook.* Ventura, CA, USA: Edu-Kinesthetics Inc.

Dennison, P., Dennison G. (1989). *Brain Gym teacher's edition.* Ventura, CA, USA: Edu-Kinesthetics Inc.

Dubrovina, I. V. (1991). *School psychology service. Theoretical and practical aspects.* Moscow, Russia: Scientific Research Institute of General and Educational Psychology.

Dubowitz, H. (1989). Prevention of child maltreatment: What is known. *Pediatrics,* 83, p. 570–577.

Doman, G. (1984). *How to multiply your baby's intelligence.* Garden City, NY, USA: Doubleday.

Elkonin, B. E. (1974). *The Psychology of the play.* Russia.

Erickson, E.H. (1980). *Identity and life cycle.* New York, NY, USA: Norton.

Feldenkrais, M. (1981). *The elusive obvious.* Cupertino, CA, USA: Meta Publications, p. 7-9.

Fidel, T. (2000). *Touch therapy.* Edinburgh, UK: Churchill Livingstone.

Fowler, M. (1994). *Attention deficit, hyperactive disorder.* Washington, D.C., USA: National Information Center for Children with Disabilities.

Gardner, H. (1993). *Multiple intelligences. The theory in practice.* New York, NY, USA: Basic Books.

Galperin, P. J. (1976). *The development of the researches on formation of intellectual operations.* Moscow, Russia.

Gesell, A. (1948). *Studies in child development.* New York, NY, USA: Harper.

Goldberg, L. R. (1993). The structure of phenotypic personality traits. *American Psychologist,* 48, p. 26-34.

Goldberg, E. (2001). *The executive brain: Frontal lobes and the civilized man,* New York, NY. USA: Oxford University Press, Inc.

Greenspan, S. I., DeGangi, G., Wieder, S. (2001). *The functional emotional assessment scale (FEAS).* Bethesda, MD, USA: Interdisciplinary Council on Developmental and Learning Disorders.

Greenspan, S. I. (1999). *Developmentally based psychotherapy.* Madison, CT, USA: International Universities Press, Inc.

Greenspan, S. I. (1997). *Infancy and early childhood.* Madison, CT, USA: International Universities Press, Inc.

Greenwald, A. G., Draine, S. C. (1998). Distinguishing unconscious from conscious cognition – Reasonable assumptions and replicable findings: Reply to Merikle and Reingold (1998) and Dosher (1998). *Journal of Experimental Psychology: General,* 127, p. 320-324.

Gross, D. R. (1987). *Psychology. The science of mind and behavior.* London, UK: Hodder and Stoughton.

Krefft, A. (2007). *Diagnosis of functions of invisible phenomena (Statistic Mathematic Analysis).* Wrocław. Poland: Oficyna Wydawnicza Politechniki Wrocławskiej

Leontiev, A. N. (1977). *The activity, consciousness, personality.* (Russ.) Moscow. Russia.

Lewis, M. (2014). *The rise of consciousness and the development of emotional life.* New York, NY: Guilford Press.

Luria, A. R. (1966). *The human brain and cognitive processes.* University of Moscow. New York & London: Transl. Harper & Row Publishers.

Masgutova, S., Akhmatova, N. (2004, 2007). *Integration of dynamic and postural reflexes into the whole body movement system.* Scientific edition: Prof. N. Akhmatova. Warsaw, Poland: MINK.

Masgutova, S., Curlee, P. (2004, 2007). *You are a winner. Trauma recovery. A new choice through natural developmental movements.* Revised Edition. Iowa, USA: 1st World Publishing.

Masgutova, S. (2011). *Infant dynamic and postural reflexes. neuro-sensory-motor reflex integration.* Revised and edited, scientific-practical manual. Orlando, FL, USA: SMEI.

Masgutova, S., Masgutov D., Akhmatova N., (2012). *MNRI® breathing reflex integration. Optimize health and cognitive resources through breathing.* (Manual) Orlando, FL, USA: SMEI.

Masgutova, S. K., Masgutov, D. (2013). *MNRI®: Reflex integration and basal ganglia. Optimal behavior and skill advancement for individuals with developmental challenges.* Orlando, FL, USA: SMEI.

Masgutova, S. K., Masgutov, D., Akhmatov, E. (2013). *MNRI®: Reflex integration disorder and PTSD recovery.* Orlando, FL, USA: SMEI.

Masgutova, S. K., Akhmatova, N. K., Lebedinskaya, O. V. (2013). Clinical-immunological assessment of therapy effect of the neuro-sensory-motor integration program of reflex patterns in airway chronic inflammatory diseases. *Event Abstract. International Congress in Milan. Front. Immunology Conference Abstract: 15th International Congress of Immunology* (ICI). doi: 10.3389/conf.fimmu.2013.02.00865. Published Online: 22 Aug 2013.

Maslow, A. H. (1943). A theory of human motivation. *Psychological Review,* 50(4), p. 370–96. Retrieved from http://psychclassics.yorku.ca/Maslow/motivation.htm.

Maslow, A. (1954). *Motivation and personality.* New York, NY: Harper.

Miller, G. A. (1966). *Psychology – The science of mental life.* Harmondworth, UK: Penguin.

Montessori, M. (1995). *The absorbent mind.* New York, NY, USA: Henry Holt and Company. (Original work was published in 1949)

O'Dell, N., Cook, P. (1997). *Stopping hyperactivity. A new solution. A unique and proven program of crawling exercises for overcoming hyper-activity.* Garden City Park, NY, USA: Avery Publishing Group, Inc.

Ornstien, R. E. (1977). *The psychology of consciousness.* New York, NY, USA: Harcourt Brace Jovanovich.

Pavlov, I. P. (1960). *Conditioned reflexes: An investigation of the physiological activity of the cerebral cortex.* (Anrep G.V., D. Sc. Trans., 1960). New York, NY, USA: Dover Publications Inc. (Original work was published in 1927 by Oxford University Press)

Pearce, J. (1985). *Magical child matures.* New York, NY, USA: Dutton.

Pearce, J. (2002). *Evolution's end. Claiming the potential of our intelligence.* San-Francisco, CA, USA: Harper.

Piaget, J. (1976). *The grasp of consciousness: Action and concept in the young child.* Cambridge, UK: Harvard University Press.

Pilecki, W., Kipiński, L., Szawrowicz-Pełka, T., Kałka, D., Masgutova, S. (2013). Spectral brain mapping in children with cerebral palsy treated by the Masgutova neurosensorimotor reflex integration method. *Journal of the Neurological Sciences:* Vol. 333, Issue (supplement 1), e550.

Pilecki, W., Masgutova, S., Kowalewska, J., Masgutov, D., Akhmatova, N., Sobieszczanska, M., Kalka, D. (2012). The Impact of Rehabilitation Carried Out Using the Masgutova Neurosensorimotor Reflex Integration Method in Children with Cerebral Palsy on the Results of Brain Stem Auditory Potential Examinations. *Advances in Clinical and Experimental Medicine.* Official Organ of the Wroclaw Medical University. 2012 (3). 2012, 21. http://www.advances.am.wroc.pl/pdf/2012/21/3/363.pdf.

Rutter, M. (1980). Attachment and the development of social relationships. *Scientific Foundations of Developmental Psychiatry.* Ed. Rutter, M. London, UK: Heinmann Medical.

Rutter, M., Silberg, J. (2002). Gene-environment interplay in relation to emotional and behavioral disturbance. *Annual Review of Psychology,* 53: p. 463 - 490.

Sadowska, L. (ed.). (2001). Neuro kinesiological diagnosis and treatment of children with impaired psychomotor development, *Ed. Akademia Wychowania Fizycznego,* Wroclaw, Poland: ss. 71, 83 - 84, 94-95.

Saint-Anne, D. (1986). *The neuro-motor and psycho-affective development of the infant.* New York, NY, USA: Elsevier.

Sechenov, I. M. (1995). *Physiology of behavior. Scientific works.* Ed. M.G. Yaroshevsky. Moscow. Russia. (Original work was published in 1863)

Semionova, K. A. (1999). *Rehabilitation treatment of the CP patients in residual state.* Moscow. Russia.

Sherrington, C. (1947). *The integrative action of the nervous system.* Press, London., UK: Cambridge University.

Tolstych, N. N. (1991). School psychology service. Theoretical and practical aspects. *Scientific Research Institute of General and Educational Psychology.* Moscow. Russia.

Ukhtomsky, A. A. (1950-1952). *The study of dominance.* Collected works in 6 books. Book 6. Leningrad, Russia.

Vojta, V. (1989). *Die Posturale Ontogenese als Basis der Entwicklungsstorungen.* Monatsschr. Kinderheilkd.

Vygotsky, L. (1986). *The child psychology. The problems of child development.* In 6 Books. Book – 4. Moscow. Russia: Pedagogika.

Watson, J. B. (1924). *Behaviorism.* New York, NY, USA: J.B. Lippincott.

www. wikipedia.org/wiki

www.MasgutovaMethod.com

Pictures

Chernobyl disaster –picture from Internet/Google: *Chernobyl: Capping a Catastrophe,* By Henry Fountain. Photographs by W. Daniels. *The New York Times*, April 27, 2014.

Historic Train Wreck in the Urals of Russia; from: Masgutova S., Curlee P. (2004, 2007).

Sandy Hook tragedy in 2013; by S. Desilets.

Philippines Island - earthquake and Typhoon Haiyan in 2013; by M. Busma.

Scientists pictures: http://en.wikipedia.org/wiki/.

The MNRI® Specialist Team sends well wishes to all the children and families that gave us the opportunity and privilege to work with them and gather the data needed for development of the MNRI® theory and practical tools.

Neurophysiological Foundation of the MNRI® Reflex Integration Program

Svetlana Masgutova, Ph.D, USA; Denis Masgutov, ISMI Director & Program Co-Creator, Poland; Nelli Akhmatova, MD, Ph.D, Russia; Elvin Akhmatov, Ph.D Student, USA

"All acts of conscious and unconscious life are reflexes in their origin." – I. M. Sechenov (1863/1995), world-known Russian physiologist

Svetlana Masgutova, Ph.D

Denis Masgutov

Nelli Akhmatova, MD, Ph.D

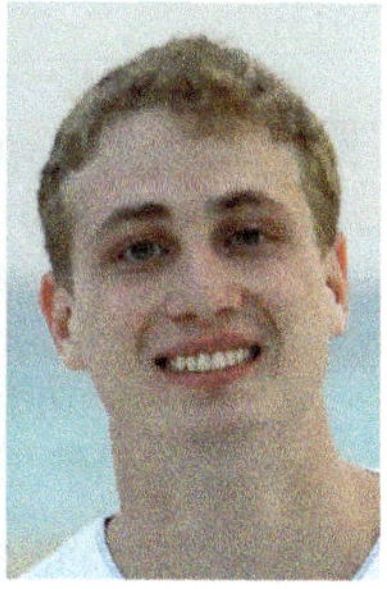
Elvin Akhmatov

In the fields of neurology and neurophysiology reflexes have been described in terms of their biological role, the character of the stimuli that trigger them, and the appearance of their motor responses in concrete behavior. I. M. Sechenov (1935, 1863/1961) and I. P. Pavlov (1927/1995) described a reflex as a biological static-dynamic unit: 1) pre-determining development of the nervous system, being unconditioned (non-changeable and strictly programmed by nature), and 2) simultaneously affected by experience and learning, creating *conditioned and specific* responses.

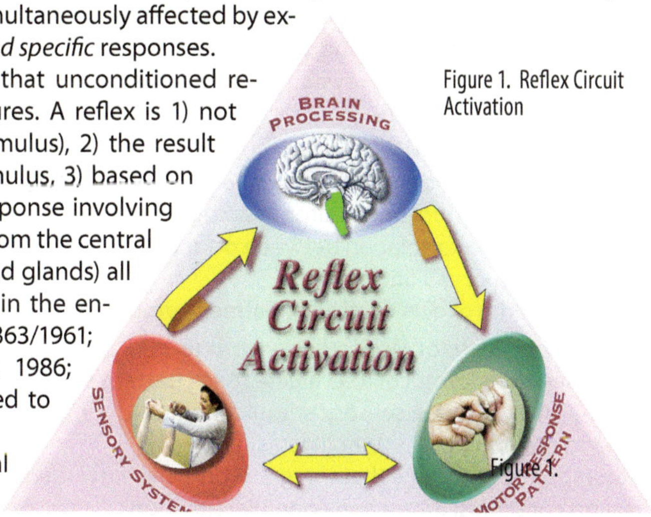

Figure 1. Reflex Circuit Activation

Research in the last century has shown that unconditioned reflexes are characterized by five distinct features. A reflex is 1) not an independent response (depends on a stimulus), 2) the result of feedback (of the nervous system) to a stimulus, 3) based on the excitation of the nervous system, 4) a response involving motor neurons that transmit nerve impulses from the central nervous system (CNS) to effectors (muscles and glands) all over the body, and 5) a response to change in the environment (Pavlov, 1927/1995; Sechenov, 1863/1961; Sherrington, 1947; Magnus, 1925; Vygotsky, 1986; Bernstein, 1997; Asratian, 1963, 1983) designed to maintain the organism's homeostasis.

The MNRI® program follows traditional neurophysiology – classical, historical (Pav-

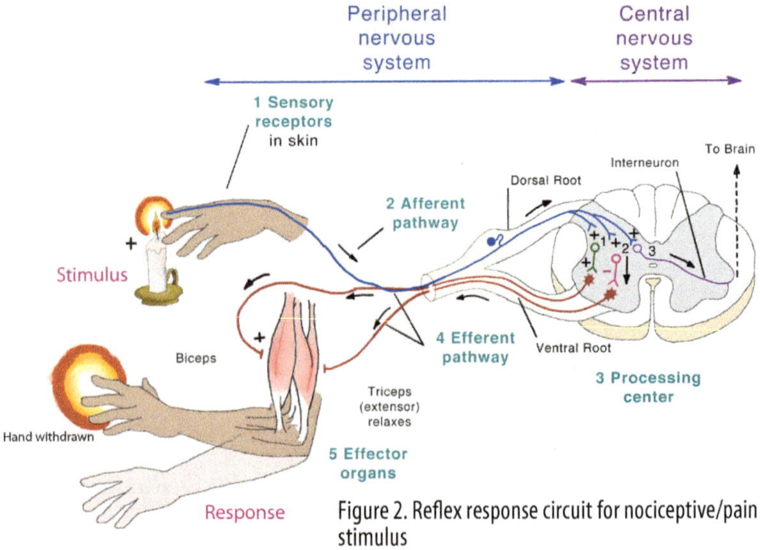

Figure 2. Reflex response circuit for nociceptive/pain stimulus

lov, 1927/1995; Sechenov, 1863/1995, 1961; Sherrington, 1947, and others), modern (Haines, 2002; Lundy-Ekman, 2002; Barker, Barasi, Neal, 2008; Barashniev, 2001, and others), and psychology (Sechenov, 1863/1995, 1961; Ukhtomsky, 1950-1952; Asratian, 1983; Luria, 1966, 1978; Anokhin, 1973, Konorski, 1969; Bernstein, 1997; Alexandrov, Sergienko, 2003; Goldberg, 2001, and others) in viewing primary reflexes as genetic/epigenetic inherent automatic responses of the CNS to adequate and specific stimuli (tactile, visual, auditory, vestibular, proprioceptive, or olfactory) in the form of adequate and specific motor/postural/glandular/pupillary/tympanic membrane reactions activating the organism's protection and survival strategies through the HPA-stress axis (hypothalamus-pituitary-adrenal triangle) (Masgutova, Akhmatov, 2013; Masgutova, Masgutov, 2013).

The Masgutova Method® uses the concept of reflex integration and a set of MNRI® programs focused on the restoration and maturation of primary motor patterns, reflexes and coordination systems to achieve optimal performance of natural mechanisms, neurodevelopmental processes, brain functioning, and sensory-motor integration. Activation of reflex patterns awakens natural genetic motor resources and self-regenerating programs, strengthens motor memory and sensory-motor coherence, and increases resilience. With this overall improvement in functioning, individuals are able to move toward realization of their full motor, social, emotional, and cognitive potential. (www.MasgutovaMethod.com).

The MNRI® Therapeutic Modality (Masgutova Neuro-sensory-motor Reflex Integration)

The Conceptual Basis

Each primary reflex is a unit serving the nervous system in the first order for protection and survival, and secondly, as a foundation for development. The reflex system involves complex supporting and opposing interrelationships among its motor patterns. In MNRI® this system is treated as a whole, much like the visual and auditory systems. Our experience shows that focusing on neurosensorimotor reflex integration in this way leads to treatment strategies and techniques that support neuromotor development and enhance function. Thus our focus is on the reflex system as a whole and the developmental process of its emergence, activation, maturation, and integration.

The Knowledge Base of MNRI® includes

- a deep understanding of motor development; its levels, stages, mechanisms, processes, and strategies and its physical structure and neurophysiology
- research on the correlation between poor performance of infant reflex patterns with a variety of developmental deficits in over 3,560 children
- analysis of the effects of MNRI® treatment programs on over 30,000 children.

Practical Applications of MNRI® theory include

- an assessment protocol to evaluate reflex pattern development in children and adults
- techniques that use sensory stimulation and the motor system to correct poorly functioning, dysfunctional, and pathological motor patterns
- therapeutic programs that integrate primary movements, controlled motor systems, and higher (intentional and conscious) motor and cognitive functions
- courses, training programs, and publications for professionals and parents.

A Reflex as a Unit of the Nervous System

The fundamental understanding of a reflex as a physical response directed by the brain was introduced by I. M. Sechenov (1995, 1961) and I. P. Pavlov (1927/1995). Stimulation from the sensory system in the form of impulses passes through neural pathways to muscles and glands, the response organs of an organism. Networks of neurons, referred to by Pavlov and Sechenov as reflex arcs, allow nerve impulses generated in response to a stimulus to transmit the data that cause a reflex (Pavlov, 1927/1995; Sechenov, 1995, 1961).

The reflex arc consists of:

1. The afferent nervous system (receptors and sensory nerve fibers bringing neural impulses to the brain). A vast system of sensory receptors detects stimuli, their strength and other qualities, and changes that occur continually in them both externally and inside the body. Our receptors monitor such things as touch, vibration, temperature, light, and sound from the external environment. Internal receptors detect functioning and changes in heart rate, pressure, pH, oxygen and carbon dioxide content, and levels of electrolytes. All of this information gathered together forms sensory input.

2. Certain brain centers whose task is to process sensory or proprioceptive information.

3. The efferent nervous system (efferent nerve fibers bringing neural impulses in the form of commands to the muscles and organs/glands to effect a response to the stimulus) (Pavlov, 1927/1995; Sechenov, 1995, 1961).

Since the times when I. Pavlov and I. Sechenov developed their theories of nervous system function, new data on neurology has validated and enriched their work.

The first part of the reflex arc is sensory input converted into electrical signals or nerve impulses that are transmitted to the peripheral nerve system and brain. These signals reach the spinal cord and medulla oblongata, pons or thalamus, and then the cortex, creating sensations, images, and thoughts. Output or decisions are made by our nervous system each moment based on an evaluation of sensory input and its meaning for the safety of the organism. The nervous system responds by sending signals to effector organs, causing muscles to contract or lengthen, or glands to produce appropriate secretions. The neuro-chemical basis for regulation of movement and posture is in the balance of neurotransmitters in synapses connecting sensory and motor neurons. Links between sensory and motor nerve systems derive from these unified mainly predetermined genetic and epigenetic 'codes' or reflex arcs.

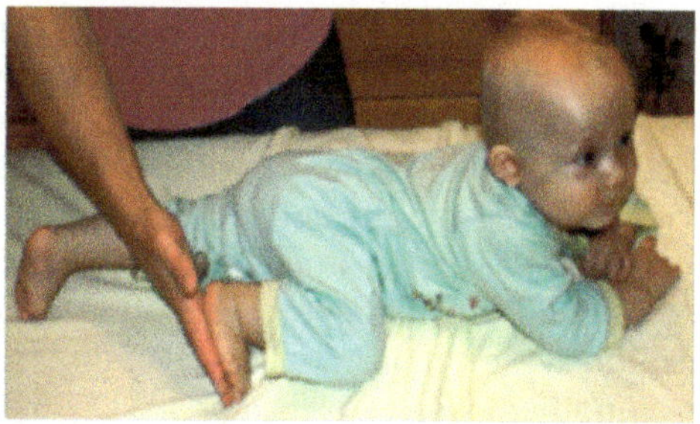

Figure 3. Baby showing Bauer Crawling.

Reflex Definition

Thus our work is based theoretically on viewing a reflex as a unit of the nervous system presenting an unconditioned physical response to a sensory stimulus (Pavlov, 1927/1995; Sechenov, 1995, 1961; Sherrington, 1947), supported by a network of neural arcs and circuits linking sensory organs, processing centers, and muscles or glands (Pavlov, 1927/1995; Sechenov, 1995, 1961). Reflexes are part of our genetic and epigenetic inheritance. They function physiologically as codes and structures created by nature to ensure protection and neurosensorimotor development. Work with reflexes touches the underlying nerve net system of the body, a net similar to the infrastructure of a leaf or a building. Most importantly, reflexes serve a dual purpose as a foundation for 'learning' that 1) ensures survival and 2) supports gradual physical, emotional, and cognitive development (Masgutova, 2010).

Unconditioned and Conditioned Reflexes

Through their unconditioned inborn 'codes' of sensory-motor reactions, reflexes allow the organism to survive in the presence of stress or unexpected danger. Nature has transferred them from generation to generation as a rigid unchangeable bedrock of protection, always dependable and ready to help us survive (Pavlov, 1927/1960; Sechenov, 1961, 1863/1995; Sherrington, 1947; Ukhtomsky, 1950-1952; Bernstein, 1997; Anokhin, 1973; and Alexandrov, 2003). Some reflexes are simple (myotatic or knee-jerk reflexes), others are more complicated (Thomas Automatic Gait, Bauer Crawling). Still others are social involving parental, territorial, ritual, and group behaviors and exist only in relationships among humans: emotional resonance (compassion, sympathy for another person) for example, or social hierarchy where an individual becomes a marriage partner, parent, child, territorial host, newcomer, leader, or follower.

Scientists throughout the 20th century (Pavlov, 1903, 1927/1960; Sechenov, 1961, 1863/1995; Sherrington, 1947; Magnus, 1925; Ukhtomsky, 1950-1952; Bernstein, 1997; Anokhin, 1973; Luria, 1972, 1976; and Konorski, 1969) recognized that reflex activity involved both 'higher' and 'lower' nervous system activity. I. P. Pavlov (1903, 1927/1960) and I. M. Sechenov (1961, 1863/1995) found that lower nervous system activity determining the work of unconditioned reflexes is governed by the brainstem (medulla spinalis, medulla oblongata, pons, and midbrain) and cerebellum.

Conditioned reflexes, on the other hand, involve the higher nervous system. Inborn reflex motor patterns become conditioned with the development of motivation and intentional actions. A conditioned reflex is a variant pattern of an unconditioned (basic) reflex. It serves as the neurophysiological foundation for development of conscious motor control and future specialized executive functions of the left and right hemispheres and cortex of the cerebrum (frontal, temporal, parietal, and occipital lobes), areas of the brain that generate complex reasoning and support socialization. Conditioned reflex activity normally depends on higher nervous system functioning (Pavlov, 1927/1960; Sechenov, 1863/1995; Ukhtomsky, 1950-1952; Bernstein, 1997; Anokhin, 1973; Konorski, 1969).

At the beginning of treatment, our repatterning work is aimed at restoration, development, and maturation of dysfunctional or retained unconditioned reflexes. MNRI® targets reflex patterns normally active in infants of 0 to 4 months. Restorative work with these reflexes provides 'neurophysiological order' in specific reflex circuits in the lower brain. Only with a reliable foundation at this level of unconditioned reflexes, as the essential protective and survival strategies of the brainstem, can further development proceed smoothly. When this foundation is absent or unreliable, its functions are taken over by higher brain centers, which are then unavailable for intentional motor activities, socialization, and complex reasoning (Masgutova, Akhmatova, 2008).

Classification

Pavlov (1927/1960) subdivided unconditioned reflexes into three groups: simple, complex, and most complex. The most complex reflexes were subdivided into two categories: individual (alimentary, active or passive, defensive, aggressive, freedom-seeking, exploring, or playing) which enable an individual to survive, and species (sexual, parental, and territorial) which ensure the survival of the species.

V. P. Simonov (1991) classified unconditioned reflexes into three groups depending on needs and motivation. In neuro-physiology unconditioned reflexes are classified according to the location of their receptors in the body: deep in the body (proprioceptive: tendinous, periosteal, articular) or on the surface (dermal or mucosal) (Vojta, 1989). Hitting a tendon or periosteum with a hammer causes tendinous and periosteal reflexes to respond. The trigger for the proprioceptive reactions (tendinous and periosteal) in the limbs is their special 'conducive' position (muscle relaxation, middle postural position) (Barashnev, 2001; Haines, 2002).

Another investigator, J. Konorski (1969), classifies all inborn activity according to its biological role: self-preserving, defensive, and orientation reflexes.

Investigators who study infant development, such as pediatricians and psychologists (Barashnev, 2001), group reflexes into specific in utero (Sucking Reflex, Tonic Labyrinthine , Trunk Extension, Leg Cross Flexion-Extension, and others), infant (Bauer Crawling, Thomas Automatic Gait, and others), early childhood (Babinski, Spinal Perez, Landau, and others), and lifelong (Core Tendon Guard, Grounding, Stability) (Masgutova, Akhmatova, 2004, 2007; Masgutova, Akhmatova, 2008). It is important to note that all of them begin their

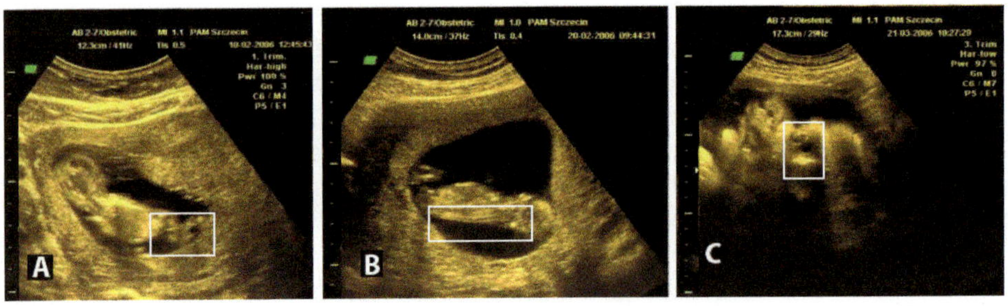

Figure 4. Ultrasound 3-D scan picture of A) 13 week unborn baby activating Leg Cross Flexion-Extension, B) Trunk Extension Reflex - vertical spine lengthening, C) Eye tracking (From materials by Prof. E. Ronin-Walknowska and Dr. S. Masgutova, 2006). Primary utero reflexes are the basis for the brain development and sensory-motor integration, which we use later postnatally for future neurodevelopment of everyday functioning and learning.

Table 1. Classification of the Dynamic and Postural Reflexes.

Type / Response / Level	Central/Core Reflexes		Peripheral Reflexes	
	Dynamic	Postural	Dynamic	Postural
Simple	Spinal Galant Spinal Perez	Babkin Palmomental Fear Paralysis	Babinski Phillipson Leg Withdrawal Foot Grasp	Peripheral Limb Reflexes
Complex	Hands Pulling Leg Cross Flexion-Extension Thomas Automatic Gait Bauer Crawling Moro Segmental Rolling Spinning Flying and Landing Locomotion	Trunk Extension Symmetrical Tonic Neck Asymmetrical Tonic Neck Tonic Labyrinthine Landau Pavlov Orientation – "What is this?" Core Tendon Guard	Hands Grasp Sequential Fingers Opening Sequential Fingers Closing	Hands Supporting

development in utero and can be seen clearly from 10-12 weeks of gestation.

Reflex Classification in the MNRI® Model

Several criteria are used in MNRI® to define reflexes according to their role in protection or survival, the complexity of their motor pattern, the level of their neurophysiological circuits and their brain 'strategy' (Masgutova, 2002-2007; Masgutova, 1990; Masgutova, Akhmatova, 2004, 2007; Masgutova, 2011). This classification also serves for practical orientation in MNRI® re-patterning of dysfunctional and retained reflexes:

Simple: single action, unidirectional, and processed mainly on the level of the spinal cord

Complex: multi-component, requiring more than one action, multi-sequential, and processed mainly on the level of the brainstem

Central: whole body activation. Central or core reflexes require whole body movement. (e.g. Spinal Galant, Spinal Perez, Bauer Crawling, Thomas Automatic Gait)

Peripheral: involve the limbs and extremities (e.g. Hands Grasp, Babinski)

Dynamic: active continuing motor response

Postural: bringing the body into a static posture

(See above TABLE: Classification of the Dynamic and Postural Reflexes)

Dynamic reflexes present a natural sequence of motor reactions to specific sensory stimulus. They are governed by the sympathetic nervous system, which controls processes of excitation that increase and conduct motor activity, leading to fight or flight responses (Masgutova, 1990) when necessary (e.g. Moro, Thomas Automatic Gait, Hands Grasp Reflex). Postural reflexes, governed by the parasympathetic nervous system, activate one or more movements in order to bring the body into a specific static position. Their task is regulation in the frame of inhibition of motor activity in order to stop or pause, leading to the freeze response when necessary and/or to activating sensory perception and processing. When matured, these reflexes are the basis for postural control and equilibrium (e.g. Asymmetrical Tonic Neck, Trunk Extension, Landau). The brain strategy here is: quiet the body to engage vision and audition.

Problem Solving with MNRI®

The human brain has its own chronology, mechanisms, and processes for development of its main strategies: protection and survival (brain stem), behavior, habits, emotional response (interbrain or diencephalon), and cognition, conscious control, and self-awareness (cortex). Infant reflex activity and neurological maturation normally coincide with the development of the brainstem. Reflex circuits connecting sensory and motor neurons (α- and γ motor neurons) and muscle fibers must be well established for initiating movement and posture both for protection and survival and for normal functioning.

Dysfunctions and delays in motor development can usually be traced back to some form of stress or insult to the nervous system that interfered with the emergence, activation, maturation, and integration of the reflex system. MNRI® research and therapy aim at re-establishing or re-routing reflex circuits to replicate what would normally exist at the earliest stage of unconditioned neonatal development. Here, in the first 0 to 4 months of life, is the key for 'reanimation' of healthy protective brain strategies and true maturation and integration of automatisms and reflexes that will eventually support mental activity. With an understanding of developmental problems as Reflex Integration Disorders, MNRI® addresses causes in the neuro-sensory-motor system, thus drawing the attention of professionals and scientists to the importance of rehabilitation at the level of the extrapyramidal nerve net system and lower motor neurons.

MNRI® uses 'regressive progressive' repatterning techniques that 'remind' the brain-body system of its genetic neurosensorimotor programs. 'Regressive' refers to the way a child positioned in a particular posture is given a precise tactile or proprioceptive stimulus and helped actively or passively to replicate the exact motor pattern that would be elicited reflexively by that stimulus in a neonate. In going back (regressive) to the 'pure' innate form of the reflex, the brain recognizes and resonates with an inner template for growth and development (progressive) supplied by its genetic inheritance. With repetition, thanks to neurological plasticity, the relevant nerve net system can then rebuild according to its original code and subsequently mature and integrate.

The Infant Nervous System

The most important function of the infant's CNS at birth is to 'produce' motor activity in the form of primitive motor reactions and inborn reflexive responses. These primitive motor responses of infancy characteristically disappear in time. From L. S. Vygotsky's point of view they are of an atavistic character (generated from the lower level of the CNS – spinal cord) and can be compared with the old phylogenetic stages of CNS development. He distinguishes this early activity from infant reflex patterns, which originate in the brain (brain stem, interbrain) and are the basis for development of higher controlling and reasoning brain centers (Vygotsky, 1986).

In older psycho-physiological literature we often find assertions that infant reflex motor patterns disappear as a result of inhibition and extinguishing by the higher centers of the nervous system. In reality we see that infant motor responses are not lost, but reappear during stress and in association with pathological conditions, even in adulthood.

In theory and practice, the Masgutova Method® is in agreement with Vygotsky. The very early most primitive motor reactions do fade as the infant develops. But the first infant sensory-motor reflex patterns and movements do not disappear; they continue to work in connection with higher nervous system structures. They hand over some of their functions and become subjugated to younger and newer brain centers (Vygotsky, 1986). Rather than being inhibited or extinguished, infant sensory-motor reflex activity integrates with, and becomes part of the higher, controlled sensory-motor systems.

Early infant behavior is governed mainly by old sub-cortical and intercerebral centers. These lower, older brain areas grow up earlier than others and are already mature at the moment of birth, as they are essential for the evolution of all organic life. In utero and in early infancy when the cerebral cortex is still immature and connections between cortical and sub-cortical brain centers are not yet established, the lower brain operates more independently. As the cortex matures, sub-cortical centers are gradually subordinated to its regulation, inhibition, and control.

L. S. Vygotsky (1986) pointed out three special traits of basic nervous system functioning in newborn and infant sensory-motor development:

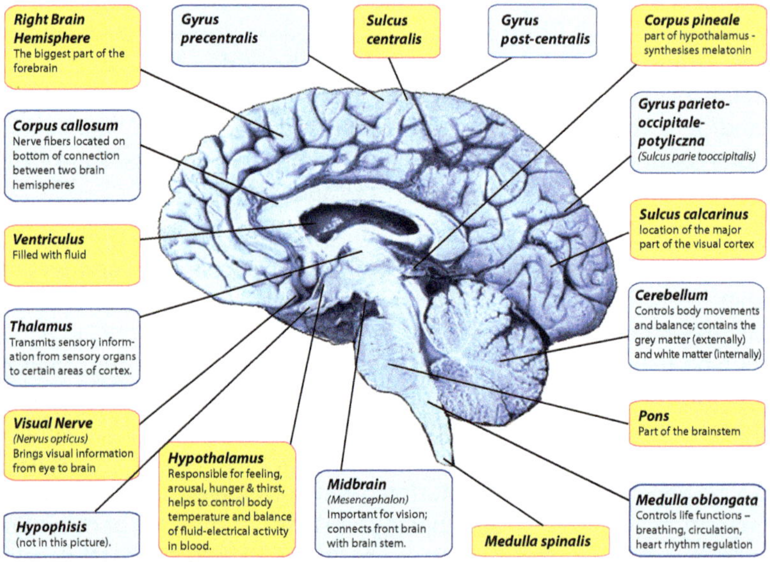

Figure 5. The parts of the brain.

1. 'Transfer' of functions to higher centers. The lower centers are unable to maintain their earlier type of functioning as the infant's learning increases and behavior becomes more complicated. The brainstem centers then transfer the greatest part of their functions to the cortical centers, thus supporting the progression of neurological development. Under control by higher centers movement becomes more nuanced, better adapted to varying conditions and more intentional.

2. 'Preservation' of lower centers. The oldest lower centers (brain stem) and their arcs do not disappear, but become subordinate to and inextricably entwined in cooperation with the higher centers (cortex). The union of lower and higher centers is so tight that it is difficult to separate them. However, the original circuits remain intact and can be observed in cases of injury to the nervous system.

3. 'Emancipation' of lower centers. In the earliest stages of newborn life the lower centers are 'emancipated' – free from control by the, as yet undeveloped, higher brain. This is the time during which unconditioned reflex activity is normal. Later, with conditioning by experience and learning and with the growth of higher brain centers, those previously automatic involuntary motor patterns are subordinated to the cortex and come under conscious control. If, however, the cortex is functioning poorly or 'separated' from the brainstem because of stress, shock, illness or injury, then the nervous system doesn't stop functioning, but has to rely on the subcortical and brainstem centers. The brainstem is then 'emancipated' again and pushed to function independently, but also maintains its primary protective functions. This is why, in cases of stress and pathology, the involuntary unconditioned reflexes become less subordinated to the cortex (less integrated with voluntary control) and more re-active, causing poor control of behavior, impulsiveness, and impaired decision-making. Thus a victim of a brain lesion may present an active Babinski response very like that of a newborn.

From L. S. Vygotsky's point of view, the reflex system as the oldest phylogenetic stage of CNS development, is the basis for development of the whole nerve system and cortex. Only neurologically matured and integrated reflexes can cooperate effectively with higher brain centers for controlled motor functions, cognitive processes, and conscious decision-making (Vygotsky, 1986).

When children present motor reactivity, impulsive behavior, and other delays; immature reflex development is often responsible. Attempting to inhibit or extinguish such involuntary reflexive motor activity seems like a reasonable approach. However, results have not been positive and the practice is no longer a standard intervention. An example, using the STNR (Symmetrical Tonic Neck Reflex), is presented below.

A retained STNR holds a child at the mercy of reciprocal reactivity linking the arms, legs, and neck. A boy with a hyperactive STNR, for example, is unable to sit still and focus. If he is asked to read in a sitting position and bends his head forward, then the automatism of the reflex causes his elbows to flex and his legs to extend. He may even impulsively stand up. When the teacher then asks him to sit, the opposite occurs – his lower limbs flex, but now his head and arms extend, taking his book out beyond his range of focus. Reflex inhibition might develop and strengthen his higher nerve system to control and suppress these reactive/impulsive reflex responses. This might even help him to adopt more appropriate habits and behavior. But would it assure the neurological maturation of a retained, delayed reflex? No. And would there be negative consequences? Yes. The child would become exhausted from constantly fighting to control the reflex motor activity. Sitting at a table through family dinners with both elbows and knees bent would be torture for him. He could never be an outfielder in baseball because reaching for a high ball would cause his knees to bend. In football he might be successful as a tackle or guard, but he'd be a poor passer or receiver.

In our experience we have seen that many habits and skills learned through reflex inhibition break down under stress and can even put a child at risk. A child, for example, whose Moro Reflex is hyperactive, may be excessively sensitive to vestibular stimulation, feel insecure in activities that involve movement, and be reluctant to participate in athletics. Normally when loss of stability in space is threatened, the Moro Reflex introduces a startle response, creates appropriate fear, and organizes a protective posture for possible falling down. An abnormally strong startle response due to vestibular hypersensitivity may also result in excessive timidity and lack of confidence, preventing a child from taking the normal risks necessary for new learning. While desensitization may calm down (inhibit or extinguish) the automatic response, it will not guarantee true maturation of the Moro neuro-circuitry. Just the contrary: in a situation of real danger an inhibited reflex will not fulfill its basic function and the child could actually sustain injury from falling with no fear and no protection from Moro's phase 2 motor pattern which should curl the body into a rolling ball to limit the harmful impact of a fall. Also, because a dysfunctional Moro Reflex is often mixed with reactivity in the Fear Paralysis Reflex circuit, failure to mature and integrate can also lead to dysfunction in both reflexes, including a predisposition to emotional fear and on-going challenges with defensiveness, avoidant behavior, and poor ability to learn.

Intervention using MNRI® would facilitate maturation and integration of any reflex that has failed to develop or has become dysfunctional. The goal would be:

1) activation of all components of the reflex circuit (receptors, sensory and motor pathways, synapses connecting the sensory and motor axons, and muscle fibers)

2) activation of the reflex circuit memory (sufficient repetition to establish the proper nerve pathways)

3) neurological maturation of the reflex circuit – myelinated axons and efficient neurotransmitter function

4) linking automaticity in the reflex response with conscious processes (mature inner control).

Reflex Circuit Memory

MNRI® re-patterning procedures propose sufficient repetitions of sensory-motor exercises approximating the sensory stimulation and unconditioned motor response of the reflex in order to activate long-term motor memory for stable nerve pathway function. This training aims at awakening the genetic motor memory encoded in the automatism of a reflex pattern and eliciting an automatic response to the stimulus. MNRI® uses its knowledge of neurophysiology to create techniques that support memory processes through activating motor neurons (alpha and gamma) with precise isometric and isotonic exercise of the particular muscles involved in the reflex pattern. In normal infant neurodevelopment, 10-15 activations of the reflex circuits are sufficient to myelinate the axons of the sensory and motor neurons (Pavlov, 1927/1960; Sechenov, 1961; Konorski, 1969). In the case of dysfunction and pathology this process can take several minutes to years depending on the extent of damage to the nervous system.

Neurological maturation of a reflex circuit means that the sensory and motor axons are myelinated (the peak of neural development) and all components of sensory processing and motor pattern expression are functioning properly. Myelination assures proper electrical conductivity in axons as well as production of the neurotransmitters at synaptic terminals. Neurological maturation of a reflex circuit creates several benefits for the nervous system and the body:

• *Freedom from automatisms.* To track the lateral movement of an object, for example, we should be able to move our eyes independently of our head. When the Head Righting Reflex is immature, its automatism links eye tracking with turning the head and shoulders; resulting in a limited field of vision and slower, less efficient visual processing. A matured, integrated Head Righting Reflex allows differentiation of eye movement from movement of the head and shoulders.

• *Reliable decoding of sensory input.* When danger is present the neurologically mature nervous system will activate its reflexes appropriately for protection and survival.

• *Regulation of the alarm state and support for cognition and communication.* When conditions are safe, the mature system will use reflex patterns, as Vygotsky (1986) says, for inner control over behavior and as the basis for emotional processing, cognition, and communication.

Automaticity in Reflex Responses and Conscious Processes

An unconditioned reflex is automatic, unconscious, and involuntary. This automaticity must mature, moving from the reactivity of infancy to a state in which a reflex activates automatically in response to real danger, and yet does not react inappropriately to stimuli that are not dangerous. Conditioned reflexes incorporate the effects of experience and learning, connecting the basic reflex (lower level of the nerve system) with higher brain structures. This is how internal control develops. Children with mature reflexes can maintain their posture, move, and use reflex motor patterns and their variants without having to think about it, so they are free to learn, interact, and grow naturally and easily. A child with immature and hyperactive reflexes must consciously try hard to initiate and control many functions that should activate and be controlled automatically. So when reflexes are delayed, hypo/hyperactive, or non-integrated, they interfere with cortical processing and impede development. Consequences of abnormal reflex development include hyper- or hypotonic muscles, metabolic diseases, neurotransmitter function deficits, vestibular dysfunction, poor brain plasticity, aberrant motor development, difficulties with auditory and visual processing, poor sensory-motor integration, delayed language, and poor social, emotional, and cognitive development.

Conclusion

Vygotsky's concept (1986) of infant reflexes as the basis for higher, controlled systems of motor skills and abilities that support cognitive functions shows clearly that reflex integration is the key to solving the problems of children and adults who experience dysfunction, delay, or regression in their motor, emotional, and cognitive functioning. By triggering the sensory system and replicating appropriate motor responses, MNRI® techniques reestablish or build connections between sensory and motor neurons.

MNRI® therapeutic modality aims at activation a reflex circuit by introducing an exemplary motor pattern, linking its components and sequences to replicate the exact biomechanics that support its original function. Rather than struggle with dysfunction and pathology, we instead invite the nerve system to explore and re-explore reflex motor patterns in accordance with their natural course of development (see the Leg Cross Flexion-Extension article in this book).

When reflex function is pathological, dysfunctional, immature, or lacks integration at any age, MNRI® techniques can re-route, re-connect or build new neural pathways and facilitate neurological maturation of its circuitry. Reflex integration exercises lead to overall maturation, including synaptogenesis, myelination, and brain plasticity. Their purpose is to provide sensory-motor coordination in a reflex circuit, regulation of the work of the sensory and motor neurons for easy postural control and motor development, and a solid foundation for intentional motor skills, inner control, emotional maturation, productive behavior, and cognitive abilities. The realization of one's highest potential necessarily rests on the foundation of reflex maturation and integration.

References

Alexandrov, Y. I., Sergienko, E. A. (2003). The psychological and the physiological: Continuality and/or discreteness? *Psychological Journal* (Ru). 24 (6), p. 98-107.

Anokhin, P. K. (1973). Biology and neurophysiology of the conditioned reflex and its role in adaptive behavior. *International Series of Monographs in Cerebrovisceral and Behavioral Physiology and Conditioned Reflexes.* USA.

Anokhin, P. K. (1968). *Biology and neurophysiology of a conditioned reflex.* (Russ.). Moscow, Russia: Medicina.

Barker, R., Barasi, S., Neal, M. (2008). *Neuroscience at a glance.* Oxford, UK: Blackwell Publishing.

Asratian, E. A. (1963). Tonic conditioned reflexes as the form of whole brain activity. *Journal of High Nerve System Activity,* v.15, n.5, p. 718-788.

Asratian, E. A. (1983). *Selected works. Reflexive theory of high nerve system activity.* Moscow, Russia: Science.

Barashniev, J. I. (2001). *Prenatal neurology.* (Russ.). Moscow, Russia: Triada-X.

Bernstein, N. (1997). *Bio-mechanics and physiology of the movement.* Moscow-Voronezh, Russia.

Goldberg, E. (2001). *The executive brain: Frontal lobes and the civilized man.* New York, NY, USA: Oxford University Press, Inc.

Haines, D. E. (2002). *Fundamental neuroscience for basic and clinical application.* 4th ed. Philidelphia, PA, USA: Elsevier Saunders.

Konorski, J. (1969). *Integrating Activity of the Brain.* Warsaw, PL: PZWL.

Luria, A. R. (1966). *The human brain and cognitive processes.* University of Moscow. Transl: New York & London: Harper & Row Publishers.

Luria, A. R. (1978). Functional brain organization. // In Book: *Biological-scientific basis of psychology.* / Ed. By A. A. Smirnov, A. R. Luria, V. D. Niebilistsin. (Russ.). Moscow, Russia: Pedagogika. p. 109-139.

Lundy-Ekman, L. (2002). *Neuroscience. Fundamentals for rehabilitation.* 2nd Edition. Forest Grove, OR, USA: Pacific University.

Magnus, R. (1925). Animal posture. *Croonian Lecture.* University of Utrecht. June, 16, 1925. p 1-15. www.jstor.com

Masgutova, S. (1990). Psychological Rehabilitation of Children – Victims of the Railway Catastrophe. *//Psychological Problems Journal.* No.1, 1990, Moscow, Russia. (Russ.)

Masgutova, S. K. (2002-2007). *Neurosensorimotor Reflex Integration.* Unpublished Lectures and Presentations. Warsaw, PL: MINK.

Masgutova, S. K., Akhmatova, N. K. (2004, 2007). *Integration of dynamic and postural reflexes into the whole body movement system.* Scientific edition: N. Akhmatova, Ph.D. (Pl.). Warsaw, PL: MINK.

Masgutova, S. K., Akhmatova, N. K. (2008). *Repatterning and integration of dysfunctional and pathological reflexes.* Orlando, FL, USA: SMEI.

Masgutova, S. (2011). *Infant dynamic and postural reflexes. Neuro-sensory-motor reflex integration.* Revised and edited, scientific-practical manual. Orlando, FL, USA: SMEI.

Masgutova, S., Akhmatova N., Masgutov D., Akhmatov E. (2013). *MNRI®: Reflex Integration for Stress-Hormone Regulation and Trauma Recovery.* Orlando, FL, USA: SMEI. 98 p.

Masgutova, S. K., Masgutov, D. (2013). *MNRI®: Reflex integration and basal ganglia. Optimal behavior and skill advancement for individuals with developmental challenges.* Orlando, FL, USA: SMEI.

Masgutova, S., Masgutov, D., Akhmatov, E. (2013). *MNRI®: Reflex Integration and PTSD Recovery. Advancement for Children and Adults Expe-*

riencing *Post-Traumatic Stress*. Orlando, FL, USA: SMEI. 98 p.

Pavlov, I. P. (1960). *Conditioned reflexes: An investigation of the physiological activity of the cerebral cortex.* (Anrep G.V., D. Sc. Trans., 1960). New York, NY, USA: Dover Publications Inc. (Original work was published in 1927 by Oxford University Press.)

Ronin-Walknowska, E., Masgutova, S. (2006). *Utero Movements and Reflexes.* DVD. Poland: ISMI.

Sechenov, I. M. (1935). *Reflexes of the Brain. Selected Works.* Volume 3. Moscow: State Publishing House for Biological and Medical Literature. p. 263–336.

Sechenov, I. M. (1995). *Physiology of behavior. Scientific works.* Ed. M. G. Yaroshevsky. Moscow, Russia. p. 26-130. (Original work was published in 1863)

Sherrington, C. (1947). *The integrative action of the nervous system.* Press, London., UK: Cambridge University.

Simonov, P. V. (1991). *Motivated Brain: A Neurophysiological Analysis of Human Behavior.* 1st Edition, USA: Taylor and Francis.

Ukhtomsky, A. A. (1950-1952). *The study of dominance //* Collected works in 6 books. Book 6. Leningrad, Russia.

Vojta, V. (1989). *Die Posturale Ontogenese als Basis der Entwicklungsstorungen.* Monatsschr. Kinderheilkd., 141, 639-642.

Vygotsky, L. (1986). *The child psychology. The problems of child development.* In 6 Books. Book – 4. Moscow. Russia: Pedagogika.

We extend our greatest gratitude to the genius of the scientists who created the grounds for understanding the complexity of a reflex as a phenomenon of nature, and giving us a pathway for its application in solving challenges in neurodevelopment and stressful conditions. – Authors

Neurotypical Development and Reflex Integration Disorder

Patty Shackleford, Ph.D., MNRI® Instructor, Melrose, FL, USA

MNRI® in Theory

The MNRI® program is frequently called the 'missing link' by professionals from around the world. It synthesizes information on reflex integration, and places it in the frame of both higher and lower nervous system activity (Pavlov, 1927/1995; Sechenov, 1961; Luria, 1969). In place of the traditional theory of reflexes as primitive responses that become cortically inhibited with normal development (Ayres, 1971, 1975; Vojta, 1972, 1981; Semionova, 1999; Goddard, 2002), MNRI® proposes the concept of reflexes as not only protective responses to stress and danger, but also the neuro-physiological foundation for higher level physical, emotional, and cognitive development. This concept allows reexamination in research of the relationship between primary sensor-motor development and physical, emotional, and cognitive skills. The key element in MNRI® is the understanding of a reflex as a genetic motor code and unit for both primary and conscious sensory-motor system development.

Patty Shackleford, Ph. D

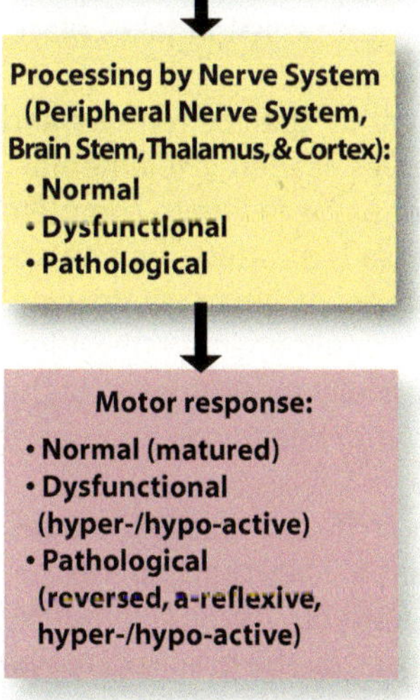

MNRI® focuses on the invitation for healthy neurological development of the infant, child, or adult using exercises rebuilding the sensory-motor reflexes (codes) rather than on diagnosis and treatment of neurological disorders and diseases. The following concepts form the theoretical core of the MNRI® approach:

- sensorimotor integration is an essential factor in human development
- motor development involves physical growth and cognitive differentiation necessary for the attainment of skills and learning
- the motor system develops on a foundation of integrated sensory perception and brain processing
- natural learning is achieved by the integration of the sensory-motor, emotional, and cognitive spheres
- a fully functional neurosensorimotor system integrates survival mechanisms and motor, emotional, and cognitive spheres.

The neurophysiology of a healthy nervous system is such that each reflex must integrate on the sensorimotor level; a specific sensory stimulus must cause a corresponding motor response. This precise link between the sensory and motor aspects of a reflex circuit through brain processing is genetically based. Each reflex consists of a three-part circuit that permits a motor response to a specific stimulus:

Part I – Sensory Stimulation, Part II – Brain Processing, and Part III – Motor Response

In the case of poorly integrated reflexes linked to dysfunctions or developmental delays, errors somewhere in the three-part circuit are likely. If the stimulus is not recognized by the sensory apparatus or transmitted efficiently by the afferent nerve system, then the brain misinterprets the input. If the efferent nerve system incorrectly channels a command to the motor system, an inappropriate reflex pattern will develop. Then maturation and integration of that reflex with controlled movements and skills will be unreliable, especially in the presence of learning challenges and stressors.

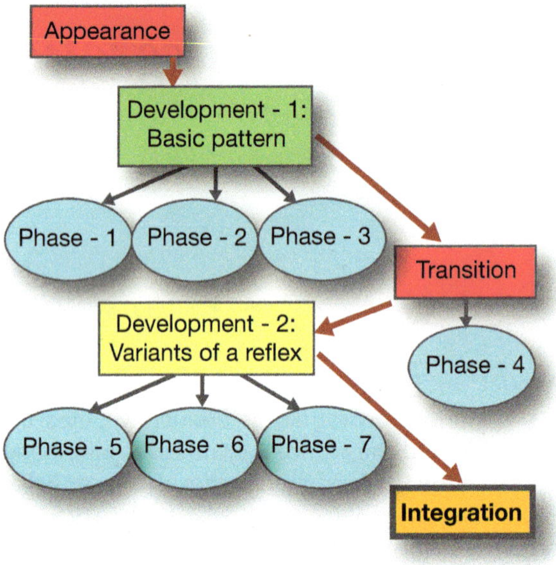

In the dynamics of healthy integration each reflex emerges at a specific time and develops its own basic patterns during the first three of seven phases (Masgutova, Akhmatova, 2004) (see figure at right). The transitional fourth phase prepares for the emergence of variants that appear during the fifth, sixth, and seventh phases. Each phase has its particular purpose. For example, a basic pattern (phases 1-3) codes the three part sensory-motor circuit and creates the nerve network that produces an appropriate response to a specific stimulus. By other words, natural training of the coordination of sensory and motor neurons takes place. The time of development of phases 1 to 3 corresponds to the time of development of the brain stem functions of physiological activity and survival. The transition phase - 4 prepares for the emergence of variant patterns, and chronologically corresponds to the time of starting more active work of the diencephalon functions (emotional communication and memory). During the last three phases, the maturation of a reflex is completed by highly developed neural networks – coordination of sensory and motor neuron work, proper electrical conductivity and biochemical neurotransmitters, and a balance of excitatory and inhibitory processes. Reflex patterns progress from the basic or unconditioned level (responsible for protection and survival) through a higher level of a conditioned response incorporating elements of learned experience and intentional action moving to maturation. With complete integration of all components of the reflex circuit and formation of myelination of the axons, proper transmission takes place, which affects the whole nervous system growth and myelination of its neural net system.

Successful completion of the phases described above ensures the development of specific reflexes and integrates them with motor skills and cognitive abilities, a combination that improves academic (elementary reading, drawing, writing, calculating) and other skills. A delay in reflex development or the failure to complete any of these phases always affects the formation of future skills. If an essential phase is omitted, then an appropriately matured neural network is not created. As a result, dysfunction or compensations may develop, but these are not true reflex patterns, and they are unreliable during periods of stress or unexpected transition. In summary, reflex development, maturation, and integration must progress successfully through each of the seven phases. This concept differs from the widely held limiting metaphysical view that healthy development requires reflex inhibition. (Masgutova, 2010; Akhmatova, Masgutova, 2006)

These concepts are based on recent neurological and neurophysiological research on the levels, stages, mechanisms, processes, strategies, and structure of primary sensory-motor integration. They draw on the theories of brain integration proposed by the outstanding scientists A. Luria (1963), A. Anokhin (1973), L. Vygotsky (1986), and A. Amosov (1978) and on the concept of neurological reflexes as responses by the brain-body system to external and internal stimuli (Pavlov, 1927/1960; Sechenov, 1863/1995; Simonov, 1991). MNRI® emphasizes the importance of primary motor development for intentional movements, motor skills such as eating, self-care, drawing, and writing; and visual and auditory processing abilities in the overall development of the motor system in children as described by numerous investigators (Vygotsky, 1986; Piaget, 1976; Rubinstein, 1984; Bozovich, 1968). The approach has been enriched by incorporating the concepts of movement development presented by N. Bernstein (1947), F. Lesfaft (1989), N. Leontiev (1974), V. Vojta (1989), B. Bobath (1972), L. Sadowska (2001), M. Feldenkries (1991), and P. Dennison and G. Dennison (1972) and theories of sensorimotor integration according to J. Ayres (1971, 1975).

Dr. Masgutova's Master's Thesis, *Unconditioned Reflexes, Unconscious Processes and their Effect on Personality Attitudes* (1980) reflected her study of these reflexes and gave her great insight into the impact reflexes have on human development.

MNRI® in Practice

MNRI® techniques are designed to restore neurosensorimotor development; to integrate primary motor system and skills; and voluntary motor skills. They activate innate reflex patterns; develop sensorimotor systems; improve sensory processing and promote cognitive and emotional growth. MNRI® also enhances cognitive abilities in highly functioning individuals. This approach encourages joyful learning in children and adults as they improve their neurosensorimotor abilities. The unique results from the pilot studies suggest that the Neurosensorimotor Reflex Integration processes have been used worldwide to change the future of challenged children and adults for positive development. Dr. Masgutova's previous research and clinical work using Neurosensorimotor Reflex Integration was devoted to unconditioned motor reflexes present in the earliest stages of child development.

Practical application of MNRI® has evolved from a combination of the theoretical basis discussed above with decades of experience and research on over 30,000 children and adults, suggesting that:

- in cases of dysfunction and delayed development natural noninvasive activation of the primary motor patterns promotes sensorimotor integration
- simple noninvasive exercises that stimulate reflex motor patterns can serve for repatterning of poorly developed, dysfunctional, or retained reflexes, and integrate into the whole sensory-motor system
- professionals, educators, and parents can assist with the transition from involuntary uncontrolled responses, emotional reactions, and spontaneous motor activity, to self-regulation and inner control.

The MNRI® program has developed 'regressive⟷progressive' repatterning procedures that 'remind' the brain-body system of its genetic neurosensorimotor programs. These procedures involve repatterning exercises to correct basic reflex patterns and/or their variants when they are dysfunctional, immature, or non-integrated. The term 'regressive⟷progressive' refers to the use of a developed or matured version of a genetically based motor pattern, combined with a precise posture and associated sensory or proprioceptive stimulation, as the foundation for future neurodevelopment. In going back (regressive) to the 'pure' innate form of the three part reflex circuit, the brain recognizes a template for growth and development (progressive) supplied by its genetic inheritance. Thanks to neurological plasticity the relevant nerve net system can then rebuild according to its original code and subsequently mature and integrate. MNRI® repatterning procedures are unique and highly effective, and their results are reproducible. They facilitate the integration of reflexes, regardless of the individual's age and neurological state. To correctly use these repatterning techniques, however, practitioners must understand the stages and specifics of human sensory and neuromotor development and be thoroughly trained in MNRI®.

All MNRI® programs use noninvasive natural techniques that can be easily learned by professionals, parents, and other adults who work with challenged individuals. Many techniques involve stimulating neuro-motor and sensory-motor points on the body, stretching the trunk and limbs, and rotating the joints. These procedures relate to the interactions among sensory or proprioceptive stimulation, reflex motor patterns and body postures. Other techniques release congestion and muscular tension throughout the body, stimulate the proprioceptive system, open communication among the muscles, tendons, and ligaments, and adjust receptors of deep touch and pressure. Reflex repatterning reconnects, strengthens, or builds neural circuitry by returning to and practicing natural reflex motor patterns and their variants in conjunction with associated sensory stimulation. The MNRI® programs address every possible aspect of a reflex and its relationship to overall sensory-motor development such as *Dynamic and Postural Reflex Integration, Lifelong Reflex Integration, Visual-Auditory*

Reflexes Integration, and others.

The MNRI® programs are used by professionals who specialize in and are certified to practice the Masgutova Method MNRI® programs. For further information on MNRI®, see the website: www.MasgutovaMethod.com.

Reflex Pattern Profile of Neurotypical Children

Dr. Masgutova's initial intent in working with children was to strengthen those with post-traumatic stress and next with neurodeficits. Dr. Masgutova has also worked with children considered to have learning challenges, to be gifted/talented, and to be neurotypical. She realized that if the Masgutova Method® was going to move forward with identifying children and adults with *Reflex Integration Disorder (RID)*, a data base was needed to see what the reflexes looked like in neurotypical children.

The research done by Dr. Masgutova from 1989-2013 on neurotypical children (2 to 19 years of age) concerned the reflexes and their parameters that display unique signs and serve as the point of orientation to measure noted deficiencies and deficits in reflex patterns of 'typical' children and adults. This information will assist in identifying children and adults with neurodeficits and learning challenges that fits within the concept of RID (S. Masgutova, 2011). The assessment procedures used for this research included assessing the 30 reflex patterns with results statistically verified with analysis by Prof. Anna Krefft (Krefft Algorythm, 2007). The results from this research (for the seventeen years of 1989-2006) on the reflex profile of neurotypical children (730 individuals) indicated that development of reflex patterns for this group of individuals was in the normal range of average – 16.77 points. The results from the data collected in 2007-2013 (780 children) showed that reflex patterns in the modern neurotypical children have changed – with the new 'normal' level of reflex development decreasing to a lower point – of 15.80 (compared to a norm of 16-17.75) (see graphs 1, 2, and 3). Comparison of these two results gave statistically significant differences and shows that the modern neurotypical group (2007-2013) range of reflex patterns have dropped in scores for Hands Pulling and ATNR, (Graph 1), Automatic Gait, Bauer Crawling, Hands Supporting, Fear Paralysis, Landau, Flying and Landing, Grounding, Head Righting (Graph 2), STNR, Spinal Galant, TLR, Foot Tendon Guard, Spinning, Locomotion, and Balancing (Graph 3). Modern children have demonstrated an increase of points in only one reflex - Pavlov Orientation – the reflex that triggers

Graph 1: Sagittal Plane

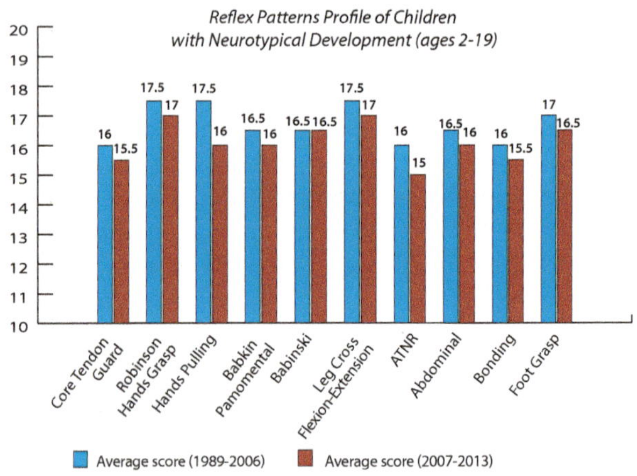

Graphs 1, 2, & 3: Motor Responses within Sagittal, Horizontal, and Dorsal Planes of the Body. 730 individuals in the year of 1989-2006, 780 individuals in 2007-2013.

Graph 2: Horizontal Plane

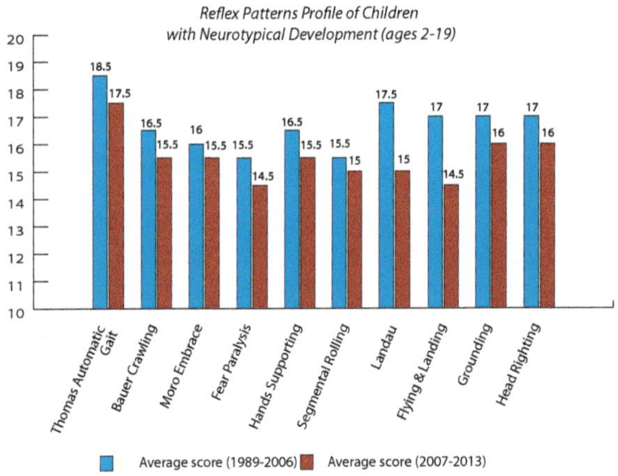

Graph 3: Dorsal Plane

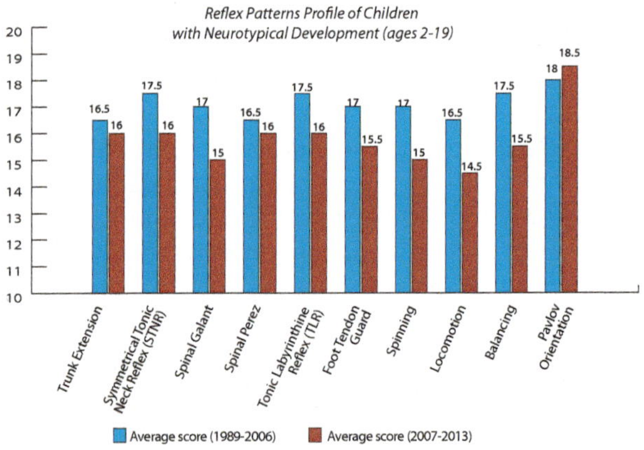

curiosity. The higher score on this reflex can explain the influence of computer technology and entertainment programs on our children. Yet this score does not guarantee the development of voluntary control and learning motivation. The statistic validity of the result is significant and equals 0.56 in the first group and 0.78 in the second, with $p < 0.001$.

Every reflex pattern has its own protective and neurodevelopmental task. A decrease in the level of development of these reflex patterns will evidently inhibit the overall physical, behavioral, emotional, and cognitive development. Comparison of reflex development in three groups demonstrate that the reflexes influencing the emotional maturation and stability, protective responses and survival, and postural control, has significantly decreased. This may be one important reason for challenges in the corresponding areas in the younger generation as reported by parents and professionals.

If changes are noted in this modern neurotypical group, consider what changes may be happening to children who are diagnosed as having neurodeficits. The downward trend in neurotypical children may be occurring due to changes in our natural environment, food supply, style of life, lack of movement, and other cultural practices such as methods used to safely raise our infants and babies.

Healthy children are naturally active, curious, and constantly moving, yet often their activity is reduced due to many factors: quiet behavior is valued and reinforced by adults in many cultures; today's increase in technology has children immobilized for hours in front of electronic screens; and, accelerated academic curriculums leave less time for recess and physical education. Even infants move less. They miss beneficial tummy time because sleeping on their back is encouraged and they spend hours each day in restrictive devices such as car seats. This kind of restriction along with overall diminished movement inhibits natural growth and development. Perhaps this is the reason many students (children and adults) today struggle with focusing, abstract information processing, and learning. Children whose movement and cognitive activity are poorly integrated face a variety of learning, behavioral, and social challenges.

Over the past few decades such profound changes have occurred in child rearing practices, education and general life style that one cannot assume all our children will grow and thrive naturally. Environmental toxins, food additives, over-reliance on medication, electromagnetic fields, electronic devices, junk food, safety measures that restrict movement and touch, developmentally inappropriate curricula, and decreased time for unstructured play have all moved childhood farther and farther away from natural development. These cultural and environmental factors can be reasons for lower results on reflex development in neurotypical children.

The unique results from the worldwide pilot studies looking at children's progress after the use of the Masgutova Neurosensorimotor Reflex Integration (MNRI®) processes indicate positive growth and development for the children with challenges and neurodeficits.

Dr. Masgutova's previous research and clinical work using MNRI® processes was devoted to unconditioned motor reflexes present in the earliest stages of human development. Every neonate has unconditioned innate reflexes that facilitate adaptation to the external environment and become the foundation of mental activity. The Masgutova Method® finds that innate primary movements and reflex patterns are expressed by every individual and are key elements of neurosensory-motor development of a human.

Many reflexes are essential for survival, especially during periods of stress and traumatic experiences (Masgutova, 2011; Akhmatova, Masgutova, 2007). These reflexes are the fundamental neurological building blocks for all learned skills and voluntary control, crucially influencing the development of the brain and many cognitive and intellectual processes as we mature. During certain events that occur in utero, at birth, or later in life, a child or adult may experience trauma that affects the ability to use those reflex patterns effectively. In addition, stressful situations may cause an individual to revert to using early infant reflexes. Experiencing stress (physical or emotional) that affects cognitive and intellectual maturation and the ability to learn is the main source of impulsive behavior and engenders a reliance on primitive reactions and reflexes, both of which lead to a regression in the formation of coordinated sensory-motor systems and movement skills. Because reactive responses and primary reflexes are designed to protect the individual and to ensure survival, they often overshadow the reasoning that occurs in the neocortex of the human brain.

In some children and adults, primary reflexes may be retained after a stressful event because of developmental motor problems, sensory processing disorders, or the poor integration of those reflexes with the intentional movements in infancy. Other survival reactions and reflexes are triggered and necessary protective and survival needs are unmet, emotions are suppressed, or everyday stressful events occur. In these individuals, primary reflexes remain active and do not integrate within their circuit. As a result, abnormal movement patterns develop and impede the development of more mature movements, skills, and intellectual processes.

Dr. Masgutova's research on over 3,350 subjects (age range 1 month–18 years) with a variety of developmental deficits, noted a correlation between the poor development of reflex patterns and the deficit(s) identified. Seventy-eight percent of those individuals (age range 4–15 years) demonstrated various difficulties such as: hyperactive or hypoactive response to sound perception, also language delays when the Asymmetrical Tonic Neck Reflex was underdeveloped. This research found that 57% of those subjects who were considered to have attention deficit disorder (ADD), attention deficit/hyperactivity disorder (ADHD), and poor transition from concrete to abstract thinking had poor Symmetrical Tonic Neck Reflex development. Fifty-eight percent of the children with an attention disorder also exhibited problems with the Spinal Galant, Spinal Perez, and Asymmetrical Tonic Neck Reflexes.

Specific Developmental Disabilities in U.S. Children Aged 3-17 Years*

Disability	Percent Change between 1997-1999 and 2006-2008
Any developmental disability	17.1%^
ADHD	33.0%^
Autism	289.5%^
Blind/unable to see at all	18.2%
Cerebral palsy	-
Moderate to profound Hearing loss	-30.9%
Learning Disability	5.5%
Intellectual disability	-1.5%
Seizures, past 12 months	9.1%
Stuttered or stammered, past 12 months	3.1%
Other developmental delay	24.7%^

*Centers for Disease Control and Prevention, National Center for Health Statistics, NHIS, 1997-2008
^Statistically significant trend over four time periods (1997-1999, 2000-2002, 2003-2005, and 2006-8)

In the USA alone, the chart at right provides information on the percentage of children of 3-17 years of age that in the 2008 census were considered as having a developmental disabilities that hindered typical growth in one of the necessary life skills. This chart also indicates the significant increase in these developmental disabilities of the last years.

Dr. Masgutova and MNRI® specialists have also gathered information on the changes that have occurred in both neurotypical children and children considered as ASD and PTSD. The graph (below-right and next page) shows the comparison of children diagnosed with Autism Spectrum Disorder (ASD) and Post-Traumatic Stress Disorder (PTSD) with neurotypical children.

The data demonstrates significant differences in all reflex patterns of modern children in the neurotypical group (first column; average level – 15.8 points) and the ASD group (second column; 9.11 points). The last group shows a dysfunctional level of development of reflex patterns (9.11 point, while the norm is 16-17.75 points), except for Core Tendon Guard, Thomas Automatic Gait, and Trunk Extension. On these graphs (graph 4, 5, 6) the level of reflex development in children with ASD before (second column) and after 8 days of the intensive MNRI® program is presented (third column), showing significant positive changes in all reflex patterns, with an average of 10.40 points (compared to 9.11 before MNRI® program). The statistic validity of the result is high and its significance equals 0.39 before and 0.51 after the MNRI® Program. The linear error in projection of change is not more than 1.77 – 2.80 percent and supports the conclusion that the MNRI® program produces statistically significant changes in reflex pattern development; $p < 0.001$.

Analysis of development of reflex patterns in the group of Post-Traumatic Stress (PTSD; 340 individuals) shows that this group's reflex patterns are of a low and very low level (11.14 points average) as the result of going through unbearable trauma and distress. Intense stress and distress activates a higher level of stress-hormones in the body and diminishes

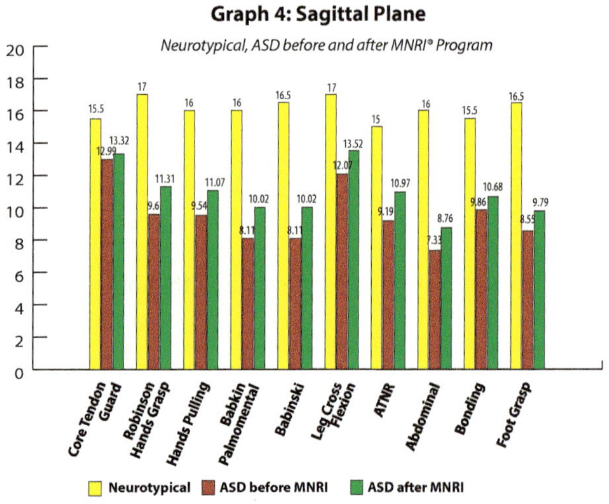

Graphs 4, 5, & 6: Motor Responses within Sagittal, Horizontal, and Dorsal Planes of the Body. Reflex Patterns Profile of Children with ASD (Age 4-20; 480 children) and Neurotypical Development (Age 2-19; 780 children).

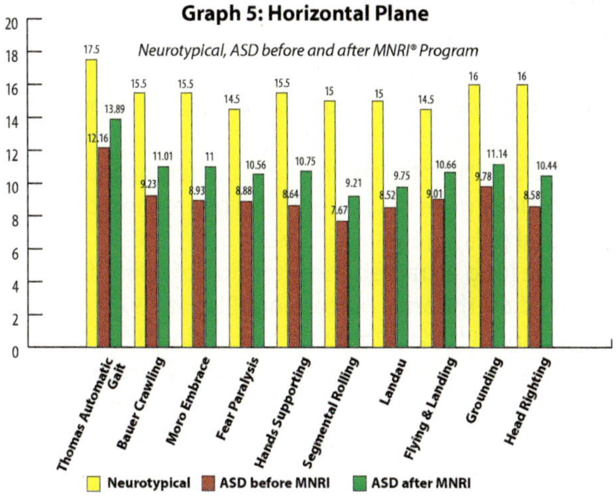

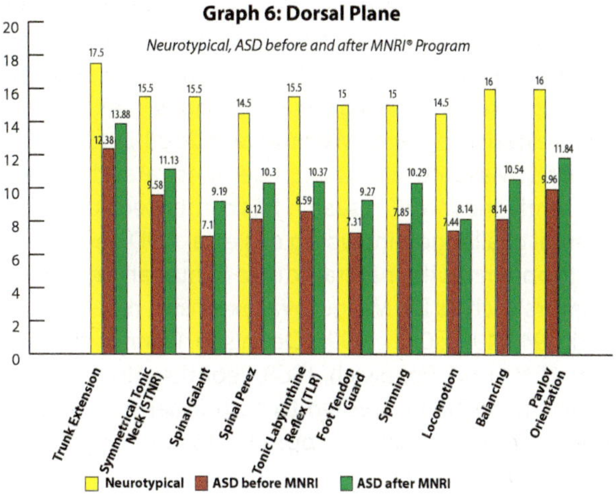

myelin of the axons in the reflex neuronal circuits and other nerve systems. The assessment of reflex patterns in individuals with PTSD clearly shows this damage to nerve circuits.

The data in the graphs 7, 8, and 9 for reflex patterns in individuals with PTSD demonstrate differences in all reflex patterns compared to reflexes in the neurotypical group (first column; average level – 15.8 points). In the PTSD group (second column) reflexes are of very low and low level on the scale (11.14 points vs. the norm of 16-17.75 points), except for Hands Grasp, Leg Cross Flexion-Extension, Thomas Automatic Gait which are closer to the level of reflexes in modern neurotypical children.

On graphs 7, 8, and 9 the changes in the level of reflex patterns in children with PTSD before (second column) and after 8 days of the intensive MNRI® program is presented (third column), and indicates a significant positive increase of points for all reflex

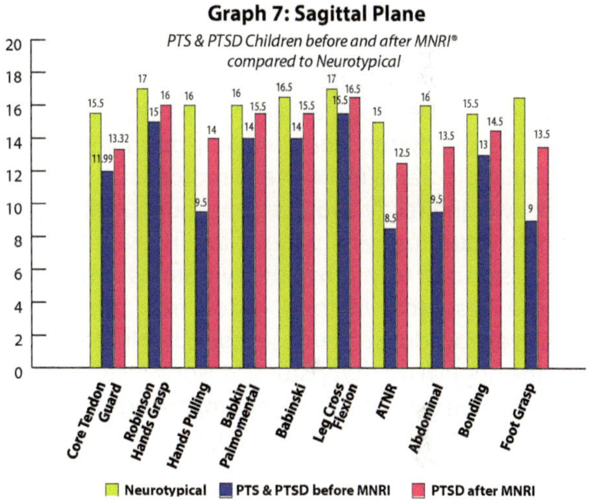

Graphs 7, 8 & 9: Motor Responses within Sagittal, Horizontal, and Dorsal Planes of the Body. Individuals with Traumatic Stress (Age 4-20; 340 children) compared to Neurotypical Development (Age 2-19; 780 children).

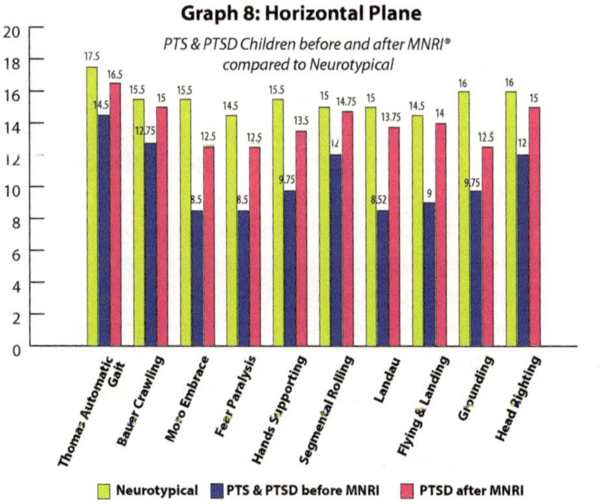

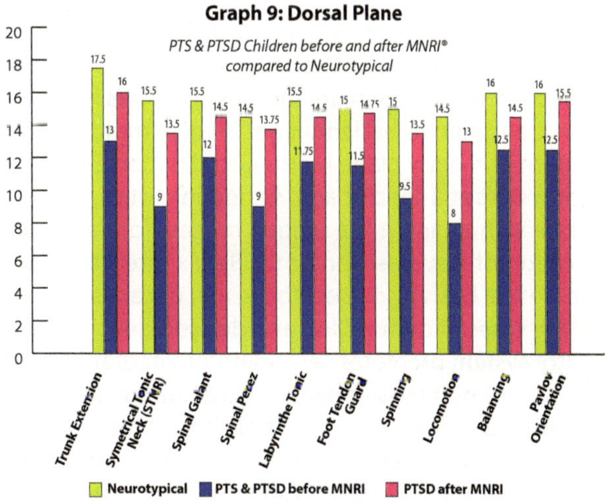

patterns (average of 14.28 points compared with 11.14 before the MNRI® program). The statistic validity of the result is high and its significance equals 0.47 before and 0.61 after the MNRI® program. The linear error in projection of change is not more than 1.67 – 2.56 percent which supports the conclusion that the MNRI® Program produces statistically significant changes in reflex pattern development; $p < 0.001$ (from lecture and presentation materials by Dr. S. Masgutova in 2008-2013).

Dr. S. Masgutova's research demonstrates the effectiveness of MNRI® for children with different challenges. Currently the MNRI® research has composed reflex profiles for 14 groups of children with neuro- and developmental challenges, including dyslexia, ADD and ADHD, cerebral palsy, genetic disorders, and others.

Problems in the natural course of human development have long attracted the attention of scientific investigators such as Dr. Masgutova. Extensive research has focused on theories of learning (Piaget, 1973; Montessori, 1949/1995; Clark, 1995; Dennison, Dennison,1972) and personality development (Vygotsky, 1986; Montessori, 1949/1995; Bozowich, 1968; Leontiev, 1977). However, in the educational and psychological literature, little research has been devoted to natural sensory-motor and motor development. As a result, few professionals use a completely natural approach to correct Reflex Integration Disorders and other neuro-motor and developmental disorders.

Currently there are many cognitive interventions aimed at mastery of specific skills by task analysis, gradual learning, and practice. Occupational and physical therapists now use sensory integration techniques to address the difficulty some children have in regulating their response to sensory input. Professionals practicing neurodevelopmental treatments emphasize the traditional idea of inhibition of retained or abnormal infant reflex patterns. Developmental optometrists use vision training to address difficulties with visual acuity that lead to delays and dysfunctions in motor-balance system and visual cognitive development. Some educational and remedial programs make use of motor activity, including natural movement. All of these interventions either directly or indirectly address some, but not all aspects of the primary reflex system. Fortunately, many professionals, educators, and parents are increasingly concerned and are interested in understanding the MNRI® processes used to improve the symptoms of Reflex Integration Disorders or poorly developed reflex patterns.

The Masgutova Method® differs from other traditional therapy systems with its concept of using a noninvasive procedure of neurosensormotor integration to rebuild the entire neural circuit of a reflex by reorganizing the original motor/posture/sensory/vestibular-proprioceptive perception systems, and the exact basic pattern, sequence, direction, strength, timing, and symmetry that would characterize an ideal initial experience of that sensory-motor reflex pattern.

The MNRI® program brings positive growth and development to typical children and effective and positive reflex integration using natural noninvasive sensory-motor techniques and movements that can be easily learned by parents, adults, and professionals who work with challenged individuals. The MNRI® techniques do not require many resources, and they complement other therapy programs such as sensorimotor integration and educational, psychosomatic, and motor sphere enhancement. The MNRI® integration exercises and techniques can be performed without equipment, require just minutes each day and can be practiced in almost any location (home, park, school, or clinic).

As the data in the above graphs demonstrate, Dr. Masgutova's research and work with over 30,000 children has found that the primary motor system has its own structure, levels of development, self-regulating mechanisms, and 'nonverbal language.' Primary motor patterns develop according to natural laws. Each reflex, motor pattern and body plane Motor Coordination System (MCS) emerges, develops, and integrates with conscious motor skills at a specific age. Although motor pattern development is based on genetic motor programs typical for all humans, it is a unique process for each individual. Dr. Masgutova's early research and practical application of Neurosensorimotor Reflex Integration techniques has evolved through the combination of the theoretical basis discussed above and decades of experience. This research suggests that:

- in cases of dysfunction and delayed development natural noninvasive activation of the primary sensory-motor system promotes sensorimotor integration and brain development
- simple noninvasive exercises that stimulate reflex patterns can integrate poorly developed, dysfunctional, or retained reflexes into the whole body sensory-motor system
- professionals, educators, and parents can assist with the transition from involuntary uncontrolled responses, emotional reactions, and spontaneous motor activity, to self-regulation and inner control.

During individual development these natural genetic programs expand, integrate, and merge with learned movements. Consciously controlled movements develop at the next level. Finally, with repeated training, these learned motor programs become the foundation for the habituated individual automatic motor program. Motor development involves:

• the formation of primary motor systems such as global archetypal motor patterns, rhythmical movements, reflexes and reflexive reactions, systems of natural motor skills, abilities based on automatic movements and reactions, and instinctive behavior

• development of basic infant reflex patterns and their variants

• integration of infant dynamic and postural reflexes with intentional and consciously controlled movements

• maturation of stages of motor development (turning from prone to supine and vice versa, sitting, standing, etc.) and the formation of transitional movements (turning around, sitting, and other motor skills needed for changing positions)

• coordination among body parts and sensory organs (left hand-right hand; hands-eyes; hands-eyes-ears; hands-legs; etc.)

• development of MCS and skills for the left and right sides of the body (involving left and right brain hemispheres), the upper and lower sections of the body (involving the cerebral cortex and midbrain), and the front and back of the body (involving the cerebral cortex and the back of the brain).

• formation of basic motor patterns used in fine motor skills such as drawing, writing, reading, playing an instrument, etc.

• development of intentional goal-oriented and controlled motor patterns, such as those used in sports, exercising, dancing, public performances, etc.

• enrichment of the pace and rhythm of motor activity

• development of lifelong reflexes regulating antigravity mechanisms and postural control (gravity, grounding, stability, and balance)

• development of adequate functioning of the muscle system within the primary motor system

• development of kinesthetic intelligence (sensory-motor aspects of consciousness, behavior, and activity)

• development of neurological links between primary motor abilities and skills and their emotional, cognitive, and behavioral expression.

Conclusions

Dr. Masgutova is the originator of MNRI® and Director of the Svetlana Masgutova Educational Institutes in Poland and the United States. Her Master's Thesis, *Unconditioned Reflexes, Unconscious Processes and their Effect on Personality Attitudes* (1980) reflected her early interest in the study of reflexes and gave her great insight into the impact reflexes have on human development. Throughout her early professional career she was given opportunities to see what happens to individuals as they experience trauma and tragedy in their lives. With her intellectual knowledge and practical experience, Dr. Masgutova has been able to bring to children and families a new paradigm on how to use the genetically given, innately developed concept of reflex development using a non-invasive, highly successful method called the Masgutova Neurosensorimotor Reflex Integration – MNRI®. The programs in this system have been used to help identify what is 'typical' for all children and how the non-integration of reflexes can cause challenges for children (and adults). This article has attempted to give research evidence of the positive growth and development that can occur when Dr. Masgutova's process for reflex integration is applied. It is hoped that the reader of this article will begin to understand the impact of a Reflex Integration Disorder on the cognitive, physical, emotional, and social development of a child.

References

Anokhin, P. K. (1973). Biology and neurophysiology of the conditioned reflex and its role in adaptive behavior. *International Series of Monographs in Cerebrovisceral and Behavioral Physiology and Conditioned Reflexes.* USA.

Amosov, N. M. (1978). *Thinking about the health.* Moscow. p. 16-29.

Ayres, J. (1971). Characteristics of types of sensory integrative functions. *The American Journal of Occupational Therapy,* 25(27). USA. p. 329-334.

Ayres, J. (1975). *Sensory integration and the child.* Los Angeles, CA, USA: Western Psychological Services.

Bernstein, N. (1997). *Bio-mechanics and physiology of the movement.* Moscow, Russ: Moscow-Voroniez.

Bobath, K. (1972). The normal postural reflex mechanism and this deviation in children with cerebral palsy. *Physiotherapy,* 57, p. 515.

Bozovich, L. I. (1997). *Basis of personality formation. Selected Psychological Works.* Moscow, Russia.

Clark, B. (1986). *Optimizing learning.* Columbus, OH, USA: Merrill Publishing Co.

Dennision, P., Dennison, G. (1972). *Brain Gym®.* Teacher's Edition. CA, USA.

Goddard, S. (2005). *Reflexes, learning and behavior: A window into the child's mind: a non-invasive approach to solving learning & behavior problems.* Eugene, OR, USA: Fern Ridge Press.

Haines, D. E. (2013). *Fundamental neuroscience for basic and clinical application.* 4th ed. Philidelphia, PA: Elsevier Saunders.

Leontiev, A. N. (1977). *The activity, consciousness, personality.* Moscow, Russia; English translation: Prentice-Hall.

Luria, A. R. (1966). *The human brain and cognitive processes.* University of Moscow. Transl: New York & London: Harper & Row Publishers.

Masgutova, S. (2011). *Infant dynamic and postural reflexes. Neuro-sensory-motor reflex integration.* Revised and edited. Scientific-practical manual. Orlando, FL, USA: SMEI. 190 p.

Masgutova, S., Akhmatova, N. (2007). *MNRI® Re-patterning and integration of dysfunctional and pathological reflexes - Svetlana Masgutova® Method of neuro-reflex integration.* Revised and changed. Orlando, FL, USA: SMEI. 82 p.

Montessori, M. (1995). *The absorbent mind.* New York, NY, USA: Henry Holt and Co. (Original work was published in 1949)

O'Reilly, R. C., Morlet, T., Cushing, S. L. (2013). *Manual of pediatric balance disorders.* San Diego, CA, USA: Plural Publishing.

Pavlov, I. P. (1960). *Conditioned reflexes: An investigation of the physiological activity of the cerebral cortex.* (Anrep G.V., D. Sc. Trans., 1960). New York, NY, USA: Dover Publications Inc. (Original work was published in 1927)

Piaget, J. (1976). *The grasp of consciousness: Action and concept in the young child.* Cambridge, MA, USA: Harvard.

Sadowska, L. (ed.). (2001). *Neurokinesiological diagnosis and treatment of children with impaired psychomotor development,* Wroclaw, Ed. AWF Wroclaw, ss. 71, 83-84, 94-95.

Sechenov, I. M. (1965). Reflexes of the brain. (Russ. tr. S. Belsky). Cambridge, MA: The M.I.T. Press. (First published in 1863.)

Sechenov, I. M. (1995). *Physiology of behavior.* Scientific Works/Ed.: M.G. Yaroshevsky. Moscow, Russia. 319 p.; p. 26-130. (Original work was published in 1863)

Semionova, K. A. (1999). *Rehabilitation treatment of the CP patients in residual state.* Moscow, Russia.

Sherrington, C. (1947). *The integrative action of the nervous system.* London, UK: Cambridge University Press.

Simonov, P. V. (1991). *Motivated Brain: A Neurophysiological Analysis of Human Behavior.* 1st Edition, USA: Taylor and Francis.

Vojta, V. (1989). *Die Posturale Ontogenese als Basis der Entwicklungsstorungen.* Monatsschr. Kinderheilkd.

Vygotsky, L. (1986). *The child psychology. The problems of child development.* In 6 Books. Book – 4. Moscow, Russ: Pedagogika.

www.MasgutovaMethod.com

Lecture and presentation materials by Dr. S. Masgutova in years 2006-2013.

My deepest gratitude goes out to all the children and families all over the world that have trusted the MNRI® processes found in the Masgutova Method®. It is only through the cooperation of these children and families that Dr. Masgutova and her MNRI® Team have been able to recognize the existence of Reflex Integration Disorder and the importance of making other helping professionals aware of this neurodevelopmental challenge. – Patty Shackleford

The Leg Cross Flexion-Extension Reflex: Biomechanics, Neurophysiology, MNRI® Assessment, and Repatterning

Elvin Akhmatov, Ph.D. Student, Orlando, FL, USA; Jakub Buraczewski, PT, MNRI® Core Specialist; Denis Masgutov, Director of ISMI, Poland

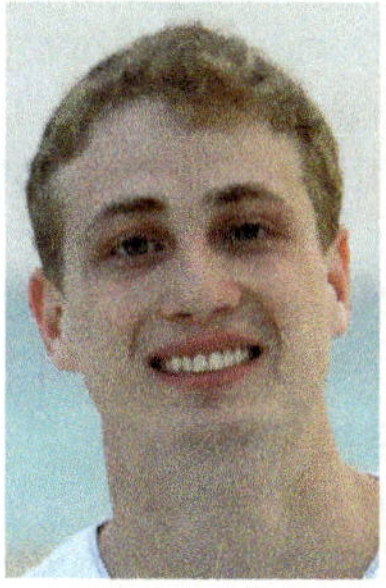
Elvin Akhmatov

Jakub Buraczewski

Denis Masgutov

Introduction

Two separate reflexes, Phillipson's Withdrawal and Leg Cross Flexion-Extension, are easily confused because they have similar motor patterns and are elicited by stimuli that can appear to be alike and usually manifest at the same time. The authors' purpose is to distinguish clearly between these two reflexes and to present detailed information on the one they refer to as the *Leg Cross Flexion-Extension Reflex*. The other reflex, often confused with Leg Cross Flexion-Extension, goes by several names: Phillipson's Withdrawal, Phillipson's Leg Flexion, Crossed Extensor, and Leg Withdrawal Reflex, among others. For clarity in this paper, the other reflex will be referred to as Phillipson's Withdrawal. On the neurophysiological level, these two reflex patterns present the work of two different nerve tracts – tactile and proprioceptive, activated and processed by different receptors.

The Leg Cross Flexion-Extension Reflex is extremely important for overall sensory-motor integration, motor programing and control. Cross-lateral or reciprocal activities such as crawling, walking, running, climbing, swimming, and jumping depend on maturation and integration of this reflex pattern. It also plays a crucial role in the 'training' of proprioceptive nerve pathways (vs. sensory-tactile) and activating the spinal-cortical tract. It is an invaluable resource for supporting neurosensorimotor-proprioceptive integration in physical rehabilitation of children and adults with developmental deficits and/or injuries affecting motor coordination and abilities. However, neither the reflex, its developmental and rehabilitative importance, nor its biomechanics and neurophysiological ascending and descending pathways are clearly or adequately defined in current literature. This study presents a new understanding of this complicated reflex and describes its biomechanics and the neurophysiological tracts of its ascending and descending pathways. The authors will also interpret MNRI® techniques for practical application that enhance precision and accuracy in the treatment of dysfunctional Leg Cross Flexion-Extension and affect other lower limb motor patterns.

Apart from the work of Masgutova and her associates, other researchers, to the knowledge of the authors, have not studied Leg Cross Flexion-Extension in children or persons with neurodeficits. In contrast, a simple search yields a wealth of information on Phillipson's Withdrawal. Two examples are 1) "The flexors in the withdrawing limb contract and the extensors relax, while in the other limb, the opposite occurs" (Rod, Trent, Tate, 1992; http://en.wikipedia.org/wiki/Withdrawal_reflex), and 2) "when a person steps on a nail, the leg that is stepping on the nail pulls away, while the other leg takes the weight of the whole body" (Ellrich, Steffens, Schomburg, 2000; Solomon, Schmidt, Adragna, 1990; http://en.wikipedia.org/wiki/Crossed_extensor_reflex). According to the authors of this article, important developmental implications of this contralateral motor pattern are missing from these descriptions and thus will be described in detail here.

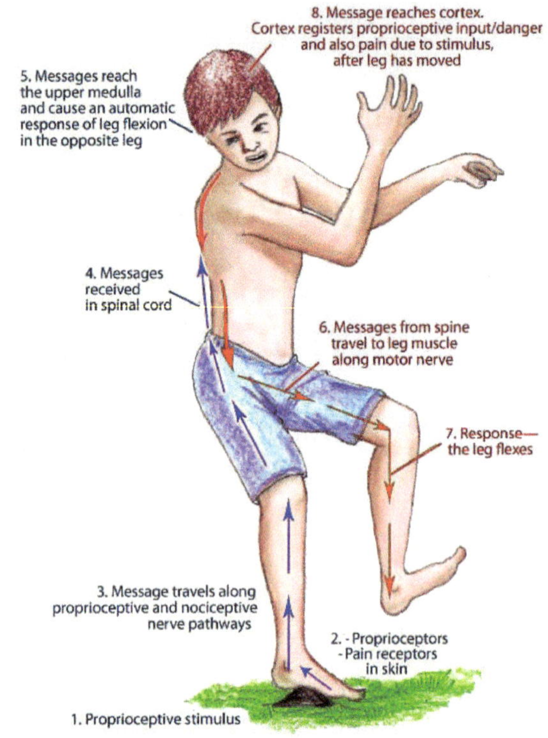

Figure 1. Leg Cross Flexion-Extension Reflex

I. The Leg Cross Flexion-Extension Reflex

Suppose you are moving along with a mature cross-lateral walking pattern when your right foot unexpectedly lands painfully on a sharp object. There are two possible scenarios: 1) your body weight is not yet on this leg and you have the chance to withdraw it, or 2) your body weight is already on this leg and if you try to withdraw it, you fall down. In scenario one, your sensory perception stimulates the tactile nerve pathway, and in scenario two the proprioceptive pathway acts in priority. In both cases the response to pain is present, and your body automatically provides shock absorption to lessen the pain by 'unweighting' the right foot with slight knee and plantar flexion and toe extension. This is a 'withdrawal'. In the case of Phillipson's Withdrawal pattern, the extension of the opposite leg is immediate.

In contrast, with Leg Cross Flexion-Extension the left (opposite) leg does not immediately extend. At this point in your gait, your left foot has just left the ground. It is impossible to withdraw your right foot from the sharp object without falling down. The slight withdrawal of your injured right foot causes a loss of equilibrium, which stimulates extension in that leg and returns that foot to the sharp object to regain balance. Thus, in order to prevent a fall, your reflex system requires that you briefly step down harder on the sharp object to support your body weight while your other leg completes its step, which it does first with flexion, then extension.

Contrary to the descriptions of Phillipson's Withdrawal, in Leg Cross Flexion-Extension, the leg where the stimulus is given extends, pushing against the stimulus rather than withdrawing. Not until your left foot touches the ground can your body weight be transferred to it, freeing you to lift the right foot away from the painful object. Pushing down hard on a sharp/unsafe object hurts more, but rapid flexion and extension of the opposite leg enables you to quickly transfer your body weight to the left foot and step away with the hurt right foot. This is the protective motor pattern of the Leg Cross Flexion-Extension Reflex. These two reflexes use different biomechanics and respond through different neural pathways.

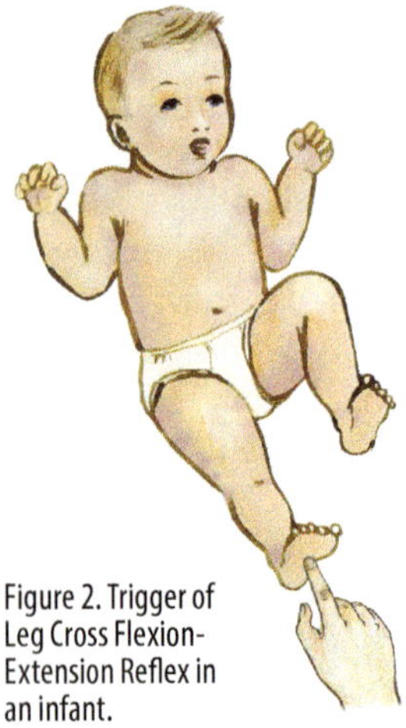

Figure 2. Trigger of Leg Cross Flexion-Extension Reflex in an infant.

In a one or two month old infant the initial phases of this sequence of movements are absent. There is no loss and regaining of equilibrium in the supine infant, though there can be some initial very slight and very brief flexion of knee and foot. Unlike the withdrawal reflex, the Leg Cross Flexion-Extension Reflex is elicited not by pain (tactile nociceptors in the skin), but by deep pressure (proprioceptors in the joints, tendons, and muscles) on the center of the plantar surface of the foot. This proprioceptive stimulus on one foot simultaneously activates automatic extension of the same side leg (pushing against the stone) and flexion in the opposite knee and hip, with slight medial adduction, followed immediately by lateral abduction, and then extension of that leg, stepping forward and to the side. This same circuitry, developed in infancy, later saves children and adults from possible injury, as described above, when they step unexpectedly on a sharp or possibly harmful object. In this complicated reflex we see how wisely nature has provided for our protection.

The circuit of this response is a complicated one activated by parallel pathways in all levels of the central nervous system, reaching the spinal cord, medulla oblongata, pons, mesencephalon, thalamus, and cortex. The input is processed in sequential and parallel ways in the spinal cord, ascending (sensory) pathways in the pons, and descending (motor) pathways in the mesencephalon, where nerve fiber crossover takes place, and in all these areas of the nerve system simultaneously (Haines, 2002; Lundy-Ekman, 2002). R. Magnus demonstrated experimentally that the processing of input and formation of the circuit of this reflex happens mainly on level of the medulla oblongata (Magnus, 1925).

The motor response is processed mainly by the spinal cord and extrapyramidal nerve net system of the brain stem. Activation of this circuit is essential for development of lower motor neuron pathways and myelination of their axons, especially on the level of the medulla oblongata.

It is very important that this reflex develop at the proper time. Normally it appears in the 12th week of gestation (Campbell, 2004; Ronin-Walknowska, Masgutova, 2006) and is active in utero and up to 1–2 months after birth (Barasznev, 2001). By 1–2 months of life it must be integrated, which means that its sensory-proprioceptive-motor circuitry is established for sequential movements of the legs, and works in a stable and timely way.

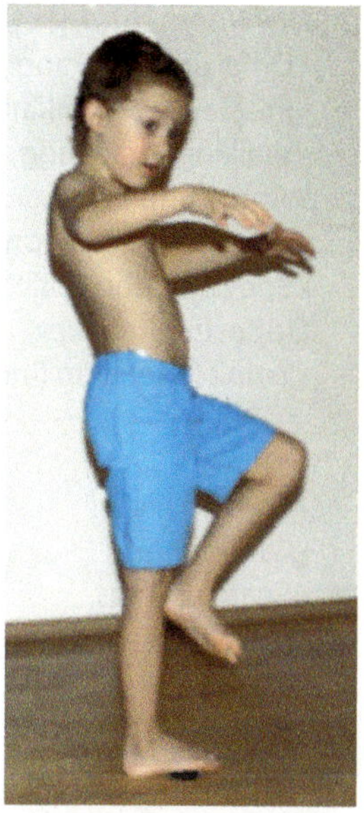

Figure 3. Stepping on an object with the whole body weight causes automatic flexion and, next, extension of the knee and hip in the other leg.

Sensory-Motor Structure

◆ *Position for eliciting the reflex:* Supine, standing, or walking

◆ *Proprioceptive stimulus:* On pressure point in the middle of the plantar surface of the foot (see stone in Figure 1). Note: In infants of 0-2 months pressure also on the lateral and medial vastus muscle fascia just above the knee (quadriceps; right leg, see Figure 4).

◆ *Proprioceptive stimulus:* Initially deep touch activates the proprioceptive receptors in the foot of a supine infant. In later development the corresponding receptors are activated (infant, child, and adult) standing, stepping, walking, and marching (see Figure 4).

◆ *Motor response:* The leg receiving the stimulus extends and opposite leg flexes in all joints; the knee moves laterally with slight adduction, then abduction, and finally extension.

Protective Meaning

• Protection from injury to the foot stepping on a potentially harmful object

• Transfer of body weight to the other leg to prevent a fall

Facilitating Reflexes

• Automatic Gait
• Amphibian
• Bauer Crawling
• ATNR (holding postural control)
• Galant
• Balancing
• Segmental Rolling
• Spinning.

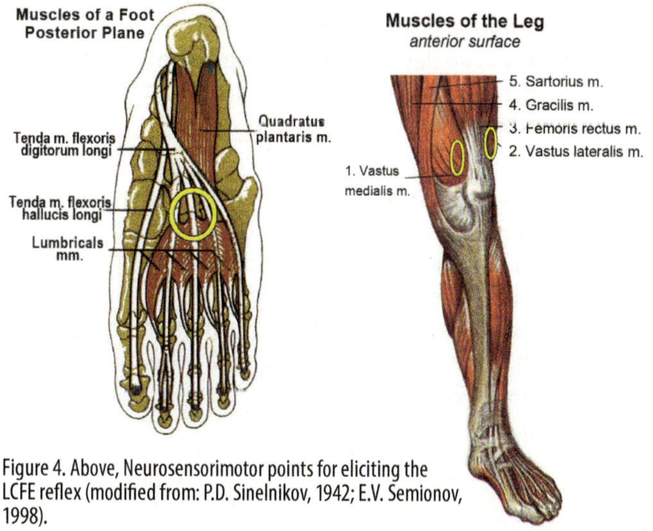

Figure 4. Above, Neurosensorimotor points for eliciting the LCFE reflex (modified from: P.D. Sinelnikov, 1942; E.V. Semionov, 1998).

Developmental Meaning

- Differentiation and coordination of legs
- Precursor to walking and running
- Walking, marching, running, swimming, skipping, jumping
- Cognitive development
- Speed of perception and thinking.

Opposing Reflexes

- Trunk Extension (including lower limbs)
- STNR
- Moro
- Spinal Perez
- Landau
- Fear Paralysis.

Figure 5, Step 1. Stepping on the object (stone)

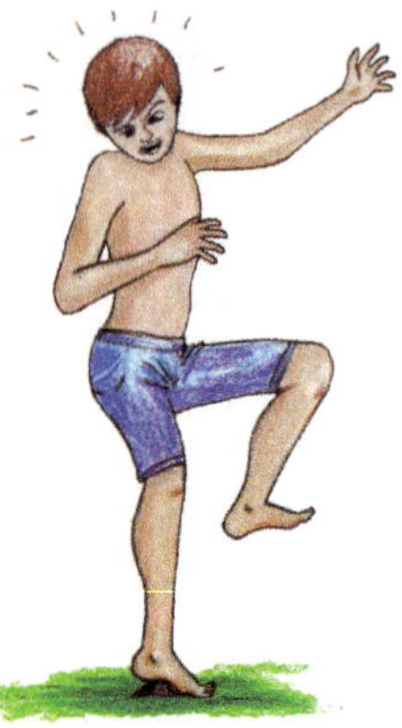

Figure 5, Step 2. Flexion of the knee and hip of opposite leg

Effects of Non-Integration

In children with neuro-motor problems, muscular hypo- and hypertension, cerebral palsy and other deficits, the Leg Cross Flexion-Extension Reflex is usually delayed and/or pathological. It is also poorly developed in children with autism and selective mutism. This is why skills requiring leg differentiation, shifting body weight from one leg to another for cross-lateral movements and balance control are all challenging for them. Additional effects of non-integration include:

- poor neural maturation of lower motor neurons
- hyper- (ADHD) or hypo-activity
- dysfunctions of the Automatic Gait and Crawling Reflexes
- postural problems: chronic inclining of the body to one side or another while standing, or instability due to poor antigravity skills
- poor ability to cross the midline and inefficient cross-motor patterns
- excessive muscle tension causing fatigue
- delay in skipping, walking, swimming, jumping, and climbing skills
- slow transition from perception to action
- poor cognitive development: comparison, classification, prioritization, calculation, and analysis
- incorrect overlapping of tactile and proprioceptive input and chaotic processing due to homologous crawling and walking patterns.

Figure 5, Step 3. Abduction of the knee laterally leading orientation of the body to side and forward

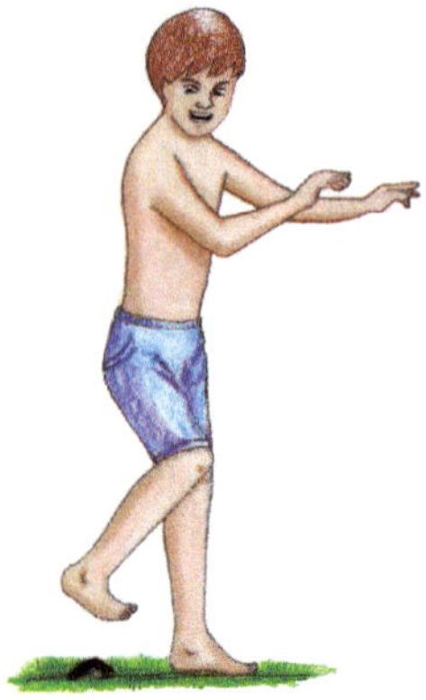

Figure 5, Step 4. Extension of the opposite leg to save the hurt foot.

II. Biomechanics of the Leg Cross Flexion-Extension Reflex

Activation and maturation of this reflex motor pattern in infancy orients the brain-body system to its later protective function when the walking child or adult encounters a potentially harmful object under foot. The biomechanics of the Leg Cross Flexion-Extension Reflex involve complex antigravity phenomena related to body weight distribution on the feet and the defensive response of reorienting body weight and re-locating it to a safer spot when gravity pulls one foot down on a painful object. Clearly the motor response can be activated from either side. For the description below, please assume that the 'stone' is under the right foot as in Figures 5, Step 1, 2, and 3.

Sequence of Movements in the Leg Cross Flexion-Extension Motor Pattern

Part 1-a. Plantar flexion of the right foot provides shock absorption—Figure 5, Step 1.

Muscle Groups	Main Muscles *(contracts)*	Opposing Muscles *(lengthens)*	Facilitating Muscles *(contracts)*
Foot *(plantar flexion for shock absorption)*	Quadratus plantae *(contracts)* Lig plantare longum *(contracts)* Flexor hallucis longus	Extensor digitorum brevis	Lumbricales Extensor hallucis longus Interosei plantares
Knee and tibial area, calf *(slight knee flexion)*	Gastrocnemius Triceps surae Popliteus Flexor digitorum longus Flexor hallucis longus	Extensor digitorum longus Peroneus longus Peroneus brevis Tibialis anterior	Soleus Fascia cruris Plantaris

Part 1-b. Extension of right leg towards stimulus (stone)—to regain equilibrium.

Muscle Groups	Main Muscles	Opposing Muscles	Facilitating Muscles
Knee and tibial area, calf *(extension)*	Gastrocnemius Tricepts surae Popliteus Flexor digitorum longus Flexor hallucis longus Tendo calcaneus (Achillis) *(lengthens)*	Extensor digitorum longus Peroneus longus Peroneus brevis Tibialis anterior *(contracts)*	Soleus Fascia cruris Plantaris *(lengthens)*
Thigh *(leg extends with slight knee extension)*	Rectus femoris Quadratus femoris (Vastus medialis and lateralis) Adductor longus Adductor minimus Gracilis *(contracts)*	Biceps femoris Adductor magnus Semimembranosus Semitendinosus *(lengthens)*	Tensor fasciae latae Extensor digitorum longus Extensor digitorum brevis Extensor hallucis longus *(contracts)*

Part 1-c. Flexion of left (opposite) hip, knee, and foot with slight medial adduction —Figure 5, Step 2.

Muscle Groups	Main Muscles	Opposing Muscles	Facilitating Muscles
Calf	Tibialis anterior Peroneus tertius Extensor digitorum longus Extensor hallucis longus *(contracts)*	Gastrocnemius Soleus Flexor hallucis longus Flexor digitorum longus Tibialis posteriori Peroneus longus Peroneus brevis *(lengethens)*	Gastrocnemius Popliteus *(contracts)*
Thigh	Quadricieps femoris Femoris rectus Adductor brevis Sartorius *(contracts)*	Semimembranosus Semitendinosus Gracilis Biceps femoris Adductor magnus *(lengthens)*	Adductor brevis Adductor longus Tensor fasciae latae *(contracts)*
Hip	Psoas major Iliacus Gluteus medius Gluteus minimus Gracilis *(contracts)*	Gluteus maximus Piriformis Lig. sacrospinale Lig. sacrotuberale *(lengthens)*	Tensor fasciae latae Iliopsoas Obturatorius internus *(contracts)*

Part 2. Lateral abduction of flexed left knee—Figure 5, Step 3.

Muscle Groups	Main Muscles *(abducts and contracts)*	Opposing Muscles	Facilitating Muscles
Thigh, hip	Gluteus maximus Gluteus medius Tensor fasciae latae Sartorius Piriformis *(abducts and contracts)*	Adductor magnus Adductor longus Adductor brevis Gracilis Pectineus *(abducting, opposing to abduction and lengthening)*	Rectus femoris Gluteus maximus Gluteus minimus *(abduct and contracts)*

Part 3. Extension of left (opposite) leg in hip, knee, and foot—Figure 5, Step 4.

Muscle Groups	Main Muscles *(lengthens)*	Opposing Muscles *(contracts)*	Facilitating Muscles *(lengthens)*
Hip	Psoas major Iliacus Gluteus medius Gluteus minimus Gracilis	Gluteus maximus Piriformis Lig. sacrospinale Lig. sacrotuberale	Tensor fasciae latae Iliopsoas Obturatorius internus
Thigh	Quadricieps femoris Femoris rectus Adductor brevis Sartorius	Semimembranosus Semitendinosus Gracilis Biceps femoris Adductor magnus	Adductor brevis Adductor longus Tensor fasciae latae
Calf	Tibialis anterior Peroneus tertius Extensor digitorum longus Extensor hallucis longus	Gastrocnemius Soleus Flexor hallucis longus Flexor digitorum longus Tibialis posteriori Peroneus longus Peroneus brevis	Gastrocnemius Popliteus

Part 4. Release of foot from stimulus: flexion of right hip, knee, and foot with slight medial adduction—Figure 5, Step 4.

Muscle Groups	Main Muscles	Opposing Muscles	Facilitating Muscles
Calf	Tibialis anterior Peroneus tertius Extensor digitorum longus Extensor hallucis longus *(contracts)*	Gastrocnemius Soleus Flexor hallucis longus Flexor digitorum longus Tibialis posteriori Peroneus longus Peroneus brevis *(lengethens)*	Gastrocnemius Popliteus *(contracts)*
Thigh	Quadricieps femoris Femoris rectus Adductor brevis Sartorius *(contracts)*	Semimembranosus Semitendinosus Gracilis Biceps femoris Adductor magnus *(lengthens)*	Adductor brevis Adductor longus Tensor fasciae latae *(contracts)*
Hip	Psoas major Iliacus Gluteus medius Gluteus minimus Gracilis *(contracts)*	Gluteus maximus Piriformis Lig. sacrospinale Lig. sacrotuberale *(lengthens)*	Tensor fasciae latae Iliopsoas Obturatorius internus *(contracts)*

With the release of the right foot from the stimulus, the protective function of the reflex is accomplished.

III. The Neuro-Sensory-Motor Pathways of the Leg Cross Flexion-Extension Reflex

Figure 6-1. Electrical currents in nerve transmission (changes from -70 mvt at rest to -55 mvt).

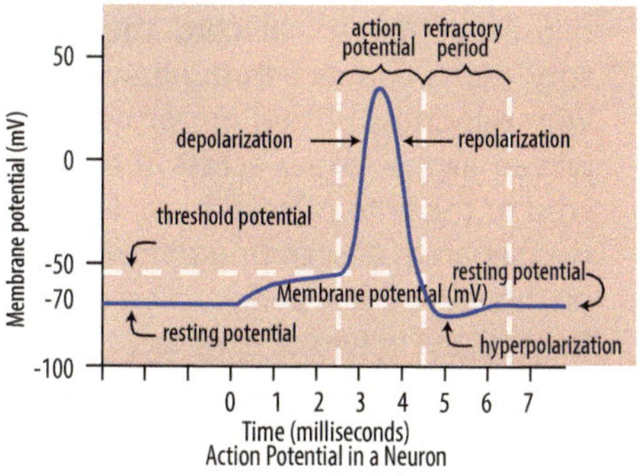

Afferent nerve pathway

A sensory-proprioceptive stimulus (on the right foot arch) generates action potentials that set off a Na+-sodium bio-chemical cascade creating electrical currents (from -70 mvt at rest to -55 mvt) to start the transmission. These currents travel via the bulbo-thalamo-cortical tract to the cortex (Figure 6-1).

Figure 6-2 presents the cellular mechanisms of action potential generation in a sensory receptor and synaptic potentials triggering conducted signals in motor and inter-neurons: A) action potential measured in a sensory neuron; B) postsynaptic triggering potential recorded in an extensor motor neuron (α-neuron); C) postsynaptic triggering potential in an interneuron (γ-neuron); D) postsynaptic inhibitory potential in a flexor motor neuron (γ-neuron) (Purves, Augustine, Fitzpatrick, et al., 2001; Haines, 2006).

The afferent tract, running from the plantar surface of the foot to the gyrus post-centralis in the motor cortex, provides sensory-proprioceptive sensation to the brain (Haines, 2006; Cohen, 1999; Barker, Barasi, Neal, 2008). It consists of three orders of neurons, working at different levels and all interacting simultaneously to process the sensory-proprioceptive input involved in this reflex:

Fig 6-2 (A, B, C, & D) Action potential generation in sensory and motor neurons.

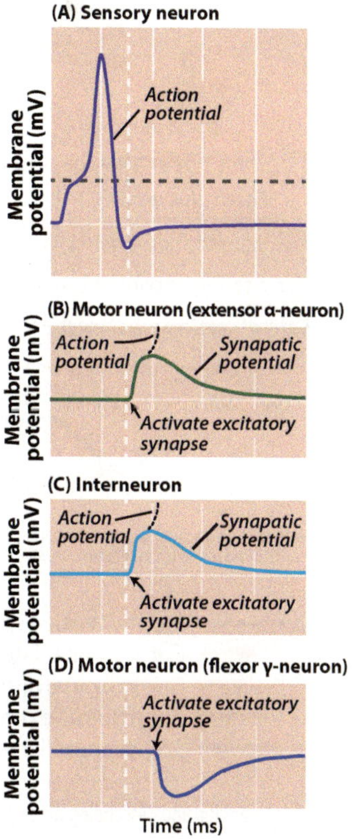

1. Axons of first order sensory neurons (connected with receptors as described above), enter the dorsal horn of the lumbar spine, travel along the fasciculus gracilis and end in the nucleus gracilis.

2. Axons of second-order sensory neurons start at the same level of the spinal cord, follow ipsilaterally to the lower olives of the medulla oblongata, and continue through the spinobulbo-thalamocortical tract to the medial lemniscus. There they transmit nerve impulses across (decussate) to the other side of the brain.

3. Third-order sensory neurons located in the thalamus direct their axons to the cortex via the thalamocortical tract, follow to the corona radiata and end in the postcentral gyrus in the motor cortex (Haines, 2006; Cohen, 1999).

Afferent nerve pathway in more detail

Ascending and processing of the sensory-proprioceptive input in the Leg Cross Flexion-Extension Reflex pattern takes place at several levels of the nerve system.

Lower limb

A sensory-proprioceptive stimulus to the tendon receptors (Golgi, Spindle cell, Meissner, Merkel, and Rufini cells) on the plantar surface of the right foot (Golgi tendons, quadratus plantae muscle) triggers receptors and dendrites of the sensory neurons, reaching their cell bodies in the spinal ganglion, continuing through their axons (proximal nerve fibers closer to spinal cord), following along the fasciculus gracilis and the nucleus gracilis and finally entering the posterior horn of the lumbar spine. Axons providing input on this level are called first-order neurons.

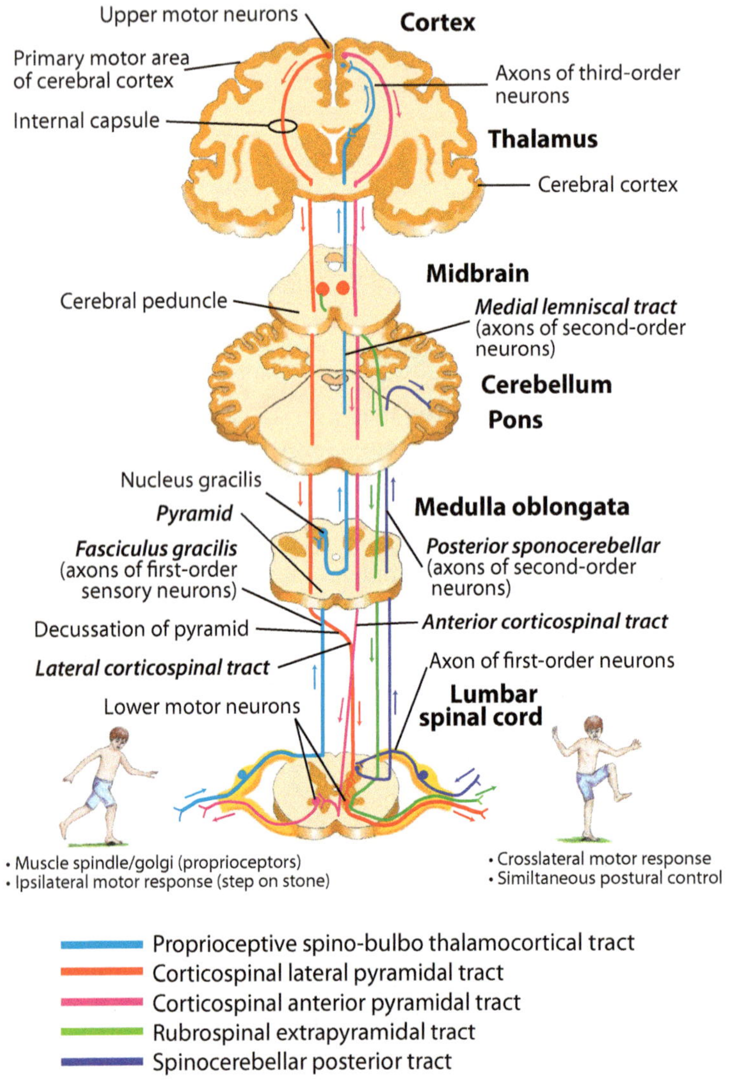

Figure 7. Ascending and descending nerve pathways of the Leg Cross Flexion-Extension Reflex pattern

Spinal Cord

First-order sensory neurons enter the dorsal horn of the lumbar spine, travel along the fasciculus gracilis, and end in the nucleus gracilis. In the middle of the spinal cord their sensory axons connect through synapses with dendrites of two motor neurons, whose axons follow across the spinal cord to the anterolateral horn, forming two separate descending pathways. So on this level, the proprioceptive input is transmitted in two directions:

- One pathway is formed by an α-motor neuron connected directly or through an excitatory interneuron with the sensory neuron. In the synapse, where the sensory and motor neurons connect, excitatory neurotransmitters such as substance P, glutamate, and CGPR (calcitonin gene-related peptide) are released to increase the action potential in the α-motor neuron and its interneuron.
- The other pathway is formed by the γ-motor neuron, connected with sensory neuron indirectly through an interneuron, whose synapse serves for the release of inhibitory dopamine and

GABA neurotransmitters. These neuro-mediators control the speed and strength of the action potential by inhibiting the impulse in order to regulate the tone in the intra- and extrafusal muscle fibers of both agonist (facilitating) and antagonist (opposing) muscles.

In synapses at the ends of axons of both α- and γ- motor neurons, where they connect with muscle fibers, the acetylcholine neurotransmitter is released to support the transmission of action potential to muscle fibers. The α-motor neuron in part 2 of Leg Cross Flexion-Extension causes the leading agonist muscle (left leg hamstring) to contract, and the γ-motor neuron inhibits and regulates the antagonist muscle (quadriceps, supporting lengthening in the same left leg). This muscle tone regulation results in flexion of the left leg. Other elements of this motor pattern occur thanks to similar interactions between sensory and motor neurons in their complex circuitry.

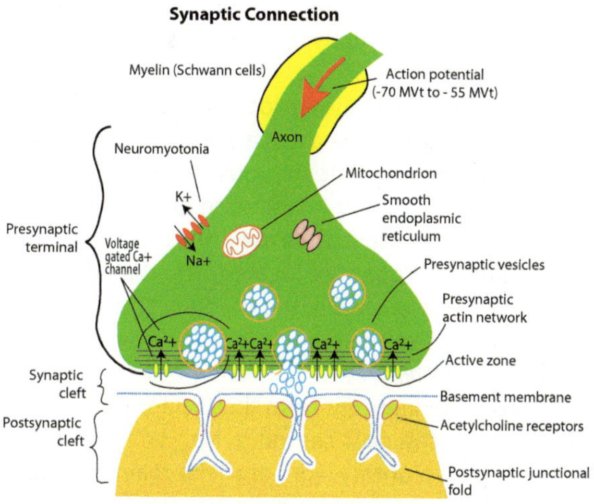

Figure 8. The mechanism of connectivity in a synapse.

Note 1: Synaptic connectivity results from currents of directed potential created biochemically by the Na+ and K+ ions (Haines, 2002; Cohen, 1999). Synaptic connectivity includes the work of an inhibitory synapse, which produces effects opposite to those of the excitatory synapse: the action of the transmitters on the receptor sites is to open those ionic channels which hyperpolarize the postsynaptic membrane. Typically these are the K+ and Cl- channels. The chemical gradients of these two ions are such that K+ diffuses outward while Cl- diffuses inward. This combination of ionic fluxes hyperpolarizes the membrane so that the internal potential becomes even more negative than the resting state. An inhibitory postsynaptic potential (IPSP) is established thanks to that. The potential is called inhibitory because the membrane potential is even farther from the excitation threshold than in the resting state (Haines, 2006; Cohen, 1999; Barker, Brasi, Neal, 2008). This process must correlate with the stimulus. The inhibitory mechanism in synaptic connectivity is a part of the nervous system functioning that contributes to:

- *regulation of sensory input and processing*
- *slowing down the flow of sensory information (especially in γ-motor neuron)*
- *control of overlapping sensory input*
- *transition of short-term memory into long-term cellular memory.*

Note 2: Dysfunction in reflex circuitry can happen in any of its parts: receptors (poor development of specific receptors and their interlinks), connection of dendrites with the axon though the cell body of a sensory neuron, in action potential (conductivity of electrical potential, Na-sodium relations, mitochondrial deficiencies), neurotransmitters at synapses (pre-synaptic, synaptic or post-synaptic processes), the motor neurons, and also effector organs (muscles and glands). Lack of adequate neural regulation in the work of a sensory neuron and its connection with a motor neuron is caused basically by imbalance between excitatory sensory and inhibitory neurotransmitters released in the interneuron, which affects the tone and coordination of opposing muscles. This kind of imbalance affects the quality of reflex circuit function, challenging the processing of input and blocking the control circuits of the reflex in the brainstem and diencephalon.

Medulla and Pons (Brainstem)

Part of the proprioceptive input in the Leg Cross Flexion-Extension Reflex travels to the medulla through axons of second-order neurons via the bulbothalamocortical tract. This afferent tract provides proprioceptive sensation to the cortex. These neurons cross contra-laterally in the lower olives of the medulla oblongata and continue to travel via the bulbo-thalamic tract to the medial lemniscus and finish in the ventroposterior thalamic nucleus. Repetitive eliciting of this reflex in our infancy and childhood allows for myelination of the axons of second-order neurons in the medulla and pons. This is the part of the brain that deals with integration of the neuro-structural aspect of a reflex and its function of protection and survival. Thus survival strategy, the ability to process danger on the sensory level and to supply grounding and stability in dangerous situations when the body is destabilized, results from the work of this reflex in the brainstem. Input from this reflex also links the brainstem (medulla, pons, and midbrain) with the cerebellum and thalamus. This level of the brain completes the processing, preparing the physical action of the Leg Cross Flexion-Extension motor pattern to provide immediate protection: stepping harder on the stone (right leg) and flexing the left leg and extending it to regain stability on safer ground. This most important part of processing Leg Cross Flexion-Extension input is automatic and entirely unconscious.

Thalamus

Third-order neurons are located in thalamus and direct their axons to the cortex along the thalamo-cortical tract. Input from the Leg Cross Flexion-Extension Reflex that reaches this level is processed here and coordinated with sympathetic nervous system responses to activate the stress axis (alarming the organism about danger and activating an increase of stress hormones). Third-order neurons form one third of the posterior limb of the internal capsule, next they follow to the corona radiata and finish in the post central gyrus, where fourth order neurons are located. The thalamus controls the circuit of the Leg Cross Flexion-Extension Reflex performed at the level of the medulla and pons. It determines the relative danger or safety of a stimulus and then regulates sympathetic activation, muscle tone and motor activity/reactivity accordingly.

Sensory Cortex

Third-order neurons from the thalamus follow along the same thalamo-cortical tract via the corona radiata and finish in the post central gyrus, where they reach the fourth order neurons. This final level of the brain controls the voluntary work of all the main muscles in the body and, according to I. Sechenov (1960), serves as the physiological basis for the psychological process of creating an 'image' of the stimulus: "Is it a stone? Is it a piece of wood, a pen?" Here the brain uses its executive functions to analyze sensory input. The whole cortex becomes involved in decoding and analyzing the input, organizing subsequent output to the muscles and making any movement more precise, goal oriented, and meaningful.

Efferent/descending nerve pathway

There are two principal groups of descending tracts: one serves for postural/gross motor coordination and control, and the other for voluntary control of fine motor activity. Descending tracts classified as postural/gross motor tracts in the Leg Cross Flexion-Extension Reflex control automatic responses of the skeletal muscles to extend one leg (right foot on the stone), and to flex, abduct, and extend the opposite leg (left).

The first group of tracts involves the purely automatic motor response and is generated from the lumbar spine, where descending innervation takes place in its different segments. The excitation of motor neurons activates the anterior leg muscles (quadriceps) for movement and depolarizes the contralateral motor-neurons of the hamstring muscles (biceps, semitendinosus, and semimembranosus), with simultaneous inhibition of motor neurons innervating antagonistic muscles creating the possibility for one leg to extend (right; led mainly by the quadriceps), and other leg to flex, then abduct and extend (left; led mainly by the biceps).

The second group of descending tracts delivers motor information from the high brain to lower motor neurons of the brainstem or spinal cord and is more involved with voluntary control. These descending tracts are formed by upper motor neurons that: a) arise in the cerebral cortex and descend to the brainstem; b) synapse with lower motor neurons and/or interneurons in the brainstem and spinal cord. They would be involved in conscious direction of the left leg forward, backward or to the side, depending where the safest place to land might be.

Control circuits of the Leg Cross Flexion-Extension Reflex operate from the basal ganglia, cortex, and cerebellum, which is responsible for planning automatic responses and adjusting to external circumstances on different levels of the nerve system.

Descending motor pathways through different levels of the nerve system

Motor Cortex

Descending motor pathways in Leg Cross Flexion-Extension present sequential and simultaneous work of the whole pyramidal and extrapyramidal nerve systems. The pyramidal system involves the corticospinal tract (lateral and anterior-pyramidal tracts), which governs rapid voluntary movements at the distal ends of limbs. They originate in the primary (Broadmann's area 4 – motor cortex) and secondary (area 6 – premotor cortex) motor cortex and the parietal lobe (areas 1, 2, 3 – somatosensory cortex). This tract decussates in the pyramids and descends as the lateral corticospinal tract to its destination: an inter- or α-motor neuron. Branches of this tract are present in: the cerebral cortex, basal nuclei, red and olivary nuclei, and the reticular formation.

This part of the descending pathway involves the pyramidal upper motor neurons, called first order motor neurons because they are unable to leave the central nervous system. Multi-synaptic, they connect with lower motor neurons to send messages to the muscles. Upper motor neuron lesions can lead to dysfunctional Leg Cross Flexion-Extension responses, as happens in individuals with CP or brain damage.

Note: 80% of the pyramidal cells are located on the precentral gyrus in the frontal lobe (motor strip), 20% of the pyramidal tract fibers originate in the postcentral gyrus of the parietal lobe (areas 1, 2, and 3). Pyramidal tract fibers begin their descent from the cortex (corona radiata) before forming the internal capsule. They are direct and monosynaptic; their axons do not synapse with other cells until they reach their final destination in the brainstem or spinal cord. Because they connect directly from the cortex to the lower motor neurons, messages can be transmitted to the periphery very rapidly.

Subcortical and Extrapyramidal tracts (diencephalon and brainstem)

The efferent output from the cortex travels via the frontopontine tract and next along the rubrospinal tract. These extrapyramidal tracts include:

- tectospinal tracts
- vestibulospinal tracts
- rubrospinal tracts
- anterior, medial, and lateral reticulospinal tracts

The tectospinal tract is involved in the control of neck muscles. It originates in the midbrain (in the brainstem) and ends in the spinal nerves, thus governing the lowered head position in the Leg Cross Flexion-Extension Reflex pattern.

The vestibulospinal tract runs from the vestibular nuclei (lower pons and medulla) to the spinal nerves. It is involved in balance. (Note that all of these tracts receive input from the cerebellum.) In Leg Cross Flexion-Extension this tract serves to keep the body in equilibrium until the legs react contra-laterally.

The rubrospinal tract passes through the red nucleus, carrying messages from the cerebellum to the spinal nerves. Information in this part of the tract flows from the superior cerebellar peduncle to the red nucleus and to the spinal nerves, allowing for somatic motor or skeletal muscle control and regulation of muscle tone for posture and voluntary movement. In Leg Cross Flexion-Extension this tract will control the direction and tone of opposing muscles to make possible this extremely complex motor response.

The reticulospinal tract runs from the reticular nuclei of the pons and medulla to the spinal nerves. It is involved in somatic (voluntary) motor control, like the rubrospinal tract, and also plays an important role in the control of autonomic (involuntary) functions in the Leg Cross Flexion-Extension Reflex such as the level of stress hormones released by the sympathetic/parasympathetic system, and regulation of breathing, heart rhythm, blood pressure).

Thalamus

The efferent fibers from the cortex and subcortical centers (including control circuits of the Leg Cross Flexion-Extension Reflex) of extrapyramidal system follow along the frontopontine, vestibulospinal, and rubrospinal tracts. After their cross-over (Foreli decussation) in the red nucleus they reach γ-motor neurons.

Control circuits of the Leg Cross Flexion-Extension Reflex reach the cortex, basal ganglia, and cerebellum which:

- regulate the excitation or inhibition of the lower motor neurons
- partially determine intensity of muscle contraction
- adjust the motor activity according to sensory-proprioceptive input.

Midbrain

Our study leads us to suppose that in the midbrain (tegmentum, pre-tectum, tectum) the descending pathway of the Leg Cross Flexion-Extension Reflex provides automatic 'assessment' of relative danger/safety and adjusts cortical programing to find a safer spot on the ground, causing abduction in the hip joint in part 3 of the pattern to move the knee and foot forward and away from the source of harm or danger.

Pons

Superior to the medulla, the pons consists of white matter with four cranial nerves attached to it. This part of the CNS includes tracts that conduct signals from the cerebrum down to the cerebellum and medulla, and tracts that carry sensory signals up into the thalamus, thus connecting the cerebellum to the pons and midbrain. Some nuclei in the pons relay signals from the forebrain to the cerebellum; others deal primarily with respiration, bladder control, hearing, equilibrium, eye movement, facial expressions, and posture. Through these connections the pons participates in control of equilibrium (vestibular system), coordination of motor, visual and auditory systems, breathing, and possible bladder activation when the Leg Cross Flexion-Extension response is triggered.

Medulla oblongata

This level of the brain is crucial for development of the Leg Cross Flexion-Extension Reflex and the timely emergence of crawling, walking, running, jumping, marching, and swimming in children, thus we describe it in more detail. The medulla oblongata is the inferior part of the brainstem between the spinal cord and the pons. It consists of: olives, pyramids, the roots of four cranial nerves (CN XII, IX, X, XI), the inferior cerebellar peduncle and the fourth ventricle.

The anterior of the medulla has two vertical bulges called pyramids. They contain axons of the corticospinal tract projecting from the cerebral cortex to the spinal cord. In the lower part of the medulla some of these fibers cross each other at the anterior median fissure known as the decussation of the pyramids (also called low motor extrapyramidal cross-over). Other fibers, the external arcuate fibers, originate from the anterior median fissure above the decussation of the pyramids and run laterally across the surface of the pons.

Beside the pyramids are two small oval lumps called olives formed by nerve fibers that connect them to the pons and the cerebellum. They have a role in motor learning.

The fourth ventricle, containing cerebrospinal fluid, forms the dorsal surface of the superior part of the medulla, where the four cranial nerves are rooted, and extends within the inferior part.

The task of the medulla oblongata in Leg Cross Flexion-Extension is control over the very complex physiological circuits and bio-dynamics of cross-lateral responses (crawling, walking, running, marching, jumping, climbing, swimming), also muscle tone, posture, respiration, heartbeat, and blood pressure. Patients with brain damage can still have functioning bodies, as long as the medulla oblongata is working. The Leg Cross Flexion-Extension test can show the level of medulla oblongata function in both healthy and persistent immobile states.

Spinal Cord

At the level of the anterior horn of the spinal cord the output follows via the frontopontine, vestibulospinal, and rubrospinal tracts, crossing over in the area of the red nucleus (Foreli) to reach the α- and γ-motor neurons of flexor and extensor muscles of contralateral limbs.

Muscular and Motor Response

Joints are controlled by two opposing sets of muscles, extensors and flexors. Thus, when a muscle spindle is stretched and the stretch response is activated, the opposing muscle group must be inhibited to prevent it from working against the resulting contraction of the homonymous muscle. This is accomplished by an inhibitory interneuron in the spinal cord.

At this level the motor neurons carry acetylcholine neurotransmitters: α-motor neurons release excitatory acetylcholine to cause a tonic contraction of protagonist muscle fibers; at the same time γ-motor neurons release inhibitory acetylcholine to cause lengthening of the antagonist muscle fibers.

In Leg Cross Flexion-Extension, one branch of the descending pathway innervates the α-motor neuron causing the homonymous muscle (same side quadriceps) to contract, producing the leg extension movement in the reflex. The other branch innervates the inhibitory interneuron, which in turn innervates the synapses connecting to the opposing muscle (same side hamstrings). Because the interneuron is inhibitory, it prevents the opposing α-motor neuron from firing, thereby protecting the opposing muscle against over-contraction and damage. The reciprocal excitation-inhibition innervates two opposing sets of muscles in synchrony: flexors and extensors causing the needed motor responses in the legs.

Note: the reticular activating system (RAS), a group of neurons coming from the thalamus and following along the brainstem, regulates arousal and inhibitory mechanisms and provides survival/protection in stress (stone under foot). It is composed of several neuronal circuits connecting the brainstem to the cortex. It plays the role of a filter for defining a stressor and its components and regulates thalamo-cortical activity and corresponding behavioral states. When the sensory input is 'not safe' the RAS system activates automatic survival responses, regulates the flow of sensory input, posture, equilibrium, motor coordination, head movements, orientation to the vertical plane, and postural adjustments needed for defensive responses. Disorders of the RAS can lead to too much arousal and hyperactive Leg Cross Flexion-Extension, causing constant running and jumping in children and adults. Known for regulation of attention, in the Leg Cross Flexion-Extension response it allows the choice of a safe spot for the opposite foot. There are links between RAS circuits and physiological pain pathways, so when the Leg Cross Flexion-Extension and Phillipson's Withdrawal are mixed, the Leg Cross Flexion-Extension can recruit nociception (pain receptor activation) instead of proprioception and compromise the whole Leg Cross Flexion-Extension circuit, causing poor cross motor coordination for walking and running (hyperactive and hypoactive, immobility), poor balance, frequent falls, injuries, as well as emotional and behavioral challenges.

IV. Neurophysiological Tracts for Leg Cross Flexion-Extension Reflex & Phillipson's Withdrawal

To produce Phillipson's Withdrawal, pain receptors in the skin send a rapid message though the afferent nerve fibers, bringing input to the posterior horn of the spinal cord. The signal then reaches out the ventral/front side of the spinal cord to muscles in both legs causing them to cooperate simultaneously: the leg that stepped on the stone must flex its knee and hip joints to lift the foot off; the opposite leg must immediately bear the full body weight (Andersen, Sonnenborg, Arendt-Nielsen, 2001; Clair, et al., 2009, Mileva, Green, Turner, 2004; see Figure 9). The input is transmitted across from the stimulated side of the body (right foot) to the contralateral anterior horn of the spinal cord and on to synapse with interneurons, which, in turn, excite or inhibit alpha motor neurons leading to the muscles of the contralateral limb (left leg) for extension of that leg and to the same side limb for pulling the injured right foot away from danger, all within 0.5 seconds (Solomon, Schmidt, Adragna, 1990). So, the Phillipson Reflex is elicited by nociceptor stimuli (pain, free nerve end receptors), not proprioceptive as in the Leg Cross Flexion-Extension Reflex. The main area for processing of nociception is at the level of first order neurons in the spinal cord (lumbar level). The biomechanics of this reflex are the following: in the ipsilateral leg (right), the flexors contract and the extensors relax to lift it off the ground. On the contralateral side (left), the flexors relax and the extensors contract to keep the leg extended and stable, since it must suddenly support the entire weight of the body. At the same time, some signals travel up the spinal cord and cause contraction of the contralateral hip and abdominal muscles to shift the body's center of gravity to the extended left leg. To a certain extent, the coordination of all these muscles, maintenance of equilibrium and feeling of the pain are mediated by the cerebellum and cerebral cortex (Spaich, Arendt-Nielsen, Andersen, 2005). However, the motor response in Phillipson's Withdrawal is generated mainly from the spinal cord, not from the brainstem (medulla oblongata) as in Leg Cross Flexion-Extension.

The Leg Cross Flexion-Extension response is elicited by a proprioceptive stimulus and uses a different nerve tract – reaching the posterior horn of the spinal cord and transmitting the input ipsilaterally to the same side

Figure 9. Pathway for A) Phillipson's Withdrawl and B) Leg Cross Flexion-Extension

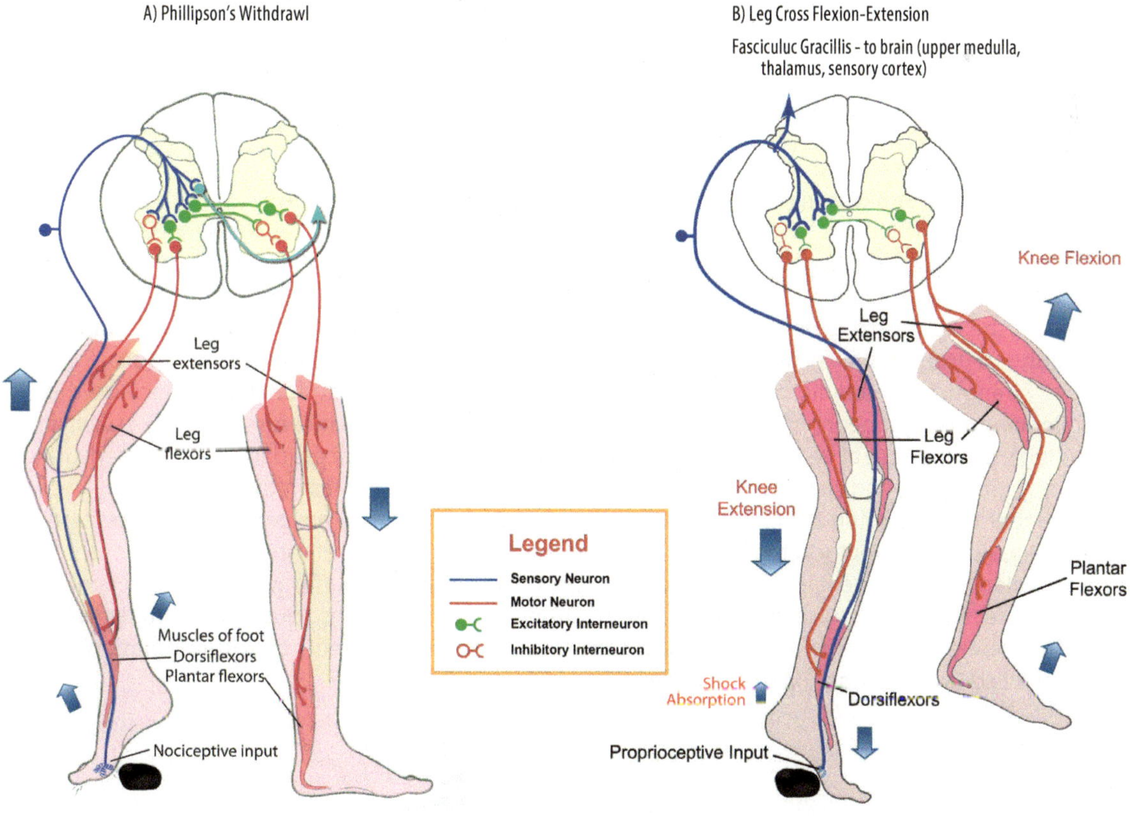

for the leg to extend (stepping even harder on the stone), and another sensory axon synapsing with second-order neurons sends the message to decussate (cross over) in the medulla oblongata (brainstem) to activate descending motor pathways in the opposite side to cause flexion, abduction and extension of the left leg so it can step away and ground the body in a safe spot. Thus, Leg Cross Flexion-Extension differs from Phillipson's Withdrawal as it is a more complex pattern, uses different ascending and descending nerve pathways and is processed mainly at a higher level of the CNS - medulla oblongata (Magnus, 1925). It is important to understand this difference as it should determine both the procedure used for repatterning and success of rehabilitation.

Conclusions

The bio-dynamics and neurophysiology of the Leg Cross Flexion-Extension Reflex schema is different from that of Phillipson Withdrawal:

• The Leg Cross Flexion-Extension Reflex is elicited by sensory-proprioceptive receptors. Organization of balance follows in first priority, and then other systems are activated (tactile/pain). The stimulus must reach the level of medulla oblongata for transmission across to the opposite leg. It also will reach the cortex level for organization of balance organization and registration of pain.

• Phillipson's Withdrawal is triggered by a nociceptor stimulus (tactile system/pain). The sensation of pain activates in priority the withdrawal away from the stimulus, and then other systems are activated for organization of balance. Contralateral stimulation of motor neurons in this reflex is reached in priority in the lumbar spine. The sensory neuron also sends signals up higher than the spinal cord to maintain balance and stabilize the body (this is why it is also called the Crossed Extension Reflex). On the level of the cortex this reflex allows the feeling and awareness of pain (Spaich, Arendt-Nielsen, Andersen, 2005).

MNRI® work with the Leg Cross Flexion-Extension Reflex activates pathways transmitting the sensory-proprioceptive stimulus, not those involved with nociception as in Phillipson's Withdrawal. This difference is essential for proper application of neuro-corrective techniques for patients.

V. Practical Application: The Leg Cross Flexion-Extension Repatterning Exercise

There are several reasons to repattern/pattern the Leg Cross Flexion-Extension Reflex pattern in children and adults:

• delay in development of the reflex pattern

• pathology in the reflex pattern caused by brain damage, cerebral palsy or a lesion in lower or upper motor neurons

• faulty connections in sensory and motor axons

• poor work of the alpha- and gamma- motor neurons (inadequate production of the GABA and dopamine, and acetylcholine neurotransmitters)

• nerve transmission mixing the Leg Cross Flexion-Extension and Phillipson's Withdrawal patterns

• negative compensation, trauma, PTSD.

In all these cases the repatterning/patterning of the physiological circuit is suggested. Training this reflex in infancy and childhood promotes myelination of second-order neuronal axons in the medulla and pons. Such training can also strengthen or rebuild these networks whenever they become dysfunctional. Because this part of the brain integrates the neuro-structural aspect of a reflex with its protective function, it is essential for survival strategies. In the case of Leg Cross Flexion-Extension, the ability to process danger on the sensory level and to supply grounding and stability when the body is destabilized, originates in the brainstem.

Processing danger and providing a survival strategy are automatic and entirely unconscious. If these functions are not accomplished efficiently in the brainstem and its midbrain, then higher more conscious levels of the brain are recruited. Development and maturation of higher-level skills then suffer when the cortex must be engaged for protection and survival and is not free to support voluntary, conscious and skillful activity. Children with developmental issues have a particularly urgent need (see Part I: Effects of Non-Integration) for repatterning of the Leg Cross Flexion-Extension Reflex.

The MNRI® Leg Cross Flexion-Extension repatterning exercise, based on its sensory-motor pathway, consists of four steps presented below (Masgutova, Akhmatova, 2004, 2007, 2010).

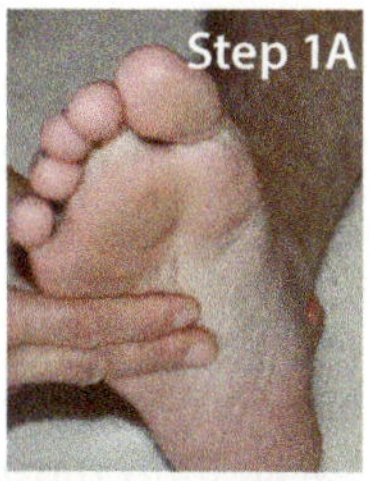

Figure 10. Points for stimuli on right foot arch and above the knee.

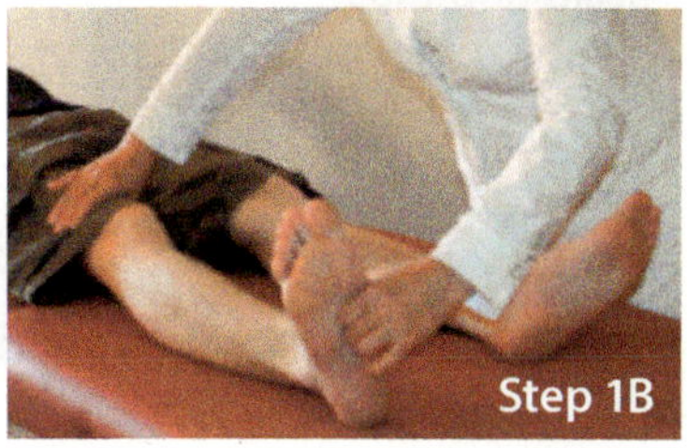

Step 1

Intervention A: Press on the center of the right foot's quadratus plantae muscle to simulate the sensory-proprioceptive stimulus of stepping on an object (See Figures 4 and 10). Hold for 5 – 7 seconds.

Reflex motor pattern: stretching and forced raising of the right arch (Sinelnikov, 1942) caused by stepping on an object.

Intervention B: Press and stretch on lower 1/6th of the quadriceps to trigger extension of the knee and hip (Figure 10). Hold for 5 – 7 seconds.

Reflex motor pattern: extension of right knee and hip, stepping harder on the object

Main muscles: Quadratus plantae, quadriceps, adductors, poplitus. Quadratus plantae, abductor hallucis, flexor digitorum longus and brevis

Opposing muscles: tensor fasciae latae, sartorius, extensor hallucis brevis, extensor digitorum brevis

Optional intervention: Perform steps A and B simultaneously.

Step 2

Intervention: Bring left knee and hip into flexion with slight medial adduction. (see Figure 11) Hold 5-7 seconds.

Reflex motor pattern: Flexion of opposite (left) knee and hip

Main leg muscles: semimembranosus, semitendedinosus, biceps femoris, gracilis, sartorius

Opposing leg muscles: gastrocnemius, popliteus

Facilitating leg muscles: quadriceps

Main hip muscles: psoas major, iliacus, rectus femoris, gluteus medius, gluteus minimus, adductor brevis

Opposing hip muscles: iliopsoas, tensor fasciae latea, pectineus, adductor longus, adductor brevis

Facilitating hip muscles: gluteus maximus, biceps femoris, semimembranosus, semitendinosus, adductor magnus.

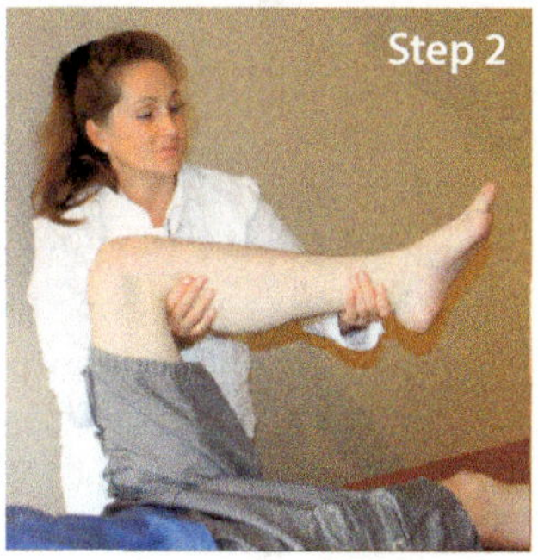

Figure 11. Bring the left knee and hip into flexion.

Step 3

Intervention: Rotate the hip, bringing the knee away from the midline. (see Figure 12) Apply traction first between the hip and knee, then between the knee and ankle. Hold each stretch 5-7 seconds.

Reflex motor pattern: Lateral rotation of hip, knee abduction

Main muscles: gluteus medius, gluteus minimus, adductor magnus, piriformis, gemeli, iliopsoas

Opposing muscles: sartorius, gracilis

Facilitating muscles: tensor fasciae latae, rectus femoris, adductor magnus, biceps femoris

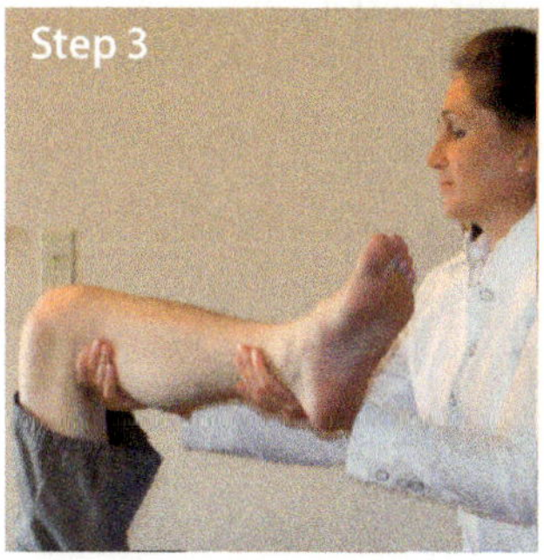

Figure 12. Rotate the hip laterally by pulling the knee to the side, stretch from hip to knee and stretch from knee to ankle.

Step 4

Intervention: Extend the flexed knee/hip and stretch the whole leg downward, holding it above the ankle joint. (see Figure 13, below)

Reflex motor pattern: extension and grounding of the flexed knee/hip to support the body's weight on a safe spot.

Main muscles: quadriceps femoris, gluteus maximus, semimembranosus, semitendinosus, adductor magnus

Opposing muscles: gluteus medius, gluteus minimus, adductor brevis

Facilitating muscles: biceps femoris, gracilis, sartorius, psoas major, iliacus, rectus femoris, gluteus medius, gluteus minimus, adductor brevis mm.

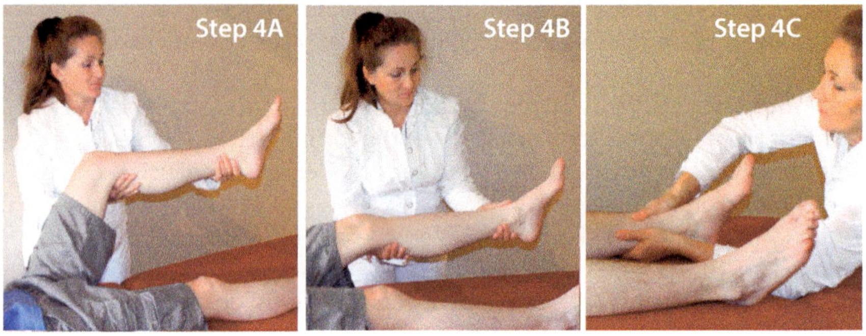

Figure 13. Extend and stretch the whole leg.

Step 5

Repeat steps 1 – 4, beginning with stimulus on the left foot.

This exercise can be done 2-3 times on one foot/opposite leg, and then on the other foot/opposite leg. Do a total of 5-7 repetitions, working slowly.

The use of this Leg Cross Flexion-Extension exercise procedure has led, during many years of the MNRI® practice, to highly significant changes in cases of severe dysfunction, such as cerebral palsy and brain damage, allowing them to learn crawling and walking. It allows individuals with autism to reset an improper hyperactive Automatic Gait pattern (toe walking, constant running, and jumping) and learn normal, matured reciprocal walking. It allows children with dyslexia to acquire differentiation functions and athletes to gain precision in their cross-lateral movements. This is a pattern which supports proper neurological pathways and CNS plasticity, and thus makes our gross movements more goal-oriented, controlled, precise, flexible, and easy.

References

Andersen, O. K., Sonnenborg, F. A., Arendt-Nielsen, L. (2001). Reflex receptive fields for human withdrawal reflexes elicited by non-painful and painful electrical stimulation of the foot sole. *Clin Neurophysiology*, 2001 Apr;112(4): p. 641-9.

Andersen, O. K., Sonnenborg, F., Matjacić, Z., Arendt-Nielsen, L. (2003). Foot-sole reflex receptive fields for human withdrawal reflexes in symmetrical standing position. *Exp Brain Res*. 2003 Oct;152(4): p. 434-43. Epub 2003 Aug 7.PMID: 12904932.

Anokhin, P. K. (1968). *Biology and neurophysiology of a conditioned reflex*. (Russ.). Moscow, Russia: Medicina.

Barashniev, J. I. (2001). *Prenatal neurology*. (Russ.). Moscow, Russia: Triada-X. 640 p.

Barker, R., Barasi, S., Neal, M. (2008). *Neuroscience at a glance*. Oxford, UK: Blackwell Publishing.

Calais-German, B. (2007). *Anatomy of movement*. Revised Edition. Seattle, USA: Eastland Press. 318 p.

Campbell, St. (2004). *Watch me grow! A unique, 3-dimensional, week look at baby's behavior and development inside the womb*. London, UK: Carroll and Brown Publishers Limited.

Clair, J. M., Okuma, Y., Misiaszek, J. E., Collins, D. F. (2009). Reflex pathways connect receptors in the human lower leg to the erector spinae muscles of the lower back. *Exp Brain Res*. 2009 Jun;196(2):217-27. Epub 2009 May 30.PMID: 19484229.

Cohen, H. (1999). *Neuroscience for Rehabilitation*. Philadelphia, PA: Lippincott, Williams & Wilkins.

Ellrich, J., Steffens, H., Schomburg, E. D. (2000). Neither a general flexor nor a withdrawal pattern of nociceptive reflexes evoked from the human foot. *Neurosci Res*. 2000, May; 37(1):79-82.

Haines, D. E. (2002). *Fundamental neuroscience for basic and clinical applications*, Fourth Edition. Philadelphia, PA, USA: Elsevier Saunders.

Knikou, M., Kay, E., Schmit, B. D. (2007). Parallel facilitatory reflex pathways from the foot and hip to flexors and extensors in the injured human spinal cord. *Exp Neurol*. 2007. Jul;206(1): 146-58. Epub 2007 May 8.PMID: 17543951.

Luria, A. R. (1969). *High Cerebrum Functions of the Human and their disorders in cases of local damages*. Moscow, Russia: Moscow State University.

Lundy-Ekman, L. (2002). *Neuroscience. Fundamentals for rehabilitation*. 2nd Edition. Forest Grove, OR, USA: Pacific University.

Magnus, R. (1925). Animal posture. *Croonian Lecture*. University of Utrecht. June, 16, 1925. P 1-15. www.jstor.com

Masgutova, S. K., Akhmatova, N. K. (2004, 2007). *Integration of dynamic and postural reflexes into the whole body movement system*. Scientific edition: Prof. N. Akhmatova. (Pl.). Warsaw, PL: MINK.

Masgutova, S. K., Akhmatova, N. K. (2010). *Repatterning and integration of dysfunctional and pathological reflexes*. (Pl.). Warsaw, PL: ISMI.

Mileva, K., Green, D.A., Turner D. L. (2004). Neuromuscular and biomechanical coupling in human cycling: modulation of cutaneous reflex responses to sural nerve stimulation. *Exp Brain Res*. 2004 Oct; 158(4): p. 450-64. Epub 2004 Jun 18.PMID: 15221175.

Penfield, W., Rasmussen, T. (1950). *The cererbral cortex of man: a clinical study of localization of function*. New York, NY, USA: Macmillian.

Pavlov, I. P. (1960). *Conditioned reflexes: An investigation of the physiological activity of the cerebral cortex*. (Anrep G.V., D. Sc. Trans., 1960). New York, NY, USA: Dover Publications Inc. (First published in 1927)

Purves, D., Augustine, G. J., Fitzpatrick, D., et al., editors. (2001). *Neuroscience*. Second Edition., Sunderland, MA, USA: Sinauer Associates.

Sechenov, I. M. (1960). *Selected physiological and psychological works* (S. Belsky, Trans., K. Koschtoyants, Ed.). Moscow, Russia: Foreign Languages Publishing House. (Original work published 1935).

Semionov, E. V. (1998). *The atlas of human anatomy*. Moscow, Russia: SEV-Press.

Sinelnikov, P. D. (1942). *Atlas of human anatomy. Book – V*. Moscow-Leningrad, Russia: Medgiz.

Solomon, Schmidt, & Adragna. (1990). *Human anatomy & physiology* (2 ed.). Atlanta, GA: Saunders College Publishing. p. 470.

Wikipedia: http://en.wikipedia.org/wiki/Crossed_extensor_reflex

Sonnenborg, F. A., Andersen O. K., Arendt-Nielsen L. (2000). Modular organization of excitatory and inhibitory reflex receptive fields elicited by electrical stimulation of the foot sole in man. *Clin Neurophysiol*. 2000 Dec;111(12):2160-9. PMID: 11090767.

Spaich, E. G., Arendt-Nielsen, L., Andersen, O. K. (2005). Repetitive painful stimulation produces an expansion of withdrawal reflex receptive fields in humans. *Artif Organs*. 2005 Mar;29(3):224-8.PMID: 15725222.

To our dear children and adults! We wish for you the proper functioning of the neurophysiological pathways of the Leg Cross Flexion-Extension Reflex circuit and for it to serve you well in sports, dancing, and your everyday life of walking, running, climbing, swimming and jumping. – Authors

Auditory Processing Disorders: Why Use MNRI®? One Audiologist's Perspective

Dr. Leah Light, Au.D., MNRI® Core Specialist & Instructor, Brainchild Institute, Hollywood, FL, USA

As an audiologist specializing in Auditory Processing Disorders (APD), I never expected to find myself involved in using manual techniques with my patient population. How I discovered MNRI® and why it has made such a difference in the lives of my patients is the topic of this article.

Dr. Leah Light

Auditory processing disorders are quite complex. We simplify things by breaking them down into a few main categories: auditory decoding, auditory integration, and prosodic deficits. Children with auditory decoding problems have 'blurry hearing.' They mis-hear one sound for another, which can lead to great confusion (e.g. 'bite' for 'bike'). Kids with auditory integration deficits have trouble getting sound information going to the right side of the brain to work together with sound information going to the left side of the brain. In essence, these children miss the big picture and are often heard saying, "I don't get it." Children with prosodic deficits have trouble understanding non-verbal speech cues such as voice intonation, facial expressions, and body language. They often make poor eye contact and have trouble understanding emotional aspects of speech (e.g. sarcasm, humor, idioms).

As a 'rookie' audiologist in the area of APD in the late 1990s, I initially set out to retrain the brain using different kinds of computer software programs and brain training games. After a few years, I became frustrated by the fact that these interventions seemed to be only modestly helpful for a handful of my kids. Why did most of the APD children not benefit? They certainly met the criteria based upon my evaluations. Maybe there was something I was missing…something that came before higher order auditory processing, which takes place in the auditory cortex at the top of the brain. Where were the fundamentals of auditory processing?

I began to take a closer look at some preliminary information in my files. Most of the children I evaluated presented with other types of problems besides auditory processing difficulties. Their case history forms showed that they had handwriting problems and eye tracking issues. They had difficulty with balance, attention problems, and poor postural control. "What does this have to do with auditory processing?" I wondered. I could not ignore the fact that the same sets of problems were staring me in the face, case after case.

I began to think a lot about what was happening at the brainstem and cerebellum, the 'ground control' centers of the brain. After all, early in evolution, our 'reptilian brain' worked only through reflexive pathways; there was no cerebrum for thinking. Many of the symptoms I read about in my patients' case histories were related to these lower brain areas, such as abnormal Acoustic Stapedial Reflexes. This reflex is a middle ear muscle reflex that is mediated by the brainstem when a loud sound is presented to the ear, thus protecting the inner ear from acoustic trauma. It acts as a buffer against loud noises. In exploring the functions of the brainstem, I learned about many other reflexes, including our earliest primary motor reflexes that are present at birth such as rooting, sucking, grasp, and the startle or Moro Reflex, to name a few.

Through an organization called Brain Gym®, I was introduced to Dr. Svetlana Masgutova in 2004. At that time, I took my first MNRI® course, Postural and Dynamic Reflex Integration. My eyes opened widely to learn what my intuition had been telling me. We need to go back to the beginning of our brain development and work forward, just as the brain develops from caudal to rostral, or from bottom to top. Every new link in the chain of development attaches to something that comes before it. At the very foundation of our neurological development is the primary reflex system, which serves as the anchor for higher levels of brain development. If there is nothing to anchor to, new learning will not 'stick.' This is what I saw happening to the children I treated with APD. I was simply aiming too high. It was like trying to teach algebra to a child who does not understand basic addition and subtraction. They just cannot get it.

I began my concentrated MNRI® journey in 2009. After five years of thinking about what I had initially learned from Dr. Masgutova, I realized that so many children with APD needed to have their primary reflexes integrated first. I became an avid 'MNRI® Groupie,' taking every course that I could attend while managing to run my neuro-audiology practice. In 2013, I completed my training as a Core Specialist in MNRI® and have begun my training as an instructor in various aspects of the method, including the Visual-Auditory Reflexes integration, where I feel I can be most useful.

In addition to being an MNRI® student, I have had the opportunity to be a researcher with Dr. Masgutova, performing audiological studies on selected children participating in MNRI® Family Conferences and analyzing physiological changes with auditory evoked brainstem responses (ABRs) and Acoustic Stapedial Reflex testing. Although the test data is still being analyzed and is presently inconclusive, it appears that most of the children tested had normal ascending (afferent) auditory system functioning through the brainstem but abnormal descending (efferent) auditory system functioning. There is evidence that the efferent or motor response system largely influences our perceptual intelligence. Motor development will be delayed or inefficient when primary reflexes are poorly integrated. Without appropriate motor responses to sensory stimulation, we may lack some of the ability to fine tune our sensory perception, ultimately affecting our higher order skills such as auditory processing and language development.

Today, I insist that MNRI® testing be a part of every auditory processing evaluation I perform. Research has shown links between the somatosensory system or touch, the auditory system, and the amygdala in the emotional center of the brain through the non-classical auditory pathways (Moller and Rollins, 2002). The phylogenetically older, non-classical auditory pathways have been shown to be active in children up until about 20 years of age. After this time, the more recently evolved, classical auditory pathways take over, providing more finely tuned information to the auditory cortex. It has been proposed that delayed maturation of the non-classical auditory pathways may be involved in some types of neurobehavioral disorders in children. This may explain why parents often report improved auditory and language skills when addressing the somatosensory system using MNRI® techniques.

In closing, I would like to add that one of the most exciting aspects of being a part of the MNRI® team of specialists is that I am joined by wonderful, caring, and dedicated professionals from many different disciplines, all bringing their expertise to our classes and meetings for a collective consciousness about human development. Being a part of the MNRI® team has been one of the most rewarding and fulfilling interventions I have been able to provide for the children and families who visit my practice. I look forward to joining Dr. Masgutova and my MNRI® teammates as we continue to make miracles happen and offer hope and new beginnings to children and families in the U.S. and around the world!

References

Moller A. R, Rollins P. (2002) The non-classical auditory pathways are involved in hearing in children but not in adults. *Neurosci Lett* 2002; 319: 41-44.

I would like to give a special thanks to Dr. Masgutova for enlightening me with her brilliant work and for helping so many. For families already benefitting from MNRI®, congratulations and may your loved ones continue to blossom. Please help us to spread the word about MNRI®. For families just beginning their MNRI® journey, your hard work and dedication will be transformed into benefits that will last a lifetime. Enjoy your rewards, and I wish you all good health, strong body and mind, and integrated hearing! – Dr. Leah Light

Reflex Integration for Post-Trauma Survival and Recovery

Svetlana Masgutova, Ph.D, SMEI, Orlando, Florida, USA

To the children and parents of Newtown, CT with whom we worked and will continue to work, and to our volunteer professionals dedicated to providing traumatic stress recovery.

Introduction

Svetlana Masgutova, Ph.D

Ten percent of American women and five percent of American men will experience Post-Traumatic stress disorder (PTSD) during their lifetimes.[1] In a national study of 4,023 adolescents ages 12 to 17 almost half reported some type of traumatic experience: sexual assault (8%), physical assault (22%), and witnessing violence (39%).[2] Neglect, domestic violence, sexual, physical and emotional abuse, terrorism, war, natural disasters, and mass shootings appear in our headlines every day. Trauma and its dire fallout is prevalent even in early childhood!

We live in harsh times and effective intervention for victims of trauma is urgently needed. Although the military and others are exploring alternative interventions such as EMDR (Francine Shapiro),[3] Somatic Experiencing (Peter Levine),[4] various forms of energy work, psychotherapy for emotional adjustment and medication for symptom relief[5] are considered the primary treatments for PTSD. The long-term experience of the author, working internationally with trauma release processes in disaster survivors, has brought her to the conclusion that the current primary treatments are not adequate and do not address the unconscious, neurosensorimotor needs of those with PTSD. While it can be helpful to address rational and emotional functioning, to work with conscious and unconscious motives, to provide pharmaceutical support and to ease tension in the muscles and tendons, these interventions do not always succeed in creating a stable on-going process that orients PTSD clients toward new perspectives and a new positive sense of life.

Both acute and chronic symptoms of trauma originate in the protective and survival mechanisms that govern the reflexive freeze and fight or flight responses. The author contends that the grip of PTSD on trauma victims can only be released by working with these non-rational non-emotional automatic responses where they originate in the nerve networks of reflex systems in the interbrain (basal ganglia, thalamus, amygdala, insula,

[1] http://psychcentral.com/lib/facts-about-ptsd/000662

[2] http://www.ptsd.va.gov/professional/pages/ptsd_in_children_and_adolescents_overview_for_professionals.asp

[3] Carlson, J., Chemtob, C.M., Rusnak, K., Hedlund, N.L, & Muraoka, M.Y. (1998). Eye movement desensitization and reprocessing (EMDR): Treatment for combat-related post-traumatic stress disorder. *Journal of Traumatic Stress,* 11, p. 3-24.

[4] Leitch, M. L., Vanslyke, J. and Allen, M, Somatic experiencing treatment with social service workers following hurricanes. *Social Work,* Volume 54, 9–18, Number 1, January 2009.

[5] http://psychcentral.com/lib/an-overview-of-treatment-of-ptsd. Accessed 12/10/13.

and limbic system) and brainstem. It is crucial for practitioners to understand and work with reflex patterns as they not only govern our actions, behavior, and thoughts in traumatic stress but are also a key resource for positive survival and neurodevelopment.

This article discusses effects of stress in normal and traumatized brains (Part 1), presents a model of the neurophysiology of stress based on reflex system functioning (Part 2), and proposes original PTSD Assessment and treatment protocols developed by the author and used effectively for over 25 years in work with trauma survivors (Parts 3 and 4). Part of a larger body of work known as Masgutova Neuro-Sensory-Motor and Reflex Integration (MNRI®), the MNRI® PTSD protocol was used with great success by the author during her many years of work with patients evacuated from traumatic events such as: the Chernobyl nuclear disaster (1986-1996), the Baku conflict (1990-1991), the earthquake in Armenia, the train crash in Ufa (1989), the Chechen War (1996-1999), conflicts in Israel (2001-2005), suicidal individuals, and survivors of fire, explosions, and abuse. MNRI® trauma recovery work uses the inherent sensory-motor links of reflex circuits to reroute brainstem neurons (lower motor neurons of the extrapyramidal nerve net system) from their anchors in an 'unbearable past trauma' to the experience of a 'safe here and now,' oriented toward positive survival in a safe healthy future. Part 5 presents in detail the results of using the MNRI® trauma protocol with 174 children and adults directly or indirectly involved in the tragic Sandy Hook School shooting in Newtown, CT on December 14, 2012.

Part 1

Stress and Trauma: the Normal Stress Response

The early work of Walter Cannon (1931) revealed that the autonomic nervous system manages two general states of function in the body as a normal course of daily life: 1) The Non-Alarm State and 2) The Alarm State. A healthy individual exposed to an isolated traumatic event will experience normal activation of neural responses designed to protect the body and enhance survival. Generated by the sympathetic nervous system, impulses from the lower brain travel to the organs and tissues, causing all the symptoms we recognize in stress: sweating or gooseflesh, dilated pupils, rapid heartbeat, shallow breathing, internal pressure, trembling, pallor, and perhaps even nausea. The release of adrenalin in the HPA-stress-axis (hypothalamus-pituitary-adrenal) pumps energy out to the limbs to support escape or self-defense (fight or flight). The enteric nervous system, popularly referred to as the 'brain in the gut,' acts to regulate digestion and elimination in ways that enhance survival.

The parasympathetic nervous system activates a relaxation response when danger has passed and one can safely return to the Non-Alarm State. In cases of overwhelming terror when no escape and no hope for survival seem possible, the parasympathetic system releases the same neurotransmitters and hormones that help the body relax, but in much larger amounts. This brings on the freeze response. In car accidents, during rape or under threat at gunpoint, protective 'freezing' enables one to collapse, faint, or dissociate from one's body.

Following a traumatic event other symptoms may persist for a few days: intense bad memories, flat effect, muscle tension, tremors, unstable gait, lack of grounding, poor balance, anxiety, and feelings of isolation. Normally, when the traumatic event is over or the source of trauma removed, a healthy individual will gradually phase out of the stressed state and move into a state mediated by the parasympathetic nervous system, recovering fully within a few days or weeks. This cycle of transition from Non-Alarm State to Alarm State and back to Non-Alarm State is governed by the autonomic nervous system with its sympathetic, enteric, and parasympathetic subsystems.

Dysfunctional Sympathetic- and Parasympathetic-Dominant States

Hans Selye's (1974) work showed that when the sympathetic nervous system dominates and the parasympathetic nervous system does not restore normal function with regular frequency, not only can the body sustain internal physiological damage, but its overall ability to function effectively and maintain emotional and behavioral stability begins to diminish. If this occurs during infancy when primary motor reflexes are developing, the underlying neuro-sensory-motor function can be challenged, making it difficult for reflexes to mature and integrate. Even after all the primary infant motor reflexes have integrated, if excessive sympathetic dominance persists, then low-level trauma, normally managed effectively by the body's near-term survival system, can trigger reflex patterns to resurface. In other words, prolonged intermittent or chronic stress can compromise maturation and integration of primary infant motor reflexes and, even in adults, can cause previously integrated primary infant reflexes to resurface.

Under conditions of sympathetic dominance, an otherwise normally functioning sensory system may become hypo- or hypersensitive or stop working altogether. Because tactile sensitivity is the first link in the three part reflex circuit connecting a sensory stimulus with brain processing and motor response, dysfunction here will lead directly to dysfunction in the reflex circuit.

Because the central nervous system regulates and directs internal and external responses based on the input it receives, skewed input will result in a skewed response. Outgoing directives will then produce motor responses that are dysfunctional (hypo- or hyperactive) or pathological (a-reflexive, incorrect, or reversed). To the outside world, an individual with a challenged sensory system can appear emotionally and behaviorally dysregulated, when in fact the quality of their emotional or behavioral response to a stimulus accurately reflects the hyper or hyposensitivity of the challenged sensory system.

An under reactive sensory system can cause the parasympathetic system to over engage the Non-Alarm state, resulting in a very high tolerance threshold and failure to react to a small or even moderate amount of stimulation. While the purpose of the parasympathetic system is to allow restoration and growth to ensure long-term survival, if the body does not understand when sensory input is dangerous or life threatening, it will not engage sufficient mobilization or immobilization strategies to ensure survival. In other words, people with under reactive sensory systems can be at risk for harm in unsafe and life threatening situations and are more likely to engage in dangerous activities.

Over reactivity, in contrast, can cause the sympathetic nervous system to engage its Alarm State mobilization system excessively, resulting in timidity and potentially depriving the individual of restorative and developmental benefits governed by the parasympathetic system. A scratch can feel like a deep cut, a light touch like an aggressive push, an everyday sound like nails scratching on a chalkboard, or simple eye contact like a glaring stare. In either case, the sensory system misinterprets the environment and sends distorted sensory signals to the central nervous system.

When Stress Becomes Trauma

Our bodies and nervous systems are designed to tolerate the normal stressors of life. In a healthy individual the autonomic nervous system engages adequate mobilization of protection to ensure near-term survival. Its sympathetic and parasympathetic subsystems function in a symbiotic fashion, regulating variations of Alarm and Non-Alarm states, with one more or less dominant depending on the level of danger or stress presented by external conditions.

With the experience of an overwhelming level of stress beyond the ability of the body and nervous system to cope, we enter the realm of trauma. Then the natural manifestations of a normal stress response may be replaced by more serious symptoms. In acute stress disorder one may experience panic reactions, chronic anxiety, confusion, dissociation, lack of trust, insomnia, poor sleep patterns, and difficulty with basic self-care, work, and relationships. Post-traumatic stress disorder (PTSD) usually involves painful flashbacks of an unbearable event, hyper-vigilance, and emotional numbness, as well. Individuals who lack resilience because of previous conditions can experience PTSD combined with other socio-psychological and psychiatric disorders. For example, co-morbid PTSD can exist together with depression, addictions, or other anxiety disorders. Complex PTSD is more likely to manifest in individuals who have experienced long term trauma such as prolonged childhood abuse. They are often diagnosed with personality or dissociative disorders and have extreme behavioral, emotional, and/or mental difficulties (Herman, 1992). In this article, the term post-traumatic stress will be used to indicate the presence of symptoms in the aftermath of a traumatic event in cases where an official diagnosis of PTSD or DESNOS (Disorder of Extreme Stress Not Otherwise Specified) has not been made.[6]

[6] http://psychcentral.com/lib/types-of-ptsd/000666 accessed 12/10/13

Part 2

Stress and Trauma in the Reflex System: The MNRI® Model

Another way to describe the cycle of stress response and return to normalcy (HPA-stress-axis) is through the activity of our neuro-sensory-motor reflex circuits and their survival functions. The mechanisms of protection and survival triggered by any stress or trauma are genetically present in our bodies and realized through automatic responses and instincts, which include motor and postural reflexes.

Reflexes are automatic encoded programs of the nervous system responding to specific sensory stimuli. Their three part neural circuits involve a chain reaction from sensory stimulation through brain processing to motor response. Given to every one of us inherently, from generation to generation (also epigenetically), they are our keys to survival. Automatic, unconsciousness, and constant, they guarantee stability for the nervous system in stress and distress (Van der Kolk, Roth, Pelcovitz, Sunday, Spinazzola, 2005).

Beginning in early gestation and continuing through the first two years of life, each individual goes through developmental stages of emergence, activation, maturation, and integration of sensory-motor patterns associated with infant reflexes. The entire reflex system is crucial for future motor, sensory, emotional, and cognitive development (Pavlov, 1927/1960). Here, however, we are primarily concerned with its function in assuring protection and survival. Three reflexes in particular call for special attention: Tendon Guard, Moro, and Fear Paralysis. These reflex patterns seem to be the most affected by stress and are most vulnerable and dysfunctional in individuals with traumatic stress and PTSD.

The Tendon Guard Reflex (TGR) is a generalized response of the brain-body system to stress. It manifests at three levels: Foot Tendon Guard, Core Tendon Guard, and Upper Tendon Guard (at the neck). The TGR is a life long reflex present in infants, children, and adults, elicited when stress enters the environment. Because the Core Tendon Guard Reflex (CTGR; see Figure 1) integrates and initiates all other Tendon Guard Responses, it will be the focus in this text.

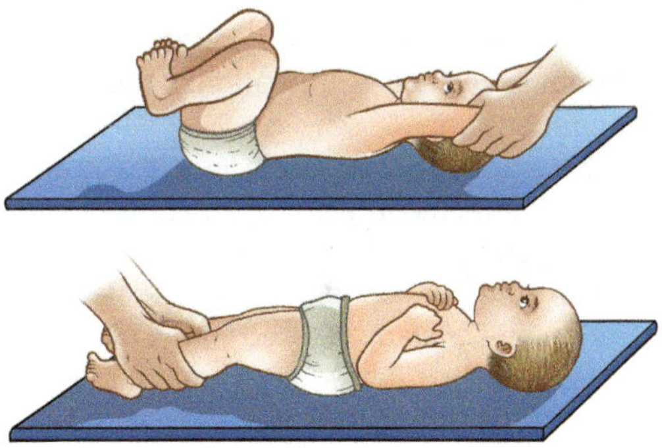

Figure 1. Core Tendon Guard Reflex

Protective task: The CTGR brings information to the brainstem/RAS/thalamus about possible danger, discomfort, or risk. Lower motor neurons of the extrapyramidal nerve net system activate the whole organism through the HPA-stress-axis, providing protection and survival by contracting flexors or extensors in the body. Depending on the level of stress, either a freeze or fight or flight response will result. The HPA-stress-axis governs the production of adrenalin, cortisol, and noradrenalin, depending on how much survival activity is needed to adapt to a physical change or injury. Using the biomechanics of the Tonic Labyrinthine Reflex (flexion-extension responses to gravity), and working through the body's Golgi system and through its links with the HPA-stress-axis, the CTGR regulates the level and intensity of excitation and inhibition needed to keep the body safe. Under intense stress, if the CTGR cannot cope alone, then the Fear Paralysis or Moro and/or other reflexes are triggered to provide increased emergency protection.

The pattern of flexion in the freeze response is characterized by automatic withdrawal of the core, holding breath, muscle contraction, visual convergence, and inhibition of movement. In contrast, the pattern of extension in the fight or flight response manifests in backward trunk extension, visual divergence, and readiness for action. When freezing enables one to hide effectively from a danger, when flight enables one to escape, or when fighting back protects one from harm, then the protective response has done its job and only lasts for a short time.

Data gathered on survivors of trauma and individuals diagnosed with PTSD who received treatment with MNRI® show that the CTGR remains affected for 4 to 12 months following a traumatic event in 100% of the cases. Their hyperactive response to the CTGR test indicates over-production of stress hormones by the HPA stress axis and a chronic state of alarm in their Golgi system. Excessive amounts of stress hormones can destroy myelin, particularly in the lower motor neurons (Sechenov, 1995), brainstem (Bremner, et al. 1995; Bremner,

2001), and cerebellum (Sherin, Nemeroff, 2011), affecting automatic life functions such as circulation, blood pressure, breathing, and digestion, as well as reflexes and instincts.

The Moro Embrace Reflex is a protective response to a sudden vestibular or proprioceptive experience of loss of stability or change in the position of the head relative to gravity. (See Figure 2.) Through its neural connections with the sympathetic nervous system it is linked to the fight or flight response. The earliest phase emerges at nine weeks in utero. The first two developmental phases of this reflex are mature at birth and remain active up to the third or fourth month, when it integrates into the whole body motor and proprioceptive systems. In adults this reflex activates movements of the limbs from core to periphery and periphery to core when a stable body position is lost.

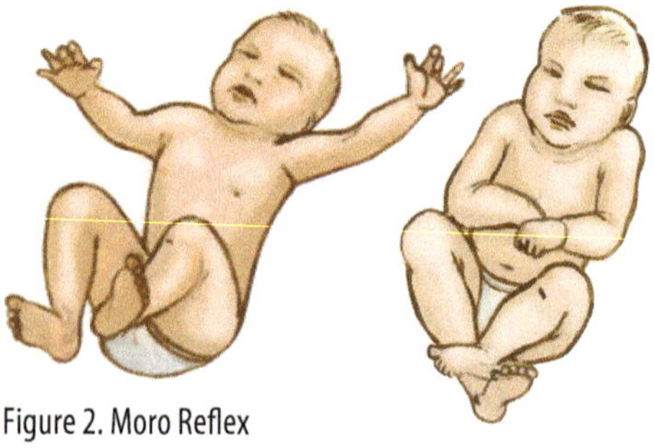

Figure 2. Moro Reflex

Protective task: this reflex trains our flexion-extension mechanisms to protect the body. Rapid flexion or extension can prevent a fall by helping us to regain equilibrium. Curling into a ball (flexion) can prevent injury if we do fall. A sudden backward or downward movement activates the posterior-anterior semicircular canals in the labyrinths, sending the HPA-stress-axis alarm system a command to produce stress hormones to strengthen the body in case of a fall. Phase 1 is characterized by disorganization or chaos. It signals the loss of stability. The infant opens the core of their body, extends arms and fingers, inhales rapidly, and holds their breath. In phase 2 order is reestablished as the whole body flexes and withdraws, activates protection and regains wholeness. In phase 3 the infant accepts comfort by embracing a person or soothing object, releases stress, and becomes ready to communicate. When fully matured, the Moro Reflex pattern supports the development of bonding and trust, integration of thought and movement, self-regulation, organization, concentration, and comprehension.

Data from MNRI® experiences with the effects of post-traumatic stress and PTSD show that the Moro Reflex pattern is affected in 100% of cases. Its response to both the MNRI® testing procedures and triggering in real life remains hyperactive for 3 to 14 months after a traumatic event and can continue longer if no stress-release work is done. This again is evidence of the chronic state of alarm created by the overworking HPA-stress-axis.

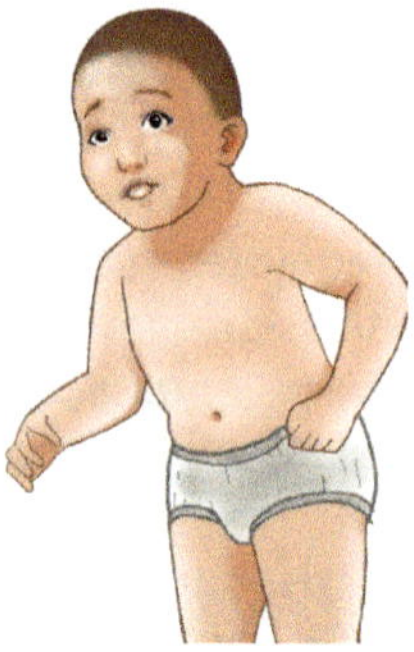

Figure 3. Fear Paralysis Reflex

Fear Paralysis Reflex (FPR) known in adults as the startle response, is active from birth throughout life (Baldacara, Jackowski, Andreoli, Mello, 2011). Although it is often confused with the infant Moro Reflex, FPR differs from the Moro in both stimulus and response. The FPR is triggered, not by vestibular, but by unexpected, sudden, intense tactile, auditory, or visual input. A further difference is that Fear Paralysis is linked to the freeze response, unlike the Moro, which is linked to fight or flight. (See Figure 3.)

FPR appears at the 9th week of gestation in flexion of the core as a withdrawal response. At the 28th week, this reaction appears as core withdrawal with abduction of limbs and bringing the head into forward flexion and an upright relationship to gravity. At birth, the reflex is already well developed. At 10 days, the infant clearly demonstrates this response to sudden sound, unexpected and unpleasant touch, and flashing light. Normally it is integrated into the whole body movement system by age three and eventually becomes the adult startle response.

Protective task: this reflex activates a strong sympathetic response followed immediately by a parasympathetic one, which organizes inhibitory processes of the nervous system, freezing the whole organism and its movements (Kaada 1986). The core withdraws (abdominal rectus muscles contract) with slight lateral limb abduction and breath is held on inhalation. This pattern may be accompanied by slight core flexion, forward tilt of the head, slight flexion in the elbow joints, extension at knees, blinking (corneal reflex), a gasp or deep sigh, and momentary fright (freeze response). It serves to economize resources and preserve them in case of longer lasting stress. Fear Paralysis not only activates its characteristic motor pattern, but also affects respiration and heartbeat, preparing the infant to coordinate physiological and emotional fear in one response for effective protection of the whole body. The freeze response also mitigates bodily sensations of discomfort, injury, and pain. All these manifestations dissipate quickly in a healthy person when the stimulus is removed. In individuals with post-traumatic stress and PTSD, this reflex is dysfunctional due to 'transmarginal inhibition' (Asratian, 1983), caused by unbearably intense or long lasting danger, pain, fright, or loss.

The effect of prolonged dysfunctional sympathetic and parasympathetic states is very clear in the way victims of every type of post-traumatic stress and PTSD demonstrate lack of integration in their Tendon Guard, Moro, and Fear Paralysis Reflexes. Symptoms of PTSD such as nervous blinking, staring, poor balance, anxiety, phobias, insomnia, fatigue, lack of trust, avoidant behaviors, depression, dependency, vulnerability to emotional overwhelm, freezing, dissociation from one's body, inability to cope with stress, and lack of resilience are all characteristic of dysfunctional Fear Paralysis and Moro Reflexes. These behavioral issues can be accompanied as well by symptoms that undermine physical health: motion sickness, digestive issues, overworked adrenal glands, and weakened immune systems.

Negative Protection

MNRI® practitioners refer to the consequences of pathology, dysfunction, and hypo- or hyperactivity in reflex expression as negative protection. If a sympathetic-dominant state of hyper-arousal persists over time, muscle tension may remain in the body, resulting in excessive reactivity. Tactile, visual, and auditory systems may become hypersensitive and one may experience the feelings, sights, and sounds of everyday life as invasive and intolerable. When the conditions of normal life are stressful one may tend to withdraw or to enlist defensive compensations. In contrast, if the system has been totally overwhelmed, a parasympathetic-dominant freeze response may persist, resulting in low muscle tone and inactivity. Individuals in this state may tend toward passivity, have difficulty resuming the activities of normal life and be unable to defend themselves in dangerous situations. Excessive or insufficient protection becomes negative protection which is present when hyperactive reflexes interfere with self-regulation, voluntary control, cognitive function, and intentional behavior, and when hypoactive reflexes provide insufficient protection from harm or danger.

Among the elements of the reflex system most critical for overall protection is tactile sensitivity. Because the external stimulus for so many reflexes is tactile, proper regulation of tactile sensitivity is essential for the brain receptors, whose job is to process incoming signals and organize appropriate motor responses. Skewed input leads to skewed output, again a form of negative protection.

Stress Anchored in a Reflex Pattern

Another way that trauma can affect reflex system functioning is through damage to innate neural pathways at the moment of stress. Suppose, for example, that an infant burns their hand by grasping a hot object and the grasp motor pattern becomes anchored to an association with pain. Thereafter, the brain avoids pain by refusing to elicit this motor response, thus arresting the training and maturation of the Grasp Reflex motor pattern and delaying the next stages of manual skill development. Accidental trauma has formed a vicious cycle: grasp motor pattern ==> memory of burn ==> brain interpretation (grasp causes pain) ==> refusal to respond to a safe stimulus that should elicit the grasp response ==> protection against grasp and poor ability to grasp (dysregulation in the balance of excitatory and inhibitory mechanisms) ==> poor muscle tone regulation in palm, fingers and arms ==> poorly developed qualities of grasp movement ==> poorly developed manual skills ==> poor academic performance (drawing, writing, typing) ==> poor self-esteem ==> fear of judgment from others concerning poor skills ==> tension and fear ==> extra effort ==> exhaustion ==> loss of motivation ==> refusal to participate in fine motor hands-on tasks ==> failure to meet expectations ==> negative relationships with teachers and parents, self-punishment and depression, behavioral challenges ==> regression (crying, infantile behavior, complaining, looking for sympathy, hiding the problem, blaming others or objects), and aggression.

A dysfunctional reflex can also be elicited as the result of being anchored to a positive outcome. Suppose an automatic Moro response (core withdrawal, holding breath, visual convergence, narrowing of visual and auditory field of attention) enabled a child to survive a life-threatening fall in the past thanks to the following sequence: sudden loss of stable body position in space (vestibular-proprioceptive stimulus), active Moro Reflex pattern while falling, core withdrawal, body maintained a round shape, child rolled and sustained no injury. The brain-body system anchored the Moro Reflex, in another vicious cycle, as ==> chronic core withdrawal (leaning too far forward), excessive visual convergence and narrow visual and auditory attention span ==> poor postural, visual, and auditory control ==> protective brain response ==> fear ==> greater tension in core withdrawal ==> limited visual and auditory perception ==> emotional instability ==> fear based behavior ==> effort to control emotions ==> guarded style of communication ==> trying hard ==> exhaustion ==> fear based self-judgment and negative attitude ==> fear based relationships ==> poor self-realization. In this kind of response, Moro is mixed with Fear Paralysis and leads to confusion in brain processing, causing fear or panic as a habitual response to stressors, large or small.

What is common to both of the above examples is that the repetitive firing of the reflex elicits protection and survival responses in a dysfunctional way. Whether the reflex pattern resonates with a negative experience that led to harm or a positive experience that enhanced survival, these responses function on the physiological level of survival and connect with the emotional response of fear or a conscious thought of fear. This is why work with reflexes requires highly trained specialists to analyze the complexities of problematic reflex pattern expression and to choose appropriate corrective tools.

Post-Traumatic Stress and Dysfunction in the Reflex System

Unbearable or prolonged post-traumatic stress affects all three parts of the reflex circuit (Sechenov, 1960; Asratian, 1983; see Leg Circuit drawing):

1) compromised transmission of sensory stimuli: receptor thresholds, electrical conductivity in the sensory neuron, and release of neurotransmitters (glutamate, substance-P, gene-related peptide) at the end of the neuron

2) inefficient or faulty processing in the synapses: release of sensory neurotransmitters to the motor neuron and its interneuron, electrical-chemical relations between sensory and motor neurons, and initial transmission in alpha and gamma motor neurons that should take place on five levels of the nerve system: spinal cord, medulla, midbrain, thalamus and cortex. Conductivity in the interneuron can also be affected, resulting in a poorly regulated release of inhibitory neurotransmitters (dopamine and GABA) at the beginning of the gamma-motor neuron.

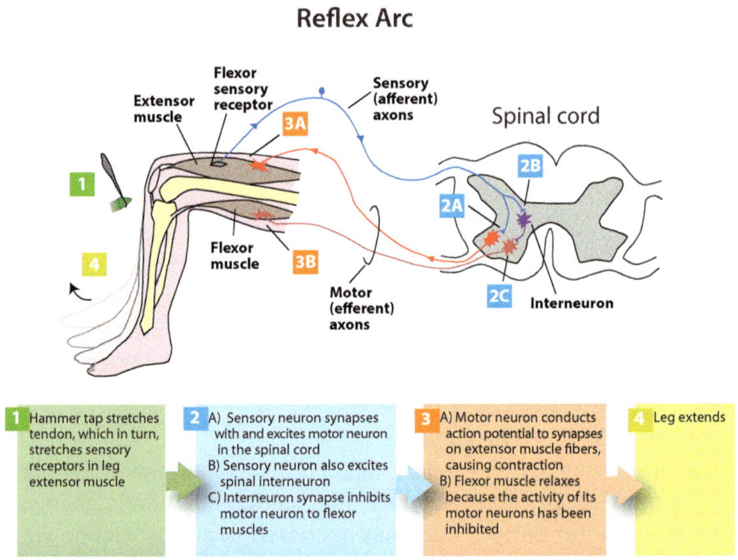

Figure 4. Quadriceps stretch (myotatic) reflex arc.

3) faulty motor neuron function: electrical conductivity in the excitatory alpha-motor neuron connecting with muscle fibers to contract them (agonist muscle), and in the inhibitory gamma-motor neuron connecting with muscle fibers to lengthen them (antagonist muscle), resulting in poor coordination of the motor or postural response.

Studies using BAER (brainstem auditory evoked response), EEGR testing, and brain imaging (Pilecki, Masgutova, Kowalewska, Masgutov, Akhmatova, Sobieszczanska, Kalka, 2012) with individuals with PTSD led us to posit a correlation among reflex dysfunctions, disharmonious brain wave spectrum imaging (Pilecki, Masgutova, Kowalewska, Masgutov, Akhmatova, Sobieszczanska, Kalka, 2012) and immune system function (Masgutova, Akhmatova, Lebedinskaya, 2013). We conclude that the most critical aspect of reflex circuit dysfunction occurs in interneuron connections where metabolic dysfunctions resulting from lack of dopamine and GABA undermine regulation throughout the reflex circuit. Poor regulation in the reflex circuit affects the whole nervous system – sympathetic/parasympathetic, lower motor neurons, the RAS (which should differentiate between safe and unsafe incoming stimuli), the midbrain's territorial and self-preservation instincts, and the thalamus. The amygdala and insula are then unable to accurately decode input and, because their first priority is protection, they trigger the alarm state in the whole organism. This situation ends in a vicious cycle by triggering the HPA-stress-axis, which produces an excess of adrenalin and cortisol that further prolongs the alarm state. Under such chronic stress, the protective aspect of a reflex circuit becomes hyperactive, overworks the HPA-stress-axis even further, and exhausts the endocrine, immune (Masgutova, Akhmatova, Lebedinskaya, 2013), and nerve systems.

Part 3

MNRI® Assessment

The MNRI® Reflex Parameters Assessment (Masgutova, 2011; see also the article in this book on Assessments) identifies the level of maturity and integration of thirty reflexes and then serves as a baseline for a program designed to improve reflex circuit functioning. A trained MNRI® practitioner will administer specific stimuli in multiple tests for each reflex to determine its level on a scale of 0 – 20 ranging from deep pathology to above average maturity and integration. Testing a reflex as an unconditioned reaction is possible only in infants and young children. In older children and adults, reflexes themselves are not checked. The MNRI® Assessment measures an older child's or adult's imitation of an 'ideal' inherent genetically encoded motor response to a particular stimulus. Techniques used to test the Moro Reflex are given below as an example. In every day life, the Moro response might be elicited by sudden vestibular and/or proprioceptive stimulation such as in a fall, a boat ride in rough waves, or a traffic accident. In the testing situation, a professional supplies the trigger.

Moro Assessment (to ensure safety should be done only by trained personnel):

1. *Inclining the head 30 degrees backwards.* The patient is lying on their back. Their head is carefully lifted forward, then suddenly tilted backwards 30 degrees. This unexpected vestibular stimulation can cause a Moro response in the limbs or whole body if the reflex is hyperactive.

2. *Straightening Legs.* The patient is supine with legs bent at the knees and feet on the table or floor. The therapist rapidly straightens their legs, supporting them with hands under the knees. This test checks the response to proprioceptive stimulation.

3. *Moro-2 Posture for protection* (for mobile patients). The patient is supine. The therapist asks them to move into the Moro 2 pattern: flexing the core, bringing all limbs into the body center, making fists, and exhaling. Next the practitioner holds all four limbs from the center in a way that limits flexion and asks the patient to resist this pressure with their core, arms, and legs (checking the patient's ability to flex core and limbs simultaneously) using 20-25% of their strength. The practitioner notices the strength of the response and whether all components of the Moro 2 motor pattern are correct.

4. *Moro-1 Posture for Core opening* (for mobile patients). The patient is supine in flexion as in #3. Now the practitioner holds all limbs from the outside in a way that prevents extension and asks the patient to resist this pressure by attempting to open the core, arms, and legs (checking the patient's ability to extend core and limbs). The practitioner notices the strength of the response and whether all components of the Moro 1 motor pattern are correct.

Five Assessment Parameters Reflex basic patterns are evaluated in each of five parameters:

- ***Sensory-motor circuit:*** Circuit integrity is evaluated in terms of levels of sensitivity to the stimulus (normal, hypersensitive, hyposensitive, or a-reflexive) and levels of activity in the response (normal, hyperactive, hypoactive or a-reflexive).
- ***Sequence and direction:*** Evaluation is based on the fidelity of the motor response to the inherent genetically encoded sequence and pattern associated with a specific stimulus. The response can be correct, incorrect, mixed, or a-reflexive.
- ***Timing and speed:*** The motor response must begin in a fraction of a second from the moment of sensory stimulation. It must also complete its pattern quickly enough to fulfill its protective function. A delayed or slow reflex response could result in injury. The response can be normal, too fast, too slow, or a-reflexive.
- ***Intensity:*** The intensity of a reflex is the amount of physical energy supplied by a system of muscles and ligaments in response to a stimulus. Muscle tone regulation governing the response should reflect the intensity of the stimulus. The response can be normal, hyper-tense, hypo-tense, or a-reflexive.
- ***Symmetry:*** Symmetry can be seen in the bi-lateral pattern, direction, timing, and intensity of the reaction. The response can be symmetrical, asymmetrical, mixed or a-reflexive.

Each reflex is analyzed with regard to these five parameters, with a possible score of 4 points for each parameter, yielding a scale of 0 – 20 points. According to the algorithm created by Anna Krefft (2007),16 – 17.75

points represents the norm and below 10-11.75 the reflex pattern is at a dysfunctional level. Below is a detailed example of scoring for the Fear Paralysis Reflex (FPR).

Fear Paralysis Assessment in Detail The subject is lying supine on a massage table. Scores are based on observation of the subject's response to stimulation of the tactile system by light taps on five different points of the body and stimulation of the auditory system by taps on the table (after a demonstration and explanation of the test). Taps are administered twice, once with eyes open and once with eyes closed, in the following locations:

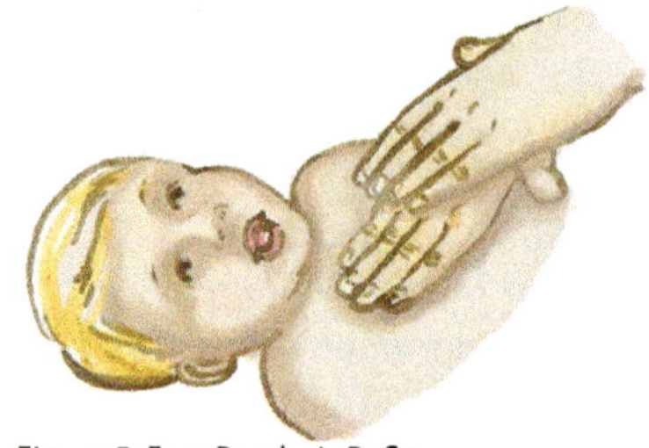
Figure 5. Fear Paralysis Reflex

1) middle of the sternum (See Figure 5.)
2) on the table surface on both sides (auditory FPR stimulus)
3) insertion of the deltoid muscles on both arms simultaneously
4) insertion of the triceps muscles on both arms simultaneously
5) sides of both hips
6) middle of the arches of both feet

Scoring the five parameters of responses to Fear Paralysis stimuli: In each case the specialist observes four features for each parameter and assigns one point (+) for a mature or normal response and zero points (–) for a response indicating pathology, dysfunction, or low level of development:

Sensory-motor circuit: Four features (total possible score - 4 points):

1) Response to tap on the middle of the sternum. Breathing: a) no change, normal breathing (+), b) rapid inhalation and/or holding breath, fear (-)

2) Response to tap in 1) above. Blinking (corneal reflex): a) no blinking (+), b) hyperactive blinking, stress, fear (-)

3) Response to tap in the area of insertion of the deltoid muscles on both arms simultaneously. Breathing: a) no change, normal breathing (+), b) rapid inhalation or holding breath, fear (-).

4) Response to tap in 3) above. Dilation of the pupils: a) no change (+), b) dilation of pupils, anxiety, shock (-). Note: in hyperactive FPR the latency of response for dilation of pupils is 1 second, in tendency for shock/fear in 1-1½ seconds, and in panic attack response 2 seconds.

Sequence and direction: Four features (total possible score - 4 points):

1) Response to tap on the middle of the sternum. Muscle tone regulation: a) little or no increase of tone in the body (+), b) visible hyperactive motor response by strong startle reaction (-).

2) Response to simultaneous taps at the insertion of the deltoid muscles on both arms. a) little or no increase of tone in the body (+), b) visible hyperactive motor response by strong startle reaction (-).

3) Response to taps on the table surface – next to the head (auditory FPR stimulus). Muscle tone regulation: a) little or no increase of tone in the body (+), b) visible hyperactive motor response by withdrawal startle reaction or by Moro motor pattern instead of FPR, stress, fear (-).

4) Response to taps as in 3) above with subject in Moro 1 position. Muscle tone regulation: a) little or no increase of tone in the body (+), b) visible hyperactive motor response by withdrawal startle reaction or by Moro motor pattern, stress, fear (-).

Intensity of response (muscle tone regulation): Four features (total possible score - 4 points):

1) Response to tap on the middle of the sternum. Muscle tone: a) slight muscle tone increase seen in stronger force of next exhalation (+), b) strong muscle tone increase in whole body, sympathetic response such as flushing or sweating (-).

2) Response to simultaneous taps at the insertion of the deltoid muscles on both arms. Muscle tone: a) slight increase in muscle tone seen in stronger force of next exhalation (+), b) strong increase in muscle tone in upper or whole body, sympathetic response such as flushing or sweating (-).

3) Response to simultaneous taps at the insertion of the triceps muscles on both arms. Muscle tone: a) slight muscle tone increase is seen in a stronger force of next exhalation (+), b) strong muscle tone increase in upper or whole body, sympathetic response such as flushing or sweating (-).

4) Response to simultaneous taps on sides of both hips. Muscle tone: a) slight muscle tone increase is seen in a stronger force of next exhalation (+), b) strong muscle tone increase in upper or whole body, sympathetic response such as flushing or sweating (-).

Timing and speed: *Four features (total possible score - 4 points):*

1) Response to tap on sternum. Latency time before response: a) normal period of latency (30 m/sec) before slight increase in muscle tone seen in next prolonged exhale (+), b) sympathetic response: rapid increase in muscle tone, flushing, sweating or dissociating from body; or excessively slow or no increase of muscle tone in the body (-).

2) Response to simultaneous taps at the insertion of the deltoid muscles on both arms. Latency time before response: a) normal period of latency (30 m/sec) before slight increase in muscle tone seen in next prolonged exhale (+), b) sympathetic response: rapid increase in muscle tone, flushing, sweating or dissociating from the body; or excessively slow or no increase of muscle tone in the body (-).

3) Response to simultaneous taps at the insertion of the triceps muscles on both arms. Latency time before response: a) normal period of latency (30 m/sec) before slight increase in muscle tone seen in next prolonged exhale (+), b) sympathetic response: rapid increase in muscle tone, flushing, sweating or dissociating from body, or excessively slow or no increase of muscle tone in the body (-).

4) Response to simultaneous taps in the middle of the arches of both feet. Latency time before response: a) normal period of latency (30 m/sec) before slight increase in muscle tone seen in next prolonged exhale (+), b) sympathetic response: rapid increase in muscle tone, flushing, sweating or dissociating from body; or excessively slow or no increase of muscle tone in the body (-).

Symmetry: *Four features (total possible score - 4 points):*

1) Response to simultaneous taps at the insertion of the deltoid muscles on both arms. Symmetry of the response: a) slight bilaterally symmetrical increase in muscle tone (+), b) no symmetry in muscle tone increase and movements (-).

2) Response to simultaneous taps at the insertion of the triceps muscles on both arms. Symmetry of the response: a) slight bilaterally symmetrical increase in muscle tone (+), b) no symmetry in muscle tone increase and movements (-).

3) Response to simultaneous taps on both sides of the hips. Symmetry of the response: a) slight bilaterally symmetrical increase in muscle tone (+), b) no symmetry in muscle tone increase and movements (-).

4) Response to simultaneous taps in the middle of the arches of both feet. Symmetry of the response: a) slight bilaterally symmetrical increase in muscle tone (+), b) no symmetry in muscle tone increase and movements (-).

Note: each location is tapped only once. Observations relating to the five different parameters and their features are based on the response to each single tap.

This is an example of the precise observation, evaluation, and scoring of one reflex motor pattern. Data from such an Assessment is processed by a computerized New MNRI® Assessment Program (Akhmatov, 2013). The final result is a thorough analysis of reflex pattern development in the subject, showing strengths, weaknesses, and tendencies in nine clusters of motor, sensory, emotional, and cognitive function:

1) Upper Limbs Reflex Patterns
2) Tonic Reflex Patterns
3) Righting Reflex Patterns
4) Lower Limb Reflex Patterns
5) Gross Motor Reflex Patterns
6) Oral-Facial, Visual, and Auditory Reflex Patterns
7) Reflex Patterns Supporting Protection and Survival
8) Reflex Patterns Supporting Curiosity and Cognition
9) Reflex Patterns Supporting Emotional Stability and Maturation

In over 35 years of clinical observation by the author and our MNRI® team of specialists dealing with Post-Traumatic stress and PTSD, these individuals have consistently demonstrated extreme hyper-reactivity in their responses to all features in all parameters of the Fear Paralysis Assessment. Interestingly, they present hyperactive pupil dilating with an unusually long latency period of 1½ to 2 seconds between the stimulus and the response. This delay, caused by *transmarginal inhibition* (Sechenov, 1935/1960; Asratian, 1963), is related to the numbing effects of panic attacks, shock, and breathless fear; conditions in which the nervous system responds more slowly. Parts of the nervous system taking longer to organize feedback are then in conflict with the hyper-excitation of survival functions in the reflex system.

Part 4

MNRI® Therapy: the MNRI® PTSD Protocol

MNRI® reflex integration techniques and exercises developed by the author for post-traumatic stress and PTSD were originally used for non-verbal work with children in states of panic, shock, and numbness, who could neither reason nor verbalize their emotions and feelings. Some of the children had even lost normal speech (for example a teenager who began to stutter after witnessing killings in the Baku conflict, a non-verbal child with CP who suffered in a fire, a child with Asperger syndrome run over by a truck, and babies and children with Down syndrome burned by vandals in fires during the war in Chechnya). These techniques were also used for immobile individuals (one wounded in the Ufa train explosion, one with MS who witnessed the killing of another person). For over 25 years this combination of exercises has proved in real life to be an effective therapeutic tool with victims of various natural and human-caused disasters, tragedies, and catastrophes. Later they were organized into a protocol with a sequence of specific steps. The MNRI® PTSD Protocol has achieved exceptional results with both children and adults in verbal (with a psychotherapeutic component) and non-verbal versions.

Steps of the MNRI® PTSD Protocol:

Setting a Goal: The goal: in the subject's own words, the goal would express something similar to: "To let go of fear and negative thoughts of physically or emotionally painful events.... I am open to courage and ready to be a winner. I am open to good health and growth under any circumstance." The subject is asked to notice where they feel the goal in their body, to touch that place and rate the goal on a scale of 0-10 (0-1 poor, and 9-10 high). How comfortable is it to contemplate the goal? How congruent is it with their current internal state? How realistic is it?

Subjective experience of the goal: the professional asks, "How does your goal make you feel? (positive, negative, trust or no trust that the goal is possible to realize)". Then, further questions concerning overall functions: "When you are tuned into your goal and feel an emotional response to it, can you breathe.... see (use your vision).... listen and hear.... think clearly.... act rationally... pause... feel your emotions congruently? These parameters are then used by the subject to create a baseline point of orientation for comparing their current state to the following two states: "I am still in the traumatic event, then and there," and "I am present, I have reached the here and now." The subject's position close to or far away from those two states can be observed at the beginning, through this process, and at the end of the session.

The subject can also choose to combine MNRI® reflex integration with emotional release and strengthening (led by a professional psychologist or psychotherapist trained in MNRI® PTSD techniques) to transform negative sensory-motor anchors into positive ones or to create new positive anchors. The strategy of the verbal emotional work (not discussed in this report), if included, must follow harmoniously with the steps of the non-verbal techniques.

Non-verbal MNRI® PTSD Protocol:

1. Release of Negative Protection. The task of this group of exercises is to diminish improper triggering of too many reflex patterns causing negative protection and poor coping.

1) Tactile Stroking: to activate generalized tactility, to bring a subject still experiencing post-traumatic stress back to the 'here and now,' to awaken them from shock, and to activate bonding so they can feel protected and safe.

2) Fear Paralysis: to let go of extreme fear, anxiety, worry, stress, and distress; to activate tolerance mechanisms for sensory perception, to re-orient them from the perception of neutral stimuli as dangerous. MNRI® research shows that this work supports the release of fear and anxiety through its possible effect on the thalamus, amygdala and insula (Masgutova, S., Masgutov, D., Akhmatov, 2013).

3) Moro: to strengthen the subject's own inner-resources for protection, to bring them back from physical and emotional withdrawal, and to open them to the surrounding stimuli and the world on a proprioceptive level. Integration of Moro also works through the thalamus, amygdala, and insula to support the release of fear and anxiety and enable the patient to see alternatives.

Note: These two exercises follow each other only in this format, where it is important to differentiate two distinctly separate defense mechanisms: Fear Paralysis based on the freeze response and Moro based on fight or flight. In post-traumatic stress and PTSD, these two responses overlap each other, confusing the sympathetic system, basal ganglia, amygdala and its insula, and causing challenges in decoding and interpretation of real or imagined dangerous stimuli.

2. Differentiation. The motor patterns of these reflexes belong to the group of asymmetrical or cross-lateral movements more associated with conscious control, rational thinking, and involvement of the cerebral cortex than the homologous, symmetrical startle responses described above. The task of these differentiation techniques is to activate neurophysiological circuits that allow differentiation of body parts on the kinesthetic level. Physical differentiation then becomes the foundation for higher levels of differentiation: past from present, present from future, here from there, and traumatic memory from positive future possibilities.

4) Leg Cross Flexion-Extension: for activation of proprioceptive pathways that govern cross-lateral lower limb motor patterns, supporting differentiation on the kinesthetic level, in this case starting with the lower limbs and hips.

5) Sequential Arms Opening: for release of the tendency to protect the core with flexed arms as in the Moro 2 position. The upper part of the body tends to be mostly under conscious control. When individuals are consciously over-focused on past trauma, hyperactive protection makes them unable to release and open their arms. Opening the arms also serves as a metaphor for opening to relationships, to new experiences, to life, and to a meaningful future.

6) Eye Tracking: for horizontal and near-far tracking, convergence-divergence, and peripheral vision and to reestablish regulation in areas of the midbrain responsible for territorial instinct and self-preservation (tegmentum, tectum, and pre-tectum) that become hyperactive during stress. Eye tracking strengthens binocular vision for tracking, focus recovery, and other visual processing unrelated to vigilance and scanning for danger.

Optional: Hands Pulling: to reestablish the coordination of core flexion with visual convergence and trunk extension with visual divergence in a visual system that has become hyper vigilant and unable to alternate appropriately between central and peripheral focus.

3. Grounding and Stability. This group of reflex techniques activates the functions of the midbrain, cerebellum, and cortex to bring back the physical feeling of stability with body weight distributed evenly on the ground. A subject can then be grounded not only physically, but also grounded in the present moment and in one's self.

7) Hands Grasp: for confidence in one's ability to 'hang on' to what is important, to protect oneself, and to feel safe in one's personal space

8) Foot Grasp: for feeling grounded and safe in one's personal space

9) Grounding: same as above

10) Foot Metatarsi/Big Toe Rotation: for proper body weight distribution on the ground and the feeling of stability

4. Adrenalin-Cortisol Release. The task of these techniques is to detoxify the overloaded HPA-stress-axis of excessive stress hormones

11) Babinski: stimulates lower motor neuron and redirects their activity away from hypersensitive tactile receptors. More efficient proprioceptive processing then leads to more accurate decoding of sensations, better grounding, and better differentiation of pain from touch. This raises the pain threshold, helps to release the negative anchoring of pain to touch and limits secondary benefit from pain. Stimulation of the Babinski sensory motor points also serves as a painkiller.

12) Perez: activates micro-movement in the spine and supports normal lordotic curvature to enhance circulation of cerebra-spinal fluid, thus helping to release toxins and stored stress hormones from the body and calming the HPA-stress-axis.

13) Abdominal Sleep Posture: works with HPA-stress-axis to relax the body and calm the emotions.

5. Brain Activation. (Masgutova, S., Masgutov, D., Akhmatov, 2013) The task of these reflex patterns is to activate fast brain waves (alpha and beta), supporting active consciousness, easier focusing, and intentional choice making.

14) Galant: releases stress, activates alpha and beta waves.

15) Hands Supporting: activates positive self-protection and alpha and beta waves.

16) Trunk Extension: awakens the body, enhances concentration and focus, and activates alpha and beta waves.

17) Foot Tendon Guard: awakens the Golgi system throughout the body, releases muscle tension, and activates alpha and beta waves, thus enhancing focus and awareness.

6. Completion:

18) Tactile Stroking: Strengthening the feelings of presence, bonding, and belonging and the awareness of a renewed capacity for real life and real interactions.

Completion of work with the Goal

Revisit the goal: "Where do you feel the stress in your body now that treatment is completed?" The subject is asked to rate the goal again on a scale of 0 – 10 as they did before treatment, "How comfortable is it to contemplate the goal? How congruent is it with their current internal state? How realistic is it?"

Subjective experience of the goal: the professional asks, "What does your goal make you feel now? (positive, negative, trust or no trust that the goal is possible to realize)". Then, further questions concerning overall functions: "When you are tuned into your goal and feel an emotional response to it, can you breathe…. see (use your vision)…. listen and hear…. think clearly…. act rationally… pause… feel your emotions congruently?" The subject compares these responses to those given to the same questions before treatment. Finally, the subject evaluates any changes in their position relative to the following two states: "I am still in the traumatic event, then and there," and "I am present, I have reached the here and now."

Part 5

Assessment, Clinical Observations, and Analysis of the Effectiveness of the MNRI® PTSD Protocol for Individuals with Post-Traumatic Stress and PTSD

Reflex Pattern Profiles in Neurotypical Children (Control Group)

To identify deviations from normal reflex development and expression in children and adults experiencing traumatic stress, PTSD and other disorders, MNRI® uses the Exemplary Reflex Pattern Profile created through research on 30 reflex patterns in children with neurotypical development (1989-2013). This exemplary profile serves as the control group. The MNRI® Reflex Parameters Assessment (for 30 reflex patterns) was evaluated and statistically verified using Prof. Anna Krefft's Algorythm (NewKrefft Method, 2007). It demonstrated that the norm on a scale of 0 to 20 is 16-17.75 points. The large bank of MNRI® reflex assessment data gathered over a period of twenty-four years made possible a comparison of two groups of neurotypical children ages 2 to 19, one earlier (Group A, 1989-2006, 780 children) and one later (Group B, 2007-2013, 730 children). Children in the earlier group

Table 1-A, B, C. Reflex pattern profiles in children with neurotypical development (age 2-19).

Group A: Motor Responses within Sagittal Plane of the Body
730 individuals (1990-1996), 780 individuals (1997-2013)

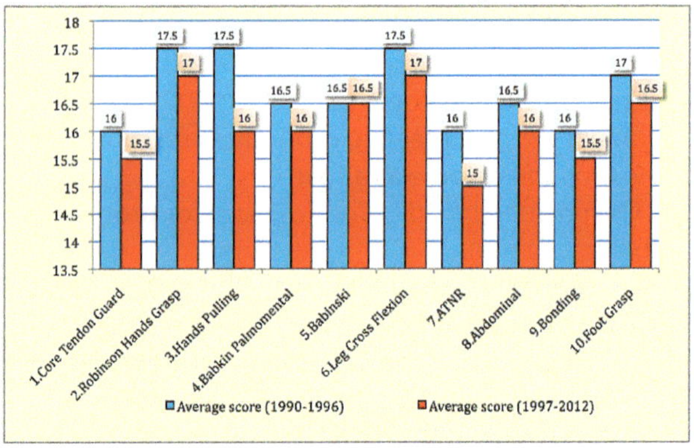

Group B: Motor Responses within Horizontal Plane of the Body
730 individuals (1990-1996), 780 individuals (1997-2013)

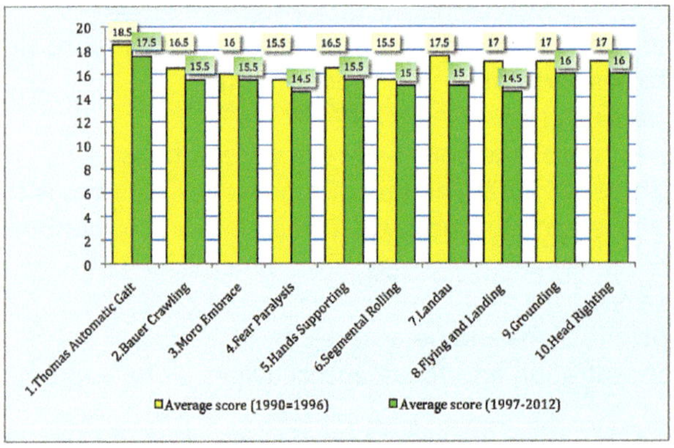

Group C: Motor Responses within Dorsal Plane of the Body:
730 individuals (1990-1996), 780 individuals (1997-2013)

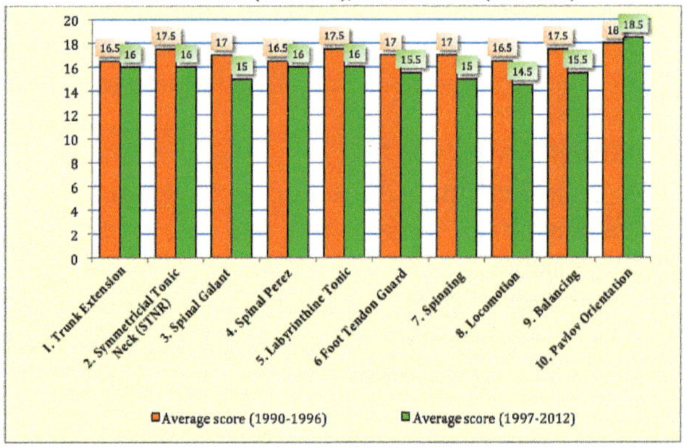

achieved average scores indicating normal reflex pattern development (16.77 points). In Group B (2007-2013) the average score was only 15.80, below the norm of 16-17.75 and below the average of 16.77 in Group A (S. Masgutova, 2008-2013, lectures). The statistical validity of both results is significant and equals 0.56 in the earlier group and 0.78 in the later, with $p<0.001$. This data is presented in the two bar graphs above, with 30 reflexes grouped according to body planes and motor coordination systems (sagittal, horizontal and dorsal).

The significantly lower scores were in reflexes directly responsible for protection: Fear Paralysis (14.5 in Group B vs. 15.5 in Group A), Core Tendon Guard (15.5 vs. 16.0) and Moro (15.5 vs. 16.0). Group B's scores for other reflex patterns influencing postural control, emotional maturation, and stability were also below the norm and below the average of Group A: Hands Supporting (15.5 vs. 16.5), ATNR (15 vs. 16), Bonding (15.5 vs.16) and Spinal Galant (15 vs. 17). Lower levels of development in these reflex patterns will inevitably lead to increased activation of defensive mechanisms in the organism and inhibit overall sensory-motor development, affecting motor ability, behavior, and cognition. These results suggest that lower levels of maturation in the reflex system may be one important reason for greater challenges in the development of protective mechanisms and less resilience under stress found in the younger generation by professionals dealing with problems of PTSD and personality growth (Masgutova S., Masgutov D., 2014).

Reflex Pattern Profiles in Children with Post-Traumatic Stress

Post-trauma MNRI® Assessment data from 340 children (1989-2013) show extremely challenged protective reflexes at levels 4 to 7 points below the norm: CTGR - 11.99, Moro and Fear Paralysis - 8.5, Hands Supporting – 9.75, and ATNR – 8.5 points. The same overall tendency is evident for other reflex patterns as well. The three bar graphs below compare neurotypical children from Group B and children with post traumatic stress before and after treatment with MNRI®.

After 1-3 treatment sessions with the MNRI® PTSD Protocol these children demonstrated significant improvement in all reflex patterns: CTGR improved from 11.99 to 12.32 points, Moro and Fear Paralysis from 8.5 to 12.5 points, Hands Supporting 9.75 to 13.5, and ATNR from 8.5 to 12.5 points. All these reflexes moved from below to above the boundary between dysfunctional and functional. The same tendency for positive change is seen in other reflexes as well.

Table 2-A, B, C. Reflex patterns profile in of children with post-traumatic stress (age 4-20; 340 individuals) compared to children with neurotypical development (age 2-19; 780 individuals).

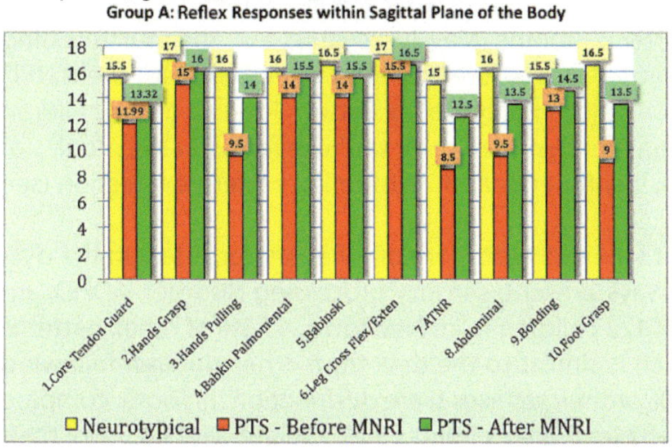

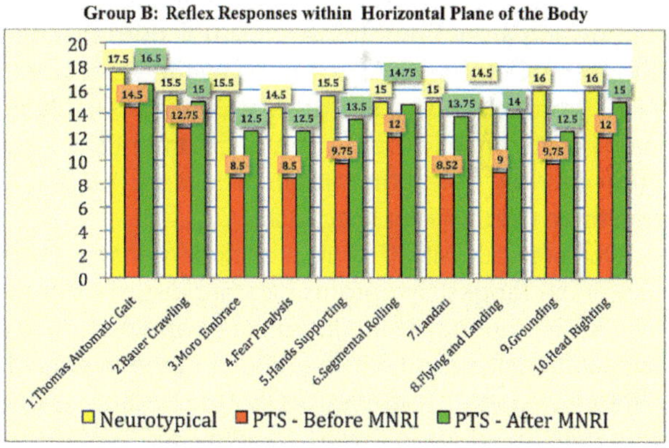

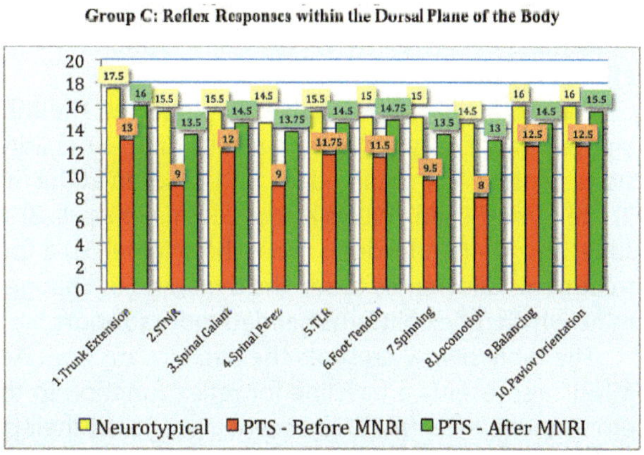

Reflex Pattern Profiles in Children from Newtown

The Sandy Hook Elementary School shooting in Newtown, CT on December 12, 2013 affected the whole community. Adults, children, and everyone in the community were in shock and suffered great emotional stress; living and suffering with the unbearable images of 20 innocent children, 6 teachers,

Table 3. Reflex patterns profile of Newtown children with Post-Traumatic Stress (122 individuals; age 4-19).

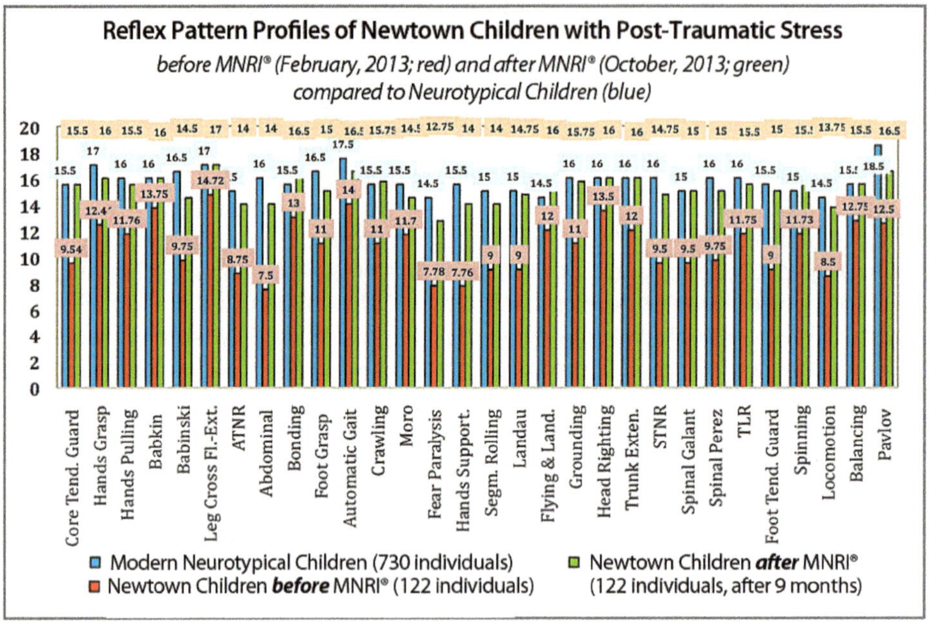

and a school principal massacred in their pleasant New England town on an ordinary school day. Our first visit took place one and a half months after the tragedy by invitation from a local family. A team of seven volunteer MNRI® specialists arrived to offer the MNRI® PTSD protocol to anyone interested. Altogether 264 MNRI® therapy sessions were given, 194 for children and 70 for adults. This method was new for Newtowners. Our professional decision was to use a non-verbal process, limiting any possibility of re-traumatizing our clients by dwelling on their tragic experience. We welcomed the opportunity to bring our method to a community in need.

The MNRI® Reflex Parameters Assessment was administered to 174 individuals in Newtown who had experienced direct and non-direct traumatic stress, among them 122 children ages 4-19 and 52 adults ages 20-63. Four successive assessments were done, the first testing was 1½ months after the event (February 2013), second testing after 3 months (April, 2013), third testing after 6 months (May-June, 2013), and the fourth testing 9 months (October, 2013) after the event. Tests were carried out before and after treatment with the MNRI® PTSD Protocol. This report presents assessment data for children only. Reflex Pattern Profiles of the children from Newtown in February before treatment show overall tendencies similar to those found in the profiles of other children from other traumatic events (1989-2013): low functional state of reflex pattern expression (see Newtown Graph #1 for specifics).

Our previous MNRI® clinical work demonstrates that when 35% or more of reflex patterns test in the dysfunctional ranges, symptoms of other disorders are always present: sensory processing disorder, ADHD, autism, CP, emotional instability, anxiety, and others. In all 122 children from Newtown, 43.3% of reflex patterns tested at dysfunctional levels following the event, which is similar to the data from other children following different traumatic experiences (340 individuals – 46.6% of their reflexes were dysfunctional). (Note: compare with data for adults following trauma - 36.7% of their reflexes were dysfunctional.) Clearly, with 43.3% of their reflexes testing in the dysfunctional ranges, the children who directly or indirectly experienced the shooting at Sandy Hook Elementary School were in a state of physical and emotional shock and fear severe enough to effect their daily functioning and well being.

Trained MNRI® professionals administered the MNRI® PTSD Protocol based on the Assessment data. This analysis specifically focuses on results of 46-51 children who received multiple assessments and therapy sessions. Scores were obtained for 1) 51 children at the initial intervention in February, 1½ months after the event, 2) the same 51 individuals a second time in April, 3) 46 individuals (from the original 51) a third time in May-June, and 4) 46 individuals (from the original 51) a fourth time in October. The sequence of assessments and treatments allowed us to see the dynamics of change in reflex profiles of the Newtown children and to determine which reflex patterns needed more support.

The table below presents the overall data from Assessments of these children. The column under 'Before MNRI®' represents a baseline for reflex function in this group 1½ months after the shooting. The three columns under 'After MNRI' represent averages of their performance before and after the MNRI® sessions offered at those times. Thus the column for April 2013 shows the improvement that held from the initial session in February combined with immediate gains from the current session. Likewise, the columns for May-June and October combine gains maintained from the previous MNRI® treatments with new gains from the current sessions.

Table 4 (at right) shows very clearly the dynamic of change in reflex motor patterns in these children over the treatment period.

Table 4. Dynamic of positive changes in the reflex profiles of Newtown children with Post-Traumatic Stress 1½, 3, 6, 9 months after MNRI intervention compared to children who experienced other traumatic experiences.

#	Table 4. Reflex Patterns (X1-X30)	Newtown Children with post-traumatic stress (46-52 children out of 184)				Children who experienced other traumatic events (340 individuals)		Control group: Neuro-typical (730 individuals)
		Before MNRI®	After MNRI®					
		52 children in 1½ months	52 children in 3 months	46 children in 6 months	46 children in 9 months	Before MNRI®	After MNRI®	
X1	Core Tendon Guard	9.54±0.78	13.5±1.03	14.75±0.71	15.5±0.45	11.99±0.96	13.32±0.78	15.5±0.45
X2	Robinson Hands Grasp	12.46±0.84	14.62±1.8	16±1.12	16±0.74	15±0.87	16±1.12	17±0.65
X3	Hands Pulling	11.76±0.7	14±7.76	15±0.71	15.5±0.71	9.5±0.45	14±0.79	16±0.54
X4	Babkin Palmomental	13.75±0.96	14±0.85	15.5±0.97	16±0.74	14±0.87	15.5±0.96	16±0.78
X5	Babinski	9.75±0.9	12.65±0.81	14±0.72	14.5±0.84	14±0.8	15.5±0.77	16.5±0.8
X6	Leg Cross Flex.-Ext.	14.72±0.79	15.75±0.76	16.5±0.96	17±0.44	15.5±0.78	16.5±1.14	17±0.87
X7	Asymmet. Tonic Neck (ATNR)	8.75±0.94	12.54±0.7	13.75±0.79	14±0.69	8.5±0.8	12.5±1.1	15±0.67
X8	Abdominal	7.5±0.72	11.5±0.82	13±0.99	14±0.69	9.5±0.87	13.5±0.75	16±1.04
X9	Bonding	13±0.85	15±0.77	16±0.87	16.5±1.08	13±0.82	14.5±0.96	15.5±0.65
X10	Foot Grasp	11±1.09	13.52±0.95	14±0.53	15±0.71	9±0.47	13.5±0.75	16.5±0.77
X11	Automatic Gait	14.03±077	15.75±0.83	16.5±0.5	16.64±0.4	14±0.8	16.5±1.04	17.5±0.78
X12	Bauer Crawling	11±1.3	13.03±1.2	13.5±1.14	15.75±0.52	12.7±0.76	15±0.77	15.5±0.58
X13	Moro Embrace	11.73±0.92	13.03±1.2	13.75±1.18	14.5±1.1	8.5±0.49	12.5±0.68	15.5±0.61
X14	Fear Paralysis	7.75±0.8	9.5±0.8	11.93±1.3	12.75±1.36	8.5±0.78	12.5±1.14	14.5±0.45
X15	Hands Supporting	7.75±0.85	10.46±1.4	13±0.87	14±0.69	9.75±0.68	13.5±0.86	15.5±0.74
X16	Segmental Rolling	9±0.73	11±1.08	13±0.87	14±0.69	12±0.45	14.75±0.87	15±0.82
X17	Landau	9±0.73	10.75±1.3	12.5±1.39	14.75±0.76	8.52±0.87	13.75±1.03	15±0.75
X18	Flying and Landing	12±1.04	13±1.02	15.75±0.75	16±0.74	9±0.92	14±1.14	14.5±0.74
X19	Grounding	11±1.08	12±0.9	14±0.98	15.75±0.52	9.75±0.78	12.5±0.82	16±0.78
X20	Head Righting	13.5±0.91	14.5±0.68	15.75±0.45	16±0.74	12±0.74	15±0.87	16±0.79
X21	Trunk Extension	12±0.89	14±0.76	15.75±0.71	16±0.74	13±0.88	16±0.69	16±0.65
X22	Symmetr. Tonic Neck (STNR)	9.5±0.8	11±1.18	13.12±1.12	14.75±1.31	9±0.96	13.5±0.78	16±0.64
X23	Spinal Galant	9.5±0.8	14±0.76	14.5±0.71	15±0.75	12±0.71	14.5±1.14	15±0.87
X24	Spinal Perez	9.75±0.9	12±0.9	13±0.97	15±0.73	9±0.78	13.75±1.07	16±0.74
X25	Tonic Labyrinthine (TLR)	11.75±0.92	14±0.78	14.5±0.72	15.5±0.77	11.75±0.8	14.5±1.05	16±0.86
X26	Foot Tendon Guard	9±0.73	11±1.18	12.67±0.96	15±0.74	11.5±0.82	14.75±0.58	15.5±0.73
X27	Spinning	11.73±0.93	14.5±0.68	15±0.79	15.5±0.66	9.5±0.81	13.5±0.83	15±0.93
X28	Locomotion	8.5±0.64	8.75±0.94	9.75±0.99	13.75±1.14	8±0.96	13±0.78	14.5±0.78
X29	Balancing	12.75±0.7	13±0.78	14±0.87	15.5±0.77	12.5±0.87	14.5±0.78	15.5±0.73
X30	Pavlov Orientation	12.5±0.8	14.5±0.64	15.5±0.53	16.5±0.55	12.5±0.91	15.5±0.83	18.5±0.67
	Total Dysfunctional patterns:	43.3%	6.6%	3.3%	0%	46.6%	0%	0%

Legend:

blue	6-9.75 points	Dysfunctional
purple	10-11.75	Border between dysfunctional and functional
pink	12-13.75	Very low functional level
orange	14-15.75	Low functional level
yellow	16-17.75	Normal functioning

By April, after only two sessions of MNRI®:

1) 40% (12 reflex patterns; 14-15.75; orange) moved higher on the functional scale though still at a low level: Robinson Hands Grasp, Hands Pulling, Babkin Palmomental, Leg Cross-Flexion-Extension, Bonding, Automatic Gait, Head Righting, Trunk Extension, Galant, TLR, Spinning, and Pavlov Orientation.

2) 33.3% (10 reflex patterns; 12-13.75 points; pink) improved from dysfunctional to a very low functional level: Core Tendon Guard, Babinski, ATNR, Foot Grasp, Bauer Crawling, Moro, Spinal Perez, Flying and Landing, Grounding, and Balancing.

3) 20% (6 reflex patterns; 10-11.75; purple) moved to the marginal level between dysfunctional and functional: Abdominal Sleep Posture, Hands Supporting, Segmental Rolling, Landau, STNR, and Foot Tendon Guard.

4) only 6.7% (2 reflex patterns) were below 10 points (blue), remaining dysfunctional – Fear Paralysis and Locomotion (vs: 43.3% [13 reflex patterns] before MNRI® therapy).

The high number of reflex patterns reaching the functional level so quickly, along with observation of the children's behavior and communication, showed that they were moving toward a more positive future perspective. This was evident in their goals and their conversations about sports, creativity, games, and other activities. Before their first session in February many of the children presented a posture of withdrawal (core flexed and eyes down when speaking). They were visually and auditorily hyper-vigilant and very dependent on adults. Already following that initial treatment, it was gratifying to see them extended, growing taller (improved Trunk Extension Reflex), courageously making eye contact and smiling and joking at the end of the sessions. Similar changes were evident in the adults as well and were welcomed by the children. One boy told his father, "Dad, it is so great to hear you laugh again!"

Table 5, right. Dynamic of reflex pattern profiles of Newtown children (122 individuals) with Post-Traumatic Stress according to specific skills.

Summary of Data

Newtown Table 5 (right) summarizes the Assessment parameter results relative to clusters of reflex motor patterns governing specific functions.

Improvement in the direction and symmetry parameters was remarkably fast and stable, suggesting that MNRI® repatterning of the reflex circuit components was highly effective. The direction parameter is particularly significant in the MNRI® model. Regaining the norm in the direction of a reflex motor pattern means, theoretically, that self-regulation in the inter- and motor neurons (alpha and gamma) and links between sensory and motor neurons have been restored. In other words, the integrity of the reflex circuit has been reestablished. This is the central construct in Masgutova reflex re-patterning and integration. Assessment scores give us dry statistical data. Practically, the achievement of reflex pattern integration is seen in improved ability to focus, to be present, to regulate fear and emotional responses, and to experience inner peace, well-being, and joy. Such positive changes were clearly evident in the appearance and behavior of the Newtown children.

Three bar graphs at right show the gradual increase in scores of all 30 reflex patterns (X1-X30; 10 patterns in each graph) to the functional level. Blue columns on the left show the baseline scores from Assessments done in February 2013, 1½ months following the event, before any MNRI® treatment. Red (April), green (May/June), and yellow (October) show increasing improvement after each intervention.

Protective reflex patterns in neuro-typical children and children who experienced trauma

The Core Tendon Guard Reflex in

Table 5. **Reflex Pattern Profiles in Children with Post-Traumatic Stress Before MNRI® (Age 4-19; 122 children, Newtown)**
Assessment Parameter Results Clustered According to Specific Skills

#	Reflex Patterns/ Clusters	Parameters of Reflex Patterns – totals in %														
		Re-Activity			Direction of movement			Intensity (muscle tone)			Latency			Symmetry		
		Norm	Hyper	Hypo	+	-	+/-	Norm	Hyper	Hypo	Norm	Hyper	Hypo	+	-	+/-
1	Upper Limb	10.7	83.6	5.7	71.3	5.7	23.0	13.9	79.5	6.6	9.0	81.1	9.8	4.1	84.4	11.5
2	Tonic	13.1	80.3	6.6	43.4	26.2	30.3	14.8	59.0	26.2	13.9	61.5	24.6	3.3	94.3	2.5
3	Righting	17.2	77.0	5.7	47.5	10.7	41.8	15.6	79.5	4.9	27.9	68.9	3.3	18.9	71.3	9.8
4	Lower Limb	24.6	73	2.5	42.6	10.7	46.7	26.2	71.3	2.5	27.0	70.5	2.5	16.4	63.9	19.7
5	Gross Motor	38.5	27.9	33.6	64.8	5.7	29.5	26.2	68.9	4.9	30.3	48.4	21.3	18.0	78.7	3.3
6	Oral, Facial, Visual and Auditory	34.4	32.0	33.6	36.9	13.1	50.0	26.2	47.5	26.2	32.0	49.2	18.9	9.8	85.2	4.9
7	Protection and Survival	9.8	86.9	3.3	36.8	40.2	23.0	1.6	95.1	3.3	4.9	91.8	3.3	23.0	54.1	23.0
8	Curiosity/Cognition	62.3	29.5	8.2	62.3	23.0	14.8	62.3	31.1	6.6	60.7	29.5	9.8	21.3	68.9	9.8
9	Emotional Stability	14.8	85.2	0.0	37.7	42.6	19.7	13.1	83.6	3.3	13.1	72.1	14.8	23.8	53.3	23.0
	Average total:	24.7	64.2	11.0	49.3	19.8	31.0	22.2	68.4	9.4	24.3	63.7	12.0	15.4	72.7	11.9

Table 6-A, B, C, below. Dynamic of reflex pattern profiles of Newtown children (122 individuals) with Post-Traumatic Stress before and after the MNRI® program (in graph expression).

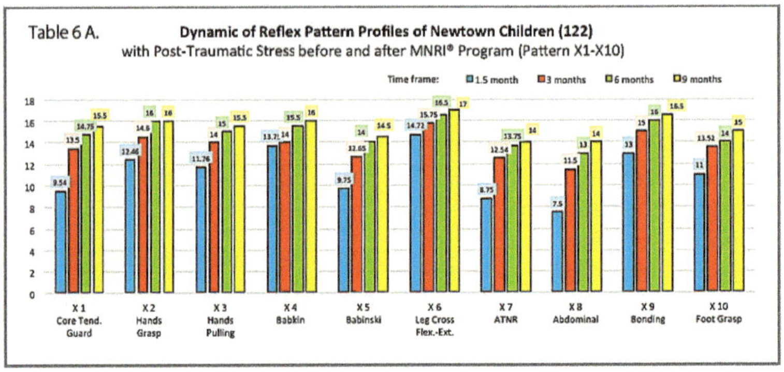

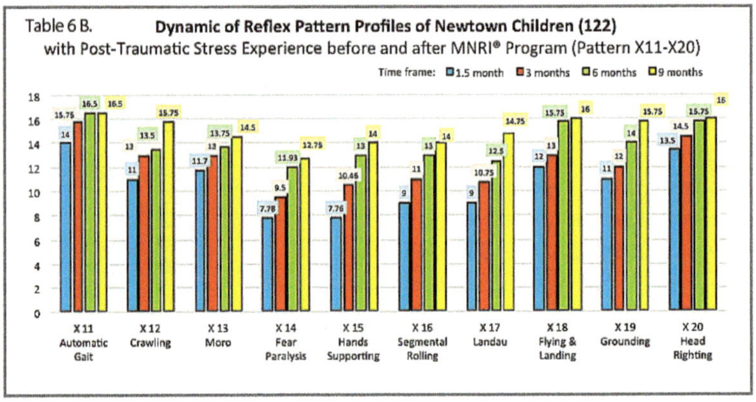

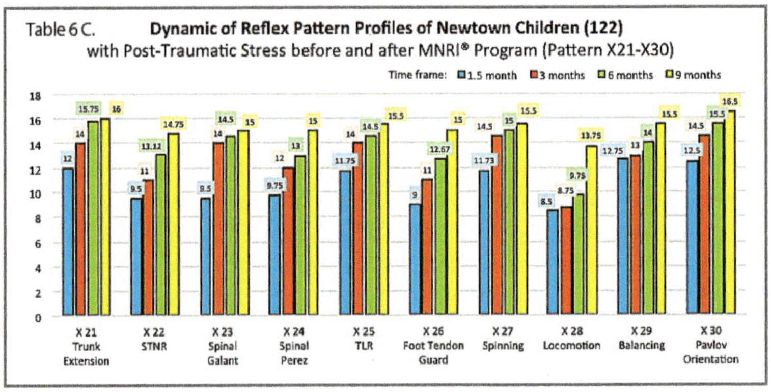

the Newtown children before MNRI® was lower than in the 1989-2006 data from other children with post-traumatic stress (9.54 vs. 11.99). The statistical significance of the Assessment results equaled 0.51 for earlier group and 0.75 for the Newtown children (with p < 0.001). Moro in the Newtown children was higher (11.7 vs 8.5); statistical significance equaled 0.57 for the earlier group and 0.78 for the Newtown children (with p <0.001). Results for Fear Paralysis were similar for the two groups (7.78 vs. 8.5) with no statistical significance: 0.54 for the earlier group and 0.51 for the Newtown children. With some variation, data from our extensive worldwide collection, including Newtown, looking at reflex integration in traumatized children before and after the use of MNRI® indicate statistically significant positive growth and development for all.

Table 7, below. Dynamic of protective reflex patterns of Newtown children (122 individuals) with Post-Traumatic Stress before and after the MNRI® program compared to other children (in graph expression).

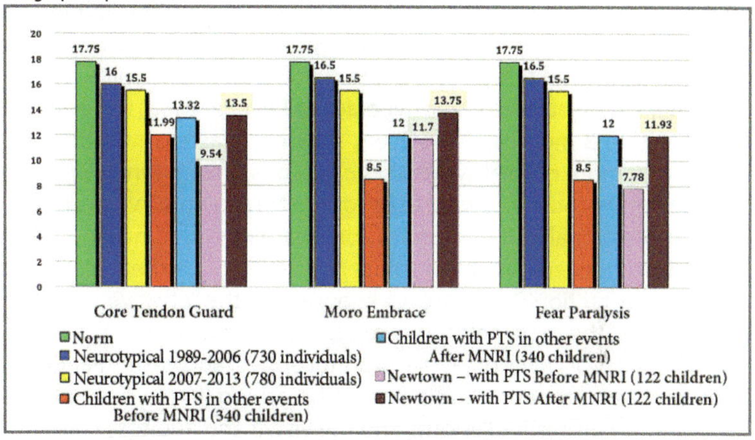

Warning signals

Along with the excellent progress cited above, some warning signals are also apparent.

It is noteworthy that 9 months after the event, complete recovery had still not occurred for such reflex patterns as Fear Paralysis (12.75 points), Locomotion (13.75), Hands Supporting (9.75), ATNR (14), Abdominal Sleep Posture (14), Segmental Rolling (14), Babinski (14.5), and Landau, (14.5). Multiple interpretations for this can apply: 1) they were less integrated developmentally before the traumatic event, 2) these reflex patterns are more vulnerable when intense or long term stress takes place and 3) more regular and longer term work is needed for full post trauma recovery. All of these results indicate that early professional intervention is essential in trauma recovery work with children.

Figure 6, below. Parameters of reflex pattern expression in Newtown children (122 individuals) - A) Feb. 2013, 1½ months after the event and B) Oct. 2013, 9 months after the event.

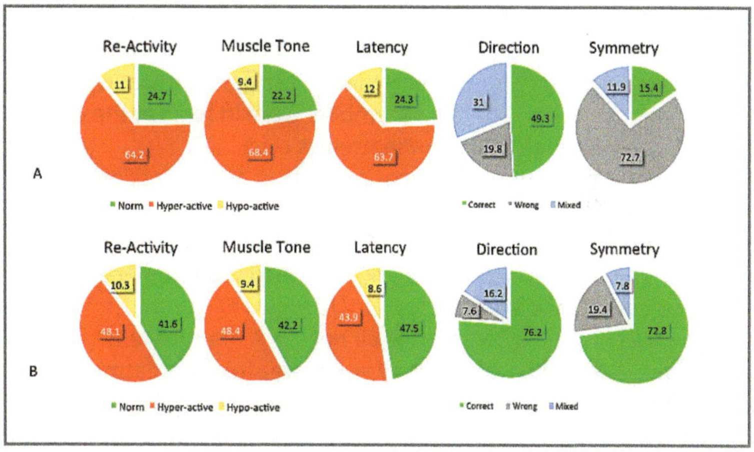

The pie charts at right show that re-activity, muscle tone, and latency/timing, which were hypersensitive and/or hyperactive immediately post-trauma, retain this tendency for as long as 9 months, even in children whose demeanor seems healthy and positive. This shows the importance of continuing the neuro-sensory-motor trauma recovery work beyond the time when the trauma might seem to be resolved, a year or even longer.

Early Intervention

Over 30% of children who came for MNRI® treatments 3 months or more after the shooting already showed signs of PTSD: hyperactive sympathetic responses in tests, sweaty palms, perspiring all over, dilating pupils, holding their breath, poor eye tracking, and startling or shaking/tremor during Fear Paralysis, Moro or Hands Supporting stimulation. They also reported tight muscles, insomnia, poor focusing at school, distracting memories, desire to escape others, and frequent crying or holding back tears. Some were already on medication because of more serious emotional issues. In contrast, children who came for reflex re-patterning before 3 months showed less hypersensitivity in protective reflexes, less rush of fear, more presence, clearer communication and reasoning, and more normal play and joking. This, along with the warning signals cited above, highlights the urgent need for early intervention.

Part 6

Discussion and Comments

The low levels of reflex integration observed in children and adults experiencing post-traumatic stress can lead to PTSD. When 35% or more of the reflexes we test are dysfunctional, we consistently find symptoms of other disorders. According to our research an average of 33.7% of these reflexes are dysfunctional in individuals diagnosed with PTSD. Thus the need for professional facilitation of reflex integration is urgent for anyone at risk for PTSD. In the Newtown children who experienced no work at the level of reflex systems in the interbrain and brainstem within 3 months of the event, slower dynamics of recovery with negative effects on the sympathetic nervous system (chronic hyperactive Fear Paralysis) were already present. The children who experienced MNRI® within 3 months of the event showed a faster return to functional or normal levels of reflex pattern expression, more normal sympathetic nervous system function, and a faster return to normal life.

Many other tools that were used for the Newtown survivors made a big difference, of course. The outpouring of professional and non-professional care provided healing, comfort, and support. Yet one missing piece must be acknowledged: the absolute need to address the unconscious involuntary physiology of positive survival and protection. The more we know about the reflexive responses involved in our protective and survival mechanisms, the more helpful we can be. It starts with 'low' brain technology, dealing with the HPA-stress-axis on the non-verbal level of tactile, proprioceptive, and vestibular processing. We must work with reflex patterns (units and keys for neurodevelopment), breathing (to physically take away the 'internal pressure' that all of our participants complained about), and the visual and auditory senses.

Although this article is limited to reflex patterns in children and does not address those of adults (our next planned article), it is interesting to mention the difference in expression in one particular reflex pattern. The children demonstrated hypo-activity for Hands Supporting, while their parents showed a hyperactive response mixed with Bonding. Instead of extending their arms straight they adducted as though encircling a child; it was quite obvious. Seeing this amazing manifestation of protection, survival, and need to protect the young was poignant. Our team members were in tears at the depth of the parents' impulse to save their dear children even though in shock themselves.

Feedback and support from participants in MNRI® sessions also touched us deeply:

"You are all angels." – A Sandy Hook father who was one of the first on the scene.

"I feel my feet on the ground and don't feel so lost." – A mom.

Deep sighs – A dad who hadn't been able to breathe since the event.

"I can't even tell you how happy I am I decided to try this. Thank you for coming!" – A mom.

"I feel like I can do what I need to do now." – A grandparent.

"Come back soon!" – a small boy as he left after his treatment session.

"Mom, they made my body feel so much better." – A boy.

"I'm a believer. This can help anyone. My blood pressure has been through the roof for the last couple months. I went to see my doctor yesterday right over there and my blood pressure was normal. I didn't believe it. I had them take it a second time." – A man arriving for his second appointment.

"I've been wanting to jump out of my skin ever since it happened. Now I'm comfortable." – A mom.

The concepts and statistics given here are not just neutral information for the author and the MNRI® team. These numbers represent the fallout from unbearable trauma experienced and lived through by the Newtowners who witnessed the unspeakable. This tragedy touched all of us personally and our goal for this report is to offer a method that can lead survivors of such dark events away from pain and despair, away from senseless loss and on toward inner peace, hopeful recovery, and a positive future. Our method is designed to promote neurosensorimotor reflex integration. Our success is measured not only by points in Assessments, but more importantly by deep restorative sleep, pain-free bodies, health, joy in the simple pleasures of life, confidence, resilience, and optimism.

Special thanks to:

- *Chris and Kathleen Yeaton and their children for their interest in our program, and for extending their invitation and support by introducing MNRI® to their community*
- *Gene and Sharon Vetrano, Newtown residents, who donated office space, a reception area, and 4 treatment rooms*
- *Scarlett Lewis for her bold and loving support of the Masgutova Method® in her video made for our MNRI® presentation at the US Uniformed Services University of the Health Sciences, USU Navy Base, USA*
- *All the children, parents, grandparents, teachers and other Newtown residents for trusting us and thereby coming to experience the freedom and inner success that comes with the MNRI® PTSD Program*
- *The MNRI® Team of Specialists who eagerly volunteered their time on our different visits to Newtown*
- *Shannon Desilets for her dedication, leadership, endless hours of coordinating, communicating, scheduling, and offering a skillful liaison with the Newtown community*
- *Wendy Elwell for finding helpful contacts*
- *Tricia Borsh and the Bridge to Healing Foundation for her donations to support this work*
- *Sponsors, donators, and organizers*
- *Southwest Airline pilot and his wife, John and Avis Harrison, who donated miles along with other pilots who helped us obtain flight tickets*
- *The Danbury Holiday Inn for providing lodging*
- *The La Quinta Inn, Danbury for providing lodging*
- *The Resiliency Center of Newtown for providing wonderful space for our work*
- *Bertucci's Restaurant*
- *Chef Bob's Catering*
- *The Best Italian Bakery in Boston for great cannolis*
- *The bakery in Newtown that provided breakfast bagels*
- *And all others wishing great results for Newtowners*

– MNRI® Team

References

Akhmatov, E. (2015). Computerized MNRI® assessment in the paradigm of modern technology. In book: *Reflexes: Portal to neurodevelopment and learning.* S. Masgutova, D. Masgutov. Orlando, FL, USA:SMEI.

Amosov, N. M. (1978). *Thinking about the health.* Moscow. p. 16-29.

Anokhin, P. K. (1968). *Biology and neurophysiology of a conditioned reflex.* Moscow. Russia: Medicina.

Anokhin, P. K. (1973). Biology and neurophysiology of the conditioned reflex and its role in adaptive behavior. *International Series of Monographs in Cerebrovisceral and Behavioral Physiology and Conditioned Reflexes.* USA.

Asratian, E. A. (1963). Tonic conditioned reflexes as the form of whole brain activity. *Journal of High Nerve System Activity,* v.15, n.5, p. 718-788.

Asratian, E. A. (1983). *Selected works. Reflexive theory of high nerve system activity.* Moscow, Russia: Science. 328 p.

Badalian, L. O. (1984). *Child neurology.* Moscow, Russia: Medicina.

Baldacara, I., Jackowski, A., Andreoli, S. B., Mello, M. F. (2011). Reduced cerebellar left hemisphere and vermal volume in adults with PTSD from a community sample. *Journal of Psychiatric Research* 08/2011; 45(12):1627-33. DOI: 10.1016/j.jpsychires.

Barashniev, J. I. (2001). *Prenatal neurology.* Moscow, Russia: Triada-X. 640 p.

Bernstein, N. A. (1947). *The construction of the movement.* Moscow, Russia.

Bernstein, N. (1997). *Bio-mechanics and physiology of the movement.* Moscow, Russia: Moscow-Voroniez.

Bremner, J. D., et al. (1995). MRI-based measurements of hippocampal volume in combat-related post-traumatic stress disorder. American Journal of Psychiatry, 152, 973-978.

Bremner, J. D. (2001). Hypotheses and controversies related to effects of stress on the hippocampus: An argument for stress-induced damage to the hippocampus in patients with post-traumatic stress disorder. *Hippocampus,* 11, 75-81.

Cannon, W. B. (1931). Again the James-Lange and the thalamic theories of emotion. *Psychological Review.* 38(4), Jul 1931, 281-295. doi: 10.1037/h0072957.

Herman, J. (1992). *Trauma and recovery: from domestic abuse to political terror.* London, UK: Pandora (an imprint of Rivers Oram Press).

Kaada, B. (1986). Sudden Infant Death Syndrome – The Possible Role of 'the Fear Paralysis Reflex'. Norwegian University Press, Oslo. 1986. 56 p.

Krefft, A. (2007). *Diagnosis of functions of invisible phenomena* (Statistic Mathematic Analysis). Wrocław, PL: Oficyna Wydawnicza Politechniki Wrocławskiej.

Leontiev, A. N. (1977). *The Activity, consciousness, personality.* Moscow, Russia.

Luria, A. R. (1966). *Human brain and psychological processes.* New York & London: University of Moscow, Harper & Row Publishers. 587 p.

Luria, A. R. (1969). *High cerebrum functions of the human and their disorders in cases of local damages.* Moscow. Russia: Moscow State University.

Masgutova, S. (2011). *Infant dynamic and postural reflexes. Neuro-sensory-motor reflex integration.* Revised and edited. Scientific-practical manual. Orlando, FL, USA: SMEI.

Masgutova, S., Curlee, P. (2004, 2007). *You are a winner. trauma recovery. A new choice through natural developmental movements.* Revised Edition. Iowa, USA: 1st World Publishing. 105 p.

Masgutova, S. , Akhmatova, N. , Lebedinskaya, O. (2013). *Clinical-immunological assessment of therapy effect of the neuro-sensory- motor integration program of reflex patterns in airway chronic inflammatory diseases.* Event Abstract. International Congress in Milan. Front. Immunol. Conference Abstract: 15th International Congress of Immunology (ICI). doi: 10.3389/conf.fimmu.2013.02.00865.

Masgutova, S., Masgutov, D., Akhmatov, E. (2013). *MNRI®: Reflex Integration Disorder and PTSD Recovery.* (Workshop manual). Orlando, FL, USA: SMEI. 67 p.

Pavlov, I. P. (1960). *Conditioned reflexes: An investigation of the physiological activity of the cerebral cortex.* (Anrep G.V., D. Sc. Trans., 1960). New York, NY, USA: Dover Publications Inc. (Original work was published in 1927)

Pilecki, W., Masgutova, S., Kowalewska, J., Masgutov, D., Akhmatova, N., Sobieszczanska, M., Kalka, D. (2012). The impact of rehabilitation carried out using the masgutova neurosensorimotor reflex integration method in children with cerebral palsy on the results of brain stem auditory potential examinations. In Journal: *Advances in Clinical and Experimental Medicine.* Official Organ Wroclaw Medical University. 2012 (3). 2012, 21, 3, 363–371. http://www.advances. am.wroc.pl/pdf/2012/21/3/363.pdf.

Sadowska, L. (2001). *Neurokinesiological diagnosis and therapy of children with mental and motor deficits.* AWF. Wroclaw. PL.

Sechenov, I. M. (1995). *Physiology of behavior.* Scientific Works Ed. Moscow: Russia: M.G. Yaroshevsky. (Original work was published in 1863)

Sechenov, I. M. (1960). *Selected physiological and psychological works* (S. Belsky, Trans., K. Koschtoyants, Ed.). Moscow, Russia: Foreign Languages Publishing House. (Original work published 1935)

Sechenov, I. M. (1965). *Reflexes of the brain.* (Russ. tr. S. Belsky). Cambridge, MA: The M.I.T. Press. (First published in 1863.)

Selye, H. (1974). *Stress without distress.* Philadelphia, PA, USA: J. B. Lippincott Co.

Sherin, J. E., Nemeroff, C. B. (2011). Post-traumatic stress disorder: the neurobiological impact of psychological trauma. *Dialogues Clin Neurosci.* Sept. 2011; 13(3); 263-278.

Sherrington, C. (1947). *The integrative action of the nervous system.* London, UK: Cambridge University Press.

Van der Kolk, B., Roth, S., Pelcovitz, D., Sunday, S., Spinazzola, J. (2005). Disorders of extreme stress: the empirical foundation of a complex adaptation to trauma. *J Trauma Stress.* 18(5): p. 389–399. [PubMed].

Our deepest thanks and gratitude to all of the children and adults of Newtown for their trust in us and for their appreciation and enthusiasm in response to our PTSD Protocol. The genuine friendship of the children with their school peers and teachers was remarkable to see! We wish you all the goodness in the world. – The MNRI® Team

MNRI® for Children with Cerebral Palsy

Svetlana Masgutova, Ph.D., Warsaw, Poland – Orlando, FL, USA

This abstract presents a study published in 2008 in the article, *Masgutova Method of Reflex Integration for Children with Cerebral Palsy*, edited by Susan Wenberg, M.A., D.C., and Mary Rentschler, M.A. at www.MasgutovaMethod.com.

Svetlana Masgutova, Ph.D

Cerebral Palsy (CP) and Primary Motor System Development

"Infantile Cerebral Palsy" (S. Freud), the group of diseases concerned with motor disorders as the result of brain damage or dysfunction of certain brain centers, is usually acquired during the first years of life, at the time when the system of primary movement patterns is developing. Primary movements are genetically programmed not only for protection and survival, but also for the development of the conscious movement system. Developmentally their role is to support:

- maturation of the nervous system (synaptogenesis, myelination, and brain plasticity)
- brain functions (cognitive development, emotional maturation)
- sensory-motor integration.

Dysfunctions of motor development and sensory-motor integration in the child with CP reflect the type of neurological insult and the developmental stage of the infant/child when the neurological insult occurred. The developmental stages can be divided into prenatal (in utero); natal (during the birth process); or postnatal (during the first years of life). Each developmental stage is vulnerable to specific neurological insults.

Prenatal palsy can be caused by infection, intoxication of the fetus, or compromised health of the pregnant mother. Many primary motor patterns and reflexes develop in utero, such as Trunk Extension, Automatic Gait, Hands Grasp, Swallowing, and Sucking. Prenatal brain damage will cause poor expression of these reflexes and adversely impact the next stage of development.

Natal palsy is generally caused by a neurological insult during birth: a consequence of premature birth, sudden delivery, narrow pelvis of the mother, use of forceps during delivery, etc. Natal trauma can negatively affect the activation of primary motor patterns and reflexes characteristic of a normal, successful birth. In such cases the expression of these genetic programs will be abnormal. The primary movements and reflexes of childbirth, such as Head Righting, Spinal Perez, Tonic Labyrinthine in Extension, Crawling, Sequential Side Rotation, Spinning, and Sucking may be dysfunctional.

Postnatal palsy is often caused by an infection with encephalitic symptoms. Other causes include childhood cranial injury, central nervous system injury, and poor systemic health. Infant motor patterns and reflexes will be stressed by postnatal CP, and may develop dysfunctionally or pathologically. Dynamic reflexes

(Grasp, Hands Pulling, Automatic Gait), positional reflexes (Asymmetrical Tonic Neck, Symmetrical Tonic Neck, Babkin Palmomental, Tonic Labyrinthine in Flexion and Extension), and postural reflexes (Trunk Extension, Spinal Perez, and Galant) are most commonly affected.

Impaired motor function associated with central nervous system damage in all types of CP may be due to:

- poor sensory perception of stimuli (tactile, visual, auditory, olfactory, vestibular, proprioceptive)
- dysfunctional or pathological processing of sensory stimuli, such as poor recognition or invalid decoding
- inadequate motor response due to abnormal muscle tone or musculoskeletal problems such as orthopedic injury or abnormality.

Motor development is a primary expression of coordinated neurological function in infant and early childhood development. In the child with CP, impaired motor function contributes to disorders in auditory and visual perception, memory, speech, and self-organization. These, together with motor problems, will influence the future development of all other spheres: physical (insufficient muscle bulk, structural weakness or abnormalities), emotional (anxiety), cognitive (sensory processing disorder, learning disabilities), and personality (lack of confidence, timidity).

Both genetically encoded primary motor patterns and consciously learned and controlled motor functions are impaired in the child with CP. In MNRI® our focus is on techniques that support primary motor patterns. We define these as genetically programmed reactions that range in complexity from simple reflex responses occurring at the spinal cord level, to more complex survival-based response patterns involving brainstem activity, to sensory-motor coordination systems (hand-eye, eye-ear, etc.) and finally, to more complex activities (visual tracking of a moving object, articulation for pronouncing sounds, etc.).

Because they are natural resources that support synaptogenesis, myelination, and optimal brain function, primary motor patterns are crucial for future development. Each step of development is based on kinesthetic memory (genetic and individual), which allows us to internalize all types of movements. Studies show kinesthetic memory to be damaged in 40-60% of children with CP (N.N. Danilova, A.L. Krilova, 1997). With motor and cognitive development limited by poor kinesthetic memory, even the simplest motor skills are very difficult for the child with CP to explore and anchor into memory. In the MNRI® model, neurosensorimotor reflex integration awakens latent brainstem genetic motor memory, so that it may serve as a resource for neurodevelopment.

Characteristics of Neurotypical Early Childhood Motor Development

To understand the uniqueness of early motor development in children with CP, it is helpful to compare their reflex pattern formation to that of a healthy child. This comparative analysis offers a profound understanding of the dysfunctional features characteristic of CP and allows us to design therapeutic interventions using developmental techniques that influence the sensory-motor links of the reflex circuit.

During the first 1½ -2 months of life, a number of reflexes develop in sequence, including the Tonic Reflexes: Tonic Labyrinthine, Symmetrical and Asymmetrical Tonic Neck, Head Righting and Trunk Extension Reflexes, and pelvic-trunk movement patterns. The individual reflexes are difficult to detect as they normally mature into other reflex patterns, and are naturally integrated into the movement system by three months of life. They are complex, resulting from spontaneous activity in the brainstem, vestibular-cerebellar centers, motor cortex centers, nuclei of the vision centers and the corpus striatum. Coordination of these areas determines the control of muscle tone and muscle activity.

During the first year, a healthy child develops still more reflexes and primary movements including antigravity reactions and supporting motor patterns (Gravity, Grounding, and Stability Reflexes; Automatic Gait, Crawling, Spinal Galant and Perez Reflexes, Grasp, Oral Automatisms, etc.)

A primary reflex, Labyrinthine Head Righting, develops as an antigravity function, allowing the supine infant to raise his head by the age of two months. It triggers him to lift his head when pulled by the hands, or when lying prone. Head Righting is mainly controlled by the labyrinthine nuclei and medulla oblongata. In children with CP, this reflex is not typically expressed until the fifth month or later.

Thanks to the development of this reflex pattern, a healthy six-month-old infant placed on his stomach is able to support his upper body on his forearms, to lengthen his trunk in an arc, using appropriate muscle contraction, and to flex his legs above ground. Later in development the child is able to turn over, to get on all fours, to crawl, and to sit without support. The Head Righting Reflex is the basis for all these movements. It is controlled by vestibular-cerebellar structures in the medulla spinalis, medulla oblongata, and the reticular formation of the brainstem.

In the second year of life the combination of the Tonic Labyrinthine and the Symmetrical and Asymmetrical Tonic Neck Reflexes helps the child learn how to control its body position at rest and how to move through space.

Characteristics of Early Motor Development in a Child with CP

All reflex patterns described above for healthy infants are dysfunctional or pathological in children with CP, and their expression is delayed by eight years or more. In CP with severe brain damage there is minimal development of the righting reflexes (Child Neurology, 2000); K. Bobath, 1972; L.O. Badalian, 1984; R. Michalowicz, 1993; C.H. Delacato, 1974; G. Doman, 1984; V. Vojta, 1989; K.B. Nelson, J.H. Elleberg, 1979; K.A. Semionova, 1999; D. E. Haines, 2002; L. Sadowska, 1998; S. Masgutova, N. Akhmatova, 2004; S. Masgutova, 2007).

Our observation and understanding of reflex patterns in children with CP has allowed us to identify dysfunction or pathology in a number of reflex patterns such as the Tonic Labyrinthine in Flexion and Extension, Robinson Grasp, Babkin Palmomental, Leg Cross Flexion-Extension, Asymmetrical Tonic Neck, Thomas Automatic Gate, Bauer Crawling, Moro, Hands Supporting, Segmental Rolling, Symmetrical Tonic Neck, Spinal Galant and Perez, Spinning, and Pavlov Orientation.

Pathological reflexes and movement patterns in children with CP are especially related to a lack of timely development, maturation and integration of tonic reflexes. Tonic reflex patterns, triggered by movement or changes in body position or in the relationship of the head to gravity, strengthen the functional links between the vestibular system and the musculoskeletal system; thus, supporting appropriate muscle tone, proprioception, posture, and motor control. The vestibular system reflexes are linked with the work of the extra-ocular and spinal muscles, creating the ground next for righting reflexes. These neurological links are modulated and matured via tonic reflexes.

Several important reflexes whose expression in children with CP is abnormal, are noted below:

1. Tonic Labyrinthine Reflex (TLR). A prone child with CP, regardless of age, demonstrates abnormally high tension in the upper and lower limbs, flexors, and abdominal muscles. The child can not raise his head, straighten his core, or extend his arms and legs – movements that are typical for a 3 -6 month old healthy child. If the child with CP is supine, the tone of extensors in the limbs increases. In severe cases, spasticity is present. A pathological TLR adversely affects the way the child sits, turns over, and stands up. It also can affect the tongue muscles, blocking articulation and preventing stimulation and development of Broca's area (speech center) and subcortical links in the brain. A pathological TLR also causes poor head position and abnormal function of the oculo-motor abductors. As a result, the child has a limited view of the objects around him/her, leading to poor development of vision. When sitting or standing, joints are overly flexed and the child has difficulty extending his limbs.

2. Symmetrical Tonic Neck Reflex. When the child with CP is put into position to test the Symmetrical Tonic Neck Reflex the head is flexed on the chest. The child's response is increased tone in the flexors of the arms and extensors of the legs. When the head is tilted backwards, tone in the limbs reverses: it increases in the leg flexors and arm extensors. This abnormal reaction causes problems with the development of postural control, inhibits the formation of the links between postural control and binocular vision and near-far accommodation, and blocks the development of superior and inferior eye muscles.

3. Asymmetrical Tonic Neck Reflex. If we turn the head of a healthy newborn lying supine to the side, the arm and leg on the same side will extend, and the opposite arm and leg flex. In a child with CP, often the limbs flex on the same side to which the head turns, or there is a global hyperactive response. This abnormal reaction is a protective response that does not support sensory awareness or orientation to the environment.

4. Segmental Rolling Reflex. This motor pattern is of crucial importance for infant development. If not developed and integrated, the child cannot rotate the shoulders independently of the pelvis. This is seen in the child with CP who cannot separate movements of different body parts. Abnormalities in the development of the Segmental Rolling Reflex prevent free balancing of the trunk while walking: the individual will have poor balance and limited ability to make subtle adjustments in response to gravity. Clinical observations include unstable and inappropriate leaning to the side and poor control of posture and movement.

5. Hands Grasp Reflex. Another reflex linked to tone and tonic function is the Robinson's Hand Grasp Reflex.

It should mature near the end of the first week, integrate with Hands Pulling by the second month, and mature to allow for easy expression of manual skills by the end of the first year. However, the Robinson's Hand Grasp Reflex can become fixed and reactive in the child with CP so that after grasping an object, he cannot relax the palm and open the fingers to release it. Children with CP who have low muscle tone will not be able to explore the grasp motor pattern as a possibility for flexing the fingers and palm into a fist for holding, or for the development of manual skills. Manual skills represented in the area of the sensory-motor cortex are strongly correlated with speech centers: Wernicke's area (auditory center for recognition of human speech), Broca's area (articulation of phonemes and sounds), and subcortical areas that integrate speech functions (W. Penfield, T. Rasmussen, 1950). Thus poor development of the Robinson's Hand Grasp Reflex may negatively influence speech formation in the child with CP.

In normal infants the tonic reflexes discussed above begin to integrate in the second month of life. In children with CP adequate reflex responses never occur. The abnormal patterns that emerge instead, involve muscular hyper-contraction and pathological muscle synergies, which limit joint mobility and morphological development, as well as movement. The expression of these pathological patterns is strikingly evident in the first year of life.

In the second year of life, hyperkinesis (unintentional movements) may be diagnosed. Hyperkinesis will involve the tongue, causing the lower lip to protrude. Abnormally high tension in the trunk and limb muscles and poor coordination in the hands may also be noted.

Abnormal muscle synergies underlie all the stereotypical pathological motor patterns of children with CP. As noted above, these abnormal synergies, expressed as abnormal tone, posture, and movement, are linked to abnormal tonic reflex patterns. In summary, the complex process of natural reflex development in children with CP is impaired. Depending on the location of the brain damage, different pathological pattern schemes are formed in the cortex and in the subcortical areas.

MNRI® Assessment of Reflex Motor Pattern Expression

Table 1 (below) shows 24 reflexes grouped according to body planes and motor coordination systems. Motor patterns for reflexes 1-8 relate to a median/axis plane dividing the body between right and left, referred to in anatomy as the sagittal plane. The sagittal body plane involves differentiation and coordination of movements of the right and left sides of the body. Motor patterns for reflexes 9-16 activate the upper and lower parts of the body, related to a horizontal plane. Reflexes 17-24 relate to a dorsal plane dividing the body between front and back.

A group of 480 children with CP were evaluated for reflex pattern integration using the MNRI® Assessment protocol. Table 1 shows the number and percentage of children from the group whose scores fell in the pathological or deeply dysfunctional range for 24 reflex motor patterns.

Table 1. MNRI® Assessment Results for a Group of 480 Children Diagnosed with CP

MCS		Deeply Dysfunctional and Pathological Reflex Patterns in Children with Cerebral Palsy	Total Number of Children: 480	
			Number	Percentage
Right/Left SAGGITAL	1	Robinson Grasp	270	56.3
	2	Hands Pulling	152	31.7
	3	Babkin Palmomental	269	56.0
	4	Babinski	360	75.0
	5	Leg Cross Flexion-Extension	279	58.1
	6	Asymmetrical Tonic Neck	286	59.6
	7	Abdominal	260	54.2
	8	Bonding	168	35.0
Up/Down HORIZONTAL	9	Thomas Automatic Gait	360	75.0
	10	Bauer Crawling	310	64.6
	11	Moro	286	59.6
	12	Fear Paralysis	312	65.0
	13	Hands Supporting	317	66.0
	14	Segmental Rolling	334	69.6
	15	Landau	427	89.0
	16	Flying and Landing	410	85.4
Front/Back FRONTAL	17	Trunk Extension	319	66.5
	18	Symmetrical Tonic Neck	312	65.0
	19	Spinal Galant	320	66.7
	20	Spinal Perez	325	67.7
	21	Tonic Labyrinthine	370	77.1
	22	Foot Tendon Guard	327	68.1
	23	Spinning	344	71.7
	24	Pavlov Orientation "What is this?"	176	36.7

For norms and levels of the MNRI® Assessment protocol, see the chapter, *MNRI® Assessment for Determining the Level of Reflex Development.*

As Table 1 shows, a very high percentage of individuals in this group demonstrated dysfunction and pathology in their expression of reflex motor patterns. In a neurotypical population all of the reflex patterns would be functional, in the usual distribution around the norm.

Masgutova Neurosensorimotor Reflex Integration (MNRI®)

MNRI® therapy uses an approach of correction and coordination of sensory and motor components of the dysfunctional reflex pattern circuits. 'Re-patterning' exercises and techniques (re-education, re-coding/re-setting of schemes) 'revive traces of genetic motor memory and reactivate the physiological circuit.' These interventions stimulate the correct expression of genetically encoded resources, such as programs of self-regulation, self-defense, and stress release. 'Re-patterning' aimed at reactivation of links between the neurostuctural aspect of a reflex pattern and its protective task (Sechenov,1863/1995, Ukhtomsky, 1950-1952) stimulates synaptic growth and myelination (Lundy-Ekman, 2002). MNRI® therapy procedures use specific protocols developed by Dr. S. Masgutova and her co-authors. These include *Neuro-Structural Reflex Integration* (2003), *Tactile Therapy* (2005), *Dynamic and Postural Reflex Pattern Integration* (2004), *Visual and Auditory Reflex Integration* (2007), *Breathing Reflex Integration* (2012), and others.

Historically, children with CP are the largest patient group in our work. Our research on their primary motor patterns has helped us identify specific dysfunctional reflexes and design individualized programs that restore function as reflex expression becomes more normalized. The study presented below demonstrates the effectiveness of MNRI® as an intervention for these children.

Application of the MNRI® Program to Children with CP

The study involved a group of 42 children with CP ranging in age from 2 to 8 years. They attended either a 14-day therapeutic clinic at the International Masgutova Institute of Movement Development and Reflex Integration in Warsaw, Poland (31 children) or a 10-day MNRI® Family Educational Conference sponsored by the Svetlana Masgutova Educational Institute, USA (11 children). MNRI® Assessments were done on each child before and after 10 or 14 days of a MNRI® treatment conference or clinic.

Each program consisted of 6 -7 hours per day of treatment that included active structural motor therapy and passive procedures.

Table 2 presents examples of a statistically significant validation of the synthesized function z = f(x). Function z = f(x) is the synthesized function of changes in reflex pattern development (z) and presents synthesized information of chosen diagnostic parameters (x). Each parameter (x) shows the level of development of a specific reflex pattern as measured according to the criteria of MNRI® reflex pattern assessment; this function allows us to measure each parameter (x) for each individual child.

Table 2. Changes in Reflex Pattern Development (Dorsal Plane)

Nr of an Individual	Assessment	X17	X18	X19	X20	X21	X22	X23	X24	Z (Coefficient of the change)
Child 1	Before MNRI®	5	5	7	5	4	5	4	8	**0.351**
	After MNRI®	8	8	9	7	7	7	5	9	**0.472**
Child 2	Before MNRI®	14	11	5	3	6	3	6	5	**0.382**
	After MNRI®	15	13	8	4	15	5	9	13	**0.611**

Function z = F(x) is the synthesized function of changes in reflex pattern development (z) and presents synthesized information of chosen diagnosis parameters (x).

The columns in Table 2 show Assessment results for specific reflex patterns as follows: X17 – Trunk Extension, X18 – Symmetrical Tonic Neck, X19 – Spinal Galant, X20 – Spinal Perez, X 21 – TLR, X22- Foot Tendon Guard, X 23 – Spinning, and X24 – Pavlov Orientation. The 'before' and 'after' numbers represent points on a continuum of reflex development ranging from 0 (deep pathology) to 20 (high level of integration), with 15/16 representing the norm and 10 representing the boundary between dysfunctional and functional.

In Figure 1 we see an example of the changes in development of reflex patterns within the dorsal body plane of child 1. Mathematical statistical analysis demonstrates the statistical importance and validity of

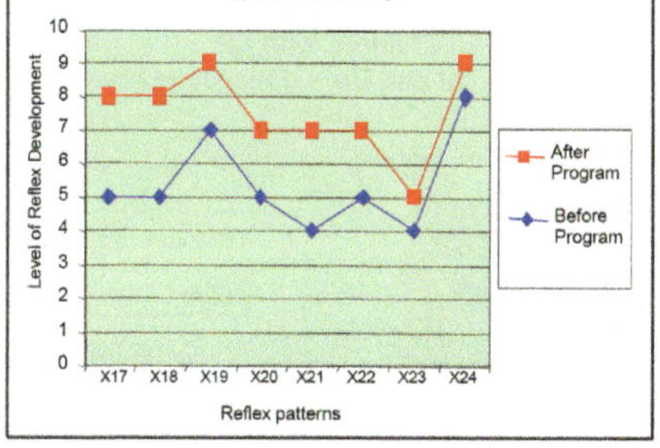

Fig. 1. Child 1:Changes in Reflex Pattern Development (Dorsal Plane)

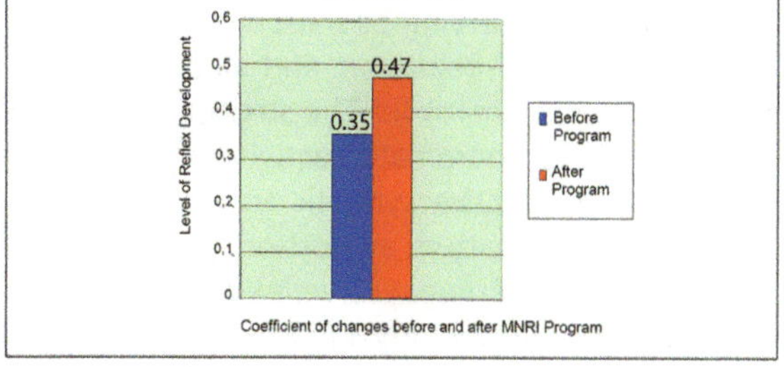

Fig. 2. Coefficient of Change in Reflex Pattern Development (Dorsal Plane)

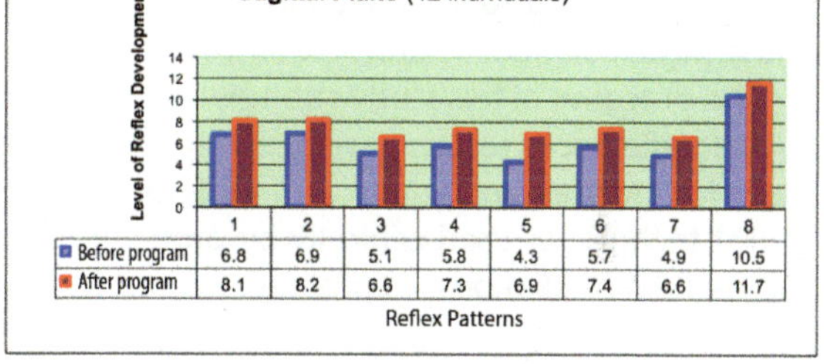

Figure 3. Changes in Level of Reflex Pattern Development in Children with CP Sagittal Plane (42 individuals)

	1	2	3	4	5	6	7	8
Before program	6.8	6.9	5.1	5.8	4.3	5.7	4.9	10.5
After program	8.1	8.2	6.6	7.3	6.9	7.4	6.6	11.7

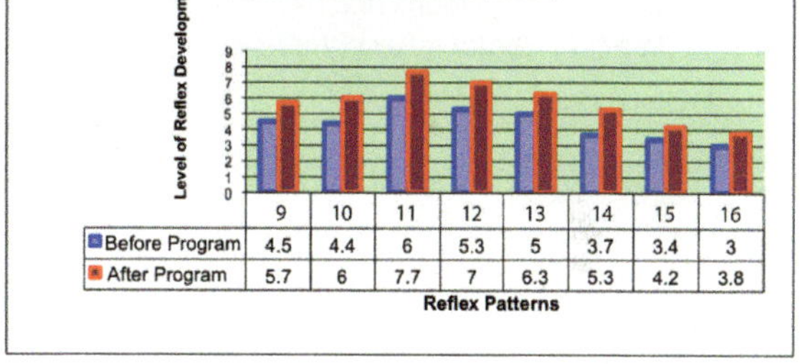

Figure 4. Changes in Level of Reflex Pattern Development in Children with CP Horizontal Plane (42 individuals)

	9	10	11	12	13	14	15	16
Before Program	4.5	4.4	6	5.3	5	3.7	3.4	3
After Program	5.7	6	7.7	7	6.3	5.3	4.2	3.8

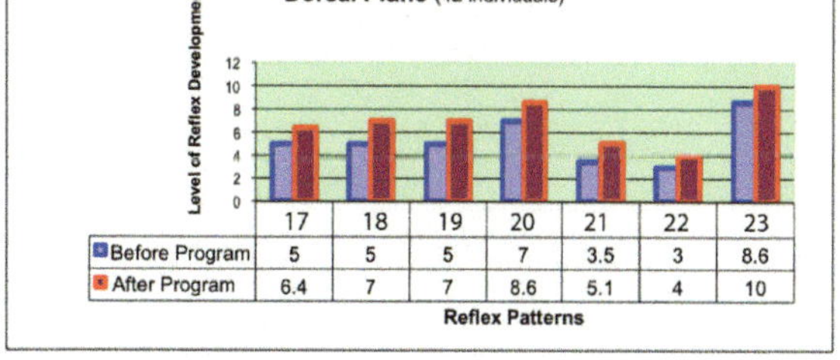

Figure 5. Changes in Level of Reflex Pattern Development in Children with CP Dorsal Plane (42 individuals)

	17	18	19	20	21	22	23
Before Program	5	5	5	7	3.5	3	8.6
After Program	6.4	7	7	8.6	5.1	4	10

these changes within each reflex pattern and also for the whole group of reflex patterns involving the dorsal body plane/MCS (Fig. 2). This type of analysis was used for all diagnostic parameters of reflex patterns for each child, with comparison of results before and after the MNRI® Program, for all three body/motor planes: sagittal, horizontal, and dorsal. Figure 2 shows an example of the coefficient of changes (z) by Prof. A. Krefft (2007), which demonstrates the statistical validity of the changes in expression of reflex patterns within this group.

Bar Graphs in Figures 3, 4, and 5 show examples of dynamic changes in the expression of reflex patterns for all three body/motor planes after 14 or 10 days of utilizing the MNRI® Program.

Figure 3. Changes in Level of Reflex Development: Sagittal plane. Again MNRI® Reflex Assessments after the MNRI® program indicated solid gains in reflexes belonging to all three body planes/MCS. The average rate of improvement in the sagittal group was 1.6 points (from 6.25 points before the MNRI® program to 7.85 after); in the horizontal group 1.69 (from 4.41 before to 6.10 after); and in the dorsal group 1.6 (from 5.3 before to 6.9 after). Clinical observations also confirm improvement in reflex patterns demonstrated by decreased pathological muscle tone, improved muscle tone regulation, decreased hyperactivity in protective responses (Fear Paralysis and Moro) and improvement in righting reflexes (Trunk Extension and Head Righting), gross motor coordination (Crawling, Automatic Gait, Perez, and Galant), manual skills (Hands Grasp, Hands Pulling and Hands Supporting), and balance.

Figure 6. (next page) Coefficient of Change in Reflex Pattern Development in Children with CP: Sagittal, Horizontal and Dorsal Planes. Mathematical statistical analysis shows remarkable improvement in reflex pattern expression in the group of children with CP after attending the MNRI® program. The coefficient of change is 0.43 before the MNRI® Program and 0.56 after, indicating statistically significant change. (Fig. 6).

The lineal error in projection of changes is not more than 1.87 – 2.82 %, again supporting the conclusion that the MNRI® program resulted in statistically significant changes in reflex pattern expression in these children.

Summary

The powerful resources of innate reflex patterns are present in everyone, serving as a foundation for motor development, sensory integration, and all our higher-level cognitive and emotional skills. The strength of this foundation depends on the coordinated activity of sensory and motor neurons and effective synaptic transmission, all essential for correct reflex pattern expression. Stress/distress, disease and injury can interfere with reflex development and functioning. Fortunately, at any age we can reawaken, repattern, or rebuild them. This is the purpose of MNRI®.

Figure 6.
Coefficient of Change in Reflex Pattern Development in Children with CP
Sagittal, Horizontal and Dorsal Planes (42 individuals)

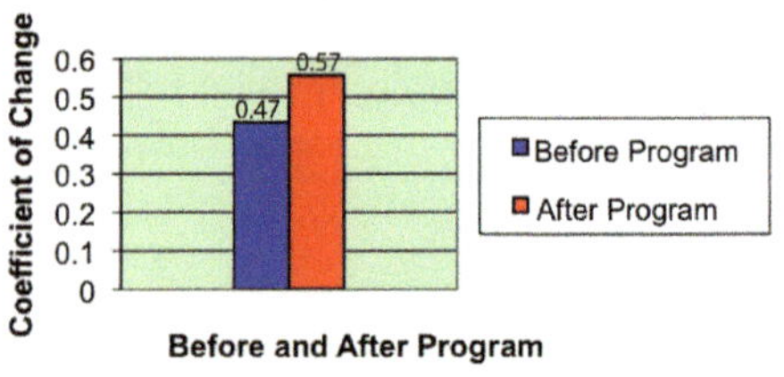

Forty-two children with CP demonstrated significant improvement in expression of primary sensory-motor patterns associated with infant Dynamic And Postural Reflexes following 10 or 14 days of MNRI® therapeutic programs. Statistical analysis supports the effectiveness of the MNRI® diagnostic protocol and validates the results of the therapeutic part of the program.

The outcome of this study strongly supports the early application of corrective procedures directed at the level of reflexes. Its results have been replicated over and over again by families who have sought MNRI® treatment for children and adults with CP and many other developmental challenges. The natural, non-invasive MNRI® techniques used clinically by professionals and applied by parents in home programs have brought successful remediation and progress toward functional health to thousands of challenged individuals.

I thank and congratulate these children for doing all that was possible to improve their sensory-motor reflex patterns and skills and for their teaching us to value life through their smiles and amazing motivation! – Dr. Masgutova

References

Badalian, L. O. (1984). *Child neurology.* Moscow. (Russ.). Russia: Medicina.

Barashniev, J. I. (2001). *Prenatal neurology.* Moscow. (Russ.). Russia: Triada-X. 640 p.

Bernstein, N. (1997). *Bio-mechanics and physiology of movement* (Russ.). Moscow, Russia: Moscow-Voroniez.

Bobath, K. (1972). The normal postural reflex mechanism and this deviation in children with cerebral palsy. *Physiotherapy,* 57, 515.

Bobath, B. (1963). Treatment principles and planning in cerebral palsy. /*Physiotherapy* April: 1–3.

Caesar, P., Dubovittz, V. (1985). Neonatal clinical neurological assessment. In Harel S. Anastasion N.J. (eds): *The At-Risk Infant. Psycho/Social/Medical Aspects.* Paul H. Brookes Baltimore. N.J. USA. 197-229.

Child neurology. (2000). Publisher: Wrocław, PL: Wydawnictwo Medyczne - Urban & Partner.

Danilova, N. N., Krilova, A. L. (1997). *Physiology of the high nerve system activity.* Moscow, Russia: Uchebnaja Literatura.

Dargasses, S-A. (1986). *The neuro-motor and psycho-affective development of the infant.* Amsterdam, NY: USA.

Delacato, C. H. (1974). *The diagnosis and treatment of speech and reading problems.* Garden City, NY, USA: Doubleday.

Doman, G. (1984). *How to multiply your baby's intelligence.* Garden City, NY: USA. Doubleday.

Freud, S. (1894). *The neuro-psychoses of defence.* SE, Part 3.

Gross, D. R. (1987). *Psychology. The science of mind and behavior.* London. UK: Hodder and Stoughton.

Haines, D. E. (2002). *Fundamental neuroscience,* Second Edition. Edinburgh, UK.

Kolár, P. (1996). Importance of developmental kinesiology for manual medicine, *Czech Journal of Rehabilitation and Physical Therapy.*

Krefft, A. (2007). *Diagnostic functions of non-observable phenomenon.* Wroclaw, PL: Oficyna Wydawnicza Politechniki Wroclawskiej.

Lewit, K. (2001). *Manual therapy in rehabilitation of motor diseases.* Ed. by: Jerzy Stodolny ZL Natura, Kielce 2001, wyd. 3. – Polish version

Lundy-Ekman, L. (2002). *Neuroscience. Fundamentals for rehabilitation.* 2nd Edition. Forest Grove, OR, USA: Pacific University.

Luria, A. R. (1969). *High cerebrum functions of the human and their disorders in cases of local damages.* Moscow. Russia: Moscow St. Univ.

Masgutova, S., Akhmatova, N. (2004, 2007). *Integration of dynamic and postural reflexes into the whole body movement system.* Scientific edition: Prof. N. Akhmatova. Warsaw, PL: MINK.

Masgutova, S. (2005). Reflexes as the basis of the nerve system development and formation of the motor patterns in infancy. / Materials of International Conference: *Modern Methods of Stimulation of Motor and Language Development. International Kinesio-Rehabilitation Camp for Children with Challenges of Dr. S. Masgutova Institute,* Warsaw, PL: MINK. p. 14-36.

Masgutova, S. (2007). *Integration of infant dynamic and postural reflex patterns: Masgutova Neuro-Sensory-Motor and Reflex Integration – MNRI® method for children and adults.* Third ed. Revised,. Illustrated. Orlando, FL, USA: SMEI. 250 p.

Masgutova, S. (2007). *Neuro-structural reflex integration therapy.* Revised Edition. Warsaw, PL: MISM. 120 p.

Masgutova, S. with Kowal, J., Mazur, G., Masgutov, D. (2005). *NeuroKinesiology tactile therapy™ by Dr. S. Masgutova,* Scientific Edition: Dr. A. Regner, J. Szymczak. 133 p. Warszawa, PL: MINK.

Masgutova, S., Akhmatova, N. (2004). *Reflexes re-patterning exercises. In: Integration of dynamic and postural reflexes into the whole body movement system.* Scientific edition: Prof. N. Akhmatova. Warsaw, PL: MINK.

Masgutova, S. (2007). *Neuro-Sensory-Motor development: Visual and auditory reflexes integration. Facilitation program of development and learning for children and adults.* Scientific Edition: Prof. N. Akhmatova. Warsaw, PL: MISM. 86 p.

Michalowicz, R. (1993). *Cerebral palsy in child.* Warsaw, PL: PZWL.

Nelson, K. B., Elleberg, J. H. (1979). Neonatal signs as predictors of cerebral palsy. *Pediatrics,* 64-68, 225-231.

Penfield, W., Rasmussen, T. (1950). *The cerebral cortex of man: A clinical study of localization of function.* New York, NY: Macmillian.

Pavlov, I. P. (1960). *Conditioned reflexes: An investigation of the physiological activity of the cerebral cortex,* (Anrep G.V., D. Sc. Trans., 1960). NY, USA: Dover Publications Inc. (Original work was published in 1927)

Piaget, J. (1976). *The grasp of consciousness: action and concept in the young child.* Cambridge, MA, USA: Harvard Press.

Rubinstein, C. L. (1946). *The general psychology basis.* (Rus.) Moscow.

Sadowska, L. (1998). *Neurological basis of the early kinesiology diagnosis and rehabilitation using Vojta Method.* Wrocław, Poland: Medycyna Manualna, 2.

Sadowska, L. (2001). Neurokinesiological diagnosis and therapy of children with motor-cognitive developmental dysfunctions. In Book: *Neurokinesiological Diagnosis and Therapy of Children with Mental and Motor Deficits.* Wroclaw, PL: AWF.

Sechenov, I. M. (1995). *Physiology of behavior. Scientific works/* Edd.: M.G. Jaroshevsky Moscow, (Russ.) Russia. (Original was published in 1863)

Semionova, K. A. (1999). *Rehabilitation treatment of the CP patients in residual state.* Moscow, (Russ.) Russia.

Ukhtomsky, A. A. (1950-1952). *The study of dominance //* Collected works in 6 books. Book 6. Leningrad, Russia.

Vojta, V. (1989). *Die Posturale Ontogenese als Basis der Entwicklungsstorungen.* Monatsschr. Kinderheilkd., 141, 639-642.

Vygotsky, L. S. (1986). *The child psychology. The problems of child development.* In 6 Books. Book – 4. Moscow, Russ: Pedagogika.

The Impact of the MNRI® Program on the Brain Stem Auditory Potential in Children with Cerebral Palsy

Prof. Witold Pilecki; Svetlana Masgutova, Ph.D; Joanna Kowalewska; Denis Masgutov; Nelli Akhmatova, MD, Ph.D; Prof. Malgorzata Sobieszczanska, MD; Prof. Dariusz Kalka, Ph.D, Warsaw-Wroclaw, Poland

This report is an adaptation of a published study by Prof. W. Pilecki (Pathophysiology Department, Wroclaw Medical Academy of Poland) in collaboration with S. Masgutova, J. Kowalewska, D. Masgutov, Dr. N. Akhmatova, Dr. M. Sobieszczanska, and Prof. D. Kalka. The original information for this article can be found in: Impact of Masgutova Neurosensorimotor Reflex Integration (MNRI®) on Children with Cerebral Palsy as Measured by Brain Stem Auditory Evoked Potentials Testing. *Advances in Clinical and Experimental Medicine*, Official Journal of Wroclaw Medical Academy 2012 (3). 2012, 21, 3, 363–371. ISSN 1899–5276. http://www.advances.am.wroc.pl/pdf/2012/21/3/363.pdf.

Prof. Witold Pilecki

Svetlana Masgutova, Ph.D

Joanna Kowalewska

Denis Masgutov

Prof. Malgorzata Sobieszczanska

Nelli Akhmatova, Ph.D

Prof. Dariusz Kalka

Abstract

Brain stem auditory evoked potentials tests (BAEP) were administered to an experimental group of 17 children with cerebral palsy, ages 1.3 to 5.9 years, before and after a brief Masgutova Neurosensorimotor Reflex Integration (MNRI®) intervention. The authors sought to obtain objective validation of positive changes as a result of the use of MNRI® in children with neurological deficits. Results showed significant improvement in nerve transmission efficiency in those children immediately following the intervention. Thirty healthy children of equivalent ages served as a control group. The Interpeak Latency I – V subtest of the BAEP is of particular interest because of its potential use for intermediate assessment and diagnosis of neurodevelopmental motor delays and abnormalities.

Introduction

The purpose of this study was to use brain stem auditory evoked potentials (BAEP) tests to obtain an objective measurement of changes produced by Masgutova Neurosensorimotor Reflex Integration (MNRI®) in children with cerebral palsy. Evaluation of neurological deficits and disorders is usually based on observation of a patient's responses to stimulation of primary reflex motor patterns and performance in motor tasks. Because this type of assessment depends to such a large extent on the skills, possibly the psychological attitude as well, of the person performing the evaluation, results cannot be entirely objective. Brain stem evoked potentials testing relies on the registration of excitation waves appearing in the sensory pathways of the central nervous system due to activation of the appropriate receptor of an adequate stimulus. Testing results are independent of both the patient's and examiner's will or state of consciousness and are given as numerical values, which accounts for their replicability and objectivity.

MNRI® is a complex body of knowledge and therapeutic application oriented on primary motor reflex patterns. As L. Vygotsky and J. Piaget recognized, these motor patterns have an important role in motor, emotional and higher cognitive functions as well as in natural inherent brain stem survival mechanisms. Because professionals are only beginning to learn about this original and broadly applicable intervention, the Masgutova Educational Institute was eager for objective research-based validation of its efficacy.

Materials and Method

The study involved an experimental group of 17 children with CP, nine girls and eight boys, ages 1.3–5.9 years (mean= 3.8 years, SD = 1.3) and a control group of 30 healthy children of equivalent ages. Testing on the control group provided a baseline of normal competence for comparison with pre- and post-testing of the experimental group. The study consisted of three parts, carried out in one continuous session lasting about one hour:

1. BAEP pre-tests on experimental and control groups
2. Therapeutic intervention: one session of MNRI®
3. BAEP post-tests on the experimental group

Part 1 BAEP Pre-testing

The BAEP tests involved bilateral administration of a stimulus in the form of a click of 70 decibels at a frequency of 10 Hertz. The children in both groups received the stimulus through earphones and their responses were registered on a computer attached to electrodes positioned on their scalps. Each testing session lasted for about 15 minutes: 3 minutes to apply the electrodes, 2 minutes to register background noise, 5 to 7 minutes to present and measure responses to clicks, 2 minutes to register background after the clicks and 1 to 1½ minutes to remove the electrodes.

The most commonly used tests of brain stem evoked potentials involve auditory (BAEP) and visual (VEP) nerve transmission. A particularly useful test parameter in the evaluation of children with neurodevelopmental deficits is the Interpeak Latency I-V (IPL I-V). This subtest directly points at transmission efficiency in the brain stem segment of the auditory pathway and specifies in milliseconds (ms) the time that elapses between electrogenesis of the I segment (auditory nerve), and electrogenesis of the V segment (inferior colliculi of the mesencephalon).

Although developmental deficits resulting from injury are usually generalized in the brain, not limited to specific structures, the evoked potentials test directly assesses only one particularly sensitive area of the brain that is accessible to it. However, researchers have found that conditions and changes in this area correlate with conditions and changes in other inaccessible areas, including motor pathways and thus also overall motor development. Majnemer, et al., have used brain stem evoked potentials for clinical Assessment and prognosis in children with CP and Down syndrome and in infants at risk for perinatal CNS damage (Majnemer, Rosenblatt, Riley, 1990; Jiang, Wu, 1990, and Pilecki, 2002). Pilecki defined this use of EP tests to evaluate and project the level of transmission efficiency in other brain areas as "intermediate assessment" (Pilecki, et al, 2002).

Part 2 MNRI® Reflex Integration Exercises

The therapeutic intervention was a modification of a typical MNRI® session. It consisted of six consecutive reflex pattern activation exercises:

- Foot Tendon Guard Reflex (automatic dorsiflexion of the foot)

- Hands Supporting Reflex (parachute reflex)
- Leg Cross Flexion-Extension Reflex
- Spinal Galant Reflex
- Asymmetrical Tonic Neck Reflex
- diaphragm mobilization for the Breathing Reflex pattern.

These exercises involved hands-on stimulation of specific sensory-motor points on the limbs and trunk and passive manipulation of limbs and trunk in patterns associated with each reflex's motor response, all performed by the same trained MNRI® specialist. The children were lying on massage tables. For each of the above reflexes the exercise consisted of seven repetitions of a basic repatterning procedure, as described by S. Masgutova in her scientific-practical manual, *Infant Dynamic and Postural Reflexes: Neuro-sensory-motor Reflex Integration*.

Part 3 BAEP Post-testing

The previous BAEP test of IPL I-V was repeated. Due to heightened anxiety about the electrodes, two children in the experimental group were unable to participate in testing. Thus post-testing results were obtained on 15 out of the 17 children's experimental group. Post-testing took about 15 minutes, following the same procedure as the pre-testing.

Absolute IPL I-V Values

Raw data resulting from the brain stem response to an auditory stimulation (click) travels through electrodes placed on the scalp to a computer software program. The software then generates a score in milliseconds to express the amount of time that passes between the click and the brain stem response to it. This is the absolute IPL I-V Value.

The Interpeak Latency (IPL) parameter was selected for analysis because researchers frequently use it as an intermediate measure of CNS integrity. Although the absolute IPL I-V value is a widely accepted parameter, it has a significant disadvantage resulting from the fact that it is not constant in the first year of a child's life. It shrinks from about 5.10 millimeters at birth to approximately 4.05 millimeters by 12 months. Because of this unusual variability the standard deviation (SD) for scores must be extremely large, making statistical comparison of groups very difficult, even when their mean scores diverge.

In this study analysis of the absolute IPL I-V results was performed based on standards established in Dr. Pilecki's laboratory. He interprets values falling within ± 2 SD of the mean as normal. Lower than the mean by more than 2 SD then indicates shortened or accelerated transmission. Higher than the mean by more than 2 SD indicates elongated or slowed transmission (in this case lower is better, indicating more efficient, faster transmission). The unusual designation of ± 2 SD as normal was due to the extreme variability in scores for the younger age groups, as discussed above.

Results

Because testing was done separately for right and left sides, the number of results was double the number of children. Scores above 4.25 ms indicate slower transmission, a performance poorer than the norm. Scores below 3.85 ms indicate faster transmission, a performance better than the norm. Table 1 shows the number and percentage of students whose scores fell within the norm, slower (worse) than the norm and faster (better) than the norm.

IPL I-V values for right and left sides	Experimental group		Control Group
	Before MNRI® N=34	After MNRI® N=30	N=60
Within Norm (3.85 - 4.25ms)	4 (11.8%)	10 (33.3%)	60 (100%)
Extended (>4.25ms)	30 (88.2%)	18 (60.0%)	0
Shortened (<3.85ms)	0	2 (6.7%)	0

Table I. Absolute IPL I-V values within, above and below normal before and after MNRI®.

All the children in the control group scored within the range of the norm determined by Dr. Pilecki's laboratory (i.e. between the 3.85 ms and 4.25 ms.). Pre-testing scores for all children in the experimental group deviated from those of the control group and from the laboratory determined norm, all slower than the norm. As would be expected given their diagnoses, the extended time intervals indicated slower less efficient nerve transmission in the brain stem segment of the auditory pathway.

Following the brief MNRI® session the percentage of results indicating achievement within the norm increased from 11.8% to 33.3%. Results indicating slowed transmission (those in which the IPL I-V value was elongated) decreased from 88.2% to 60%. One child's post MNRI® IPL I-V value decreased to a speed even faster than the control group mean, better than the laboratory standard. Statistical analysis using the Wilcoxon test yielded p = 0.001 for the right-sided responses and p = 0.004 for left-sided responses. The Student's T-test for Pairs yielded p-values for the respective sides of 0.001 and 0.002. These results can be considered statistically significant with a high degree of confidence.

Experimental Group Pre-Test Absolute IPL I-V Results Compared to Control Group

In the experimental group, both the mean aIPL I-V values and the standard deviation differ considerably from those of the control group. The experimental group values were 1.24 ms higher for the right- and 1.09 ms higher for the left-sided responses. Variability within this group was also far greater, as seen in the larger standard deviation. The differences were statistically significant at the level of p <0.001.

Absolute IPL I-V value	Experimental group before MNRI		Control group	
	Right side	Left side	Right side	Left side
Mean	5.29 ms	5.14 ms	4.05 ms	4.05 ms
SD	1.39 ms	1.18 ms	0.10 ms	0.09 ms
Min.	4.24 ms	4.16 ms	3.85 ms	3.85 ms
Max.	9.20 ms	8.40 ms	4.25 ms	4.23 ms

Table II. Absolute IPL I-V values in experimental (before MNRI®) and control groups.

The mean absolute IPL I-V values in the experimental group, whose average age was 3.8 years, were worse (right 5.29 ms, left 5.14 ms) than in full-term newborns (5.10 ms). At the same time dispersion of the results was also enormous, from within the laboratory norm (minimum scores of 4.24 and 4.16) to a result (9.20) exceeding the average value of the control group (4.05) and the laboratory norm by nearly 50 standard deviations.

Pre- and Post-test Absolute IPL I-V Results in the Experimental Group

Mean values in the experimental group decreased after MNRI®. The right side improved (got faster) by .45 ms and the left side by .32 ms. Statistical analysis using the Wilcoxon test for the right and left sides was p = 0.001 and p = 0.004 respectively and the Student's t-test for pairs yielded p-values for the respective sides of 0.001 and 0.002, all significant with a high degree of confidence.

Absolute IPL I-V value	Before MNRI		After MNRI	
	Right side	Left side	Right side	Left side
Mean	5.29 ms	5.14 ms	4.84 ms	4.82 ms
SD	1.39 ms	1.18 ms	1.41 ms	1.14 ms
Min.	4.24 ms	4.16 ms	3.76 ms	3.84 ms
Max.	9.20 ms	8.40 ms	8.96 ms	8.24 ms

Table III. aIPL I-V values in experimental group before and after MNRI®.

Variation in transmission speeds remained very high, although it decreased: the minimum (fastest) response got faster by .48 ms on the right side and .32 ms on the left, and the maximum (slowest) got faster by .24 ms on the right and .16 ms on the left. Such changes are understood as improvement in transmission efficiency in the brain stem segment of the auditory pathway.

Relative IPL I-V Values

The extreme absolute IPL I-V score variability in the first year of life and the resulting large size of the standard deviation, as discussed above, makes comparison of groups very difficult, even when their mean scores diverge. To reduce this inconvenience Pilecki developed an additional parameter derived from his earlier research with test results from 411 children of adjusted ages from -8 to +78 weeks. This early research enabled him to give mean values week by week for age groups with differing SD values. In contrast to the absolute IPL I-V value, the relative IPL I-V results from a sophisticated mathematical equation that relates raw data from children with severe deficits to absolute scores of neurotypical children, standardized for ages week by week from ages -8 to +78 weeks. Scores are then expressed not in milliseconds, but in units that directly reflect their relationship to the standard deviation for the corresponding age group. Thus a relative IPL V–I of 5 or 12 indicates a result 5 or 12 SD below the mean for children of the same age. Use of the relative IPL V–I value is a new feature in this study.

The benefit of these new values is that they give a clearer measure of the degree of change from test to test and for very short age intervals. For professionals unfamiliar with evoked potentials, understanding the severity of deficit or degree of change shown in absolute BAEP test results can be challenging, because the standards are not widely known. By indicating how many standard deviations a result deviates from the mean of 1 standardized by age for the neurotypical population, the relative IPL I-V value enables average physicians to interpret BAEP test results more easily.

In addition to giving scores that directly reflect a relationship to the norm, the mathematical formula that generates relative IPL I-V scores also amplifies their numbers, so that differences are more clearly apparent.

Thus in both groups the maximum values reach highs near 50 units, deviating from the mean of 1 determined for the standard group by up to 50 SDs. Improvement after MNRI® was also very great: 4.5 units for the right side and 3.2 for the left. Though not shown in this report, variability among individual children's scores and in their changes from pre- to post-testing was likewise very high. The largest improvement in relative IPL I-V value after MNRI® for an individual child was 15.2 units (from 14.7 before MNRI® to -0.5 after). In this child the initial deficit of 14.7 standard deviations below the norm jumped to 0.5 of one standard deviation above the norm! Other individual results were less spectacular: some children improved only slightly and in two cases there was a slight deterioration.

Discussion

The data obtained from pre- and post-testing showed clear improvement in transmission efficiency in the brain stem segment of the auditory pathway immediately following a brief MNRI® session. The fact that all other testing conditions remained constant in both rounds and the MNRI® intervention was the only variable suggests very strongly that MNRI® caused the improvement.

Relative IPL I-V value	Before MNRI		After MNRI	
	Right side	Left side	Right side	Left side
Mean	12.4	10.9	7.9	7.7
SD	14.0	11.8	14.1	11.4
Min.	1.9	1.1	-2.9	-2.1
Max.	51.5	43.5	49.1	41.9

Table IV. Relative IPL I-V values in experimental group before and after MNRI®.

Variability in the outcome from child to child was not surprising, as the experimental group included children in whom CP differed in both etiology and severity. The fact that statistical analysis showed these positive results to be highly significant is all the more remarkable given the fact that post-testing followed a one-time-only and very brief therapeutic intervention.

The question of the sustainability of this improvement remains open. However, such an immediate low result, especially following only one very brief therapeutic intervention, can be seen as a positive sign. In the authors' experience any such clear indication of plasticity suggests that improvement is possible and that regularly performed exercises should enhance sustainability. Fortunately, the children participating in this study are under continuous observation and further studies are in process.

Another open question of great interest to those concerned with motor development, involves the relationship between improvement in the auditory pathway and improvement in mobility. Many authors have pointed out this possibility, claiming that brain stem auditory potentials are of great diagnostic and prognostic importance when it comes to motor development (Pilecki, 2003; Pilecki, Szawrowicz, Jagielski, Janocha, Borodulin-Nadzieja, 2002; Yilmaz, Degirmenci, Akdas, Kulekci, Ciprut, Yuksel, Yildiz, Karadeniz, Say, 2001; Scalais, 1998; Pike, Marlow, 2000; Majnemer, Rosenblatt, 1995). Dr. Pilecki also found a close relationship between mobility and the IPL I-V parameter (Pilecki, Szawrowicz, Jagielski, Janocha, Borodulin-Nadzieja, 2002). He described BAEP test results as an "intermediate diagnosis." Although the term is of his authorship, the rule he applied is widely accepted among researchers, who use the BAEP test both for measuring motor development and for its prognostic value (Yilmaz, Degirmenci, Akdas, Kulekci, Ciprut, Yuksel, Yildiz, Karadeniz, Say, 2001; Majnemer, Rosenblatt, 1995).

Further support for the use of this 'intermediate assessment' comes from other research by Pilecki, in which he found that in healthy children, the ability to walk independently is associated with the achievement of a certain transmission speed in the brain stem segment of the auditory pathway. He demonstrated that this relationship was more than a coincidental simultaneous convergence of two developmental processes. In the children he observed with slowed brain stem transmission, manifested by elongation of the IPL I-V, the development of independent walking was also delayed. Those children only learned to walk when the IPL I-V value decreased to about 4.25 ms (Pilecki, Szawrowicz, Jagielski, Janocha, Borodulin-Nadzieja, 2002).

The authors look forward to further research (e.g. based on wavelet analysis or single responses), using the equipment capabilities and the professional expertise of the Department of Pathophysiology of Medical University of Wroclaw, to expand their knowledge and understanding of these promising possibilities (Kipiński, Maciejowski, 2010).

Conclusions

1. BAEP pre- and post-testing showed significant improvement in transmission speed in the brain stem segment of the auditory pathway in an experimental group of 15 children with CP following a brief therapeutic session of MNRI®, as measured by changes in absolute and relative IPL I-V values.

2. Statistical analysis using the Wilcoxon test and the Student's t-test for pairs yielded p-values ranging from 0.004 to 0.001, indicating that results were significant at high level of confidence.

3. BAEP test results strongly suggest that MNRI® exercises can have a rapid and significant positive effect on brain stem functions in children with neurodevelopmental delays and abnormalities.

4. The possibility that BAEP results reflect changes in other areas of the brain, including motor pathways lying in the direct proximity to the brain stem segment of the auditory pathway, opens the prospect of potentially very wide use for that test.

References

Bobath, C. K. (1958). Neurophysiological bases of authors' method of treatment of cerebral palsy. *Acta Neurol Psychiatr Belg,* 58, 469–474.

Chiappa, K. H. (1997). *Evoked potentials in clinical medicine.* Third edition. New York, NY, USA: Raven Press.

The Doman-Delacato method. (1968). *J Iowa Med Soc,* 58, 507–509.

Jiang, Z. D., Wu, Y. Y., Liu, X. Y. (1990). Early development of brain stem auditory evoked potentials in Down's syndrome. *Early Hum Dev* 1990, 23, 41–51.

Kipinski, L., Maciejowski, A. (2010). Analysis of brain evoked potentials based on digital registered single-trial responses. *Adv Clin Exp Med,* 19, 289–299.

Lilienfeld, S. O., Ammirati, R., Landfield, K. (2009). Giving debiasing away: Can psychological research on correcting cognitive errors promote human welfare? *Persp Psycholog Sci,* 4, 390–398

Majnemer, A., Rosenblatt, B., Riley, P. (1990). Prognostic significance of multimodality evoked response testing in high-risk newborns. *Pediatr Neurol* 1990, 6, 367–374.

Majnemer, A., Rosenblatt, B. (1995). Prediction of outcome at school entry in neonatal intensive care unit survivours, with use of clinical and electrophysiological techniques. *J Pediatrics*, 127, 823–830.

Majnemer, A., Rosenblatt, B. (1996). Evoked potentials as predictors of outcome in neonatal intensive care unit survivors: review of the literature. *Pediatr. Neurol* 1996, 14, 189–195.

Masgutova, S. (2009). *Integration of infant dynamic and postural reflexes: Neuro-sensory-motor Reflex Integration® method for children and adults.* Orlando, FL, USA: SMEI.

Masgutova, S. (2011). *Integration of infant dynamic and postural reflexes: Neuro-sensory-motor reflex integration method for children and adults.* Orlando, FL, USA: SMEI. 257 p.

Masgutova, S. (2009). *MNRI® neurosensorimotor development: Visual and auditory reflexes integration. Facilitation program of development and learning for children and adults.* Orlando, FL, USA: SMEI.

Masgutova, S. K., Akhmatova, N. K., Kiselevsky, M. V. (2008). Clinic-immunological assessment of effect of the therapy program of neuro-sensory-motor integration of reflex patterns at chronic inflammatory diseases of respiratory system. *Russ Immun J,* 2, 454–463.

Oswald, M. E., Grosjean, S. (2004). Confirmation bias. in: a handbook on fallacies and biases in thinking, judgement and memory. *Cognitive Illusions. Eds.:* Pohl RF, Hove and NY, USA: Psychology Press, 79–96.

Piaget, J. (1976). *The grasp of consciousness: Action and concept in the young child.* Cambridge, MA: USA: Harvard.

Pike, A. A., Marlow, N. (2000). The role of cortical evoked responses in predicting neuromotor outcome in very preterm infants. *Early Hum Dev,* 57, 123–135.

Pilecki, W., Masgutova, S. M., Kowalewska, J., Masgutov, D., Akhmatova, N., Sobieszczanska, M., Kalka, D. (2012). The Impact of the rehabilitation carried out by the MNRI® - Masgutova Neurosensorimotor Reflex Integration Method in children with cerebral palsy on the results of the brain stem auditory potential examinations. *Adv Clin Exp Med.* 21 (3): p. 363-71.

Pilecki, W. (2002). Prognoza rozwoju ruchowego u dzieci z poważnie zaburzoną transmisją w pniu mozgu obserwowaną w badaniu BAEP. Wydawnictwo WTN, Wrocław. pp. 125–130.

Pilecki, W. (2002). *Ocena transmisji w obwodowym odcinku drogi słuchowej u dzieci z moALzgowym porażeniem dziecięcym* (MPDz). Wrocław, PL: Wydawnictwo WTN. 107–112.

Pilecki, W. (2003). Multimodalne potencjały wywołane w prognozie rozwoju ruchowego dzieci. In: *Komputerowe wspomaganie badań naukowych.* Wydawnictwo WTN, Wroclaw – Polanica ZdroALj, 75–80.

Pilecki, W. (2004). Względna wartosc IPL I-V – nowy parametr przydatny w badaniu sluchowych potencjaloALw pniowych u dzieci. In: *Komputerowe wspomaganie badań naukowych.* Wydawnictwo WTN, Wrocław – Polanica ZdroALj, 93–98.

Pilecki, W., Szawrowicz, T., Jagielski, J., Janocha, A., Borodulin-Nadzieja, L. (2002). Zaleznosc między prędkością transmisji w pniowej części drogi słuchowej a rozwojem ruchowym dziecka. Wydawnictwo WTN, Wrocław. pp.119–124.

Scalais, E. (1998). Multimodality evoked potentials as a prognostic tool in term asphyxiated newborns. *Electroencephalogr Clin Neurophysiol*, 3, 137–144.

Vojta, V. (1964). Current methods of rehabilitation in the treatment of motor disorders in perinatal encephalopathy. *Cesk Neurol*, 27, 81–86.

Vygotsky, L. S. (1986). *The child psychology. The problems of child development.* In 6 Books. Book – 4. Moscow, RU: Pedagogika.

Yilmaz, Y., Degirmenci, S., Akdas, F., Kulekci, S., Ciprut, A., Yuksel, S., Yildiz, F., Karadeniz, L., Say, A. (2001). Prognostic value of auditory brain stem response for neurologic outcome in patients with neonatal indirect hyperbilirubinemia. *J Child Neurol*, 16, 772–775.

We thank the children for their participation in our research and for their courage in using the evoked brain potential equipment. We wish you new levels of success in your crawling, walking, and swimming! – Authors

EEG Mapping Shows Changes in Brainwave Spectrum in Children with Cerebral Palsy During and After Masgutova Neurosensorimotor Reflex Integration Therapy

Prof. Witold Pilecki, Ph.D; Anna Pilecka-Kalamarz; Prof. Dariusz Kalka, Ph.D; Lech Kipinski, Ph.D; Svetlana Masgutova, Ph.D

Prof. Witold Pilecki

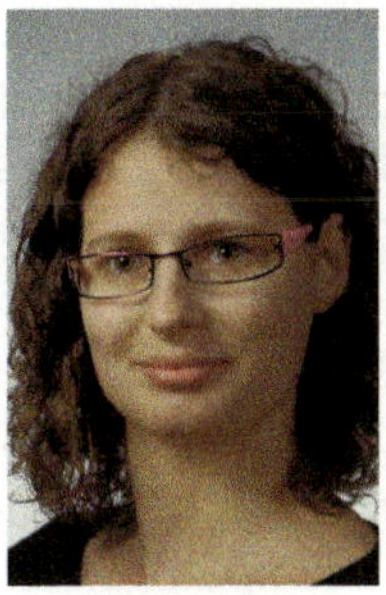
Anna Pilecka-Kalamarz

Prof. Dariusz Kalka

Lech Kipinski, Ph.D

Svetlana Masgutova, Ph.D

Abstract

The rehabilitation of children with impaired motor function due to a damaged central nervous system can be performed using a variety of methods. One of them is Masgutova Neurosensorimotor Reflex Integration (MNRI®), often used with children suffering from cerebral palsy. The objective of this report is to demonstrate the effectiveness of MNRI® therapeutic methods by means of mapping the brain. The paper presents the case of a 13-year-old boy with cerebral palsy. EEGs were performed using ASA Lab's ANT Software BV data acquisition system. Nine measuring sessions were performed before, during, and following an MNRI® session of seven specific therapeutic exercises. Spontaneous EEG activity was recorded in a 10-20 electrode system using 32 leads. The recording automatically underwent an elimination of artifacts, after which the software produced qualitative descriptions as well as mappings of the EEG frequencies. The fast Fourier transform (FFT) algorithm and ASA Lab's mapping software were used in the computer analysis. Spectral maps were compared in relative and absolute scale for the measured signals obtained before, during, and after MNRI® rehabilitation. Pre- and post treatment results showed changes in the spatial distribution and relative amplitude of alpha and rapid beta activity taking place under the influence of the therapy. It is concluded that neurosensorimotor reflex integration (MNRI®) stimulated the child's central nervous system by activating the cortical centers of his brain, thereby causing modification of spontaneous activity.

Reprinted with official permission. This article is a variation of a publication by Pilecki, W., Kipiński, L., Pilecka-Kalamarz, A., Kałka, D., and Masgutova, S. (2013). Spectral brain mapping in children with cerebral palsy treated by the Masgutova Neurosensorimotor Reflex Integration method. *Journal of the Neurological Sciences*, 21, 3, s. 363–371.

Introduction

Childhood cerebral palsy (CP, paralysis cerebralis infantum), also known as Little's disease, is a syndrome of various disorders of movement and posture resulting from permanent, non-progressive brain damage in the early stages of development (Zablocki, 1998). Childhood cerebral palsy can be defined as a set of chronic and non-progressive central nervous system disorders resulting from damage to the brain during pregnancy (20%), during the perinatal stage (60%) or in the first years of life (20%). Causes of CP include: abdominal injuries to the mother, chronic disease during pregnancy, malformations, fetal hypoxia, infection, the influence of ionizing radiation, drugs or toxins including cigarette smoking and alcohol consumption, perinatal trauma, prematurity, brain injury, hypoxia after birth, severe neonatal jaundice, and neuro-infections (Zablocki, 1998).

Four main forms of the disease can be distinguished, depending on the area of damage and the symptoms: spastic (pyramidal) and dyskinetic (extrapyramidal) forms, and atactic (brain) and mixed forms, which are the most numerous (Michalowicz, 2001).

One of the most important tasks in treating a child with CP is finding an effective method of rehabilitation. Among the many interventions for children with locomotor and nervous system disabilities are those of: Vojta, Bobath, Doman-Delacato and the Wroclaw Improvement System (Vojta, 1964; Bobath, 1958; The Doman-Delacato method, 1968). An alternative for these is the relatively new method of Masgutova Neurosensorimotor Reflex Integration (MNRI®), described in various works (Masgutova, 2005; Masgutova, Akhmatova, 2004; Masgutova, Regner, 2009).

An assessment of the impact of different methods of rehabilitation on the central nervous system can be made in an objective manner using electrophysiological procedures. Such tests are carried out by the Medical University's Department of Pathophysiology in Wroclaw, under the direction of Professor Witold Pilecki. With regard to the effect of MNRI® on childhood cerebral palsy, examinations of brainstem auditory evoked potentials have already been used, as presented in other research (Pilecki, Masgutova S., Kowalewska, Masgutov D., Akhmatova, Poręba, Sobieszczanska, Koleda, Pilecka, Kalka, 2012). In the present study we used a different electoencephalographic technique – EEG mapping, introduced to the study of the brain by Duffy et al (Duffy, Burchfiel, Lombroso, 1979).

Purpose

The object of this study is to demonstrate the effectiveness of MNRI® using mapping of brain wave frequencies performed in a computer system, based on multi-channel EEGs recorded in a child with cerebral palsy.

Material

A 13-year-old boy diagnosed with mixed form CP, with a distinct component of the spastic form, but without other chronic diseases, was treated with exercises from the MNRI® protocol. Choosing a child with chronic spasticity as part of the MPD was important because this is the most common form associated with damage to the cerebral cortex, and the EEG mapping allows one to observe the spatial distribution of bioelectrical activity in precisely that part of the brain.

Rehabilitation involved the modification of a typical MNRI® therapeutic process and included 7 consecutive progressive exercises chosen in such a way as to affect various parts of the body and to activate various motor functions. MNRI® repatterning exercises relating to the following exercises were used:

1. Foot Tendon Guard Reflex (automatic dorsal flexion of the foot) – right limb
2. Foot Tendon Guard Reflex (automatic dorsal flexion of the foot) – left limb
3. Leg Cross Flexion-Extension Reflex – right limb124
4. Leg Cross Flexion-Extension Reflex – left limb
5. Hands Supporting (parachute) reflex – both sides
6. Asymmetric Tonic Neck Reflex
7. Spinal Galant
8. Breathing Reflex for mobilizing the diaphragm.

The time spent conducting the exercises was only about 45 minutes, whereas typically a therapeutic session would last more than an hour. This change was necessary due to technicalities involved in conducting EEG measurements after each exercise. Three separate EEG recordings were done: phase 1: before exercises, phase 2: after each of the seven exercises, and phase 3: a final measurement a few minutes after the end of the session.

For the measurements, we used an ASA (Advanced Source Analysis) measurement and diagnostic system from the Dutch firm A.N.T., which consists of the ASA-Lab's computerized data acquisition system combined with an electroencephalographic amplifier from TSMI Refa-8 (*ASA User Manual* version 4.6, 2008). Thirty-two silver-plated cupped electrodes were placed against the child's scalp, in accordance with the international standard of 10-20 leads (Fp1, Fpz, Fp2, AFz, Fz, F3, F4, Fc1, Fc2, Fc5, Fc6, F7, F8, C3, Cz, C4, Cp1, Cp2, Cp5, Cp6, T7, T8, P7, P8, P3, Pz, P4, POz, O1, Oz, O2) and configured in a unipolar system with the reference electrode placed on the left ear lobe and grounded on the forehead. During the measurements, the impedance was held between the skin and the electrode < 5k Ω. The sampling frequency was 625Hz.

Method

The study was carried out in four phases:

Phase 1: EEG measurement before rehabilitation

Phase 2: EEG measurements during rehabilitation (immediately following each of the 7 MNRI® exercises.)

Phase 3: EEG measurement a few minutes after completion of the MNRI® session

Phase 4: Calculations, production of brain mapping images and analysis of results

The calculations were made off-line on the ASA computer system (*ASA User Manual* version 4.6, 2008). The preprocessing of signals was applied through low-pass filtration with a Butterworth filter 0.53 Hz, providing automatic detection of artifacts outside the range of amplitude -150 - 150μV. The elimination of interference was necessary because artifacts of locomotive origin are a significant problem during the rehabilitation of motor functions from an electrophysiological point of view. Only selected portions of the recording containing undisturbed EEG activity were chosen for the calculations by opting for the fragmentation of signals on windows 1s. Every 1s window containing artifacts detected by the software was eliminated (Fig. 1). A compromise

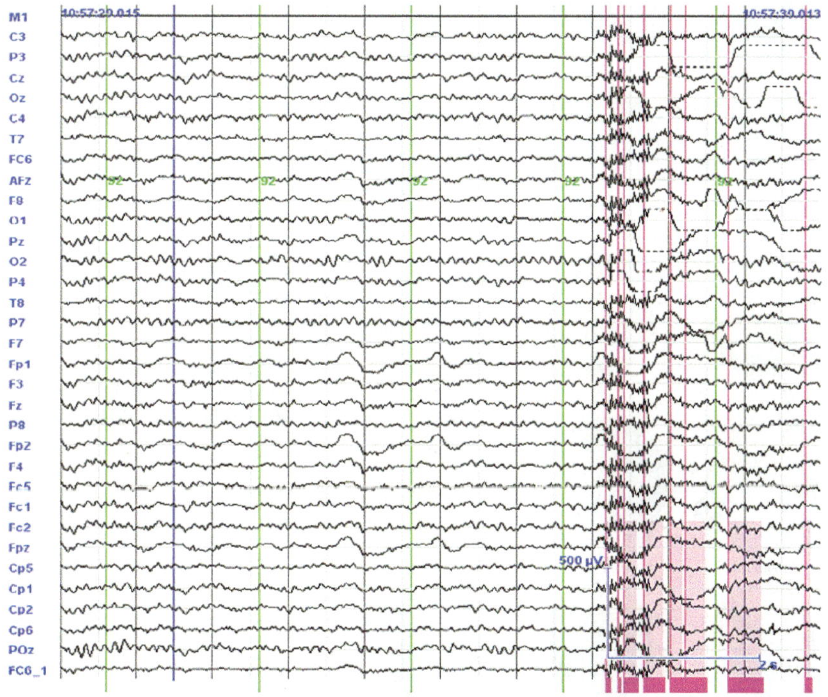

Figure 1: Detection of artifacts in the ASA program. Interference marked on the right side was automatically cut from EEG recordings before performing calculations.

between the frequency resolution of the calculations, and records sufficiently clean of artifacts was agreed on.

Each EEG signal's spectrum was calculated for the purpose of demonstrating the frequencies generated by the brain's neural structures in each stage of the study. For this purpose we used the so-called Discrete Fourier Transform (DFT) as determined by introducing the ASA algorithm of Fast Fourier Transform (Fast Fourier Transform, FFT) into the system (Brigham, Oran, 1988). The average spectrums were calculated in windows 1s. The spectrum frequencies of EEG signals for each lead obtained in this way, after the application of the appropriate interpolation techniques, allowed for the creation of a spectral power map of EEG signals in selected frequency bands, and thus imaging their spatial distribution in each phase of the study.

We analyzed the EEG spectrum from 0.5 Hz to 30Hz, divided into five sub-ranges: 0.5 - 3.5 Hz (delta), 3.5 - 7.5 Hz (theta), 7.5 - 12, 5 Hz (alpha), 12.5 - 20 Hz (free beta, beta 1) and 20 - 30 Hz (fast beta, beta 2). For these sub-bands we performed spectral distribution maps in 3D, using interpolation techniques and the standard model of the head as found in the functions library of the ASA software. It was decided to create maps in two scales: absolute [μV2/Hz] and relative [%]. Absolute maps allow one to compare the spatial distribution of signals in the different ranges of the energy spectrum in terms of amplitude (the same as in research by the electroen-

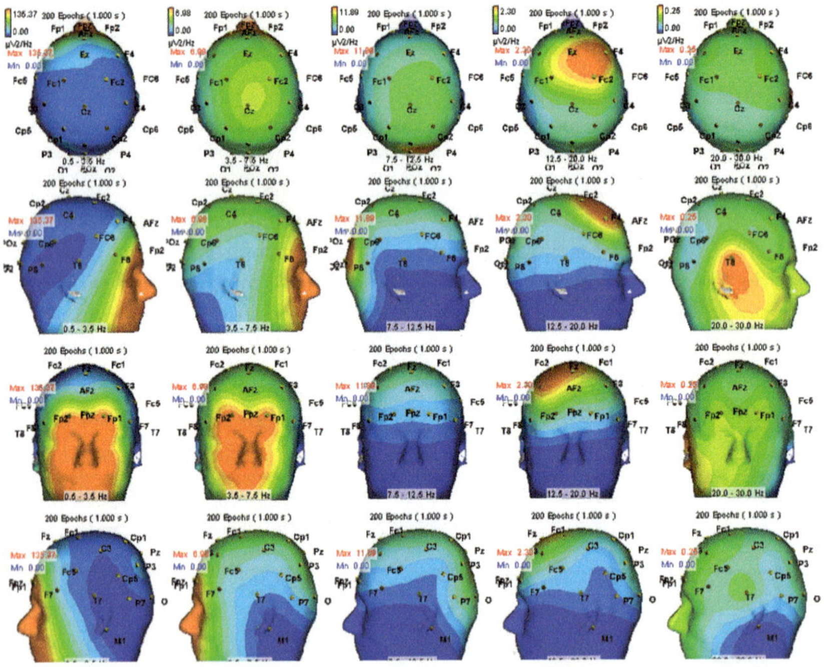

Figure 2: Maps of the spatial distribution of the power spectrum expressed in absolute units [μV2/Hz] for resting EEG performed at the outset of the experiment. In subsequent columns different projections are shown in 3D using a standard model of the head for the frequencies: delta, theta, alpha, beta 1, and beta 2.

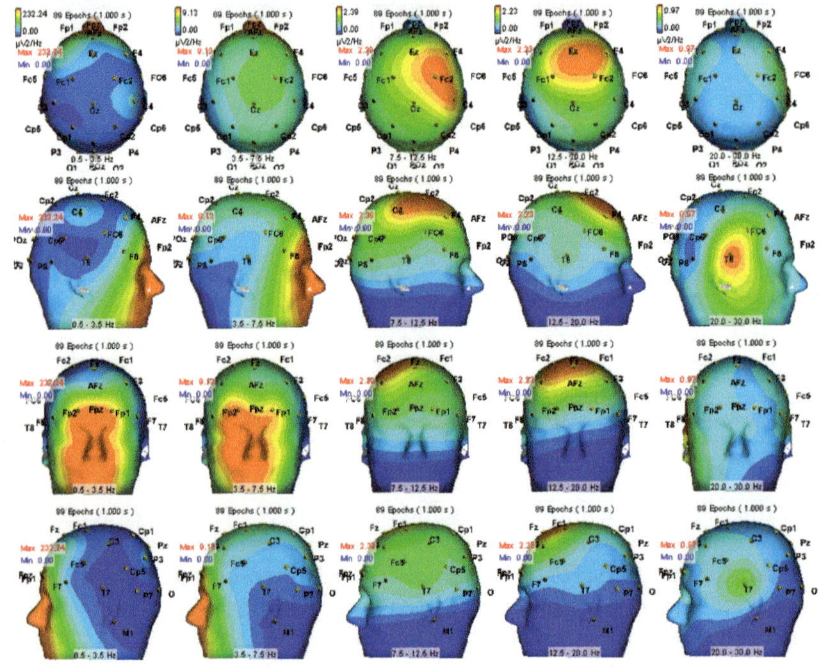

Figure 3: Maps of the spatial distribution of the power spectrum expressed in absolute units [μV2/Hz] for one of the EEG recordings made during phase 2 (after MNRI® work involving the limbs). Subsequent columns show the different projections in 3D using a standard model of the head for the frequencies: delta, theta, alpha, beta 1, and beta 2.

cephalograph) and it shows areas that are mostly intensely activated in the sub-ranges of the spectrum. In contrast, representing the transmitted energy of the signal in a given frequency range relative to the energy of the entire spectrum allows one to determine which areas are primarily responsible for EEG activity in a particular sub-range. Images are presented in an automated scale using the RGB color scale.

Results

A qualitative analysis of EEG records from phases 1, 2, and 3 (before, during, and after rehabilitation by MNRI®), show certain differences among results. A symmetrical record was achieved, with no seizure activity, showing a mixture of slow delta-theta waves mainly in the frontal and parietal leads, alpha waves mainly above the occipital leads, and a correct halting response. In addition, in the records taken during phase 2, rehabilitation, a greater admixture of fast beta waves is found in fronto-temporo-parietal leads.

Through EEG frequency mapping we obtained images of the distribution of the spectral power of the EEG signal for each of the 9 registrations. Due to the short scope of this study we present only selected images, presenting the results for recordings done in phase 1 (Figs. 2 and 5), in phase 2 after the Galant Reflex exercise (Figs. 3 and 6) and in phase 3, after the whole rehabilitation session (Figs. 4 and 7). Figs. 2 to 4 present maps in absolute scale [μV2/Hz] and Figs. 5 to 7 are drawn in relative scale [%]. Each series of images was analyzed separately.

Mapping of the recorded output in an absolute scale shows the dominance of low frequencies in the upper frontal leads, and the lowest amplitude in the vicinity of the temporo-occipital leads. Alpha activity is correctly located mainly in the occipital region. Fast frequencies from beta waves are of the highest amplitude in the fronto-parietal leads. This information leads to presupposition that the quality of focusing, presence, and awareness in the patient can increase.

During rehabilitation, the minor periodic voltage fluctuations, delta and theta frequencies did not change

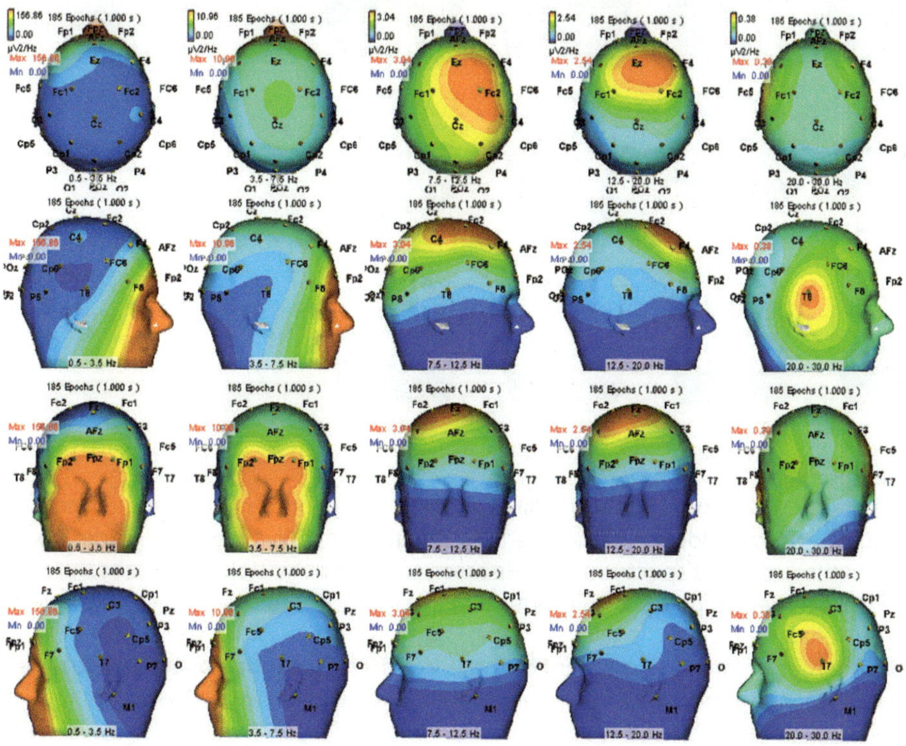

Figure 4: Maps of the spatial distribution of the power spectrum expressed in absolute units [μV2/Hz] for an EEG performed after MNRI® rehabilitation Subsequent columns show the various projections in 3D using a standard model of the head for the frequencies: delta, theta, alpha, beta 1, and beta 2.

significantly throughout all three phases.

Changes did occur in the alpha frequency range (7.5-12.5 Hz). Alpha activity initially observed (phase 1) in the occipital leads disappeared at the beginning of rehabilitation (phase 2). Subsequent phases of the study show clearly that there is a lack of alpha activity in this area, and at the same time the alpha activity did appear above the central leads (Fig. 3 middle column), mainly on the left side, although it had a relatively not high amplitude. This effect proved to be permanent, as it held until phase 3, the end of the experiment (Fig. 4 middle column). The result again shows a possibility of more proper organization of brain wave spectrum in the patient and a basis for improvement of the abilities to focus, increase awareness and critical thinking.

EEG frequencies in the range free beta (12.5 - 20Hz) did not change in spatial distribution or in average absolute values. Certain changes were observed for fast beta (20 - 30Hz): this stronger voltage appeared during phase 2, rehabilitation, (expiring after its completion) in the temporal leads of the dominant hemisphere (Figs. 3 and 4, last column).

Mapping in the relative scale brings additional information. As in the absolute scale, the dominance of delta activity asserts itself in the frontal leads; however, the theta function has the highest percentage (17%) in the central region, while low in the frontal leads. Clearly, this is related to the extremely high dominance of lower brain waves (delta and theta) in this area, approaching up to 89% (Fig. 5, first column). A second strongly expressed frequency in phase 1 is alpha activity above the occipital leads, filling 48% of the signal spectrum at

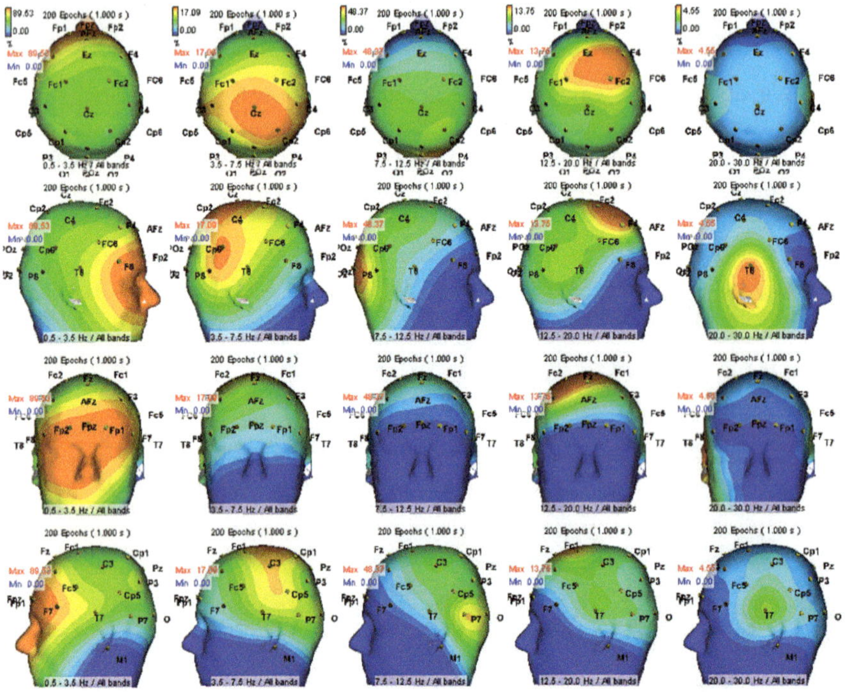

Figure 5: Maps of the spatial distribution of the power spectrum, expressed in the relative scale as the ratio of power in a given range of frequency to the power of the entire spectrum. Illustrations were made for resting EEG recorded in phase 1. Subsequent columns show various projections in 3D using a standard model of the head for the frequencies: delta, theta, alpha, beta 1, and beta 2.

this location (Fig. 5, middle column).

The percentage of free beta (12.5 - 20Hz) is strongly (13%) expressed in the frontal central leads. In contrast, the fastest EEG output frequency (beta), is localized in the temporal regions, mainly on the right side. It is also associated with low activity in the rest of the spectrum in these areas, which is not visible on the absolute scale maps. The percentage of the total spectral power of EEG output in this (beta) frequency range is the lowest of all and reaches just over 4% in channel T8.

During phase 2, rehabilitation, as before, no significant changes in either distribution or amplitude of delta and theta activity were shown. The disappearance of alpha activity in the occipital leads was confirmed. This frequency range strongly (several dozen %) saturates spectrum signals measured in the temporo-parietal electrodes, which is a new discovery. In contrast to the absolute scale maps, relative scale maps show a tendency of occipital alpha activity to revert after the exercises. This discovery suggests that MNRI® treatments temporarily arrest/inhibit alpha activity in its physiological location, while stimulating it in regions operationally linked to the motor cortex system as it works in a healthy brain (Fig. 6, middle column). It is interesting that, at the same time, it does not return to its phase 1 level immediately after phase 2. The absolute scale maps show this, as do

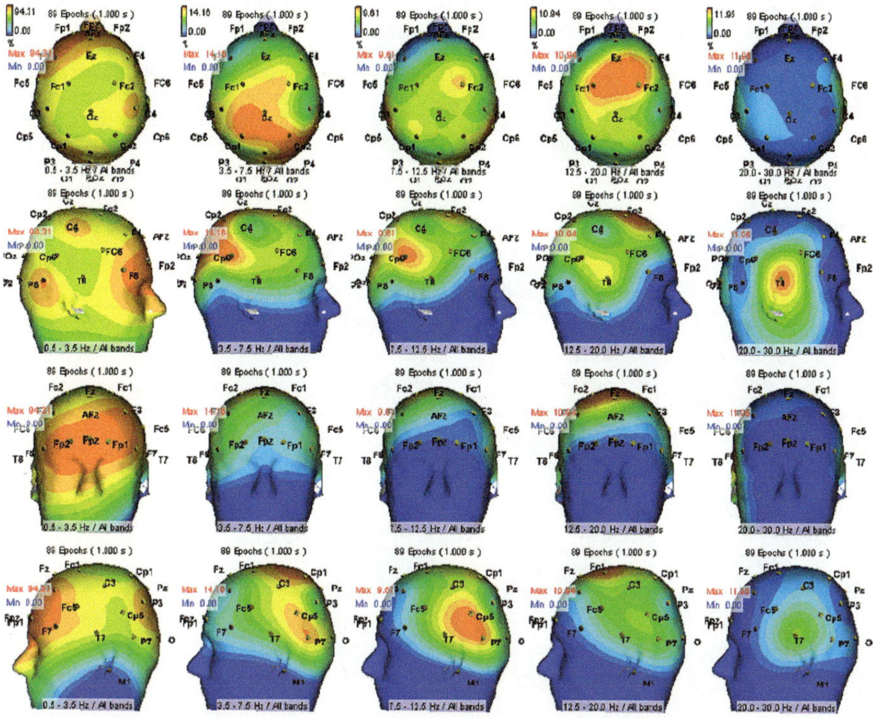

Figure 6: Maps of the spatial distribution of the power spectrum, expressed in a relative scale, as the ratio of power representing a given range of frequency to the power of the entire spectrum signal. The illustrations were made for one of the EEG recordings taken during phase 2 (after stimulation of the limbs). Subsequent columns show various projections in 3D using a standard model of the head for the frequencies: delta, theta, alpha, beta 1, and beta 2.

the maps in Fig. 7 where alpha activity represents a lower percentage of the spectrum in phase 3, after MNRI®, than in phase 1, before. This can mean a tendency that MNRI® can reach a stable transition in the work of the brain waves.

Free beta activity of 12.5-20Hz does not change its spatial properties or relative amplitude during the test. However, maps in relative scale confirm the increase of rapid beta activity (20-30Hz) in the temporal leads (up to several percent), expiring after completion of MNRI® exercises focused on the extremities (Figure 6, and Figure 7, last column).

Conclusions

We conclude that rehabilitation using MNRI® in a child patient with cerebral palsy caused a reorganization of spontaneous electrical activity in his brain. This is observed in the alpha frequency range as well as in fast beta activity. Increased activity in these frequency bands during rehabilitation appears in the parietal and temporal locations. This may be connected with the positive and stable therapeutic effect of this method on motor disorders originating in the central nervous system. Our observations are consistent with the findings of other authors who have noticed changes in EEG mapping in children with CP during experiments related to

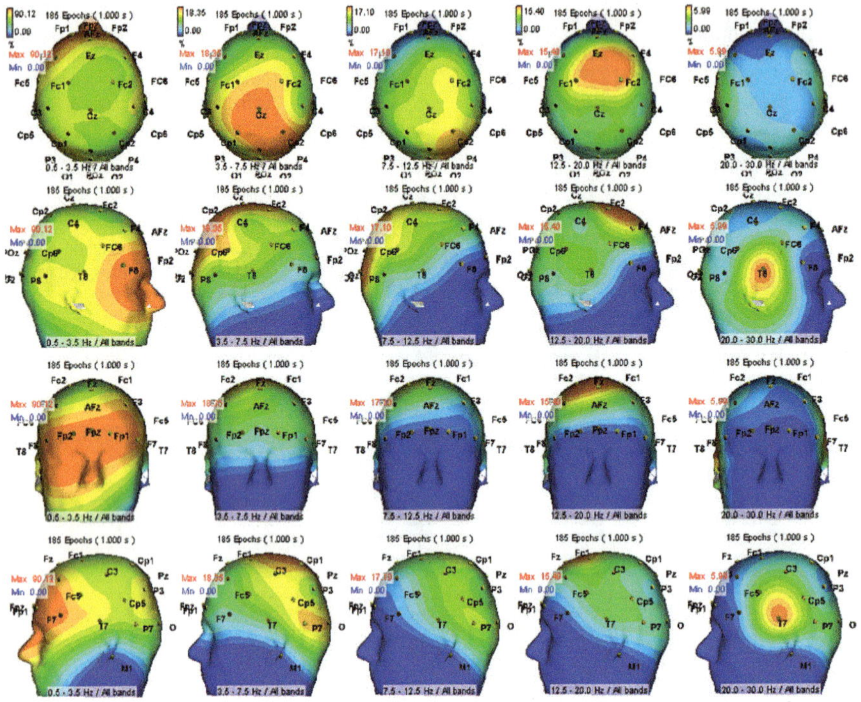

Figure 7: Maps of the spatial distribution of the power spectrum, expressed in a relative scale, as the ratio of power representing a given frequency range to the power of the entire spectrum of the signal. Illustrations were made for the final EEG (phase 3), performed after completion of the entire MNRI® session. Subsequent columns show the various projections in 3D using a standard model of the head for the frequencies: delta, theta, alpha, beta 1, and beta 2.

the motor system (Shin, Lee, Hwanh, You, Im, 2012).

It should be made clear that although the results presented in this analysis are positive, they apply only to a single case. Taking into account inter-individual variability and the influence of many external factors on EEG results, drawing objective conclusions about the physiological effects of MNRI® would require testing a homogeneous group of children with childhood cerebral palsy, and comparing mapping results to the results obtained in a control group of healthy children. A future publication by our team will be devoted to this issue.

References

ASA User Manual version 4.6, A.N.T. BV, Enschede (2008).

Bobath, C. K. (1958). Neurophysiological bases of authors' method of treatment of cerebral palsy. *Acta Neurol Psychiatr Belg*, 1958, 58, s. 469–474.

Brigham, E., Oran, E. (1988). *The Fast Fourier transform and its applications.* Englewood Cliffs, NJ, USA: Prentice Hall.

The Doman-Delacato method. (1968). *Journal of the Iowa Medical Society*, 58, 507–509.

Duffy, F. H., Burchfiel, J. L., Lombroso, C. T. (1979). Brain electrical activity mapping (BEAM): A method for extending the clinical utility of EEG and evoked potential data. *Annals of Neurology*, 5(4), s. 309 – 321.

Masgutova, S., Masgutov, D. (2005). *The integration of facial reflexes by Svetlana Masgutova. Working techniques supporting the development of motor skills and speech.* Warszawa, PL: MINK.

Masgutova, S. (2005). *Reflexes as a basis for the development of the nervous system and the development of movement patterns in infancy in Modern methods of stimulating the development of motor and speech.* Krynica Górska, Warszawa, PL: MINK. s. 14–36.

Masgutova, S., Akhmatova, N. (2004). *The integration of dynamic and postural reflexes of the entire mobility system.* Warszawa, PL: MINK.

Masgutova, S., Regner, A. (2009). *The development of child's speech in the light of sensory integration.* Wrocław, PL: Continuo.

Michalowicz, R. (Ed.) (2001). *Childhood cerebral palsy.* Wydawnictwo Lekarskie PZWL. Warszawa, Pl.

Pilecki, W., Masgutova, S., Kowalewska, J., Masgutov, D., Akhmatova, N., Poręba, M., Sobieszczanska, M., Kolęda, P., Pilecka, A., Kalka, D. (2012). The Impact of Rehabilitation Carried out Using the Masgutova Neurosensorimotor Reflex Integration Method in Children with Cerebral Palsy on the Results of Brain Stem Auditory Potential Examinations. *Advances in Clinical and Experimental Medicine*, 21, 3, s. 363–371.

Sadowska, L. (2003). Comprehensive diagnostics and stimulation of the development of children with congenital and acquired dysfunctions of the central nervous system according to the Wroclaw model of improvement. *Psychosocial Problems of Child Development: Diagnostic and Therapeutic Aspects.* Adam Marszałek, Toruń 2003, s. 81–97.

Shin, Y. K., Lee, D. R., Hwanh, H. J., You, S. J., Im, C. H. (2012). A novel EEG-based brain mapping to determine cortical activation patterns in normal children and children with cerebral palsy during motor imagery tasks. *NeuroRehabilitation*, 31(4), s. 349 – 35.

Vojta, V. (1964). Current methods of rehabilitation in the treatment of motor disorders in perinatal encephalopathy. *Cesk Neurol*,1964, 27, s. 81–86.

Zablocki, K. J. (1998). *Childhood cerebral palsy in theory and therapy.* Wydawnictwo Akademickie "Żak". Warszawa, Pl.

We thank all the families and the children participating in and supporting our brain research and for scientifically showing such great results with the MNRI® procedures! We appreciate your patience during all test procedures and for tolerating the unpleasant electrodes on your heads as well as the auditory and visual stimuli. You were the best and you helped uncover a new level of science! – Authors

Neuro-Motor Facilitation in Children with Cerebral Palsy Using the Masgutova Method®

Karolina Niezgodzka, PT, MNRI® Core Specialist, Warsaw;
Prof. Ludwika Sadowska, MD; Joanna Kowalewska, PT, MNRI® Instructor;
Anna Maria Choinska, MD., Ph.D, Wroclaw, Poland

This abstract presents, in brief, the study, *Effects of Activation of the Neuromotor System in Children with CNS Damage using MNRI®* by K. Niezgodzka, L. Sadowsksa, J. Kowalewska, and A-M. Choińska as reported at International Conference of the Day of Handicapped: *Healthy Children – Healthy Europe* (Poland) and received an award for second place.

Karolina Niezgodzka

Prof. Ludwika Sadowska, MD

Joanna Kowalewska

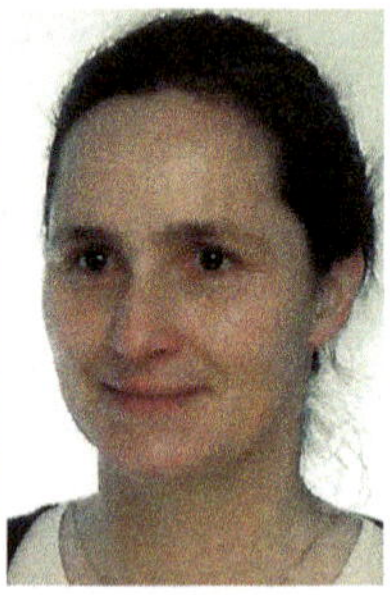
Anna Maria Choinska, MD, Ph.D

Introduction

Damage to the central nervous system (CNS) is characterized by a combination of symptoms of varying etiology and complex clinical symptomatology, including disorders of the CNS as determined using the Vojta Method (reflex locomotion), to various forms of cerebral palsy (CP) as described by the International Classification of GMCs (Palisano, Hanna, Rosenbaum, et al., 2000; Palisano, Rosenbaum, Walter, et al., 1997; Sadowska, 2001).

In the psychomotor assessment of children with damage to the CNS, functional tests are used to verify the reactions and reflex mechanisms of posture, rectification, and balance. Currently there are many diagnostic and therapeutic concepts used in treating children with brain damage such as NDT-Bobath, Vojta, Pető, Ayres, Castillo-Morales, and Masgutova Neurosensorimotor Reflex Integration (MNRI®). MNRI® evaluates and also provides integrating exercises for reflexes, a genetic pattern of physiological functions created by nature as a code and structure insuring neurosensorimotor development based on the clinical functioning, neurological status, and biological age of the child (Sadowska, 2004, 2012; Kaluzna, 2004).

The basis of the MNRI® method is the fact that a child is born equipped with a complex system of natural resources, reflexes, and instincts. Reflexes and primary motor responses in infancy, early childhood, and later in life pave the way for the connection between sensory and motor neurons creating reflex circuits and connecting with the motor system, brain, and body functions. These initial reactions and reflexes occur in children as automatic sensory-motor patterns.

Reflexes fulfill the role of activating the nervous system, a defensive mechanism triggering survival strategies in situations where the hindbrain focuses on reactions of survival, blocking the conscious reasoning process in the new cortex. There are several reasons why we need reflexes, such as the development of the nervous system, physiological survival, and the development of a protective defense system (Masgutova, Akhmatova, 2005). Reflexes can exhibit various dysfunctions such as poor coordination in sensory and motor processing and lack of development and maturation with negative affects on overall neuro development and learning.

The MNRI® method puts forth the idea of a 'conversation' with pathological reflex patterns in a positive way, not activating their defense programs, offering therapy that works with the genetically programmed static and dynamic reflexes of the mind-body system. MNRI® specialists create modeling exercises or repatterning procedures showing the brain how it can return to a positive system of functioning normally (Masgutova, Sadowska, 2008; Masgutova, 2006). The MNRI® programs used in the therapy of these pathological reflex patterns consist of the following: Neuro-structural integration of reflex patterns, integration of Archetype Movements, Neurotactile therapy, Oral-Facial and Visual-Auditory Reflexes, and the re-patterning of reflexes. (S. Masgutova, D. Masgutov, 2005; Masgutova, Regner, 2009).

An Assessment of 30 dynamic reflex and postural patterns is completed before an individual program is developed in order to determine which reflexes are disturbed, or non-integrated, and develop a plan direction of corrective, integration exercises. For every malfunctioning reflex there are specific exercises aimed at the coordination and integration between the sensory and motor neurons (Masgutova, 2004; Masgutova, Akhmatova, 2005). This work is important for children who fall in developmentally delayed at-risk groups and those diagnosed with neurological damage in the operation of reflex arcs similar to children with CP (Sadowska, 2012).

Table 1 Study groups of children with CP based on neurological symptom, age, and gender.

Groups of Children with CP based on their neurological symptoms, age and gender									
CP Groups / Gender and age		GR I Spastic hemiparesis (Hemiparesis) n = 19		GR II Spastic paresis of four limbs (Tetraplegia) n = 8		GR III Cerebellar CP syndrome (Ataxia and Hypotonia) n = 22		GR IV Extrapyramidal CP (Dyskinesia accept Dystonia) n = 11	
Age, Number and %		n	%	N	%	n	%	N	%
Boys n=33	0- 2 year old	3	5	0	0	4	6.7	2	3.3
	2.1- 4	5	8.3	3	5	2	3.3	3	5
	4.1- 6	0	0	0	0	3	5	0	0
	6.1-8	0	0	0	0	1	1.7	1	1.7
	Total	8	13,3	3	5	10	16,7	6	10
Girls n=27	0-2	2	3,3	1	1,7	4	6,7	2	3,3
	2.1- 4	8	13,3	3	5	4	6,7	1	1,7
	4.1 – 6	1	1.7	0	0	2	3.3	1	1.7
	6.1- 8	0	0	1	1.7	2	3.3	1	1.7
	Total	11	18.3	5	8.3	12	20	5	8.3
	Total n = 60 (100%)	19	31.6	8	13.3	22	36.7	11	18.3

Children who have brain injuries described as CP are found to exhibit involuntary psychomotor retardation and excessive muscle tension or laxity. The negative impact of CP on the development of motility is an impairment or delay of the equivalent reaction mechanisms and postural muscular tone control. The description and criteria of clinical evaluations of four age groups with large motility disorders of children with brain damage are determined by the international classification system Gross Motor Function Classification System (GMFCS) Test. This is an approximate five-level scale specifying increasing levels of movement disorders to assess the overall current level of large motor skills and their limitations and possibilities (see Table 2). Generally, when the GMFCS scale has a high point level, the intensity of the restriction of movement activity will be noted (Palisano, Hanna, Rosenbaum, et al., 2000; Palisano, Rosenbaum, Walter, et al., 1997).

Purpose of This Study

This study will determine the effectiveness of neuromotor rehabilitation on 23 chosen reflex patterns of children with CP in infancy, preschool, and early school using the MNRI® program and the evaluation of motor skills according to the International classification system GMFCS.

Materials and Method of Research

Characteristics of the study group

The research study was performed on 60 children aged 0 – 8 years, who were medically diagnosed with CP, in the form of: 1) spastic paresis of varying severity (hemiparesis), 2) spastic paresis of four limbs (tetraplegia), 3) mild cerebellar syndrome with mental disorders (ataxia and hypotonia), and 4) extrapyramidal syndrome (dyskinesia, not including dystonia). The children all attended MNRI® camps organized by the Dr. Svetlana Masgutova International Institute (SMEI) in Mielno, Poland at the Sirena Rehabilitation Center. Participants of the two-week camp were administered rehabilitation for 6 hours a day by MNRI® therapists with basic training as PT, OT, SP, and other fields of special corrective education. Parents/caregivers of the children who were involved in the therapy and this study gave their consent for the children to participate in this study. Evaluation and statistical analysis of the results were carried out before and after the completion of the MNRI® therapy program. Statistical information for the children with regard to clinical diagnosis, age, and gender is illustrated in Table 1.

Description of research methods

The level of development of reflex patterns was diagnosed within a scale of 0 – 20 points according to the Masgutova system. The Assessment takes into account the accuracy of the sensory-motor reflex pattern, the direction of the reaction, strength (regulation of muscle tone), latency (time of onset, duration and termination of the reaction), and the symmetry of the response to a stimulus (Masgutova, Akhmatova, 2010). Reflex Assessment criteria with the level of pathology, dysfunction, and norm, and their patterns is discussed in the chapter, *MNRI® Assessment for Determining the Level of Reflex Development* in this book.

The children in this study were also evaluated using the GMFCS system to determine their level of motor disability. The GMFCS was administered at the beginning and end of the MNRI® camp experience. The criteria and description of the GMFCS scale is on Table 2. A two-way analysis of variance (ANOVA) was calculated on participants' ratings on the MNRI® Reflex Assessment and the GMFCS to the statistical verification tests ANOVA and Chi-2McNemary using the level of $p < 0.05$ as statistically significant.

Table 2: Gross Motor Function Classification System
Criteria and Description of Scale

GMFCS Level I:
- **Can walk indoors and outdoors and climb stairs without using hands for support**
- **Can perform usual activities such as running and jumping**
- **Has decreased speed, balance and coordination**

GMFCS Level II:
- **Has the ability to walk indoors and outdoors and climb stairs with a railing**
- **Has difficulty with uneven surfaces, inclines or in crowds**
- **Has only minimal ability to run or jump**

GMFCS Level III:
- **Walks with assistive mobility devices indoors and outdoors on level surfaces**
- **May be able to climb stairs using a railing**
- **May propel a manual wheelchair and need assistance for long distances or uneven surfaces**

GMFCS Level IV:
- **Walking ability severely limited even with assistive devices**
- **Uses wheelchairs most of the time and may propel own power wheelchair**
- **Standing transfers, with or without assistance**

GMFCS Level V:
- **Has physical impairments that restrict voluntary control of movement**
- **Ability to maintain head and neck position against gravity restricted**
- **Impaired in all areas of motor function**
- **Cannot sit or stand independently, even with adaptive equipment**
- **Cannot independently walk but may be able to use powered mobility**

(Palisano, Rosenbaum, Walter, Russell, Wood, Galuppi, 1997)

Results and Discussion

Analysis of levels of integration of specific reflexes before and after therapy

Figure 1 presents the results of the average levels of integration of 23 reflex patterns before therapy (in red/orange), and after the MNRI® camp (in green). The lowest level of integration before treatment was characterized by reflexes: X19 - Flying and Landing (average value of 4.15), X13 – Bauer Crawling (average value of 4.50), and X9 - Leg Cross Flexion-Extension (average value of 4.89). The highest level of integration before treatment were characterized by reflexes: X27 - Pavlov Orientation (average value of 8.48), X15 – Fear Paralysis (average value of 7.65), X14 - Moro (average value of 7.65), and X6 – Hands Pulling (average value of 7.37). All reflex pattern test results show a higher level of integration after therapy at the MNRI® camp. Detailed comparative analysis of statistical tests (t-test for dependent samples) showed statistically significant differences at level $p < 0.05$.

Interestingly, the change in the level of reflex integration in the children of the study is different depending on the form of CP, as illustrated in Table 3. Analysis of changes in the level of the integration of reflexes in children, depending on the clinical form of paralysis showed that children with symptoms of extrapyramidal syndromes achieved, on average, the best improvement in the level of the development of reflexes, followed by children of hemiparesis, tetraplegia, and ataxia (cerebellar form).

Figure. 1

Average Level of Reflex Development Before and After MNRI® Therapy

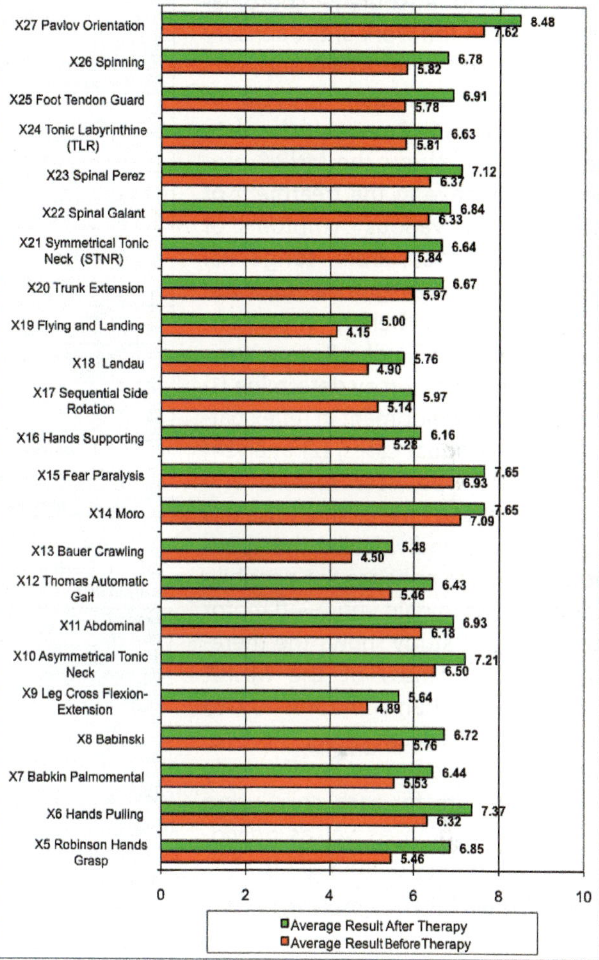

Testing motor disability level in children in GMFCS scale

Analysis of motor disability in the GMFCS scale in the study group, before (Test 1) and after (Test 2) MNRI® treatment during the rehabilitation camp, was carried out on the percentile basis, as illustrated in Table 4.

Children examined before therapy (Study 1 - Test-1) attained skills from levels III-V, which indicates low psychomotor development, meaning a majority, as 39 children (65%) reached the fifth level, 16 children (27 %) achieved level IV and only 5 children (8%) achieved a level III on the GMFCS scale. The level of motor skills was also examined at the end of the rehabilitation camp. Statistical comparison shows significant differences $p < 0.05$, which are indicated in Table 4 in bold with an asterisk*.

The age group 0-2 years from level V, with limited self-control of reflexes and maintenance of antigravity posture of the torso and head, showed improvement entering compartment IV, acquiring skill rotation from back to stomach. After treatment, three children aged 2-4 years with level V motor skills, passed to level IV, and three children from level IV to level III, already achieving the ability to sit with help and to move about on the floor. They also started demonstrating an ability for trunk righting posture with the help of people and equipment.

Statistical analysis shows significant differences in the level of motor disability in test groups I and II. However, within the age group 4-6 years of seven surveyed, four present level IV, that is, sitting with the help of equipment and moving with the help of people and Instruments for a short distance in trunk righting but having difficulty maintaining balance on an uneven floor. The remaining three children, in terms of motor skills display level V, that is, impaired movement including a limited ability of self-control and maintenance of an anti-gravity posture. After treatment in this age group, one child achieved abilities that reclassified him from level V to level IV.

In the oldest group, children aged 6-8 years, in the six surveyed before treatment two qualified for the III skill level, which translates to the ability to walk on flat surfaces indoors and outdoors with the help of additional equipment, and three presented level IV, whose skills remained at level of six years of age. One child presented level V of development. Improvement after therapy was observed in one child, recorded in the change of classification from level IV to level III, which can be considered as a great therapeutic success.

Discussion

The MNRI® therapeutic program is aimed at improving sensorimotor and cognitive function within the concept of reflex integration in children with various forms of CP. The MNRI® program proposes unique opportunities to encourage coordination between the work of the sensory and motor neurons which supports the neuro-rehabilitation process of children with CP. Based on the data gained in this study, the MNRI® program is effective in the treatment of children with various symptoms of brain dysfunction, both congenital and acquired (Masgutova, Sadowska, 2004). Physiotherapy specialists, knowing the profile of reflex patterns in children with CP may determine (1) the level of development in reflex patterns of individual cases, (2) analyze/find the links between reflexes and motor skills, and (3) effectively use this information to individualize therapeutic programs to enhance the possibilities of a child to gain improved motor function. This study shows the therapeutic effectiveness of the MNRI® method in children with CP with two weeks of intensive program. These results can be used to more effectively plan a program for each child using the MNRI® program as well as individualize it for different types of CP.

Table 3. Differences in the level of reflex development before and after MNRI therapy in children with hemiplegia, tetraplegia (=quadriplegia) and ataxic CP and extrapyramidal CP (athetosis).

#	Reflex	Reflex #	Average difference			
		X	Hemiplegia n=19	Tetraplegia n=8	Ataxia n=22	Extrapyramidal CP n=11
1	Robinson Hands Grasp	X5	0.79	1.69	1.43	2.14
2	Hands Pulling	X6	0.79	1.50	0.93	1.41
3	Babkin Palmomental	X7	0.71	1.00	0.82	1.36
4	Babinski	X8	0.89	1.19	0.77	1.27
5	Leg Cross Flexion-Extension	X9	1.03	0.94	0.48	0.68
6	Asymmetrical Tonic Neck Reflex (ATNR)	X10	0.71	0.75	0.50	1.09
7	Abdominal Sleep Posture	X11	0.82	0.88	0.57	0.91
8	Thomas Automatic Gait	X12	1.18	0.88	0.68	1.27
9	Bauer Crawling	X13	1.29	0.81	0.64	1.27
10	Moro	X14	0.63	0.56	0.25	1.05
11	Fear Paralysis	X15	0.79	0.63	0.34	1.45
12	Hands Supporting	X16	1.21	0.88	0.41	1.23
13	Sequencial Rolling	X17	0.97	1.11	0.30	1.45
14	Landau	X18	0.74	0.88	0.77	1.23
15	Flying and Landing	X19	0.84	1.19	0.68	0.95
16	Trunk Extension	X20	0.61	0.69	0.66	0.95
17	Symmetrical Tonic Neck	X21	1.00	0.56	0.68	0.86
18	Spinal Galant	X22	0.50	0.63	0.68	0.09
19	Spinal Perez	X23	0.84	0.50	0.64	1.00
20	Labyrinthine–tonic	X24	0.92	0.31	0.80	1.05
21	Foot Tendon Guard	X25	1.16	0.50	0.82	2.18
22	Spinning	X26	0.95	1.06	0.84	1.14
23	Pavlov Orientation	X27	1.34	0.81	0.45	0.86
24	**Average Difference for all reflexes**		**0,90**	**0,87**	**0.66**	**1.17**

Table 4. Level of disability according to the GMFCS scale before and after treatment compared to Chi-²McNemary test.

Level of disability according to the GMFCS scale	I GR. (0-2 y) n=18*		II GR. (2. 1-4 y) n=29*		III GR. (4. 1-6 y) n=7		IV GR. (6. 1-8 y.) n=6	
	Test-1	Test-2	Test-1	Test-2	Test-1	Test-2	Test-1	Test-2
I	0	0	0	0	0	0	0	0
II	0	0	0	0	0	0	0	0
III	0	0	3	6	0	0	2	3
IV	3	5	6	6	4	5	3	2
V	15	13	20	17	3	2	1	1
Total	18	18	29	29	7	7	6	6

Above, GMFCS scale results before and after MNRI® treatment show the percentage of children placed in each of four groups, based on their test performance compared to Chi-2McNemary test (asterisks indicate statistically significant results according to McNemar's Chi-square test).

Summary and Conclusions

The accuracy of the diagnosis determines the effectiveness of the rehabilitation process for children with brain damage at birth. This study explores the use of tests for assessing the functional motor skills (psychomotor test GMFCS scale) and the MNRI® Reflex Assessment (targeting the level of maturation of 23 reflex patterns). The neurological state of children with brain damage is quite diversified and diagnosis is difficult, both in terms of determining the level of reflex development, as well as motor skills related to chronological age of a child. Both the Assessment of the level of reflex development of the 23 studied reflex patterns as well as the test of the level of motor disability according to the GMFCS scale are complicated and demand the special training of a professional. This study demonstrates that when appropriate protocols are administered, the possibility of a positive outcome of the therapy process is probable and falls in a statistically significant range ($p > 0.05$). The research presented in this paper offers the following conclusions:

1. Children with CP are characterized by a pathological or low level of neurological development of reflexes and impaired skills in spontaneous motor activity.

2. Improvement in the reflex development in children with CP after 2 weeks of therapy following the typical program offered at the MNRI® camp was statistically significant ($p < 0.05$). Children with extrapyramidal CP symptoms achieved the greatest improvement followed by children with hemiplegia, quadriplegia, and ataxia.

3. The younger age group of children (under 4 years) in MNRI® treatment favorably correlated with the level of reflex integration and a reduction of motor disability at a higher statistically significant level of $p < 0.05$.

Abstract

The aim of this study was to determine the degree of development of 23 neurological reflexes in spontaneous motor activity according to the MNRI® method and the level of motor disability, measured by the GMFCS scale in 60 children with different clinical forms of cerebral palsy, aged 0-8 years. This study group of children was divided into four age groups (I- from 0 to 2 years, II- 2-4 years, III- 4-6 years, IV- 6-9 years) undergoing improvement according to the MNRI® program offered at a two-week MNRI® camp. The results presented in the tables and graphs, which were verified statistically significant at the $p > 0.05$ level, demonstrated the efficacy of the MNRI® therapy, which was manifested by an increase in the average level of reflex development, while still remaining within the limits of a pathology or deep dysfunction. In each tested age group, after therapy, children were found to demonstrate a significant change in their level of motor disability on the GMFCS scale.

References

Kaluzna, A. (2004). Principles of diagnosis and treatment of developmental disorders of sensory integration in children [in] edited by L. Sadowska *Neurophysiological methods of rehabilitation of children with developmental disorders* ed. Wroclaw, PL: AWF p. 142-162.

Masgutova, S. (2004). *Children with challenges: Motor Development and reflexes integration, neurosensorimotor and reflex integration approach*, Warsaw, PL: MINK. pp. 54-58.

Masgutova, S., Sadowska, L. (2008). Reflexes integration in children with Down syndrome. The results of diagnostics and therapy. [in] edited by J. Patkiewicz: *Down syndrome - Progress in treatment, rehabilitation and education.* Wroclaw,PL: PTWK. 2008, p. 35 -56.

Masgutova, S. (2006). Integration of reflexes of newborns in the general mobility arrangement and effectiveness of learning. [in] edited by A. Regner and S. Masgutova: *Modern methods of stimulation of motor development and learning.* Conference Materials, April 29, 2006, Warsaw-Szczecin, PL: MINK. pp. 5-28.

Masgutova, S., Akhmatova, N. (2005). *The integration of dynamic and postural reflexes with the motor system of the whole body*, Ed. II, Warsaw, PL: MINK. pp. 14 – 42.

Masgutova, S., Kowal, J., Mazur, G., Masgutov, D. (2005). *Neurokinesiological tactile therapy by Dr. Svetlana Masgutova.* Warsaw, PL: MINK. pp.23 -33.

Masgutova, S., Masgutov, D. (2005). *The integration of facial reflexes using Svetlana Masgutova method. Techniques supporting the development of motor skills and speech.* Warsaw, PL: MINK. pp.14 -16.

Masgutova, S., Regner, A. (2009). *The development of the child's speech in the light of sensory integration,* Ed. Wroclaw, PL: Continuo. p.131–138.

Masgutova, S., Sadowska, L. (2004). The use of educational kinesiology in children with learning difficulties in the content of development of early motor dynamisms [in] edited by J. Patkiewicz: *Quality of life of children and disabled young people in Poland and in the European union*, Wroclaw, PTWK. pp. 97-107.

Palisano, R. J., Hanna, S. E., Rosenbaum, P. L., et al. (2000). Validation of a model of gross motor function for children with cerebral palsy. *Phys. Ther.,* 80 (10), pp. 974-985.

Palisano, R., Rosenbaum, P., Walter, S., et al. (1997). Development and reliability of a system to classify gross motor function in children with cerebral palsy. *Dev Med Child Neurol,* 39 (4), pp. 214-223.

Pilecki, W., Masgutova, S., Kowalevska, J., et al. (2012). The influence of rehabilitation using masgutova neurosensory reflex integration in children with cerebral palsy on the results of the study of auditory evoked potentials. // *Advances in Clinical and Experimental Medicine.* 21 (3), pp. 363 -371.

Sadowska, L. (ed.). (2001). *Neurokinesiological diagnosis and treatment of children with impaired psychomotor development*, Wroclaw, Ed. Wroclaw, PL: AWF. ss.71, 83 - 84, 94-95.

Sadowska, L. (2004). Neurophysiological methods of rehabilitation of children with developmental disorders, Wroclaw, Ed. *Physical Education*, p.10 -30.

Sadowska, L. (2012). Childhood cerebral palsy - definition, comprehensive diagnostics. [in] ed. L. Sadowska, M. Dzicwulska: *Neurophysiological basis for diagnosis and treatment of children with developmental disorders.* Warsaw, PL: Ed. WSzM, pp. 85-92.

Wood, W., Rosenbaum, P. (2000). The gross motor classification system for cerebral palsy: A study of reliability and stability over time. *Dev Med & Child Neur,* p. 42, 292-296.

We express our endless gratitude to these children and their parents for the participation in our research. We congratulate our children for their great improvement in sensory-motor development and wish them good health and new achievements! Your loving hearts and smiles are the best example to us on how to create joy in this life. – Authors

Reflex Integration of Children with Down Syndrome: MNRI® Assessment and Therapy Results

Prof. L. Ludwika Sadowska, MD; Svetlana Masgutova, Ph.D; Joanna Kowalewska, MA, PT, MNRI® Instructor; Denis Masgutov, Director of the ISMI; Henryk Filipowski, Ph.D., Wroclaw-Warsaw, Poland

This article presents the study of the assessment and therapy results of 38 children with Down syndrome using MNRI® (based on the presentation by L. Sadowska, S. Masgutova, J. Kowalewska (2009), *Reflex Integration of Children with Down Syndrome; Diagnosis and Therapy Results Using Masgutova Neurosensorimotor Reflex Integration Program – MNRI®. International Scientific-Practical Conference - XXXII National Day of Child Rehabilitation: Down Syndrome. Treatment, Rehabilitation and Education. 17-18.10.2008*). Editing assistance was provided by David Miller (USA).

Prof. Ludwika Sadowska, MD

Svetlana Masgutova, Ph.D

Joanna Kowalewska

Denis Masgutov

Henryk Filipowski, Ph.D

Objective of Article

The objective of this article is to offer an appropriate support tool for new solutions based on the 'reason and cause' of the problems concerning the deficits and challenges of motor development in individuals with Down syndrome. Also, to document statistical research to verify that the functions of children with Down syndrome are improved and better facilitated by the use of the Masgutova Method® program, MNRI® processes. This article presents the results of MNRI® with 38 individuals with Down syndrome.

Key words: Children with Down syndrome, reflex integration, Masgutova Method®, MNRI® - Masgutova Neurosensorimotor Reflex Integration program, A. Krefft algorithm.

Introduction

Down syndrome is a genetically originated developmental pathology involving multi-organ and multi-level human development dysfunction, occurring in 1 for 700 born neonates with increasing frequency in the last decades, due to advancement in medical sciences. Genetic disorders determine the set of symptoms of moderate or deep mental retardation, which affect development of different spheres of the child: psychosomatic, intellectual, emotional, social behavior, communication, and personality development. Down syndrome is a multiform disorder and any therapy program used should provide multidirectional support, according to the child's general genetic pathological mechanisms and their individual development. The current study demonstrates significant improvement of functioning of reflex patterns in children with Down syndrome as verified by mathematical statistical analysis. Presenting the model of the evaluation of reflex pattern development according to the mathematical statistical analysis by synthesized Z function by A. Krefft allows for an objective approach and scientific analysis of the effectiveness of the MNRI® program designed for 8 days at Therapeutic Rehabilitation Camps in Poland, Canada, and the US.

Specifics of the Development of Children with Down Syndrome

Down syndrome is a genetically determined, permanent, and incurable development disorder. However, the primary motor coordination disorders and acquired patterns can be improved or corrected (Sadowska, Masgutova, Kowalewska, 2009). Repatterning offers one such possibility. A priority in the purpose of MNRI® therapy is improvement of functioning of reflex patterns and their integration in a general child's static-motor system, facilitation of brain functioning plasticity, plasticity of neurodevelopment, and activation of the emotional, social, and cognitive processes.

Infant reflex patterns of a child with Down syndrome are usually developed poorly and it takes more effort and time to trigger and activate their functioning. The development of reflexes of these children is delayed. The reasons for this are some specific physical and somatic features as well as specific functioning of their central nervous system. Most children after birth have poor muscle tone control (hypo-tonicity), general flaccidity, lower muscle strength, and excessive/hyper motor rotation range in their joints. Characteristic physical features for Down syndrome can cause developmental dysfunctions and deficits in reflex patterns integration as well as a child's motor and cognitive functioning.

About the MNRI® Program

The MNRI® processes are designed for individuals with neuro-developmental disorders and aimed at improvement of their physical and cognitive functioning. The Masgutova Method® concept of neurosensorimotor reflex integration is based on the idea of awakening the latent brainstem genetic motor memory, so that it may serve as a resource for overall neuro-development.

The purpose of MNRI® processes is to support the reflex patterns' integration within the work of sensory and motor systems and brain functioning to facilitate a greater physiological foundation for appropriate human development. Human development, whether normal or abnormal, is continuous. Stages of maturation and the emergence of reflex patterns should not be thought of as static points in development, but as a glimpse of one moment in a dynamic process. The Masgutova Method® defines specific reflexes and stages of development through the Assessment procedure and therapeutic intervention.

Nature gave every human motor reflexes as the response to sensory and proprioceptive/vestibular stimuli. These responses serve as a basis for neural and defensive mechanisms. Reflexes appear in the prenatal period and after birth. Thereafter, each reflex develops and matures through phases of integration supporting controlled motor coordination and higher sensory-psychical functions.

The concept of reflex pattern integration proposes the means for improvement through the use of primary motor patterns. These patterns present natural, genetically programmed, inner human resources. Integration of these reflex patterns takes place within three parts of the reflex circuit:

• sensory stimulation and identification/decoding of the stimulus by the nervous system and sense organs and transferring the stimulation through the afferent pathway to the brain

• brain information processing and recognition of the stimulus; filtering the information for brain strategy organization for protection and survival or for processing on the cognitive level

• motor reaction to a sensory stimulus, as a brain response coming through the efferent pathway.

The Role of the MNRI® Program for Neurodevelopment

A reflex as an automatic response presents inherent, genetically programmed motor activity to a sensory or proprioceptive stimulus. Every human, regardless of their health condition, has a system of primary movements and reflexes. In a definition of a reflex, MNRI® uses the physiological concept of conditional and unconditional activity of the nervous system.

Knowledge of reflexes, their structure, and developmental dynamic is very important. Functioning of reflexes is related to fight or flight and also freezing reactions. Those reactions are the basis of the body's protective and defensive mechanisms. In the case where these reflexes are poorly developed at their proper time or are immature, persistent, and inadequately active, they may cause deficits and pathologies in the motor system (genetically given or learned) at certain stages, and in the formation of cognitive abilities – processes of perception, focusing/attention, and thinking. Knowledge of this allows an understanding of the links between gross and fine motor and thinking processes, particularly to:

- choose a method of intervention for integration of the spheres to support development of voluntary control
- facilitate motor skills and their coordination with intellectual processes
- support formation of defensive reflexes for survival in stress or in a state of neurodeficit.

Properly matured and functioning defensive reactions determine proper neurodevelopment. The delay of reflex development restricts our behavior and thinking, and negatively influences the coordination of the sensory-motor-brain processing system at various developmental stages. In times of stress, events may cause the individual system to return to primary reactions: postural and motor patterns – reflexes.

The reflex naturally influences the formation of cognitive processes to a certain extent. It also forms the basis for individual motor development on a higher level of human functioning. Intervention therapy aimed at the correction of the reflex circuit components can bring improved results for the release of dysfunctions and improve various spheres of human functioning: motor development; functioning level for skills and abilities; and, emotional and intellectual abilities.

The information in this article offers an option for appropriate support tools to give new solutions based on the reason and cause of problems concerning deficits and challenges in motor development for individuals with Down syndrome.

Reflex integration Assessments help to explain specific disorders in the functioning of a reflex circuit. A detailed Assessment of a reflex pattern and its level of development and integration evaluates the following parameters: reflex pattern components: sensory perception, brain processing of the sensory stimulus, and motor response (the individual reactions for specific stimuli), latency (time after the stimuli influence, time of duration of the response; and other dynamic features), movement direction in the reflex pattern, strength of reaction, and locomotor or positional symmetry.

In order to determine the level of reflex pattern development, the MNRI® Assessment uses the criteria in points from 0–4 for each of five parameters (see the chapter, *MNRI® Assessment for Determining the Level of Reflex Development* in this book). If a reflex is in full integration – a maximum score of 20 points is possible. If the reflex is not fully developed and integrated, the score can fall in the following ranges: a) pathological and dysfunctional development from 0–10, and b) partly appropriate or completely developed and integrated state of patterns from 10–20.

The MNRI® processes demonstrate the basics of assessment and corrective procedures for the intervention of different motor patterns to support the following:

- developmental mechanisms presenting the proper defense reactions (Tendon Guard, Tonic, Antigravity, Grounding, and Hands Supporting Reflex patterns)
- physical development (Spinal Galant and Spinal Perez Reflex patterns)
- body posture, locomotion control, and coordination (postural reflex patterns)
- cognition processes (Asymmetrical Tonic Neck, Pavlov Orientation, Auditory and Visual Reflex patterns)
- listening and memory (Asymmetrical Tonic Neck, Pavlov Orientation, Auditory and Visual Reflex patterns)
- proprioceptive system development (Tonic Labyrinthine, Symmetrical Tonic Neck, Trunk Extension Reflex patterns)
- body posture control related to binocular seeing and binaural hearing (Auditory and Visual Reflex patterns)
- perception for reading and writing (Robinson Hands Grasp, Hands Pulling, and Hands Supporting Reflex patterns)
- the motor aspects of writing and drawing skills.

Assessment procedures of the integration/non-integration of a reflex pattern are based on:

• features of defensive reflexes, as an automatic, unconditioned reaction for a certain stimulus. These reflexes stay in relationship with an individual's age, sensorimotor reflex integration maturity, dynamics of development, and reflex integration.

• occurrence of characteristic reflex components in a range of sensorimotor pattern properties, including: proper duration time, reaction strength, proper direction, and response symmetry. The matter of reflex integration is a key program idea.

Neurophysiology dictates that each reflex must integrate on the sensory-motor level. A specific sensory stimulation will trigger a corresponding motor/gland response. The neural link between the sensory and motor aspects of a reflex is genetically based, having evolved over thousands of years. Defined sensual or proprioceptive stimulation will produce a proper glandular response or motor reaction. This type of specific link, throughout the nervous system and between the sensory apparatus in a reflex circuit, is genetically conditioned.

If the sensory stimulus is not properly recognized by the sensory apparatus, it will result in an incorrect interpretation by the brain. Also, if the efferent nervous system conducts the command in an incorrect way, the development of the reflex pattern will be pathological, which means the development of the reflex and its integration with controlled movements will be delayed and inadequate. This fact is especially true in stressful situations or situations where the person is learning something new.

Every reflex appears at a certain time and develops its own basic pattern in three phases. The fourth phase of development is a period of change that prepares for the development of variants during the fifth, sixth, and seventh phases. Each phase has its own role. For example, a basic pattern is responsible for coding the sensory-motor circuit. It creates the nerve network for specific stimuli in order to establish appropriate physiological functioning and protection. The task of reflex pattern development in successive phases is the basis of the creation of reflex pattern integration with new skills and movement capabilities. This serves as a foundation for new skills, including accomplishments at school like reading, drawing, writing, and math. Maturation of the nervous system involves the inter-connection of reflex circuits. The role of these latter phases is to expand the development of a reflex in order to create the groundwork for reflex integration with motor skills and abilities. This supports the development of academic skills such as elementary reading, drawing, writing, and calculating. Delayed reflex development, or the omission of any phase, adversely affects the formation of future skills. The result is evident in the next level of development; the reflex will not develop an appropriately matured, neural network. Thus there will be dysfunctions or compensations rather than ideal patterns. The altered patterns are less reliable in situations of stress or unexpected transition. It is critical for a reflex to evolve through each phase for full development, maturation, and integration. This concept is unique, and should be distinguished from the traditional understanding based on the inhibition of a reflex. As it was mentioned before, attitudes differ considerably from traditional repression (inhibition) of a remaining, partial reflex.

Evidence and Methods of Research

In this article, the results of the MNRI® Assessment and intervention for 38 children ranging from 6 months to 10 years of age, diagnosed with Down syndrome, are demonstrated. The children were participants of the MNRI® program aimed at the improvement of dysfunctional and pathological reflex patterns according to the following MNRI® programs: Neuro-Structural Reflex Integration, Tactile Integration, Dynamic and Postural Reflex Pattern Integration, Lifelong Reflex Integration, Proprioceptive and Cognitive Integration, Visual and Auditory Reflexes Integration, Oral-Facial Reflex Integration, Dance Therapy, Art Creation and Reflex Integration, and Archetype Movement Integration. These programs were facilitated by a group of professionals trained in the MNRI® program during organized therapeutic camps, and individual therapy with children with developmental deficits. Statistical analysis was carried out before and after 8 days of MNRI® programming, at the neurosensorimotor rehabilitation camps/clinics organized by the International Dr. Svetlana Masgutova Institute (Warsaw – Poland; San Francisco, New Jersey, and Florida – USA, and Vancouver – Canada).

An evaluation was conducted of 24 reflex patterns in three groups referring to the following body movement planes: sagittal, horizontal, and dorsal. Every reflex pattern was evaluated on a scale of 0–20 with regard to five parameters: reflex pattern, direction of movement, strength of reaction, timing of reaction, and symmetry. The testing within these parameters was done for each reflex pattern individually. The rating 10 on the scale determines the transition state in development of the pattern between its pathological/dysfunctional and normal state, whereas numbers from 15–20 represent normal functioning.

Statistical Methods

Measures. The primary outcome of interest was change in the reflex patterns of the 38 children diagnosed with Down syndrome. Each of 38 study participants received a pretest at the beginning of an 8-day conference and a post-test at the end. Initial evaluations of motor and cognitive patterns considered the child's age, neurologic abnormalities, and status of inborn reflex patterns. This entailed assessing 24 reflex patterns (coded X_1-X_{24}) using five criteria: reflex pattern (or sensory-motor circuit), direction of a response (or movement), strength of reaction, time of reaction, and symmetry. Rating for each parameter was assigned on a scale of 0-4, with 4 indicating full display of a parameter, and 0 indicating the parameter's absence. This results in a maximum score of 20 for each pattern. Summary scores of 0-9 reflect varying degrees of abnormal development, 10 to 11.75 are intermediate, and 11-20 represent varying degrees of partially or fully integrated reflex patterns. Scores 16-17.75 represent the norm. Reflex patterns were further categorized according to body movement planes. There were eight reflex patterns each corresponding to a plane of body movement: sagittal (medial-lateral), horizontal (superior-inferior), and dorsal (anterior-posterior) (Masgutova, 2011).

Statistical analysis. Results of the reflex pattern assessment completed on 38 children diagnosed with Down syndrome were analyzed based on the multivariable function $z = f(x)$ of directly non observable phenomena. (Krefft, 2007). Briefly, this function estimates the level of the reflex pattern integration Z as a function of the entailed grading reflex patterns $X_1, X_2, \ldots, X_{24}$, with the assumption that this is a linear function. So, variable Z_S (sagittal body plane) collects the information from the first 8 reflex patterns X1, X2, ..,X8, variable Z_H (horizontal body plane) describes the information from the second set of 8 reflex patterns $X_9, X_{10}, .., X_{16}$ and variable Z_D (dorsal body plane) integrates the last 8 reflex patterns $X_{17}, X_{18}, .., X_{24}$. The level of the reflex pattern integration Z_C is estimated by all measured reflex patterns $X_1, X_2, \ldots, X_{24}$. Mean values of Z_C, Z_S, Z_H, and Z_D were compared before and after 8 days of an MNRI® program using an ANOVA test (IBM SPSS Statistics Grad Pack 22.0). Results were considered statistically significant where $p < 0.05$.

A Brief Description of the Reflex Patterns of 38 Children Diagnosed with Down Syndrome

The main feature of dysfunctional/pathological reflexes in children diagnosed with Down syndrome is the delay of their Tonic Reflex patterns. They inhibit formation of many other static postural dynamic reflexes, such as:

Tonic Labyrinthine Reflex (TLR). In a normally functioning infant/child, the prone position (lying down on their stomach) triggers the tone of flexors of the neck, abdominal muscles, and legs. When supine (on back) TLR also triggers the tone of extensors of the upper and lower limbs and also the trunk. TLR is active during the first 3-6 months of life and later must integrate with head and trunk righting and flexion and rotational movements. The development of this reflex in a child with Down syndrome is very poor, and typically extremely hypoactive. All of this blocks the muscle tone and postural control development resulting in a delay of the ability to roll over from their back to stomach and to support the trunk on their elbows in a prone and sitting-up position. This reflex also affects the tongue muscle tone regulation. With children with Down syndrome, the tongue is usually placed too much in the frontal part of the oral cavity, disturbing articulation formation. The delay of control of head righting affects the tone of extra-ocular muscles and causes dysfunctions in eye leveling (horizontal and vertical) limiting perception of surrounding objects. In the vertical position, they have hyper-rotation in their joints, and difficulties with muscular-proprioceptive control and spatial orientation.

Asymmetrical Tonic Neck Reflex (ATNR) is triggered by spontaneous turning of the head to the side. ATNR appears when the limbs extend on the side of the body on which the head is turned, and the limbs of the opposite side flex. This reflex is active up to 4-6 months of age and integrates with the intentional turning of the head around 6-7 months. Thanks to ATNR, the motor abilities of the left and the right body sides are developed. ATNR most actively helps with the exploration of the body midline (sagittal plane) functions. It stimulates development of the monaural and then, binaural hearing. During the next stage, this reflex helps the development of the STNR also. During 'the asymmetric period' of development, babies spend 80% of their time in the ATNR position. For a child with Down syndrome, poor muscle tone regulation and hypo-tonicity leads to limitation of ATNR pattern use – thus leading to a lack of head-core-trunk motor control, hand-eye coordination, and spatial orientation. This reflex especially creates the basis for left hemisphere functions and speech development – for the Wernicke Center to start decoding human speech (emotional and cognitive aspects), and for the Broca Center to code and express the child's first sounds and syllables at 6-9 months of life). As the child with Down syndrome demonstrates a considerable delay in development of this reflex, their phonemic auditory discrimination, speech formation, and comprehension processes are delayed.

The Symmetrical Tonic Neck Reflex (STNR) is stimulated by straightening the head of an infant lying on their abdomen which triggers the arms to straighten and the legs to flex automatically. It also occurs while flexing the head down to the chest, triggering the arms to flex and the legs to extend. This reflex develops actively around 6 months of age and integrates up to 10 months preparing the baby for the first creeping movements, and then to be able to stand and walk. A child with Down syndrome demonstrates significant delay in development of this reflex, as well as of muscular hypo-tonicity and postural control resulting in an inability to sit down and crawl on all fours. All this affects the formation of visual perception mechanisms (horizontal and vertical eye tracking, close-far vision, and visual adjustment to gravity changes in the body). The STNR of a child with Down syndrome is poorly developed. This leads to hyper-tonicity of the neck flexors (a tendency for a low or upraised head position). It negatively affects horizontal eye leveling and two-dimensional perception (e.g. for drawing, reading, and writing).

The intensive development of tonic reflexes of healthy children begins in the second month of life. For children with Down syndrome, tonic reflexes stay dysfunctional or pathological for a much longer time period. Their muscular hypo-tonicity creates excessive mobility of their joints, inhibiting the development of motor coordination and motor control.

Antigravity and Rotational Reflexes

The Segmental Rolling Reflex is triggered by the turning of the body, starting from the top (head ==> shoulder ==> pelvis ==> knee) when an infant is rolling over from back to front. This happens, thanks to the ability to rotate isolated horizontal parts of the body in the range of the neck, shoulders, and pelvis. This reflex develops actively from the fourth month and integrates at up to 10-11 months of age preparing the child for rolling over both ways, from back to front and vice versa. The development of this reflex for a child with Down syndrome is also delayed. They are not capable of rotating in a standing position or while walking, which challenges their equilibrium. The lack of proper development of this reflex causes a homolateral manner of walking for the child instead of cross-lateral walking.

Hands Supporting Reflex is an automatic response of the arms and hands to extend forward in the direction of the ground at the moment an individual loses the stable position of their body in space. This reflex integrates up to 6 months of life. It influences the development of the control of the upper part of the body, grounding, and stability. It also supports formation of 'hand-eye' coordination, close-far vision, self-defense abilities, and orientation in their own personal and interpersonal space. This reflex is delayed for a child with Down syndrome limiting the exploration of their personal space, demonstrating immaturity, infantile features, lack of ability to build communication, and causes dependency on other people. This reflex pattern is used in MNRI® as a corrective exercise for improving one's ability to protect their own personal boundaries.

The Hands Supporting Reflex of a child with Down syndrome typically stays dysfunctional or pathological. This results in muscular hypotonicity with excessive mobility of joints inhibiting the extension mechanisms, gravity, grounding, and stability, as well as motor coordination and postural control.

The Bauer Crawling Reflex is activated by touch on the plantar surface of the feet of an infant lying down on their stomach with their head and trunk in the midline. This triggers the head to move up and initiates automatic cross-lateral crawling movements by pushing off by their feet. This reflex integrates at up to 4 months of life. For a child with Down syndrome, this creeping reflex develops later causing a delay in crawling on all fours, standing and walking, as well as rolling over on their stomach and back. The Bauer Reflex influences development of reciprocal motor coordination. Creeping and crawling on all fours for the child with Down syndrome is usually homologous (appears to 'jump like a frog'). This can negatively affect motor coordination as well as their pace for perception, thinking, multitasking, and learning.

Spinal Reflexes

Galant Reflex: the linear stimulation along one side of the spine when a child is lying on their side causes the lateral flexion of the trunk on the same side. Sometimes the arm and leg abduction/extension occurs additionally. This reflex is active from the third to the ninth month. For a child with Down syndrome, this reflex is often hyper-sensitive and hyperactive. Together with muscular hypotonicity, this causes overall hypersensitivity to touch for the body, as well as poor antigravity mechanisms, and thus lack of control of spine movements.

The Perez Reflex is a reaction to stroking the fingers along the spinal column in the direction from the sacrum to the neck causing the trunk and head to straighten in a posterior and inferior direction, resulting in the upper limbs abducting with flexion. This reflex is active from birth to 2-3 months of age. For a child with Down syndrome, this reflex is often oversensitive and hyperactive. In combination with hypotonic muscles, it leads to overall hyper- or hypo-sensitivity to touch and physical weakness, as well as poor control of spine (lordosis-kyphosis) movements, and reactivity in movements and behavior.

The poor development of spinal reflexes negatively affects focusing, short and long term memory, and thinking processes (resulting in a lack of cause-and-effect thinking). Children with Down syndrome show a lack of persistence in focusing, motor reactivity, limited visual span, poor internal control, poor gross-motor coordination and walking patterns, as well as emotional instability, and fear and phobias. Hyperactive Galant and Perez Reflexes can also cause hypersensitivity to tight clothes or belts.

Manual Reflexes

Hands Grasp Reflex is a response to touch of the palm (at the base of the fingers) by closing the palm and fingers and holding them tightly while the stimulus is present. This reflex normally is reactive during the first week of life, then follows through seven developmental phases (phase 2 is a basically completely closed palm and phase 4 is the use of a 'hanging grasp' pattern), and integrates with manual abilities and skills at the end of the first year of life. For a child with Down syndrome, this reflex is hypoactive and, usually, strongly delayed. It influences the development of hand-eye and hand-mouth coordination, manual skills (for holding objects, manipulating them, taking and giving, and catching), and fine motor control of fingers and palms (for drawing, writing, playing with toys and objects). Delay of this reflex causes an inability or refusal to do activities requiring precision and fine motor control. It also affects the development of speech and communication.

Visual and Auditory Reflexes

Visual and Auditory Reflexes affect horizontal eye tracking, convergence/divergence, ocular-vestibular, vestibular-ocular, ocular-kinetic; stapedius-acoustic, and sound localization (directional) skills.

Eye Tracking Reflexes in a child with Down syndrome are usually hypoactive and immature. Chaotic eye movements, poor eye mobility, and problems with sight make visual focusing impossible or cause the ignoring of visual objects. Immature eye tracking leads to the use of too much head righting, which means the lack of freedom of all horizontal planes of the body, and rigidity of the body within the sagittal body plane.

Vestibular-Ocular Reflexes (horizontal and vertical) support stability of visual images created due to the movement of their eyes in the opposite direction of head movement. For a child with Down syndrome, we observe rigidity, the lack of equilibrium, hyper- or hypo-active physiological nystagmus, and poor vestibular-visual coordination leading to visual and proprioreceptive disorientation.

Convergence/Divergence Reflex responses for bringing the eyeballs close together (to see close objects) and apart (for distant objects) often are delayed in children in this group. Bringing their eyeballs together is especially challenging because of the genetic physical features of their face (more flat), and location of their eyes (more distant and lateral) which affects their visual cognition (focusing, decoding, and analysis). Divergence can cause too much separation of the eyes and an inability for proper peripheral vision. Convergence/divergence affect the visual perception and cognition of the world. In children with Down syndrome these reflexes are delayed in development, leading to poor eye mobility, and poor visual analysis and comprehension, to a certain extent.

The Stapedius Acoustic Reflex also affects sound localization/direction. The Stapedius Acoustic Reflex (triggered by the contraction of the stapedius muscle in the middle ear) is an involuntary response to auditory stimuli of high frequency. At the time of exposure to the high frequency sounds, this reflex reduces the vibration of the cochlea and is used for sound perception and processing by the brain. The Stapedius Reflex is usually stimulated only by intense sounds; its activation at low sound frequency can lead to auditory dysfunctions – lack or loss of hearing, poor auditory decoding, or discrimination. Children with Down syndrome often have a hypo- or hypersensitive Stapedius Reflex leading to their 'ignoring' an auditory stimulus, possible self-stimulation (closing their ears, talking to themself, or constant vocalizing), overall poor muscle-tendoneous tone, and excessive loosening of the jaw.

Pavlov Cognitive Orientation Reflex is caused by a new bright or interesting stimulus or its qualities. It is seen in a variety of reactions such as, head and eye movement toward the stimulus/source of information, change in breathing rhythm, increase in muscle tone, and cognitive reactivity (cortex activation, dominance of beta and alpha high frequency brain waves). The sensory component of this reflex leads to an increase of visual-auditory focusing, decoding and analysis, and a decrease in the sensory threshold ('protection'). The novelty, which triggers sensory-cognitive activity within the Pavlov Reflex in a child with Down syndrome is poorly developed in infancy and thus causes a delay in whole cognitive development at its later stages. The demands to learn at home and at school often cause stress and the child may refuse to participate in learning activities, have low learning motivation, and defensive reactions toward learning.

Oral-Facial Reflexes

The Sucking Reflex is triggered by tactile stimulation in the oral cavity, especially the tongue and upper hard palate closer to the soft palate. The response is a sequential and rhythmical retraction of the tongue and cheek muscles and muscles around the soft palate and deep mouth cavity followed by swallowing and breathing (inhale-exhale) regulation. This reflex was originally connected with a survival function (the nutritional need). It gives the feeling of comfort and relaxation. The Sucking Reflex of a child with Down syndrome is often poor as they have lack of muscle tone and poor coordination for swallowing and breathing. This leads to difficulty for the child in eating and drinking well, as well as regulation of satiation. This reflex can also affect the feeling of being content and bonding, which, in turn, influences communication.

The Babkin Palmomental Reflex is triggered by pressing the middle of the palms resulting in the mouth opening and the head flexing and moving down and forward. This reflex develops actively during the first months of life and integrates with 'hand-mouth' coordination around the fourth month. A child with Down syndrome usually has a very poorly developed and pathological Babkin Reflex. Dysfunctional oral-facial morphology results in pathologies of this reflex (open mouth, poor tongue posture and coordination, and increased salivation). This reflex influences the development of other reflex patterns concerned with eating, articulation, and speech formation. Dysfunction in this reflex and the developmental pathology of a child with Down syndrome is significant.

Our reflex Assessments of children with Down syndrome show a multi-dysfunctional nature of development in combination of poor reflex pattern and genetic disorders, which, at the same time, differs from child to child. This is why the knowledge of an overall reflex profile and individual differences in them can help to create an individualized MNRI® program for a child with Down syndrome. MNRI® techniques are directed toward correction of disorders, improvement of sensory-motor dissonance, neuro-facilitation of reflex circuits, integration of tactile, visual, auditory, vestibular, and proprioceptive senses.

Down syndrome is a permanent and incurable disorder, but the primitive and voluntary motor activity can be improved significantly. The work of 'remodeling' reflex patterns is one of such possibilities. As Down syndrome is a polymorphic disorder, the therapy program should be of multi-directional character and provided in accordance with genetic pathological mechanisms including individual developmental specifics for each child. A priority goal of MNRI® therapy for a child with Down syndrome is to improve their overall neurodevelopment, emotional, social, and cognitive functioning through reflex integration.

Results and Discussion

The results of Assessments of the level of development/integration of each of the 24 reflex patterns (in points from 0–20), presented as the qualities X_1–X_{24} of children with Down syndrome at the beginning and after the MNRI® program are shown in Table 1.

Table 1 shows the results of changes in the level of reflex pattern integration of 38 children diagnosed with Down syndrome after completing treatment at a 8 day MNRI® camp. The information in this table demonstrates a positive change for all reflexes. This data, as given in the examples, suggests the results of statistically important validation of the synthesized function z = f(x) and a high degree of effectiveness for the MNRI® process applied to children with Down syndrome during this time of intervention. Each parameter (x) shows the level of development of the specific pre and post Assessment of the given child.

Analysis of the level of development/integration of all 24 reflex patterns was carried out, and it has been described using function Z_C and three functions for systems in planes A, B, and C. (A – Sagittal, B - Horizontal,

C - Dorsal). Results are shown on graphs and statistical verification in tables, based on both before and after the MNRI® program.

The level of changes in development of all the examined reflex patterns is illustrated in Figure 1. There we see the results of synthesized information of all the diagnostic qualities (X_1–X_{24}). In Figure 1, the 24 reflex patterns of children with Down syndrome within different planes of body symmetry and movements are presented as follows:

A) Sagittal: 1) Robinsons Hand Grasp, 2) Hands Pulling, 3) Babkin Palmomental, 4) Babinski, 5) Leg Cross Flexion-Extension, 6) Asymmetrical Tonic Neck, 7) Abdominal, 8) Bonding

B) Horizontal: 9) Thomas Automatic Gait, 10) Bauer Crawling, 11) Moro, 12) Fear Paralysis, 13) Hands Supporting, 14) Segmental Rolling, 15) Landau, 16) Flying and Landing

C) Dorsal: 17) Trunk Extension, 18) Symmetrical Tonic Neck, 19) Spinal Galant, 20) Spinal Perez, 21) Tonic Labyrinthine, 22) Foot Tendon Guard, 23) Spinning, 24) Pavlov Orientation.

Figure 1 illustrates the pathology and dysfunction of the 24 reflex patterns and shows the synthesized function Z. It illustrates the level of pattern development before and after the MNRI® program. After the program we still observe dysfunction and pathology in 15 reflex patterns. However, we also see significant positive changes and improvement in motor development overall. The comparison of the results before and after MNRI® intervention shows a significant difference at the level of $p < 0.001$.

See Table 2 for the Z_C synthesized function of the level of development/integration of reflex patterns in all body movement planes described as qualities from X_1–X_{24}. Children with Down syndrome demonstrate poor maturation or immaturity. They also have dysfunction and pathology of a wide variety of reflex patterns. Particularly, they show developmental deficits for such reflex patterns as: Babkin Palmomental, Robinson

Diagnostic Quality/ Feature	Body Movement Plane	Reflex Pattern	Assessment Before Program	Assessment After Program
X1	A	Robinson Hands Grasp	6.4	8.2
X2	A	Hands Pulling	8.4	9.2
X3	A	Babkin Palmomental	4.5	5.4
X4	A	Babinski	6.7	7.3
X5	A	Leg Cross Flexion-Extension	5.5	6.9
X6	A	Asymmetrical Tonic Neck	6.7	7.4
X7	A	Abdominal	8.2	10.4
X8	A	Bonding	12	13.2
X9	B	Thomas Automatic Gait	8.8	9.3
X10	B	Bauer Crawling	6.5	9.4
X11	B	Moro Embrace	11	13.2
X12	B	Fear Paralysis	12	13.4
X13	B	Hands Supporting	8.3	8.9
X14	B	Segmental Rolling	7.2	8.1
X15	B	Landau	6.4	7.2
X16	B	Flying and Landing	5.2	5.8
X17	C	Trunk Extension	8.2	8.8
X18	C	Symmetrical Tonic Neck	6.3	8.4
X19	C	Spinal Galant	8.3	11.2
X20	C	Spinal Perez	11	12.3
X21	C	Tonic Labyrinthine	9.2	11.2
X22	C	Foot Tendon Guard	8.1	10
X23	C	Spinning	8.3	12.4
X24	C	Pavlov Orientation	6.6	6.8

Table 1. Changes in the level of average reflex pattern integration for 38 children with Down syndrome before and after completing treatment at MNRI® camps.

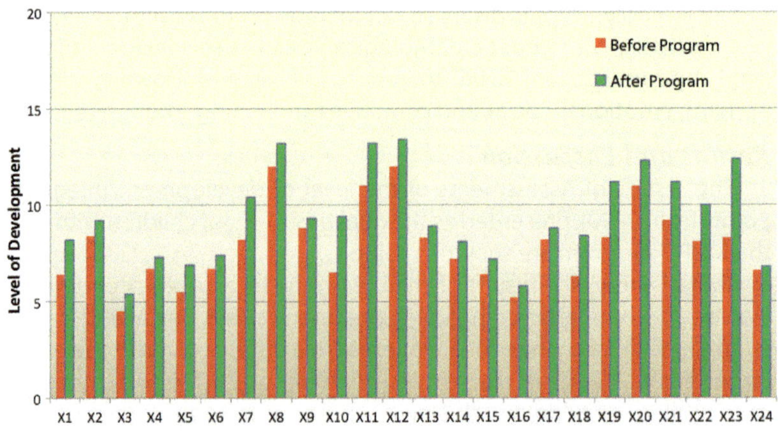

Figure 1. Graph of diagnostic results of the level of development/integration of each of the 24 reflex patterns (in points from 0–20) of 38 children with Down syndrome at the beginning and after the MNRI® program.

Quality Group	Zc			ANOVA Test
	Average	N	Stand. Error/deviation	
Before	0.3924		0.1884	p < 0.001
After	0.6038		0.1790	
Total	0.4981		0.2113	

Table 2. Statistical verification of the value of average functions and synthesized Z_C described as qualities from X_1–X_{24} in research of 38 children diagnosed with Down syndrome before and after 8 days of MNRI® intervention.

Hands Grasp, Sequential Fingers Opening and Closing, Hands Pulling, Leg Cross Flexion-Extension, Symmetrical Tonic Neck, Bauer Crawling, Moro, Thomas Automatic Gait, Segmental Rolling and Spinning, Asymmetrical Tonic Neck, Spinal Galant, Spinal Perez, Fear Paralysis, and Pavlov Orientation. The reflexes of a child with Down syndrome emerge and develop with delay. They are disharmonious in their development. The sensory system of some reflex patterns is hypersensitive and others are hyposensitive. This is seen in reflex responses, accordingly, as hyperactivity and hypoactivity. Poor and inadequate development of reflexes in these children is the result of genetic deficits inhibiting and blocking their motor and cognitive development.

The comparative analysis of the level of development/integration of reflex patterns in this group of 38 children with Down syndrome occurring before and after the MNRI® program is presented in the following three figures. A reflex Assessment was done for each of the 38 children. The detailed analysis of the level of development/integration of the reflex patterns in the three body movement planes is presented in Figures 2–4 with statistical results given in Tables 3–5.

Figure 2 shows the dynamic of the changes in the development of reflex patterns grouped within the Sagittal Plane. The resulting statistical analyses are presented in Table 3. Z_S represents the synthesized function of the level of reflex integration within plane A – Sagittal described as diagnostic qualities from X_1–X_8.

Figure 3 shows the dynamic of the changes in the development of reflex patterns grouped within the Horizontal Plane. The resulting statistical analyses are presented in Table 4. Z_H represents the synthesized function of the level of reflex integration within plane B – Horizontal described as diagnostic qualities from X_9–X_{16}.

Reflex patterns found in the Dorsal body movement plane are noted with a high degree of integration and indicate an improvement of psychomotor skills verified as statistically important. This result is illustrated in Figure 4 and Table 5. Z_D represent the synthesized function of the level of development and integration of the reflex patterns in plane C – Dorsal described as diagnostic qualities from X_{17}–X_{24}.

The statistical analysis shows a high validity of the results that the MNRI® intervention demonstrates improved reflex pattern expression in children with Down syndrome after 8 days of therapy. The results indicated in the previous graphs and charts demonstrate the improvement in development of reflex patterns in the control research. The average value of the synthesized function Z_C for the whole group of children with Down syndrome shows the change in the coefficient from 0.3924 before program, to 0.6038 after 8 days of MNRI® techniques. Lineal function with possible deviation of the Method is 2.82%. This result indicates a high level of accuracy and validity of the MNRI® processes diagnosis and therapy.

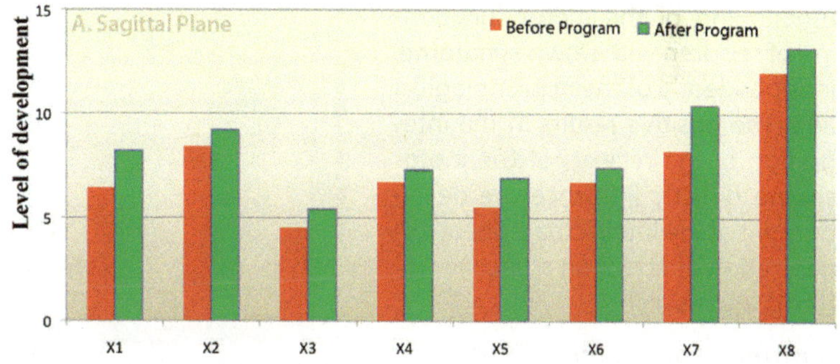

Figure 2. The dynamic of change in the development of the eight reflex patterns of these 38 children with Down syndrome in body movement plane A – Sagittal: 1) Robinson Hands Grasp, 2) Hands Pulling, 3) Babkin Palmomental, 4) Babinski, 5) Leg Cross Flexion-Extension, 6) Asymmetrical Tonic Neck, 7) Abdominal, 8) Bonding.

Quality Group	Zs			ANOVA Test
	Average	N	Stand. Error/deviation	
Before	0.4083	38	0.1890	p<0.001
After	0.5912	38	0.1940	
Total	0.4998	76	0.2113	

Table 3. Statistical verification of the value of average functions and synthesized Z_S described as qualities and features from X_1–X_8 in research of the 38 children diagnosed with Down syndrome before and after MNRI® intervention.

The Masgutova Method® program supports optimal function of the motor, tactile, visual, and auditory systems. This program differs from others in that it proposes neuro-sensory-motor integration of reflex patterns instead of inhibition. The program demonstrates the possibility of integration of reflexes (natural genetic motor programs) with consciously learned and controlled movements, skills, and abilities. MNRI® integrating techniques and exercises are aimed at facilitation and maturation of "neurological pathways" corresponding to specific reflex patterns (Pavlov, 1960; Sechenov, 1927/1960). The program proposes non-invasive gentle movements and playful exercises which can be learned by parents of challenged children, adults, and professionals who work with challenged individuals. These techniques require few external resources and can be used in conjunction with other therapies. The results as highlighted in the above tables and charts demonstrate the significant power of the MNRI® processes which assist in the sensorimotor integration of dysfunctional reflex patterns and facilitate the maturation of these reflex patterns for natural development of the body-mind system for self-regulation of motor patterns. The neuro-facilitation of the genetic developmental motor system supports improvements in the balanced and higher developmental functioning of the motor functioning of children with Down syndrome, independently of their chronological age. The positive results in the integration of the reflexes of the motor sphere directly influence the development of the intellectual sphere (organization of attention span, memorizing, and thinking).

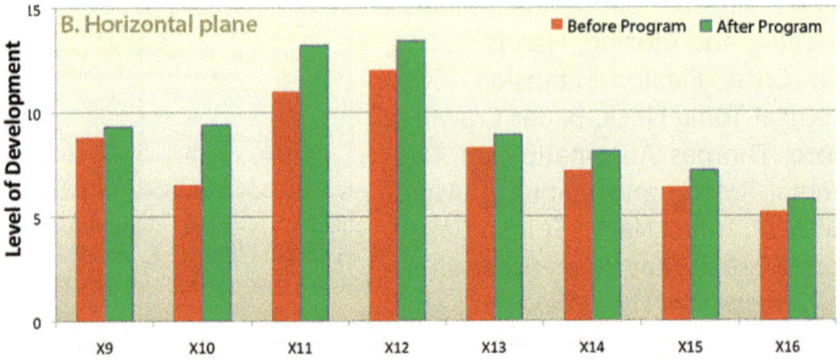

Figure 3. The dynamic of change in the development of eight reflex patterns of 38 children with Down syndrome in plane B - Horizontal: 1) Thomas Automatic Gait, 2) Bauer Crawling, 3) Moro, 4) Fear Paralysis, 5) Hands Supporting, 6) Segmental Rolling, 7) Landau, 8) Flying and Landing.

Quality	Z_H			ANOVA test
Group	Average	N	Standard error/deviation	
Before	0.3861	38	0.1792	p<0.001
After	0.6088	38	0.1754	
Total	0.4974	76	0.2088	

Table 4. Statistical verification of the value of average functions and synthesized Z_H described as qualities from X_9–X_{16} in research of the 38 children with Down syndrome before and after MNRI® intervention.

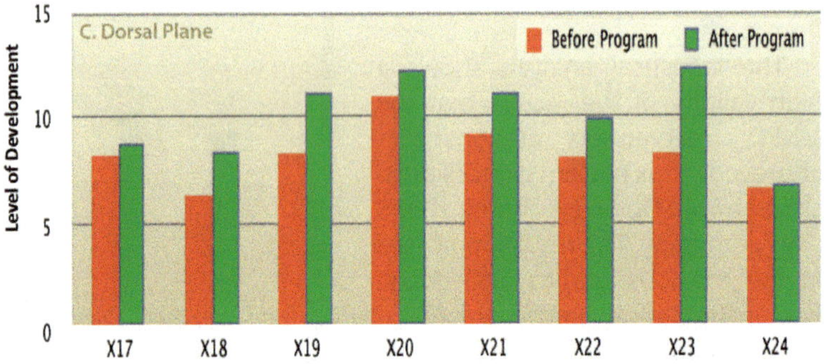

Figure 4. Dynamics of changes in development of eight reflex patterns of 38 children with Down syndrome in plane C - Dorsal: 1) Trunk Extension, 2) Symmetrical Tonic Neck, 3) Spinal Galant, 4) Spinal Perez, 5) Tonic Labyrinthine, 6) Foot Tendon Guard, 7) Spinning, 8) Pavlov Orientation.

Quality	Z_D			ANOVA Test
Group	Average	N	Standard error/deviation	
Before	0.4135	38	0.2076	p < 0.001
After	0.5849	38	0.1853	
Total	0.4992	76	0.2136	

Table 5. Statistical verification of the value of average functions and synthesized Z_D described as qualities from X_{17}–X_{24} in research of 38 children diagnosed with Down syndrome before and after the MNRI® intervention.

Summary and Conclusion

The Masgutova Method® is designed to facilitate the growth and potential of children or adults with challenges. This article provides the results of MNRI® Assessments and corrective therapy with children with Down syndrome. The children were participants of an 8 day MNRI® Intensive intervention aimed at the improvement of dysfunctional and pathological reflex patterns.

The interpretation of these results, based on scientific analysis given within this article indicates improvement in the development of reflex patterns of the children diagnosed with Down syndrome. The average value of the synthesized function Z_C for the whole group of 38 children with Down syndrome demonstrates the change in coefficient from 0.3924 (before the therapy) to 0.6038 (after 8 days of MNRI® process interventions). Lineal function with possible error/deviation of MNRI® is 2.82%. These results indicate a high level of accuracy, validity, and effectiveness of the MNRI® processes with this specific group of Down syndrome children. The results of the MNRI® program with these children show the importance of appropriate corrective procedures directed at the level of motor reflexes.

The aim of rehabilitating children with Down syndrome using MNRI® corrective techniques and exercises is to activate the proprioception, tactility, and hearing-vision system for optimal motor function. This concept differs from others in the way that it draws on neurosensory-motor integration of reflex patterns instead of inhibiting them when they are preserved and active. MNRI® also draws on the integration of reflexes and their basic motor patterns (genetic 'programs' with learned and controlled movements, habits, and skills). The exercises and corrective movement activities are formation and maturation oriented, which means facilitation of impulse transmissions along nerve pathways and coordination of reflex patterns. The program includes exercises that facilitate reflex patterns are noninvasive, delicate, and child-friendly, as well as, fun movement activities. They are easy to learn by parents, caregivers, and professionals working with children with Down syndrome.

The level of reflex integration is presented with a synthetic function Z based on a mathematical model, which allows for an objective evaluation and analysis of the efficacy of this corrective 8 day MNRI® rehabilitation program used with children with Down syndrome. The parents' reports and the therapists' observations were positive, noting functional improvements of the children. Statistical tools and mathematical analysis according to Prof. Anna Krefft allowed for an objective evaluation of the outcome and provided the following:

Children diagnosed with Down syndrome need early intervention based on the MNRI® processes to start as soon as possible for the correction of dysfunctional and pathological reflexes which otherwise serve as the basis for the Down syndrome child's development of postural control, overall postural and motor development, fine motor coordination, speech and thinking processes, and learning skills.

Children having Down syndrome need a total program of intervention at the neurosensorimotor integration level for the correction of primary motor system and reflex patterns that are poorly developed. MNRI® processes are composed for that purpose, and were realized by MNRI® professionals during the 8 days of direct intervention with these children.

The results of mathematical and statistical analyses prove the high level of validity of diagnosis techniques with the Masgutova Method®, as well as, the intervention processes and procedures aimed at the correction of deficits in the Down syndrome children's primary motor system and cognitive functioning.

The use of the Masgutova Method® opens new possibilities for triggering natural, genetically given motor memory resources of children with Down syndrome and supports, through the use of the neurosensorimotor reflex integration procedures, reflex circuit functioning, and consciously learned motor system.

The Masgutova Method® offers a strong support for creation of new developmental possibilities and programs for Down syndrome children that can be applied in conjunction with other therapies. The Masgutova Method® demonstrates the primary importance of addressing reflex patterns to support the development and function of motor and cognitive systems. Statistical analysis has supported the effectiveness of the MNRI® diagnostic protocol and validates the results of the therapeutic part of the program.

References

Bernstein, N. (1997). *Bio-mechanics and physiology of movement* (Russ.). Moscow, Russia: Moscow-Voroniez.

Bobath, K. & Bobath, B. (1984). The neurodevelopmental treatment. In D. Scrutton (Ed.). *Management of the Motor Disorders of Children with Cerebral Palsy* (pp. 6-18). Oxford, England: Blackwell Scientific Publications Ltd.

Delacato, C. H. (1974). *The diagnosis and treatment of speech and reading problems.* Garden City Park, NY, USA: Avery.

Doman, G. (1984). *How to multiply your baby's intelligence.* Garden City, NY, USA: Doubleday.

Haines, D. E. (2002). *Fundamental neuroscience.* Second Edition. New York, NY, USA: Churchill Livingstone.

Krefft, A. (2007). *Diagnosis functions of non observable phenomena.* (Statistic Mathematic Analysis). Wroclaw, Poland: Oficyna Wydawnicza Politechniki Wroclawskiej.

Hooker, D. (1952). *The prenatal origin of behavior.* Laurence, KS, USA: University of Kansas Press.

Masgutova, S. K. (2006b). *Infant reflex integration in movement development.* In S. K. Masgutova, A. Regner, & M. D. E. Ronin-Walknowska

(Eds.), Materials of Polish National Conference: Modern Methods of Stimulation of Movement Development and Learning in Children with Difficulties in Learning, ADHD and Autism (pp. 5-28). Warsaw, Poland: Maternity Department, Pomorska Medical Academy, Szczecin .

Masgutova, S. K. (2007). *Neuro-sensory-motor development: Visual and auditory reflexes integration. Facilitation program of development and learning for children and adults.* Warsaw, Poland: MISM.

Masgutova, S. K. (2011). *Infant dynamic and postural reflexes. Neuro-sensory-motor reflex integration.* Warsaw, Poland: SMEI.

Masgutova, S. K., Akhmatova, N. K., & Lebedinskava, O. V. (2013). *Clinical-immunological assessment of therapy effect of the neuro-sensory-motor integration program of reflex patterns in airway chronic inflammatory diseases.* Front. Immunol. Conference Abstract: 15th International Congress of Immunology (ICI). http://www.frontiersin.org/10.3389/conf.fimmu.2013.02.00865/event_abstract?sname=15th_International_Congress_of_Immunology_(ICI)

Masgutova, S. K., Akhmatova, N. K., & Kiselevsky, M. V. (2008). Clinic-immunological assessment of effect of the therapy program of neuro-sensory-motor integration of reflex patterns at chronic inflammatory diseases of respiratory system. *Russian Immunology Journal*, 2, 454-463.

Masgutova, S. K., Masgutov, D. (2012). *MNRI® Oral-facial reflex integration in children and adults with neurodeficits* (Polish version). Warsaw, Poland: ISMI.

Montessori, M. (1995). *The absorbent mind.* New York, NY, USA: Henry Holt and Company.

Pavlov I. P. (1960). *Conditioned reflexes: An investigation of the physiological activity of the cerebral cortex.* (Anrep G.V., D. Sc. Trans., 1960). New York, NY, USA: Dover Publications Inc. (Original work was published in 1927)

Piaget, J. (1976). *The grasp of consciousness: action and concept in the young child.* Cambridge, MA, USA: Harvard Press.

Pilecki, W., Masgutova, S., Kowalewska, J., Masgutov, D., Akhmatova, N., Poreba, M. et al. (2012). The impact of rehabilitation carried out using the Masgutova Neurosensorimotor Reflex Integration method in children with cerebral palsy on the results of brain stem auditory potential examinations. *Advances in Clinical and Experimental Medicine,* 21, 363-371.

Sadowska, L. (2008). Wroclaw Rehabilitation Model (WRM) in early diagnosis and therapy in children with Down syndrome. In B. Kaczmarek (Ed.), *Facilitation of Development of Children with Down Syndrome - Theory and Practice* (pp. 195-224). Krakow, Poland: Impuls Kraków.

Sadowska, L., Masgutova, S., Kowalewska, J. (2009). *Reflex integration of children with Down syndrome; Diagnosis and therapy results using Masgutova neurosensorimotor reflex integration program – MNRI®.* International Scientific-Practical Conference - XXXII National Day of Child Rehabilitation: Down Syndrome. Treatment, Rehabilitation and Education. Article based on Presentation on Oct. 17-18, 2008.

Sechenov, I. M. (2014). *Selected physiological and psychological works* (Russ.). Moscow, Russia: Foreign Languages Publishing House. (Original work published 1935)

Schmid, F. (1987), *Das Down-syndrom,* Munsterdorf, Germany: Verlag Hansen & Hansen.

Sherman, S. L., Allen, E. G., Bean, L. H., & Freeman, S. B. (2007). Epidemiology of Down syndrome. *Mental Retardation and Developmental Disabilities Research Reviews,* 13, 221-227.

Sherrington, C. S. (1947). *The integrative action of the nervous system.* London, UK: Cambridge University Press.

Souchay, C., Guillery-Girard, B., Pauly-Takacs, K., Wojcik, D. Z., Eustache, F. (2013). Subjective experience of episodic memory and metacognition: A neurodevelopmental approach. *Frontiers in Behavioral Neuroscience, 7,* 212.

Vygotsky, L. S. (1986). *The child psychology. The problems of child development.* In 6 Books. Book – 4. Pedagogika. Moscow. (Russ.). Russian.

Wisniewska, K. E. (1990). Down syndrome children often have brain with maturation delay of growth and cortical dysgenesis. *American Journal of Medical Genetics Supplement,*7, 274-281

Weijerman, M. E. & de Winter, J. P. (2010). Clinical practice. The care of children with Down syndrome. *European Journal of Pediatrics,* 169, 1445-1452.

Wiseman, F. K., Alford, K. A., Tybulewicz, V. L., Fisher, E. M. (2009). Down syndrome – recent progress and future prospects. *Hum. Mol. Genet.*, 18, R75-R83.

Dear children, thank you for your participation in our research, you were so great! We are so happy for your personal growth and development! Be healthy and reach even better results in your learning and life! – Authors

The Role of STNR for Visual Reflexes and Primary Visual Skills in Children with Dyslexia

Svetlana Masgutova, Ph.D., Warsaw, Poland – Orlando, FL, USA

Svetlana Masgutova, Ph.D

This article is based on preliminary research and an article by S. Masgutova titled, *The Effect of Combining Vision Training with Reflex Movement Patterns and Related Postures on the Development of Primary Visual Skills in Children with Dyslexia* presented at the Neuro-Optometric Rehabilitation Association International Conference in 2009 where it was awarded a Founding Fathers Neuro-Optometric Achievement Medal. This article is published at www.MasgutovaMethod.com.

Visual and motor reflex patterns in babies serve as the neurophysiological basis for future academic skills in school children. Visual skills such as binocular vision, eye tracking, head-righting, visual prioritizing and focusing, convergence/divergence, figure-ground discrimination, and others build on the integration of overall neurosensorimotor development with manual skills, balance, stability, grounding, and the vestibular system. Timely maturation and integration of these systems enable children to avoid both delays in sensory and motor skills and lags in academic achievement.

Specialists working with various MNRI® reflex integration programs have consistently found a correlation between primary visual skills and reflex patterns among children with learning challenges. Initial research on the influence of visual and motor reflexes on visual skills by MNRI® researchers resulted in quantification of these correlations in a group of 370 (7-10 years old) children diagnosed with dyslexia. A further project measured the response among 240 of those children with severe dyslexia and the poorest visual skills to MNRI® exercises. The purpose was to determine whether combining vision training with motor reflex patterns and related developmental postures (as practiced in MNRI®) would influence the development of visual skills in these children. This report describes and comments on findings from those two investigations.

Correlation between primary visual skills and developmental postures among children with dyslexia

The following responses and reflex patterns associated with primary visual skills were assessed:

1) binocular vision
2) eye tracking (smooth pursuit movements used to track a slow moving object):
   a) on a horizontal plane
   b) on a vertical plane
   c) in a circle
   d) coordination of eye tracking with breathing
3) oculovestibular reflex pattern (head righting, eyes open)

4) optokinetic reflex pattern (head righting, eyes open/closed)

5) convergence-divergence

6) focusing

7) visual figure-ground discrimination

8) visual analysis/synthesizing (visual following from details to whole and vice versa)

9) staring (freezing)

10) corneal (eye-blinking)

11) visual Fear Paralysis (response to sudden unexpected intensive visual stimulus by pupil dilating, rapid eye blinking, holding breath, muscle overtension, and screaming).

Each reaction or reflex pattern was evaluated according to the following main parameters:

1) pattern (corresponding the response to the specific sensory or sensory-proprioceptive stimuli)

2) direction of the motion or sequence of the responses/postures

3) stability of the response (smooth movements of eyes, proper range of eye movements)

4) timing (in tracking, the ability to maintain a slow moving object in the center of the field of view with the eyes neither lagging behind nor moving ahead; in oculovestibular and optokinetic reflexes, speed of head righting adjustment and number of saccades)

5) symmetry of the response in both eyes.

The following additional parameters of the reflex pattern were also evaluated:

1) efficiency of eye movements in terms of completion of the task, range of visual field, and ability to maintain binocularity in the mid-field

2) muscle tone (easy tracking, no tension, appropriate blinking reflex, no staring reflex, easy breathing).

The initial research group consisted of 370 children (7-10 years old) diagnosed with dyslexia. Specialists measured their performance of the visual skills described above and correlated them with developmental postures in the following way. The children performed visual tasks in each of three postures: **Posture 1,** seated in a chair; **Posture 2**, lying down prone supported by their elbows. **Posture 2** represents a very early developmental stage in which an infant first becomes able to overcome gravity by raising his head and maintaining it in a vertical position. **Posture 3**, sitting in the Symmetrical Tonic Neck Reflex position, on their knees and heels with head up, arms extended and hands flat on the floor slightly in front of the knees.

The testing results showed that the level of development of their primary visual skills correlated with the three body positions. All the children demonstrated variations in visual skills depending on their body position, with poorer results in the two developmental postures. Eye tracking on the horizontal plane, for example, was less controlled and more difficult in **Postures 2** and **3**, which represent earlier developmental stages. This sensitivity of visual skill to body position became the basis for further research.

Table 1. Level of primary visual skill development demonstrated by 370 children with dyslexia in three different postures.

LEVEL	POSTURE-1				POSTURE-2				POSTURE-3 (STNR)			
	No	%	No	%	No	%	No	%	No	%	No	%
Good	39	10.54	118	31.89	17	4.59	77	20.81	23	6.22	91	24.59
Average	79	21.35			60	16.22			68	18.38		
Poor	156	42.16	252	68.10	169	45.68	293	79.19	164	44.32	279	75.40
Very poor	96	25.95			124	33.51			115	31.08		

The group of 370 children with dyslexia demonstrated the following (see Table 1):

• Overall inadequate primary visual skill development: 68.10% showed poor or very poor results in **Posture 1** (sitting); 79.19% in **Posture 2** (prone on elbows); and 75.40 % in **Posture 3** (STNR pattern).

• The children tended to achieve worse results when they performed visual tasks in postures associated with reflex patterns from earlier stages of development. Whereas 68.10% demonstrated poor or very poor skill levels in **Posture 1** (sitting), that percentage grew to 79.19% in **Posture 2** (prone on elbows) and 75.40% in **Posture 3** (STNR pattern).

Based on long-term and current clinical observations, MNRI® practitioners interpreted these results as follows:

1) Visual skills in children with dyslexia are insufficient and poorly developed. Deficiencies in different visual reflex patterns cause major difficulties in their reading and writing (reading speed, silent reading, reading

comprehension, spelling, hand writing, composition) and the deficiencies become worse depending on the position of their bodies. The children demonstrated better results in **Posture 1** (sitting). This can be due to the growth of ability to control visual skills with increasing age due to training of the visual system in their daily lives (conscious control, learned skills).

2) The high level of poor and very poor results in **Postures 2** (prone) and **3** (STNR) can indicate the lack of sufficient development of visual skills at earlier stages of neurosensori-motor development. Children with dyslexia seem to use their visual skills better when they can control them (often with considerable effort), but cannot use these skills automatically and spontaneously. Using more controlled than spontaneous eye activity indicates poor neurological maturation of automatic visual function. This behavior, requiring more effort, can cause overloading of the visual system, visual inhibition, poor coordination of visual activity, and fatigue. The visual skills of such children are too unstable, too poorly developed, and require too much conscious control to adequately support their learning process. **Fig. 1** presents the data from **Table 1** in chart form.

Figure 1. (below) Level of Visual Responses for Children with Dyslexia in Different Postures: 1) sitting, 2) prone supported on elbows, and 3) STNR.

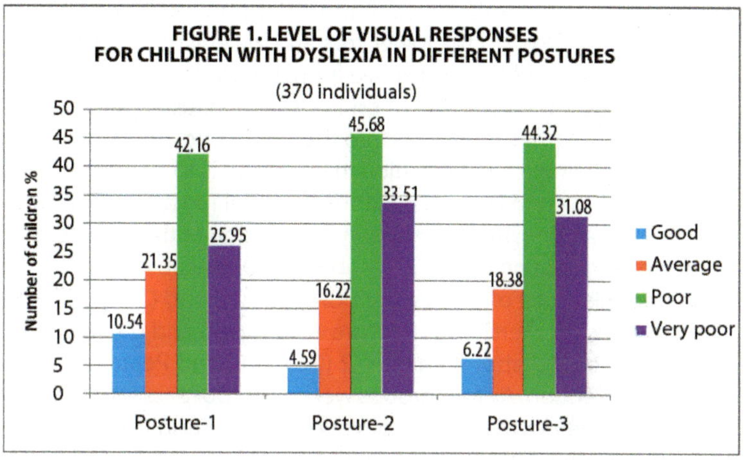

Results shown in Table 1 became the baseline for the next stage of research.

Response of control and research groups to special vision training exercises

This stage of research involved 240 children: 150 (62.5%) with poor and 90 (37.5%) with very poor results on the primary visual skills assessment shown in Table 1. They were divided into the following four groups of 60 children each:

1) **Control group:** received no intervention.

2) **Research group A:** performed vision exercises (from MNRI® Visual and Auditory Reflexes Integration by S. Masgutova, 2008) in **Posture 1** (sitting) with no use of the postural reflex patterns.

3) **Research group B**: performed the same vision exercises in **Posture 2** (prone supported on elbows) as practiced in the MNRI® program.

4) **Research group C:** performed the same vision exercises in **Posture 3** (STNR) as practiced in the MNRI® Program.

Posture 1:

Posture 1 is sitting down, keeping the head and spine straight upright, and supported with the palms and elbows, while being in a chair with arms. It is a stage of infant motor development in which development of vertical alignment postural control and antigravity mechanism occurs. This position presents a later motor stage which is explored after being on the stomach and crawling and sitting on knees. It is important for head righting and all visual skills: binocular and distant vision, convergence-divergence, near-far accommodation, and others. The skill of sitting down should start developing at the sixth month. It affects the formation of future walking and coordination of visual and gross motor abilities.

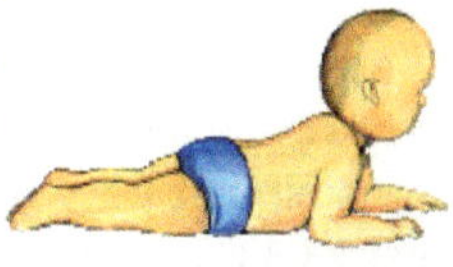

Posture 2

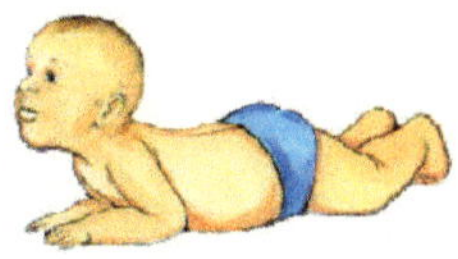

Posture 2:

Posture 2 is prone supported on elbows and represents a stage of motor development that infants must achieve. The main task of this pattern is the development of antigravity mechanisms that enable the baby to raise her head and keep its vertical alignment. This position in turn develops the important skills of head righting, eye leveling, binocular vision, eye tracking, convergence, focusing, near-far accommodation, and visual evaluation of the horizontal and vertical planes. Eye tracking is one of the main automatic responses that trigger the midbrain and RAS system in the brainstem to organize the protection strategy for visual information processing. All of these crucial skills should be developed by the end of the second month. Later on they will affect the development of the infant and child's early ability to use visual skills when they sit on their knees and heels in the STNR position in extension.

Posture 3

Posture 3:

Posture 3 or the STNR pattern is the sitting-on-the-knees-and-heels posture in which head extension causes arm extension and the tendency for leg flexion. The gravity center shifts toward the lower core and lower limbs as the child quiets motor activity to engage vision and audition. It is active from the age of six months. At about the sixth to seventh month, the baby begins to crawl on her stomach, using an asymmetrical style of motion. The matured STNR then integrates into the whole body movement system at about the tenth month. The task of this STNR position in extension is to bring the above visual reflex patterns together with posture to enable automatic visual function to become a consciously regulated skill. The integration of vision and posture is necessary for creation of the code for future automatic postural control in response to information coming from the vestibular, oculovestibular, and optokinetic systems.

During this stage of the research **Groups A, B and C** were given the MNRI® Visual Reflex training. **Group A** performed vision exercises in **Posture 1. Groups B** and **C** performed the same exercises in the developmental postures described above, **Posture 2 (Group B)** and **Posture 3 (Group C).** The control group experienced no further intervention. All four groups were given post-assessments after one and four months. All of the children had initially demonstrated poor or very poor results in the assessment of their visual skills.

Control group. This group was given neither visual nor postural exercises. In pre-tests, 61.67% of these children performed at a poor level and 38.33% at a very poor level. In post-tests, after one month they did not demonstrate significant changes (60.0% poor and 36.67% very poor, with an average negative change of 3.33%). Results in four months again did not show significant change (53.33% poor and 40.0% very poor, with an average negative change of 6.67%; see Table 2).

Table 2. CONTROL GROUP
(60 individuals, 7-10 years old)

Level of visual skills	Initial results				Results in one month				Results in four months			
	No	%	No	%	No	%	No	%	No	%	No	%
Good	0	0	0	0	0	0	2	3.33	0	0	4	6.67
Average	0	0			2	3.33			4	6.67		
Poor	37	61.67	60	100	36	60	58	96.67	32	53.33	56	93.33
Very poor	23	38.33			22	36.67			24	40.0		

Research group A (Posture 1: Sitting). Children in this group were given MNRI® Visual Reflex training with no developmental or STNR postural component. They were not given any postural exercises. Their performance in post-tests given one month later showed that 20% of them moved to good or average levels and 80% still demonstrated poor or very poor levels; in four months the number of children achieving good or average levels grew to 26.67% and the number remaining at poor or very poor levels was 73.33%. These results did not show any significant change in visual skills after four months in comparison with results after one month (see Table 3).

Table 3. RESEARCH GROUP A: POSTURE 1 – Sitting down
(60 individuals, 7-10 year old age)

Level of visual skills	Initial results				Results in one month				Results in four months			
	No	%	No	%	No	%	No	%	No	%	No	%
Good	0	0	0	0	3	5	12	20	4	6.66	16	26.67
Average	0	0			9	15.0			12	20.0		
Poor	37	61.67	60	100	29	48.33	48	80.0	28	46.67	44	73.33
Very poor	23	38.33			19	31.67			16	26.67		

The graph at right (Figure 2) shows the dynamic of the change in primary visual skills of 60 children with dyslexia in **Posture 1 (sitting)**: initial results, results in one month after doing visual exercises, and results in 4 months with homework/visual exercises given two times a week.

FIGURE 2. DYNAMICS OF CHANGE IN VISUAL RESPONSES FOR CHILDREN WITH DYSLEXIA IN POSTURE - 1

(60 children of 7-10 year old age)

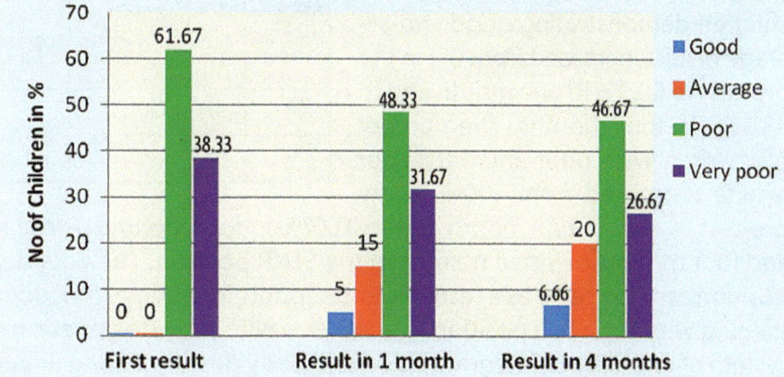

Research group B (Posture 2 - Prone supported on elbows). Children in this group were given the vision exercises in the developmental posture, prone supported on elbows. The number of children demonstrating good and average results increased from 0% in the pre-test to 50.0% in one month and to 60.0% in four months (See **Table 4**). The number of children with poor and very poor results decreased from 100% in the pre-test to 50% in one month and to 40% in four months. These results show significant change after one month of visual training in **Posture 2 (prone on elbows).** The data demonstrate that the developmental position of prone on elbows can serve as a resource and support for optimal functioning of visual skills. MNRI® combines vision training with the developmental position as a way to activate genetic motor memory and remind the body-brain system of its ability to integrate vision and body posture during infancy. Results of the post-test in four months showed that optimized visual skills remained stable and in fact tended to improve further. The results also suggest that body position can be used as a key for future development of visual skills. The combination of visual work and developmental posture can be a valuable corrective tool for children with visual challenges and learning difficulties.

Table 4. RESEARCH GROUP B: Posture 2 - Prone supporting on elbows (60 individuals, 7-10 years old)

Level of visual skills	First results				Results in one month				Results in four months			
	No	%	No	%	No	%	No	%	No	%	No	%
Good	0	0	0	0	10	16.67	30.0	50.0	13	21.67	36.0	60.0
Average	0	0			20	33.33			23	38.33		
Poor	37	61.67	60	100	15	25.0	30.0	50.0	14	23.33	24.0	40.0
Very poor	23	38.33			15	25.0			10	16.67		

The graph at right (Figure 3) shows the dynamic of the development of primary visual skills of 60 children with dyslexia in **Posture 2 (prone on elbows):** initial results, results in one month after doing visual exercises, and results after 4 months with homework/visual exercises given two times a week.

FIGURE 3. DYNAMICS OF CHANGE IN VISUAL RESPONSES FOR CHILDREN WITH DYSLEXIA IN POSTURE - 2

(60 children of 7-10 year old age)

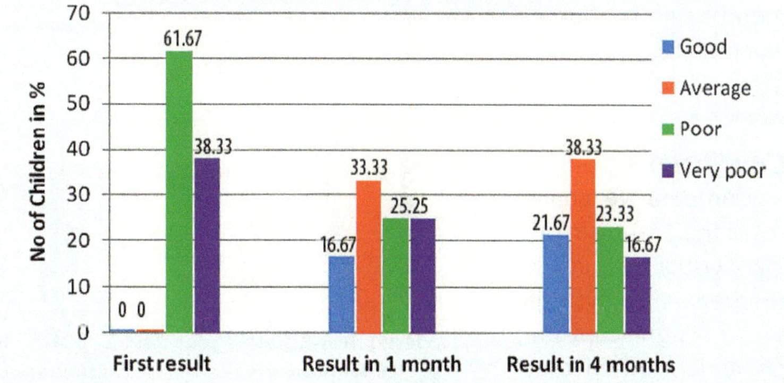

Group C (Posture 3 – STNR position; see Table 5) Children in this group were given the vision exercises in **Posture 3 (STNR).** The number of children demonstrating good and average results increased from 0% in the pre-test to 60.0% in one month and to 68.33% in four months. The number of children with poor and very poor results decreased from 100% in the pre-test to 40% in one month and to 31.67% in four months. These results show significant change after one and four months of visual training in the STNR position. The data demonstrate that the STNR developmental position can also serve as a resource and support for optimal functioning of visual skills. MNRI® combines vision training with the STNR position as another way to activate genetic motor memory and remind the body brain system of its ability to integrate vision and body posture during infancy. Results of the post test in four months were similar to those of Group B indicating optimized visual skills remained stable and also tended to improve. These results again suggest that body position can be used as a key for future development of visual skills.

Level of visual skills	Table 5. RESEARCH GROUP C: Posture -3 – STNR (60 individuals, 7-10 year old age)											
	First results				Results in one month				Results in four months			
	No	%	No	%	No	%	No	%	No	%	No	%
Good	0	0	0	0	12	20.0	36	60.0	15	25.0	41	68.33
Average	0	0			24	40.0			26	43.33		
Poor	37	61.67	60	100	13	21.67	24	40.0	11	18.33	19	31.67
Very poor	23	38.33			11	18.33			8	13.33		

The graph at right (Figure 4.) shows the dynamic of the development of primary visual skills of 60 children with dyslexia in **Posture 3, STNR:** initial results, results in one month after doing visual exercises, and results after 4 months with homework/visual exercises given two times a week.

Figure 5, right, is a Summary of Research Group Findings which shows the dynamic of the development of primary visual skills of 180 children with dyslexia in Research Groups: A (sitting); B (prone); and C (STNR); as demonstrated in: Test 1, initial results; Test 2, one month after doing visual exercises; and Test 3, after four months with homework/visual exercises given two times a week.

FIGURE 4. DYNAMICS OF CHANGE IN VISUAL RESPONSES FOR CHILDREN WITH DYSLEXIA IN POSTURE -3 (STNR)

(60 children of 7-10 year old age)

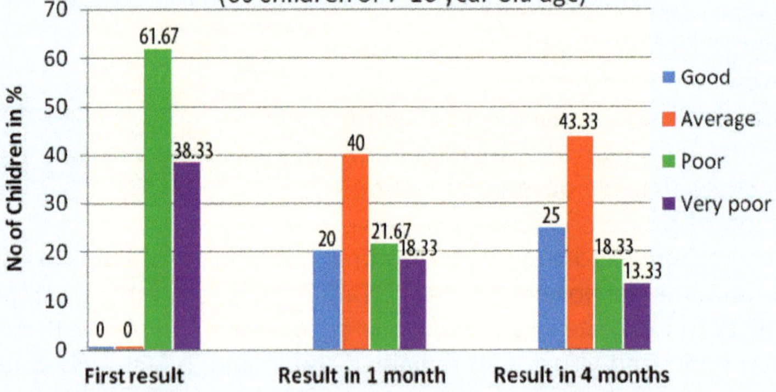

FIGURE 5. DYNAMICS OF CHANGE IN VISUAL RESPONSES FOR CHILDREN WITH DYSLEXIA

(180 individuals - 60 in each Research Group)

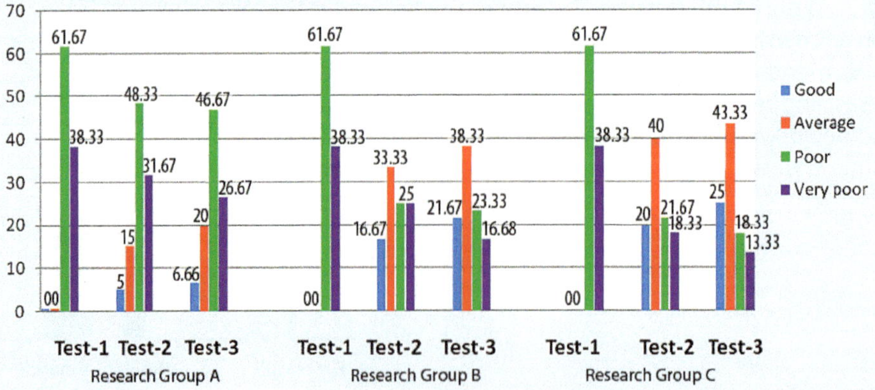

Conclusions

Comparative analyses of the change in primary visual skills in 180 children with dyslexia showed significant improvement when they performed visual exercises in **Postures 2 (prone supported on elbows)** and **3** (STNR position). Children training in **Posture 1 (sitting)** showed lower results than those in the other two groups after both one and four months. Children with dyslexia may need a much longer time for visual training in the more stable sitting position. Post-testing showed improvement in 120 children with dyslexia in less time, with less effort and more stable results thanks to the use of the developmental resources of postural reflex patterns inherent in **Postures 2 (prone on elbows)** and **3 (STNR position).** The researchers interpret this improvement as resulting from activation of the 'tasks' of the reflex patterns. Developmentally these positions provide for the formation of the visual skill of estimation of horizontal and vertical planes and lines in the surroundings, and for awakening and strengthening the mechanisms linking vision with balance, stability, grounding, and the vestibular system. At the ages of two and four months, these reflexes support the formation of the visual-vestibular-proprioceptive code, which then supports the development of visual skills: binocular vision, tracking along the horizontal and vertical lines/planes, perception of two-dimensional planes and three-dimensional perspective with easy visual transition from one to the other, near-far accommodation, and figure/ground discrimination (important for future attention span). The results of this investigation suggest that combining vision training with reflex patterns and related developmental postures (as practiced in MNRI®) reactivate the developmental potentials of visual, oculovestibular, and optokinetic reflex patterns that become the neurophysiological basis for academic achievement.

References

Brandt, T., Dieterich, M. (1999). *The vestibular cortex: Its locations, functions, and disorders.* Annals of the New York Academy of Science. US.

Bryan, J. (1995). *Your amazing brain. A fascinating see-through view of how our brain works.* Forward by Daria Luisi, New York, NY, USA: Victoria House Publishing Ltd.

Delgado-Garcia, J. M. (2000). *Why move eyes if we can move the head?* Brain Research Bulletin. USA.

Dieterich, M., Brandt T. (2000). *Brain activation studies on visual-vestibular and ocular motor interaction. Current Opinion in Neurology.* USA.

Haines, D. E. (2002). *Fundamental neuroscience.* 2nd Edition. The Curtis Center Independence Square West. Philadelphia, Pennsylvania. USA.

Lindsay, P., Donald, N. (1972). *Human information processing: An introduction to psychology.* New York & London: Academic Press.

Masgutova, S. (2009). *The effect of combining vision training with reflex movement patterns and related postures on the development of primary visual skills in children with dyslexia.* www.MasgutovaMethod.com.

Masgutova, S. (2009). *MNRI® Visual and auditory reflexes integration.* Orlando, FL, USA: SMEI.

Masgutova, S. (2011). *Infant dynamic and postural reflexes. Neurosensorimotor Reflex Integration.* Revised and edited, scientific-practical manual. Orlando, FL, USA: SMEI.

Masgutova, S. & Masgutov, D. (2001). *Creative drawing and integrating movement.* Warsaw, Poland: ISMI.

Montgomery, T. M. (1997-2007). *Anatomy, physiology and pathology of the human eye.* US. Retrieved August 2007, from www.tedmontgomery.com

O'Reilly, R., Morlet, T., Cushing, S. (2013). *Manual of pediatric balance disorders.* San Diego, CA, USA: Plural Publishing Inc.

Pavlov, I. P. (1960). *Conditioned reflexes: An investigation of the physiological activity of the cerebral cortex.* (Anrep G.V., D. Sc. Trans., 1960). New York, NY, USA: Dover Publications Inc.

Piaget, J. (1954). *The construction of reality in the child.* (M. Cook, Trans.). New York, NY, USA: Ballantine Books.

Vygotsky, L. (1986). *The Child Psychology.* The Problems of Child Development. In 6 Books. Book – 4. Moscow, Russia: Pedagogika.

Our MNRI® Research Team on visual reflexes thanks all our children for their excellent cooperation in allowing us to find that the main key for visual exercises is the STNR! Many children and adults can now benefit from this news! – The MNRI® Team in Poland

An Investigation of the Effects of MNRI® Techniques on the Educational Performance of Kindergarten Students

Lisa Ortego, MS, CCC-SLP; Emily Pelican, MS, CCC-SLP; Laurie Callaba, PT; Tina Marks, MS, CCC-SLP, MNRI® Core-in-Training Specialists, Louisiana, USA

Abstract

Three speech-language pathologists and one physical therapist who work in several private and public schools in southwest Louisiana undertook this project. After taking the *MNRI® Maximizing Brain Potentials* program, we, the authors, became interested in the possibility of implementing this MNRI® program in a few of the schools we worked in and to research what improved academic progress may occur in the childrens' learning. The research was planned in three different kindergarten classes at two different schools. The goal of this research was to find out if the MNRI® techniques would affect the education performance of kindergarten students as measured by their performance on the Dynamic Indicators of Basic Early Literacy Skills (DIBELS) thus increasing overall reading skills. Due to the limited time each therapist had to spend in each class, we decided that the participating kindergarten classes would be provided the *MNRI® Archetype Movement* DVD (8 standing poses/exercises). The kindergarten teachers were requested to use the DVD daily in the designated classrooms.

Clockwise from left: Tina Marks, Laurie Callaba, Lisa Ortego, and Emily Pelican

Introduction

Four different therapists working in several private and public schools in southwestern Louisiana undertook this project. An analysis of the results of the childrens' school skills and educational work for the last several years in these schools showed that there was a puzzling lack of academic progress demonstrated by many students while receiving traditional reading intervention techniques. The schools included a mixture of low and middle-income families. The progress for these students was measured according to the results on their early literacy skills assessments (indicators on the DIBELS). Prior to the start of school in August 2013, the 4 therapists and 40 teachers from Reeves, LA and surrounding areas attended the *MNRI® Maximizing Brain Potentials* course taught by Dr. Patty Shackleford. After completing this program, they investigated the possibilities of implementing this MNRI® program and researching if any academic progress would occur in the identified students' academic progress as a result of implementing the MNRI® Archetype Movements. The research was planned in two different schools, Reeves High School and Kinder Elementary School, in three of their kindergarten classes. It was determined that a comparison of the results in overall reading skills of children could be noted as measured by their performance with DIBELS before and after implementing the MNRI® program. Due to the limited time each therapist had to spend in each class, it was decided that two of the kindergarten classes would use the DVD showing the MNRI® Archetype Movements (8 standing poses/exercises) daily. The teachers in these two kindergarten classes had completed the *MNRI® Maximizing Brain Potentials course*. The third kindergarten class was used as a control group. The teacher in this class did not take the *MNRI® Maximizing Brain Potentials* course nor was she using the MNRI® Archetype Movements with her class.

For clarity in this study, Reeves High School Kindergarten Class is labeled **Class # 1** (with MNRI® intervention), a kindergarten class in Kinder Elementary School is labeled **Class # 2** (also with MNRI® intervention), and a second kindergarten class in Kinder Elementary School is labeled **Control Class # 3** (with NO MNRI® intervention) as our control group.

Assessment Tool and Therapeutic Techniques

"The Dynamic Indicators of Basic Early Literacy Skills (DIBELS) are a set of procedures and measures for assessing the acquisition of early literacy skills from kindergarten through the sixth grade... [It's] fluency measures [are] used to regularly monitor the development of early literacy and early reading skills." The DIBELS is *"comprised of seven measures to function as indicators of phonemic awareness, alphabetic principle, accuracy and fluency with connected text, reading comprehension, and vocabulary...[It] was designed for use in identifying children experiencing difficulty in acquisition of basic early literacy skills."* (Good, pg. 1.) The Reading Coach and Reading Interventionist administered the DIBELS three times a school year in August, January, and May. Once data was analyzed, intervention groups were formed based on the scores. The students were then grouped according to their score: core (benchmark with no support needed), strategic, and intensive. The strategic and intensive groups received direct intervention services. These students had their progress monitored by the teacher every two to three weeks based on student need and performance.

MNRI® Archetype Motor Patterns – A child is born with a complex of natural motor resources – reflexes and primary movements. These movements present the basis for the genetic motor 'programs' of human development to create the nerve net system and myelination, determining brain maturation and functioning as a child grows (Masgutova, 2011; Teitelbaum, P., Teitelbaum, O., et. al. 2002; Rentschler, 2013). The education of a child uses all these resources as physiological indirect support of cognitive functions (Vygotsky, 1986).

Learning, planning, and the development of inner control are some of the most important functions of the brain's cortex. Successful development of these skills ultimately depends upon the physiological maturation of the brainstem (myelination of the extrapyramidal nerve net) through early movements and sensory motor integration which are the foundation of human development. Movement, behavioral, and learning challenges often result when reflex patterns do not develop properly, are immature, and/or poorly integrated in comparison with typical patterns of development. In addition, primary reflex patterns may not function appropriately following significant life stressors and/or trauma that children and adults may experience. In fact, data collected from thousands of MNRI® Assessments over the past 30+ years has demonstrated that, as the number of non-integrated primary motor reflex patterns in neurotypical children increase, the number and severity of learning challenges correspondingly increases (Masgutova, S., Masgutov, D., 2013).

In the primary motor system we recognize Archetype Movements as the most global and general motor 'constructions' determining bio-mechanical functions and strategies of motor activity. A child during development and spontaneous learning incorporates these innate bio-mechanic aspects of movements with primary cognition including focusing, curiosity, memorizing, comparing, classifying, analysis, visualization, and abstract thinking. The specific patterns of these innate bio-mechanics are identified as Archetype Movement patterns: These include:

Intentional Movements

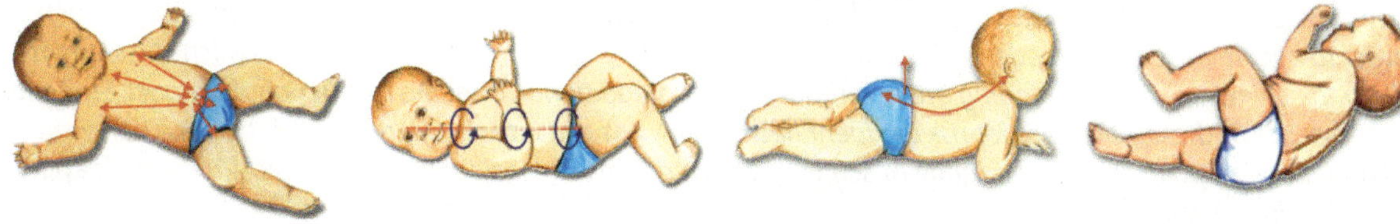

Core-Limbs Flexion-Extension — Mouth – Spine rotation — Trunk Extension — Lateral Spine Flexion-Extension

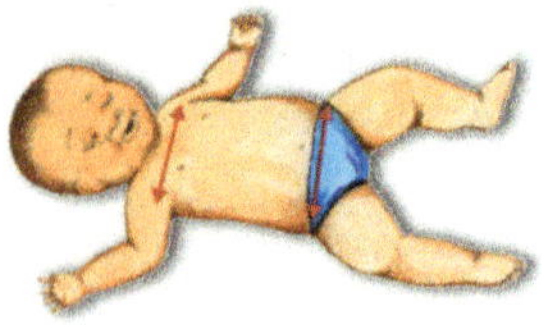

Homologous movement

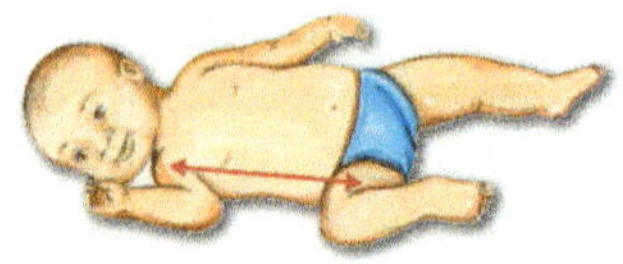

Homolateral movement

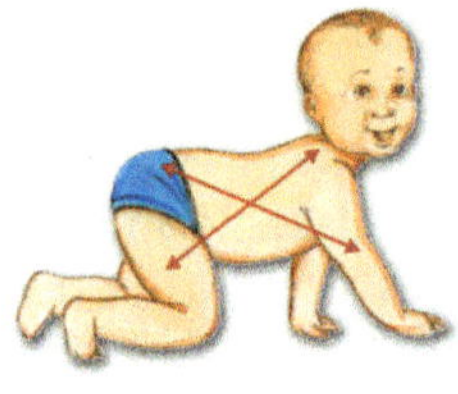

Cross-Lateral Movement

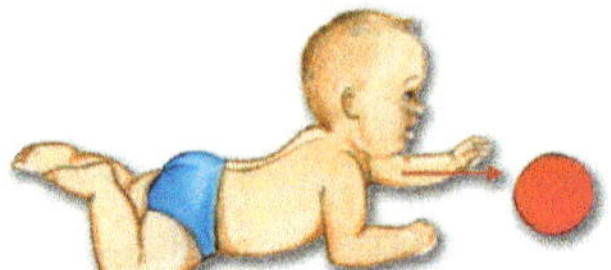

Intentional/Differentiated Movement

- Core-Limbs Flexion-Extension – Six Ended Star
- Horizontal Spine Rotation – mouth-spine rotation
- Vertical Spine Compression – trunk extension
- Lateral Spine Flexion-Extension – lateral leaning (to sides)
- Homologues Movements – bilateral activities
- Homolateral Movements – one-sided activities
- Cross-Lateral Movements – opposite limbs movements
- Intentional Movements – goal-oriented/consciously controlled activities

Each of the Archetype Motor patterns creates the basis for the similar motor construction of various reflex patterns.

Method

Participants. The participants of this study include kindergarten children in two schools – Reeves High School (Class # 1) in Reeves, LA and Kinder Elementary School (Classes # 2 and 3) in Kinder, LA. Both schools are in Allen Parish, Louisiana. The total number of students involved in this study initially was 64. As happens in any classroom, several children moved in and out of classes due to being new to the area or moving out of the area. Therefore, only the students who began the school year and remained all year long were used in the study. The final break out was 15 in Class # 1 and 40 in Classes # 2 and 3 for a total of 55 students.

Materials. In this study, each student was administered the DIBELS which are a set of procedures and measures for assessing the acquisition of early literacy skills from kindergarten through the sixth grade. This assessment is administered three times each year in August, January, and May:

- the Beginning of the Year DIBELS assessment consisted of the First Sound Fluency (FSF) and Letter Naming Fluency (LNF) (a Composite Score is obtained from these two scores)
- the Middle of the Year DIBELS assessment consisted of the First Sound Fluency (FSF), Letter Naming Fluency (LNF), Phoneme Segmentation Fluency (PSF), and Nonsense Word Fluency (NWF) (a Composite Score was obtained from these four scores)
- the End of the Year DIBELS assessment consists of the following measures: Letter Naming Fluency (LNF), Phoneme Segmentation Fluency (PSF), and Nonsense Word Fluency (NWF) (a Composite Score was obtained from these three scores).

For research purposes, we tracked FSF and PSF for Middle and End of the Year Composite Scores.

"The DIBELS First Sound Fluency (FSF) is a standardized, individually administered assessment that provides a measure of phonemic awareness skills in the beginning and middle of kindergarten" (http://dibels.uoregon.edu/market).

"The DIBELS Phoneme Segmentation Fluency (PSF) measure is a standardized, individually administered test of phonological awareness." (Kaminski & Good, 1996). The PSF measure assesses a student's ability to segment three- and four-phoneme words into their individual phonemes fluently. (http://dibels.uoregon.edu/market).

The following chart indicates the kindergarten benchmark goals and cut off points for risk based on the numeric scores at the beginning, middle, and end of the year testing (http://dibels.org/papers/DIBELSNext-BenchmarkGoals.pdf). *"The DIBELS benchmark goals are used to identify students who need additional instruction in order to become successful readers... The cut off points for risk indicate a level of skill below which the student is unlikely to achieve subsequent reading goals without receiving additional, targeted instructional support. Students with scores below the cut off point for risk are identified as likely to need intensive support… Between a benchmark goal and a cut off point for risk is a range of scores… referred to as strategic support."* (http://dibels.org/papers/DIBELSNextBenchmarkGoals.pdf.)

Kindergarten: Benchmark Goals and Cut Point for Risk

Measure	Score Level	Likely Need For Support	Beginning of Year	Middle of Year	End of Year
DIBELS Composite Score	At or Above Benchmark	Likely to Need Core Support	26 +	122 +	119 +
	Below Benchmark	Likely to Need Strategic Support	13 – 25	85 – 121	89 – 118
	Well Below Benchmark	Likely to Need Intensive Support	0 – 12	0 - 84	0 – 88
FSF (First Sound Fluency)	At or Above Benchmark	Likely to Need Core Support	10 +	30 +	
	Below Benchmark	Likely to Need Strategic Support	5 - 9	20 – 29	
	Well Below Benchmark	Likely to Need Intensive Support	0 - 4	0 – 19	
PSF (Phoneme Segmentation Fluency)	At or Above Benchmark	Likely to Need Core Support		20 +	40 +
	Below Benchmark	Likely to Need Strategic Support		10 - 19	25 – 39
	Well Below Benchmark	Likely to Need Intensive Support		0 – 9	0 – 24
NWF-CLS (Nonsense Word Fluency-Correct Letter Sounds)	At or Above Benchmark	Likely to Need Core Support		17 +	28 +
	Below Benchmark	Likely to Need Strategic Support		8 – 16	15 – 27
	Well Below Benchmark	Likely to Need Intensive Support		0 – 7	0 - 14

The benchmark goal is the number provided in the "At or Above Benchmark" row. The cut off points for risk indicate a level of skill below which the student is unlikely to achieve subsequent reading goals without receiving additional, targeted instructional support. (Source: https://dibels.org/papers/DIBELSNextBenchmarkGoals.pdf)

The kindergarten teacher in Class # 1 and 2 were instructed in MNRI® standing Archetype Movements and reflex dance techniques and given the *MNRI® Archetype Movement* DVD with the instructions to use it daily with the entire total class with the exception of the Control Class #3. They did not use any MNRI® techniques.

Procedure

Class # 1 started the school year with 17 students. One student moved in the fall and two other students transferred into the class around midyear. One of these students then transferred to another school system prior to the DIBELS End of the Year testing. Results of the study are provided on the 15 students who remained from the beginning of the year through the end of the year.

In Class # 1, the two students who transferred into the school were not receiving MNRI® techniques at their previous school. It should also be noted that this class does not have any special education students. Two students repeated kindergarten and two students were screened for special education services. Of the two screened for special education services, only one repeated kindergarten. However, by the time the evaluation referral process should have occurred, the referrals for that evaluation were no longer valid due to gains made through use of MNRI® techniques. The teacher used the following MNRI® techniques daily for whole group instruction: *MNRI® Archetype Movements* DVD in the morning, Hands Supporting Reflex against the wall during bathroom breaks. ATNR dance reaching for sight words on the word wall, and cross tap for counting, ABCs, and any other rote activity MNRI® could be incorporated into. During small group activities, she used ATNR from repatterning, STNR (daily at the beginning of the year), and Archetype Movements before reading. Around midyear she added all phases of Robinson Hand Grasp and manual skills for fingers. At mid year, the speech therapist (an MNRI® Core Specialist-in-Training), pulled one student weekly using specific techniques addressing negative transitions – severe anger and coping issues. The teacher also pulled two other struggling students who were demonstrating weaknesses in basic gross and fine motor skills.

Class # 2 began the school year with 23 students. One student transferred in mid fall from another kindergarten class within the same school to this class at the parent's request. At both the middle and end of year DIBELS testing 24 students were in the class. This class had three special education students, and two additional students receiving speech and language therapy. From August through December 2013, this class did not use any MNRI® techniques in the classroom, in the special education setting, or in the speech and language therapy. Results of the study are provided on the 22 students who remained from the beginning of the year through the end of the year. In January the teacher began using the following MNRI® techniques daily for whole group instruction: *MNRI® Archetype Movement* DVD in the morning, Hands Supporting against the wall during bathroom breaks, ATNR dance and Spine Rotations for sight words on the word wall, Robinson Hand Grasp dance, cross tap for counting, ABCs, and any other academic activity MNRI® could be incorporated into. The teacher also used Archetype Movements during small group activities. In October, one of the regular education students in this class was screened by the school's occupational therapist (OT) to evaluate concerns in fine motor control. This student was found to be a couple months delayed but did not meet the state requirements to qualify for OT services through the school system, so the OT recommended a 6 month follow-up.

When the six months follow-up evaluation was completed in April 2014, the student was rescreened and was found to have gained a year in fine motor skills and was not in need of OT services. It does need to be noted that one of the special education students, about whom the teacher reported great gains in academic skills in the classroom, declined in the scores on the end of year District Common Assessment: a 9% decrease in Math and a 33% decrease in English Language Arts. This special education student was reported to have had a seizure at the beginning of May 2014. The seizure activity may have caused neurological deficits and may be a reason for the decline on the assessment. Doctors are evaluating other areas of decline noted in this student: regression in developmental skills including gross motor, fine motor, cognition and speech. Due to his medical issues, he was removed from the study.

Control Class #3, our control group, began the school year with 24 students. Three students moved and one student transferred into the class by midyear. At the middle of the year DIBELS testing in January 2014, there were 21 students in the class. By the end of the school year, 3 students moved and 2 that had moved at mid-year transferred back into the class. The end of the year DIBELS testing in May included 21 students. However, results of the study are provided on the 18 students who remained from the beginning of the year through the end of the year. In this class there were two special education students with one repeating kindergarten and two students who received intervention for speech and language. One of the intervention students was recommended for a full evaluation for special education placement. This class received no MNRI® techniques or additional exercises in their curriculum.

It should be noted that all students in the study received scores on all assessments and composites. The scores of zero are actual scores that they received rather than lack of participation on that particular assessment/composite.

Results & Discussion

First Sound Fluency (FSF)

FSF measures the phonemic awareness skills in kindergarten students. It was only assessed on the Beginning and Middle of the Year Assessments. It should be noted that the benchmark score on the Beginning of the Year Assessment was 10 and the Middle of the Year Assessment benchmark score was 30.

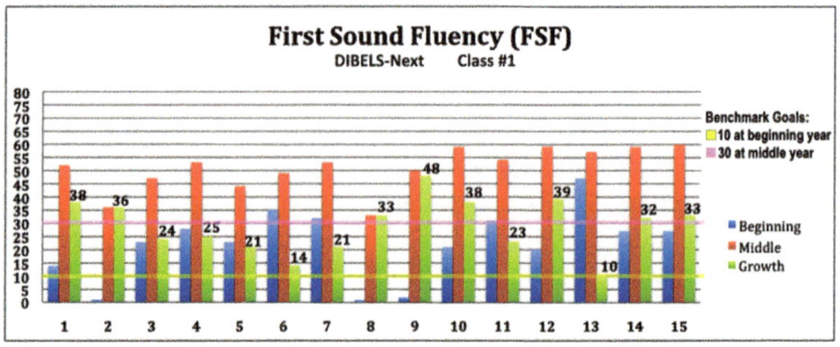

(See graph at right.) In **Class # 1**, on the Beginning of the Year Assessment for FSF, 3 students (20%) scored in the at risk group which indicated intensive interventions, 0 students scored some risk, and 12 students (80%) scored core or low risk which indicated no need for interventions. On the Middle of the Year Assessment for FSF, 0 students (0%) scored in the at risk group which indicated intensive interventions, 0 (0%) students scored some risk, and 15 students (100%) scored core or low risk which indicated no need for interventions. The average growth between the Beginning of the Year Assessment and Middle of the Year Assessment was 29 points. The greatest point gain was 48 points and the lowest point gain was 10 points. Despite the lowest point gain of 10 points, this student still scored 37 points higher than the Beginning of the Year Assessment benchmark criteria and 27 points higher than the Middle Assessment benchmark criteria. The student who made the greatest gain scored 2 points on the Beginning of the Year Assessment criteria (8 points below benchmark criteria) and by the Middle of the Year Assessment scored 50 points (20 points above benchmark criteria).

(See graph at right.) In **Class #2,** on the Beginning of the Year Assessment for FSF, 0 students (0%) scored in the at risk group which indicated intensive interventions, 0 students scored some risk, and 22 students (100%) scored core or low risk which indicated no need for interventions. On the Middle of the Year Assessment for FSF, 0 students (0%) scored in the at risk group which indicated intensive interventions, 0 (0%) students scored some risk, and 22 students (100%) scored core or low risk which indicated no need for interventions. The average growth between the Beginning and Middle of the Year Assessments was 28 points. The greatest point gain was 48 points and the lowest point gain was 12 points.

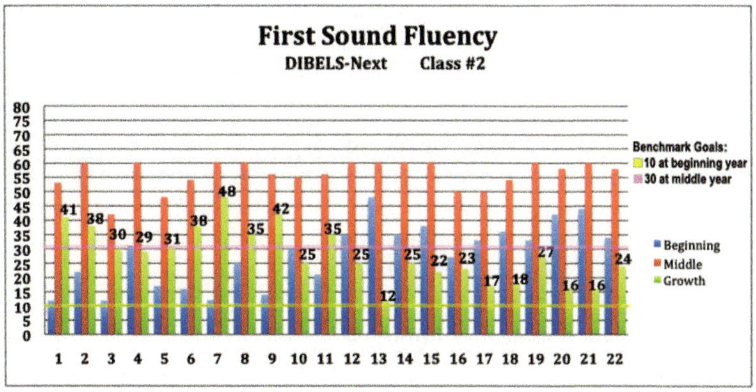

(See graph at right.) In **Control Class #3** (control group) on the Beginning of the Year Assessment for FSF, 7 students (39%) scored in the at risk group which indicated intensive interventions, 1 student (6%) scored some risk, and 10 students (55%) scored core or low risk which indicated no need for interventions. On the Middle of the Year Assessment for FSF, 0 students (0%) scored in the at risk group which indicated intensive interventions, 1 (6%) student scored some risk, and 17 students (94%) scored core or low risk which indicated no need for interventions. The average growth between the Beginning and Middle of the Year Assessments was 34.4 points. The greatest point gain was 54 points and the lowest point gain was 20 points.

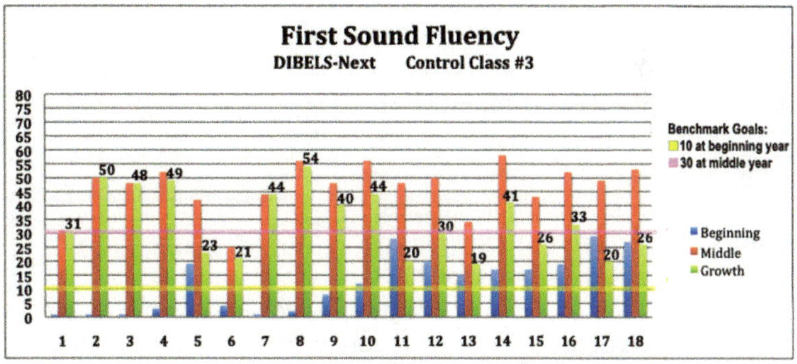

Phoneme Segmentation Fluency (PSF)

Phoneme Segmentation Fluency (PSF) assesses a student's ability to segment three and four phoneme words into their individual phonemes fluently within one minute. The benchmark scores for the Middle of the Year Assessment was 20 and End of the Year Assessment was 40. (PSF was only administered during the Middle and End of the Year Assessments.)

(See the graph at right.) In **Class # 1,** on the Middle of the Year Assessment for PSF, 0 students (0%) scored in the at risk group which indicated intensive interventions, 0 (0%) students scored some risk, and 15 students (100%) scored core or low risk, which indicated no need for interventions. On the End of the Year Assessment for PSF, 0 students (0%) scored in the at risk group which indicated intensive interventions, 0 (0%) students scored some risk, and 15 students (100%) scored core or low risk which indicated no need for interventions. The average growth between the Middle and End of the Year Assessment was 15.3 points. The greatest point gain was 46 points and the lowest point gain was -3 points. Despite the lowest gain being -3 points, this student still scored 18 points higher than the core/low risk score.

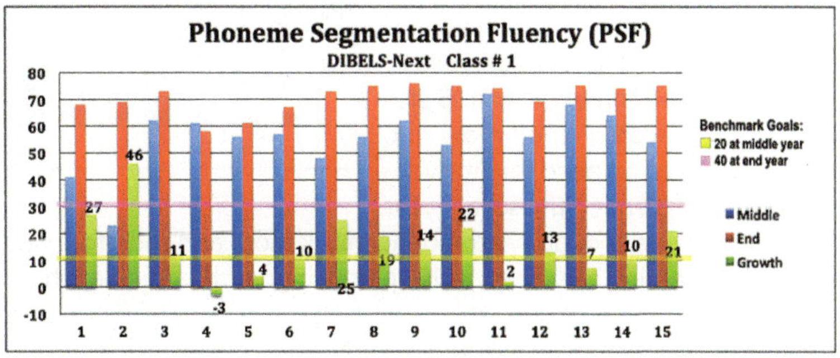

(See graph at right.) In **Class #2**, on the Middle of the Year Assessment for PSF, 0 students (0%) scored in the at risk group which indicated intensive interventions, 0 (0%) students scored some risk, and 22 students (100%) scored core or low risk which indicated no need for interventions. On the End of the Year Assessment for PSF, 0 students (0%) scored in the at risk group which indicated intensive interventions, 0 (0%) students scored some risk, and 22 students (100%) scored core or low risk, which indicated no need for interventions. The average growth between the Middle and End of the Year Assessment was 7 points. The greatest point gain was 38 points and the lowest point gain was -9 points. This student who had the lowest point gain still scored 16 points above the core/low risk score.

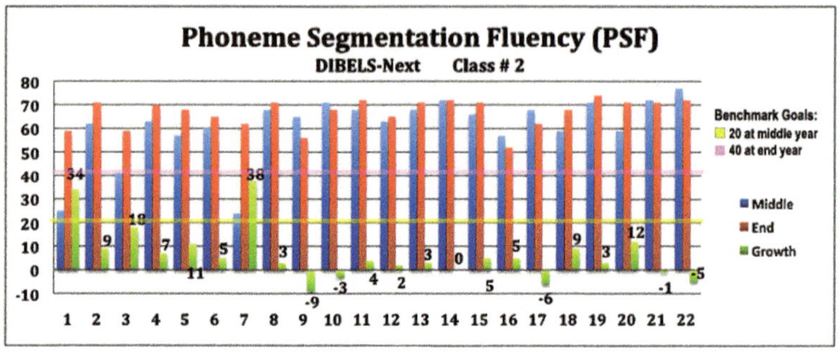

(See graph at right.) In **Control Class #3** (control group), on the Middle of the Year Assessment for PSF, 0 students (0%) scored in the at risk group which indicated intensive interventions, 1 (6%) student scored some risk, and 17 students (94%) scored core or low risk which indicated no need for interventions. On the End of the Year Assessment for PSF, 0 students (0%) scored in the at risk group which indicated intensive interventions, 1 (6%) student scored some risk, and 17 students (94%) scored core or low risk which indicated no need for interventions. The average growth between the Middle and End of the Year Assessment was 15.1 points. The greatest point gain was 49 points and the lowest point gain was -4 points.

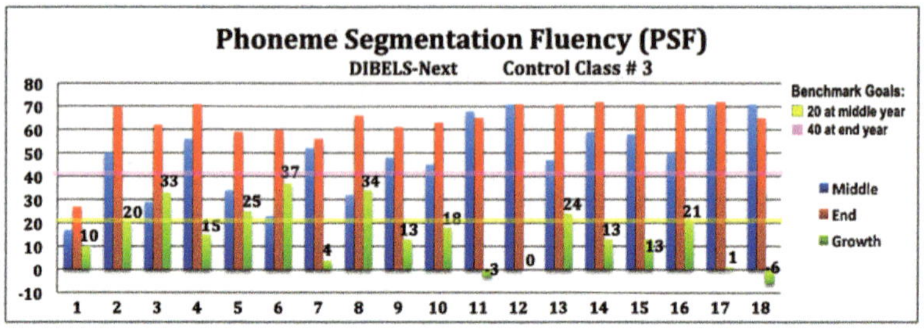

Composite Scores

The Composite Scores from the Beginning and End of the Year assessments were compared. Benchmark score on the Beginning of the Year assessment Composite was 26 or more and the End of the Year assessment Composite benchmark score was 119 or more.

(See graph at right.) In **Class #1**, on the Beginning of the Year Assessment Composite, 2 students (13.3%) scored in the at risk group which indicated intensive interventions, 2 (13.3%) student scored some risk, and 11 students (73.3%) scored core or low risk which indicated no need for interventions. On the End of the Year Assessment Composite, 0 students (0%) scored in the at risk group which indicated intensive interventions, 0 (0%) students scored some risk, and 15 students (100%) scored core or low risk, which indicated no need for interventions. This class ended the school year with all students at or above the benchmark score and therefore not needing interventions. The highest End of the Year Composite score was 119 points higher than the benchmark score.

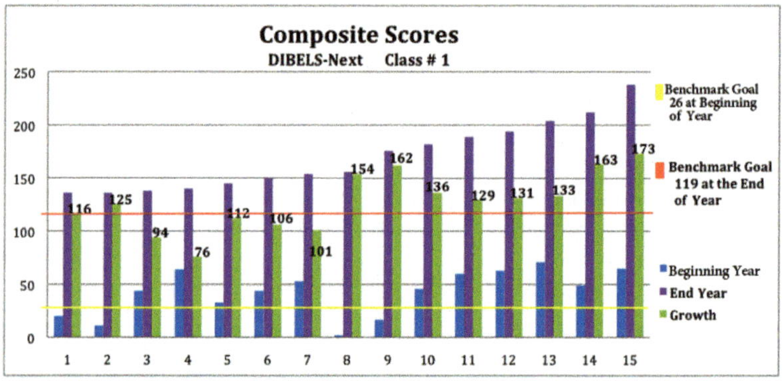

(See the graph at right.) In **Class #2**, on the Beginning of the Year Assessment Composite, 0 students (0%) scored in the at risk group which indicated intensive interventions, 1 (4.5%) students scored some risk, and 21 students (95.5%) scored core or low risk which indicated no need for interventions. On the End of the Year Assessment Composite, 0 students (0%) scored in the at risk group which indicated intensive interventions, 0 (0%) students scored some risk, and 22 students (100%) scored core or low risk, which indicated no need for interventions. Both the Beginning and End of the Year Assessments indicated that all students scored at or above the benchmark scores. The highest End of the Year Composite score was 148 points higher than the benchmark score.

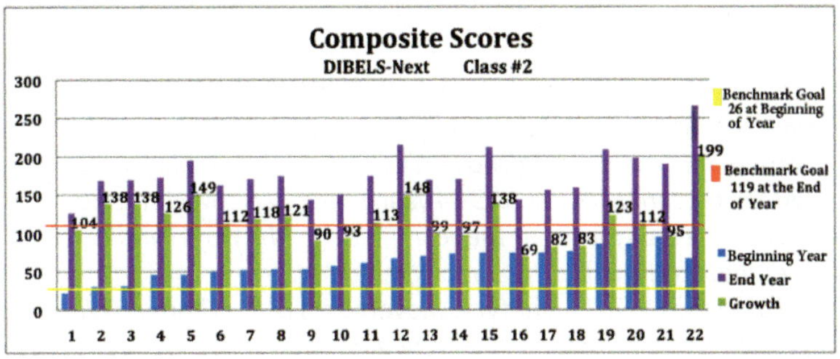

(See the graph at right.) In **Control Class #3** (control group), on the Beginning of the Year Assessment Composite, 3 students (16.7%) scored in the at risk group which indicated intensive interventions, 3 (16.7%) student scored some risk, and 12 students (66.6%) scored core or low risk which indicated no need for interventions. On the End of the Year Assessment Composite, 1 student (6%) scored in the at risk group which indicated intensive interventions, 1 (6%) student scored some risk, and 16 students (88%) scored core or low risk, which indicated no need for interventions. The lowest Composite of the Year score was 36 points below the benchmark score. The highest End of the Year Composite score was 118 points higher than the benchmark score. This control class was the only class in the study to have students in interventions due to their Composite scores.

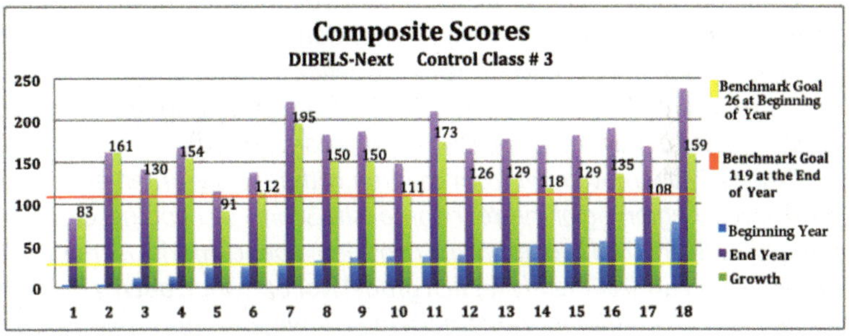

Conclusion

Based on the results of the preliminary testing, our research indicates that using daily MNRI® Archetype techniques had a positive effect on the education performance of kindergarten students as measured by their performance on DIBELS, thus increasing their overall reading skills:

• the Beginning of the Year Assessments showed Class # 1 as being ranked with the lowest reading scores in the parish but, based on the End of the Year Composite Score, it was then ranked as having the highest reading scores in the parish.

• Class # 1 showed a significant decrease in students scoring at the intensive intervention level and in students needing strategic and intensive interventions.

• Class #2 also showed the same trend as the Class # 1. There was a decrease in students needing intensive and strategic interventions. A difference between the Class #1 and Class #2 was that Class #2 started the school year as a stronger overall class as compared to Class #1.

• Control Class #3 overall had more students needing intensive and strategic interventions and also had the smallest percentages of students scoring at the core support level.

Our research concludes that use of MNRI® Archetype techniques in Classes # 1 and 2 resulted in a significant increase in students scoring at core support as measured by the DIBELS. These classes also showed a significant decrease in intensive and strategic interventions. Therefore, it is believed that the MNRI® techniques had a positive effect on the academic skills of these two kindergarten classes as measured by the DIBELS.

Back in the classroom, how were the faces and lives of the students behind the research affected by this research? One kindergarten teacher began who began the school year (in August) using MNRI® techniques offered this perspective of its impact on her class:

"As a kindergarten teacher facing the 'horrors' of Common Core in the classroom and the rigor and high expectations of my young students, I was blessed to have the Masgutova Method® introduced to me by Tina Marks, SLP and MNRI® Core Specialist. I began using the Archetype DVD in March of my first year, and saw such great results in that classroom that I was excited and ready to see what a whole year of implementation could do.

At the beginning of the 2013-14 year, I attended a Maximizing Brain Potential Class *taught by Dr. Patty Shackleford and felt ready to tackle the year. These are the ways I implemented the method in my classroom (Class # 1), and the results I received from that implementation:*

• *I used the* MNRI® Archetype Movements *DVD (Thomas DVD) every morning as the announcements were being made. Later, as the students were able to do the movements without my help, I would do it during my carpet time, and throughout the day, with student leaders leading it instead of the DVD. As the student leaders led, I walked around, correcting some postures, and encouraging some lazy bones.*

Results: Less discipline issues on the floor; better attention and retention of the concepts taught all day, not just during carpet time.

• *My reading intervention group consisted of the students who had the lowest reading scores based on the DIBELS assessment. This intensive reading intervention consisted of 30-45 minutes of daily instruction. With the permission of my principal and Literacy Coach, I did reflex integration with the group for the first 15 minutes. The reflexes we worked on were ATNR and STNR (on the floor as a group); Robinson's Hand Grasp and Sequential Hands Opening and Closing (using a musical rhyme Diane Whiteside created).*

Results: I saw movement from children with intensive intervention scores on the DIBELS test come up to benchmark level. (Beginning of the Year Composite Scores of 12 points or less are considered intensive and End of the Year Composite Scores of 119 or more are considered benchmark.) All my students, including the 6 at intensive level in my group, scored on-level, ready for first grade! Not only that, but my low-income class scored the highest on the DIBELS test in the whole parish for kindergarten! Also, we saw impressive handwriting improvement from all students.

Other reflexes used during the day were Hands Supporting on the wall by the bathroom and with partners at their desks; mouth-spine rotation (Archetype Movement) before a test or while studying sight words and math facts; Babinski foot walking (walking on the outside of the feet when going outside), and whatever else I could create to fit into the day."

– April Dunnehoo, Kindergarten Teacher

References

Good, R. H., Kaminski, R. A. (2014). *What are DIBELS?* Dynamic Measurement Group. https://dibels.org/dibels.html

Masgutova, S. (2011). *Infant dynamic and postural reflexes. Neuro-sensory-motor reflex integration.* Revised and edited. Scientific-practical manual. Orlando, FL, USA: SMEI. 190 p.

Masgutova, S., Masgutov, D. (2012). *MNRI® archetype movement integration. A blueprint for movement and cognitive development.* Orlando, FL, USA: SMEI 92 p.

Masgutova, S., Masgutov, D. (2013). *Maximizing brain potentials in the classroom with MNRI®.* Orlando,FL, USA: SMEI.

Rentschler, M. (2013) *Dynamic & postural reflex integration - The heal. (class outline)* published online at www.e-bookspdf.org.

Teitelbaum, P., Teitelbaum, O. B., Fryman, J. and Maurer, R. G. (2002). *Infantile reflexes gone astray in autism.* (available online at www.moveplaythrive.com.)

Vygotsky, L. *The child psychology. The problems of child development.* In 6 Books. Book – 4. Moscow, Russ: Pedagogika.

https://dibels.uoregon.edu/market/assessment/resources/benchmarkgoals.php#3kindergarten

https://dibels.uoregon.edu/market/assessment/measures/psf.php

https://dibels.uoregon.edu/market/assessment/measures/isf.php

https://dibels.uoregon.edu/market/assessment/measures/fsf.php

https://dibels.uoregon.edu/market/assessment/measures/lnf.php

Dear children, Congratulations for demonstrating your excellent results in early intervention of sensory-motor patterns and reflex integration. We also send our gratitude to the teachers and parents for all their support and participation in our research! – MNRI® Team

Immunological Study of Effectiveness of MNRI® Program on Respiratory Diseases

Svetlana Masgutova, Ph.D, Warsaw, Poland; Nelli Akhmatova, Ph.D, MD; Irina Shubina, Ph.D, MD; Prof. Michael Kiselevskij, MD, Moscow, Russia

Svetlana Masgutova, Ph.D

Nelli Akhmatova, Ph.D

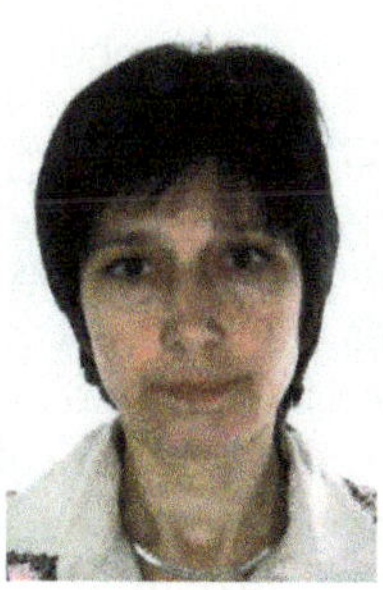
Irina Shubina, Ph.D

Prof. Michael Kiselevskij

This article presents an evaluation of the immunological effectiveness of the MNRI® program for patients with respiratory disease (see article by S. Masgutova, N. Akhmatova, I. Shubina, M. Kiselevskij, 2009). *Immunological Assessment of Effectiveness of Reflex Integration Therapeutic Program for Chronic Inflammatory Respiratory Disease. Immunology Journal.* Moscow, Russia. (Translation at www.MasgutovaMethod.com).

Keywords: Masgutova neurosensorimotor reflex integration, MNRI, recurrent obstructive bronchitis, lymphocytes, cytokines, natural killer cells

Introduction

Chronic inflammatory disease of the respiratory system and recurrent exacerbations are mainly caused by damage to the immune regulation mechanisms. Over the last years an increase in the incidence rate of this disease, with concomitant obstructive bronchitis syndrome, has been registered and the total percentage of inflammatory respiratory diseases continues to grow. Also, the frequency of cases requiring hospitalization has increased (Martinez, Curtis, Albert, 2008).

In this study treatment of chronic inflammatory diseases of the respiratory tract in children and adults was carried out by combining a complementary therapy using MNRI®, based on activation of the primary motor system in addition to the traditional treatment. Parameters of the immune status and cortisol level in children suffering from recurrent obstructive bronchitis (ROB) were studied to assist efficiency of the MNRI® therapy.

The study showed that combining MNRI® with conventional treatment corrects some disorders of the immune system mechanisms normalizing the number of lymphocytes (CD3, CD4, CD8) and NK-cells, the metabolic function of leukocytes, and the level of regulatory and anti-inflammatory cytokines. The therapy based on reflex patterns integration, improves the insufficient effects of standard therapy on immune system cells and strengthens polarization of the immune response of Th-1 type. These results from including MNRI® therapy included a decrease in the incidence of viral inflammatory diseases, positive healing dynamics, and prolonged remission in chronic respiratory diseases.

Materials and Methods

The study included the examination data of children and adults with chronic inflammatory and atopic diseases (chronic bronchitis, bronchial asthma) with concomitant obstructive bronchitis syndrome.

Observation cohorts involved 196 children from 2 to 13 years and 94 adults 20 to 60 years, who first underwent in-patient and out-patient treatment in clinical hospitals in Poland and Russia, then underwent MNRI®. Diagnosis was made in the hospital on the basis of complaints, objective examination, laboratory and x-ray analysis, according to classification of clinical forms of bronchitis and lung diseases of children and adults. The results were submitted by parents and patients on a voluntary basis and they signed contracts to use their personal data for research purposes. A control children's cohort included children of the same age group (examination of these children was performed according to parents' consent). Control for the immune-modulating effect of the MNRI® program was a children's group (15 children) suffering from ROB and receiving conventional treatment only.

The main parameters of the immune system were studied on the documentation of children's history of disease prior to therapy (initial background, group 2), after completing conventional therapy (group 3), and MNRI® program (group 4). Blood tests from patients of groups 3 and 4 were taken on the same day (last day of MNRI® therapy).

Neutrophil phagocytosis activity was evaluated by the following criteria: 1) percentage of phagocyting cells (% P); 2) absolute number of phagocytes; 3) phagocytic number (PN), average number of intracellular latex particles (the result of division of total phagocuted particles by cell number involved in phagocytosis); 4) phagocytic index (PI), the number of latex particles counted for one 'true' phagocyte. Latex particle size was of 1.3– 1.5 mm.

Flow Cytometry Analysis (FACS-Analysis) and Antibodies

Subpopulations of lymphocytes were determined by flow cytometry and monoclonal antibodies (mAb) (Caltag Laboratories, USA) against corresponding antigens. The cells were washed by cold phosphate-salt buffer (PBS) and stained by FITC and PE conjugated antibodies according to the manufacturer's instruction. The suspension was washed twice by cold PBS. The results were measured by flow cytometer FacsCalibur (Becton Dickinson, USA). The gate of cell population was fixed by front and side light scattering and cell size. The results counted 5000 cells in the gate. Statistics were calculated by using WINMDI 2.8 software.

Cortisol level (nMl/L) was determined in blood plasma by IFA with «Cortisol Kit», DPC «Immule », USA.

Statistical data evaluation was made by Student's t-criterion, Vilkokson's criterion with standard statistical program Windows (StatSoft 6.0).

Results and Discussion

Clinical assessment of the effectiveness of MNRI® therapy was performed in patients with chronic bronchitis (CB) in less than one year before treatment. MNRI® therapy of patients was started immediately or within one or two weeks following completion of basic therapy (antibiotics). Besides the patient group receiving alternative therapy (22 patients), a control group was formed including the same number of patients and the same disease severity, who received conventional therapy only.

MNRI® therapy resulted in a significant decrease of the period and severity of the disease. Frequency and duration of CB exacerbations in the study group (MNRI®) was substantially lower than in the control; no exacerbations were registered in 15 patients during one year after the MNRI® course, while three patients of the control group developed exacerbations. Before MNRI® therapy, 13 patients developed three or more exacerbations a year, and after a year of the MNRI® study group there was only one patient. But there were nine patients in the control group receiving standard therapy who developed exacerbations. The index of average days of sick leave registered during a year before and after the conducted therapy, decreased from 65.1±3.4 to 6.6±3.24 in the studied group (MNRI®) and from 55.8±2.53 to 46.8±2.58 in the control group (standard therapy), i.e. in 9.7 and 1.2 times, respectively.

The study of the effectiveness of MNRI® therapy also focused on bronchial asthma (BA). Development of bronchial asthma is closely connected with infection and sensitivity to bacterial antigens. These factors play the leading role in pathogenesis of not only infection-dependent bronchial asthma, but also a number of its combined and transitive forms. This was the basis for the use of MNRI® therapy for patients with infection-dependent and mixed bronchial asthma. Clinical effectiveness of the therapy was assessed during a period of not less than 1 year. A significant effect was registered in the patients of this group in a reduced number and severity of recurrent disease, decrease in the use of antibiotics and hormones. A positive effect in 19 out 25 patients with BA and two patients demonstrated no marked effect (see article). Twelve patients of 25 of this group had a severe course of the disease. After MNRI® therapy completion, the course of the disease became easier. A positive effect in the control group was registered in only 16% cases versus 76% in the MNRI® group.

The effectiveness of MNRI® therapy was studied in 35 children with BA aged 2–13 years. The therapy was performed immediately or within 1-2 weeks following the basic treatment approved in the Allergology Department (Kuvatov RCH, Ufa, Russia). MNRI® therapy effectiveness was assessed within 12 months after the last treatment. The results demonstrate positive effects of this alternative therapy on the mixed form of bronchial asthma in children. After the MNRI® course most children did not have any disease recurrences during 3 months, the rest of the patients also had some attacks with decreased frequency, duration, and severity of the disease. The disease attacks occurring in the later follow-up period were less frequent, and not as long and severe, as compared to those in the period before MNRI® therapy. Positive effects of the therapy were registered in 85.7% patients (versus 34.3% in the control group).

Along with the positive dynamics in the course of bronchial asthma, a 3.7-fold decrease in ARVD (acute respiratory virus infection) and 1.8- fold decrease in bronchitis incidence were registered after MNRI® therapy and consequently the number of antibiotics taken was reduced as well.

The following data present examination results of 10 children, aged 3–6 years suffering from ROB with the background of acute respiratory virus infection. At the moment of examination these children did not have sufficient basis for bronchial asthma diagnosis.

Analysis of the disease etiology showed that 94% of obstruction recurrences were induced by ARVI, which in 72% was complicated by bacterial infections. The tests in children with ROB showed a statistical significant decrease in the level of phagocyting neutrophils (% P) and their absorbing activity. At the same time a reliable increase of phagocytosis indexes was noted after MNRI® therapy in comparison with standard therapy. The children with ROB showed a decrease in bactericidal activity of leukocytes as a result of inhibition of oxygen-dependent microbicidal mechanisms. Leukocyte stimulation also decreased as compared to that of healthy children indicating a inhibition of immune system reactivity. The leukocyte index was reverting to normal values after the completed therapies.

A study of the subpopulation structure in children with ROB revealed the decreased number of D3+, CD4+, D8+, NK, NKT- cells. This data also indicates an increased number of cells expressing molecules of early (CD25) and late (HLA-DR) activation. An increase in B-lymphocyte and T-regulatory numbers was noted. After completing standard therapy, cell phenotype began to normalize gradually though an enhanced immune system reactivity was still noted, which disappeared after MNRI® therapy. The levels of CD3, CD4, CD8, CD25 lymphocytes and NK-cells were already reliably registered in groups 3 and 4.

Cortisol levels in the morning and evening blood plasma samples of children with ROB was statistically significantly lower than that of the control group. Conducting immunocorrecting treatment involving both standard and MNRI® therapy led to a gradual shift towards increased cortisol levels, however the cortisol level of the morning sample after standard therapy was consistently lower than that of the sample taken after adding the MNRI® therapy. It is well-known that the anti-inflammatory and immunomodulating effect of cortisol in a physiological concentration is associated with lipocortine synthesis, which blocks formation of lipooxi- and ciclooxigenases products of arachidonic acid. Lipocortine phosphorilated form promotes intensive synthesis of IgE – a suppressive factor and therefore inhibits IgE synthesis. The pathological chemical phase of ROB leads to an extensive release of LTB4 that plays a leading role in formation of bronchus obstruction (Sechenov, 1935/1960). Possibly the initially low cortisol concentration in this disease may inefficiently regulate formation of LTB4, IgE and other effectors, thus potentiating Th2 response.

This fact is confirmed by the decrease in the production rate of pro-inflammatory cytokine IL-1, regulatory IFN- and IL-12, and the enhanced rate of antiinflammatory cytokines IL-4, IL-10 in peripheral blood mononuclear leukocyte culture (PBML) in children with ROB. In spite of the positive dynamics of cytokine levels after standard therapy, their concentration in the culture medium of leukocytes was reliably different from the control group. When MNRI® therapy was added, the level of regulatory and anti-inflammatory cytokines normalized.

Thus, children with ROB in the exacerbation phase provoked by viral infection develop failure in immunological reaction. Significant changes were noted in innate (inhibition of phagocytic neutrophil function, decrease in the number of NK and NKT-cells) and adaptive immunity (decrease of T and B-lymphocyte number), levels of pro-inflammatory (IL–1) and regulatory (IFN, IL-12) cytokines. The results suggest that the immune response in children with this disease is polarized towards the Th2 pathway, since IL-4 products are Th2-lymphocytes, while IL-12 polarizes immune response towards the Th2 pathway.

The low cortisol level did not allow adequate neuroendocrine system response to the pathogen. Standard therapy had practically no effect on the number of D3, CD4, and NK and T-regulatory cells. A tendency to immunophenotype normalization was noted when additional MNRI® therapy was performed. Standard treatment exacerbated depression of parameters of neutrophil phagocytosis, pro-inflammatory IL- production and regulatory cytokines, and had no effect on cortisol level.

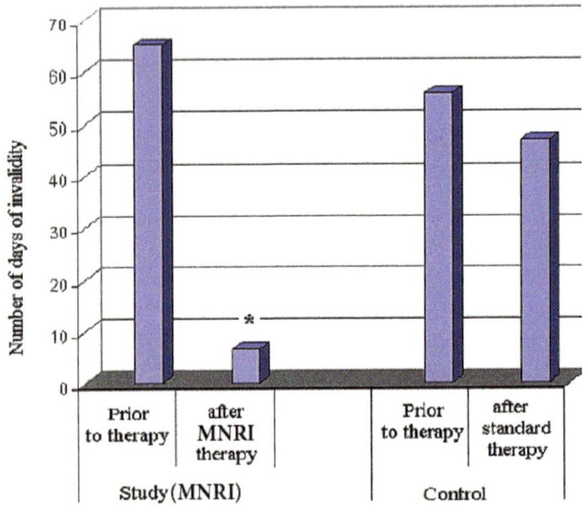

Figure 1. Effect of MNRI® therapy in CB of adult patients.*-<0.01 in comparison with control group (prior to therapy)

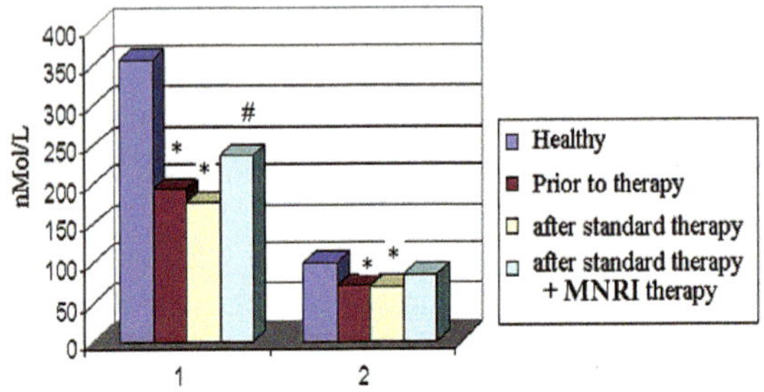

Figure 2. A level of cortisol in plasma of blood at children with ROB. 1 - morning test, 2 - evening test. *- <0.01 in comparison with control group (prior to therapy); # - P <0.01 in comparison with group after standard therapy.

Therefore, after conventional therapy despite clinical improvement, children with ROB still had insufficient cortisol symptoms and simultaneous inhibition of T-cell immunity.

After the addition of MNRI® to the standard therapy of ROB the results showed a reliable increase in the absolute number, of segmented neutrophils, both in relation to the initial number and to the standard therapy in the group number. MNRI® with standard drug therapy leads to an increase in the absorbing activity of neutrophils and normalization of leukocyte metabolic function, raising the stimulation index of NST-test to the normal value (from 0.97 to 1.3). In addition, a statistically significant increase of the number of cells expressing differentiation antigens and NK-cells (CD16) was registered. NK-cells are the key effectors of innate immunity and play an important biological role in mechanisms of immune surveillance targeting tumor cells; in destruction of the virus and parasite infected cells, in regulation and differentiation of bone marrow cells (eliminating rapidly proliferating hemopoetic = hematopoetic cells); and in graft-versus-host reaction.

In addition to its immunoregulating effect, the therapeutic MNRI® program increased cortisol levels up to physiological concentrations. IL-1 produced by mononuclear leukocytes enhanced the secretion of glucocorticoids as a consequence. MNRI® therapy may determine adrenal cortex regulation by hypophysis ACTH by IL-1 production. Thus MNRI® therapy enhanced the Th-1 pathway immune response and its positive effect on immune system cells surpassed that of standard therapy.

Follow-up observation for one year showed that ARVI incidence in children receiving MNRI® decreased by 2.8 times (from 4.51±1.1 to 1.6±0.62) and the frequency of exacerbation decreased by 2.5 times (from 6.4±2.35 to 2.52±0.33). ARVI incidence and exacerbation frequency in children receiving standard treatment only remained at the previous rate. Similar results were noted in patients with CB and BA after addition of MNRI® to the basic therapy the disability index in children decreased, the frequency and severity of the disease reduced, and, ARVI and bronchitis incidence decreased.

We can conclude that including the MNRI® therapeutic program in treatment of children with recurring obstructive bronchitis and adults with BA and CB leads to correction of the impaired mechanisms of the immune system. MNRI® therapy leads to a decreased incidence of ARVI disease, of ARVI disease incidence, an increase in the positive dynamics of the course of chronic respiratory diseases and an extension in the remission of the disease.

References

Andronova, T. M., Chumakova, M. M. (2002). Theoretical problems of use of an immunomodulator Licopid in clinical practice. (Russ.) // *Ter Arkh* 2002;74(1): 70–2.

Anokhin, P. K. (1968). *Biology and neurophysiology of a conditioned reflex.* (Russ.). Moscow, Russia: Medicina.

Anokhin, P. K. (1973). *Biology and neurophysiology of the conditioned reflex and its role in adaptive behavior.* // International series of monographs in cerebrovisceral and behavioral physiology and conditioned reflexes. Rockville, MD: Elsevier.

Bartoli, M. L., Di Franco, A., Vagaggini, B., Bacci, E., Cianchetti, S., Dente, F. L., Tonelli, M., Paggiaro, P. L. (2009). Biological markers in induced sputum of patients with different phenotypes of chronic airway obstruction. // *Respiration* 2009;77(3):265–72.

Bergeron, C., Boulet, L. P. (2006). Structural changes in airway diseases: characteristics, mechanisms, consequences, and pharmacologic modulation. // *Chest* Apr. 2006;129(4):1068–87.

Classification of clinical forms of bronchitis and lung diseases in children. *Russian Bulletin of Perinatology and Pediatrics* 1996;2:52–56.

Egorova, N. B., Semenov, B. F., Kurbatova, E. A., Semenova, I. B., Kuzmina, L. A., Efremova, V. N., Skhodova, S. A., Krasnoproshina, L. I., Beliaev, A. L., Burtseva, E. I., Gruber, I. M. (2001). Immune response following administration of multicomponent vaccine VP-4 in elderly patients. (Russ.). //*Zh Mikrobiol Epidemiol Immunobiol* 2001 Jul–Aug;(4):21–3.

Erel, F., Gulec, M., Kartal, O., Caliskaner, Z., Ozturk, S., Yaman, H., Kurt, Y., Gocgeldic, E., Ors, F., Karaayvaz, M. (2007). Serum leptin levels and lipid profiles in patients with allergic rhinitis and mild asthma. // *Allergol Immunopathol (Madr)* 2007 Nov–Dec;35(6):232–8.

Gordienko, S. M. (1983). Comparative evaluation of the results of the nitroblue tetrazolium reduction test in microscopic and spectrophotometric variants of the method with various tetrazolium salts. (Russ.). *Lab Delo* 1983;(2):21–4.

Holtzman, M. J., Byers, D. E., Benoit, L. A., Battaile, J. T., You, Y., Agapov, E., Park, C., Grayson, M. H., Kim, E. Y., Patel, A. C. (2009). Immune pathways for translating viral infection into chronic airway disease. // *Adv Immunol* 2009;102: 245–76.

Kurbatova, E. A., Semenov, B. F., Egorova, N. B., Efremova, V. N., Gruber, I. M., Semenova, I. B., Kushchenko, O. M. (2004). Immunostimulating activity of polycomponent vaccine "Immunovac-VP-4" and Grippol after their combined administration. (Russ.). // *Zh Mikrobiol Epidemiol Immunobiol* 2004 May-June(3):34–9.

Leow, M. K., Kwek, D. S., Ng, A. W., Ong, K. C., Kaw, G. J., Lee, L. S. (2005). Hypocortisolism in survivors of severe acute respiratory syndrome (SARS). // *Clin Endocrinol (Oxf).* 2005 Aug;63(2):197–202.

Löfdahl, C. G., Postma, D. S., Laitinen, L. A., Ohlsson, S. V., Pauwels, R. A., Pride, N. B. (1998). The European Respiratory Society study on chronic obstructive pulmonary disease (EUROSCOP): recruitment methods and strategies.// Respir Med 1998 Mar;92(3):467–72.

Macdonald, H. R. (2007). NKT cells: in the beginning. // *Eur J Immunol* 2007;37(suppl 1):S111–5.

Martinez, F. J., Curtis, J. L., Albert, R. (2008). Role of macrolide therapy in chronic obstructive pulmonary disease. // *International Journal of Chronic Obstruct Pulmon Dis* 2008;3(3):331–50.

Masgutova, S., Akhmatova, N., Shubina, I., Kiselevskij, M. (2009). Immunological assessment of effectiveness of reflex integration therapeutic program for chronic inflammatory respiratory disease. *Immunology Journal.* Moscow, Russia. (Translation at www.MasgutovaMethod.com).

Musajanova, R., Pathogenetic approaches to the treatment of children with chronic pneumonia. *Medical and Health Science Journal,* MHSJ 2011V/ 7, pp. 87-89.

Rozinova, N. N. (2007). Clinical forms of inflammatory lung diseases in children: diagnosis criteria and principles of therapy. *"Problem Patient" Journal Archives* 1995;2007(10):18-2

Rozinova, N. N. (1992). *Classification of Clinical Forms. In World Health Organization: International statistical classification of diseases and related health problems,* 10th revision. Geneva: World Health Organization.

Sechenov I. M. (1961). *Reflexes of the brain.* (Russ.). Moscow, Russia. pp. 263–336. (Original work was published in 1863)

Sechenov I. M. (1995). *Physiology of behavior.* Scientific Works/Ed.: M.G. Yaroshevsky. Moscow, Russia. 319 p.; p. 26-90.

Semenov, B. F., Zverev, V. V. (2007). Concept of inducing of rapid immunological protection against pathogens. (Russ.). // *Zh Mikrobiol Epidemiol Immunobiol* 2007 Jul–Aug;(4):93–100.

Terawaki, K., Yokomizo, T., Nagase, T., Toda, A., Taniguchi, M., Hashizume, K., Yagi, T., Shimizu, T. (2005). Absence of leukotriene B4 receptor 1 confers resistance to airway hyperresponsiveness and Th2-type immune responses. *J Immunol* 2005;175:4217–25.

Virella, G., Goust, J. M., Fudenberg, H. H. (1990). *Introduction to medical immunology,* 2nd ed. New York, NY, USA: Marcel Dekker.

Zakharov, P., Rosinova, N. (2001). Outcomes of chronic inflammatory disease of the lungs in children on the basis of long-term follow-up findings. *Russian Journal of Perinatology and Pediatrics* (Russ.). No.3, pp.30-35.

World Health Organization: *International statistical classification of diseases and related health problems,* 10th revision. Geneva: World Health Organization; 1992.

Zhang, D. X., Gauthier, K. M., Falck, J. R., Siddam, A., Campbell, W. B. (2007). Steroid-producing cells regulate arterial tone of adrenal cortical arteries. // *Endocrinology* 2007;148:3569–76.

Dear children, We wish you a strong immune system and healthy growth! We also thank all the adults who helped us in this research and wish them good health, less germs and viruses, and an active healthy lifestyle to keep their immune system strong! – Authors

Clinical-Immulogical Assessment of Therapy Effect of Neuro-Sensory-Motor Integration Program of Reflex Patterns in Chronic Airway Inflammatory Diseases

Svetlana Masgutova, Ph.D, Orlando, FL, USA; Nelli Akhmatova, Ph.D, Moscow, Russia; Prof. Olga Lebedinskaya, MD, Perm, Russia

Svetlana Masgutova, Ph.D

Nelli Akhmatova, MD, Ph.D

Prof. Olga Lebedinskaya, MD

Airway chronic inflammatory diseases and recurrent exacerbations are mainly caused by damage to the immune regulation mechanisms. Treatment of chronic inflammatory diseases of the respiratory tract in children and adults was carried out by combining complimentary therapy using the method of neuro-sensory-motor reflex patterns integration (Masgutova Neurosensorimotor Reflex Integration – MNRI®), which works with primary motor system activation and conventional treatment. The parameters of the immune status and the cortisol level in children suffering from of a recurrent obstructive bronchitis are studied, as well as the assessment of therapy efficiency based on MNRI®. It is established that combining the therapy using primary motor system/reflex integration with the conventional treatment allows correction of the immune system mechanisms' damage to normalize the amount of T lymphocytes (CD3, CD4, CD8) and NK-cells, the metabolic function of leukocytes, as well as the level of regulatory and anti-inflammatory cytokines. The MNRI® therapy improves the effect of standard therapy on immune system cells and strengthens the polarization of immune response to Th-1 type. It decreases the frequency of airway viral inflammatory diseases, leads to positive dynamics of chronic respiratory diseases, and prolongs their remission.

The above research was carried out at the I. I. Mechnikov Research Institute for Vaccines and Serum, Russia, and E. A. Wagner Perm State Medical Academy, Russia. Its abstract was presented at the 15th International Congress of Immunology (ICI), in Milan, Italy, on August 22-27, 2013. The topic: Immune-mediated disease pathogenesis.

Citation: Masgutova, S.K., Akhmatova, N.K., and Lebedinskaya, O.V. (2013). Clinical-immunological assessment of therapy effect of the neuro-sensory-motor integration program of reflex patterns in airway chronic inflammatory diseases. *Front. Immunol.* Conference Abstract: 15th International Congress of Immunology (ICI). doi: 10.3389/conf.fimmu.2013.02.00865

We wish all the participants of our research excellent health in the future! – Authors

Effect of Reflex Integration Techniques on Dynamic of Congenital and Adaptive Immunity in Herpes-Associated Patients with Multiforme Erythema

Ekaterina Sorokina, MD, Ph.D; Nelli Akhmatova, MD, Ph.D, Russia;
Elvin Akhmatov, Ph.D Student, Orlando, FL, USA

Ekaterina Sorokina, MD Nelli Akhmatova, MD Elvin Akhmatov

Abstract

The purpose of this research was to study features of the functioning of congenital and adaptive links of immunity for patients with herpes-associated multiforme erythema (HAEM) and an assessment of efficiency in its treatment with the Masgutova Neurosensorimotor Reflex Integration (MNRI®) program. Patients with HAEM before treatment observed a decrease in the number of CD4+, CD8+, CD25+, CD95+ against an increase of CD16+/32+. When studying patients' blood serum, their cytokines profiles were initially observed as decreased in the IL-2, IFN-γ level and increased in IL-10. Patients on MLPB observed a high level of an expression of TLR2, TLR4 and TLR3, TLR9. During treatment with MNRI®, a consecutive increase of lymphocytes of CD4+, CD8+, CD25+ against decrease in CD16+ was observed. MNRI® therapy promoted the IL-2 and IFN-γ level increase in blood serum to normal values, raised induced production of IFN-γ, and lowered induced production of IL-4, IL-5. As a result of the use of MNRI® therapy, the effect of strengthening of an expression of TLR2,3,9 was noted. Inclusion of MNRI® therapy of patients with HAEM had an immunocorrective effect, promoting the turning on of mechanisms of congenital immunity. This research was carried out at the Laboratory for Studying the Mechanisms of Immunity Regulation, I. I. Mechnikov Scientific Research Institute of Vaccines and Serums at the Russian Academy of Medical Sciences in Moscow, Russia.

Key words: neuro-sensory-motor reflex patterns integration (MNRI®), herpes-associated erythema multiforme (HAEM), TLRs, cytokines, lymphocytes.

Introduction

Herpes-associated erythema multiforme is the most frequent form and occurs in 50-93% of patients with exudative erythema multiforme (EEM) (Mashkilleyson, Alihanov, 1983; Alihanov, 1986; Samgin, et all, 1990; Wetter, Davis, 2010; Drago, Romagnoli, Loi, Rebora, 1992). Erythema multiforme relapses in at least 30% of patients.

A considerable number of causes for the infectious-allergic form of erythema have been detected, and clinical manifestations of infectious herpes are found in about half of patients. Relapses of EEM are often accompanied by reactivation of HSV1 and 2. However, in the period of erythema relapse, clinical signs of HSV infection may not be found (Lamoreux, Sternbach, 2006).

Selecting a method of therapy to treat erythema exudative multiforme (EEM) is determined by a trigger factor as well. The trigger role of herpes simplex virus in the pathogenesis of exudative erythema multiforme gives objective reasons to proscribe acyclic nucleosides to such patients (Ivanov, et all, 2002; Laeger, 2012; Lemak, Duvic, Bean, 1986; Huff, 1988; Weston, Morelli, 1997).

Therefore, the success in using immunomodulatory drugs in EEM therapy show that finding new combinations of methods of therapy is feasible. In the literature of the field, one can find statements that the monotherapy of EEM by acyclic nucleosides is not effective enough (Samgin, Khaldin, Khaldina, 2004; Samgin, Khaldin, 2000).

In many cases, psycho-emotional stress has a role in precipitating or triggering a herpes simplex relapse which comes prior to HAEM development. Because of what has been said, it appears important to study the connection between the immune system and the nervous system, and to search for new combined methods for adequate therapy to treat herpes-associated erythema multiforme, while taking into account the peculiarities of innate and adaptive functioning of the immune system.

This article uses the MNRI® (Masgutova Neurosensorimotor Reflex Integration®) therapeutic program developed by Dr. S. Masgutova, which includes diagnostic and therapeutic procedures (Masgutova, Akhmatova, 2005).

The purpose of this study is to investigate the effect of the MNRI® therapeutic program on the innate and adaptive immune systems functions, and the peculiarity of cells cooperation in the immune response at the patients with HAEM.

Materials and Methods

In this research, two groups of patients were involved: Control Group-A – 15 healthy individuals (8 males and 7 females, age range of 18-54), and Research Group (B and C) - patients with HAEM - 21 patients.

Research group of patients with HAEM. Twenty-one patients were chosen for research (age range of 25-40 years old). The duration of the HAEM disease varied from two to ten years with a frequency of relapse of one to two times per year. The laboratory DNA diagnostics identified the herpesviridae virus in these 21 patients: 11 patients were identified with HSV type 1 in salivary glands secretion and in foci, four patients had HSV-2 in the discharge of the urogenital tract, three patients were identified with EBV and HHV-6, two patients with EBV and HSV-1, and one patient with EBV DNA. The identification of viruses correlated with the specific titer of antibodies IgM and IgG to HSV-1 and HSV-2, and IgM to capsid antigen EBV (VCA), IgG to capsid protein, IgG to the early antigen (EA) and nuclear antigen (EBNA) while analyzing blood serum by ELISA.

The research group consisted of two other sub-groups: B) combined MNRI® and immunomodulation treatment – 16 patients (7 males and 9 females), that received the MNRI® therapeutic program and also standard medical immunomodulation treatment, and C) immunomodulation treatment– 15 patients (8 males and 7 females) that went through only standard medical immunomodulation treatment, whose results served as the criteria for comparison of the data in these two sub-groups and analysis of the effect of the MNRI® program. The 8 patients that participated in Group-B were from the international Svetlana Masgutova Educational Institute (Warsaw, Poland). The other 8 patients of Group-B and patients receiving immunomodulation treatment from Group-C (15 people) and also individuals of the Control Group (15 healthy individuals) were patients of the laboratory of immunity regulation of I. I. Mechnikov Research Institute of Vaccines and Serums (Moscow).

Determining the level of cytokines in the serum/plasma and blood of patients was determined by a flow cytometer Cytomix FC- 500 (Beckman Coulter, USA) with the help of the test system FlowCytomix Thl/Th2 11 plex (e-Bioscience, USA) according to the manufacturer's instructions.

Lymphocyte subpopulation structure and TLRs were determined by using monoclonal antibodies according to the manufacturer's instructions (Beckman Coulter, USA). The content of cells was determined by flow cytometer Cytomix FC-500 (Beckman Coulter, USA).

Combined Therapy of the Patients with HAEM

Results of treatments of patients with HAEM in Group-B (16 patients) who received the MNRI® and immunomodulation treatment were compared with results of Group-C (15 patients) who received immunomodulation treatment only. The Control Group-A of clinically healthy individuals (15 individuals) served as the criteria of the effect of both therapeutic programs. Patients with HAEM in Group-B received MNRI® therapy sessions for 14 days (6-8 hours daily). The patients were observed for one year and were evaluated before and after the MNRI® sessions. The groups were comparable in age, sex, and severity of the disease.

Standard immunomodulation treatment (Group-C; 15 patients) included etiotropic treatment in the form of antiviral chemotherapeutic remedy-acyclic nucleosides (acyclovir, Zovirax, viroleks, valacyclovir, and famciclovir). The etiotropic treatment was administered taking into account clinical and laboratory activation indicators of viral infections.

Results and Discussion

All observed patients were evaluated and measured for immunological function before treatments began.

The immunological research revealed some changes in the lymphocyte subpopulation reference ranges in the group of patients (see Table 1), which were taken before the treatment. In the whole group the percentage of T-helper CD4 + was on the lower level of normal (38.6%). They also revealed a decrease in the relative number of CD8+ regulatory lymphocytes to 22.2%. In the content of the subpopulation of natural killer-cells, CD16 + changes were observed in the form of increasing the percentage and absolute number up to 19.6% and 544.2 × 106 cells/ml. Lymphocytes with CD95 + marker were initially reduced to 34.6%. In peripheral blood lymphocytes of patients, a very low activation level of molecules CD25 + (a receptor for IL-2) was noticed, showing 2.4%.

The cytokine profile in the blood serum of patients at the beginning of the study showed a reduction of the proinflammatory cytokine IL-2 (9.1 ± 0.2%), IFN-γ (18.4 ± 1.1%) with increased anti-inflammatory cytokine IL-10 (51.8 ± 14.6%) in comparison with the healthy subjects (see Table 2).

These cytokines were stabilized after the MNRI® method was performed, except IL-10. Its level remained elevated, probably as a controlling mechanism of hyperactivation of the immune system.

The failure of IFN-γ in the system of the observed patients showed a lower level in the serum with the reduction of induced production. With normal levels of serum IL-5 and IL-12, the patients with induction showed increased production of IL-5 and IL-12 (up to 267.8±55.2 and 6559.3±1798.2 pg/ml, accordingly) that may indicate that there are stored immunocompetent cells to induce switching B lymphocytes to IgA synthesis and activation of antigen-presenting cells.

The next stage of our research was the study of the Toll-like receptors (TLRs) of peripheral blood mononuclear leukocytes (PBMC) in patients with HAEM (see Table 3), our research revealed the following features: in the cells of the patients we initially observed higher levels of TLR3 (exceeding 2 times higher) and TLR9 (exceeding 2.3 times higher) expression, than in a healthy person.

It was established that TLR3 recognizes double-stranded RNA and molecular structure of viruses, but does not conduct a signal from single-stranded RNA or double-stranded DNA. TLR9 is involved in the recognition of DNA-containing viruses. TLR9 by interconnecting with nucleic acid sequences of bacteria or DNA viruses, induces the production of proinflammatory cytokines and interferon type I, which play an important role in the immunological defense. It can be assumed that prolonged viral replication in the observed patients may increase the number of cells with TLR3 and TLR9 expression (Akhmatova, Kiselevsky, 2008; Takeuchi, Kaufmann, Grote, et al., 2000).

Results of Treatments

Anamnestic data analysis showed that all of the patients previously received therapy with temporary clinical effect or without any effect at all. Twenty-one patients were prescribed with antiviral chemotherapeutic agents: acyclic nucleosides (acyclovir, Zovirax, viroleks, valacyclovir, and famciclovir). Eleven patients showed insignificant temporary improvement of blanching foci, reduction of frequency, and clinical severity of HAEM relapses. Patients also received vitamin A and B-group vitamins. Despite ongoing earlier etiopathogenic therapy the condition of HAEM was torpid. Our study revealed that the role of the trigger factor in all patients was the reactivation of the viral infection. Additionally, all patients showed impairment of the innate immune system, which was the basis for including the MNRI® method into the combined therapy. The goal of including this combined therapy was to correct immunological disorders and subsequent removal of the trigger factor.

As a result of applying the MNRI® therapy, the duration and severity of the disease significantly decreased (14 patients), which coincided with the dynamics of laboratory indicators of the herpes virus infection activation. Two patients also demonstrated similar result, but not on a significant level. In the group of traditional medical treatment, only 6 patients demonstrated similar results with lower level of significance.

During the MNRI® test we observed, for example, a consistent significant increase of CD4+ lymphocyte population up to 42.1%, CD8+ up to 27%, and CD25+ up to 3.7% with the CD16+ reduced to a normal range (18.1%) in 14 patients out of 16, and in Group-C non-significant increase of the same lymphocyte population in 5 patients out of 15.

The evaluation of the dynamics of the humoral immunity showed a decrease of IgA levels to normal levels after MNRI® therapy in the same 14 patients. The level of circulating immune complexes, which was significantly higher than normal prior the treatment, tended to decrease.

Studying the serum levels of cytokines after the therapy revealed that MNRI® therapy enhanced the lowered baseline level of IL-2 in serum and IFN-γ to a normal level in the same patients of Group-B. Major producers of IL-2 are CD4 T-lymphocytes. IL-2 causes two principal physiological effects: to induce antigen-dependent proliferation of all types of T-cells and to promote the differentiation of certain functional lymphocyte subpopulations: Cytotoxic lymphocyte and regulatory T cells. The principal biological effects of a IL-2 signal was stimulation of T- cell and NK-cell proliferation. It is known that the immune defense against viruses is formed with the participation of many mechanisms of innate and adaptive immune systems and is implemented by using four main factors: interferon type I, natural killer cells, cytotoxic T cells, and neutralizing antibodies. Therefore, the tendency of increasing IL-2 during immunotherapy serves to enhance antiviral protection, and thereby helps to eliminate the trigger factor of the erythema development in the observed patients. IFN-γ (known as an immune interferon) is produced by immune T- lymphocytes subpopulations Th1, CD8+ CTLs, and NK- cells. IFN-γ is the most potent activator of

Table 1. Influence of the MNRI therapy, percentage lymphocyte subpopulations in peripheral blood of patients with HAEM

Indicator	Norm, Unit	Before Treatment	Group 1	Group 2
			After MNRI	After Basic Therapy
$CD3^+$	60-76%	63,1±3,1	66,1±3,8*	53,7±4,9*
	1100-1700 x10^6 cell/McL	1748,3±331,7	1201,4±127,4	1036±112,6
$CD3^+/CD4^+$	38-46%	38,6±2,8*	42,1±2,3*	35,2±3,5*
	700-1100 x10^6 cell/McL	1015,6±172,6	1039±74,9	689,8±71,3
CD25+	8-12 %	2,4±0,6*	3,7±0,6*	2,9±0,9*
$CD3+/CD8^+$	31-40%	22,2±2,1	27±2,6	23,1±5,8
	500-1000 x10^6 cell/McL	608,7±170,7	524,5±84,4	612,4±81,5
$CD16^+/CD32^+$	10-19%	19,6±2,3*	18,1±2,3*	11,9±1,6*
	200-400 x10^6 cell/McL	544,2±147,2	352,1±62,1	262,5±41,7
$CD72^+$	11-16%	11,1±2,1*	16,1±2,6*	13,6±2,2
	200-400 x10^6 cell/McL	325±75,5	346,3±40	319,8±31,2
$HLADR^+$	5-20%	13,7±3,1*	18,7±1,7*	12,0±1,3*
	100-500x10^6 cell/McL	416,7±80,9	401,8±40,7	249,3±49,7
CD95+	41-63%	34,6±4,2*	30,3±2,9*	28,9±6,4*
	100-500x10^6 cell/McL	540±44,5	603,7±77,9	566,3±52,3
CD21+	11-16%	12,2±2,8*	16,8±2*	11,2±1,2
	100-500x10^6 cell/McL	378,4±88,6	327,1±34,6	347,1±62,8

Note. * p <0.05 - significance of differences between the indices before and after the treatment (Mann-Whitney U-test).

TABLE 2. The effect of MNRI on cytokine levels in the serum of patients with HAEM

Cytokine	Healthy individuals	Cytokine level in the patients with HAEM, pg/ml		
		Before Therapy	MNRI	Basic Therapy
IL-1b	10,1±4,7	9,8±0,4	12±1,5	10,1±3,4
IL-2	18,3±6,5	9,1±0,2*↓	14,9±3,2*	11,6±1,7*
IL-4	9,2±0,1	10,3±1,2	12,4±1,6	9,7±1,7
IL-5	11,4±0,1	9,4±0,3*	11,3±2,1	12,5±0,9*
IL-8	9,8±1,2	10,9±0,5	14,3±1,8	N/D
IL-10	20,2±6,8	51,8±14,6*↑	71,6±23,3* ↑	36,2±11,8
IL-12	6,3±0,2	7,8±0,9	10,3±1,3	5,0±1,9*↓
IFN-γ	23,6±5,9	18,4±1,1* ↓	32,1±8,6*↑	21,4±4,2*
TNF-β	8,8±0,2	10,2±1	11,7±0,8	N/D
TNF-α	8,5±0,1*	9,5±0,2	10,2±0,6*↑	N/D

Note. * p<0.05 - reliability of differences relative to the control group (healthy individuals);
#- p<0.05 - reliability of differences between the study groups (Mann-Whitney U-test).

TABLE 3. The effect of MNRI therapy on the expression level of TLRs in PBMC in patients with HAEM

	Number Of patients	Number of cells with expression,%					
		TLR2	TLR4	TLR3	TLR9	TLR7	TLR8
Before treatment	10	17,1±4,5*	23,4±5,8	34,2±6,7*	28,1±8,7*	49,9±8,1	52,4±4,9
MNRI Method	11	26,8±6,1*	26±5,2*	41,4±3,1*	42,3±3,6*	52,4±6,2*	68,2±3,7
Basic therapy	10	7,6±2,1*	12,3±1,5 *	21,5±6,1*	29,3±6,7	n/e	n/e
Healthy individuas	15	4,8±0,6	4,2±0,3	15,4±3,4	11,2±1,3	n/e	n/e

Note. The accuracy of the difference between the rates before and after the treatment:
* p <0.05, ** p <0.01, n/i{n/e}, and - not examined.

macrophages. It also activates NK- cells, and induces the MHC I and MHC II protein cells expression, thereby facilitating antigen presentation (including virus) to T- lymphocytes, which leads to the formation of the antiviral immune response. IFN-γ which is produced by CD8+ cytotoxic T-lymphocytes contributes to the antiviral action of these cells.

The MNRI® therapy increased IFN-γ production to the normal level (452.7 pg/ml) and reduced initial elevated (high) levels of induced IL-4 and IL-5 production.

The observed patients initially had nine times higher levels of induced IL-5 production. IL-5 is produced by a subpopulation of immune CD4+ T- lymphocytes and mast cells, and promotes the differentiation and activation of eosinophils. IL-5 is described as a differentiating factor of B-lymphocytes. The increased IL-5 levels in the observed patients (showing a high rate of determined herpes virus infection), on one hand, may have a positive effect. This is accomplished by increasing levels of IgA which helps to increase mucosal defense. On the other hand, a significant increase in IL-5 levels can contribute to the formation of an allergic reaction. Considering that HAEM is a manifestation of a hypersensitivity reaction, reduction of the induced production of IL-5 during the course of therapy has a beneficial efficacy of the MNRI® therapy.

The study of the expression levels of TLRs the PBMC of the patients, as a result of MNRI® therapy, noted the enhanced expression of TLR2, 3.9 (to 26.8%, 41.4%, and 42.3 % accordingly). The very noticeable increase of the TLRs expression was TLR9 by a multiple of 1.5. MNRI® therapy did not affect the initially elevated (5 times) level of TLR4 expression. Therefore, the application of MNRI® therapy on the patients with HAEM contributed to the inclusion of the innate immunity.

After 12 months of monitoring the patients with HAEM, our research revealed a reduction of 2.7 times in the HAEM relapse rate and, in general, a reduction of clinical symptoms during relapses in Group-B (MNRI® combined with traditional medical treatment). The traditional medical treatment reduced the relapse rate by only 1.5 times. As a result of MNRI® therapy, we saw positive changes in the clinical and laboratory indicators of HSV infection activation, which is reflected in the reduction of herpes simplex relapse and reduction of laboratory indicators of active viruses.

In summary, based on our data, the use of the MNRI® therapeutic program allows correction of damaged mechanisms of the immune system and activation of the mechanisms of innate immunity in patients with HAEM. MNRI® therapy also helps to reduce the severity and duration of herpes simplex relapses by acting to influence the trigger factor and exerting a positive influence on the HAEM pathogenesis.

References

Akhmatova, N. K., Kiselevsky, M. V. (2008). *Innate immunity: anti-tumor and anti-infective.* Moscow, Rus: Practical Medicine.

Alihanov, A. M. (1986). *Improving the treatment of patients with exudative erythema multiforme based on their clinical and immunological studies.* MD, Ph.D Dissertation. Moscow 1986, p. 128.

Drago, F., Romagnoli, M., Loi, A., Rebora, A. (1992). Epstein-Barr virus-related persistent erythema multiforme in chronic fatigue syndrome. *Arch. Dermatol.* 1992; 128(2):p. 217–222.

Huff, J. C. (1988). Acyclovir for recurrent erythema multiforme caused by herpes simplex. *J Am Acad Dermatol.* 1988 Jan.18 (1 Pt 2):197-9.

Ivanov, O. L., Khaldin, A. A., Samgin, M. A. and others. (2002). *Rational choice of therapy for herpes simplex.* Tutorial. Moscow. 2002, p. 20.

Lamoreux, M. R., Sternbach, M. R. (2006). Erythema Multiforme. *Am Fam Physician.* 2006, 1, 74(11), 1883-1888.

Lemak, M. A., Duvic, M., Bean, S. F. (1986). Oral acyclovir for the prevention of herpes-associated erythema multiforme. *J Am Acad Dermatol.* 1986 Jul;15(1):50-4.

Masgutova, S. K., Akhmatova, N. K. (2005). *Integration of dynamic and postural reflexes into whole body movement system.* Warsaw, PL. MINK. 207 p.

Mashkilleyson, A. L., Alihanov, A. M. (1983). Analysis of 100 cases of exudative erythema multiforme. *Journal of Dermatology and Venereology,* 11, p. 4-8.

Samgin, M. A., Khaldin, A. A., Khaldina, M. V. (2004). Therapeutic possibilities of Valtrex (valaciclovir) on herpes simplex and herpes-associated erythema multiforme (a generalization of the five-year experience). *The Journal of Dermatology and Venereology.* 200, 4, p. 52 -54.

Samgin, M. A., Khaldin, A. A. (2000). Valtrex in the therapy of patients with exudative erythema multiforme, associated with herpes simplex. *Russian Journal of Skin and Venereal Diseases.* 2000, 2, p. 34 -36

Samgin, M. A., Ivanov, O. L., Kuzheleva, S. A. and co-authors. (1990). The role of herpes simplex virus in the development of exudative erythema multiforme. *Bulletin of Dermatology and Venereology.* 1990, 13, p. 71-74.

Takeuchi, O., Kaufmann, A., Grote, K. et al. (2000). Cutting edge: preferentially the R-stereoisomer of the mycoplasmal lipopeptide macrophage-activating lipopeptide-2 activates immune cells through a toll-like receptor 2- and MyD88-dependent signaling pathway *J. Immunity.* 2000. – v. 164(2). – p. 554-557.

Vestergard, G. D, Henningsen, E. (2012). Recurrent erythema multiforme triggered by herpes simplex virus. *Ugeskr Laeger.* 174 (23):1611-2.

Weston, W. L., Morelli, J. G. (1997). Herpes simplex virus-associated erythema multiforme in prepubertal children. *Arch Pediatr Adolesc Med.* 1997 Oct; 151(10):1014-6.

Wetter, D. A., Davis, M. D. (2010). Recurrent erythema multiforme: Clinical characteristics, etiologic associations, and treatment in a series of 48 patients at Mayo Clinic, 2000 to 2007. *J. Am. Acad. Dermatol.* 62(1):p. 45-53.

We scientists thank all the adults who helped us in this research. We wish them good health along with a strong active and healthy lifestyle to keep their immune system activated! – Authors

Oral-Facial Reflex Integration: Research Results and Case Study

Svetlana Masgutova, Ph.D.; Anna Regner, Ph.D. Dean, Medical Academy, Wroclaw; Denis Masgutov, Director, ISMI, Poland; Oksana Lutcenko, University Lecturer, Moscow, Russia

Svetlana Masgutova, Ph.D

Anna Regner, Ph.D

Denis Masgutov

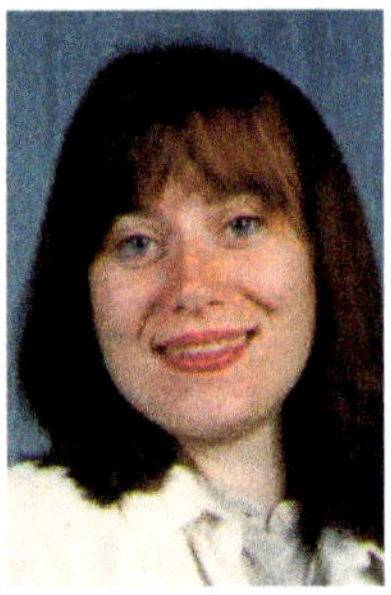

Oksana Lutcenko

This article is based on the publication, Masgutova S., Regner A., Masgutov, D. (2008). *Language Development: Using Sensorimotor Integration Approach.* Continuo, MISM, Wroclaw,PL 154 p. Updated version, 2013.

Introduction

Our goal is to evaluate the effect of MNRI® Oral-Facial Reflex Integration techniques on motor-postural control of oral and articulation skills and abilities in children with global sensorimotor deficits as compared with traditional speech pathology treatment.

One of the most important aspects of human social and cognitive development is the development of communication through language and speech. Limitations in verbal communication are caused by various neurophysiological, anatomical, psychological, or social deficiencies. The most common reasons are neurophysiological and anatomical developmental deficits and abnormal development of Oral-Facial Reflex patterns. Early intervention addressing these aspects allows us to prevent the cause of the dysfunctions and facilitate the development of neurostructural and functional aspects in the oral-facial area including its fine motor control formation.

Oral-facial structure is an integral part of the human body; it works very intensively and participates in many complex functions such as breathing, swallowing, sucking, chewing, pronouncing human speech sounds (decoding/coding), emotional expression, and visual-auditory perception. Oral-Facial Reflexes play an important protective task providing the state and feeling of safety and comfort, thus stimulating developmental mechanisms of Oral-Facial Reflex patterns and speech apparatus.

The Oral-Facial Reflex patterns mentioned above emerge and start developing during the prenatal time, then mature and integrate during infancy and early childhood. Some reflexes displayed as automatisms stay active throughout a person's life (chewing and swallowing) and function along with consciously controlled oral-motor skills.

Oral-Facial Reflexes need to be matured in order to create a proper neurophysiological foundation for such functions, as eating (sucking, biting, chewing, and swallowing reflexes, coordination of those with breathing), articulation and speech, emotional expression and communication, and also self-realization and self-worth.

Tasks and Methods

The main task of this research and study was to determine the effectiveness of the MNRI® Oral-Facial Reflex Integration program and create a program and procedural guidelines for therapists and also parents executing a Home Program. The study included reflex patterns Assessment, clinical observations, clinical questionnaire/survey for parents, and the analysis of bioelectrical function of the brain using electroencephalograph-recorder (EEGR). The novelty of this study was comparative analysis of the reflex Assessment and of EEGR results.

The reflex Assessment was an evaluation of the level of maturation of the Oral-Facial and other sensory-motor reflex patterns (such as Sucking, Babkin Palmomental, pucker, swallowing, biting, chewing, smiling, Asymmetrical Tonic Neck – ATNR, Hand Grasp, Hands Pulling, and some others) and was based on A. Krefft mathematical statistic algorithm (NewKrefft, 2007) using the following criteria points: 0-3.75 – pathological, 4-9.75 – dysfunctional, 10-11.75 – transitional from dysfunctional to functional, 12-15.75 – very low and low level, 16-17.75 – norm.

The EEGR (electroencephalography containing typical features of the EEG and brain mapping) that we used allowed a non-invasive diagnostic method showing bioelectrical brain activity. It registered the electrical potential of the brain – its waves, rhythms, fazes - specifically the neurons of the cortex. Brain waves at neural oscillations (electrode locations) in normal physiological conditions present regular electric activity between 0.5- 35 Hz., and the brain wave spectrum frequencies between 1-100 Hz and amplitude between 5-200 µV (micro Volt) were used as the criteria for the electroencephalographic reading of the source of the brain activity, and showed if the brain wave spectrum was functional or pathological.

The structural elements of bioelectrical activity of brain waves for a human in a normal state are:

• alpha (α) - frequency 8-13 Hz. Amplitude between 30-100 µV. (Alpha brain waves are well seen during absence of visual stimulus, activation of visual perception decreases alpha waves.)

• beta (β) - frequency 12-60 Hz. Amplitude below 30 µV (Located mostly in frontal region, active during active concentration or thinking. It can be activated by pathological changes and chemical substances like benzodiazepines.)

• theta (θ) - frequency 4-80 Hz (Activation of Theta can be observed during hypnosis.)

• delta (δ) - Frequency below 4Hz (High amplitude. Observed during the third and fourth non-REM stage of sleep.)

• gamma (γ) - frequency 26-100 Hz.

As is known in the case of any pathology, e.g., damaged cells or inappropriate chemical connectivity and transfer, the speed of wave transition is typically delayed or accelerated, and their amplitude increases or decreases with a changed shape and configuration.

Recordings of the EEG tests were performed by a team of trained scientists including specialists in the use of EEGR, psychologists, and medical doctors of neurology.

The standard recording of the electroencephalography was performed with the use of 19 electrodes belonging to a 10-20 system recommended by the International Federation of Clinical Neurophysiology (IFCN). One electrode monitors the heart recorded by the electrocardiogram (EKG), other electrodes record movements of the eyeballs, and the electro-occulogram (EOG) allows for automatic blocking/inhibition of the impact of the eye movements on the results of electroencephalogram (EEG).

The entire procedure was recorded mathematically. New generation, small size electroencephalograph equipment (EEGR-19/26 produced in the Medicom MTD Research Institute in Russia) was used to provide optimal comfort for the patient during the therapeutic treatment. Physiological responses were checked and recorded with the eyes open and closed before, during, and after therapy in order to identify the best therapeutic approaches for the patient. For analysis of the data, only elements of the recording with a minimal number of artifacts and computer errors were selected.

Study Participants

A group of 1,080 children between 2-7 years old were the subjects of our study, 80% of them were diagnosed with various speech delays related to:

- speech neurodevelopmental deficits and disorders (aphasia, dysarthria, anarthria, dysphasia)
- neurosensorimotor integration disorders (like in Autism Spectrum Disorder, Selective Mutism, and others)
- cognitive delays (e.g. oligophasia)
- genetic problems (Down syndrome, Prader-Willi syndrome, Apert's syndrome, and others)
- social, communication, and environmental conditions (delayed cognitive and speech development).

The work with this study group was done as a collective effort by the early intervention Clinical Neuro-Speech Development post-graduate faculty of the Medical Academy of Wroclaw and clinic work and rehabilitation camps of the Svetlana Masgusova Educational Institute during the years of 2006-2008.

In the second part of the study using the EEGR, 60 children from this group participated with scientists of the Medicom MTD Institute of Russia by invitation of the SMEI. The research was provided by its full extent for 38 children with delays in speech development and poor verbal communication. All 1,080 children participated in the therapeutic part of the study of MNRI® Oral-Facial Reflex Integration mentioned above but the comparative study of the reflex Assessment and EEGR results was accomplished with 38 of those children.

Discussion of Results

The reflex Assessment in children with various speech deficits and delays demonstrated a poor level of Oral-Facial and related reflex patterns (see Table 1).

The analysis of the results of the reflex Assessment before and after the MNRI® Oral-Facial Reflex Integration intervention presented in this graph shows that reflex developmental dynamics changed for a significant number of children, particularly: pathological reflexes were characteristic – for 8.13%, dysfunctional – for 60.43%, low and very low – 25.8%, norm – 4.55%, and after the intervention: pathological were demonstrated by 3.28% of children (a decrease of over 50%), dysfunctional – 29.14% (a decrease of over 50%), low and very low - 52.46% (an increase of over 100%), norm – 15.12% (an increase of over 200%). These results show that the pathological and dysfunctional reflex patterns decreased and moved to next higher level - low and very low; also the reflex patterns showing in the norm increased significantly after the intervention.

Therapeutic application of the MNRI® Oral-Facial Reflex integration (14 days of intensive rehabilitation by professionals at their therapy clinics and International Dr. Svetlana Masgutova Institute camps in Poland) demonstrated a significant level of changes in oral-motor and speech/communication skills.

Table 1. Results of reflex Assessment in children with various speech deficits and delays.

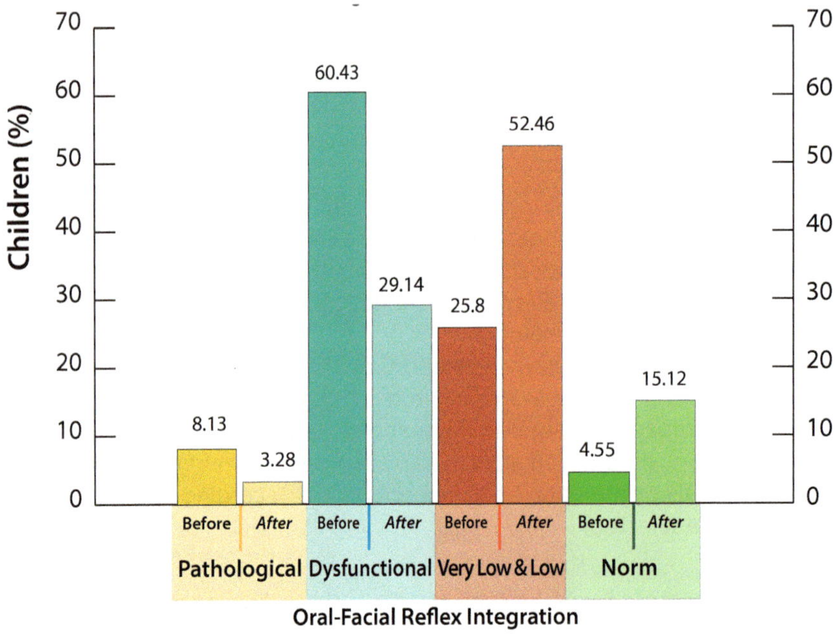

An evaluation of the progress in every child was done with parents and specialists with a clinical questionnaire/survey every 6 months for two years before and after the two-week rehabilitation camp, followed up and monitored further by the parents and local specialists.

An analysis of submitted questionnaires show that positive changes for children on average for every rehabilitation session and camp happened in the following areas:

- morphological structure and function of oral-facial apparatus – in 45% of participated children
- eating function (chewing, swallowing) - 90%
- breathing function - 86%

- integration of reflex patterns and increased control of oral-facial motor functions - 87%
- integration of reflex patterns and increased control of motor coordination – 56%
- improvement in speech articulation/pronunciation - 48%
- increase in available vocabulary - 39%
- improvement of sensory processing – 67%
- improvement in verbal and non-verbal communication – 65%
- improvement in communication and motivation level – 78%
- positive changes in self regulation mechanisms and functions – 27%
- improvement in cognition, ability to focus, and memory – 54%
- increase in emotional maturity and stability - 36%
- improvement in behavior and habitual behaviors - 43%.

The study relating to the analysis of bioelectrical function of the brain using the EEGR recorder was focused on testing before and after the MNRI® rehabilitation sessions and was oriented specifically on the analysis of:

- brain waves correlating to the age and development of a child demonstrating maturity of the brain
- individual brain wave patterns in cases of developmental pathology/dysfunction
- normalization and correction of brain waves after the application of every single MNRI® exercise
- causes of dysfunctions in brain waves in children with speech neurodevelopmental deficits and delays.

Overall analysis of the brain wave spectrum by the EEGR showed that 76.3% of the children (in a group of 38) had average or severe dysfunctions in brain wave spectrum. Low frequency brain waves prevailed in our children cohorts over their fast frequency (in 29 or 76.3% of the children), which showed a possible low maturity of brain development.

Results of the statistical analysis processed for the 38 children before and after the MNRI® Oral-Facial Reflex sessions demonstrated positive tendencies for certain improvements of Oral-Facial Reflexes under the influence of this intervention. Results of the MNRI® effect were estimated by scientists for: 8 children (21.1%) as the results being excellent; 19 (50%) as the results being average; 2 (5.2%) as the results being less effective; and 9 children (23.7%) showed the same results of their brain wave spectrum as when initially tested and their waves were estimated as harmonious. Overall, the MNRI® intervention was considered to be highly effective.

Below is an example of one case study with the use of EEG diagnostic recording presented to verify the effectiveness of the MNRI® Oral-Facial Reflex Integration Program.

Example of the EEGR study of a 6 year-old child (male) with diagnosis of cerebral palsy is presented below.

Assessments used:

Level of speech development has been assessed with the ATA scale created for patients with aphasia (scale authored by Paluch, Drewniak-Wolosz, Milosz, 2005) and with the Dysarthria Scale for pediatric patients developed by Polish scientists (Grabiasa, Gustaw, Mirecka, 2006). Using both scales the following elements were assessed:

1) vocabulary
2) development of articulation/pronunciation
3) language development and competency (comprehension and expression)
4) verbal and non verbal communication.

In all above parameters, the results were very low for the object of the study. Vocabulary was at a very low level of development, very poor articulation and pronunciation, poor comprehension and expression, and a low level of development of verbal and non verbal communication. The child represented a deep dysfunction without an ability for audible speech. All communication took place with non linguistic signals of various sounds and cries. His cognitive development was assessed with a Raven Weksler test showing a low level of development. This data also correlated with results of the reflex Assessment.

Based on the initial results of the EEG recording of this participant, Theta rhythm was shown to be recorded in the occipital region with dominant frequency below 8 Hz (result demonstrating lack of proper brain development in relation to the age of the child).

Theta is a component of brain wave spectrum with frequency 4-7 Hz and various amplitude. It appears in children and adults during their sleep, resting state, and vigilance. Low amplitude of theta and high amplitude

Fig 1. Topography of the brain waves spectrum - Delta, Theta, Alpha, Beta. Initial scan figure is presented below.

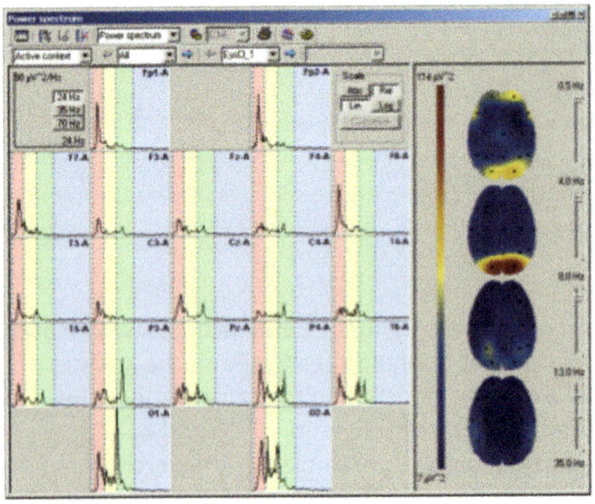

Fig 2. Improvement in topography of the brain waves spectrum in the region of right occipital area EEG electrode (02) - Delta, Theta, Beta. Initial scan.

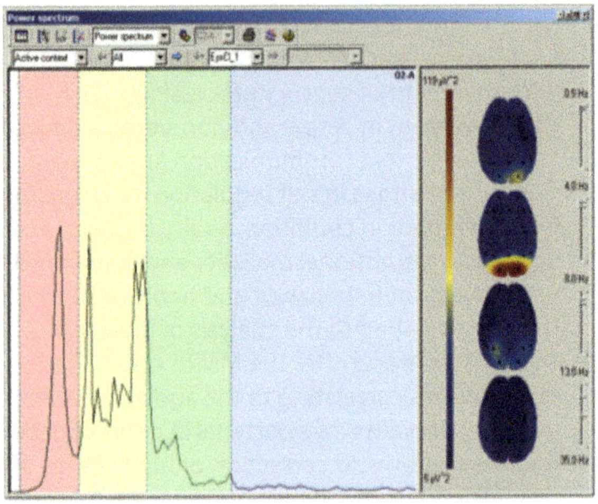

After application of oral-facial techniques of MNRI®, the EEG recording shows tendencies of better brain functioning.

Fig. 3. Topography of the brain waves spectrum- Delta, Theta, Beta. Initial results.

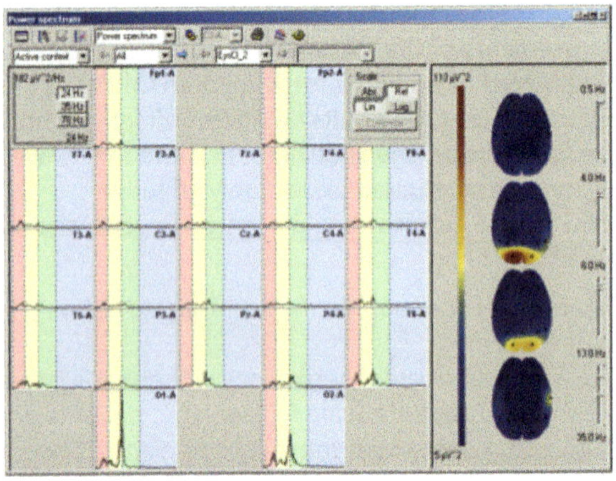

Fig 4. Increasing topography of the brain waves spectrum recorded with EEG (02) electrode in the right occiput area- Delta, Theta, Beta. Basic results.

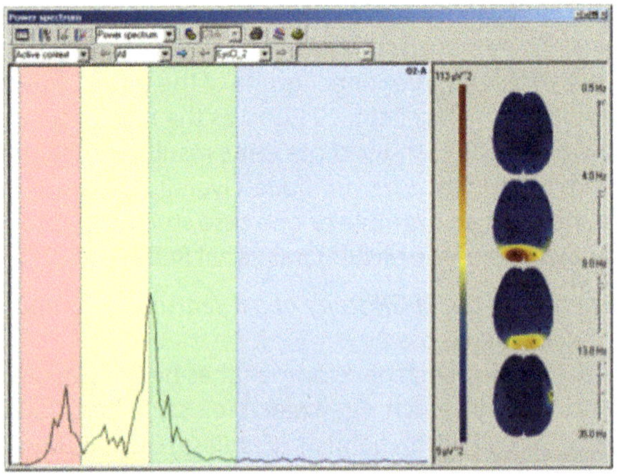

of beta waves are an indication of a good ability to focus and be aware of the world around them.

Initial Results

After application of MNRI® Oral-Facial techniques, the EEG recording shows increased tendencies in basic brain frequencies (changes in the Alpha brain wave spectrum). We see maximal activity of the Theta and Alpha waves in the occipital region of the brain.

The presented figure above shows a magnified form of EEGR recording, specifically the recording from (02) electrode in the right side of the occiput.

EEGR recordings of bioelectrical brain activity taken before and after applying MNRI® Oral-Facial Reflex Integration therapy demonstrated changes in the frequency of the Alpha brain wave activity and, thus, the effectiveness of its techniques supporting the improvement of brain wave spectrums and brain maturation mechanisms.

The results of long term observations and our experience in working with children with special needs allow us to create an individual therapeutic plan for continuous home and clinical programs.

All therapeutic procedures should be based on individual needs and the developmental abilities of a child, taking under consideration:

Table 2. Results before therapy. Alpha brain waves spectrum signal from electrode in the occipital region. Initial results: Alpha brain waves (recorded by electrode 02) frequency 24.19 Hz.

Tables - Relative powers values [%]

S | →S

APV | RPV | DFV | AWFV | EFB | RSC | FSC | ASC

Name	Delta	Theta	Alpha	Beta	Sum
02-A	24.61	46.40	24.19	4.77	99.97
01-A	20.56	39.76	33.92	5.72	99.96
P4-A	27.08	30.66	35.80	6.42	99.96
P3-A	19.16	26.25	45.32	9.20	99.93
C4-A	15.74	35.32	35.70	13.13	99.90
C3-A	34.92	33.84	19.57	11.55	99.88
F4-A	25.58	31.06	27.71	15.56	99.91
F3-A	31.80	23.53	29.24	15.33	99.90
Fp2-A	42.37	30.43	18.25	8.78	99.83
Fp1-A	41.12	28.13	19.17	11.46	99.88
T6-A	21.80	26.64	34.23	17.26	99.94
T5-A	23.55	26.18	24.66	25.52	99.90
T4-A	19.79	34.95	21.79	22.96	99.49
T3-A	20.99	28.94	17.75	31.53	99.21
F8-A	36.23	32.18	17.33	14.12	99.86
F7-A	43.98	32.89	14.06	8.84	99.78

Table 3. EEGR results after MNRI® techniques. Increasing Alpha brain waves to frequency of 47.16 Hz (recorded by electrode 02).

ables - Relative powers values [%]

S | →S

APV | RPV | DFV | AWFV | EFB | RSC | FSC | ASC

Name	Delta	Theta	Alpha	Beta	Sum
02-A	18.05	28.15	47.16	6.60	99.96
01-A	13.37	30.55	50.19	5.84	99.96
P4-A	15.96	29.72	44.00	10.25	99.93
P3-A	17.22	31.93	33.96	16.80	99.91
C4-A	16.08	38.01	29.60	16.17	99.86
C3-A	26.85	29.64	29.13	14.24	99.87
F4-A	21.70	32.04	29.47	16.68	99.88
F3-A	27.41	22.84	30.21	19.40	99.86
Fp2-A	30.22	31.66	25.07	12.84	99.80
Fp1-A	27.12	30.61	27.89	14.19	99.81
T6-A	22.91	28.98	30.15	17.90	99.93
T5-A	20.19	31.18	21.58	26.97	99.91
T4-A	9.86	21.07	20.73	47.01	98.66
T3-A	20.95	19.87	19.58	38.75	99.15
F8-A	25.70	25.34	28.48	20.11	99.62
F7-A	34.34	30.43	21.87	12.99	99.64

- neurological conditions
- age
- therapeutic possibilities related to developmental deficits (Autism Spectrum Disorder, ADD, ADHD, etc.)
- the child's motivation and ability to receive/participate in therapy
- other individual conditions: social, emotional, communicative.

General recommendations. The most appropriate treatment schedule consists of 60 minute sessions 1-3 times a week. For children with ADHD, Autism Spectrum Disorder, sensory integration issues, proprioceptive system issues, phobias as well as other symptoms including hypersensitivity or hyperactivity, we recommend a 60 minute session one time per week.

For infants we recommend two 60-minute sessions weekly, provided they tolerate tactile, visual, and auditory stimulation.

In some cases in particular, even 4 sessions per week are recommended for children with strong hyposensitivity, agenesis of corpus callosum (a congenital disorder in which there is a complete or partial absence of the corpus callosum), delayed cognitive development, delay in speech development, Down syndrome, FAS (fetal alcohol syndrome), or other genetic disorders.

Parents play an active and essential role in supporting positive changes in a child. Parent workshops and educational conferences are offered as part of our services.

Summary

The results of these studies add to our understanding leading to the development of the Neurosensorimotor Reflex Integration Program that has proven effective in treating people suffering from oral-facial dysfunction. If left untreated, such dysfunction disrupts proper motor control of the face and neck, potentially leading to problems with speech development.

The role of the MNRI® Oral-Facial Reflex integration programs lies in directing development and correction of reflex patterns towards their integration with movements and conscious habits of facial motor control. The process of integration allows for the development of adequate regulation in motor activity of the face: proper fluidity of movements, accuracy of righting responses, rhythm, strength, and speed of responses. Facial integration also strongly impacts development of various cognitive functions.

In instances where facial reflexes are not integrated we can frequently see negative compensations, overactivation and over-recruitment of the muscles with increased muscle tension, poor endurance, and fatiguing quickly, abnormalities in symmetry of the face and the whole body, increased conscious control, motor hyperactivity or hypoactivity, poor general physical endurance, emotional instability, and poor coordination of cognitive functions.

The MNRI® Oral-Facial Reflex integration program is based on an understanding of reflexes and their dynamics. Additionally they are based on studies and clinical observations of working with children and adults with a variety of developmental challenges such as cerebral palsy, various genetic disorders, Autism Spectrum Disorder, ADD, ADHD, cognitive disabilities, communication and speech delay, and other developmental dysfunctions.

MNRI®, as presented by the novel approach of Oral-Facial Reflexes classification, derives from work of classic neurophysiologists I. P. Pavlov, I. M. Sechenov, A. Lurie, and N. Berstein.

From this body of work, we know that the neurophysiological bases for activity of facial reflexes are underpinned by complex functions of the entire nervous system. Facial reflexes are based on central nervous system activity in the regions of both brain hemispheres and the cortex, midbrain, brain stem, cranial nerves, and peripheral somatic and autonomic nervous systems.

The level of maturation and integration of the reflex patterns will present information concerning the level of motor development of oral and facial functions, while at the same time non-integrated reflexes will prevent and inhibit proper function and motor control.

MNRI® has proven to provide positive results in development of motor and postural control on the oral-facial area. Techniques used in the MNRI® programs serve as a basis for morphological and functional changes influencing development of articulation, pronunciation, and speech.

Early intervention for children in a risk group allows for the activation of innate neurodevelopmental processes. Each child is given a chance to be physically more capable, cognitively more active, and emotionally more mature. All techniques are aimed at improving daily functions, to enable greater ease in learning and daily functioning.

References

Grabiasa, S., Gustaw, K., Mirecka, U. (2006). *Dysarthria Scale*. University of Lublin, Poland

Krefft, A. (2007). *Diagnosis functions of non observable phenomena*. (Statistic Mathematic Analysis). Wroclaw, Poland: Oficyna Wydawnicza Politechniki Wroclawskiej. Pl.

Masgutova, S., Regner, A. (2008). *Language development: Using sensorimotor integration approach*. Wroclaw, PL: Continuo, MISM, 154 p.

Masgutova, S. K., Akhmatova, N. K. (2005). *Integration of dynamic and postural reflexes into whole body movement system*. Warsaw, MINK. Pl. 207 p.

Masgutova, S. K. (2006). *Infant reflex integration in movement development*. In S. K. Masgutova, A. Regner, & M. D. E. Ronin-Walknowska (Eds.), Materials of Polish National Conference: Modern Methods of Stimulation of Movement Development and Learning in Children with Difficulties in Learning, ADHD and Autism (pp. 5-28). Warsaw, Poland: Maternity Department, Pomorska Medical Academy, Szczecin, Pl.

Masgutova, S. K. & Masgutov, D. (2012). *MNRI® Oral-facial reflex integration in children and adults with neurodeficits* (Polish version). Warsaw, Poland: ISMI.

Paluch, A.,Drewniak-Wolosz, E., Milosz, L. (2005). *AFA scale. How to measure development of speech in children with aphasia*. Krakow, Poland

Pilecki, W., Masgutova, S., Kowalewska, J., Masgutov, D., Akhmatova, N., Poreba, M. et al. (2012). The impact of rehabilitation carried out using the Masgutova Neurosensorimotor Reflex Integration method in children with cerebral palsy on the results of brain stem auditory potential examinations. *Advances in Clinical and Experimental Medicine*, 21, 363-371.

http://www.brainworksneurotherapy.com/what-are-brainwaves

http://www.medicom-mtd.com/en/products/catalog_neuromyan.html - Medicom MTD Research Institute

http://www.castillomoralesvereinigung.de/Castillodata/English/CM.html

We thank all our children, their parents, and specialists for participation in our research, and wish our young patients maturation of their oral-facial reflex patterns and successful development of speech and language! – Authors

Reflex Integration Disorder as a New Treatment Paradigm for Children with Autism

Svetlana Masgutova, Ph.D., Poland-USA; Denis Masgutov, Poland, MNRI® Program Creators

Svetlana Masgutova, Ph.D Denis Masgutov

Introduction to this Study

The MNRI® program is based on clinical observations and research on assessed reflex delays found in children and adults, especially those with neurodeficits. Many individuals with Autism Spectrum Disorder (ASD) are lacking in skills related to early motor milestones. The MNRI® program supports maturation in the neuro-sensory-motor system through specific strategies and techniques that access the innate and natural resources of reflex patterns. (Masgutova, 2011; Masgutova, Akhmatova, 2008, 2012).

The goal of this article is to introduce the basic principles of the MNRI® program and its application to the ASD population, particularly:

• Assessment of reflex patterns to evaluate the level of reflex maturation and functionality in children diagnosed with ASD (see *MNRI® Assessment for Determining the Level of Reflex Development* in this book)

• Reflex development profiles as a guide for use with each child to measure increased performance (at cognitive, physical, emotional, and behavioral levels)

• The effectiveness of MNRI® for increasing neurosensorimotor integration and restoring reflex pattern expression in a child with ASD.

The MNRI® program is based on knowledge and experience of neurodevelopment through the use of reflex patterns to develop physical and cognitive skills as well as behavioral and emotional regulation. Reflexes are genetically determined motor-behavioral patterns that must be integrated *by every child* into consciously controlled sensory-motor abilities and skills (Sechenov, 1863/1995; Sherrington, 1947; Vygotsky, 1934/1986; Myles, Huggins, Rome-Lake, et al., 2003; Masgutova, 2011). They provide an individual with the neurological foundation to process sensory input, program and control motor and behavioral actions, enhance memory and learning, and develop appropriate language and communication skills.

Statistical analysis has allowed us to calculate the number of reflex patterns that were dysfunctional/pathological or in the norm in large numbers of subjects. We have concluded that when 35% or more of reflexes were dysfunctional, we were dealing with a phenomenon of 'Reflex Integration Disorder' (RID; S. Masgutova, 2011). Analysis of reflex profiles of 3,700 children with ASD revealed that RID was typical for children with autism as 86.7% of their reflex patterns were assessed to be dysfunctional. Knowledge of the state of a person's reflex functioning can be key in choosing an effective strategy of intervention.

Autism Study Group Description

ASD disorders develop during early childhood (Prelock, 2006; Piña-Garza, 2001, Cashin, Sci, 2006; Hall, 2012; Lemer, 2008). Current demographic information indicates that the number of identified cases of ASD are rapidly increasing in many countries, especially in technologically developed countries (Lemer, 2008; Rutter, Silberg, 2002; O'Reylly, 2013; Gutstein, 2009; Randall, Parker, 1999). The US Center for Disease Control research indicates that, in some states in the US, one out of every 68 children (one out of 42 boys) is diagnosed with ASD, a 30% increase in comparison with year 2012 (*IACC Strategic Plan for Autism Spectrum Disorder* (ASD) Research, 2013).

Dysfunctional reflex pattern expression is evident in two separate groups of children diagnosed with autism. There are those whose reflexes were immature and deeply dysfunctional from birth, and another group that seemed to develop normally but regressed suddenly into autism at age 2 or 3. Their reflexes may have been immature and thus possibly their nerve systems were not resilient enough to handle the stress that they were exposed to. Or, their reflexes couldn't mature because of asynchronicity in cortical and brain stem function development, resulting in a predisposition for neurodevelopmental disorders beginning at or after 2 years of age. A mild problem initially unrecognized can get worse with age.

To gather the data, MNRI® Reflex Assessments were administered to 3,700 children diagnosed with ASD over the last 15 years (1997-2014). The assessments, of 30 separate reflexes, were given prior to and after 6 hours a day of MNRI® therapeutic techniques, given for 8 days (total of 48 hours). Overall results of the MNRI® Reflex Assessments showed that all of these children (3,700 individuals) presented dysfunctional reflex patterns. The average percentage of dysfunctional reflex patterns found in the children was: 86.6% in those diagnosed with severe autism, 72.4% in those with moderate autism, and 63.3% in those with high functioning autism. The results of this study indicate a clear correlation of ASD with a poor level of reflex functioning.

A smaller, more focused research project focused on ASD included 480 children, aged 2-19, from Poland, USA, and Canada. The MNRI® Reflex Assessment was again administered before and after 6 hours a day of MNRI® therapeutic techniques, given over 8 days (total of 48 hours). The level of development of reflex patterns in these children with ASD improved with the MNRI® program and positive changes were found in their overall development.

The children with ASD who participated in these studies all demonstrated delays in three important areas:

- deficits in social development
- significant speech and communication delays
- behavioral rituals (such as stimming) and abnormal habits.

The typical characteristics of development in the ASD population are well documented (Lemer, 2008; O'Reylly, 2013; Gutstein, 2009; Randall, Parker, 1999).

- tactile and/or auditory hyper-sensitivity
- lack of or poor eye contact
- alimentary behavior disorders leading to a preference for a limited number of foods
- poorly developed self-defense reactions
- lack of curiosity about the world around them
- tendency to focus on the negative
- poor muscle tone
- gaps in the development of kinesthetic memory and self-consciousness
- inability to imitate and follow instructions.

MNRI® research provides a key addition to this list:

- dysfunctional and pathological reflex system development. (Masgutova, Akhmatova, 2008, 2012):

Many children with ASD skip the development of genetically given phases of unconditioned reflexes (Lemer, 2008; Masgutova, Akhmatova, 2008, 2012). These children may even appear to grow well during the first months and years of their lives. Some develop very well in the areas of gross motor coordination, intelligence, and language (and may even develop language earlier than their peers). Parents often report that they unexpectedly lose these abilities and skills, often after some significant stress. In all cases, a global delay in the formation of unconditional reflexes and the next progressive development into conditioned reflex patterns is observed. This significant delay in development of unconditioned reflex patterns may be one of the early signs of a predisposition for ASD. It is very helpful for reflex integration therapy to be offered when such a delay in development of reflex patterns is first observed. It is the authors' premise that reflex integration should be used as a tool for early intervention for all children demonstrating developmental challenges. All children with ASD, whatever their age, can benefit from correction of motor patterns and underdeveloped or dysfunctional reflexes (Lemer, 2008; Masgutova, Akhmatova, 2008, 2012).

Aspects of Developmental Deficits in Children with ASD

Muscular disharmony and lack of muscle tone regulation beginning in infancy results in impulsive reactions that often turn into permanent behaviors as the child with ASD grows older. Impulsivity leads to other challenges, including a lack of goal orientation and inner control, hyperactivity, chaotic and disorganized behavior, inability to focus or follow instructions, impatience and irritability.

Poor muscle tone regulation further includes difficulty with motor programming, motor planning, and motor control, thus poor motor coordination. (Masgutova, 2011; Masgutova, Akhmatova, 2008, 2012). This poor regulation is caused by a lack of integration in the excitation and inhibition processes in the reflex circuits, including incorrect connections between alpha and gamma motor neurons. Typically, muscle tone disharmony in children with ASD presents as:

• hypertonic muscles in the dorsal part of the body (along the spinal column – thoracic longus, trapezius) together with hypotonic abdominal muscles and diaphragm which affect postural control. In an effort to release tension, the child with ASD frequently displays impulsive movements involving balance mechanisms.

• hypotonic extensors in the upper limbs (bracioradialis, biceps), in the back of the neck and arms, and poorly developed muscles in the palms and fingers. This muscle tone disharmony creates a tendency to keep their arms flexed and results in poor dexterity.

• hypertonic lower limb muscles (soleus, gastrocnemius, Achilles tendon, quadriceps, hamstrings) which results in toe walking, as well as frequent running and jumping.

• low tone in oral-facial muscles, which causes a tight jaw leading to decreased mobility. These tight jaw joints inhibit muscles responsible for chewing, swallowing, and articulation.

A deficit in muscle tone regulation in children and adults with ASD also increases their tendency for hyperventilation, incomplete, rapid, and shallow breathing, and holding their breath during times of stress. The tendency for hyperventilation in individuals with chronic diseases is 4 to 7 times higher than the general population and, in individuals with ASD, it can be as much as 7 to 12 times higher.

Imbalance in postural control is another challenge caused by poor muscle tone regulation. Children with ASD may also have undeveloped body structure which results in their slow overall growth (due to a growth factor hormone deficit). The deficit in muscle tone regulation can indirectly affect intellectual development because the child with ASD often does not notice important external stimuli and escapes into their own world, thus being caught in a cycle of repetition of the same stimuli over and over. Improper muscle tone regulation, shallow breathing, and slowed growth may be some of the causes for delayed development in their physical, intellectual, and emotional systems.

Challenges with postural control. Many children with ASD habitually lean forward with their heads turned downward. This posture indicates tight spinal muscles, poor abdominal and neck muscles, and tight calf muscles. Their gravity center shifts disproportionately from the back to the front, making them incapable of maintaining normal balance. In order to compensate, the brain sends an order to tighten and/or lock the tendons in order to maintain a standing position. This posture then over-activates the defensive reactions of fight or flight, which leads in turn to more impulsive movements (Masgutova, Akhmatova, 2008, 2012).

Challenges with motor coordination. Children with ASD demonstrate a lack of stability in their standing and walking patterns. Incoherent coordination within the sagittal (right-left, righting movements), horizontal (rotational, lateral extension-flexion), and dorsal planes (flexion-extension, postural control) results in impulsive, awkward or slow motor activity. Although their head righting is sometimes better developed, children with ASD have difficulty with active movements because their rotational movements are rather poorly formed. For many this leads to toe walking, jumping in the same spot with over-activation of balancing mechanisms, and running around. All this hyperactivity creates repetitious behavior that further increases their responses of either excitation or freezing. Such dissonance in a child's excitation-inhibition processes, along with chaotic motor responses, must be the target for work by parents and professionals. The re-patterning exercises given in the MNRI® programs offer a method to calm these responses. Participation in active sports such as walking, running, skiing, swimming, and weight training can also help them to learn motor pattern archetypes and positive use of their reflexes (Masgutova, Akhmatova, 2008, 2012). The creators of the MNRI® program observe that gross muscle tone regulation also influences the development of fine motor coordination and assists with the improvement of manual dexterity, visual and auditory perception, and integration of these systems.

Challenges in visual and auditory systems. The eyes of children with ASD present as restless or immobile with dilated pupils. When their eyes appear to freeze, they typically provide a narrow limited visual span, poor visual attention and focus, and hyperactive peripheral vision. Many children with ASD demonstrate a dependence or addiction to computers, tablets, or cell phones, often with compulsive repetition of the same object or program over and over again. The individual becomes over focused, which over stimulates their static balance (vestibular system). The reactive Pupillary Reflex may become hypersensitive, over-stimulating the sympathetic system, with either hyperactive or hypoactive motor activity. Children with ASD tend to cope poorly with this chaos in their visual processing.

Auditory Reflexes and processing are also usually problematic in children with ASD (Masgutova, 2011). The majority of children with ASD have inadequate sound orientation, prioritizing, focusing and binaural perception, along with poorly developed stapedius acoustic, ATNR, and Auditory Fear Paralysis reflex patterns. This is why so many children with ASD are hypersensitive to surrounding sounds. Their auditory systems cannot prioritize important sounds, therefore they hear all incoming sound input at the same time, causing over activation of the auditory system and the resulting reactive behaviors. These children attempt to protect their auditory system by covering their ears, trying to cut down external sounds. Other behaviors often noticed with hypersensitive hearing are hitting themselves around the ears and/or rapid head turning or shaking. Problems and delays in development of auditory reflex patterns lead to poor auditory decoding/coding, sound differentiation and analysis. Dysfunctional auditory reflexes are at the root of the delays with expressive language and speech patterns experienced by so many children with ASD.

Challenges in reflex patterns development. Over the last 25 years, MNRI® has collected data from direct observation and research on children with ASD (Masgutova, 2011; Masgutova, Akhmatova, 2008, 2012). This data suggests that children with ASD have many non-integrated reflexes, each child presenting their own individual pattern of immaturity or dysfunction. Some examples are presented below.

• *Tonic Labyrinthine Reflex* is significantly delayed in children with ASD. Neurotypical infants finish the development of this reflex between 4-5 months of life (Barashniev, 2001; Masgutova, 2011). The delay in tonic reflex patterns inhibits the formation of other static (postural) and dynamic reflexes such as the Asymmetric and Symmetric Tonic Neck and Head Righting.

• *Head Righting Reflex* develops and the unconditioned pattern matures between 1½-2 months of life in a typical infant (Barashniev, 2001; Masgutova, 2011). The delay in development for children with ASD results in dysfunction of 'head-core-trunk' coordination and Trunk Extension.

• *Hands Grasp Reflex* in the neurotypical infant has matured by the end of the fourth month of life (Barashniev, 2001). The delay in maturation in children with ASD is noticeable in their inability to fully close their fingers into a fist and in an improper demonstration of phase 2 (fist) and phase 4 (hanging grasp) patterns. This delay interferes with their ability to use their hands for proper protection. The pincer and tripod grips, which support fine motor skills such as handwriting and using silverware, are also significantly delayed.

• *Hands Pulling Reflex* in a neurotypical infant has matured by the end of the second month of life (Barashniev, 2001; Masgutova, 2011). A delay in development interferes with the appropriate 'elbows-head-core' sequence of flexion, which is essential for the natural biomechanics of this reflex. A delay of this pattern negatively affects core flexion-visual convergence, trunk extension-visual divergence coordination, divergent/distance vision, and 3-D vision.

• *Asymmetric Tonic Neck Reflex (ATNR)* in neurotypical infants integrates at between 6-7 months (Masgutova, 2011). Delays in this reflex occur frequently in children with ASD, causing poor spatial orientation, poor development of monaural and later binaural auditory perception/processing, and over-all speech delays.

• *Trunk Extension Reflex* is fully integrated between 7-8 months of life in the neurotypical infant (Barashniev, 2001; Campbell, 2004; Masgutova, 2011). A delay in maturation for children with ASD results in diminished cognitive skills, toe walking, over focus on details, and limited imagination and lack of curiosity. Trunk Extension becomes hyperactive in the majority of children with ASD and is expressed with excessive jumping and running behaviors.

• *Crawling Reflex* was found to be not integrated in 83% of the children with ASD. Many of these children were early walkers who did not crawl in infancy. Due to poor development of this reflex pattern, children with ASD are delayed in their cross-lateral gross motor skills, motor programming, planning and control, and in their ability to multitask.

• *Automatic Gait Reflex* in children with ASD is demonstrated as either a hyperactive or hypoactive-hypotonic reflex and leads to a tendency to freeze (S. Campbell, 2004; S. Masgutova, 2011).

• *Hands Supporting Reflex* is usually delayed in children with ASD and does not develop without specific training. When this reflex pattern is undeveloped, children have limited means of protection on a biochemical level in using their arms for support, most importantly to break their fall when they stumble. (S. Masgutova, 2011).

• *Bonding* in typical infants is noted from the first months of life. Almost every child tested presented signs of inadequate bonding. When it is not developed, there are delays in visual focusing on the face/eyes of their mother and other adults, limited development of a smile, inability to label objects in their environment, and limited emotional communication (S. Masgutova, 2011). Lack of bonding, of course, is directly connected to attachment disorders.

• *Moro and Fear Paralysis Reflexes* are patterns that organize the protective freeze response based on sensory stimulation (Fear Paralysis), and the fight or flight response based on vestibular or proprioceptive stimulation (Moro). Clinical observations indicate that these two reflex patterns are usually confused in children with ASD. This confusion leads to a lack of differentiation of the sympathetic (excitation) and parasympathetic (inhibition) nerve systems, which can lead to extreme hyperactivity or freezing (an absent state, indifference), panic attacks, and possibly a tendency for aggressive behaviors. An immature Moro Reflex is usually not functional in its second protective phase (closing the core of the body), resulting in an insufficient positive self-protection (Masgutova, 2011).

• *Segmental Rolling and Spinning* are reflex patterns which allow the infant to roll over and move across space by rolling. When these rotational reflex patterns are delayed or minimally expressed in a rigid way, the child with ASD uses their axis body plane for movement and posture much more than the horizontal plane. Lack of horizontal rotation results in toe walking, excessive jumping, and/or running around (Masgutova, 2011).

• *Vestibular-Ocular Reflexes* (horizontal and vertical) should support the stability of visual images. Many children with ASD have poorly developed or dysfunctional visual reflexes. A child with visual reflexes which are not integrated will present with poor equilibrium, rigidity, hyper- or hypo-active physiological nystagmus, poor vestibular-visual coordination and visual and proprioceptive disorientation (Masgutova, 2011).

• *Convergence/Divergence Reflexes* work to bring the eyes together for near focus and apart for distance. Often delayed in children with ASD, these reflex patterns affect visual focusing, decoding, analysis and, in turn, visual cognition. The delay also affects the appropriate functioning of peripheral vision, gaze, and tracking. These delays lead to poor voluntary control of eye mobility, poor visual analysis, reading, and visual comprehension (Masgutova, 2011).

• ***Acoustic Stapedial Reflex and sound prioritization.*** The Acoustic Stapedial Reflex triggers the eardrum with a sharp high frequency sound, causing contraction of the stapedius muscle in the middle ear. Children with ASD often exhibit a hyperactive Stapedial Reflex, leading to a hypersensitive auditory system, which interferes with their ability to prioritize sounds. This leads to unpleasant overlapping and over processing of sounds (Myles, et al., 2003; Pilecki, et all, 2012). Protective behaviors of ignoring or inhibiting auditory stimuli are then invoked to calm down the auditory system. Other compensatory behaviors may include self-stimulation, putting hands over or fingers in the ears, hitting their own ears, head, jaws and/or chin, vocalizing a constant sound, or talking to themselves (Masgutova, 2011; Lemer, 2008).

• ***Swallowing and Chewing Reflex patterns*** are often underdeveloped and either hypo- or hyperactive in a child with ASD (Campbell, 2004; Masgutova, 2011). Shallow breathing and/or hyperventilating accompany these patterns. Delays in the Swallowing and Chewing Reflex patterns create difficulties in eating, digestion, and appropriate development of articulation and speech (Lemer, 2008; Masgutova, Akhmatova, 2008, 2012). The Sucking Reflex also may not be integrated and can hinder the integration of the Swallowing and Chewing Reflexes because of low muscle tone in the oral cavity and tongue muscles during infancy.

The MNRI® program offers a system to organize sensory-motor and muscle coherence through re-education of non-integrated reflexes, which helps the child to rediscover how their body should naturally sense, move, and react. This also assists in relieving any inhibited and hyperactive patterns of the child with ASD. Calming or eliminating impulsive and defensive responses, in turn, will positively affect the development of kinesthetic consciousness, postural and motor programming, and control.

Results and Discussion

The MNRI® reflex Assessment was administered to 3,700 children with ASD over the last 15 years (1997-2014) and to children in a smaller, more focused research project which included 480 children, aged 2-19, with ASD from Poland, USA, and Canada. The MNRI® Reflex Assessment was administered before and after the MNRI® intervention (see Table 1 and Figures 1 - A, B, C).

Table 1 describes:

1) The level of reflex pattern development of children with ASD (480 individuals) compared to children with neurotypical development (780). This data indicates a significant delay in children with ASD with 86.7% of their reflex patterns at the dysfunctional level; 13.3% at a functional level but of very low development; and no reflex patterns in the normal range. Children with neurotypical development demonstrated 0% in the dysfunctional range; 63.33% at a functional level but of low development; and 36.67% at the normal level indicating mature development.

2) The level of development of reflex patterns in the children improved with the MNRI® program and positive changes were found in their overall development. Changes in their level of reflex pattern integration after completing six hours a day of treatment at 8-day MNRI® Family Conferences (compare pre- and post Assessment data) demonstrate the positive result of the MNRI® intervention (see Figures 1 – A, B, and C). The results were validated statistically using data on a level of synthesized function z = f(x) (by NewKrefft algorithm, A. Krefft, 2007) as well as the nonparametric comparison of two variables by Wilcoxon Matched Pairs test ($p>0.001$; Statistical Program). The statistical evaluation of the data demonstrates significant change in all reflex patterns and a high degree of effectiveness of the MNRI® program.

The majority of the reflex patterns (86.7%) evaluated by the Reflex Parameter Assessment fell at the dysfunctional level (below the transitional point of 10-11.75), which in the MNRI® model indicates RID (Reflex Integration Disorder); 35% of these dysfunctional reflex patterns were linked to officially diagnosed disorders according to clinical observation of over 30,000 children by MNRI® specialists. After MNRI® treatment the level of development of reflex

Reflex #	Reflex pattern	Children with ASD (480 individuals)		Neurotypical Control Group (780 individuals)
		Before Program	After Program	
1	Core Tendon Guard	12.99	13.32	15.5
2	Robinson Grasp	9.6	11.31	15
3	Hands Pulling	9.54	11.07	17
4	Babkin Palmomental	8.11	10.02	16
5	Babinski	8.11	10.02	16.5
6	Leg Cross Flexion Reflex	12.07	13.52	17
7	Asymmetrical Tonic Neck	9.19	10.97	15
8	Abdominal	7.33	8.76	16
9	Bonding	9.86	10.68	15.5
10	Foot Grasp	8.55	9.79	16.5
11	Thomas Automatic Gait	12.16	13.89	17.5
12	Bauer Crawling	9.23	11.01	15.5
13	Moro Embrace	8.93	11	15.5
14	Fear Paralysis	8.88	10.65	14.5
15	Hand Supporting	8.64	10.75	15.5
16	Segmental Rolling	7.67	9.21	15
17	Landau	8.52	9.75	15
18	Flying and Landing	9.01	10.66	14.5
19	Grounding	9.78	11.14	16
20	Head Righting	8.58	10.44	16
21	Trunk Extension	12.38	13.88	17.5
22	Symmetrical Tonic Neck	9.58	11.13	15.5
23	Spinal Galant	7.1	9.19	15.5
24	Spinal Perez	8.12	10.3	14.5
25	Tonic Labyrinthine	8.59	10.37	15.5
26	Foot Tendon Guard	7.31	9.27	15
27	Spinning	7.85	10.29	15
28	Locomotion	7.44	8.14	14.5
29	Balancing	8.14	10.54	16
30	Pavlov Orientation	9.96	11.84	16
Level of reflex development (criteria)	Dysfunctional reflexes (6-9.75points)	86.7% (26 patterns)	23.33% (7)	0
	On boundary of dysfunctional and functional state (10-11.75)	0	63.33% (19)	0
	Functional but very low level of development (12-13.75)	13.3% (4)	13.3% (4)	0
	Functional though lower level of development (14-15.75)	0	0	63.33% (19)
	In norm (16-17.99)	0	0	36.67 (11)

Table 1. The level of reflex pattern development in 480 children with ASD compared to a group of children with neurotypical development and dynamic changes in reflex patterns of children with ASD before and after completing treatment at MNRI® conferences/camps.

At right, Figures 1– A, B, C. Reflex profile patterns for children with ASD (age 2-19; 480 children) and for those with Neurotypical Development (Age4-20; 780 children).
Figure 1-A. Reflex group 1: Sagittal Plane of the Body
Figure 1-B. Reflex group 2: Horizontal Plane of the Body
Figure 1-C. Reflex group 3: Dorsal Plane of the Body

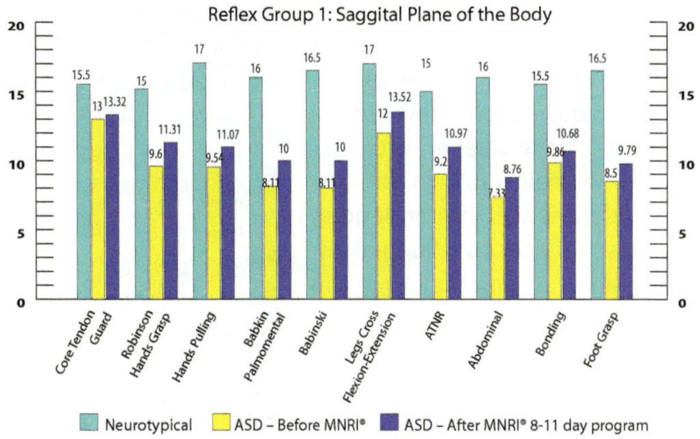

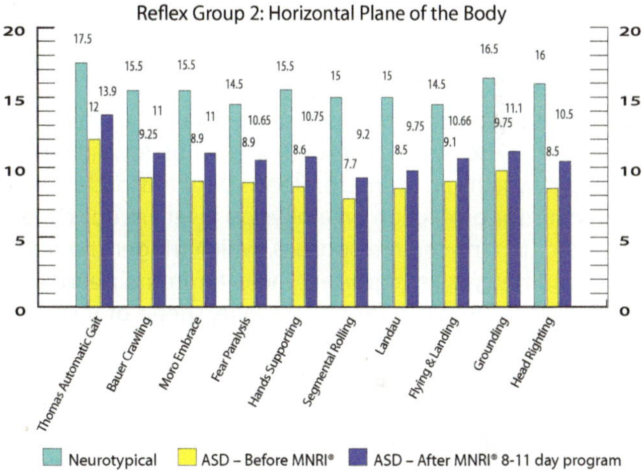

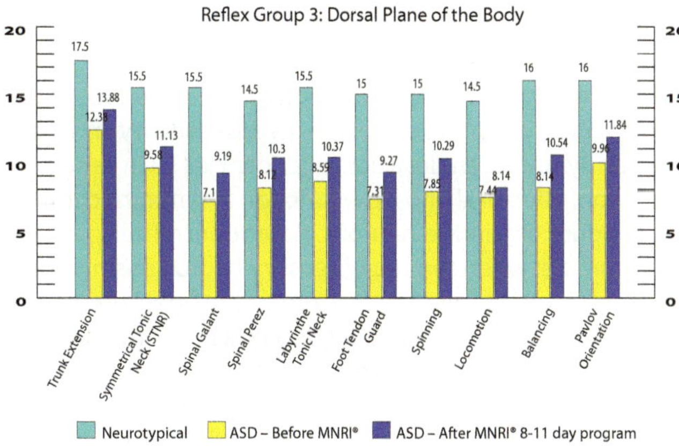

patterns improved significantly for 86% of assessed reflexes (26 reflex patterns out of 30; the coefficient of change is valid - 0.49 before the MNRI® program and 0.61 after the intervention at the $p < 0.001$).

When using the MNRI® programs to re-educate non-integrated reflexes of a child with ASD, the time for correction may be longer than for children facing less severe developmental challenges. Work oriented at facilitation of these reflexes should begin at an earlier age, when automatic and voluntary motor performance can improve in a shorter amount of time. As the child with ASD exhibits a multi-dysfunctional disorder, the therapy program should provide a multi-directional approach based on the individual pathological characteristics of the child. A priority of MNRI® therapy for a child with ASD is to improve their neuro-development, emotional, social, and cognitive functioning through reflex integration and neuroplasticity. Some of the MNRI® programs that address these needs are *MNRI® Repatterning of Dysfunctional Dynamic and Postural Reflex Integration, MNRI® Neuro-Structural Reflex Integration, MNRI® Neuro-Tactile Reflex Integration, MNRI® Archetype Movement and Lifelong Reflex Integration, MNRI® Birth & Post-Birth Reflex Integration, MNRI® Oral-Facial Reflex Integration, MNRI® Proprioceptive & Cognitive Reflex Integration, MNRI® Visual & Auditory Reflex Integration,* and other programs.

The MNRI® Reflex Assessment provides a reflex profile for children with ASD and the opportunity to use their unique individual differences in reflex patterns to create an individualized MNRI® program. The individualized program offers a blueprint for the correction of sensory-motor dissonance and neuro-facilitation of reflex circuits and parameters. It is oriented toward the integration of tactile, visual, auditory, vestibular, and proprioceptive systems for proper functioning.

A more in-depth analysis of reflexes within their five parameters: circuit, direction, intensity, timing and symmetry has shown more detailed aspects of RID in children with ASD (see Table 2 and Figure 1-A). Reflex patterns have been grouped into nine main clusters for this in-depth analysis of tendencies for these reflexes to influence different areas of functioning in the child with ASD.

#	Reflex Patterns/ Clusters	Parameters of Reflex Patterns – total in %														
		Sensory-Motor Circuit			Direction of Movement			Intensity			Latency			Symmetry		
		Norm	Hyper	Hypo	+	-	+/-	Norm	Hyper	Hypo	Norm	Hyper	Hypo	+	-	+/-
1	*Upper Limb*	10.2	13.3	76.5	16.3	7.5	8.7	18.6	33.3	48.1	18.5	34.4	47.1	2.5	91.3	6.2
2	*Tonic*	16.8	15.7	67.5	20.4	66.7	12.9	21.3	30.6	48.1	21.3	31	47.7	5	91.7	3.3
3	*Righting*	17.3	14.4	68.3	56.3	11.7	32	20.2	30.6	49.2	20.2	30.6	49.2	7.2	88.8	4
4	*Lower Limb*	16.8	81.3	1.9	52.1	15.8	32.1	15.8	52.1	32.1	15.8	52.1	32.1	5.6	89.8	4.6
5	*Gross Motor*	1.5	28.3	70.2	60.4	17.9	21.7	21	30.6	48.3	20	31.7	48.3	6.5	90.4	3.1
6	*Oral-Facial/Visual and Auditory*	0.0	20.4	79.6	25	47.7	27.3	22.9	47.7	29.4	20.4	47.7	31.9	6.9	90.2	2.9
7	*Protection and Survival*	0.0	70.0	30.0	18.8	72.9	8.3	0.0	75	25	0.0	78.7	21.3	3.1	94.6	2.3
8	*Curiosity/Cognition supporting*	20.6	44.2	35.2	62.9	16.7	20.4	12	26.1	61.9	12.1	40.2	47.7	5.8	90.4	3.8
9	*Emotional Stability, Maturation supporting*	0.0	11	89	16.5	56	27.5	16.5	27.5	56	15.6	28.3	56.1	5	89.4	5.6
	Average Total:	9.2	33.2	57.6	36.5	42.3	21.2	16.5	39.3	44.2	16	41.6	42.4	5.3	90.7	4

Table 2. The parameter of reflex patterns in the Reflex Profile of 480 children with ASD; MNRI® pre-assessment values (in percentages).

In Table 2, the red colored areas signify difference/dominance levels, whereas yellow areas show it to be even with other parameters. This table provides the data for each of the five parameters evaluated for each reflex. Each reflex has been assigned to a Reflex Pattern Cluster. Interestingly, notice the dissonance in sensitivity, for example, which is hypo in the upper limb reflexes, while hyper in the lower limbs. Also motor reflexes (upper limbs, lower limbs, head righting) are hypo while protective responses are hyper, which can be interpreted as poor motor development that triggers more defense mechanisms since the means for protection on a physical level are not developed or sufficient. Also interesting is the lop-sided symmetry parameter in responses, which trigger the right and left sides of the body with 100% asymmetry. Note: symmetry is an important indicator in the neurological evaluation of reflex responses. Asymmetry can show as asynchronicity for muscle tone, speed of response, and uneven biomechanics (different in the right and left sides of the body).

Assessment of reflexes and their dynamics in nine clusters (according to their effect on different sensory systems, behavioral and emotional patterns, their protective role, and affect on cognitive functions) can help to identify some general tendencies of reflex patterns that impact different aspects of an individual's functioning. Our long-term work with children with ASD and research on their reflex profile assessments allows us to hypothesize about the possible neurophysiological causes of RID. They often have:

1) poor prenatal development of proprioceptive tactility (deep touch, pressure, and stretching receptors) and, in infancy also, improper differentiation of tactile from proprioceptive sensory stimuli

2) hypersensitivity in the auditory and/or visual system and also dissonance in the sensitivity of theses two systems

3) confusion between nociceptive and proprioceptive ascending pathways (this increases sensitivity and over-protection, or the opposite – hyposensitivity and inability to protect personal boundaries, including one's own body, and even destructive tendencies such as self-aggression)

4) confusion and improper function in different components of the reflex circuit: sensory receptors, electrical conductivity in the neurons, neurotransmitter production, possibly also metabolic processes in the cells. For example: when too many sensory neurotransmitters (substance-P or glutamate) are present as the result of hypersensitivity and too few interneuron neurotransmitters are released into gamma motor neurons (GABA and dopamine). Or, there could be a lack of coordination between alpha and gamma motor neurons and their connection with muscle fibers (too few or too many acetylcholine neurotransmitters), affecting regulation of muscle tone and the movement biomechanics.

5) improper function in the stress-axis of the organism caused by some traumatic experience (acute or chronic stress) Our MNRI® Parents Support Group (over 1,600 parents and professionals) sent a questionnaire to parents to identify factors causing post-traumatic experiences in children with neurodeficits. In a group of 870 parents having children with ASD, 86.9% stated that their children (ages 3½-19) had a history of traumatic experience, which turned into post-traumatic stress.

A comparison of changes in reflex parameters before and after MNRI® intervention shows the effect of MNRI® especially in improvement of: circuit (restoration of sensory-motor links), direction (the coordinated work of the alpha and gamma motor neurons to create first proper tone followed by appropriate movement), and symmetry (processing on the extrapyramidal and then lower motor neuron level – in this order; not predominantly in the cortex).

The data above show that children diagnosed with ASD need early intervention based on MNRI® reflex integration processes as early as possible for timely facilitation of their postural and motor development, fine motor coordination, receptive and expressive language, thinking and learning skills. MNRI® re-patterning exercises address neurologically developmental mechanisms, resulting in:

1) physical growth (often in height and weight)
2) activation of natural mechanisms of development
3) reduction in stress from learning
4) development of positive attitude and motivation
5) motor, emotional and cognitive development

Conclusion

Children with ASD demonstrate significant delays in reflex pattern development, which typically is linked to their muscle tone deregulation and hyperactive protective behavior. ASD is a serious disorder characterized by immature and poorly functioning reflex motor patterns. Lack of reflex integration in infancy or within the first two years of life affects neurodevelopment in a negative way and can interfere with the development of conscious motor-cognitive skills and cause all of the features of ASD. As the alarming number of children diagnosed with ASD continues to grow, this crucial problem needs proper attention from professionals and scientists.

MNRI® assessments of children with ASD show that most of their reflex patterns (86.7%) are at the dysfunctional level, which in this model indicates Reflex Integration Disorder. Among these dysfunctional reflex patterns 36% are linked to official diagnoses of ASD. MNRI® research provides statistically valid data in formatting the hypothesis that RID is an important underlying cause of ASD.

MNRI® programs used for children with ASD show significant improvement in development of their reflex patterns, leading to corrective effects in motor coordination, muscle tone regulation and bringing a sense of calm and safety to their system. Reflex re-patterning exercises for children with ASD have provided integration of their sensory-motor-muscular system, building harmony and coherence in the foundation of their neurodevelopment.

After MNRI® treatment the level of development of reflex patterns improved significantly in 86% of assessed reflexes (26 reflex patterns out of 30; the coefficient of change is valid - 0.49 before the MNRI® program and 0.61 after the intervention at the $p < 0.001$). Statistical evaluation of the data demonstrates significant changes in all reflex patterns and a high degree of effectiveness of the MNRI® program applied to children with ASD during 8 days of 6 hour daily intervention.

A comparison of changes in reflex parameters before and after the MNRI® intervention showed specifically positive change occurred in: reflex circuit (restoration of sensory-motor links in a reflex circuit), direction (the coordinated work of the motor neurons to create the proper tone and movement), and symmetry in the response. These changes can be interpreted as a positive achievement in 'rerouting' and bringing the reflex pattern back to its 'genetically given place' in the nervous system profoundly supporting a tendency toward normal development, integration of all systems. In other words, the MNRI® program 'teaches and retrains' the brain to improve its own sensory-motor links and functions on the level of primary unconditioned reflex responses.

MNRI® Assessment provides unique reflex profiles of children with ASD and can orient professionals in creating individualized therapy programs depending on their specific developmental challenges: tendencies for fear and phobias, and aggressive behavior, obsessive compulsive problems, routine and repetitive behavior, poor learning motivation, chronic school failure and dyslexia, hyperactivity (ADD and ADHD), and delayed intellectual and emotional development, regardless of their diagnosis.

MNRI® reflex repatterning exercises and other programs for children with ASD open up the possibility of easier learning, creativity, and the development of their own unique personality.

References

Barashniev, J. (2001). *Prenatal Neurology.* (Russ.). Moscow, Russia: Triada-X. 640 p.

Campbell, S. (2004). *Watch me grow! A unique, 3-dimensional, week look at baby's behavior and development inside the womb.* London, UK: Carroll and Brown Publishers Limited.

Cashin, A., Sci, D. (2006). Two terms—One meaning: the conundrum of contemporary nomenclature in autism. *J Child Adolesc. Psychiatr Nurs* 19 (3): 137–44.

Fowler, M. (1994). *Attention deficit, hyperactive disorder.* Washington, D.C., USA: National Information Center for Children with Disabilities.

Gross, D. (1987). *Psychology. The science of mind and behavior.* London, UK: Hodder and Stoughton.

Gutstein, S. (2009). *The RDI book. Forging new pathways for autism, Asperger's and PDD with relationship development intervention® program.* Houston, Texas, USA: Connections Center Publishing.

Hall, L. (2012). *Autism spectrum disorders: From theory to practice* (2nd Edition). Boston, MA, USA: Pearson.

Interagency Autism Coordinating Committee (IACC). (April 2014). *IACC strategic plan for autism spectrum disorder (ASD) research —2013 update.* April 2014. Retrieved from the U.S. Department of Health and Human Services Interagency Autism Coordinating Committee website: http://iacc.hhs.gov/strategic-plan/2013/index.shtml.

Lemer, P. (2008). *Envisioning a bright future. Interventions that work for children and adults with autism spectrum disorders.* Santa Ana, CA, USA: OEP Foundation, p. 71-115, 290-296.

Luria, A. R. (1978). Functional brain organization. // In Book: *Biological-scientific basis of psychology.* / Ed. By A. A. Smirnov, A. R. Luria, V. D. Niebilistsin. (Russ.). Moscow, Russia: Pedagogika. p. 109-139.

Masgutova, S. (2011). *Infant dynamic and postural reflexes. Neuro-sensory-motor reflex integration.* Revised and edited, scientific-practical manual. Orlando, FL, USA: SMEI.

Masgutova, S., Akhmatova, N. (2008, 2012). *MNRI® Children with challenges* – Revised and Updated. Orlando, FL, USA: SMEI. 154 p.

Myles, B., Huggins, A., Rome-Lake, M., et al. (2003). Written language profile of children and youth with Asperger syndrome: From research to practice. *Education and Training in Mental Retardation and Developmental Disabilities* 38 (4): p. 362-369.

O'Dell, N., Cook, P. (1997). *Stopping hyperactivity. A new solution. A unique and proven program of crawling exercises for overcoming hyperactivity.* Garden City Park, NY, USA: Doubleday.

O'Reilly, R., Morlet, T., Cushing S. (2013). *Manual of Pediatric Balance Disorders.* San Diego, CA, USA: Plural Publishing Inc.

Pilecki, W., Masgutova, S., Kowalewska, J., Masgutov, D., Akhmatova, N., Sobieszczanska, M., Kalka, D. (2012). The impact of rehabilitation carried out using the Masgutova neurosensorimotor reflex integration method in children with cerebral palsy on the results of brain stem auditory potential examinations. // *Advances in Clinical and Experimental Medicine.* Official Organ of the Wroclaw Medical University. 2012 (3). 2012, 21. http://www.advances.am.wroc.pl/pdf/2012/21/3/363.pdf.

Piña-Garza, J. (2013). *Fenichel's clinical pediatric neurology: A signs and symptoms approach.* Atlanta, GA, USA: Elsevier.

Prelock, P. A. (2006). *Autism spectrum disorders: Issues in assessment and intervention.* Austin, TX: Pro-Ed.

Randall, P., Parker, J. (1999). *Supporting the families with autism.* Chichester, England: Willey.

Rutter, M., Silberg, J. (2002). Gene-environment interplay in relation to emotional and behavioral disturbance. // *Annual Review of Psychology;* 2002; 53: p. 463 - 490.

Sechenov, I. (1863/1995). *Physiology of behavior. Scientific works/* Edd.: Moscow. (Russ.). Russia: M. G. Jaroshevsky.

Sherrington, C. (1947). *The integrative action of the nervous system.* London, UK: Cambridge University Press.

Vygotsky, L. (1986). *The child psychology. The problems of child development.* In 6 Books. Book – 4. Moscow, Russia: Pedagogika.

Vygotsky L. (1934). *Thinking and Speech.* (Edited and translated in part by Eugenia Hanfmann and Gertrude Vakar, and in part by Norris Minnick). The M.I.T. Press (1962).

We thank all the children who helped us reach our conclusions by participating in this study. We wish you increased focus, learning, and friendships through continued reflex integration. – Authors

A Search for Excellence in Gifted Children with Reflex Integration

Svetlana Masgutova, Ph.D, Orlando, FL, USA, Denis Masgutov, Poland, MNRI® Program Creators

Introduction to MNRI® for Gifted Children

Svetlana Masgutova, Ph.D Denis Masgutov

The Masgutova Neurosensorimotor Reflex Integration (MNRI®) program presents a whole-personality approach which is based on the knowledge and experience of neurodevelopment through the use of reflexes. Reflexes are our genetic motor patterns (Pavlov, 1927/1960; Sechenov, 1863/1995; Sherrington, 1947), which optimize the development of our physical and cognitive spheres and the regulation of our behavior and emotions (Masgutova, 2011).

The MNRI® program has been applied to gifted children for two important reasons: this program is aimed at the support of natural genetic sensory-motor memory and motor-cognitive processes and, although gifted children display unique abilities and talents in specific areas, they often also have challenges in areas of sensory-motor integration, motor-physical development, emotional stability and maturation, as well as, personality growth and communication (Masgutova, Ozerova, 1995).

Key words: Masgutova Neurosensorimotor Reflex Integration (MNRI®) program, gifted and talented, Reflex Profile.

Goals

The results of the MNRI® Reflex Parameter Assessment were completed for 270 gifted children. The results from this study were compiled to:

1) evaluate the levels of reflex pattern maturation and function in gifted children and create their Reflex Profile

2) study the effect of the MNRI® reflex integration therapy program on the neurodevelopment of these children based on restoration of their primary reflex patterns.

Definition

Reflexes are automatic responses of the nerve system to a specific stimulus of certain modality (tactile, visual, auditory, vestibular, proprioceptive, olfactory) expressed through specific posture or motor (or glands secretion or pupillary) responses (Sechenov, 1863/1995; Pavlov, 1927/1960) that activate neurodevelopment and also trigger protective mechanisms simultaneously (Masgutova, 2011).

Reflexes, the first and earliest genetically determined motor-behavioral patterns, must be integrated into higher controlled sensory-motor abilities and skills to provide for processing or decoding of the input, programming and control of motor and behavioral actions, memory, learning, language and communication (S. Masgutova, 2011). Thus matured reflex patterns create the neurophysiological foundation of development and learning, while poor development of the patterns leads to a lack of sensory-motor integration, reactive movements and responses, impulsive behavior, lack of postural control, and poor motor programming and coordination that can negatively affect motivation and achievement.

Based on the National Association for Gifted Children's definition for giftedness, a child must demonstrate "outstanding levels of aptitude or an exceptional ability to reason and learn, and/or competence (documented performance or achievement in the top 10% or better) in one or more domains: mathematics, music, language, and/or a set of sensorimotor skills – painting, dance, or sports" (www.nagc.org, 2010). There is no typical definition of giftedness that is accepted universally. Often this term is used associated with terms such as: 'intelligent', 'talented', 'genius', and 'creative' (Chambliss, 2010; Fulkerson, Horvich, 1998; Eide and Eide, 2004; Silverman, 2002). The term 'giftedness' still differs from these terms though it includes some features of each. It has multiple meanings and specific nuances depending on the area of creativity/intelligence, and the typical and individual features of the child with exceptional abilities (Gagné, 1985; Renzulli, 1978; Fulkerson, Horvich, 1998; Silverman, 2011). Overall accepted definitions in psychology are as follows:

- Giftedness is determined as an innate predisposition for productive exploration and successful learning of some human ability/skill. Note, that to be gifted does not mean being able to do the relevant activity successfully. Giftedness means that a person can easily master specific types of activity/skills and achieve considerable success in it based on this predisposition.
- Talent is displayed when a person possesses advanced skills and not just innate predisposition. A talented person is an individual who has demonstrated or has performed in real action/life an outstanding ability to perform with considerable success with a specific type of activity/work. The results of such performance are at a higher level and exceed the average performance of the work done by many other people.
- Genius typically refers to a level of performance or achievement in an activity or work when a person is not only talented, but has already achieved outstanding or phenomenally recognized success in one specific or several areas of activity in life.

Giftedness can be typical for more children and adults (potentially almost every person can have some giftedness as well as people with high levels of abilities). However, fewer individuals demonstrate extraordinary talents. Rare individuals will demonstrate genius abilities (Chambliss, 2010; Gagné, 1985; Renzulli, 1978; Fulkerson, Horvich, 1998; Eide and Eide, 2004; Silverman, 2002).

Exceptional Abilities and Developmental Challenges in Gifted Children

So giftedness as the basis for development of extraordinary skills can be seen in children as "exceptional performance on tests and/or other measures of ability or as a rapid rate of learning, compared to other students of the same age, or in actual achievement in a domain" (www.nagc.org, 2010; Gagné, 1985; Renzulli, 1978). As a child matures to adolescence, high levels of *achievement and motivation* in the domain become the primary characteristic of their giftedness. The formation of ability or talent can take a life span and depends on a deeper level of exploration and experience.

Various factors can either enhance or inhibit the development and expression of the abilities of gifted children. Some gifted children with exceptional aptitudes may not demonstrate outstanding levels of achievement due to different circumstances. (Gagné, 1985; Renzulli, 1978). Among these are: environmental/educational (family and school), social and personality (limited opportunities to learn as a result of cultural barriers, social demands and beliefs, family expectations, personality growth, motivation), health (physical state and possible learning disabilities), genetic predispositions (strength of the nerve system and resilience, diseases, etc.), and also neurophysiological (nerve system functions, brain integrative and plasticity mechanisms, myelination of the reflex circuits), (Chambliss, 2010; Gagné, 1985; Renzulli, 1978; Fulkerson, Horvich, 1998; Eide and Eide, 2004; Silverman, 2011), and neurodevelopmental including overall maturity of automaticity, sensory motor integration and integrity of reflex patterns (Masgutova, Ozerova, 1995).

Gifted children may develop asynchronously when their intelligence formation is ahead of their physical growth and sensory-motor integration, thus their social-emotional and cognitive abilities may often develop unevenly (Chambliss, 2010; Gagné, 1985; Renzulli, 1978; Fulkerson, Horvich, 1998; Eide and Eide, 2004; Silverman, 2011; Nikitin, Nikitina, 2004; Masgutova, Ozerova, 1995). This asynchronicity and dichotomy between potential for and demonstrated achievement should be the focus of early intervention. Focusing on integrated neurodevelopment can be one of best solutions added into the design of school programs and services for gifted students.

The MNRI® program offers a unique possibility to support the natural resources for this group of children and adults using reflex patterns as the earliest and most fundamental brain integrative patterns (Masgutova, 2011). The program serves to accomplish the missing link in neurodevelopment for this group of children. It involves:

- the MNRI® Reflex Parameter Assessment tool for determining the level of reflex development
- knowledge of the Reflex Profile of Gifted Children based on research of their level of reflex development
- a MNRI® Reflex Integration Procedure to support neurosensorimotor coordination.

The children who participated in this study were determined or described by school psychologists as *gifted children*, whose main descriptions were equal to those described above. Here we present typical, very general tendencies of the gifted children's neurodevelopment, not differentiating between the differences that we can observe in a wider spectrum of their traits or focusing on their developmental challenges.

Typical challenging features of gifted children described by various authors (Chambliss, 2010; Gagné, 1985; Renzulli, 1978; Nikitin, Nikitina, 2004;) in combination with immature reflex patterns are (Masgutova, 2011; Masgutova, Ozerova, 1995):

- sensory hyper-sensitivity (tactile, auditory, tactile) and, at the same time, ignoring some important stimuli
- poor muscle tone regulation (hyper- or hypo-tension)
- delay in physical development as a result of ignoring movement-driven activities
- poor motor coordination, chaotic movements, and reactivity in actions and behavior (especially in situations of stress
- poor postural control and equilibrium
- gaps in the development of kinesthetic-motor memory
- lack of ability to imitate and follow motor instructions
- difficulties with communication and socialization as a result of avoiding interaction with others, lack of interpersonal boundaries, ignoring the importance of communication, and 'absence' during conversation (being preoccupied with new ideas or images)
- negative or critical viewpoints of others' thinking, logic, and creativity (considering this as uninteresting, incompetent, and boring)
- impatience and poor ability to wait for others' turn and irritation at their slow speed
- poorly developed self-defense reactions
- over-focus and too much specialization on their specific topics of interest or curiosity
- tendency to focus on negative emotional experiences
- emotional sensitivity and overwhelm (crying and becoming easily frustrated)
- fears and phobias
- poor adaptive habits

Some Aspects of Developmental Challenges

Below are some typical challenges of gifted children in physical and neuro-development based on our long-term clinical observations when working within the MNRI® intervention (Masgutova, 2011).

Challenges in muscle tone regulation. Difficulties with coordination of motor programming, planning, and control causing a delay in motor-postural control development. This also is pre-determined by a lack of proper muscle tone regulation, and, particularly, muscular hypotonia of one group of muscles and hypertonia of another group of muscles in infancy and early childhood. This is due to an imbalance of the excitation and inhibition processes in the reflex circuits. Typically, a lack of coordinated muscle tone in gifted children is presented by:

- hypertonic muscles in the dorsal part of the body (along the spinal column – thoracic longus, trapezius) together with hypotonic latissimus dorsi, abdominal muscles, and diaphragm affecting postural control and a need to release tension by constant and impulsive movements, challenging the equilibrium mechanisms
- hypertonic upper limb muscles (bracioradialis, biceps) together with the hypotonic extensors (servical longus, levator scapular, scalene, upper trapezius) in the back of neck and arms, and also a lack of proper muscle tone in the forearms, which causes a tendency to keep elbows flexed
- overworking lower limb muscles (soleus, gastrocnemius, Achilles tendon, quadriceps mm., hamstring) resulting in tight muscles and, for some children, a tendency to walk on the toes
- poorly toned muscles of the oral-facial area (buccinators, masseter, major zygoma mm.) and very tight muscles around the jaw joint (lateral and medial external and internal pterygoid mm.) causing difficulty with jaw relaxation, chewing, and swallowing.

Another reason for the imbalance in muscle tone is shallow and rapid breathing with a tendency for hyperventilation, especially in moments of stress.

Improper muscle tone regulation can cause serious imbalances in postural control. Our clinical observations demonstrate that gifted children may also have an undeveloped body structure and lack of flexibility for rotational movements, postural control, and overall mobility due to muscle tone regulation issues.

The disharmony of the muscle system can indirectly result in a tendency to escape from external stimuli, over focusing, and withdrawing into one's 'own world'. That is why a child may find himself in the vicious circle of an asynchronous delay in the development of their physical, motivational, and emotional spheres. This muscular disharmony leads to improper impulsive reactions which may with age turn into permanent disharmony and cause specific challenges in their behavior, resulting in a lack of goal orientation, poor inner control, hyperactivity, being chaotic, an inability to focus on and follow instructions, conflicting behavior, irritability, impatience, impulsiveness, ignoring stimuli, and other issues. Therefore, a correction of muscle tone is necessary.

Challenges with postural control. Many gifted children demonstrate the following body posture characteristics: leaning too far forward, in dissonance with too tight spinal muscles, poor neck and abdominal muscles, and tight calf muscles. This often is linked to an immature vestibular system and visual over-convergence. The child is challenged to keep a natural balance with all senses integrated. Their gravity center, which is shifted disproportionately to the front and the muscles are tight in area of their spine and lower limbs, activates the over-defensive response of fight or flight. This leads to an increase of impulsive movements. It is important to organize activities with an emphasis on sensory-motor and muscular-motor coordination and control and development of kinesthetic consciousness. Impulsiveness and over-defensive responses can cause emotional sensitivity and instability in a child. The more the body is inclined forward, the more tightened are the muscles of the back, causing breathing to be inhibited and hyperactive.

Challenges with motor coordination. Gifted children can be more inclined toward poor stability while standing and walking with uncoordinated movements that are delayed or reactive when compared to the development of neurotypical children.

Chronologically, gifted children often do not reach the same motor development milestones as their neurotypical peers. This development may either be delayed or exhibit early development which can undermine the correct develop of skills. Parents of gifted children report that the head righting pattern forms in their child at 3-4 months of development (a delay of 1-2 months, while it typically takes place at 1½-2 months), rolling over at 7-10 months (delay of 2-6 months, while it normally takes place at 4 months), delayed in skipping (at 2½-4 years), jumping (4-6 years), starting with one-two years of delay. In contrast, some other motor milestones may be achieved earlier than for neurotypical children, like an ability to stand up can take place at 6 months (typically it occurs at 8-11 months); many of these children skip crawling and start walking at 7 months (occurring typically at 10-14 months); and some start sitting at 4 months (occurring typically at 6 months).

The coordination of movements within the sagittal plane (right-left axis plane), horizontal plane (rotational, lateral extension-flexion), and dorsal plane (flexion-extension, postural control) is not harmonious which limits movements that happen in a 3-dimensional plane. Many gifted children have comparatively better developed righting movements and delayed rotational movements resulting in poor postural control and motor coordination. The result of all this for many children in this group is the same – hyperactivity, fast walking with over-activation of the balancing patterns, running around or the opposite when muscles are too tense (they often refuse exercise), that form certain features and habits in behavior such as over-excitation or freezing.

A dissonance in their excitation-inhibition processes and chaotic motor or freeze responses should be the target of therapy by parents and professionals. Specially oriented sports activities and regular gymnastics, are important. These sports can include active walking, running, skiing, swimming, weight training, soccer, and other types of exercises. This is a good opportunity to also work on communication skills and abilities, for coordinating movements with other participants to learn skills such as estimating distance, direction, and amount of applied strength in playing a game (such as playing with a ball).

Fine motor coordination is better developed in gifted children than gross motor coordination. Still, the tone regulation in muscles that supply gross motor coordination can influence the development and ease of fine motor coordination supporting manual dexterity, visual and auditory perception, and integration of these systems.

Oral-motor (fine motor) coordination and the formation of speech is also better developed in gifted children than cross motor coordination. This allows them to advance quickly with their abilities in language and abstract thinking. The MNRI® program based on analysis of clinical observations proposes certain strategies and procedures to target maturation of the gross and fine motor systems and any milestones that were skipped by a child.

Challenges in visual and auditory systems. Gifted children tend to have restless or hyperactive eye movements as a result of their advanced capability to explore many visual objects simultaneously, and also over focusing on details that they are interested in. However, this results in immobilization of their eye movements, and dilation of the pupils leading to eye freezing effect. Thus, gifted children can have a narrower, limited visual span and visual over-focusing (for screens, i-Pads, i-Phones, and computers) with hypoactive peripheral vision. This will often cause an impaired ability for protective visual responses (poor estimation of 'danger in the area', bumping into furniture and other people, or poor estimation of distance, especially in times of stress), and poor vestibular balance. For some specific visual or auditory objects, they can be over focused and sitting for too long watching something of interest, until their brain becomes exhausted, over stimulating their static balance (vestibular system). Many of these children demonstrate an obsessive interest in computers (with traits of addiction). Their pupillary reflex can be over reactive showing hypersensitivity (too much excitation of the sympathetic system). In such cases, impulsive traits can follow, activating territorial behavior (caused by visual reactivity). In some cases, it can be the opposite, exhaustion can lead to hypoactive work of the vision system.

The auditory system and its processing mechanisms are usually well developed in gifted children, still their protective reflexes (stapedius, ATNR and Auditory Fear Paralysis) can be too sensitive causing overall auditory sensitivity. They have rather well developed sound orientation and monaural and binaural perception. However, their auditory system can be overloaded by difficulty in sound prioritizing (a tendency to hear several sounds/texts simultaneously) and their 'inner dialog' is overloaded. Over-activation of the auditory system can cause reactive behavior and results in attempts to protect their own auditory system (ignoring stimuli, escaping conversations, closing ears). Their exceptional ability for productive processing (decoding/coding, sound differentiation, sound source orientation, and auditory analysis) and unique aptitude for language and speech can work together with hypersensitivity to sounds and cause them to become easily irritated, noted in their behavior and communication.

Challenges in reflex patterns development. Some reflexes in gifted children are immature. In each case, this immaturity or lack of integration is very individual. Usually these children skip development of some unconditioned reflexes (from 0 to 4 months of life). Below is the individual description of reflex development based on over 25 years of clinic observation of unconditioned reflex patterns in over 270 gifted children:

- Tonic Labyrinthine Reflex can be delayed by several months and mature to 6-12 months of life (in norm – the unconditioned pattern develops between 4-5 months of life; Barashniev, 2001; *Child Neurology,* 2000; Florintino, 1973; Masgutova, 2011). The delay of their Tonic Reflex patterns inhibits formation of their static (postural) and dynamic reflexes (i.e., STNR, Head Righting).
- Head Righting Reflex starts developing after 2-4 months of age or later (in norm – the unconditioned pat-

tern matures between 1½-2 months of life; Barashniev, 2001; Florintino, 1973; Masgutova, 2011). A delay in this pattern results in a delay of 'head-core-trunk' coordination, STNR, and Trunk Extension. Later, Trunk Extension becomes hyperactive in many gifted children (leading to hyperactive muscles and tension in their body).

• Hands Grasp Reflex starts developing timely (after birth) and matures earlier, up to 3 months (in norm – this reflex develops up to the end of the 4th month of life; life; Barashniev, 2001; *Child Neurology*, 2000; Florintino, 1973; Masgutova, 2011). There are also many gifted children that are delayed with the unconditioned pattern and, later, just skip the development of the earlier phases (phase 2 and 3) and develop phase 4 (the 'hanging grasp' with fingers more actively closed and palms more open). This delay causes poor ability for proper protection (when falling down) in contrast to the conditioned patterns (using a pincer and tripod grip) that support hand manipulation skills (eating, using silverware, and handwriting) which happen to develop faster than in other children.

• Trunk Extension Reflex develops closer to the end of the 6-7th month (in norm – infants start demonstrating this response in the 1st month and it must be matured by the end of the 7-9th months of life; Barashniev, 2001; *Child Neurology,* 2000; Florintino, 1973; Masgutova, 2011). So gifted children are advanced in the development of this pattern compared to peers with neurotypical development, though they demonstrate hyperactive responses because of hyper muscle tone and too active functioning of the excitatory processes of the nerve system. The tendency for hyperactive Trunk Extension also inhibits the development of crawling abilities.

• Crawling Reflex develops in norm from prenatal time (12 weeks of gestation) and integrates up to 4th months of life (Campbell, 2004; Barashniev, 2001; *Child Neurology,* 2000; Masgutova, 2011). Many gifted children do not crawl in infancy and therefore didn't develop this pattern, preferring to start walking earlier than their neurotypical peers. Thus, these children are not well coordinated for reciprocal (cross-lateral) gross motor coordination, motor programming, planning and control, and motor multitasking. Reflex Assessments reveal that many of these children develop their corresponding skills through conscious learning based on activity in the higher brain (instead of matured automaticity). Unfortunately, development achieved by 'skipping' phases is most affected by stress. Therefore, motor coordination is more affected in these children during times of stress on the neurological system.

• Automatic Gait Reflex develops early in gifted children and is often hyperactive (in norm – infants start demonstrating this response in the 1st month and it must be matured by 2 months of their life; Barashniev, 2001; *Child Neurology,* 2000; Florintino, 1973; Masgutova, 2011). However, in children with muscle hypotonia, it is more often hypoactive (with a tendency for freezing and lack of interest for active gross movements). This pattern is linked with visual over focusing on one side, it supports the development of inner direction and orientation on goals; on the other, it leads to over focusing.

• Hands Pulling Reflex can be delayed and forms between 4-5 months of life or later (in norm, development occurs up to end of 2nd month of life; Barashniev, 2001; Florintino, 1973; Masgutova, 2011). The sequence of flexion ('elbows-head-core') is essential for the biomechanics of this reflex. When the sequence is incorrect, it negatively affects the 'core flexion-visual convergence' and 'trunk extension-visual divergence' coordination. Thus, vision and Hands Supporting support each other or, they support each other less than is required, resulting in stronger abilities to form 2-dimentional visual images than 3-D. It should be noted, however, that 3-D vision appears to develop later, thanks to their giftedness.

• Hands Supporting Reflex in children of this group is typically not developed (in norm – infants start demonstrating this response in the 4 months and it must be matured by 6 months of their life; Barashniev, 2001; Child Neurology, 2000; Florintino, 1973; Masgutova, 2011). We don't see development of this specific reflex without specially oriented training. Thus, gifted children often have limited means of protection on a biochemical level (in case of falling down and using their hands for support when needed).

• Asymmetric Tonic Neck Reflex (ATNR), within its unconditioned pattern (integrating normally at 6-7 months; Barashniev, 2001; Child Neurology, 2000; Florintino, 1973; Masgutova, 2011), usually develops timely in gifted children but tends to be hypersensitive and hyperactive. This causes too much parallel processing of sounds resulting in increased tension in the auditory system. The tension results often in a tendency to ignore some important stimuli. The delay of this reflex causes poor spatial sound orientation, challenges in transitioning from monaural to binaural auditory perception, struggling to process information (led more through conscious control), missing some specific sounds, and irritation and impatience in their communication.

• Moro and Fear Paralysis (FPR) Reflex patterns serve for protection against sensory stressors (FPR) and proprioceptive stressors (Moro) (in norm – infants demonstrate this response at 2-4 months of life and it must mature to the 4 months and the FPR integrates to 2 years of life; Vygotsky, 1934/1986; Barashniev, 2001; Florintino, 1973; Masgutova, 2011). These two reflexes are often confused in many gifted children thus leading to a lack of differentiated function in the sympathetic (excitation) and parasympathetic (inhibition) nervous systems causing hyperactivity (Moro) or 'freezing' affect (FPR, an 'absent state' or indifference). Also, fear and panic, with a tendency for irritation or anger is sometimes observed. Some of these children have phobias from being too sensitive to stimuli that have caused negative emotional responses in their life (such as fear of animals or birds, fear of natural phenomena such as the night, earthquakes, wind, diseases, microbes/viruses, or death). Because of their over focusing, they can take fear to a deeper inner place, making it challenging to lessen their fear anchors.

• Segmental Rolling (in norm integrates at the 3-4 months of life; Barashniev, 2001; Sadowska, Zaleski, 2008; Masgutova, 2011) and Spinning Reflex pattern (in norm integrates at the 4 months of life; Barashniev, 2001; Sadowska, Zaleski, Florintino, 1973; 2008; Masgutova, 2011) are delayed, expressed poorly, and are rigid (the child activates the axis body plane for movements and posture much more for activation of movement in the horizontal plane). This causes a lack of rotational movements and inflexibility resulting in too much running, and, in some children, walking on their toes and jumping while walking.

• Vestibular-Ocular Reflexes (horizontal and vertical; in norm integrate at the 2 and 6 months of life; Masgutova, 2008; Florintino, 1973; O'Reilly, Morlet, Cushing, 2013) support stability of moving visual images and are less developed in many gifted children leading to a lack of equilibrium causing rigidity, hyper- or hypo-active physiological nystagmus, poor vestibular-visual coordination, ocular-proprioceptive disorientation, too much convergence activity, and a lack of peripheral vision development. Both focused and peripheral vision usually develop later, thanks to their fast learning. However, as with other reflexes developed after 'skipping early phases' of development, the vestibular-ocular reflexes in gifted children are often less resilient during times of stress.

• Convergence/Divergence Reflex responses (bringing eyeballs close together to see close objects and apart for distant objects; in norm integrate at 1½ – 2 months of life; Florintino, 1973; Masgutova, 2008; O'Reilly, Morlet, Cushing, 2013) basically develop in gifted children at the typical time which helps their cognition of the objects of the world (focusing, decoding, analysis). Their highly advanced intelligence makes their visual work significantly richer and unique. Nevertheless, the work of divergence is not balanced and causes some delay of their peripheral vision. This can trigger protective responses in the brain and a chaotic gaze, lack of smooth eye movements, and more voluntary control for eye mobility, visual analysis, and comprehension (on a conscious level) than automaticity.

Acoustic Stapedial Reflex and sound prioritization. The Acoustic Stapedial Reflex (triggering the eardrum with a sharp, high frequency sound, causing contraction in the stapedius muscle of the middle ear; in norm integrate at the 2-4 months of life; Masgutova, 2008; O'Reilly, Morlet, Cushing, 2013) in gifted children is often hyperactive. The reflex is developed in a timely manner and helps to differentiate phonemes and other sounds precisely. Thus, early articulation and speech development occurs, but the auditory system is hypersensitive. Gifted children also have a tendency for over-focusing, which inhibits their ability to prioritize sounds as to what is important or not important) overloading their processing. This leads to protective behaviors of ignoring or inhibiting the auditory stimuli (confusion with Abdominal Sleep Posture Reflex). They can easily lose their interest in working longer. The result of this is over-focusing, incomplete auditory processing, becoming tired, and jumping to the next stimuli or topic which makes them appear to be hyperactive.

• *Breathing Reflex patterns* (in norm integrate at the 2-6 months of life; Masgutova, Masgutov, Akhmatova, 2012) can be hypo- or hyperactive in gifted children causing challenges in motor development. Their diaphragm can be too tight or hypotonic resulting in hyperventilation and shallow inhaling, causing disharmony in their breathing cycle. This is especially true when they are trying hard or under stress (physical, emotional, or cognitive).

The above brief descriptions of some of the basic reflex patterns show that there is a need for effective, practical options for addressing the neurodevelopmental challenges of gifted children. MNRI® allows us to treat these problems using the reflex re-patterning concept of a reflex is a unit of the nerve system given genetically. Therefore, working with it guarantees activation of the sensory-motor genetic memory creating profound results in the integration of the sensory, motor, and muscular systems.

MNRI® is a novel and non-invasive method of support for neurodevelopment and functions of the central nervous system, including neurodeficits. More than 30,000 children have benefited from this program over the last 20 years. MNRI® initially evaluates children's reflex patterns using the MNRI® Parameter Assessments of 30 reflexes (see the Assessment article in this book). This is followed by an individualized program for correction (repatterning) and integration of reflex patterns into the child's central nervous system function through sensory motor circuits. Repatterning dysfunctional reflex patterns results in improved gross and fine motor coordination and skills, as well as postural control (Bernstein, 1997). This leads to improved behavior because children learn to regulate involuntary unconscious spontaneous motor activity and emotional responses (Bobath, Bobath, 1984; Masgutova, Akhmatova, 2007). Improved cognition also occurs (Piaget, 1936/1976; Montessori, 1949/1995).

Results and Conclusions

This article provides the results of the MNRI® Assessment and repatterning interventions for 270 gifted children aged from 5 to 19. The children were participants of the following MNRI® programs: Neuro-Structural Reflex Integration, Tactile Integration, Dynamic and Postural Reflex Pattern Integration, Lifelong Reflex Integration, Proprioceptive and Cognitive Integration, Visual and Auditory Reflexes Integration, Oral-Facial Reflex Integration, Art Creation and Reflex Integration, and Archetype Movement Integration. These programs were facilitated by a group of professionals trained in the MNRI® program during organized therapeutic conferences, and individual therapy with children with developmental deficits. Statistical analysis was carried out before and after 8 days of MNRI® programs at the neurosensorimotor rehabilitation camps/conferences organized by the International Dr. Svetlana Masgutova Institute.

An evaluation was conducted for patients of their 30 reflex patterns which were divided into three groups coordinating with the following body movement planes: sagittal, horizontal, and dorsal (Krefft, 2007). Every reflex pattern was evaluated on a scale of 0–20 with regard to five parameters: reflex pattern, direction of movement, strength of reaction, timing of reaction, and symmetry (Masgutova, 2011). The testing within these parameters was done for each reflex pattern individually. A rating of 10 on the scale determines the reflex to be at the transition state in

Table 1. Reflex #	*Reflex pattern*	*Gifted Children (270 individuals; age 5-19)*		*Neurotypical Control Group (780 individuals; age 2-19)*
		Before MNRI®	*After MNRI®*	
1	Core Tendon Guard	14.54	15.56	15.5
2	Robinson Grasp	16	17	17
3	Hands Pulling	15.5	16.76	16
4	Babkin Palmomental	15.5	16.34	16
5	Babinski	14.5	15.72	16.5
6	Leg Cross Flexion	16.5	17.5	17
7	Asymmetrical Tonic Neck	15.87	17.86	15
8	Abdominal	14.5	15.76	16
9	Bonding	15	16.5	15.5
10	Foot Grasp	15.5	16.86	16.5
11	Thomas Automatic Gait	14.5	16.5	17.5
12	Bauer Crawling	14.5	16.76	15.5
13	Moro Embrace	14.54	16.65	15.5
14	Fear Paralysis	14.65	15.24	14.5
15	Hand Supporting	14.5	16.5	15.5
16	Segmental Rolling	14.52	16.54	15
17	Landau	14.34	15.52	15
18	Flying and Landing	14.64	16.5	14.5
19	Grounding	14.25	16.5	16
20	Head Righting	16.56	17.34	16
21	Trunk Extension	15.76	17	17.5
22	Symmetrical Tonic Neck	16.62	17.56	15.5
23	Spinal Galant	14.74	16	15.5
24	Spinal Perez	14.37	15.65	14.5
25	Tonic Labyrinthine	15.02	15.74	15.5
26	Foot Tendon Guard	14.45	16.25	15
27	Spinning	15.52	16.36	15
28	Locomotion	14.25	16.78	14.5
29	Balancing	15	16.5	16
30	Pavlov Orientation	17.5	18.75	16

Table 1. Changes in the average level of development of reflex patterns for 270 gifted children before and after completing treatment at MNRI® camps.

development of the pattern between its pathological/dysfunctional and normal state, whereas numbers from 15–20 represent normal functioning (Krefft, 2007). The *Assessments* article in this book shows the points scale determining the criteria of evaluation of pathology, dysfunction, norm, and integrated levels of reflexes and their patterns, and the levels of integration of the basic pattern and its variants with intentional and controlled movements (Masgutova, 2011).

The MNRI® Assessment of 270 gifted children shows that some typical reflex patterns are immature at the age when they should have accomplished their development. At the same time this differs from child to child. The knowledge of Reflex Profiles of gifted children and individual differences in patterns can help to create an individualized MNRI® program for gifted children directed at facilitating improvements on sensory motor dissonance, neuro-facilitation of reflex circuits, and all parameters and integration of tactile, visual, auditory, vestibular, proprioceptive protective and breathing systems. A priority focus of the MNRI® program for gifted children is to support their neurodevelopment, as well as emotional, social, and cognitive functions through reflex integration and neuroplasticity (Syka, Merzenich, 2003; Sadowska, Zaleski, 2008).

Reflex Assessment of gifted children demonstrates that many of their reflex patterns have not accomplished their maturation though these children have grown seemingly well, thanks to their giftedness, intelligence, and language skills (developing earlier than their peers). Still, their physical abilities are delayed, to some extent, and this can be seen in their hyperactive protective mechanisms and lack of flexibility in stress and distress (as also was reported by their parents). The delay in development of

Figure 1-A. Saggital Plane of the Body

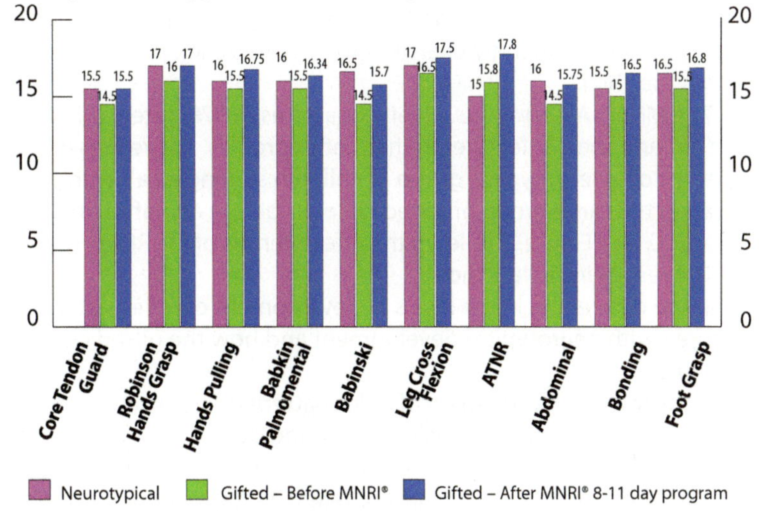

Figure 1-B. Horizontal Plane of the Body

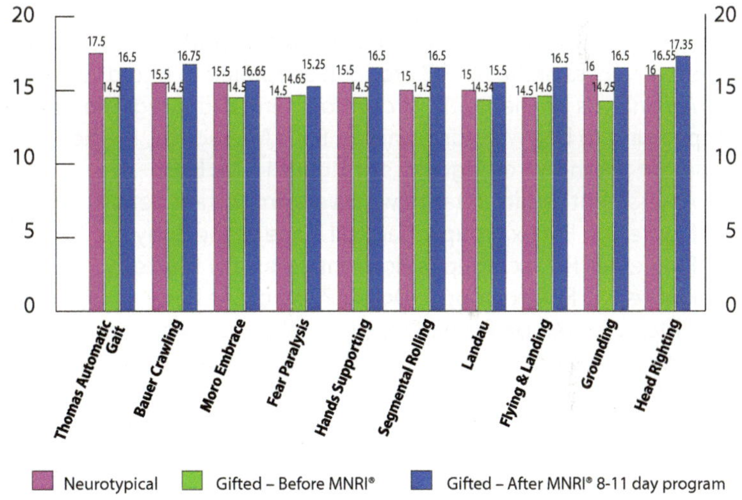

Figure 1-C. Dorsal Plane of the Body

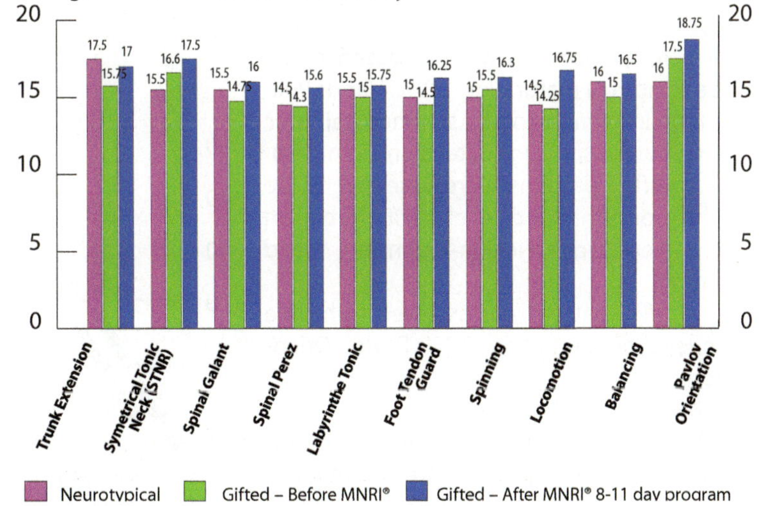

unconditioned phases for some reflex patterns can be one of the early signs for a predisposition for asynchronisity. This is why reflex integration therapy must be an early intervention tool for every child in the world as it utilizes natural, friendly exercises to activate reflex patterns and help them to mature using the potential within the child.

The MNRI® Assessments of reflex patterns in 270 gifted children were performed to learn the overall tendencies and specific features in their reflex profiles. The results of the Assessments were compared with reflex patterns of a neurotypical group of children serving as a control group. Reflex patterns also have been compared in the same group of gifted children before and after an 8-day MNRI® Conference (see Table 1, Figures 1 - A, B, C, and Figure 2) to learn the effectiveness of MNRI® on gifted children.

The results in Table 1 show:

1) the differences in the levels of development of reflex patterns in gifted children in comparison to those children with neurotypical development and how the gifted children show some lack of maturation of reflex patterns.

2) the level of development of reflex patterns in gifted children improved (after MNRI® treatment) which can lead to positive changes in their physical and overall development, and as management of stress and survival.

The results of changes in the level of reflex pattern integration of the 270 gifted children after completing treatment at an 8-day MNRI® Conference/Camp (compare pre- and post assessment data) demonstrates a positive change in all reflexes after the MNRI® intervention. The results of statistical analysis were validated by synthesized function $z = f(x)$ (by NewKrefft algorithm, A. Krefft, 2007) and also by nonparametric comparison of two variables by Wilcoxon Matched Pairs test ($P<0.001$; Statistica Program), which proves the high degree of effectiveness of the MNRI® program as applied to gifted children during 8 days/6 hour daily intervention.

An analysis of the levels of development and integration of 30 reflex patterns within three groups corresponding to body-motor planes A, B, C (A – sagittal, B – horizontal, C – dorsal) before and after the MNRI® program are shown on Figures (see Figure 1 – A, B, C).

Assessment results data shows several important details:

Figure 1-A. Reflex group 1: Sagittal Plane of the Body

Figure 1-B. Reflex group 2: Horizontal Plane of the Body

Figure 1-C. Reflex group 3: Dorsal Plane of the Body

Gifted children performed reflex patterns (see Table 1 above) on the following level of development:

• 20 reflex patterns were assessed to be lower than normal maturation (66.67%; Core Tendon Guard, Babinski, Abdominal Sleep Posture, Bonding, Thomas Automatic Gait, Bauer Crawling, Moro, Fear Paralysis, Hands Supporting, Segmental Rolling, Landau, Flying and Landing, Grounding, Symmetrical Tonic Neck, Spinal Galant, Spinal Perez, Foot Tendon Guard, Balancing, Tonic Labyrinthine, Locomotion)

• 6 reflex patterns were assessed to be close to normal maturation (20%; Hands Pulling, Babkin Palmomental, Asymmetrical Tonic Neck, Foot Grasp, Trunk Extension, Spinning)

• 3 reflex patterns were assessed to be at the norm/matured level (10%; Robinson Grasp, Leg Cross Flexion-Extension, Head Righting)

• 1 reflex pattern was assessed to have a high level of integration (3.3%; Pavlov Orientation).

These results show that:

1) all reflex patterns are on functional level

2) 13.3% of reflex patterns are on a mature/

Figure 2. Gifted Children: Reflexes before and after MNRI®

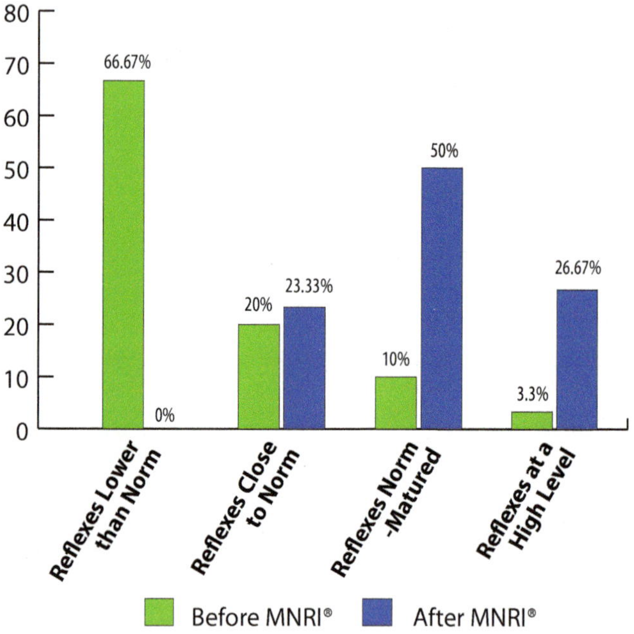

integrated level

3) 20% of reflex patterns are close to matured

4) 66.67% are still not matured, though they are functional (Figure 1, 2, 3).

5) reflex profiles in gifted children show the need of the support for neurointegration of reflex patterns

6) immature reflex patterns cause, to some extent, asynchronicity in the neurodevelopment of gifted children.

After the MNRI® program, the level of development of reflex patterns improved significantly, and specifically for the following reflex patterns (see Figure 2):

• 0 reflex patterns were assessed to be lower than norm maturation (0%)

• 7 reflex patterns were assessed to be close to norm (23.33%; Core Tendon Guard, Babinski, Abdominal Sleep Posture, Fear Paralysis, Landau, Tonic Labyrinthine, and Spinal Perez)

• 15 reflex patterns were assessed to be at the norm/matured level (50%; Bonding, Thomas Automatic Gait, Bauer Crawling, Moro, Hands Supporting, Segmental Rolling, Flying and Landing, Grounding, Spinal Galant, Foot Tendon Guard, Balancing, Hands Pulling, Babkin Palmomental, Foot Grasp, and Spinning)

• 8 reflex patterns were assessed to be at a high level of integration (26.67%; Pavlov Orientation, Symmetrical Tonic Neck, Locomotion, Asymmetrical Tonic Neck, Trunk Extension, Robinson Grasp, Leg Cross Flexion, Head Righting).

These results of the Post-Assessment show that:

1) all reflex patterns (100%) improved.

2) the majority of reflex patterns moved to a level of complete maturation and integration – 76.67% patterns.

3) 26.67% of reflex patterns out of the above matured reflexes (76.67%) moved to a high level of integration.

4) Reflex Profiles of gifted children after MNRI® intervention program show significant and fast improvement of reflex patterns (for 76.67%) meaning their ability to learn quickly also takes place on the level of automaticity. The coefficient of change is significance and thus, valid - 0.47 before the MNRI® program and 0.62 after; $p < 0,001$).

5) the MNRI® program supports the maturation of reflex patterns and can solve the problem of asynchronicity in neurodevelopment of gifted children to a significant extent.

This new data is unique and shows the importance and possibility of facilitation of neurodevelopment of gifted children through reflex integration. This allows one to presuppose that integration of reflexes can correct asynchronicity.

Conclusions

The MNRI® program was specifically applied in this research to 270 children identified as gifted ranging from 5 to 19 years of age. The study included both a pre and post MNRI® Assessment for children attending 8 day MNRI® Conferences. The study was conducted to create a Reflex Profile of gifted children based on evaluation of the level of reflex pattern maturation and function. It also studied the effects of the MNRI® therapy program on their neuro-reflex integration.

Assessments of reflex patterns in gifted children show specifics of their overall Reflex Profile, which can alert persons in helping professions in creating facilitation programs for their neurodevelopment. It demonstrates that many of their reflex patterns have not accomplished maturation phase, which can be interpreted as a lack of neurosensorimotor integration. It was noted that gifted children often exhibit asynchronicity among their intelligence formation, physical growth, and sensory-motor integration. This frequently results in their social, emotional, and cognitive abilities developing at different times and levels.

Several clinical observations were noted in this population within the MNRI® program. They were found to exhibit challenges in muscle tone regulation, postural control, motor coordination, and reflex pattern development as well as difficulties with their visual and auditory systems. Their motor coordination was better developed than gross motor coordination in this group.

The initial Assessment revealed that some of their reflexes were considered to be immature. It was found that 20 reflex patterns were assessed to be lower than the normal maturation (66.67%: Core Tendon Guard, Babinski, Abdominal Sleep Posture, Bonding, Thomas Automatic Gait, Bauer Crawling, Moro, Fear Paralysis, Hands Supporting, Segmental Rolling, Landau, Flying and Landing, Grounding, Symmetrical Tonic Neck, Spinal

Galant, Spinal Perez, Foot Tendon Guard, Balancing, Tonic Labyrinthine, Locomotion). Six reflex patterns were assessed to be close to norm in maturation (20%: Hands Pulling, Babkin Palmomental, Asymmetrical Tonic Neck, Foot Grasp, Trunk Extension, Spinning). Three reflex patterns were assessed to be at the norm/matured level (10%: Robinson Grasp, Leg Cross Flexion-Extension, Head Righting). Only one reflex was assessed to have a high level of integration (3.3%: Pavlov Orientation).

After treatment with MNRI®, Assessments showed there was 100% improvement of all reflex patterns. It was found that no reflex patterns (0%) were now assessed to be lower than the normal maturation. The post Assessment further indicated only 7 reflex patterns were assessed to be close to norm (23.33%: Core Tendon Guard, Babinski, Abdominal Sleep Posture, Fear Paralysis, Landau, Tonic Labyrinthine, Spinal Perez). Now, there were 15 reflex patterns assessed to be at the norm/matured level (50%: Bonding, Thomas Automatic Gait, Bauer Crawling, Moro, Hands Supporting, Segmental Rolling, Flying and Landing, Grounding, Spinal Galant, Foot Tendon Guard, Balancing, Hands Pulling, Babkin Palmomental, Foot Grasp, Spinning). To the benefit of these children, 8 reflexes were now assessed to be at a high level of integration (26.67%: Pavlov Orientation, Symmetrical Tonic Neck, Locomotion, Asymmetrical Tonic Neck, Trunk Extension, Robinson Grasp, Leg Cross Flexion-Extension, Head Righting). Therefore, this study indicates that the MNRI® program can correct the asynchronicity of reflex integration of children identified as gifted and talented.

References

Barashniev, J. (2001). *Prenatal Neurology.* (Russ.). Moscow, Russia: Triada-X. 640.

Bernstein, N. (1997). *Bio-mechanics and physiology of the movement.* Moscow, Russ: Moscow-Voroniez.

Bobath, K., Bobath, B. (1984). *The neurodevelopmental treatment.* In: Scrutton D, editor. *Management of the Motor Disorders of Children with Cerebral Palsy.* Oxford, UK: Blackwell Scientific Publications Ltd. p 6–18.

Campbell, S. (2004). *Watch me grow! A unique, 3-dimensional, week look at baby's behavior and development inside the womb.* London, UK: Carroll and Brown Publishers Limited.

Chambliss, A. (2010). *Instructional needs of gifted and talented students. A review of literature.* www.nagc.org.

Child Neurology (2000). Poland, Urban & Partner.

Florintino, M. R. (1973). *Reflex testing methods for evaluating C.N.S. development.* 2nd Edition. Springfield, USA: Charls C Thomas Publisher.

Fulkerson, J., Horvich, M. (1998). Talent development: Two perspectives. // *Phi Delta Kappan,* 79(10), 756.

Eide, B., Eide, B. (2004). *Understanding gifted children - Brains on fire.*

Gagné, F. (1985). Giftedness and talent: Reexamining a reexamination of the definitions. // *Gifted Child Quarterly,* 29, 103-112).

Gross, D. (1987). *Psychology. The science of mind and behavior.* London, UK: Hodder and Stoughton.

Krefft, A. (2007). *Diagnosis of functions of invisible phenomena (Statistic Mathematic Analysis).* Wroclaw, PL: Oficyna Wydawnicza Politechniki Wroclawskiej.

Masgutova, S., Ozerova, S. (1995). The aspects of the psychological work with gifted children having high level of anxiety. // in Book: *Actual problems of psychological assisstance in education.* Material of Regional Conference, June 28-30, 1995 Orechovo-Zouevo, Moscow, (Russ.). p. 36-41, Russia.

Masgutova, S. (2009). *MNRI® Visual and auditory reflexes integration.* Orlando, FL, USA: SMEI.

Masgutova, S. (2011). *Infant dynamic and postural reflexes. Neuro-sensory-motor reflex integration.* Revised and edited, scientific-practical manual. Orlando, FL, USA: SMEI.

Masgutova, S., Masgutov, D., Akhmatova, N. (2012). *MNRI: breathing reflex integration. Optimize health and cognitive resources through breathing.* Orlando, FL, USA: SMEI.

Montessori, M. (1995). *The absorbent mind.* New York. Henry Holt and Company. USA. No Child Left Behind Act, P.L. 107-110 (First published in 1949, Title IX, Part A, Definition 22) (2002); 20 USC 7801(22) (2004). (First published in 1949).

NAGC's position paper, *Redefining giftedness for a new century: Shifting the paradigm.* State Definitions of Gifted and Talented.

Naglieri nonverbal ability test or the *Test of nonverbal intelligence.*

National Association for Gifted Children. (2010). *Redefining giftedness for a new century: Shifting the paradigm.* Position Statement, Washington, DC, www.nagc.org.

Nikitin, B. P., Nikitina, L. A. (2004). *We and our children.* (Russ.). Moscow, Russia: Medicina.

O'Reilly, R., Morlet, T., Cushing, S. (2013). *Manual of Pediatric Balance Disorders.* San Diego, CA, USA: Plural Publishing Inc.

Pavlov, I. P. (1960). *Conditioned reflexes: An investigation of the physiological activity of the cerebral cortex.* (Anrep G.V., D. Sc. Trans., 1960). New York, NY, USA: Dover Publications Inc. (First published in 1927).

Piaget, J. (1936). *Origins of intelligence in the child.* London: Routledge & Kegan Paul.

Piaget, J. (1954). *The construction of reality in the child.* (M. Cook, Trans.). New York, NY, USA: Ballantine Books.

Renzulli, J. S. (1978). What makes giftedness? Re-examining a definition. *Phi Delta Kappa,* 60, 180-181.

Sadowska, L., Zaleski, A. (2008). *Diagnosis and Therapy of the Psychological and Motor.* Silverman L. (2012). Upside Down Brilliance. website.

Spheres in Development Rehabilitation. // in: Sadowska, L. (Ed.). *Neurophysiological correction methods for children with developmental disorders.* AWF, Wroclaw 9-49.

Sechenov, I. (1995). *Physiology of behavior. Scientific works/* Edd.: Moscow. (Russ.). Russia: M. G. Jaroshevsky. (First published in 1863.)

Sherrington, C. (1947). *The integrative action of the nervous system.* London, UK: Cambridge University Press.

Silverman, L. (2002). *Upside-down brilliance: The visual-spatial learner.* Denver, CO, USA: DeLeon.

Syka, J., Merzenich, M. (2003). *Plasticity and signal representation in the auditory system.* New York, NY, US: Springer Publishing Co.

Vygotsky, L. (1986). *The Child Psychology. The Problems of Child Development.* In 6 Books. Book – 4. Moscow, Russia: Pedagogika.

Vygotsky L. (1962). *Thinking and speech.* (Edited and translated in part by Eugenia Hanfmann and Gertrude Vakar, and in part by Norris Minnick). Boston, MA: The M.I.T. Press (First published in 1934).

Wescher intelligence scale for children, 4th Edition. Read NAGC's position paper Use of the WISC-IV in Gifted Education

http://www.nagc.org/resources-publications/resources/my-child-gifted#sthash.dRnoctpF.dpuf

http://www.nagc.org/resources-publications/gifted-education-practices/identification/tests-assessments#sthash.V2W7Hnfm.dpuf

Dear children, Thank you for your participation in our work and for helping us to understand that the reflex integration is also important for gifted people because it can facilitate motivation and a desire for better and more active expression of our talents. – Authors.

MNRI® as a System

Sally Averkamp, MNRI® Parent, Madison, WI, USA

When I was introduced to the Masgutova Reflex Integration® program in 2012, I had a 22 year old son who had physically recovered from autism but was left with an extremely low self-image and substantial cognitive delay. I first read everything I could find on the MNRI® website. To learn the techniques for use on my son, I ordered one of the manuals on reflex integration. I quickly realized that this work was not intuitive to me; I had a background in art and graphics, not science and anatomy. So I looked for a certified specialist in my area and found MNRI® Core Specialist Peg Johnson who works in the south-central Wisconsin area. As she worked with my son each week, she also taught me techniques to do at home. It was not easy but slowly, I saw my son's very difficult problems with his feet and hands improve, and with this, cognitive changes also occurred. I knew then that we were onto something big.

Sally Averkamp

But success made me greedy for more and faster success so we enrolled in a Conference – four intensive days where my son received six hours of MNRI® each day, in different MNRI® areas. First, we received an Assessment which was very enlightening as to my son's strengths and challenges and how we were to proceed. In just four days, he could feel the difference (and I wished we would have stayed for all 8 days). I then took my first *MNRI® Dynamic & Postural Class* and we were off to amazing growth from our Home Program. Three years later, my son is approaching typical for his age in self-image and cognitive ability. I see college in his future.

From this stumbling beginning, I now see that MNRI® is a complete system for me to plug into. The major components of this system are the child or adult needing sessions, the parents in need of education, and more professionals who want training to become MNRI® Core Specialists. In this system, support is provided for each of these groups and, with the Assessment, it forms a complete system.

1) MNRI® Assessment Clinic

The MNRI® Assessment Clinic option is designed to provide a parent who does not have easy access to a MNRI® Core Specialist resource, qualified guidance and education in creating a basic MNRI® Home Treatment program. During a typical 1½ hour Clinic session, an Assessment is given then, a certified MNRI® Core Specialist shares MNRI® Neuro-structural and Dynamic and Postural Reflex program. Toward the end of the session, the Core Specialist provides the parent with a basic checklist that helps to outline the reflex focus important for forming a productive home treatment program. While much can be learned by reading the *Dynamic and Postural Reflex Manual*, videos, or attending a MNRI® course, the practical experience and in depth knowledge that a certified MNRI® Core Specialist can offer helps parents rest assured that the MNRI® Home Treatment

The MNRI® Child/Adult Centered Program

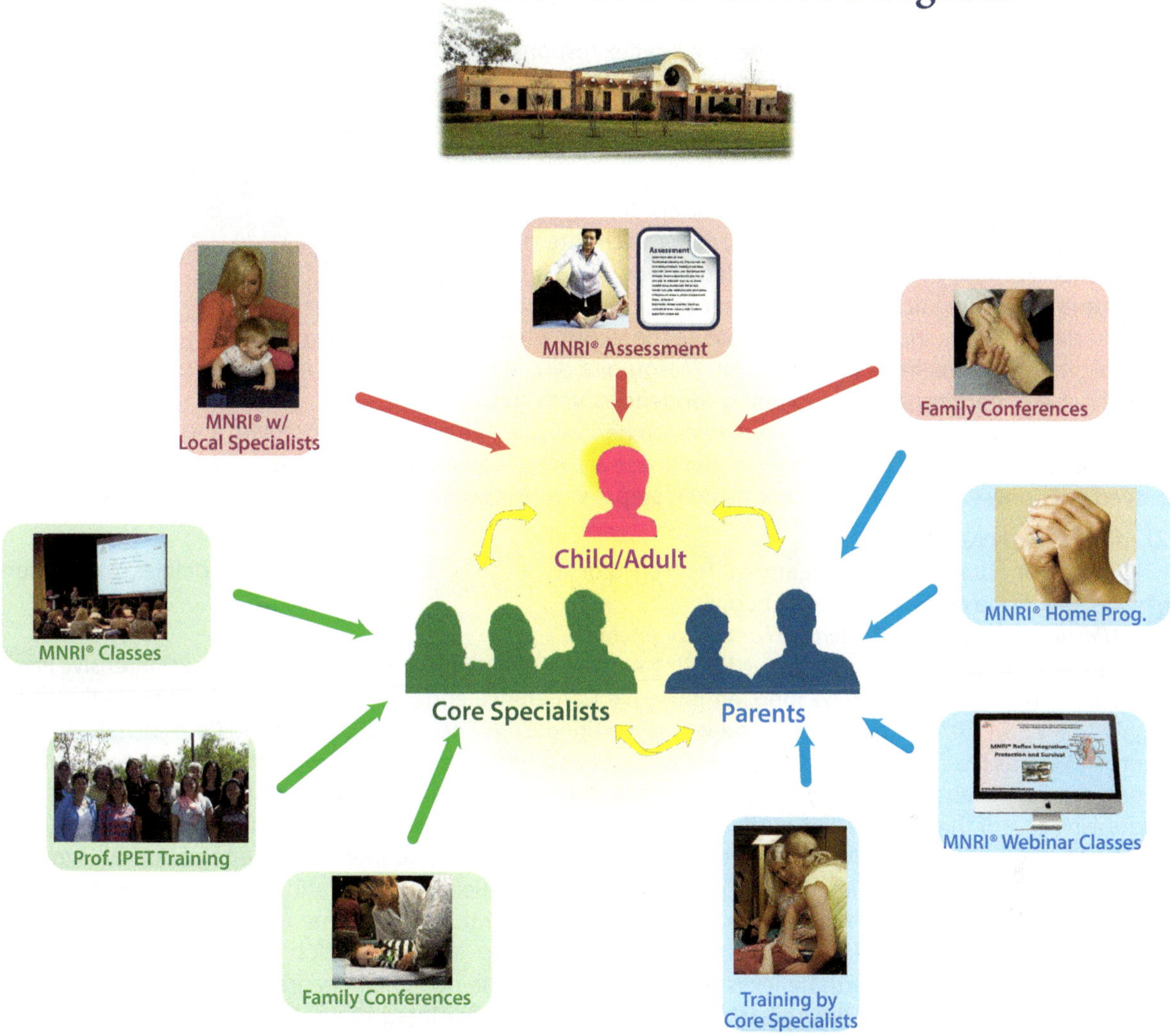

program they are following is focused on the right reflexes and techniques. Several individual training sessions are scheduled during a clinic, among which families can decide to participate in one or a number of sessions.

A basic Reflex Assessment tells if reflex patterns are developed:

- appropriate to age
- appropriate to the reaction/action of the stimuli (reflexive or compulsive level of reaction/action)
- appropriate reaction to the reflex circuit: sensory stimuli and corresponding response.

A specialized Reflex Assessment gives parameters and features to assess reflex pattern function:

Part I – The basic pattern should be:

- normal to the age of an individual
- showing a proper response for the stimulus
- have a correct and stable functioning circuit.

Part II – The variant patterns:

- in the proper connection within the programming aspect
- in the correct connection with the control aspect
- able to serve in its intentional function of motor and cognitive activity

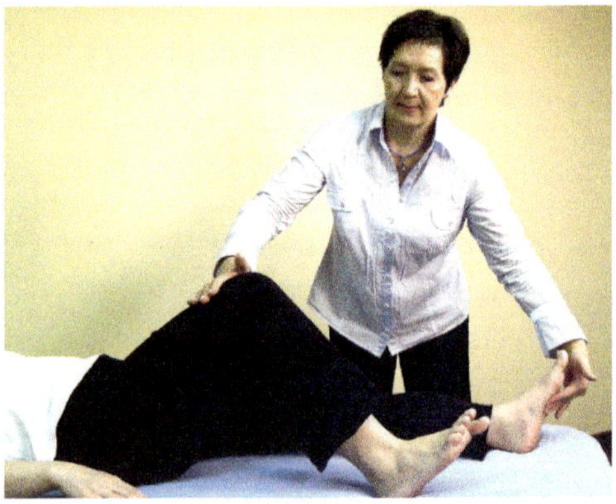
An Assessment by Dr. Masgutova checking Leg Cross Flexion-Extension.

Part III - Parameters and features should be:

• appropriate function of a reflex circuit (sensory stimuli and corresponding motor response)

• appropriate direction of movement in a reflex response

• appropriate strength of reaction/action (muscle tone regulation in pattern)

• appropriate timing of the response (time for emergence, duration, and completion)

• symmetrical in a response to the stimulus

Assessments are also given at the outset of each intensive Conference. For many years, my husband and I wondered if our daughter's eye-rolling tic was related to her seizures. (Our neurologist did not seem interested in this connection.) But when my daughter received her Assessment, Dr. Masgutova directly linked her eye tracking and eye-jumping tendencies with her seizures, could tell us specifically which nerves on her left side needed to be addressed, and she directed the Core Specialists during the Conference to give these areas special attention. (Her memory and eye-rolling improved after only two days.) So the Assessment step is very important.

2) MNRI® Success

Options for obtaining MNRI® treatment for a child or adult can be done in several ways or in a combination of ways. Below is a discussion of the options.

A) Participate in an MNRI® Intensive Family Conference

The most intense and comprehensive MNRI® treatment training option available is an MNRI® Intensive Family Conference. Typically 4-8 days long, an Intensive Conference engages Dr. Masgutova's Team – the most highly trained MNRI® professionals in the world (certified Core Specialists) and MNRI® Core-in-Training students (at various levels of MNRI® certification) to create and apply a treatment program unique to each participating child's challenges. Each day of the conference includes a series of treatment sessions, with each session addressing one of the many MNRI® modules important to the complete, holistic, MNRI® approach.

The MNRI® programs are Neuro-Structural Reflex Integration, Tactile Integration, Integration of Dynamic and Postural Reflex Patterns and Lifelong Reflexes, Oral-Facial Reflexes Integration, Visual and Auditory Reflexes Integration, Archetype Movements Integration, Upper Limbs Reflexes Integration, Children with Challenges, Dysfunctional and Pathological Reflexes Repatterning and Integration, Proprioceptive-Cognitive Integration, Birth and Post- Birth Reflexes Integration, and Breathing Reflex Integration. In addition, complementary MNRI® programs are Pet Assistant Therapy based on Reflex Integration, Art-Therapy based on Reflex Integration, Dance-Therapy based on Reflex Integration, Speech Development based on Reflex Integration, Proprioceptive-Cognitive Integration based on Reflex Integration, and Aqua-Reflex Integration.

B) MNRI® Home Program

While the MNRI® Family Educational Conference is the ultimate experience, what makes everything work in the end is the parent's ability to carry the MNRI® work out at home. An intensive Family Conference is a quick-start approach to treating a child. Gains are almost always made at conferences, many of which are maintained, but even greater gains depend on the Home Program follow-through. A parent industrious enough to learn and apply MNRI® on his/her own, through a manual, video, by attending a course or combination of courses, possesses the long-hall stamina necessary to win the race. Success in an MNRI® program depends not simply on repatterning/integration of the reflex patterns by a Core Specialist or outside resources, but also on parents able to follow through with a committed neurodevelopment program support at home, usually 5 days per week (20 min. to 2 hours, daily). Determined and steady wins the race.

There are four general ways to get training, each with different levels of engagement. The choice of where to begin is a personal one, and often depends on a combination of factors including proximity of MNRI® resources, financial situation, and time available.

a) Learn at an Intensive MNRI® Family Conference. The Intensive Conferences have a session most evenings for parents to learn techniques. In addition, the therapists often provide training that will benefit their child or adult. To guide post-conference success, a basic MNRI® Home Program is created for each child and given to parents to continue the reflex integration work at home.

b) Work with an MNRI® Authorized Specialist (see website for list) through the Home Services. The easiest way to get started with treatment is to locate an authorized MNRI® Core Specialist in your area. This Specialist begins by completing a basic Reflex Evaluation to identify your child's unique reflex challenges. Once your child's specific reflex challenges are identified, a basic MNRI® Home Program is created and the MNRI® Specialist will begin the therapy and educational process to show you how to work with your child. The more parents are able to participate in the MNRI® process, the greater the possibility for functional change in their child.

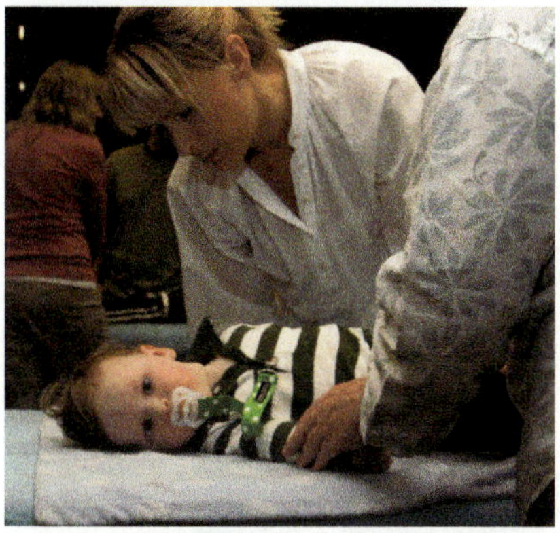

Parent receiving feedback on MNRI® procedures at an intensive Conference.

c) Parents Attend an MNRI® Course. Another option is for parents to consider attending an MNRI® educational class. Parents often join professionals at classes to learn for themselves how to apply MNRI® restorative techniques to integrate automatic motor reflexes and challenged sensory systems. Each MNRI® training class is designed to make sure that parents and professionals leave understanding the basic assessment and integration techniques to create and begin using the MNRI® program. While some parents are apprehensive about participating side by side with experienced professionals, we have observed that a parent's intimate understanding of their child's unique challenges and deep motivation to improve their child's situation, often prepares them to, not only learn, but also to productively contribute to class content. Unique parent insight can lead to questions in class that help MNRI® instructors address technique nuances that benefit both the parent and professionals. We have found that often, even when parents have access to a local MNRI® resource, they choose to attend MNRI® classes to deepen their knowledge and enrich the MNRI® therapy techniques provided by a local MNRI® Specialist. We encourage parents to take MNRI® classwork whenever possible, whether to create programs for their children or to enrich the impact of other therapy options they have had the ability to choose.

d) Parents use an MNRI® Dynamic & Postural Reflex Integration Manual or Watch Videos Online. For those who do not have access to an authorized MNRI® Core Specialist resource and simply do not have the time or means to attend a course or other MNRI® sponsored event, the *MNRI® Dynamic and Postural Reflex Integration* manual and videos are available for purchase through the MNRI® website shop. While direct instruction and guidance is always the preferred approach, benefit can be gained from simply reading the manual or watching the videos and slowly beginning the MNRI® therapeutic process at home. The MNRI® Team encourage the families that choose this option (and other families as well) to deepen and expand their MNRI® knowledge through whatever means are possible.

Reflex patterning videos available on the MNRI® website (www.MasgutovaMethod.com) include:

- the Asymmetrical Tonic Neck Reflex (ATNR) which supports the development of one-sided and asymmetrical, cross-lateral motor coordination within the right-left motor coordination system
- the Automatic Gait Reflex which prepares the child for a Matured Gait Reflex by stimulating the development and dynamics of the legs and feet, as well as muscles in the abdominal and back areas
- Abdominal Sleep Posture Reflex which is used to unconsciously find the best position to prepare to fall asleep
- the Babinski Reflex which has a direct influence on the development of gravity, grounding, stability, and balancing reactions
- the Babkin Palmomental Reflex which creates the basis for the development of the 'palm-mouth' coordination system
- the Bauer Crawling Reflex which assists the development of coordination of whole body movement
- the Bonding Reflex which supplies the physical state and emotional well-being of being safe and protected
- the Fear Paralysis Reflex which is triggered by a sensory stimulus – auditory, visual, or tactile, such as a sudden or loud sound, an unexpected flash of light, or an uncomfortable touch
- the Flying and Landing Reflex which is strongly related to gravity, balance, centering, and grounding reactions

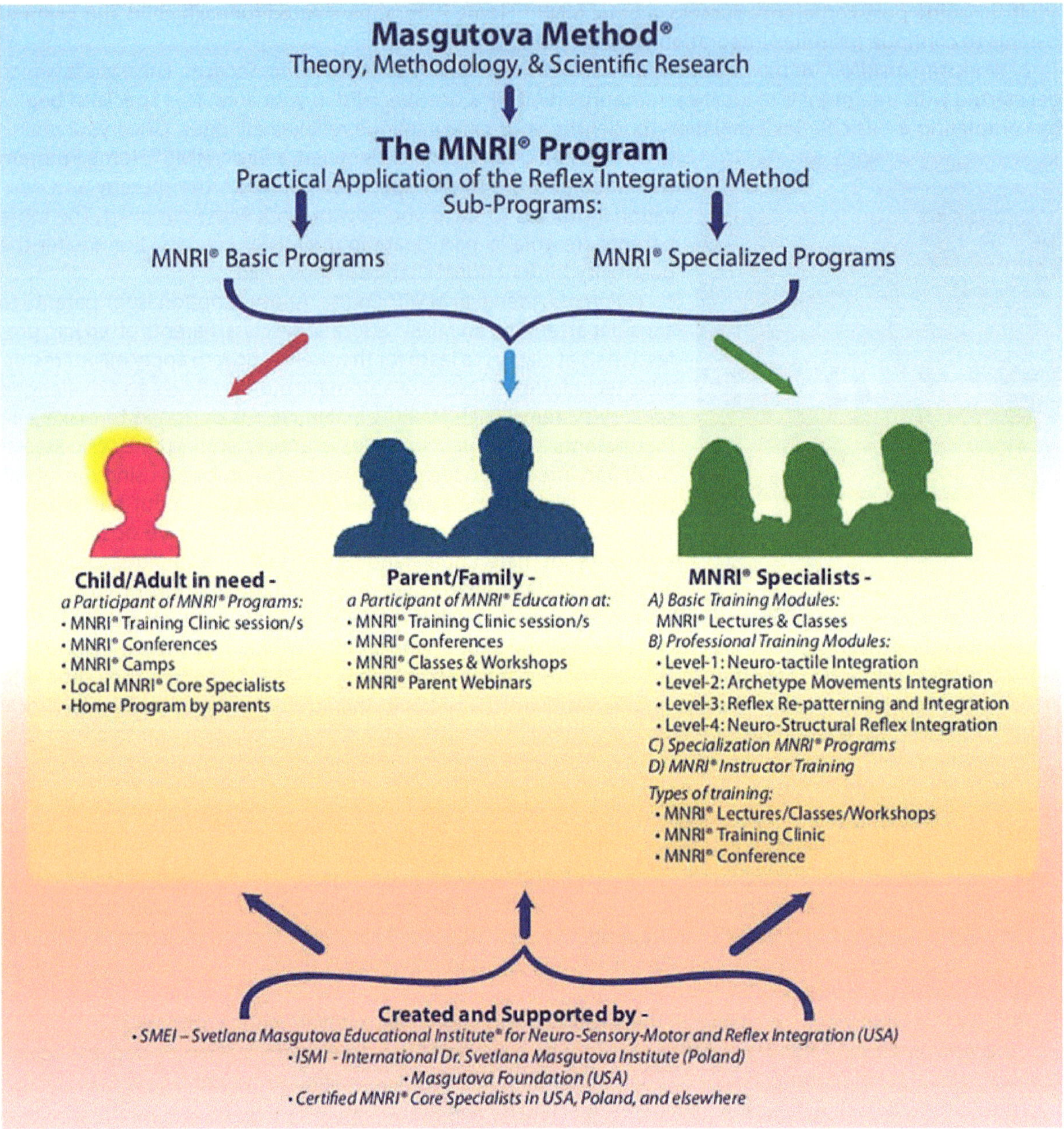

• the Foot Tendon Guard Reflex which is a protective reaction to a physical or emotional stress

• the Hands Pulling Reflex which becomes a part of the pattern of pulling and flexing the arms in an attempt to sit or stand up

• the Hands Supporting Reflex which offers protection when falling and helps provide clear boundaries in relationships

• the Landau Reflex which is called a reflex of joy and happiness; this reflex develops the motor control of the back of the body and the relationship of horizontal and vertical perspectives of the body

• the Leg Cross Flexion-Extension Reflex which influences muscle development and coordination of the legs

• the Pavlov Orientation Reflex which is a result of activity of the nervous system reaction scheme on being stimulated by something unexpected, new, or interesting

• the Robinson Hands Grasp Reflex which influences the development of gross hand motor coordination, grasping and holding objects, manipulation of bigger objects, and later, the development of fine motor coordination of fingers for drawing, writing, playing musical instruments, and using eating utensils

• the Spinal Galant Reflex which influences future physical development like spine motions, stretching flexibility, and postural control; it also the affects gravity and stability reflexes

• the Spinal Perez Reflex which is the basis for whole body coordination and development, especially for the links between the limbs and head through the core of the body, both front and back

• the Symmetrical Tonic Neck Reflex (STNR) which assists in the development of bilateral patterns of body movement and information processing in the left and right hemispheres

• the Trunk Extension Reflex which activates the cerebral-spinal fluid pumping mechanism, influencing the development of the nervous system.

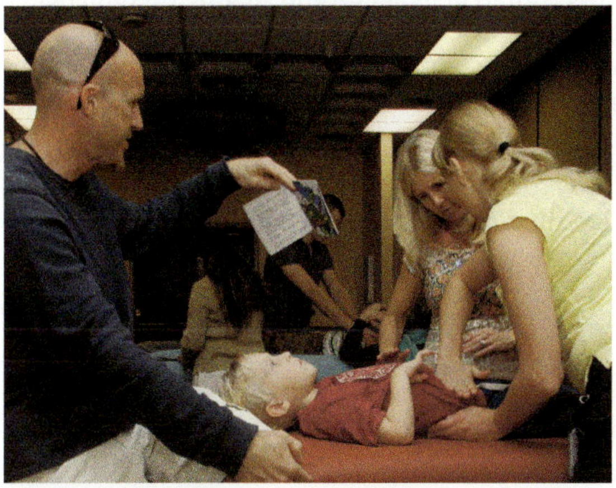

During this Intensive MNRI® Conference, the child's mother is receiving training as her child receives MNRI® treatment.

A Webinar MNRI® Parent School is provided via webinar sessions similar to on-line classes with the goal of education and support for parents, especially those who cannot reach the classes and workshops because of being in different countries or other reasons of distance. The direction of the Webinars is geared toward classes on different programs of reflex integration. Lectures help parents address many 'hot' topics important to them and for consultation on the use of MNRI® techniques as applied to dealing with specific learning challenges and neurodevelopmental deficits. Examples are *Protective Behavior in Children and Reflex Integration, Oral-Facial Reflex Integration and Expressive Speech Development, Bonding Reflex and Attachment,* and other classes.

Other simple ways to advance knowledge and experience include membership in the Children with Challenges Chat Group and networking with other parents and MNRI® professionals. Continue to check the MNRI® Resource Directory on the MasgutovaMethod website for a list of people in training and for our updated list of authorized Core Specialists. As you experience success in your Home Treatment program, share your story on our website through *Tell Us Your Story*, to help others understand how they can accomplish the same thing and to help newcomers understand what can happen just by parents just like them.

3) MNRI® Training for Core Specialists and Parents

The Masgutova Method® has been actively developed in the USA, Poland, Canada, and other countries and is approved for Continuing Education Credits and hours by these professional organizations in the USA:

- ASHA – The American Speech- Language-Hearing Association (2010)
- AOTA – The American Occupational Therapy Association (2009)
- PTBC – Physical Therapy Board of California (2009)
- NCBTMB – National Certification Board for Therapeutic Massage and Bodywork (2010).

The MNRI® Course Education Program is the entry point for all MNRI® students wishing to become an MNRI® Core Specialist. The MNRI® Core-in-Training Program consists of several classes offering specialized training, which must be completed in order to become a MNRI® Core Specialist. This training program is designed with sufficient rigor to ensure students become solidly grounded in MNRI® and integrative techniques. The classes are given above.

To enter the MNRI® Core-in-Training Program an applicant must have a minimum of 1) a BA or BS from an accredited college program or 2) completion of six semester hours of training in anatomy and physiology before the MNRI® Core Training Program is completed.

In addition, they need to have completed the MNRI® Dynamic and Postural course twice and another MNRI® course and complete at least one case study which should be ready to submit to a professional journal before the completion of the MNRI® Core-in-Training program is finished.

The program is progressive and culminates with full certification as a MNRI® Core Specialist after:

- completion of at least eleven MNRI® courses
- completion of at least four MNRI® IPET practicum courses

- completion of a total of 304 clinical hours (accumulated at designated MNRI® Family Conferences)
- passing all required practical exams
- completion of the MNRI® Core Specialist application process
- completed MNRI® Core Specialist Agreement.

All new MNRI® Core-in-Training participants must follow the IPET Training Program requirements as listed on the website at www.MasgutovaMethod.com.

MNRI® IPET (Intensive Professional Educational Training): The MNRI® Core Specialist education and corresponding title is given to those professionals who accomplish such in-depth training as Neuro-Tactile Integration, Archetype Movement Integration, Re-patterning and Integration of Reflexes, and Neurostructural Reflex Integration.

MNRI® IPET Fast Track is a new possibility for MNRI® students and specialists (from local to far away states and other countries) to study at the Orlando MNRI® Training Center to get their training certification in the extensive MNRI® programs in a shorter time.

Initiating treatment for your child or loved one is not an easy route but the important point here is that improvement and healing are possible for all children and adults, physically and cognitively, that otherwise have historically had little hope. For parents passionate about doing the most they can for their children, it is a path that is do-able for them. When MNRI® is combined with love and persistence, the potential is huge. In this book alone are just a few examples of the many hundreds of children and adults who should not be able to walk but can, should not be able to talk or hold a conversation but they do, and many, many who would otherwise not have the quality of life that they have now.

The Masgutova Method® has raised the bar of possibilities of what can be done to help those with disabilities and other challenges. So jump in! It takes work but is well worth the effort!

Parents, children, and specialists: Congratulations in creating a unique system and excellent programs to provide MNRI® education and training. – Sally Averkamp

MNRI® Assessment for Determining the Level of Reflex Development

Svetlana Masgutova, Ph.D, Orlando, FL, USA;
Denis Masgutov, Warsaw, Poland, MNRI® Creators

The MNRI® program and its techniques are designed to restore neurosensorimotor development; to integrate primary movements, reflexes, coordination systems, and skills and to enable optimal functioning, development, and learning. This program uses a problem solving approach for neurodeficits through strengthening or rerouting the physiological circuits of poorly developed reflexes. It also enhances reflex pattern functioning while maximizing the motor-cognitive abilities in highly functioning individuals. The MNRI® program is aimed at activation of genetic motor memory and, using this type of memory, it optimizes and restores a normal route for the flow of information from the environment and its processing by the child/adult's nerve system. It simultaneously addresses the processes that release negative effects of stress and over or under-protection as these reflexes are also units of our defense mechanisms used for protection and survival.

Svetlana Masgutova, Ph.D

Denis Masgutov

MNRI® Assessment of Reflex Development

With a clear understanding of reflex development and its place in the overall conceptualization of the chronology of an infant's motor system, one can identify normal or abnormal reflex motor pattern expression during infancy. However, maturation and integration of reflex patterns differs among infants at different ages as they progress through interrelated phases such as:

- reflex circuit establishment (linking all its components – ascending sensory and descending motor) during the newborn period of development
- basic pattern or unconditioned reflex development (proper functioning of their automaticity) and neurological maturation
- transition of a reflex from an unconditioned to a conditioned level (incorporating experience and learning)
- maturation of a conditioned reflex pattern (some reflexes, like eye tracking for reading, need to be cortically controlled and free from being overridden by an automatic reflex like head righting).

Early assessment and correction of abnormal reflex patterns are necessary to ensure needed support for adequate future neurodevelopment. For example, the Bauer Crawling Reflex emerges at 12 weeks of gestation and should be integrated at the end of four months, postnatally. Although children do not demonstrate active crawling at four months, the unconditioned reflex must be integrated in order for it to serve as a resource at six months to begin intentional crawling – to move the limbs and whole body in a reciprocal way for physical development and towards interesting objects. This intentional movement will soon evolve into successful crawling on all fours between 7–10 months. Crawling influences the motor development that is required for conscious reaching and other movements, the infant's orientation in space and time, and multitasking and cognitive development (Doman, 1984; Delacato, 1954; Dennison, 1986; Nikitin, Nikitina, 2004; Masgutova, 2004, 2008, 2011). A delay in the initiation of early unconditioned crawling negatively affects all other phases of the Crawling Reflex. Thus, an assessment of the Bauer Crawling Reflex, followed by noninvasive training in support of the natural sequence of this reflex development, is a key element in a child's neurosensorimotor development. Matured reflexes are an essential part of a child and adult's development. Additionally, they are displayed in specific ways at different ages – not obviously seen in our movements and behavior directly, nevertheless, they are a unit of our nerve system.

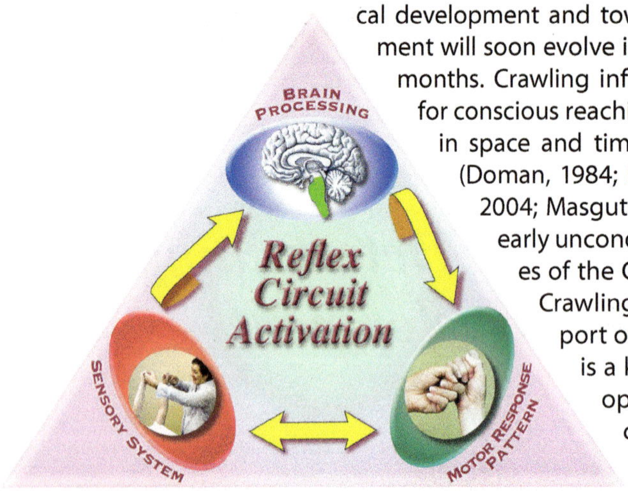

Figure 1. Depiction of a reflex circuit with the Hands Grasp Reflex.

The role of an evaluation of reflex patterns is important in cases of individuals with neurodeficits. These neurodeficits result in poor functioning circuits (the improper functioning of receptors and nerve pathways), poor electrical conductivity of nerve impulses, deficits of biochemical mediators (neurotransmitters in sensory and/or motor nerves), and an inability to regulate muscle tone and create a motor or postural responses.

Reflex Assessment Parameters

For diagnostic purposes, some investigators have classified reflex response expression as follows: normal, hypo-reflexive, hyper-reflexive, or a-reflexive (Barashnev, 2001; Cohen, 1999; Heines, 2002; Sadowska, 2001; Krasikova, 2001; Muller, 1966; Illingworth, 1978; Nelson, 1979; Akhmatova, Masgutova, 2008). Dr. Svetlana Masgutova has also created a reflex dysfunction-maturation-integration scale to assess their relative level of maturity and integration (Masgutova, 2011).

The MNRI® program uses this Assessment protocol to evaluate the level of reflex development/integration. The Assessment identifies the effects of poor reflex integration and then serves as a baseline for a program designed to improve reflex circuit functioning. An MNRI® practitioner administers specific stimuli in multiple tests for each reflex and then determines whether or not it is:

- present to an age-appropriate degree
- appropriately developed and functional
- integrated with sensorimotor functions and skills used in movement, physical development, and learning.

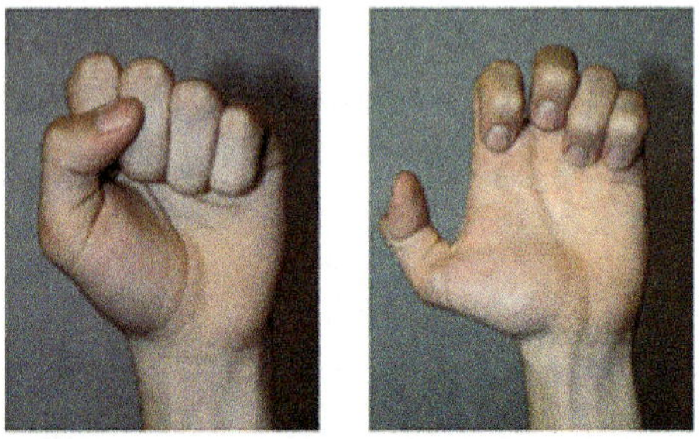

Figure 2. Example of the parameter of the Hands Grasp sensory-motor circuit and direction of the response: above: correct Grasp Reflex pattern-left (phase 2) and, transition variant-right (phase 4)

Below: incorrect Grasp Reflex pattern-left (phase 2) and, transition variant-right (phase 4)

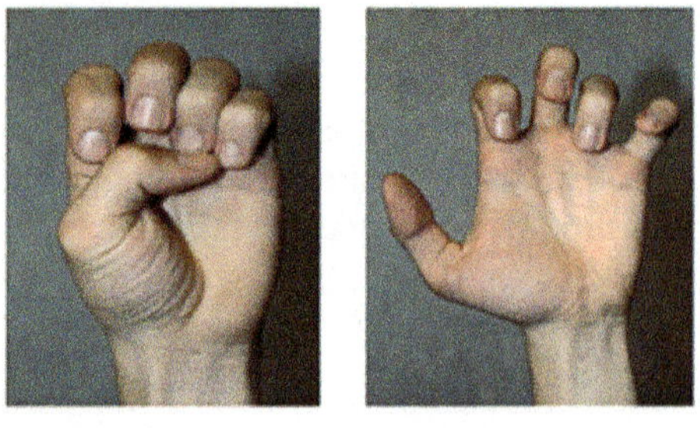

A basic reflex pattern consists of the coordinated performance of sensory-motor systems in a three-part circuit (sensory input, its processing, and motor outcome, see Figure 1). Because measuring sensory sensitivity and brain processing is not currently easily available, the Assessment procedure evaluates whether the motor response is proper or not (see Figure 2). The basic reflex pattern is evaluated in each of five parameters: integrity of the sensory-motor circuit, sequence and direction of the motor response, latency (timing and speed), intensity (muscle tone regulation), and symmetry.

• *Sensory-motor pattern of a reflex (or a circuit):* Each reflex consists of coordinated sensorimotor links within the reflex circuit (receptors, sensory neurons, synapses, and motor neurons) which are determined genetically. The motor response should be accurate or correct to the inherent encoded response pattern associated with the specific sensory stimulus (Cohen, 1999; Heines, 2002).

• *Direction of the response in a reflex:* Each reflex consists of a specific sequence of motor responses that culminate in a posture or are continued as movement in a specific direction. The muscle system coordinates these postures and movements. The direction depends on the coherent performance of alpha and gamma motor neurons.

• *Latency (timing and speed of response):* The reflex circuit (involving sensory input, brain processing, and motor response) must provide the proper speed for the nerve impulse, along the neuron axons, to achieve a motor response which should begin in a fraction of a second (107 bit/seconds) from the moment of sensory stimulation and must continue for an adequate length of time (Cohen, 1999; Heines, 2002). It must also complete its response/pattern quickly enough to fulfill its information and protective function. A delayed or slow reflex response could result in injury.

• *Intensity (strength) of a reflex response:* The intensity of a reflex is the amount of physical strength supplied by a certain system of muscles, tendons, and ligaments in response to a stimulus received by the reflex pattern. The strength of that response should reflect the intensity of the stimulus (excitation). Hyperactive, hypo-active, and a-reflexive responses are inadequate; only specific responses stand for a matured reflex pattern capable of serving as the basis for neurodevelopment and protection.

• *Symmetry:* Symmetry/asymmetry in a response can be seen in the bi-lateral response on a level of sensory-motor pattern, direction of the response, latency/timing, and also intensity of the reaction (see Figures 3 and 4).

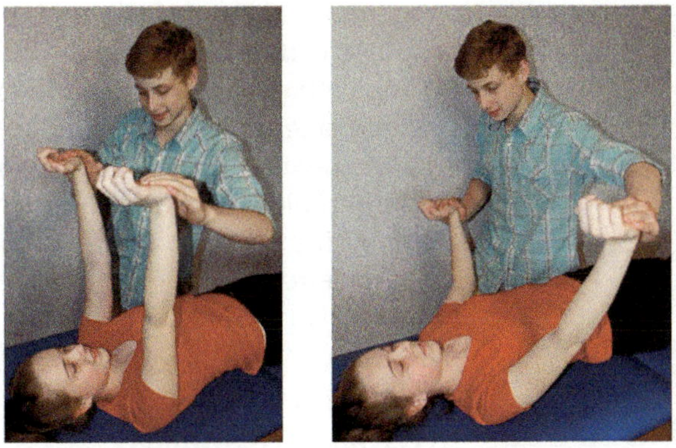

Figure 3. Direction of motion in Hands Supporting Reflex pattern: left is correct and right is incorrect.

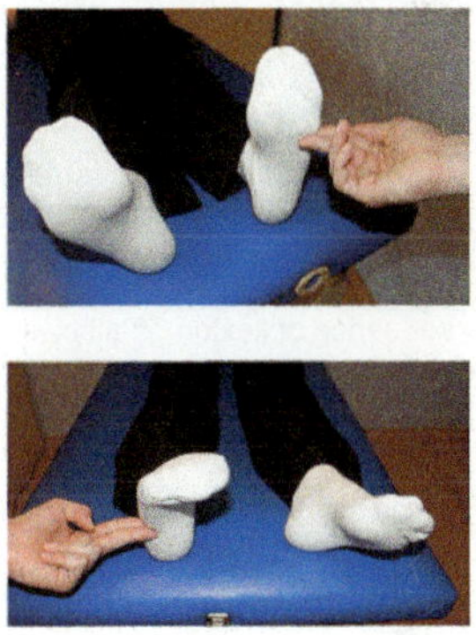

Figure 4. Asymmetry in Babinski: reaction in left foot (above) is correct, and in right foot (below) is incorrect.

Evaluating reflex pattern integration requires an in-depth knowledge of neurosensorimotor reflex development gained via specialized education and extensive clinical experience. Based on the Assessment, an individualized program is designed for the correction of dysfunctional reflex patterns and to enhance motor-cognitive development. The main goal of the individualized program is to strengthen and support physical, emotional, and cognitive growth in the individual. MNRI® practitioners demonstrate through their work that it is possible to change dysfunctional or pathological reflex patterns into higher levels of functioning that support learning. The Assessment findings are used to help children and adults gain access to the genetically given resources that support motor-cognitive functioning and maximize their developmental potential.

Because each one has its own developmental dynamic and purpose, reflex patterns often reveal the reasons for functional delays. MNRI® Assessment practices show that delayed maturation or integration of a reflex can greatly disrupt the next level of motor development and cognitive functioning. Below are examples of reflexes and the skills that may be compromised if they fail to integrate:

• Core Tendon Guard and Tonic Labyrinthine – postural control, protection mechanisms, which also influences visual-auditory processing, regulation of the HPA-stress axis (hypothalamus-pituitary gland-adrenal glands)

• Asymmetrical Tonic Neck – formation of antigravity and, within that, the exploration of acceleration and inhibition mechanisms used while our bodies move (for example, head turning to the side) which also influences monaural and binaural hearing, language and auditory processing, auditory memory, and spatial orientation

• Symmetrical Tonic Neck and Trunk Extension – postural control of the head and trunk which also influences eye tracking, visual convergence-divergence, binocular vision and binaural hearing, attention, reading and writing

• Spinal Galant – spine motion control and spinal column development which also influences cross-lateral coordination, hyperactivity, auditory processing, and focusing and awareness

• Spinal Perez – spine motion control and spine lordosis-kyphosis regulation which also influences reaching and brain detoxification

• Visual Reflexes – Horizontal, Vertical Eye Tracking which also influences Convergence-Divergence, Ocular-Vestibular, Optokinetic, Visual Fear Paralysis (eyes freezing/dilating), visual recognition, comparison, analyses and prioritizing, reading (fast and with comprehension), and writing

• Robinson Hands Grasp, Babkin Palmomental, and Hands Pulling – influence hand-eye coordination, manual skills, including writing and drawing, and expressive language

• Sequential Fingers Opening and Closing – supination and pronation movements of the wrist which also influences spatial orientation for hands, differentiation, calculation, and other mathematical skills.

Table 1. See also the description of parameters and features of the Hands Pulling Reflex pattern in the *Computerized MNRI® Assessment in the Paradigm of Modern Technology* article by E. Akhmatov and Moro and Fear Paralysis in the article, *MNRI® Reflex Integration for Post-Trauma Survival and Recovery* of this book).

Table 1. *Hands Pulling – Parameters and Features*

1. Reflex Circuit (sensory-motor aspect):

#	*Features*	*Functions* properly (+)/ not (-)	Not clear (+/-)
1.1	Motor response is in norm and corresponds to a sensory-proprioceptive stimulus (and age)		
1.2	Stimulus triggers the flexion of elbows in priority (and is coordinated with visual convergence)		
1.3	Proper sequence of flexion of: elbows, next head and then the core		
1.4	Ability to extend the spine when reaching a sitting up position (and is coordinated with visual divergence)		

2. Direction of the response in a reflex pattern:

#	*Features*	*Functions* properly (+)/ not (-)	Not clear (+/-)
2.1	Direction of response is flexion and in norm and corresponds to a stimulus (and age)		
2.2	Flexion of the elbows causes sequence of flexion of the head and core (and is coordinated with visual convergence)		
2.3	Abduction/adduction of arms functions properly		
2.4	Ability to extend the spine when reaching a sitting up position (and is coordinated with visual divergence)		

3. Intensity of the response (muscle tone regulation):

#	*Features*	*Functions* properly (+)/ not (-)	Not clear (+/-)
3.1	Strength of response is normal and corresponds to a stimulus and age		
3.2	Flexion of the elbows causes sequence of flexion of the head, first, and then, the core		
3.3	Contraction of leading muscles show appropriate motor activity		
3.4	Ability to keep strength of muscles proper during every step of sequence of flexion of elbows, head and core with ability to extend the spine when reaching a sitting up position		

Application of the MNRI® Assessment (Examples)

Assessment Procedure (Parameters and Features, see Table 1) Every reflex pattern is evaluated according to five main parameters and four features in each parameter (20 points are possible). At left is an example of parameters and features of Hands Pulling for a sample evaluation. The functioning of the features (proper or not proper) shows the level of the development of the reflex pattern and also allows the therapist to create an individual program for the individual.

Interpreting the MNRI® Reflex Parameters

In the MNRI® Assessment Reflex Parameters quantitative analysis evaluates the expression of basic reflex patterns in terms of their functionality. A functional reflex generates an automatic protective response when needed and also serves as a foundation for further neurosensorimotor development. The scoring for each reflex is based on evaluation of the five parameters described above: pattern, direction, intensity, latency (timing/speed), and symmetry. Each parameter can earn a score from 0 to 4 as a result of appropriate functioning of four features (e.g., the features of a reflex circuit in Hands Pulling are: 1) adequacy of motor response to a sensory-proprioceptive stimuli; 2) stimuli triggers the flexion of the elbow in priority (and is coordinated with visual convergence); 3) proper sequence of flexion of: elbows, next head, and then the core; 4) ability to extend the spine when reaching a sitting up position (and is coordinated with visual divergence; see Table 2). The combined number of points earned for each feature parameter (1±1±1±1) becomes the measure of functionality for that particular reflex. Each reflex earns a total score between 0 and 20 (5 parameters and 4 features in each), placing it on a continuum between severe pathology and a high level of integration.

4. Latency of the response (timing):

#	Features	Functions: properly (+)/ not (-)	Functions: Not clear (+/-)
4.1	Motor response emerges timely (normal and corresponds to a stimulus, and age)		
4.2	Proper timing of response in the sequence: 1) flexion of the elbows, 2) flexion of the head, and 3) flexion of the core		
4.3	Stability in flow of the response – response stays active till stimulus is present		
4.4	Completion of response when reaching the sitting position		

5. Symmetry in response:

#	Features	Functions: properly (+)/ not (-)	Functions: Not (+/-) clear
5.1	Circuit is displayed symmetrically for both arms		
5.2	Direction of response and sequence of flexion: 1) elbows, 2) head and 3) core is symmetrical for both arms		
5.3	Strength of motor response is symmetrical for both arms		
5.4	Latency (timing) of motor response is symmetrical for both arms		

Table 1 (continued)

Criteria for individual features of basic reflex pattern evaluation scores (see Table 2 below):

4 = Appropriate – all four features are correct
3 = Three features are correct, one is incorrect
2 = Two features are correct, two are incorrect
1 = Dysfunctional – one feature is correct, three are incorrect
0 = Pathological – all four features are incorrect

Table 2. Main Parameters of the Reflex Pattern(s) Test(s)*

Features	Pattern Before	Pattern After	Direction Before	Direction After	Strength Before	Strength After	Timing Before	Timing After	Symmetry Before	Symmetry After
1	–	–	–	+	–	–	–	+	–	–
2	–	–	–	–	–	–	–	–	+	+
3	–	+	+	+	–	+	+	+	+	+
4	+	+	+	+	–	–	+	+	–	–
Score	1	2	2	3	0	1	2	3	2	2

Table 2. Basic reflex pattern evaluation scores.

An Assessment of reflex patterns before and after receiving MNRI® intervention (usually administered every 6 months) allows us to see the progress in the reflex pattern development. An example Assessment of the main parameters of a basic reflex pattern evaluation is given in Tables 5 and 6.

The example in Table 6 shows the Assessment of the Hands Pulling Reflex pattern of a child with cerebral palsy. The total score at pre-assessment is 7 (incorrect pattern – average dysfunction of a basic reflex pattern), and at post assessment is 11 (the pattern is on the boundary of dysfunctional and normal development – elements of right pattern). Usually, this scoring is done on each side/limb individually and then the average of the points is tallied and identified according to the criteria found on the basis of research done with children in different groups of challenged and neurotypical children (see Table 3 & Figure 5). These criteria for determining the level of reflex development/functioning allow us to determine the levels of pathology, dysfunction, development, and integration.

Level of function/integration		Level of dysfunction/pathology	
Points	**Definition**	**Points**	**Definition**
20.00 and over	High level of integration of basic reflex pattern and its variants	8.00-9.75	Incorrect pattern with some correct features. Mild dysfunction.
18.00-19.75	Matured level of a basic reflex. Functions on a higher level than average	6.00-7.75	Incorrect, dysfunctional.
16.00-17.75	Average development. Basic pattern support the development of variant patterns	4.00-5.75	Incorrect. Deep dysfunction.
14.00-15.75	Functional, but at a low level of development	2.00-3.75	Incorrect pattern. Pathology.
12.00-13.75	Basic reflex pattern is functional at a very low level of development	0.00-1.75	Incorrect pattern. Severe pathology.
10.00-11.75	On the boundary between normal function and dysfunction. Elements of correct pattern		

Table 3. Criteria of development/functioning of a reflex pattern

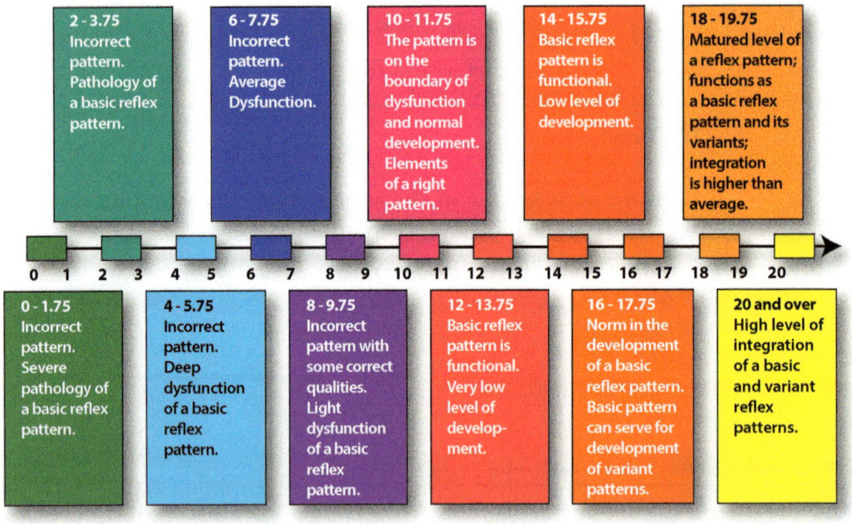

Figure 5. Levels of Reflex Pattern Integration

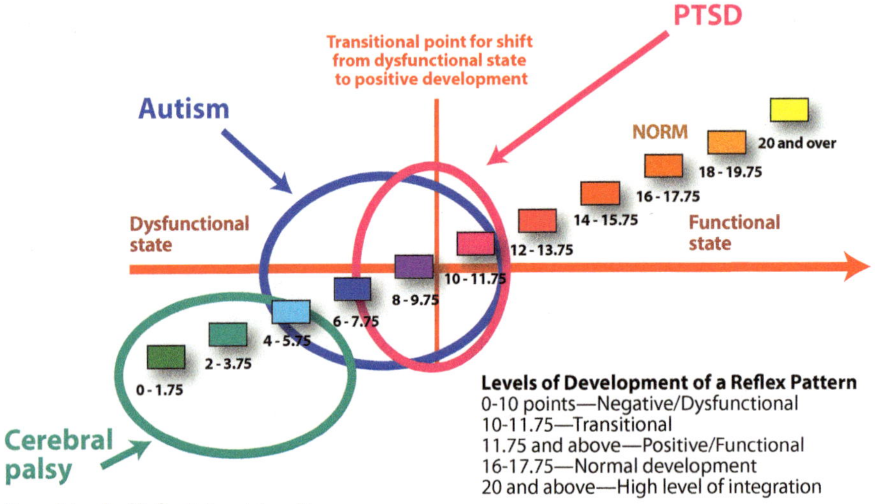

Figure 6. Levels of Reflex Pattern Integration

The MNRI® Assessment program can evaluate up to 30 or more reflexes of gross-motor/core and limb coordination which can be carried out altogether or chosen according to specific research or therapy program goals. This Assessment is approached in three related groups according to 1) the sagittal, 2) horizontal, and 3) dorsal planes of the body and movements.

Statistic analysis allows us to calculate the number of reflex patterns that are dysfunctional/pathological or in the norm; and, when determined to be 35% or more reflexes are dysfunctional, to presuppose that a phenomenon of 'Reflex Integration Disorder' is present (RID; Masgutova, 2011). Analysis of reflex profiles of children from different groups with challenges shows that RID is typical for children with autism (3,700 assessed individuals) as 86.7% of their reflex patterns are dysfunctional. Assessments of children (2,560 individuals) with post-traumatic stress show that 46.6% of their reflex patterns are dysfunctional. Knowledge of the state of a person's reflex functioning can be key in choosing a professional facilitation strategy (Masgutova, Akhmatova, 2008).

Reflexes and their dynamics are also interpreted in Assessment reports within 9 clusters according to their effect on different sensory systems, behav-

Clusters	Functional Area	Reflex Patterns
Cluster 1	Upper Limb Reflex Patterns	Hands Grasp, Hands Pulling, Babkin Palmomental, Hands Supporting, Upper Core Tendon Guard
Cluster 2	Tonic Reflex Patterns	Core Tendon Guard, TLR, ATNR, STNR, Abdominal, Spinal Galant, Spinal Perez
Cluster 3	Righting Reflex Patterns	Trunk Extension, Head Righting, Spinal Perez, Landau, Foot Tendon Guard, Hands Pulling, Hands Supporting, Leg Cross Flexion-Extension, Thomas Automatic Gait, Spinal Galant, Flying and Landing, Segmental Rolling, Spinning, Locomotion, Balancing
Cluster 4	Lower Limb Reflex Patterns	Babinski, Foot Grasp, Foot Tendon Guard, Leg Cross Flexion-Extension, Thomas Automatic Gait, Bauer Crawling, Trunk Extension, Core Tendon Guard, Grounding, Flying & Landing
Cluster 5	Gross Motor Reflex Patterns	Core Tendon Guard, STNR, Trunk Extension, Thomas Automatic Gait, Moro Embrace, Fear Paralysis, Spinal Galant, Spinal Perez, ATNR, Segmental Rolling, Spinning, Locomotion, Flying& Landing, Landau, Grounding, Hands Pulling, Hands Supporting
Cluster 6	Oral-Facial /Visual and Auditory Reflex Patterns	ATNR, STNR - Convergence-Divergence, Babkin Palmomental, Hands Grasp, Hands Pulling
Cluster 7	Protection and Survival Reflex Patterns	Core Tendon Guard, Fear Paralysis, Moro Embrace, Hands Supporting, Hands Grasp, Bonding
Cluster 8	Curiosity/Cognition Supporting Reflex Patterns	Pavlov Orientation, Core Tendon Guard, ATNR, STNR, Babkin Palmomental, Hands Grasp, Hands Pulling, Bauer Crawling, Leg Cross Flexion- Extension, Landau, Flying & Landing, Segmental Rolling, Spinning, Bonding
Cluster 9	Emotional Stability, Maturation Supporting Reflex Patterns	Bonding, Fear Paralysis, Moro Embrace, Core Tendon Guard, Pavlov Orientation, Babkin Palmomental, Hands Grasp, Hands Supporting, Landau, Flying& Landing, Segmental Rolling, Spinning

Table 4. Clusters of reflex patterns serving specific areas of an individual's functioning

#	*Reflex Title*	**Table 5.** *Parameter of Reflex Pattern*														
		Sensory-motor circuit			*Direction of movement*			*Intensity (muscle tone)*			*Latency*			*Symmetry*		
		Norm	*Hyper*	*Hypo*	+	-	+/-	*Norm*	*Hyper*	*Hypo*	*Norm*	*Hyper*	*Hypo*	+	-	+/-
1	*Hands Grasp*			√			√			√			√		√	
2	*Hands Pulling*			√	√					√			√		√	
3	*Hands Supporting*			√		√				√			√		√	
4	*Babkin Palmomental*			√			√			√			√		√	
5	*Core Tendon Guard*			√			√			√			√		√	
	Total %	0	0	100	20	20	60	0	0	100	0	0	100	0	100	0

Table 5. Cluster 1. Upper Limb reflex patterns (example of the pre-assessment of S.D., a 5-year old child with CP, hypotonia).

ioral and emotional patterns, their protective role, and the affect on cognitive functions (see Table 4). Summarizing the results within clusters can help to identify some general tendencies of reflex patterns that impact different aspects of an individual's functioning (also a statistical analysis is used according to NewKrefft algorithm for determining if the change is significant – A. Krefft, 2007; see *Computerized MNRI® Assessment in the Paradigm of Modern Technology* article by E. Akhmatov in this book). Tables 5 and 6 are examples of the information processing on a separate cluster done by a computerized system.

Figure 7 of this Assessment report shows the dynamics of changes in a cluster of reflex patterns after the MNRI® program compared to its pre-assessment. In this example, the progressive dynamic in development of each parameter is shown. Significant changes in parameters, such as the circuit, direction, and symmetry allow us to conclude that the program was effective as these reflex parameters are especially difficult to change. This also allows us to create the best individualized strategy for positive growth for a person in their future MNRI® therapy sessions.

#	Reflex Title	Table 6. Parameter of Reflex Pattern														
		Sensory-motor circuit			Direction of movement			Intensity			Latency			Symmetry		
		Norm	Hyper	Hypo	+	-	+/-	Norm	Hyper	Hypo	Norm	Hyper	Hypo	+	-	+/-
1	Hands Grasp	√			√					√			√		√	
2	Hands Pulling	√			√					√			√	√		
3	Hands Supporting			√			√			√			√		√	
4	Babkin Palmomental			√			√			√			√		√	
5	Core Tendon Guard			√	√				√			√			√	
	Total %	40	0	60	60	0	40	0	20	80	0	20	80	20	80	0

Table 6. Cluster 1. Upper Limb reflex patterns (example of Post-assessment of S.D., the same 5½ year old child with CP, hypotonia in 6 months).

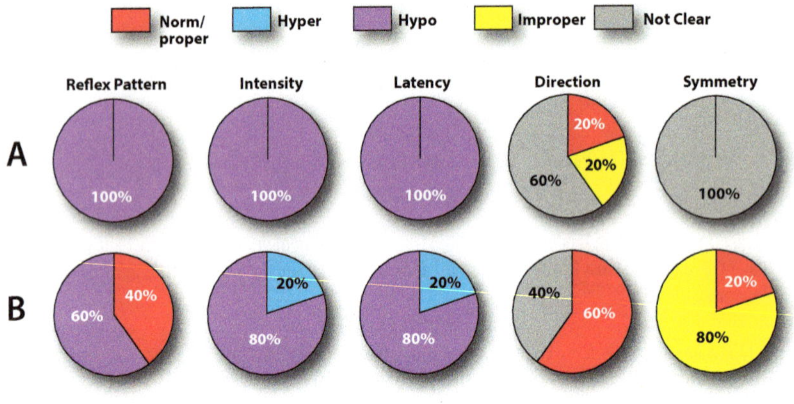

Figure 7. The information from Tables 5 & 6 given here in graph form. Table 5 is shown as (A) and Table 6 is shown as (B).

Conclusion

The MNRI® program was created for the facilitation of the growth potential in children and adults with challenges such as cerebral palsy, autism, attention deficits – ADD and ADHD, dyslexia and hyperlexia, genetic syndromes, developmental delays, fetal alcohol syndrome, PTSD, and many other challenges. An Assessment also can be used as a tool to define the progress or results of therapy sessions using MNRI® or any other method used by an individual.

The MNRI® Assessment allows for qualitative and quantitative analysis. The quantitative analysis is oriented on measuring the specific parameters, such as the reflex circuit, direction and sequence of movements in response, intensity, timing and symmetry, which altogether allows one to create a specific and individual oriented therapy program. An Assessment is suggested every 6 months for individuals going through the therapy process to evaluate changes and to change the therapeutic program or strategy, if needed. The results of MNRI® for individuals with challenges emphasizes strongly the importance of appropriate corrective procedures for dysfunctions on the level of primary movements and reflexes. They demonstrate that reflexes are the units of the sensory, motor, and proprioceptive systems influencing health, learning, and development. The MNRI® program proposes new opportunities arising from the use of natural resources, which awaken the self-regulating process through reflex integration. It is offered as a strong support for the creation of new developmental possibilities and programs for children and adults. The reflex integration methods involve natural, non-invasive movements that can be easily learned by parents of challenged children or adults and by professionals who work with challenged individuals. These techniques also do not require a lot of external resources.

Evaluation of the reflex pattern is not intended to be the basis for labeling any child or adult because of neurological symptoms, physical or mental disability, age, and other factors. It serves exclusively to identify goals for future support of the individual's natural, genetic motor programs and as the basis for further development of motor-cognitive functioning.

With the MNRI® concept, it is possible to change dysfunctional or pathological reflex patterns to function better for development and learning. The information on reflex pattern evaluations gives insight to the highest potential of the person's development for the realization of their natural innate resources and possibilities.

References

Barashniev, J. I. (2001). *Prenatal neurology.* (Russ.). Moscow, Russia: Triada-X. 640 p.

Child neurology. (2000). Wrocław, PL: Wydawnictwo Medyczne - Urban & Partner.

Cohen, H. (1999). *Neuroscience for rehabilitation.* 2nd Edition. Philadelphia, PA, USA: Lippincott Williams and Wilkins.

Delacato, C. H. (1974). *The diagnosis and treatment of speech and reading problems.* Garden City, NY, USA: Doubleday.

Dennison, P., Dennison, G. (1989). *Brain Gym® teacher's edition.* Ventura, CA, USA: Edu-Kinesthetics Inc.

Doman, G. (1984). *How to multiply your baby's intelligence.* Garden City, NY, USA: Doubleday.

Heines, D. E. (2002). *Fundamental neuroscience.* 2nd Edition. Ny, Philadelphia, PA, USA: Churchill Livingstone.

Illingworth, R. S. (1971). *The development of the infant and young child: Normal and abnormal.* 4th ed. Edinburgh: ES Livingstone.

Krasikova, I. (2001). *Child massage and gym for prophylactics and treatment of spine disorders, scoliosis and flat foot.* (Russ.). Moscow, Russia.

Krefft, A. (2007). *Diagnosis of functions of invisible phenomena* (Statistic Mathematic Analysis). Wrocław. Poland: Oficyna Wydawnicza Politechniki Wrocławskiej.

Masgutova, S. K., Akhmatova, N. K. (2008). *Repatterning and integration of dysfunctional and pathological reflexes.* Orlando, FL, USA: SMEI.

Masgutova, S., Akhmatova, N. (2004, 2007). *Integration of dynamic and postural reflexes into the whole body movement system.* Scientific edition: Prof. N. Akhmatova. Warsaw, PL: MINK.

Masgutova, S. (2011). *Infant dynamic and postural reflexes. Neuro-sensory-motor reflex integration.* Revised and edited, scientific-practical manual. Orlando, FL, USA: SMEI.

Miller, G. A. (1966). *Psychology – The science of mental life.* Harmondsworth, UK: Penguin.

Nelson, K. B., Elleberg, J. H. (1979). Neonatal Signs as Predictors of Cerebral Palsy. *Pediatrics*, 64-68, 225-231.

Nikitin, B. P., Nikitina, L. A. (2004). *We and our children.* (Russ.). Moscow, Russia: Medicina.

Sadowska, L. (2001). *Neurokinesiological diagnosis and therapy of children with mental and motor deficits.* Wroclaw, PL: AWF.

We thank all the children and specialists helping us to create a unique Reflex Integration Assessment tool that made certain difference in comprehension of the specifics of the neurodevelopment in children and adult in various groups of challenges, including PTSD and learning difficulties, and finding most sufficient corrective program for them. We believe that many other specialists will be interested in our assessment tool and professional benefits. – Authors.

Computerized MNRI® Assessment in the Paradigm of Modern Technology

Eldon Akhmatov, MA, Computer Technology Engineer
Ph.D. Student, Moscow, Russia

Overall Approach:

Eldon Akhmatov

Nowadays information technology, with its high level of development, variability in application, and enormous possibilities can allow the creation of computerized programs to serve different aspects of the information bank and processing in the helping specialties. Electronic computers offer complicated operations to process information for an entire organization's methodologies and methods, such as assessment, registration, statistical analysis, and various other applications. All these new possibilities were a dream for specialists of helping professions some 30 years ago. In this report, new technological solutions and possibilities to support the Masgutova Neurosensorimotor Reflex Integration (MNRI®) tools in the area of reflex pattern development evaluation is proposed. The MNRI® program created by Dr. Svetlana Masgutova contains two parts: 1) the Reflex Assessment oriented to the evaluation of the level of development of 30 dynamic and postural reflex patterns, and 2) therapeutic programs called reflex repatterning/integration aimed at practical application of neurosensorimotor functions for children/adults with various developmental challenges (Masgutova, 2010, 2011).

The basic goal of the MNRI® program is the support of neurodevelopmental mechanisms and processes in children or adults through facilitation of their primary motor reflex patterns while simultaneously assuring the natural and genetically given brainstem defense strategy. MNRI® programs develop the concept of integration or sensory-motor coordination of a damaged or dysfunctional reflex circuit versus the traditional concept of "inhibition" and "extinction" of reflexes that are "non-reduced" by a specific developmental age and are retained (Semionova, 1999). The MNRI® program is based on the concept by L. Vygotsky (1986) and J. Piaget (1976) concerning the role of the child's primary motor system in formation of higher cognitive functions during childhood – and on into adulthood.

MNRI® therapeutic procedures are designed for work with Reflex Integration Disorder (RID) and directed at reestablishing or 're-patterning' dysfunctional and pathological reflex patterns. They include specific techniques and motor activities similar to basic reflex patterns in order to stimulate the neural connections of the sensory (afferent) and motor (efferent) pathways of the nerve system that support the inborn mechanisms of brain functions and its neuroplasticity (Masgutova, 2005; Masgutova, Regner, 2008).

This program works with a variety of neurodeficits and learning dysfunctions, such as the Autism Spectrum Disorders, cerebral palsy (CP), genetic disorders, Post-Traumatic Stress Disorder, dyspraxia, dyscalculia, dysgraphia, etc. The program is directed toward recognition of damage to reflex circuits and creating a rapid and effective manner for MNRI® rehabilitation intervention.

The evaluation of the clinical condition of a patient with motor neuro-disorders and diseases in traditional medicine is usually based on examination of motor programming and motor control and partly of reflexes. The diagnoses, due to the vast impact of the human factor, are more subjective, and the results are often dependent on the skills of the person performing the evaluation. The attitude of the researcher, expecting to obtain a particular result, may increase the subjectivity of the evaluation. (Oswald, Grosjean, 2004; Lilienfeld, Ammirati, Landfield, 2009).

About Methods

There is no practical and timely method to gather data from the evaluation that is accepted and widely used. Because there is not a consistent method for data gathering, it is difficult to compare results obtained by different researchers in different settings.

Thus, it would be very useful to develop a method/tool to measure the state of reflex development from an objective level of the actual clinical condition of patients and as a measure of the effectiveness of the performed rehabilitation. The analysis of the MNRI® Assessment tool is based on the highest level of objectivity of the researcher, as the focus of the evaluation is very detailed and oriented on the evidence of automatic and non-controlled responses of specific stimuli.

The development of this tool/method also assists in the gathering of other necessary information for use in reporting and the development of various programs and methods used in the therapy of children with neurodevelopmental deficits and neurological disorders.

Goal of the Report:

This report presents a brief description of the computerized support for the MNRI® Reflex Assessment called NEW-MNRI® Reflex Assessment (New-MNRI-RA). This computerized version of the Assessment is a new level of the MNRI® program for the mathematical analysis, processing of the data concerning different parameters and features of the reflex patterns, summarizing the evaluation information, and allowing the formation of reports for evaluation of children/adults for the trademarked MNRI® sub-programs by Dr. S. Masgutova.

The goals for creating the computerized version of the Reflex Assessment are 1) to find new technological solutions to record multiple data, 2) to create improvements in 'the technical environment' to simplify the manual, monotonous work and 'mental overload', 3) to find ways to make working procedures easy and interesting for users, while guaranteeing precision, accuracy and errorless recording of given data, and 4) to economize the time needed to calculate, process, and summarize the gathered data.

The development of this new version of the Assessment program will allow a computerized analysis to determine possible incidence of RID in individuals, having evidenced poorly developed and/or dysfunctional state of at least 1/3 of reflex parameters.

The Objective Basis for Creating a Computerized Version of the MNRI® Reflex Assessment

1) The existing MNRI® Assessment program presents a unique tool for evaluation of reflex pattern development: its level, functional state, and specifics, for the creation of a rehabilitation program for a MNRI® specialist and a Home Program for parents or caregivers (Masgutova, 2010, 2011).

2) The validity of the Assessment tool went through testing trials several times using mathematical statistical analysis by Prof. Anna Krefft (Krefft, 2007), and the results were discussed and published in several works (Masgutova, Akhmatova, Shubina, Kiselevsky, 2008; Masgutova, Regner, 2008; Masgutova, Sadowska et al., 2008; Pilecki, Masgutova et al., 2012).

3) The validity of the Assessment tool was tested through objective studies before and after rehabilitation procedures of the neurosensorimotor reflex integration – EEGR and brainstem auditory evoked potential, and an immunology study presented in several publications (Masgutova, Regner, 2008; Pilecki, Masgutova, et al., 2012).

Based on approved validity of the MNRI® Assessment and evidences of significant therapeutic effects, the author undertook the creation of this computerized version of the New-MNRI-RA program.

Brief Description of the MNRI® Assessment Program

MNRI® Assessment procedure evaluates every reflex pattern from the point of view of its components or parameters, specific features of the reflex circuit components, and their biomechanics. The correspondence to age and neurological norm is included in a statistical and quality analysis.

Every reflex pattern is evaluated according to five main parameters and four features in each parameter (20 points are possible), which makes the evaluation tool more exact and valid in creating objectivity for the entire testing procedure and every component. The example of evaluation of reflex parameters and features can be demonstrated by description of the following evaluation of the Robinson Hands Grasp Reflex. The functioning of the features (proper or not proper) shows the actual level of development for a reflex pattern (Masgutova, 2009, 2010, 2011).

Robinson Hands Grasp

Parameter 1. Reflex Circuit (sensory-motor coordination):

Features (evaluated as 'proper' – 1 point, 'not proper' – 0 point)

1.1 Sensory sensitivity (corresponds to age and neurodevelopmental norms)

1.2 Activity of the motor response (corresponds to age and neurodevelopmental norms)

1.3 Correct biomechanics:

1) Grasp phase 2 – response comes from the middle of the palm with flexion and closing of all fingers in fist evenly;

2) Thumb is located on closed palm in between the index and middle fingers

1.4 Elbow is extended with a possibility for absorption (no overextension or over-flexion)

Parameter 2. Direction of the motor response in a reflex/reflex pattern:

Features (evaluated as 'proper' – 1 point, 'not proper' – 0 point)

1.1 Grasp from the middle of the palm with all fingers closing in fist evenly in phase 2

1.2 Thumb is located on closed palm in between the index and middle fingers

1.3 The arms can keep the position of 90 to 43 degrees as directed toward the head

1.4 Arm moves in front of the body core – elbows are extended with a possibility for absorption (there is no overextension with rotation of palms)

Parameter 3. Intensity of the motor response (muscle tone regulation):

Features (evaluated as 'proper' – 1 point, 'not proper' – 0 point)

1.1 Strength of motor response (corresponds to age and neurodevelopmental norms)

1.2 The grasp motion comes from whole palm and fingers and is strong and adequate for intensity of sensory stimulus (muscle tone regulation is proper)

1.3 Thumb, index, and middle fingers are stronger in closing the palm (than the ring and small fingers)

1.4 The palm grasp motion lets go/decreases adequately when the intensity of the sensory stimulus decreases (the person lets go/stops intense grasp accordingly when stimulus stops being active/intense).

Parameter 4. Latency of the motor response (timing):

Features (evaluated as 'proper' – 1 point, 'not proper' – 0 point)

1.1 Motor response emerges in a timely manner

1.2 The emergence of the palm grasp motion is timely and adequate to the sensory stimulus presentation

1.3 Palm grasp motion lasts for some seconds adequately to the length/presence and intensity of sensory stimulus (the person holds the grasp for a bit longer time accordingly to the prolongation of the time of stimulation)

1.4 Palm grasp motion lets go/decreases adequately when the intensity of the sensory stimulus decreases (the person lets go/stops intense grasp accordingly when stimulus stops being active/intense.

Parameter 5. Symmetry in response:

Features (evaluated as 'proper' – 1 point, 'not proper' – 0 point)

1.1 Circuit: circuit displayed is the same pattern for both sides

1.2 Direction: grasp from the middle of the palm with all fingers closing in fist evenly in phase -2 - symmetrical for both sides

1.3 Intensity: the grasp motion comes from both palms and all fingers with the same and adequate strength in phase -2 symmetrical for both sides (muscle tone regulation is proper)

1.4 Latency: the emergence of the grasp motion in both hands is timely and adequate to sensory stimulus presentation, and simultaneous and symmetrical for both sides

Analysis of the work applied by the MNRI® team to record results manually for processing of the information demonstrates the complexity, uniqueness, and manpower hours (6-7) per client. These elements take much effort and time, and significantly increase the time it takes for clients to receive the Assessment results. Thus, creating a new computerized program means structural improvement and time savings for both the recorder of the Assessment and for the information to be received by the client. The development of a new computerized program also increases the possibilities of gaining a wider picture and interpretation of obtained results which can lead to a greater understanding of the nature of dysfunctions in children and adults and help provide statistical information on the actual occurrence of RID.

Introduction to the New-MNRI®-RA

The New-MNRI®-RA program has been created for the following basic functions:

• developing a research base of clients/patients with RID

• the analysis of results of the reflex Assessment based on parameters and features of every reflex pattern

• development of reports in Text Editor (Microsoft® Word) in three formats

• construction of various therapeutic programs for conferences and clinics which can be duplicated and shared with all therapists and other professionals as needed.

As a testing tool:

• establishing the state of the development of 30 reflex patterns (with possibility to expand this number)

• scoring the reflex evaluation using the mathematical expression in points from '0' to '20' as proposed and tested by Prof. Anna Kreft (Krefft, 2007; Masgutova, Sadowska et al., 2008; Pilecki, Masgutova et al., 2012)

• gathering data on the state of the RID (as defined as at least 1/3 of dysfunctional patterns and their features of the total 30 reflex patterns) in mathematical statistic expression

• describing the dynamics of change in the condition of the reflex patterns in clients

• providing a written format for the Home Program recommendations in a user-friendly educational form for clients to continue the MNRI® therapeutic program at home

• preservation and loading of the data

• importing the entire data (or its separate components) in an Microsoft® Excel version to allow the development of needed reports based on selected criteria

• providing voice-activated control functions so one evaluator will be able to complete the Assessment, thus reducing the number of trained specialists needed.

Features of the program:

The program provides:

• intuitively clear, convenient interface, and modern design

• automatic saving to an off site server to eliminate possible damage of information data files and easy recovery

• encrypted information files that cannot be read without the software to ensure safety of the confidential information

• the possibility to control the whole system against tester/human errors.

Directions for development of the software are:

• creation of servers for maintenance of synchronous work of several parallel running Assessments and report copies of the program on various points of the globe

• integration of process equipment with the software for the purpose of all-around automation of the process of diagnostics of patients.

The New-MNRI®-RA program consists of the following components:

• system of authorization providing control of accounts of those using the system and is controlled by the level of need to use specific programs; an informational section to provide consistent gathering of necessary data parts containing the basic information on patients/clients

• the ability to use the information from the reflex Assessment to generate reports to build Home Programs and schedules, and to analyze the information on the features and parameters of reflex patterns for further research projects

• the system includes the most up-to-date voice-activated controls allowing control of the data base of patients, algorithms for encrypting, and other tools necessary for the use of gathered data.

The minimum system requirements:

- an operating system: Windows Me - Windows 8 (all Windows operation system)
- the screen resolution: 800x600
- clock frequency of the processor: 400 MHz.

Further Development of MNRI® Assessment Tool – Three Dimensional EMG and BTS Type Technology for Reflex Evaluation

The program is compatible with all modern requirements and is able to be used in all operating systems. The method of mathematical analysis according to Prof. Anna Krefft algorithm was used in order to evaluate the synthesizing of levels of reflex pattern integration, called "synthesized diagnostic function Z" according to New Krefft method (Krefft, 2007). In order to check the result on a scientific statistical level, primarily if the result relates to whole populations of individuals, statistical tests have been used.

Three models of diagnostic functions for individual groups of reflex patterns in three 'body-movement planes' have been introduced, determining the level of integration before and after the MNRI® program.

It is traditionally known that the level of pathology of physiological functioning of a child with Down syndrome, for example, is determined by a doctor on the basis of genetic analysis and, next, on the basis of evaluation of synthesized factors of diagnostic characteristics. Variables describing qualities are called synthesizing variables (complex/qualities). Traditionally, obtaining a sufficient exact diagnosis based on synthesized factors (diagnostic features) is a very challenging task, and sometimes even impossible. This method of statistic research can be used, not only in medicine, but also in economics and many sciences of the humanities.

The static model of the synthe-

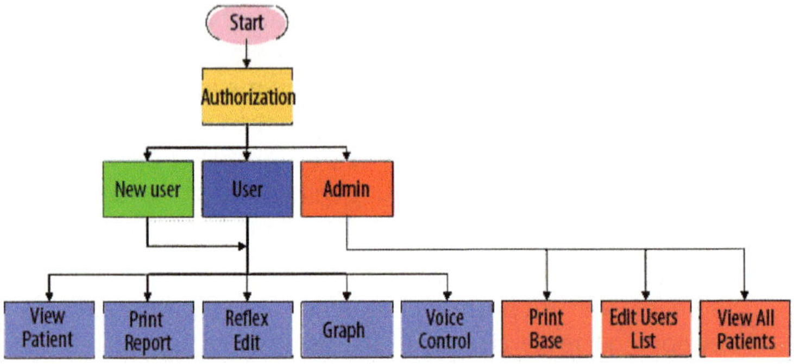

Figure 1. Structure schema of the NEW-MNRI®-RA program.

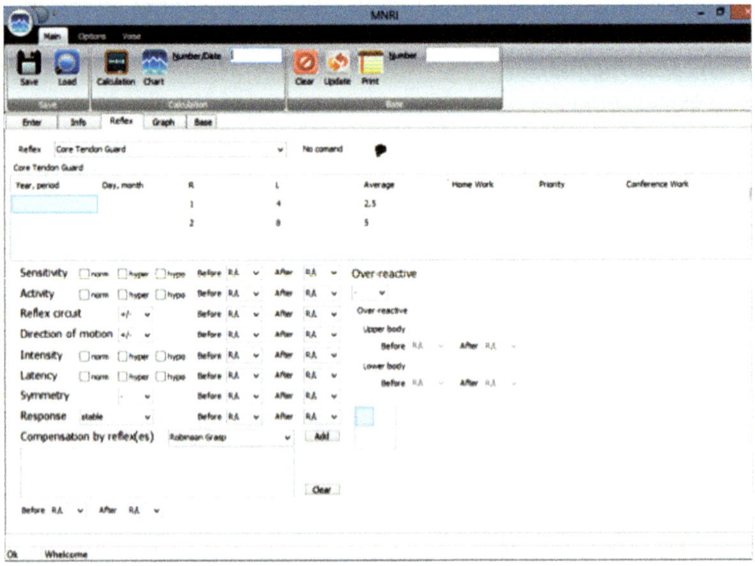

Figure 2. Reflex Assessment registration form.

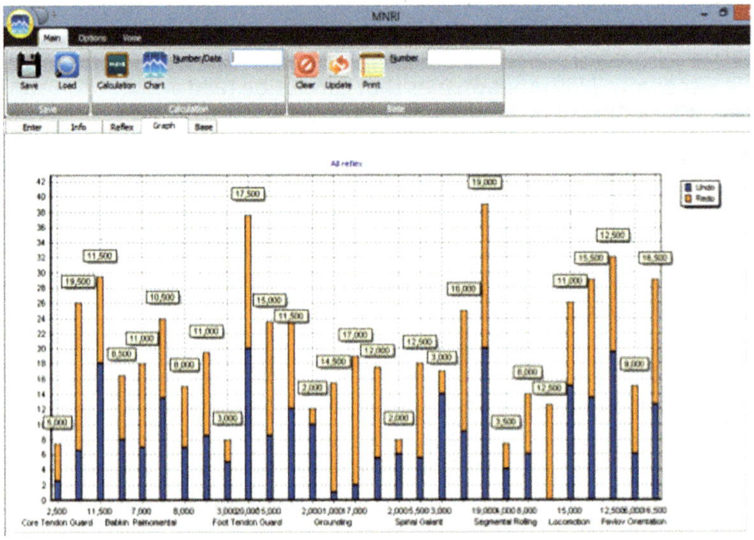

Figure 3. Reflex Assessment results in points by a graph expression.

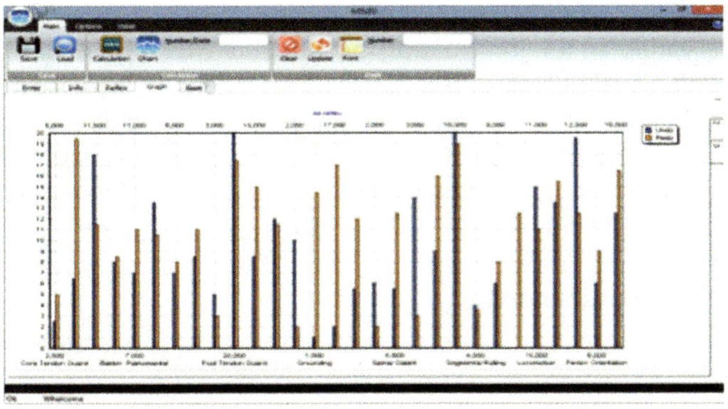

Figure 4. Reflex Assessment results in points by a graph expression (variant)

Figure 5. Authorization into the program.

sized variable that has been used in this article, is a linear model or can be adapted to be a linear one. To make the considerations easier, we are establishing that it is linear.

Assumptions like that do not influence the general (overall) deliberations. This model can be expressed in the formula, as follows:

$$Y = \beta_1 X_1 + \beta_2 X_2 + \ldots + \beta_K X_K + \beta_0 \qquad (1)$$

where: Y - is an explained variable that describes the considered synthesized variables (pathology level of a child with Down syndrome)

$X_1, X_2, \ldots, X_K$ - are explanatory variables in model (1) (diagnostic variables)

$\beta_1, \beta_2, \ldots, \beta_K$ – are model parameters (1) presenting some constant characteristics (qualities, factors) of this model.

To identify a model (1) means to determine individual values (qualities) β_i corresponding to individual parameters of this model.

If the model (1) is identified, it displays as a part of an instrument, which any moment will allow for a given set of diagnostic features $X_1, X_2, \ldots, X_K$ to determine the value of synthesized variable Y.

To set a synthesized function Z, we need to have the following experimental material:

• statistical material in the form of a matrix X dimensioned [n x k], n>k, where n is a number of simultaneous researches (tests) of diagnostic features (researches related to the same moment of correction procedures), where the researcher describes the states of synthesized variable Y, that is the variables within the model (1), the number k expresses the quantity of these variables. Generally the matrix X can be presented as follows:

$$\mathbf{X} = \begin{pmatrix} X_{11} & X_{12} & \ldots & X_{1k} \\ X_{21} & X_{22} & \ldots & X_{2k} \\ . & . & \ldots & . \\ . & . & \ldots & . \\ . & . & \ldots & . \\ X_{n1} & X_{n2} & \ldots & X_{nk} \end{pmatrix}$$

• information about so called 'directions of influence' of the corresponding individual diagnostic variables in relation to explained variable Y. For the individual variables, this information is expressed with '+' (plus) or '–' (minus). If the given variable X_j numbered 'j', j=1,2,...,k is a stimulant for a variable Y, that means an increase in variable X_j level also creates an increase in variable Y. In that case, the information about 'influence direction' in relation to variable Y is expressed with '+'. In the case where variable X_j is a negative influence to variable Y, that is, it creates a decrease of the variable Y level – the 'direction of influence' is expressed with '-'. Therefore, the information about the 'direction of influence' of the individual explanatory variables in model (1) are given in a form of a sequence with k elements in which each of them is expressed with '+' or '-'. Each of these symbols express a 'direction of influence' of variable X_j with an adequate number.

$$\boldsymbol{K} = [\underbrace{+ \; + \; - \; \dots \; +}_{k\text{ – elements}}]$$

Therefore, the experimental material [X, K] using Prof. A. Krefft's method allows us to determine a model (1).

In order to obtain a scale of value for a synthetic variable that is easy to use, variable Y is transformed into variable Z (Y→Z) :

$$Z = e^{Y} : (1 + e^{Y})^{-1}, (2)$$

Where: e – is a base of natural logarithms, Z∊ (0;1).

The closer the value of synthetic variable Z to number 1, the higher is its level.

On the base of a determined model (1) the values of three synthetic variables Z_C, Z_{LP}, Z_{GD}, Z_{PT} have been obtained for the individual research of children in terms of the chosen diagnostic variables (features). This means that every simultaneous test of diagnostic features allowed us to obtain corresponding values of synthesized variables and average values of individual functions Z_C, Z_{LG}, Z_{GD} and Z_{PT} in a group of children with Down syndrome before and after two weeks of MNRI® programs. Statistical analysis, using the ANOVA test, of children examined before and after the MNRI® program, has shown statistically significant differences on the level of $p < 0,001$.

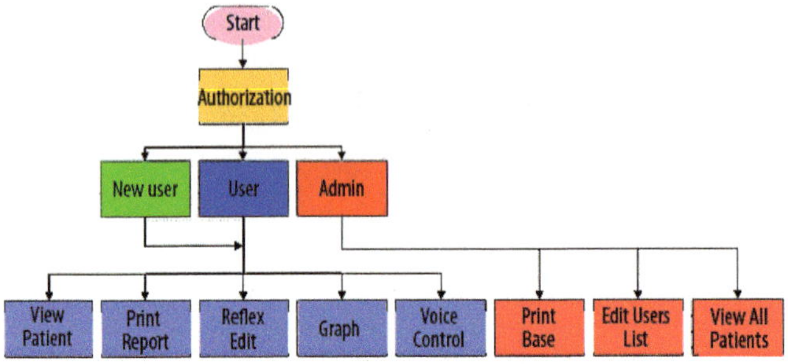

Summary

The New-MNRI®-RA Assessment program was created to direct the MNRI® evaluation and therapeutic system. This Assessment program functions at full capacity when measuring capabilities for reflex development in therapeutic-rehabilitation with individuals with various challenges and neurotypical clients. Although this program was created to become a separate educational program in the MNRI® system for the continuous education of specialists or other helping technical personnel, it can also be successfully used by other educational and therapy schools, centers, hospitals, and clinics.

References

Krefft, A. (2007). *Diagnosis of functions of invisible phenomena (Statistic Mathematic Analysis)*. Wroclaw, PL: Oficyna Wydawnicza Politechniki Wrocławskiej.

Lilienfeld, S. O., Ammirati, R., & Landfield, K. (2009). Giving debiasing away: Can psychological research on correcting cognitive errors promote human welfare? *Perspectives on Psychological Science*, 4, p. 390-398.

Masgutova, S., Regner, A. (2008). *Language development using sensory-motor integration approach*. Scientific edition: Prof. T. Galkowski, Prof. J. Mezwa, Dr. B. Dolyk. Wroclaw, PL: Continuo, MISM. 167 p.

Masgutova, S., Sadowska, L. et al. (2008, Oct. 17-18). *The MNRI® - Masgutova neurosensorimotor reflex integration program of diagnosis and therapy for children with Down syndrome*. International Scientific-Practical Conference - XXXII *National Day of Child Rehabilitation: Down syndrome. Treatment, Rehabilitation and Education*.

Masgutova, S. K., Akhmatova, N. K., Shubina, Z. H., Kiselevsky, M. V. (2008). Immunological assessment of effectiveness of reflex integration therapeutic program for chronic inflammatory respiratory diseases, *Russian Journal of Immunology*, 2 (11), 4. (Russian) p. 454-463.

Masgutova, S. (2009). *Integration of infant dynamic and postural reflexes: Neuro-sensory-motor reflex integration® method for children and adults*. Orlando, FL, USA: SMEI. 257 p.

Masgutova, S. (2010, 2011). *MNRI® dynamic and postural reflex integration*. (Revised and edited. English version.) Scientific-practical manual. Orlando, FL, USA: SMEI. 273 p.

Masgutova, S., Masgutov, D. (2004). *Dr. Svetlana Masgutova method of facial reflex integration. Techniques for facilitation of motor and speech development*. Orlando, FL, USA: SMEI. 45 p. (English version)

Masgutova, S., Akhmatova, N. (2004). *Integration of dynamic and postural reflexes into the whole body movement system*. Scientific edition: Prof. N. Akhmatova English and Polish version. Warsaw. PL: MINK.

Masgutova, S. (2005). Reflexes as the basis for the development of the nervous system and formation of motor patterns in infancy. materials of international conference: modern methods of stimulation of motor and language development. *International Kinesiology-Rehabilitation Camp for Children with Challenges*. Krynica-Zdroj, 04-05.06.2005. Warsaw, PL: MINK. p. 14-36.

Masgutova, S. (2005, 2008). *MNRI® neurosensorimotor development: visual and auditory reflexes integration. Facilitation program of development and learning for children and adults*. Warsaw, PL & Orlando, FL, USA: SMEI. 86 p.

Oswald, M. E., Grosjean, S. (2004). *"Confirmation Bias"*, in Pohl, Rüdiger F., *Cognitive illusions: A handbook on fallacies and biases in thinking, judgement and memory*, Hove, UK: Psychology Press. p.79–96.

Piaget, J. (1976). *The grasp of consciousness: action and concept in the young child*. Cambridge, MA, USA: Harvard.

Piaget, J. (1936). *Origins of intelligence in the child*. London: Routledge & Kegan Paul.

Pilecki, W., Masgutova, S., Kowalewska, J., Masgutov, D., Akhmatova, N., Sobieszczanska, M., Kalka, D. (2012). The impact of rehabilitation carried out using the masgutova neurosensorimotor reflex integration method in children with cerebral palsy on the results of brain stem auditory potential examinations. *Advances in Clinical and Experimental Medicine*. (Official Organ Wroclaw Medical University) 21, 3. p. 363–371. http://www.advances.am.wroc.pl/pdf/2012/21/3/363.pdf

Semionova, K. A. (1999). *Recovery treatment of the cerebral palsy residual state*. Moscow, Russia: Antidor.

Vygotsky, L. S. (1986). *The child psychology. The problems of child development*. In 6 Books; Book – 4. Moscow, Russ: Pedagogika (Russian).

Computerized MNRI® Reflex Assessments are oriented towards accurate and fast feedback for children and adults in need. I invite you to explore It and love it... – E. Akhmatov

Visualizing the Pathways to MNRI® Success: The Six Cs, C2C, and +iT Together Tools

Dasaratha Rama, Ph.D, Parent, Weston, Florida, USA

Dasaratha Rama

Countless parents raise children, as we raised our first child, without having to learn the intricacies of human development. With our second child we found ourselves in a completely different situation. In retrospect, it seems incredible that our first son made it to an Ivy League School without our knowing anything about the Babinski or Foot Tendon Guard Reflex patterns, the various phases of the Hand-Grasp Reflex, and many other interesting issues related to human development. Now our second son is ensuring that we are developing a strong foundation in child development from a variety of specialists!

The process of raising our second child has been like completing a gigantic jigsaw puzzle that you have to work on for many years, knowing that you only have a few pieces. You may not even have the right pieces. At least, the pieces that you have may not be the right ones for your current situation. You know the right pieces are there somewhere for you to find and fit together. To make it more challenging, different people have different pieces of the puzzle. You have to find these people, get the right puzzle pieces, and start putting the puzzle together. As with many families of children with challenges and sensory motor deficits, we have been working on this jigsaw puzzle for years. We have found many pieces and know we will probably need to find many more in the future. One of the most valuable pieces that we have found is Masgutova Neurosensorimotor Reflex Integration (MNRI®), a program created by Dr. Svetlana Masgutova, which proposes a new modality of nerve system functioning – neurosensorimotor reflex integration – to facilitate development in children and adults with different abilities.

In this article, I will not discuss MNRI® itself but share three visualization tools that we as parents can use to get the most out of MNRI® and other approaches. These tools include: (1) The +iT Together™ Canvas, (2) The Six Cs Tree, and (3) C2C™ linkages.[1] These visualization tools can be used by parents and professionals to map their MNRI® journey, and to track progress. Given the uncertainty and complexity of juggling therapies, education, and daily life, I believe that these three visualization tools can help parents chart a course and navigate their MNRI® journey with greater ease and confidence.

[1] The six Cs model was developed by Dr. Rama for The PhD Project (www.phdproject.org) but is applicable in many other contexts. The iT Together™ Canvas, the Six Cs tree, and C2C™ have been developed by Dr. Rama, and facilitate the delivery of the Six Cs model.

The +iT Together Canvas

The first visualization tool that parents can use to think about their MNRI® journey is the +iT Together canvas. The +iT Together canvas (see Figure 1) includes three main elements: (1) + represents services, experiences, resources, and ideas being added through MNRI® programs, (2) imagining (i) represents possibilities for activities and experiences that open up as the child's sensorimotor system develops, and (3) Transformation refers to changes that occur in the child's sensorimotor system, in the child's well-being, and in the family's well-being.

Figure 1 shows each of these three elements. Families can add (+) different services provided by the Svetlana Masgutova Educational Institute (SMEI) to their current mix of therapies. The '+' component includes Assessment activities, the treatment provided to clients, and the variety of concepts and techniques taught to parents for implementing their MNRI® home programs.

The 'i' (imagining possibilities) component is the less visible but powerful component of experiences such as MNRI® Family Conferences. In our experience, the value of Family Conferences goes beyond learning MNRI® techniques. One key benefit is being able to imagine new possibilities that can transform the child's and family's experiences. We have attended six family conferences since 2010. During each conference, we learned many techniques to implement in our Home Program. However, one of the most unexpected recommendations from a MNRI® Conference was to teach dance to our son. This recommendation has had a significant impact on our son and our family. Prior to the conference, I had not given this idea any thought. In fact, given our son's motor problems, I was not sure about how dance would work for him.

Our son started learning hip-hop and tap after one year of MNRI®. He also decided that he wanted to learn Bharathanatyam (classical dance from Southern India). Further, he has set himself a long-term goal of performing an 'arangetram' (a debut on-stage solo performance of a Bharathanatyam student after undertaking many years of training). He has started performing small pieces in public. He is highly motivated by dance and is very present when preparing for and presenting on stage. He also loves to attend classical dance performances in the US and in India. These new interests and activities have expanded his experiences and interactions with others. Dance has been a transformative force in our lives, one that we would not have discovered without MNRI®!

The key point is that MNRI® professionals have an important role to play in helping parents imagine possibilities. The changes in the child's sensorimotor system can be gradual and the significance of these changes may not be apparent to an untrained observer. By guiding parents to imagine possibilities, and to reflect on transformations in the child's sensorimotor system, MNRI® specialists can strengthen the pathways to parent success with MNRI® services.

Figure 1. The +iT Together Canvas for SMEI

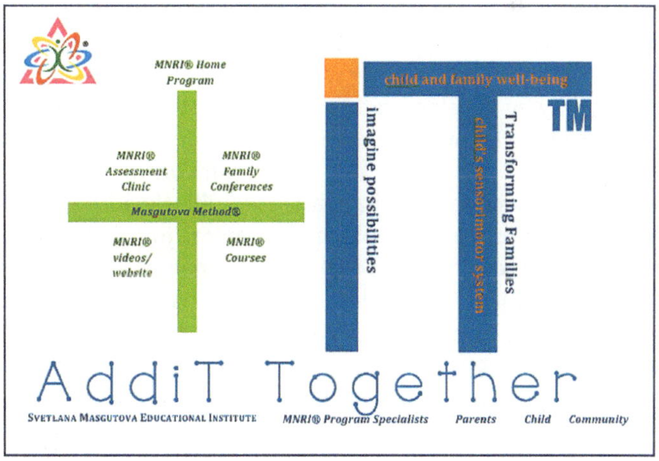

The third component of the canvas is transformation (T). Through implementing the Home Program and expanding experiences, parents can support the transformation of their child's sensorimotor system as well as the child's and family's well-being. The key point is that transformation unfolds gradually over time. Parents play a critical role in this process; MNRI® Conferences help prepare the parent for this role.

Thinking in terms of these three components can help parents make better decisions. Family Conferences are expensive. However, these conferences have been invaluable for us and have provided benefits that we may not have received by following another path (e.g., working with a local professional over a period of time). The Six Cs model discussed in the next section provides further clarification into the value of different MNRI® services.

Based on Figure 1, here are some questions to get started with the +iT Together Canvas to think about your MNRI® journey (see Table 1):

Table 1. Using the +iT Canvas

+: Which MNRI® services are you using (planning to add)?
i: What possibilities can you imagine for your child and your family? within one month: within six months: within a year:
T: What changes do you see in your child's sensorimotor system since you started using MNRI? How can these changes enhance your child's well-being? How can these changes enhance your family's well-being?

Figure 2. The Six Cs Tree - Parents

The Six Cs Tree

The Six Cs model (Figure 2) identifies six factors (Connections, Conversations, Coordination, Clarity, Choice, and Competence) that can help parents add the right mix of services, imagine possibilities for their child's development, and set the stage for their child's well-being as well as for overall family well-being.

The two green areas in the figure represent critical Six Cs variables in a given situation or for certain individuals. Figure 2 focuses on the parent's role in selecting and implementing an appropriate mix of therapies and other experiences to support their child's development. Technology has made it possible for us to learn about and use many types of therapies and resources from all over the world. At the same time, this large array of choices makes it difficult for parents to figure out which combination of therapies and experiences is best-suited for their family's needs at any given time. Parents of children with disabilities are thus faced with the challenge of making decisions under uncertainty on an ongoing basis.

Parent choice is a critical variable that affects MNRI® outcomes. Parents have to choose the appropriate mix of services (e.g., number and timing of MNRI® conferences and clinics, and video resources). Figure 2 shows how the other Cs affect a parent's ability to make effective choices. A parent with a supportive and well-informed network of Connections can engage in meaningful Conversations with others to gain Clarity and thus make better Choices. In other words, parents can cope with the challenges of ongoing decision-making under uncertainty by developing their own Six Cs. In summary, the value of the Six Cs elements can be understood in terms of the following statement:

Chance favors those who invest in their Six Cs

The Six Cs are factors that parents intuitively address while dealing with the challenges of raising a child with disabilities. For example, many parents join support groups and participate in discussion forums. The value of the Six Cs tool is to make these variables explicit so that parents can think about the impact of these variables on MNRI® outcomes. During events such as the Family Conferences, parents can get too focused on developing competence through learning concepts and techniques. They may not take advantage of the other valuable opportunities at these conferences. For example, there are many opportunities to engage professionals and parents in conversations and to develop new connections.

C2C Linkages

The concept of C2C Linkages is designed to enhance the practical use of the Six Cs model. C2C Linkages draws attention to critical relationships between the Six Cs factors in a given situation. For example, a key question while implementing MNRI® programs is:

How can parents coordinate their relationships and resources to make MNRI® service choices work for their family?

Figure 2 identifies parent choice as a critical Six Cs variable that affects successful implementation of MNRI®. Figure 3 builds on this idea and expands the focus to a key relationship between two of the Six Cs elements (Choice and Coordination). While making their choices, parents have to figure out how to coordinate their existing relationships, resources, and activities with MNRI® treatment and Home Program. In other words, the key issue is not simply about choosing a service but of how to make it work in the long run through effective coordination with existing activities and experiences. By looking ahead and thinking about this issue, parents will be better positioned to succeed in implementing MNRI®. Figure 3 thus shows Choice to Coordination as a critical C2C pathway for parents.

Figure 3. The Six Cs Tree with C2C Pathways

Figure 3 also identifies an important supporting pathway: Conversations to Clarity. Parent conversations are a critical variable that affects MNRI® outcomes. During Family Conferences, parents also get opportunities to interact with many professionals and other parents. These conversations help parents gain clarity in making choices. By becoming aware of these factors, parents can make better decisions about services.

Considering C2C pathways explicitly in service delivery is important for parents, professionals, and the SMEI organization. Parents may be overwhelmed with the demands of parenting a child with significant challenges. They may need encouragement, opportunities, and support to engage in conversations. Professionals may be more focused on delivering services than on encouraging and participating in conversations. From an SMEI perspective, the Family Conferences are already designed to support parent conversations. By articulating the Six Cs and C2C Pathways, SMEI can continue to strengthen key pathways for parent success.

Table 2. Using the Six Cs Tree and C2C Pathways tools

Six Cs How can you develop/use your Six Cs to be more successful with MNRI® programs? Connections Conversations Coordination Choice Coordination Competence
C2C: Choice to Coordination How can you coordinate your relationships and resources to make your MNRI® service choices work for your family?
C2C: Conversations to Clarity How can you sustain your participation in meaningful conversations to make MNRI® service choices work for their family?

Six Cs and C2C Pathway Tools: Child

One of the key benefits of the Six Cs Tree is that we can use these tools to think about parents, children, and professionals. Again, the specific Cs that are most significant vary depending on the child's developmental level and family situation. The example below focuses on the child's Choices – a C that we have started emphasizing more as we have seen greater progress. Other Cs may be more important earlier in the MNRI® journey. Figure 4 shows Choices to Connections as a critical child pathway.

Continuing our dance example, one of the fundamental aims of the MNRI® Family Conferences is to develop the child's ability to be present and to focus, using specialized reflex patterns. Independence, inner control, choice making, motivation, and initiative evolve through focusing and presence. As our son started exploring music and dance, he started making his own Choices. He would initiate ideas about what kinds of music he wanted to learn. For example, he initiated the idea of learning Indian dance. He initiates routinely in music. This is still an area we need to strengthen but it is emerging and we expect to see it grow in the coming years.

As our son started making choices, his capacities to interact with others strengthened considerably. The music teacher responded to his initiations. He would jump up and down in excitement when she came to teach him! Once he asked her to take him to a music store to help him buy a book on patriotic music. Seeing his interest, she further nudged his musical development. She invited him to play for her friend who is a concert pianist in Venezuela and for an adult student. Recently, our son made an interesting observation about a piece he was playing and she encouraged him to write to the author. Dance, music, and visual arts have opened the doors to making choices and developing connections.

The Six Cs tool can be used to imagine possibilities for the child in a systematic way. As the child's sensorimotor system strengthens, he is better able to make connections, engage in conversations to gain clarity, coordinate with others, make choices, and develop competence. The next section integrates the +iT Together and Six Cs tools for parent and child.

Figure 4. The Six Cs Tree - Child

Strengthening +iT Together in Families with MNRI® Family Conferences

As discussed in an earlier section, it is important to consider the 'iT' elements in understanding the benefits of the MNRI® program. In this section, I will use the +iT Together, Six Cs, and C2C Linkages tools to discuss MNRI® Family Conferences. In our experience, the family conferences are the most expensive yet most beneficial service delivery option. Given the expense, the use of the tools discussed in this article to think about the Family Conference experience can help families better use the opportunities at these Conferences.

Table 3 is a reflection template that integrates the +iT and Six Cs elements. For example, the first row of the table lists examples of different opportunities for interactions that parents can use to engage in conversations, to gain clarity into MNRI® programs, and to make choices. Together, these experiences can help parents learn MNRI® techniques and to begin imagining new possibilities for their child.

Summary and Additional Ideas

This article presented three visualization tools (+iT Together canvas, Six Cs Tree, and C2C Linkages) that can be used by parents and professionals to map and to track their MNRI® journey. The use of these tools can help parents navigate their MNRI® journey with greater ease and confidence.

Figure 5 presents an integrated +iT Canvas. This canvas includes the parent and child Six Cs tree and can be used as a one page reference that summarizes all three tools. The items at the bottom of the canvas are a useful part of the overall concept. We can think about the +iT and Six Cs components in terms of SMEI, MNRI® specialists, parents, children, and the community. For example, as the child's sensorimotor system evolves, parents are better able to leverage others in their community to work with their child.

In thinking about the Six Cs, it is useful to think about parent-professional interactions as well as the interactions between parents. For example, while we started dance because of a suggestion made by Dr. Masgutova, our experiences have created possibilities for additional conversations. Two of the parents I know have started or are planning to start a dance program for their sons. Dr. Masgutova can also point to our son and others like him to encourage other parents to pursue this option.

Finally, we live in an environment that offers many opportunities to address the significant challenges that our children face. If we can harness resources and use them effectively, we can enhance the child's sensorimotor systems, and the capacity to learn and to interact with others. The

Table 3. +iT Together Reflection Tool for Family Conference

(1) + services: MNRI® Family Camp therapy sessions; evening presentations/workshops; lunch; snack breaks
(2) +: MNRI® Home Program
(3) i: imagining Possibilities
(4) i: imagining Possibilities: Child Six Cs
(5) T: Supporting Transformation: Parent Six Cs
(6) T: Reflecting on Transformation (a) Child sensorimotor system (b) Child well-being (c) family well-being

MNRI® method offers unique opportunities to empower individuals through strengthening their sensorimotor systems. By encouraging parents to develop their own Six Cs, SMEI can empower parents to select appropriate services and to persist in their application of MNRI® techniques over an extended period of time in order to realize the potential of this approach.

More information on +iT Together and Six Cs tools can be found at www.c2cpathways.com. Follow C2C Pathways Facebook at https://www.facebook.com/pages/C2C-Pathways/1410769812558852 and on Twitter (@c2cpathways) for updates and suggestions on using these tools and resources.

Figure 5. Visualizing Pathways to Success: Family Conference

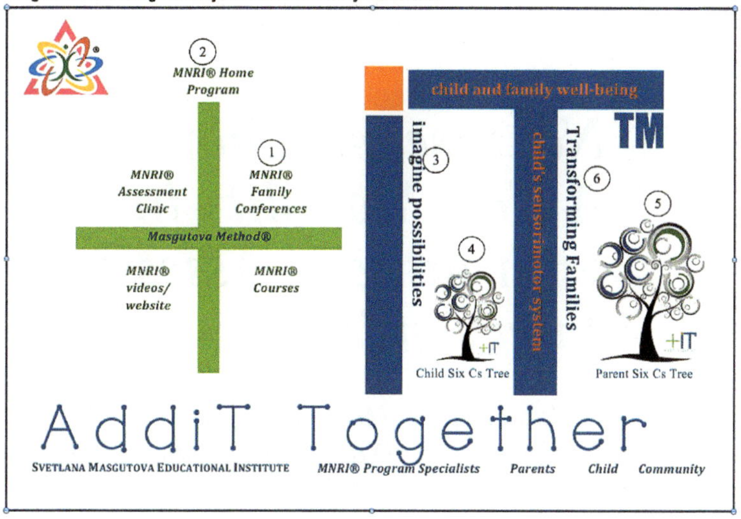

Congratulations to Ananth for your success in personality development, curiosity, and expression through profound dance. Your dancing skills are extraordinary and express the depth of sensorimotor integration and the beautiful blossoming of your heart and emotions. Our MNRI® Team expresses our congratulations to you and the wonderful support and interest of your dear parents. – Pamela Curlee

Neurosensorimotor Reflex Integration - A New Modality in Work with a Child with Autism: Assessment and Intervention Results

Renata Chodyko, MA, Defectologist, MNRI® Specialist;
Denis Masgutov, Director of ISMI, Warsaw, Poland

History and Diagnosis

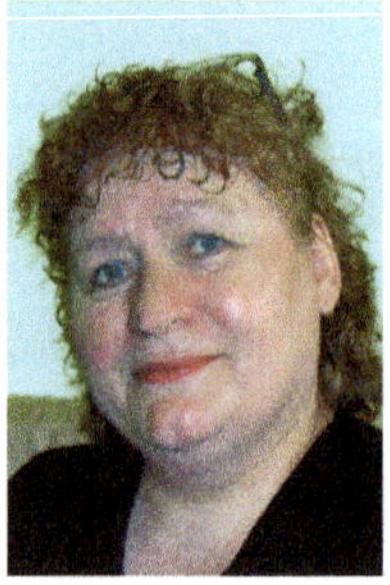
Renata Chodyko

Denis Masgutov

Erik (born October 14, 2008) was 3 years old when he started Masgutova Neurosensorimotor Reflex Integration (MNRI®). He was diagnosed with severe Autism Spectrum Disorder (ASD) in November of 2010, at two years of age. His symptoms included a lack of verbal communication, hyperactivity, lack of eye contact, severe tactile defensiveness, poor muscle tone, poor gross motor coordination, poor oral motor skills including chewing, and intense crying when encountering new environments. Erik's parents were very concerned with his behavior and his mother consulted with many specialists for prognoses about his future development.

Presenting Concerns

At the beginning of MNRI® in October of 2011, Erik demonstrated very severe symptoms of autism – refusal of any touch or interaction, anxiety, frequent crying and tantrums, and an inability to function in social situations. When anxious, he would physically go into a 'fight or flight' response, sometimes screaming for hours. As he was a completely nonverbal three year old, it was difficult to understand his needs and feelings, anticipate his anxiety, and to find ways to help him cope with different situations. His ability to sit, crawl, and stand was compromised by very low muscle tone and poor muscle bulk throughout his body. Erik had started walking very late, at close to three years. His posterior leg muscles were very tight, resulting in a strong toe-walking posture with poor grounding and stability and an tendency to turn the feet inward. His motor skill development ranged from the 4th to the 6th percentile. This was related to his severe delay in motor planning, especially in cross-lateral gross and fine motor coordination. Erik also had very significant sensory integration dysfunction, especially in the areas of auditory processing, tactile and visual perception, and vestibular-proprioceptive-motor control.

Erik's emotional development was also significantly delayed. His protective behavior and fight or flight responses were excessive: any new environment or new person approaching him was perceived as a potential danger. He never stayed in the therapy room and his sessions were done in the 'corner' of the room or in a corridor to provide a more calming environment for him when beginning the MNRI® sessions; his behavior was still one of 'flight' from touch and interaction.

MNRI® Intervention Method

Erik has participated in MNRI® reflex integration therapy for the past year and a half, since the age of 3. Erik's family brought him to Poland from Estonia to attend three MNRI® International Camps starting in October 2011. During these events he received interventions based on the Masgutova Neurosensorimotor Reflex Integration® program during 6-7 hours a day for 12 days and was given MNRI® exercises to continue the program at home. Erik's local specialists also attended MNRI® Conferences to learn this program and to continue it at home on a professional level.

The MNRI® program, based on the Masgutova Method® of Neurosensorimotor Reflex Integration, developed by Svetlana Masgutova, Ph.D, is aimed at restoration of neurophysiological and sensory-motor functioning of dysfunctional, pathological or delayed sensory-motor patterns (Masgutova, Akhmatova, 2004; Masgutova, 2006, 2012; Masgutova, Akhmatova, Kiselevsky, 2008; Masgutova, Akhmatova, Lebedinskaya, 2013). This program offers a unique approach to rehabilitation: neurosensorimotor reflex integration (Masgutova, Akhmatova, 2004; Masgutova, Masgutov, 2013). Results from work with thousands of children have proven that this modality opens new perspectives for intervention with rapid, optimum and stable results, saving valuable neurodevelopmental time for individuals with developmental deficits (Masgutova, 2012). The MNRI® program is designed to build a physiological foundation for neurodevelopment and sensorimotor integration, resulting in the improvement and maturation of physical, emotional and cognitive function (Masgutova, Akhmatova, 2004; Masgutova, 2006, 2012; Masgutova, Akhmatova, Kiselevsky, 2008; Masgutova, Akhmatova, Lebedinskaya, 2013; Pilecki, Masgutova et al., 2012). In their use of MNRI® as the primary tool in their interventions with Erik and many other children, the authors of this article draw upon four aspects of the program:

1) The real possibility to work with reflex patterns on the level of neurosensorimotor links within a genetic structure, "a reflex pattern as an inherent unit of the nerve system" (Sechenov, 1863/1960; Pavlov, 1927/1960; Luria, 1978; Badalian, 1984; Haines, 2002; *Child Neurology*, 2000) that naturally opens the door for genetic sensory-motor long-term memory (Masgutova, Akhmatova, 2004; Masgutova, 2006, 2012).

2) MNRI® works with retained reflex patterns using integration techniques rather than traditional inhibition. The concept of integration is logical from the point of view of individual psychology and human species survival: retained reflexes must not be inhibited or interrupted; they "must complete their development to be available and reliable when needed for protection" (Sechenov, 1863/1960; Pavlov, 1935/1960; Luria, 1978). Equally important, they must be properly supported in order to mature and become integrated within their circuitry, where they serve as the basis for further development. "Inhibition as part of nerve system function must mature" itself and serve not just for interrupting the immature reactive response, but also "for assuring the freedom from automatic/reactive responses" (Sechenov, 1863/1960; Pavlov, 1935/1960; Badalian, 1984; Anokhin, 1973, Masgutova, 2012).

3) True integration takes place within the following reflex components: a) sensory receptors and neurons, b) six levels of synaptic connections for nerve system processing (peripheral nerve system, spinal cord, upper medulla, midbrain, thalamus, cortex), and c) motor neurons with effector organs (muscles, glands) (Sechenov, 1836/1960; Pavlov, 1927/1960; Asratian,1974; Badalian, 1984; Anokhin, 1973). This approach leads to integration on all levels of the nerve system (Masgutova, 2012).

4) Reflex integration manifests in the coordination of the biomechanical/neuro-structural aspect of the motor response with its protective task (Masgutova, Akhmatova, 2004; Masgutova, 2006, 2012). Thus, work with reflex patterns in their automatic unconditioned phases must be targeted in intervention to reach this result.

MNRI® Assessments and Results

The MNRI® Reflex Development Assessment can easily identify normal or abnormal sensory-motor reflex pattern expression. An MNRI® practitioner administers specific stimuli in multiple tests for each reflex to determine whether it is:

- present to an age-appropriate degree
- appropriately developed and functional
- integrated with sensory-motor functions and skills used in learning and movement (Masgutova, Akhmatova, 2004; Masgutova, 2006, 2012).

Basic reflex patterns consist of coordinated movements of the sensory-motor system's three-part circuit and are evaluated in each of five parameters: integrity of the circuit, sequence and direction of movement, timing and speed, intensity (muscle tone regulation of those muscles that serve specific reflex patterns), and symmetry of the response (see the article, *MNRI® Assessment for Determining the Level of Reflex Development* in this book). Each reflex parameter is evaluated based on functional state of four distinct features as in the example below:

Table 1. Reflex Pattern: Hands Pulling Parameters
(based on: S. Masgutova, N. Akhmatova, 2004; S. Masgutova, 2006, 2012)

1		2		3		4		5	
***Circuit* (sensory-motor coordination)**		***Direction of the movement in a reflex response***		***Intensity of the motor response***		***Latency of the motor response***		***Symmetry in the reflex response***	
1.1. Sensory sensitivity corresponds to age		2.1. Motor reaction corresponds to age		3.1. Strength of motor reaction corresponds to age		4.1. Motor reaction emerges with Appropriate speed		5.1. Circuit (sensory-motor coordination)	
Proper	improper	proper	Improper	proper	improper	Proper	improper	Proper	improper
1.2. Motor reaction corresponds to age		2.2. Sequence: 1) flexion of elbows, 2) flexion of head, 3) *flexion* of the core		3.2. Flexion of elbows triggers muscles in area of: a) neck, b) abdomen		4.2. Timing in: 1) flexion of elbows, 2) flexion of head, 3) flexion of core		5.2. Direction of the movement is balanced equilaterally	
Proper	improper	proper	Improper	proper	improper	Proper	improper	Proper	improper
1.3. Sequence of flexion: 1) elbows, 2) head, 3) core		2.3. Abduction/ adduction of arms (15-20 degrees away from body sides)		3.3. Muscle tone of arms		4.3. Timing in flexion and abduction/ adduction of arms		5.3. Intensity of the motor response is balanced equilaterally	
Proper	improper	proper	Improper	proper	improper	Proper	improper	Proper	improper
1.4. Core flexion and extension coordinates with visual convergence and divergence		2.4. Core flexion and extension coordinates with visual convergence and divergence		3.4. Muscle tone of neck area		4.4. Timing in core flexion and extension coordinates with visual convergence and divergence		5.4. Latency of the motor response is balanced equilaterally	
Proper	improper	proper	Improper	proper	improper	Proper	improper	Proper	improper

Corresponding tests for 30 reflex patterns were administered and scored on a scale of 0 to 20 points (five parameters with four features each). The evaluation of test results was based on the criteria developed by Prof. Anna Krefft (Krefft, 2007), in which 10.00-11.75 points indicates that the reflex pattern is in a transitional state on the boundary of normal function and dysfunction, containing elements of both correct and incorrect features. A result below 10.00 means the pattern is dysfunctional. A score above 11.75 means the pattern is functional and represents low-average or average development.

Erik has participated in three MNRI® Reflex Assessments conducted by Svetlana Masgutova, Ph.D (at Camp 1: 10/24/2011, Camp 2: 05/24/2012, and Camp 3: 11/25/2012). The purpose of the Assessments was: to define his actual level of reflex development, to analyze the parameters of each reflex pattern, and to create a treatment program based on the results. Erik's initial reflex Assessment at Camp 1 (10/24/2011) indicated dysfunction, or non-integration, of a number of reflexes that typically should be integrated by the age of 12 months. At four years old Erik's reflexes were delayed due to poor muscle tone regulation, poor processing of sensory stimuli, poor motor responses, and his diagnosis of ASD. The specialized reflex repatterning program developed for Erik was based on this initial Assessment, which also served as a baseline for tracking his progress and modifying his treatment plan every six months.

The tables-graphs below show the dynamic of changes in Erik's reflex patterns that involve the sagittal plane of the body (see tables 1 - X_1 to X_{10}; 2 - X_{11} to X_{20}; and 3 - $X_{21}$to X_{30}) at six-month intervals.

MNRI® Reflex Assessment Results for Erik, ages 3.0, 3.7 and 4.1 yrs.

The data in Erik's initial evaluation (Test 1) demonstrates the immaturity, improper functioning, and dysfunctional state (below 10 points) of 90% of his reflex patterns. The red line marks the transition point between dysfunctional and functional. Only one reflex out of ten (10%), Core Tendon Guard (X_1) was in the positive range, though still on a very low level of development and hyperactivity.

X_n	Reflex Pattern (X_1 to X_{10})	Table 2. Level of Reflex Development in Sagittal (Right-Left) MCS (X_1 to X_{10})												
		Test	Points	Criteria in points										
				1	2	3	4	5	6	7	8	9	10	11
				0-1.75	2-3.75	4-5.75	6-7.75	8-9.75	10-11.75	12-13.75	14-15.75	16-17.75	18-19.75	20.00
1	Core Tendon Guard	Test-1	12.25											
		Test-2	13.25											
		Test-3	14.5											
2	Hands Grasp	Test-1	5.75											
		Test-2	6.5											
		Test-3	8.5											
3	Hands Pulling	Test-1	6.25											
		Test-2	7.25											
		Test-3	9.5											
4	Babkin	Test-1	5.25											
		Test-2	6.25											
		Test-3	8.5											
5	Babinski	Test-1	5.25											
		Test-2	7.25											
		Test-3	9.5											
6	Foot Grasp	Test-1	6.25											
		Test-2	8.25											
		Test-3	9.5											
7	Leg Cross Flexion-Extension	Test-1	9.25											
		Test-2	11.25											
		Test-3	13.5											
8	ATNR	Test-1	6.75											
		Test-2	8.25											
		Test-3	11.25											
9	Abdominal	Test-1	5.25											
		Test-2	8.25											
		Test-3	10.25											
10	Bonding	Test-1	5.25											
		Test-2	8.25											
		Test-3	11.25											

Coefficient	Features	Test 1 - Before Camp 1		Test 2 - After Camp 2		Test 3 - After Camp 3	
		Average	SD	Average	SD	Average	SD
Z_1	Level of integration of reflexes, including X_1 to X_{10}	0.3881	0.1790	0.5912	0.1940	0.6512	0.2940
Test ANOVA		p < 0.001				p < 0.001*	

Table 2. Reflex patterns in the group of sagittal plane ($X_1 - X_{10}$) in Erik with diagnostic features scored on a scale of 0-20 points and mathematic significance of synthesized diagnostic function Z_1-$_{10}$ measuring the level of reflex integration/development with verification by statistic tests (see the Assessments chapter in this book) before MNRI® Program at Camp 1 (10/24/2011), after MNRI® Programs at Camp 2 (05/24/2012) and Camp 3 (11/25/2012) (* P< 0.001 - Compared with 'Before MNRI® Program' at Camp 1).

Subsequent MNRI® Assessments show significant changes in Erik's level of reflex development:

- After participating in the MNRI® Program at Camp-2, improvements were noted in 80% of Erik's reflex patterns in the sagittal plane (Hands Grasp, Babkin, Babinski, Abdominal Sleep Posture, Foot Grasp, ATNR, Leg Cross Flexion-Extension, Bonding). Additionally, Leg Cross Flexion-Extension moved to the transitional point (in the direction of positive function): it was no longer dysfunctional.
- After participating in MNRI® at Camp-3, improvements were observed in 90% of Erik's reflex patterns in the sagittal plane. Forty percent of these reflexes moved to the next higher level within the dysfunctional range (Hands Grasp, Hands Pulling, Babkin, Babinski) and 30% moved to the transitional point (ATNR, Abdominal Sleep Posture, Bonding). Twenty percent of these reflexes moved to a higher level of positive dynamics in the scale (Core Tendon Guard and Leg Cross Flexion-Extension) though he still needed to work further to reach the norm.
- When Erik's scores are compared from his first camp to his third camp improvements were seen in 100% of reflex patterns involving the sagittal plane.

Reflexes of the Sagittal Plane

Special attention was given to several reflex patterns in this group (X_1 to X_{10}) for the reasons described below:

- *Core Tendon Guard Reflex,* a protective response of the whole body to stress by flexion (withdrawal - freezing) or extension (flight or flight), was hypersensitive and hyperactive in Erik. When this reflex is dysfunctional it causes a tendency for excessive withdrawal, vulnerability to stress, fears, and worries, and feeling unsafe. It is seen in the response to traction/stretching of the limbs (instead of stretching easily, resistance against stretch presents with compression as a non-productive protective response) and in poor postural control (Masgutova, Akhmatova, 2004; S. Masgutova, D. Masgutov, 2005, 2013; Masgutova, 2012). The Core Tendon Guard in Erik was asymmetrical by ¾" for upper and lower Tendon Guard.
- *Hands Grasp Reflex,* a response to stimulation of the upper part of the palm and lower level of phalanges (Badalian, 1984; Masgutova, Akhmatova, 2004; Masgutova, 2012), was hyposensitive and hypoactive in Erik. This reflex pattern was definitely not integrated within its biomechanical structure and protective function, which resulted in poor grasp and manual skills, lack of interest in manipulating objects and toys, and poor hand-eye coordination. Erik's level of development of Hands Grasp was equal to that of an 11

month old child.

- *Hands Pulling,* a response to being grasped by the wrists to activate a sitting up movement based on sequential flexion of: a) the elbows, b) head, c) core (Masgutova, Akhmatova, 2004; S. Masgutova, D. Masgutov, 2013; Masgutova, 2012), was hyposensitive and hypoactive in Erik, with a tendency for delay of head righting and elbow extension (instead of flexion). This reflex pattern was dysfunctional, resulting in poor coordination of core flexion with visual convergence, hyperactive gaze and visual staring (over-reactive 'territorial instinct'), poor hand-eye coordination, poor postural control when sitting, and poor neck muscle tone regulation. The development of the Hands Pulling Reflex was equal to that of a 10 month old child.
- *Babinski Reflex,* a response of the foot to stimulation on the external plantar surface, inward rotation, dorsi flexion of the big toe, and fanning out of the other toes used in crawling, walking, and running on uneven ground (Badalian, 1984; Masgutova, Akhmatova, 2004), was hypersensitive, hyperactive, and dysfunctional in Erik. The direction of the movement was also improper as his foot moved in the opposite direction. Delay in development of the Foot Tendon Guard and Babinski Reflexes is usually attributed to poor neurological functioning of the lower motor neurons (extrapyramidal nerve system) leading to overall delays in sensory-motor integration (Badalian, 1984).
- *Leg Cross Flexion-Extension* is a response to stimulation of the middle of the foot arch on one leg expressed by flexion of the other leg at the hip and knee, followed by extension. This pattern is used in crawling, walking, running, and climbing. Its physiological circuit accesses the upper medulla area where the cross over of nerve fibers take place, thus it affects the formation and myelination of neural connections. This pattern is very important for the organization of nerve transmission to higher brain levels (Magnus, 1925; Badalian, 1984; Masgutova, Akhmatova, 2004; Masgutova, 2012). Erik's score for Leg Cross Flexion-Extension was below the transitional point. This reflex was observed to be hyposensitive, hypoactive, and dysfunctional. The direction of the movement in this reflex was also improper and opposite which is why Erik walked on his toes with an unstable and poorly coordinated gait. The delay in this reflex leads to poor neurological functioning of the lower motor neurons (extrapyramidal nerve system) responsible for integration of early motor and sensory-motor system.

Although the above reflexes are singled out for this report, it is important to note that overall attention was given just as thoroughly to every reflex pattern.

Reflexes of the Horizontal Plane

Analysis of the next group of reflexes (X_{11} to X_{20}) shows a positive dynamic in development of reflex patterns that involve the horizontal plane of the body (X_{11} to X_{20}). (see Table 3.) Again the data from Erik's initial evaluation at Camp 1 clearly demonstrate the immaturity and dysfunctional state of his reflex development. 100% of Erik's reflex patterns that involve the horizontal plane scored below the transitional 10-11.75 point level.

MNRI® Assessments after Camps 2 and 3 show the following changes in Erik's level of reflex development:

- After Camp 2 improvements were noted in 90% of reflex patterns in the horizontal plane (all except Fear Paralysis). One reflex, Automatic Gait (X_{11}), moved up to the transition point between dysfunctional and functional. The other eight reflexes moved up to the next level of development within the dysfunctional state (Bauer Crawling, Moro, Flying and Landing, Hands Supporting, Grounding, Segmental Rolling, Landau, and Head Righting).
- After Camp 3 further improvements were observed in 90% of Erik's reflex patterns in the horizontal plane. 20% moved up to the next level within the dysfunctional state (Segmental Rolling and Landau), 20% moved up to the transitional point (Moro and Fear Paralysis), and 50% (Automatic Gait, Bauer Crawling, Hands Supporting, Grounding, Head Righting) in the horizontal body plane group achieved a functional level though he still needed to work further to reach the norm.
- When Erik's scores from his first camp to his third camp are compared, improvements are noted in 100% of reflex patterns involving the horizontal plane.

Special attention was targeted at several reflex patterns in this group (X_{11} to X_{20}) for the reasons described below:

- *Hands Supporting,* a protective response occurring when the individual extends the arms toward a horizontal or vertical surface to avoid harm from a fall or an approaching object (Badalian, 1984; Masgutova,

#	Reflex Pattern (X_{11} to X_{20})	Test	Point	Table 3. Level of reflex development in Horizontal (Upper-Lower) MCS										
				Criteria in points										
				1	2	3	4	5	6	7	8	9	10	11
				0-1.75	2-3.75	4-5.75	6-7.75	8-9.75	10-11.75	12-13.75	14-15.75	16-17.75	18-19.75	20.00
11	Automatic Gait	Test-1	6.75											
		Test-2	10											
		Test-3	12.5											
12	Bauer Crawling	Test-1	6											
		Test-2	8.25											
		Test-3	12.75											
13	Moro	Test-1	7.25											
		Test-2	8.75											
		Test-3	11											
14	Fear Paralysis	Test-1	8.75											
		Test-2	9.75											
		Test-3	11											
15	Hands Support	Test-1	6.25											
		Test-2	8.25											
		Test-3	12.5											
16	Segmental Rolling	Test-1	5.75											
		Test-2	8.75											
		Test-3	9.75											
17	Landau	Test-1	0.5											
		Test-2	4.25											
		Test-3	7.5											
18	Flying/ Landing	Test-1	5.25											
		Test-2	6.5											
		Test-3	7.75											
19	Grounding	Test-1	6.75											
		Test-2	8.75											
		Test-3	12.5											
20	Righting	Test-2	9.75											
		Test-3	13.5											

Coefficient	Features	Test 1 - Before Camp 1		Test 2 -After Camp 2		Test 3-After Camp 3	
		Average	SD	Average	SD	Average	SD
Z_{LP}	Level of integration of reflexes including X_{11} to X_{20}	0.3667	0.1670	0.4914	0.1837	0.5911	0.2738
Test ANOVA		p < 0.001				p < 0.001*	

Table 3. Reflex patterns in the group of horizontal plane (X_{11} – X_{20}) in Erik with diagnostic features scored on a scale of 0-20 points. The mathematic significance of synthesized diagnostic function Z_{11-12} measuring the level of reflex integration/development was verified by statistic tests before MNRI® Program at Camp 1 and after MNRI® Program at Camp 2 and MNRI® Program at Camp 3 (* P< 0.001 - Compared with 'Before MNRI® Program at Camp 1').

Akhmatova, 2004; Masgutova, 2012), was hyposensitive, hypoactive and dysfunctional in Erik, resulting in inability to protect himself, lack of emotional stability, poor 'hand-eye' coordination and weak postural control. The development of his Hands Supporting reflex was equal to that of a 4 month old child.

- *Fear Paralysis Reflex,* a protective startle response to unexpected sensory stimulation: tactile, visual, and auditory (Kaada, 1986), was extremely hypersensitive and hyperactive, especially for tactile and auditory stimuli.
- *Moro,* another protective response, in this case to vestibular-proprioceptive stimulation from the loss of stability in space (Vygotsky, 1986; Badalian, 1984; Masgutova, Akhmatova, 2004; Masgutova, 2012), was retained, hypersensitive, hyperactive, and had a negative effect on Erik's balancing and grounding.
- *Automatic Gait* is an automatic stepping response to movement of the gravity center forward, away from a vertical body position. This pattern is used in crawling, walking, running, jumping, and climbing (Badalian, 1984; Masgutova, Akhmatova, 2004; Masgutova, 2012). It was dysfunctional, hyposensitive, and hypoactive in Erik. The direction of the movement in this reflex was also improper causing toe walking, instability, and poor coordination. A delay in the development of this reflex leads to poor neurological functioning and affects other milestones in early motor development (Haines, 2002; Masgutova, Akhmatova, 2004; Masgutova, 2012).
- *Bauer Crawling* is a response to stimulation of the plantar surface of the feet expressed by sequential flexion-extension of hips/knees and pushing with the feet for crawling forward. This pattern is used in walking, running, climbing. Its physiological circuit is rather complicated and involves parallel processing on several neural pathways, thus helping the development of future multitasking in motor planning and control (Badalian, 1984; Doman, 1984; Masgutova, Akhmatova, 2004; Masgutova, 2012). It was dysfunctional, hyposensitive, and hypoactive in Erik. The direction of his movement in this reflex was also improper. A delay in the development of this reflex leads to poor functioning of the lower motor neurons (extrapyramidal nerve system) and cortical areas responsible for integration of early motor and cognitive development (Badalian, 1984; Doman, 1984; Masgutova, Akhmatova, 2004; Masgutova, 2012).

Reflexes of the Dorsal Plane

Data from the last group of reflexes (X_{21} to X_{30}) demonstrates the following positive dynamics in the development of Erik's reflex patterns involving the dorsal plane of the body. (See Table 4.)

Data from Erik's initial evaluation at Camp 1 demonstrates the immature and dysfunctional (below 10 points) state of 100% of his reflex patterns in the dorsal plane. The recorded level of change in his reflex development after MNRI® again indicates significant growth:

#	Reflex Pattern (X_{21} to X_{30})	Test	Point	Table 4. Level of reflex development in Dorsal (Front-Back) MCS — Criteria in points										
				1	2	3	4	5	6	7	8	9	10	11
				0-1.75	2-3.75	4-5.75	6-7.75	8-9.75	10-11.75	12-13.75	14-15.75	16-17.75	18-19.75	20.00
21	Trunk Extension	Test-1	9.25											
		Test-2	11.75											
		Test-3	14											
22	STNR	Test-1	7.25											
		Test-2	9.25											
		Test-3	12											
23	Spinal Galant	Test-1	6.25											
		Test-2	7.5											
		Test-3	10.5											
24	Spinal Perez	Test-1	6.25											
		Test-2	7.75											
		Test-3	9.75											
25	TLR	Test-1	5.25											
		Test-2	7.75											
		Test-3	11.25											
26	Foot Tendon	Test-1	5.25											
		Test-2	7.25											
		Test-3	9.75											
27	Spinning	Test-1	6.5											
		Test-2	8.25											
		Test-3	11.5											
28	Locomotion	Test-1	4.25											
		Test-2	5.25											
		Test-3	7.75											
29	Balancing	Test-1	5.25											
		Test-2	8.75											
		Test-3	11.25											
30	Pavlov Orientation	Test-1	5.25											
		Test-2	8.50											
		Test-3	11.75											

Coefficient	Features	Before Program		1-After Program		2-After Program	
		Average	SD	Average	SD	Average	SD
Z_3	Level of integration of reflexes including X_{21} to X_{30}	0.3583	0.1473	0.52924	0.1646	0.6314	0.2534
Test ANOVA		$p < 0.001$				$p < 0.001$*	

Table 4. Reflex patterns in group of dorsal plane (X_{21} – $X2_{30}$) in Erik with diagnostic features scored on a scale of 0-20 points. The mathematical significance of synthesized diagnostic function $Z_{21\text{-}30}$ measuring the level of reflex integration/development was verified by statistic tests before MNRI® at Camp 1, and after MNRI® at camp 2 and after MNRI® at Camp 3 (* P< 0.001 - Compared with 'Before MNRI® Program' at Camp 1).

- After Camp 2 70% of the reflexes in the dorsal plane (STNR, Spinal Galant, Foot Tendon Guard, Spinning, TLR, Balancing, Pavlov) moved one or more levels closer to the transitional (10-11.75) point range and 10% (Trunk Extension) reached as far as 11.75 points.
- After Camp 3 (six months later) improvements were noted in 100% of his reflex patterns in the dorsal plane. 30% (Spinal Perez, Foot Tendon Guard, Locomotion) moved to the next higher level within the dysfunctional state, 50% (Spinal Galant, TLR, Spinning, Balancing, Pavlov) moved to the transitional level (10-11.75 points), and 20% of the reflex patterns (Trunk Extension and STNR) moved to the level of positive functional development and were no longer deemed dysfunctional.
- When all camps considered together, it shows improvement in 100% of Erik's reflex patterns.
- When Erik's scores are compared from his first camp to his third camp improvements were seen in 100% of his reflex patterns involving the dorsal plane.

The synthesized diagnostic functions (Z_2, Z_3) measuring the level of changes at Camp 2 and Camp 3 are statistically significantly higher (with a p value of <.001) in each post-test (Camp 2 and Camp 3) in comparison with initial tests (Z_1, at Camp 1).

Our focus in repatterning the reflex patterns in this third group was on the following reflex patterns:

- *Foot Tendon Guard Reflex*, part of the whole body Tendon Guard response to stress, manifests in plantar and dorsi flexion of the feet, and trains our ankle joints in the flexion-extension necessary for crawling, standing, walking, and running (Mahoni, 1999; Badalian, 1984; Masgutova, Akhmatova, 2004; Masgutova, 2012). This reflex was hypersensitive, hyperactive and dysfunctional in Erik and partly responsible for his toe walking.
- *Spinal Galant*, a protective response of the whole body to a stimulus on the side of the spine by lateral flexion of that same side away from the stimulus. It is an intense fight or flight response (Badalian, 1984; Masgutova, Akhmatova, 2004). In Erik it was extremely hypersensitive and hyperactive. Dysfunction in this reflex causes poor spinal and overall postural control, poor head righting, and vulnerability to stress, fears, worries, and feelings of being in danger.
- *Spinal Perez* is also a protective response of the whole body to tactile stimulation on the spine by spinal extension with anterior pelvic tilt and hip flexion. This pattern is based on the 'fight or flight' response (Badalian, 1984; Masgutova, Akhmatova, 2004) and was very hypersensitive and hyperactive in Erik. A dysfunctional Perez Reflex has a negative effect on spinal and overall postural control, head righting, stress tolerance, emotional stability, and immune system function (Badalian, 1984; Masgutova, Akhmatova, 2004; Masgutova, Akhmatova, Kiselevsky, 2008; Masgutova, Akhmatova, Lebedinskaya, 2013).

- *Spinning* is an active response by the whole body to turning from front to back, starting by flexing and turning one leg across the midline, followed by spine rotation with support of all limbs. This pattern activates the entire body for changing its position in space, thus training the body in spatial orientation and flexibility, skills needed for crawling, walking, and all rotational movements of the spine (Sadowska, 2001; S. Masgutova, D. Masgutov, 2012). In Erik this reflex pattern was hyposensitive and hypoactive, limiting his flexibility, his kinesthetic cognition of space, and his orientation in the visual and auditory fields. A dysfunctional Spinning Reflex pattern has a negative effect on postural and motor control, head righting, flexibility, movement dynamics, stress tolerance, and cognition (Sadowska, 2001; S. Masgutova, D. Masgutov, 2012).

During application of the MNRI® Program, Erik demonstrated significant changes in all these reflex patterns (X_1-X_{30}) as seen in the tables and graphs above.

Other Assessments

Erik was given several other pre and post tests at Camp 1 including: Evoked Brain Potentials of Visual (VEP) and Auditory nerves (BAER), Wii Balance and GMFM 66 (Gross Motor Function Measure 66), and the Peabody Developmental Motor Scales (PDMS 2).

Evoked Brain Potentials Research on measurement of Evoked Brain Potentials (organized by the Masgutova Institute and provided by Prof. W. Pilecki of the Department of Pathological Physiology of the Medical Academy of Poland) was aimed at measuring the excitation wave evoked by a stimulus to Erik's brainstem with Auditory Evoked Potentials (BAEP) and Visual Evoked Potentials (VEP). These measurements show transmission efficiency and provide a basis for prognostic assessment of children at risk for central nervous system deficits (Pilecki, Masgutova, et al., 2012). Erik's VEP test results was within the norm, thus only his BAEP results are presented here.

BAEP testing consisted of 2 parameters: 1) Peak I Latency and 2) Inter Peak Latency I-V. The Interpeak Latency I-V parameter is very important for this study because it describes transmission in the brainstem segment of the auditory pathway. The results specify in milliseconds (ms) the time that elapses between electrogenesis of the I segment (auditory nerve, located in the ear), and electrogenesis of the V segment (inferior colliculi of the mesencephalon, part of the brainstem). (Pilecki, Masgutova, et al., 2012)

During normal maturation, the speed of transmission increases and the IPL value decreases. Prolongation of this parameter indicates a delay in maturation. Of high interest to those working with motor development is the fact that changes in the auditory pathway correlate with changes in a motor neuron located very close by. This makes possible an indirect diagnosis, in which we see the connection between the work of the auditory pathway and the motor neuron pathway (Pilecki, Masgutova, et al., 2012).

Table 5. Peak I and Inter Peak I-V Latency test results before and after MNRI®.

	10/24/2011, pre-test		11/7/2011, post-test		Norm
	Right	Left	Right	Left	
Peak I	1.50 ms	1.64 ms	1.62 ms	1.82 ms	≤1.75 ms
Change			+.12 ms	+.17 ms	
Diff. R - L	Right side .14 ms faster		Right side .20 ms faster		
IPL I-V	4.34 ms	5.28 ms	4.35 ms	4.50 ms	3.85 – 4.25 ms
Change			-.01ms	-.78 ms	
Diff. R - L	Right side .94 ms faster		Right side .15 ms faster		

The BAEP Test (70 dB given at a frequency of 10 Hz) was administered to Erik before (10/24/2011) and after 12 days of MNRI® exercises (11/7/2011).

BAEP Rest Results: The norm for Peak I Latency is a value no higher than 1.75 ms. Higher than 1.75 ms indicates deficient auditory acuity. The slight prolongation (1.82 ms on left side) is without clinical significance. The testing showed no significant change in Erik's auditory acuity after MNRI®.

IPL I-V Latency: Prolongation of the IPL I-V (i.e. speeds slower than the norm) is evident both before and after rehabilitation. However, lower values in the second examination show significant improvements, which indicate that transmission was much faster after MNRI®, especially on the left side, which improved by .78 ms. These results are statistically significant at the level of $p < 0.001$.

The BAEP assessment clarifies an interesting aspect of Erik's behavior, one often noticeable in other children with ASD. Eric was and still is demonstrating some hyperactive responses (running away, covering or rubbing his ears) to noises, loud voices, or sudden unexpected and high-pitched auditory stimuli. His MNRI® Assessment at Camp 1 showed that his ATNR pattern was mixed with the Abdominal Sleep Posture pattern. According to our previous research on the brain wave spectrum (Masgutova, Regner, 2007; also unpublished research by Skhomorokchov, Zakharov, et al., 2006-2007, material of the SMEI) ATNR activates rapid frequency brain waves (alpha and beta) and triggers the Auditory Stapedius Reflex, while the Abdominal Sleep Posture activates slow brain waves (delta and theta), inhibits the Stapedius Reflex, and encourages the auditory system to ignore stimuli and to rest.

In testing the ATNR at Camp 1, turning Erik's head to the left triggered an ATNR response and turning to the right triggered the Abdominal Sleep Posture reflex. Turning to the right is understood to activate primarily auditory processing in the left hemisphere of stimuli coming into the right ear. The link between the right ear and the left hemisphere is especially important for language development because of the brain's left side location of Broca's and Wernike's areas, where language is processed. In Erik's case, his Abdominal Sleep Posture caused him to shut down in the presence of stimuli coming to these areas. This was evident in his behavior as well: Erik often ignored instructions, explanations, and logical speech directed to him. Instead he liked to listen to songs and poems when they were combined with tactile stimuli.

According to our interpretation, Erik's Auditory Stapedial Reflex had been improperly activating his Abdominal Sleep Reflex instead of his ATNR. The conflict between his ATNR and Abdominal Sleep Posture Reflexes was preventing him from developing efficient binaural hearing.

This explains perfectly Erik's tendency to use his mixture of patterns to calm down his hypersensitive auditory perception by compensating with the Abdominal Sleep Posture pattern and favoring more monaural auditory perception with his left ear. Like many compensatory strategies, this was difficult, inefficient, and ultimately stressful. Thus, his need to escape auditory stimuli and avoid auditory decoding, which in turn may have inhibited his development of receptive and expressive language.

Before MNRI®, Erik's brain was responding to auditory input from his right ear .94 ms faster than from his left ear. How confusing would one's experience of the auditory world be if sounds arrived in the brain with a delay of .94 ms between reception from the right and left ears? Only twelve days later, after MNRI®, Erik's IPL – IV speeds increased by .01 ms on the right and .78 on the left. Now the left ear has improved and instead of a difference in speed of .94 ms between his two ears, the difference is only .15 ms. How much might this change help Erik to make sense of what he hears?

The results of the BAEP strongly suggest that improvements in Erik's reflexes contributed to improved functioning of his auditory processing skills. Improvements in transmission speed observed on the BAEP correspond to improvements in MNRI® Assessments of the ATNR during Camps 2 and 3. Erik's work towards integrating his ATNR, Stapedius, and other reflex patterns allowed him to begin to resolve the issue of the improper links between his ATNR and Abdominal Sleep Posture. It is to be expected that maturation and integration of these two reflexes would result in more synchronized functioning of his two ears and with it more efficient binaural hearing. This opens the possibility for Erik to improve in language decoding and expressive speech, which is now growing rapidly day by day and has surprised many specialists.

The Wii Balance Board – Center of Balance: This test is considered a valid and reliable means of assessing standing balance (Clark, Bryant, Pua, McCrory, Bennell, et al., 2010). At Camp 1 in this test Erik demonstrated a strong asymmetry in his center of balance: 63% of his body weight was distributed on his left foot and only 37% on the right. After 12 days of participating in MNRI®, Erik demonstrated a significant improvement in his center of balance: 51% of his body weight was distributed on his left foot and 49% on his right. These results show 8 months growth in ability to balance, according to the test standards.

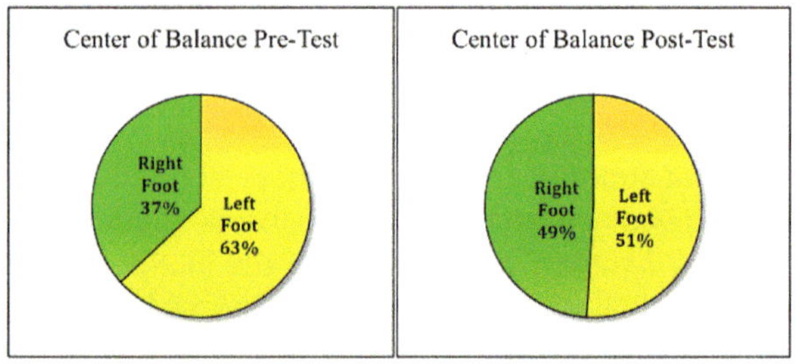

Chart 1. Wii Balance test results for Erik – Before and After 12 days of MNRI® at Camp 1 (10/24/2011 - 11/07/2011) carried out by physical therapy specialists.

The Gross Motor Function Measure-66 (GMFM-66): is a criterion based clinical assessment designed to evaluate change in gross motor skills and their progress in children with developmental delays and neurodeficits. The GMFM measures traditional gross motor milestones. The GMFM-66 is a subset of tasks taken from the GMFM-88 that are entered into a scoring program called Gross Motor Ability Estimator (GAME) in order to produce an estimate of the child's gross motor function. The GAME program has been shown to estimate gross motor skills within a 95% confidence interval with as few as 13 tasks scored; therefore all 66 tasks do not necessarily need to be tested and scored if time constraints or physical impairments prevent it. When an item is not tested it is entered into the GAME program as "not tested" and not used as part of the gross motor skill estimate score (Linder-Lucht, Othmer, Walther, et al., 2007.)

Table 6. GMFM-66 scores.

Date	*Age*	*GMFM-66 Score*	*Standard Error*	*95% Confidence Interval*	*Items Tested*	*GMFCS*	*Change in Score*
11/7/2011	4y	61.80	1.29	59.27-64.33	31	Level 2	9.71
10/24/2011	4y	52.09	1.23	49.68-54.50	26	Level 2	N/A

The mean score for a child of a similar age and GMFCS level is 60.8 (standard deviation of 7.2, n=29). The mean change in a child's score after 1 year of skilled intervention is 2.23 (standard deviation of 6.48, n=9). Erik's GMFM score improved by 9.71 points in 13 days! This is a highly significant result.

Peabody Visual Motor (Folio, Fewell, 2000; Goyen, Lui, Woods, 2008; Hartingsveldt, Cup, Oostendorp, 2006). The Peabody Developmental Motor Scale Revised II (PDMS-2) is a test of early motor abilities that was designed to assess the gross and fine motor skills of children from birth through six years of age in the areas of reflexes, static balance, locomotion, object manipulation, grasping, and visual-motor integration. The visual-motor integration subtest of the PDMS-2 was administered to assess Erik's fine motor skills.

Table 7. Results of Peabody Visual Motor test.

Visual-Motor Integration	Standard Score	Standard Score Performance	Percentile Rank
Camp 1: 10/24/11	2	Very Poor	<1%
11/8/11 (14 days later)	3	Very Poor	1%

The data in the table at right demonstrates that at the beginning of Camp 1 Erik had a standard score of 2 in the Visual Motor Integration subtest and in the post-test 12 days later he achieved a standard score of 3. Erik's visual-motor integration scores improved by one standard score and he demonstrated a four month improvement in skills.

Height Measurement

Erik's height at Camp 1 (10/24/2011) initially measured 94 cm. (37 inches), and twelve days later, 101 cm (39.76 inches), a difference of 7 cm (2.76 inches). We attribute this significant difference in height to improved structural alignments due to MNRI® work with postural reflex patterns. When he was measured again after Camp 3 a year and two months later he had grown an additional 8 cm (3.15 inches).

MNRI® Program for Erik

Based on the results of the MNRI® Assessment described above, an individualized program for Erik was established for MNRI® Camp specialists and for parents at home for the following 6 months. The purpose of the individualized program was to strengthen and support physical, emotional, and cognitive growth through work with non-integrated and dysfunctional reflex patterns. Erik's MNRI® Camp program included the following:

- MNRI® Repatterning of Dysfunctional Dynamic and Postural Reflexes
- MNRI® Neuro-Structural Reflex Integration
- MNRI® Neuro-Tactile Integration
- MNRI® Archetype Movement and Lifelong Reflex Integration
- MNRI® Birth & Post-Birth Reflex Integration
- MNRI® Oral-Facial Reflex Integration
- MNRI® Proprioceptive & Cognitive Integration
- MNRI® Visual & Auditory Reflexes Integration
- Art Creation & Reflex Integration

Parents' report on Erik's changes at Camp-1 (10/24/2011 - 11/07/2011)

"Erik attended his first camp in October 2011 at age three. He neither spoke, nor maintained eye contact. He also did not understand basic words and orders, didn't use the potty, and couldn't dress himself. He was also a very picky eater. He did not seek active contact with children or adults.

After the first 12-day MNRI® Camp in Poland, Erik started demonstrating better stability and grounding and less toe-walking. His gross motor coordination improved. He became less hypersensitive, more tolerant for touch and sounds, less defensive, and more stable emotionally. His new ability for visual contact was such a happy moment! He demonstrated better postural control and his cross-lateral movement increased significantly, enabling him to walk in a more coordinated way. He began to articulate specific sounds ("ng", "g", "k") as the result of work with the Oral-Facial Program and tried to repeat some simple syllables ("am", "ma", "ta"). He also became more curious about his surroundings and his own activities. His skill and interest in imitation increased as well, and he seems more motivated for activities that stimulate cognitive development."

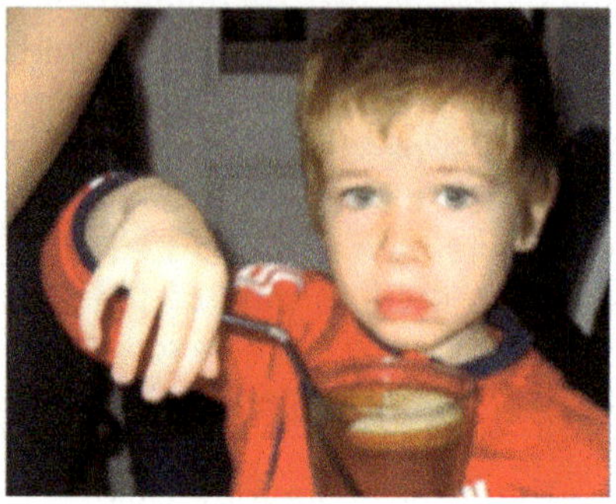
Erik at age four.

MNRI® Home Program

The Home Program that we suggested for Erik for the next six months included exercises from the following programs:

1. MNRI® Neuro-Structural Reflex Integration (30-40 min.) – 2 times a week
2. MNRI® Neuro-Tactile Integration (30-40 min.) – 2-3 times a week
3. MNRI® Archetype Movements Integration (20-25 min.) – 2-3 times a week
4. MNRI® Proprioceptive-Cognitive Integration (30 min.) – 2-3 times a week
5. MNRI® Oral-Facial/Visual-Auditory Reflex Integration (20 min.) – 2-3 times a week
6. MNRI® Repatterning and Integration of Reflex Patterns (20-40 min.)

Reflex Re-patterning in Erik's home program especially targeted ATNR, Babinski, Core Tendon Guard, Grounding, Fear Paralysis, Foot Tendon Guard, Hands Supporting, Leg Cross Flexion-Extension, Locomotion, Segmental Rolling, Spinal Galant, Spinal Perez, Spinning, STNR, and Trunk Extension. His parents were encouraged to maintain a cycle of work during 1½ months followed by rest for 10 days, repeating for 6 months until his next Assessment.

Short Home Program

An optional Short Program served for periods when time for work with Eric was limited:

1. Core Tendon Guard Reflex pattern - 2 times a week
2. Embrace Squeeze - 5 times a week
3. Spinal Gallant – 4 times a week
4. Thoracic Embrace - 3 times a week
5. Foot Tendon Guard - 5 times a week
6. Spinal Perez – 3 times a week
7. Segmental Rolling - 2 times a week
8. Spinning – 3 times a week
9. Balancing Board – 5 times a week (5-10 minutes daily)
10. Visual-Auditory exercises - 2 times a week

Erik's progress at home was fast and amazing.

Parents' Report on Erik's Changes

After following the Home Program for six months, Erik's local professionals and parents shared his progress:

"... All of the improvements from the first Camp were stable, held with time, and even became much stronger. In six and a half months of work using our reflex integration program Erik became stronger with all reflex patterns and began to speak! (His first words were "Mama", "Dad", "dog", "cow", "lion").

Parents' report on Erik's changes at Camp 2 (05/24/2012) and Camp 3 (11/25/2012)

"Now in 2013 Erik understands and follows simple and more complicated orders. He communicates in his own language, but actively uses about 100 words and sometimes 2-word sentences.

He maintains longer eye contact, is potty-trained, and with some help, also dresses himself. His menu is more varied and compared to earlier he needs less and less help with eating. He also seeks contact with other children and adults and smiles, while earlier he cried a lot and was often in a bad mood.

Within one year (at and after his third camp) of participating in MNRI® our dear Erik is showing significant changes: his gross-motor coordination is improving amazingly well, and his walking and running skills almost correspond with his age. Erik has begun to jump! He constructs simple sentences, his vocabulary and comprehension is growing very actively; he initiates communication even with new people; he sings and talks about his emotions, and explores the environment with great curiosity. Erik started to draw and he is very happy showing his drawings to our family. He uses pencils and paper as well as different chalks on the board. He always gives names to the objects he draws. His brother and parents are very happy with Erik's progress. He has become better friends with his brother.

Erik with his mother.

He still prefers a limited number of foods, but he gets served the same menu as the whole family. He started using a spoon and a fork instead of eating only with his hands as he did earlier. He still is using his hands to feel the texture of a food. His balance improved very well: he can run and has begun galloping and jumping. He still needs to improve his space orientation and control, mostly in bigger rooms. His eye contact has improved significantly, but he needs to continue working on eye tracking and expanding his active visual work in a wider visual field. His internal world is changing, and his motivation for learning new things and communication is growing, which is beyond any expectation and amazes both his parents and specialists.

Erik started attending daycare for a few hours a day and he is doing well with other children and activities during these hours. His favorite subjects are the music lessons, dancing, and singing. His speech therapist is extremely happy about the progress Erik makes from day to day. He started to initiate new games and activities during the lessons. Erik has attended four Camps now and at Camp 4 (May-June, 2013) we have ongoing and stable success and we are very pleased by all the great changes in his neurodevelopment."

A local specialist states, that "Erik has a beautiful and dedicated family working with MNRI® on his physical, emotional, cognitive, and motivation development".

Conclusions

MNRI® programs aimed at restoration of neurophysiological and sensory-motor functioning of dysfunctional or delayed reflex patterns have been highly effective for Erik, a child with severe autism. The results of MNRI® Assessments, as well as standardized tests done concurrently, show significant and statistically valid changes as a result of this therapy.

The model of MNRI® Camps combined with MNRI® Home Programs has led to stable and on-going changes in a child with autism in many areas, his sensorimotor reflex integration, physical, cognitive, speech and emotional development.

Neurosensorimotor Reflex Integration as a specific modality can be an important element in treatment of children with autism. The rapid and stable results achieved by MNRI® Specialists with Erik are remarkable and open new perspectives for intervention for children with autism that can save neurodevelopmental treatment time for many individuals with developmental deficits.

The integration of reflex motor patterns achieved through the use of MNRI® programs serves as the physiological basis for neurodevelopment and sensorimotor integration and leads to rapid overall improvement.

References

Anokhin, P. K. (1973). *Biology and neurophysiology of the conditioned reflex and its role in adaptive behavior.* International series of monographs in cerebrovisceral and behavioral physiology and conditioned reflexes. USA.

Asratian, E. A. (1974). *Ivan Petrovich Pavlov - 1849-1936. Science Academy of Russia.* Scientific-bibliography literature. Moscow, Russ: Science.

Badalian, L. O. (1984). *Child Neurology.* Moscow, Russ: Medicina.

Doman, G. (1984). *How to multiply your baby's intelligence.* Garden City, NY: USA. Doubleday.

Child neurology. (2000). (Publisher:) Wydawnictwo Medyczne - Urban & Partner. Wrocław, PL.

Clark, R. , Bryant, A. , Pua, Y. , McCrory, P. , Bennell, K. , et al. (2010). Validity and reliability of the Nintendo Wii balance board for assessment of standing balance. *Gait & Posture.* 31(3), 307-310.

Haines, D. E. (2002). *Fundamental neuroscience,* Second Edition. Edinburgh, UK.

Folio, M. R., Fewell, R. R. (2000). *Peabody developmental motor scales examiners manual* (2nd edition). Austin, TX: Pro-Ed.

Folio, M. R., Fewell R. R. (2000). *Peabody visual motor.* http://cdd.unm.edu/ecspd/pubs/pdfs/Levels%20of%20Gross_Fine%20Motor.pdf

Goyen, T-A., Lui, K., Woods, R. (2008). *Visual-motor, visual-perceptual, and fine motor outcomes in very-low-birthweight children at 5 years.* DOI: 10.1111/j.1469-8749. 1998.

Hartingsveldt, M., Cup, E., Oostendorp, R. (2006). Reliability and validity of the fine motor scale of the Peabody Developmental Motor Scales–2. *Occupational Therapy International*, v. 12, Issue 1. p. 1-13. DOI: 10.1002/oti.11.

Kaada, B. (1986). *Sudden infant death syndrome – The possible role of 'the fear paralysis reflex'.* Oslo, Norway: Norwegian University Press. 56 p.

Krefft, A. (2007). *Diagnosis of functions of invisible phenomena: statistic mathematic analysis.* Wrocław, PL: Oficyna Wydawnicza Politechniki Wrocławskiej.

Linder-Lucht, M., Othmer, V., Walther, M., Vry, J., Michaelis, U., Stein, S., Weissenmayer, H., Korinthenberg, R., Mall, V. (2007). Validation of the gross motor function measure for use in children and adolescents with traumatic brain injuries. *Pediatrics.* 2007 Oct;120(4):e880-6.

Luria, A. R. (1978). *Functional Brain Organization.* (In Book:) *Biological-scientific basis of psychology.* Ed. By A. A. Smirnov, A. R. Luria, V. D. Niebilistsin. Moscow, Russ: Pedagogika p.109-139.

Magnus, R. (1925). *Animal posture.* Croonian Lecture. Utrecht, Netherlands: University of Utrecht. p. 1-15. www.jstor.com

Mahoni, F. (1999). *Hypertonic muscle release.* Moscow, Russ: Assent.

Masgutova, S. (1997). *Creative art and Brain Gym®. Using Brain Gym® in creative art for learning, integrated being and personal growth.* Moscow, Russ: Personality Development and Art Education Research Institute of the Russian Education Academy and 'Ascent' Educational Institute of Psychological Assistance.

Masgutova, S. (1998). *The basis of psychological work with children using their pictures.* Bydgoszczy, PL: Psychological Institute Publications.

Masgutova, S. (2007). *Neuro-sensory-motor development: visual and auditory reflex integration. Facilitation program of development and learning for children and adults.* Warsaw, PL: ISMI.

Masgutova, S. (2006). Infant reflex integration in movement development. In materials of Polish National Conference: *Modern Methods of Stimulation of Movement Development and Learning in Children with Difficulties in Learning, ADHD and Autism.* 29.04. 2006. Maternity Department, Pomorska Medical Academy in Szczecin MINK. Warsaw – Szczecin, Poland.

Masgutova, S. (2009, 2011). *MNRI® visual and auditory reflexes integration: MNRI® facilitation program of development and learning for children and adults.* Revised edition. Orlando, FL, USA: SMEI.

Masgutova, S. (2011). *Infant dynamic and postural reflexes. Neuro-sensory-motor reflex integration.* Revised and edited. Scientific-practical manual. Orlando, FL, USA: SMEI.

Masgutova, S. (2012). *MNRI® master's: Theory and practice enrichment - level one.* Orlando, FL, USA: SMEI.

Masgutova, S., Akhmatova, N. (2004, 2007). *Integration of dynamic and postural reflexes into the whole body movement system.* Scientific edition: Prof. N. Akhmatova. Warsaw, PL: MINK.

Masgutova, S., Akhmatova, N. (2010). *MNRI® re-patterning and integration of dysfunctional and pathological reflexes - Svetlana Masgutova® method of neuro-reflex integration.* Second Edition. Revised. Orlando, FL, USA: SMEI.

Masgutova, S., Akhmatova, N. (2007, 2013). *Masgutova method of birth and post-birth reflex integration. Sensory-motor aspect.* Revised edition. Orlando, FL, USA: SMEI.

Masgutova, S., Akhmatova, N., Kiselevsky, M. (2008). Clinic-immunological assessment of effect of the therapy program of neuro-sensory-motor integration of reflex patterns at chronic inflammatory diseases of respiratory system. *Russian Immunology Journal,* 2(11),#4. (Russ.). 454–463.

Masgutova, S., Masgutov, D., Akhmatova, N., Akhmatov, E. (2011). *MNRI® Neuro-Structural Reflex Integration.* Revised, new research material included. Warsaw, PL: ISMI.

Masgutova, S. K., Akhmatova, N. K., Lebedinskaya, O. V. (2013). *Clinical-immunological assessment of therapy effect of the neuro-sensory-motor integration program of reflex patterns in airway chronic inflammatory diseases.* Event Abstract. International Congress in Milan. Front. Immunol. Conference Abstract: 15th International Congress of Immunology (ICI). doi: 10.3389/conf.fimmu.2013.02.00865.

Masgutova, S. with Kowal, J., Mazur, G., Masgutov, D. (2005). *Neurokinesiology tactile therapy™ by Dr. S. Masgutova*, Scientific Edition: Dr. A. Regner, J. Szymczak. Warszawa, PL: MINK.

Masgutova, S., Masgutov, D. (2005a). *Integration of lifelong reflexes.* In: The Neuro-Kinesiology Rehabilitation Camp. Education Materials for Parents and Specialists. Warsaw-Krynica-Zdroj, PL.

Masgutova, S., Masgutov, D. (2005b). *Dr. Svetlana Masgutova method of facial reflexes integration™. Techniques for facilitation of motor and speech development.* Warsaw, PL: MINK.

Masgutova, S., Masgutov, D. (2011a). *Tactile integration. Masgutova Method ® of neurosensorimotor facilitation.* Orlando, FL, USA: SMEI.

Masgutova, S., Masgutov, D. (2011b). *MNRI: Proprioceptive-cognitive integration. Optimize innate auditory-ocular processing and learning skills.* Manual. Orlando, FL, USA: SMEI.

Masgutova, S., Masgutov, D. (2012). *MNRI® archetype movement integration. A blueprint for movement and cognitive development.* Orlando, FL, USA: SMEI.

Masgutova, S., Masgutov, D. (2013). *Parents' guide to MNRI®: Neurosensorimotor reflex integration.* Orlando, FL, USA: SMEI.

Masgutova, S., Regner, A. (2007). *Language development using sensory-motor integration approach.* Warsaw, PL: Continulo.

Pavlov, I. P. (1960). *Conditioned reflexes: An investigation of the physiological activity of the cerebral cortex.* (Anrep G.V., D. Sc. Trans., 1960). NY, USA: Dover Publications Inc. (Original work was published in 1927)

Pilecki, W., Masgutova, S., Kowalewska, J., Masgutov, D., Akhmatova, N., Sobieszczanska, M., Kalka, D. (2012). The impact of rehabilitation carried out using the Masgutova neurosensorimotor reflex integration® method in children with cerebral palsy on the results of brain stem auditory potential examinations. In journal: *Advances in Clinical and Experimental Medicine.* Official Organ Wroclaw Medical University. 2012 (3). 2012, 21, 3, 363–371. http://www.advances. am.wroc.pl/pdf/2012/21/3/363.pdf.

Sadowska, L. (2001). Neurokinesiological diagnosis and therapy of children with motor-cognitive developmental dysfunctions. In Book: *Neurokinesiological Diagnosis and Therapy of Children with Mental and Motor Deficits.* Wroclaw, PL: AWF.

Skhomorokchov, A., Zakharov, S. (2006-2007). *The Technical Non-Medical Case Description of the Patients Brain Response for the MNRI® Intervention Program Registered by the Use of Encephalan-EEGR-19/26.* Non published research material – property of PL.: ISMI.

Sechenov, I. M. (1863/1995). *Reflexes of the brain.* (Russ.). Moscow, Russ: Pedagogika. p. 28-54. (Original work was published in 1863)

Vygotsky, L. S. (1986). *The child psychology. The problems of child development.* In 6 Books. Book – 4. Moscow, Russ: Pedagogika.

Other literature with no reference used in this article:

Amen, D. G. (2000). *Change Your Brain – Change your life the breakthrough program for conquering anxiety, depression, obsessiveness, anger, and impulsiveness.* New York, NY, USA: Three Rivers Press.

Ayres, J. (1975). *Sensory Integration and The Child.* Los Angeles, CA, USA: Western Psychological Services.

Bates, E. (1999). *Plasticity, Localization and Language Development. // In Changing Nervous System: Neurobehavioral Consequences of Early Brain Disorders.* Ed. S. H. Bronan. 1999: 214-53. New-York, NY, USA: Oxford University Press.

Doidge, N. (2007). *The brain that changes itself: Stories of personal triumph from frontiers of brain science.* USA: Viking, Penguin Group Inc.

Goldberg, E. (2009). *The new executive brain. Frontal lobes in a complex world.* New York, NY, USA: Oxford University Press.

Gross, D. R. (1987). *Psychology. The science of mind and behavior.* London, UK: Hodder & Stoughton.

FitzGerald, M. T., Gruener, G., Mtui, E. (2007). *Clinical neuroanatomy and neuroscience.* Amsterdam, Netherlands: Elsevier Urban and Partner. 458 p.

Khomskaja, E. D. (1987). *Neuropsychology.* Moscow. Russ: Moscow University.

Masgutova, S. (2007). *Neuro-sensory-motor development: visual and auditory reflex integration. Facilitation program of development and learning for children and adults.* Warsaw, PL: ISMI.

Merzenich, M. (2013). *Soft-wired. How the new science of brain plasticity can change your life.* San-Francisco, CA, USA: Parnassus Publishing.

Rutter, M. (1980). *Attachment and the development of social relationships.* In: Rutter, M. (ed). *Scientific Foundations of Developmental Psychiatry.* London, UK: Heinmann Medical.

We congratulate Erik in each small and bigger success and watch with great pleasure the emergence of his wonderful personality and new interest in communication. We thank his Mom, Eva, for her endless love, patience, and faith in Erik as a Winner! – Denis Masgutov & Renata Chodyko

Early Positive Results with Autism

Jo Anne Tierney, OTR/L, MNRI® Core Specialist, Mendota Heights, MN, USA

Jo Anne Tierney

Max is a young boy with autism. At the time this article was written, he was 4½ years old. Here I will describe some very nice changes seen in the areas of sensory integration, anxiety, behavior, and functional skills within just 5 months of adding the Masgutova Method of Neurosensorimotor Reflex Integration® (MNRI®) to his regular therapy program.

History of Therapeutic Intervention

At the time described in this article, Max attended an Early Childhood Special Education program where he received limited speech and occupational therapy services. He had received outpatient occupational therapy 2-3 times per week for two years, utilizing traditional therapy modalities (repetitive practice of functional skills, the Wilbarger brushing protocol, the Listening Program). During that time, some of the OT sessions involved co-treatments with a speech therapist, as it was felt that he was unable to tolerate two separate treatment sessions. He was also receiving physical therapy services once per week. At the recommendation of his physical therapist, he began therapy with an occupational therapist trained in MNRI®. At that time, 5 months ago, he began receiving MNRI® therapy once per week, with traditional therapy continuing twice per week. Approximately 3 months later, his schedule was changed to allow for MNRI® therapy twice per week, with traditional therapy once per week.

Presenting Problems

Max was referred for MNRI® therapy by his physical therapist. By her description, he was making nice progress, but she recognized that his underlying neurophysiology was not well-developed, and felt that the lack of appropriate reflex integration might be limiting his pace and level of development. She noted that motor planning and coordination were issues, as well as core strength (he was unable to play in prone or lift his head while swinging prone in a hammock swing). Significant sensory issues were noted (refusing to take off his socks at home or at therapy and meltdowns when asked to wear a helmet for bike riding). He frequently 'scripted' (rote recitation without actual interactive communication) and used his peripheral vision rather than giving good eye contact.

My initial Assessment 5 months ago indicated significant delays in his reflex development. He exhibited significant anxiety, with strong resistance to being touched. The initial reflex Assessment indicated hyperactivity of his Core Tendon Guard, Fear Paralysis, Foot Tendon Guard, and Babinski Reflexes, and very significant hypo-activity of upper limb reflexes. His Robinson Hands Grasp Reflex was very hypo-active with minimal active grasping; when he did grasp, it was a loose hanging grasp using the ends of his fingers, and minimal or no use of his thumbs. He was unable to focus in order to imitate grasp or finger patterns. His Hands Supporting Reflex was hypo-active and asymmetrical, with an incorrect pattern of abduction. Hands Pulling was also very hypo-active, with no engagement of his arms, and a strong head lag. Visual tracking was poor, with compensation by head righting (he needed to turn his head in order to follow the stimulus). With his head stabilized, he demonstrated a very narrow range for visual tracking.

MNRI® Assessment by Dr. Masgutova

On March 12, 2013, 4½ months after starting MNRI® therapy, Max underwent a complete reflex Assessment by Dr. Svetlana Masgutova. The results of this Assessment were compared to the reflex integration profile of neurotypical children, in order to create a more individualized Home Program for Max. The results of this Assessment are as follows, at right.

This comparative analysis demonstrates that Max's reflex pattern development is significantly below the norm when compared with neurotypical children. A number of Max's reflexes demonstrate an incorrect pattern of response, and are at the level of light dysfunction – these include Babinski, Foot Grasp, Hands Supporting, Segmental Rolling, Landau, Flying and Landing, STNR, Foot Tendon Guard, and Locomotion. Reflexes that demonstrate elements of a correct pattern and are at the boundary between dysfunction and normal function include Moro and Fear Paralysis Reflexes, and Abdominal Sleep Posture. The remaining assessed reflexes including Robinson Hands Grasp, Babkin, ATNR, Bonding, Bauer Crawling, Grounding, Head Righting, Spinal Galant, Spinal Pereze, Tonic Labyrinthine, Spinning, Balancing, Pavlov, Core Tendon Guard, Leg Cross Flexion-Extension, Automatic Gait, and Trunk Extension are in the functional range, but at a low or very low level of development.

Table 1. Reflex Pattern Evaluation Results: Reflex profile of Max in comparison with neurotypical children*

Reflex Pattern										
#	1	2	3	4	5	6	7	8	9	10
Reflex Group-1	Core Tendon Guard	Hands Grasp	Hands Pulling	Babkin Palmo-mental	Babinski	Leg Cross Flexion Extension	ATNR	Abdominal Sleep Posture	Bonding	Foot Grasp
Neurotypical	15.5	17	16	16	16.5	17	15	16	15.5	16.5
Max (March 2013)	14.25	12.25	15.25	12.25	9.25	13.25	11.75	10.25	11.25	9.25
Reflex Group-1	Automatic Gait	Bauer Crawling	Moro	Fear Paralysis	Hands Support	Segmental Rolling	Landau	Flying and Landing	Grounding	Head Righting
Neurotypical	17.5	15.5	15.5	14.5	15.5	15	15	14.5	16	16
Max (March 2013)	13	11.75	10.25	10	9.25	9.25	8.75	9.25	12.25	12.75
Reflex Group-1	Trunk Extension	STNR	Spinal Galant	Spinal Perez	Tonic Labyrinthine	Foot Tend	Spinning	Locomotion	Balancing	Pavlov Orientation
Neurotypical	16	16	15	16	16	15.5	15	14.5	15.5	18.5
Max (March 2013)	13.75	9.75	12.25	11.75	12.25	8.75	11.25	7.75	12.25	11.13

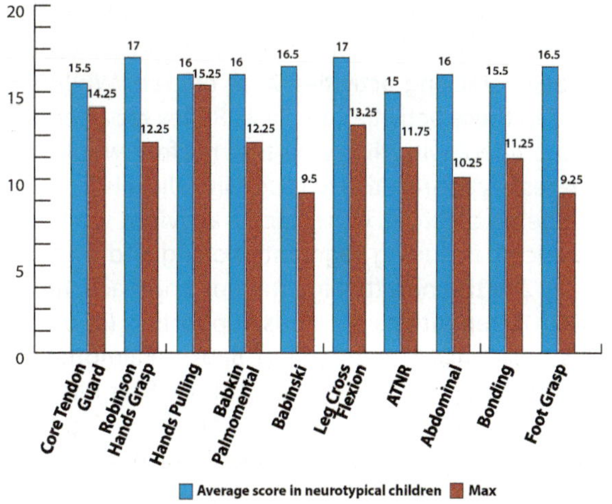

Graph 1 (above). Reflex patterns on the saggital plane of motor coordination; Max in comparison with neurotypical children

One reflex pattern – Hands Pulling – was within the norm when assessed by Dr. Masgutova. This represents a very strong and significant improvement since Max's initial Assessment with me 4.5 months earlier, described above. When one looks at the developmental purpose of this reflex, one can begin to understand how its development and integration into a child's motor system can impact so many aspects of that child's functioning. This reflex supports the development of muscle tone regulation in the arms and the development of precise movements of arms, hands, and fingers while reaching for objects. In this way, it supports the development of gross motor, fine motor, and cognitive skills. This reflex participates in the development of visual convergence-to-divergence, and has to do with engagement, curiosity, and pulling oneself into the world. The development toward integration of this and other reflexes is felt to be a significant factor in the many developmental gains that Max is experiencing.

Therapeutic Intervention Using the Masgutova Method® of Reflex Integration

At the start of MNRI® therapy 5 months ago (at which time, he was receiving traditional therapy twice per week, and MNRI® therapy just once per week), Max demonstrated very strong hyper-sensitivity to touch, as well as aversion to the therapy. Therefore, therapy began very playfully, with two of his favorite activities – riding a tricycle down the hallway, and riding on the 'horsey swing.' Calming tactile input from the MNRI® Tactile Integration protocol was introduced occasionally and briefly, at a level that he would accept, during his preferred play activities. When he arrived at the end of the hallway on the tricycle, a few Embrace Squeezes were applied before sending him off again. When he fell off the swing into a pile of pillows, Embrace Squeeze and other tactile inputs were again given, providing deep, calming pressure. Soon he was also accepting brief repatterning of the Hands Supporting Reflex because it was playful and because he knew that we would count to 10 and he would be done. Then it was back to playing.

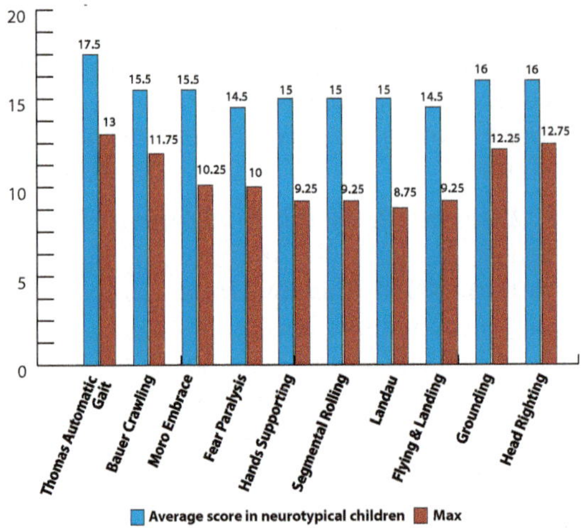

Graph 2 (above). Reflex patterns on the horizontal plane of motor coordination; Max in comparison with neurotypical children

Graph 3 (below). Reflex patterns on the dorsal plane of motor coordination; Max in comparison with neurotypical children

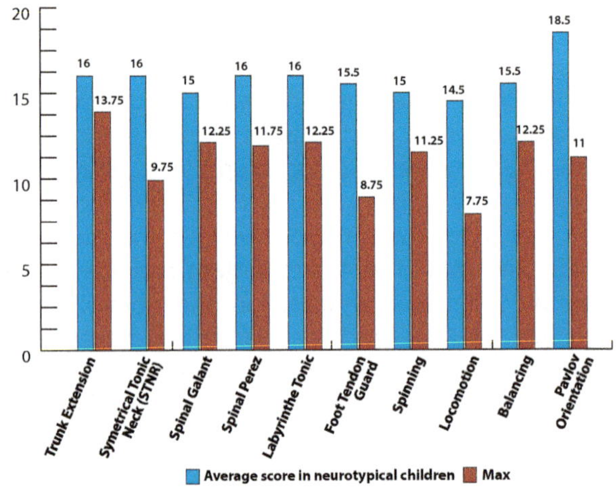

Because Max was used to doing fine motor activities during his traditional therapy sessions, pegboard and bead-stacking activities were offered. While engaged in these activities, Babinski Reflex repatterning was provided, with the hope that his feet would become less hyper-reactive to touch and allow him to tolerate being barefoot. Other tactile activities were also introduced, including segment stroking and traction. On day 2 of this new (to him) therapy approach, he accepted 10 repetitions of Hands Supporting (in one of the 3 positions), 1 repetition of Babkin-Palmomental on each hand, brief repatterning of the Fear Paralysis Reflex, and a total of 2 repetitions of Embrace Squeeze on each limb, either when distracted or between activities, briefly.

Initially, Max's therapy took place throughout the clinic – in the gym or the hallway – as he was averse to entering the small room where the therapy table was located. However, by day 2, he chose to ride the trike into the small room, where he enjoyed playing briefly with a balloon. On the 4th day of therapy, he announced to his mother that he did not want to go into the small room. However, he chose the room twice that day – once when riding the tricycle, and later when given a choice of the gym or the small room! On that day, he accepted some therapy on the table, mostly in sitting, but also in a supine position for Hands Supporting. On day 5 of therapy, we spent the first 20 minutes of the session in the small room on the therapy table. He fussed a bit, but accepted it fairly well. On the 7th day of therapy he accepted ½ hour on the therapy table, and on the 8th day he was on the therapy table for 50 minutes, until the therapist suggested going to the gym! Throughout this time, he became more and more comfortable with the therapy. Strong initial whining and fussing gradually decreased, as did his level of anxiety.

Two months following the start of MNRI® therapy (still at one time per week at this point), Max was demonstrating significantly decreased anxiety; he was becoming very playful, and started giving hugs. He was accepting therapy quite well, and was becoming very calm during Tactile Reflex Integration and Core Tendon Guard repatterning exercises. He even began to allow the therapist to work briefly on his head and face in order to address tactile, auditory, and visual hyper-sensitivity. At this point, he also allowed removal of his socks, although he was very anxious about this, and it was for a very brief period of time. Two weeks later, he allowed his socks to be off for 10 minutes; one week after that, they were off for the entire hour, plus the hour that he spent with the PT. He now automatically takes off both shoes and socks upon request at the start of the therapy session.

Progress Noted within the First 5 Months of Starting MNRI®

Although some of the items listed are changes noted during his therapy sessions, most of the following changes are reported by his mother.

Sensory Integration:

Max demonstrates decreased anxiety and a significantly greater acceptance of MNRI® therapy. Tactile processing has improved: he now allows socks off at therapy, and allows touch to his face, including around his ears and eyes, which was initially extremely hyper-sensitive and aversive to him. He has improved tolerance for wearing hats, acceptance of new foods, and now allows his toenails to be clipped (it used to be a struggle, with 2 people needed to get the job done). He had his first haircut at which there were no problems whatsoever, including using the clipper around his ears (this was previously an ordeal for him). At his first chiropractic visit recently, Max readily climbed onto the table and had no problem accepting treatment.

Max in 2014.

Body Awareness and Motor Control:

Max demonstrates improved overall awareness of his body and position in space, and improved head control when pulled to sit from supine (he previously had a strong head lag, and no engagement of his arms for pulling himself up). When transitioning from sitting to lying down, he previously rotated onto his hands, then lowered himself down; he is now able to transition from sitting to supine simply by leaning back and lying down. He is now moving his body with less rigidity and with more natural fluidity of motion, and has improved coordination for swimming, with use of his arms and legs at the same time. Max is also beginning to blow bubbles in the water, has improved ability to catch a ball and to be engaged in a game of catch, and an improved enjoyment of coloring and other fine motor activities. Max now uses his pointer finger to play with the iPad (previously he could not figure out how to use the iPad).

Self-Care Independence:

Max has spontaneously fed himself with a spoon for the first time (his parents had given up on this, but always had feeding utensils available, just in case), has begun dressing himself independently, and employs improved problem-solving when dressing (the first day he did this, his mother told him that he might be cold with the short-sleeve shirt he had chosen; he then went back to his room and put a long-sleeve shirt on under his short-sleeve shirt). He is beginning to wash himself during baths.

Social/Emotional and Communication:

Max now holds hands with his mother, comes forward for hugs and is beginning to reach with his arms for a hug (previously, he turned and backed up for hugs). He is showing greater interest in his peers. He is beginning to ask questions several times per day, and for the first time, he spontaneously asked a 'wh' question: "Where's Daddy?" He is beginning to sing songs all the way through and is following directions on a Dora game.

Max's mother decided to restructure Max's therapy schedule, in order to receive MNRI® therapy three times per week. Therapy continues to incorporate work with functional and age appropriate skills. However, it is felt that these skills will develop more easily and quickly as he develops a stronger neuro-physiological basis to support them. The reflexes are the foundational pieces for future growth and development. As these reflexes are addressed and appropriate nerve pathways develop and become myelinated, his responses will continue to move in the direction of positive protection and growth. The feeling of safety and inner peace that comes with this will allow for the drawing out of developmental resources, and will provide the neuro-physiological basis for further motor, emotional, and cognitive development, in order to reach his full potential.

Although we as MNRI® specialists see the value of this therapy for our clients, it is especially gratifying when a parent offers their observation and belief that these amazing, functional changes can be attributed to this work. It is deep and profound work, and while we know that we are making neurological changes, it is the day-to-day, functional gains that are important to our clients and to those who love them.

I would like to thank Max and his amazing family for allowing me to be a part of their lives and their journey. It never ceases to amaze me that so many areas of life can be impacted simply by addressing reflex development and integration. Max, you are such a joy – I wish you the brightest of futures! Love you Max! – Jo Anne Tierney

The Effectiveness of MNRI® for a Child with PDD-NOS and Social Language Delay

Lori Burgess, MST CCC SLP, MNRI® Core Specialist , Minneapolis, MN, USA

Introduction

Lori Burgess

Professionals involved in every aspect of development are acutely aware of the increasing prevalence of motor, emotional, and cognitive disabilities among young children. The epidemic of Autism Spectrum Disorders that began in the 1980s continues today. Parents, educators, and child development professionals are all in search of effective methods to address the problems that challenge so many of our children. This study examines the effects of Masgutova Neurosensorimotor Reflex Integration (MNRI®) on one child with diagnoses of PDD-NOS (Pervasive Developmental disorder), Mixed Developmental Delays, Mixed Receptive and Expressive Language Delay, and Social Language Delay. The goal of the therapy program was to improve the child's social language skills.

Case History/Background Information

JC was delivered by C-section weighing 2 pounds, 13 ounces and was 15 inches long. The pregnancy had been complicated by his mother's diagnosis of preeclampsia and HELLP syndrome at 29 weeks gestation. As a 31-week preemie JC remained in the NICU for 5 weeks due to immature lung development. He was placed on oxygen for the first 23-48 hours and tube fed for the first 3-4 weeks in the NICU. Breast feeding was difficult due to his poor ability to latch on.

JC's developmental milestones were delayed for both walking and talking. He did not experience frequent ear infections, nor did he have P. E. tubes. There were no reported concerns regarding his acuity for speech frequencies bilaterally. No allergies were reported. He slept well, though he had frequent nightmares. Current medications included a multivitamin and Miralax for constipation.

In 2011 at age four, JC was diagnosed with PDD-NOS, Mixed Developmental Delays, Mixed Receptive and Expressive Language Delay and Social Language Delay.

The Presenting Problem

According to a report submitted by his parents, in 2011 (at age 4) JC had verbal rituals in which he repeated phrases and sentences over and over again, lengthy emotional outbursts, and limited social pragmatic skills. JC had difficulty with auditory and visual input, was scared of new situations and transitions, and required instant gratification. He often kicked, screamed, and cried for hours at a time when his rituals were not adhered to. JC had difficulty with bowel movements, averaging once every 3 or 4 days. He was still in diapers and did not appear to be motivated for toilet training. His play skills remained parallel and not interactive. Sharing and turn taking were difficult for JC (i.e. head butting when a toy was not given to him). JC's language skills were limited to labeling, with emerging ability to ask questions such as 'what' and 'where'. He enjoyed repetitive play, looking at cars, and learning about the makes and models of cars. He did not engage in pretend play. JC enjoyed books, but showed no interest in the written word and/or sound symbol relationships. His comprehension skills were unknown due to lack of expressive speech.

JC has received the following diagnoses:

1) Pervasive Developmental disorder – Not otherwise Specified (PDD-NOS) – 299.9. This diagnosis indicates some characteristics of children on the autism spectrum but of lesser amount or intensity. Children with this diagnosis generally carry a better prognosis for long-term developmental outcomes.

2) Mixed Receptive and Expressive Language Delay – 315.32 - Difficulties in both understanding and using language as demonstrated by results of the CASL and objective information.

3) Social language delay 438.19 – Difficulty with pragmatic skills such as the initiation, maintenance, and termination of conversations, reading emotions, accepting delay of gratification, eye contact, turn taking, asking/answering questions, describing feelings, gaining attention, greeting others, etc, as determined by results of the CASL and objective information.

4) Mixed Developmental Delays – 315.9 – Unspecified developmental disorder – NOS – as determined by the neuropsychological examiner.

Known treatment history included Birth to Three services via his home school district, which provided an Early Childhood Special Education classroom and traditional speech therapy. Lack of progress was the reason for his referral to outside speech therapy.

The period under study for this report began September 7, 2011 (JC's age was 4 years 1 month) and ended August 22, 2012 (JC's was age 5). During this time he continued to receive traditional physical therapy once a week and traditional occupational therapy twice a week. Beginning on September 8, 2011, he began to receive MNRI® twice a week in place of traditional speech therapy.

Evaluation

The following instruments were used in evaluating JC:

1. Case history completed by JC's parents (as reported above)
2. The Comprehensive Assessment of Spoken Language
3. MNRI® Reflex Development Assessment
4. Review of the neuropsychological evaluation

The Comprehensive Assessment of Spoken Language (CASL):

JC's was given the CASL and his scores are reported below:

Subtest 1. Comprehension of Basic Concepts: measures the ability to comprehend words representing basic precepts and concepts. These words are nouns, pronouns, prepositions, and adjectives for concepts such as size, shape, quantity, quality, distance, direction, and position. The concepts tested are important parts of the instructions given to children during their early years in school.

Subtest 2. Antonyms: measures the examinee's ability to identify words that are opposite in meaning. Because this test is not given to 4 year olds no Pre Test score is given.

Subtest 3. Sentence Completion: measures word retrieval, knowledge, and expression within a linguistic context.

Subtest 4. Paragraph Comprehension: measures the examinees' comprehension of syntax by means of a series of spoken narratives. The syntactic complexity builds within each narrative and from narrative to narrative. This test is not intended to assess word comprehension, ability to make inferences, or general comprehension of spoken connected language. It assesses the ability to derive meaning from the syntactic structures.

Standard scores on the CASL are based on a mean of 100, with a Standard Deviation of 15, standard scores between 85 and 115 being within normal limits. Percentile Rank scores indicate the percentage of children in the standardization sample who earned either the same score or lower, therefore, a child ranking at the 35th percentile has earned a score equal to or surpassing 35% of the children in the group. Age equivalency scores indicate the age at which children in the normative sample received the same raw score. Stanines are given with the mean being 5 and a standard deviation of 2.

JC's scores demonstrate good labeling skills but he presented with a lack of interaction and ability to use words with intentional communication. (See table below.)

Informal Language Sample:

Sub-test	Raw score	Standard Score	Percentile	Stanine	Test Age Equivalent
1. Comprehension of Basic Concepts	21	103	58	5	4.3
2. Antonyms					
3. Sentence Completion	13	108	70	6	4.7
4. Syntax Construction	12	99	47	5	4.0
5. Paragraph Comprehension	3	91	27	4	3.0
6. Pragmatic Judgment	4	88	21	3	3.0

An informal language sample was obtained in a free play situation in order to analyze JC's language use in context. An analysis of intelligibility was also calculated to assess how well JC was understood by an unfamiliar (albeit trained) listener in conversation.

Number of spontaneous utterances during free play: 100

Number of utterances completely intelligible -58 – left alone to play

Number of utterances partially unintelligible – 42 – attempted reciprocal play.

JC was given a choice of toys to play with. He chose cars and immediately began to take them out of the bag. JC did not notice the therapist's hand was outstretched to receive her share of the cars. After repeated attempts to get JC to share the cars, the therapist then took the bag and started playing with the left over cars. Repeated attempts to engage JC in reciprocal play did not get any response except that, the more requests that were made, the more JC would curl up while sitting, talk/mumble louder with a higher pitched voice, and turn his back to wherever the therapist was sitting. At one point the therapist took a car JC had finished playing with. JC became more and more agitated when he realized he was not going to get the toy back right away and that the therapist was playing with the car in a manner that was not how he started to play. During this time, JC's expressive language skills appeared to deteriorate. He had the vocabulary to express himself, but he did not appear to know how to use his skills in a functional manner when stress was introduced. JC could label everything he saw, but he could not play verbally with another person. Agitation became a huge factor in his vocal quality. He began to mumble and miss correct place, manner, and voice for connected speech making him highly unintelligible. This concern was noted in all prior reports from a speech pathologist and the neuropsychological evaluation. JC qualified for speech services for social language skills.

MNRI® Reflex Development Assessment

The MNRI® Assessment protocol determines whether a child's reflex development or integration is present to an age-appropriate degree, appropriately matured and functional, and integrated with overall sensorimotor development and skills used in learning.

An informal reflex assessment was given by the author on September 7, 2011. JC demonstrated a lack of integration in the following reflexes: Fear Paralysis, Bonding, Trunk Extension, Spinal Galant, Flying and Landing, Spinal Perez, ATNR, STNR plus visual tracking, Babkin, Hands Grasp, Hands Pulling, Foot Tendon Guard, Automatic Gait, Hands Supporting, Landau, and Babinski. When reflexes fail to mature and dysfunctional reflex responses become habitual, they interfere with motor, emotional, and cognitive development. A child with this poor level of reflex integration is getting unreliable feedback from his sensory system and is at the mercy of irregular involuntary motor reactions to external and internal stimulation. Positive compensations drain energy away from assigned tasks and cause fatigue and frustration. Negative compensations can take the form of withdrawal, impulsivity, emotional meltdowns, and misbehavior. He would be at high risk for underachievement, especially in the presence of learning challenges and stress. All of this was true for JC.

The Intervention: MNRI®

The Masgutova Method® consists of specific interventions to integrate retained primary reflexes in order to awaken self-regulation of the sensory, motor and proprioceptive systems that influence learning and development. Research regarding the use of these techniques demonstrates significant improvement in motor, emotional, language and cognitive function in people diagnosed with LD, ADD, ADHD, PDD, PDDNOS, OCD, apraxia, Asperger's syndrome, ASD, Down Syndrome, Fetal Alcohol Syndrome, ODD, emotional disturbance, sensory processing disorders, Post-Traumatic Stress Disorder, cerebral palsy, genetic disorders, brain stroke, and others. This was the approach used with JC during this investigation period.

MNRI® uses 'regressive progressive' repatterning techniques that 'remind' the brain-body system of its genetic neurosensorimotor programs. These procedures involve repatterning exercises to correct basic reflex patterns and/or their variants when they are dysfunctional, immature, or unintegrated. The term 'regressive progressive' refers to the use of a developed or matured version of a genetically based motor pattern, combined with a precise posture and associated sensory or proprioceptive stimulation, as the foundation for future neurological development. In going back (regressive) to the 'pure' innate form of the reflex . . . the brain recognizes a template for growth and development (progressive) supplied by its genetic inheritance. As a result of neurological plasticity, the relevant nerve net system can then rebuild according to its original code and subsequently mature and integrate.

Following his evaluation for retained primary reflexes, an MNRI® therapy program was developed for JC. He received MNRI® therapy twice a week on those reflexes for approximately 1 year. JC's parents were given a Home Program consisting of exercises for Bonding, Hands Supporting, Spinal Galant, and Visual Convergence-Divergence, and Eye Tracking.

Two examples of the chosen reflexes and exercises are as follows:

The Asymmetrical Tonic Neck Reflex (ATNR) was chosen due to its developmental effect on hemispheric differentiation and specialization for hearing, auditory, and visual integration, attention, and memory; and its importance as the foundation for binaural hearing and binocular vision. The trigger for this reflex is reflexive and passive head turning to the side. In response, the limbs extend on the same side of the body to which the head is turned, while the limbs on the opposite side flex.

ATNR activation: The patient is supine, head turned toward straight limbs; opposite limbs are flexed. Activation is done by gently pulling on the straightened arm and leg (arm at a 30° angle). The exercise is repeated for the opposite side. During this exercise the acoustic reflex can also be activated by providing different stimulation for the left and right auditory fields.

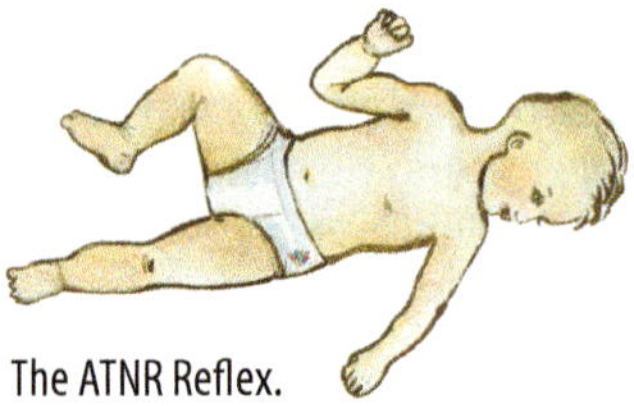

The ATNR Reflex.

The Symmetrical Tonic Neck Reflex (STNR) was chosen due to its developmental effect on binocular vision and near-far accommodation. In healthy infants it provides natural vision training to expand distance and peripheral vision, coordinate proprioceptive, vestibular and visual perception, and develop spatial concepts as well as visual tracking skills for horizontal plane, vertical plane, horizontal arc, and convergence-divergence. These two tonic neck reflexes are designed to help the eyes and ears team together, which was extremely difficult for JC.

In the STNR an upward (extension) or downward (flexion) movement of the head causes a reciprocal movement of the limbs. Head flexion (chin to chest) causes arms to flex and a tendency for the legs to extend. Head extension (chin up) causes arms to extend and a tendency for the legs to flex.

STNR-2 activation: the patient moves into and holds the STNR Sphinx position for 7–8 seconds, exhaling while the therapist triggers the neck flexors and extensors by holding on the front and back of the head. The patient presses back with a very delicate strength. The exercise is repeated several times allowing for rest in between activations. Additionally, the patient performs horizontal eye tracking and convergence-divergence training in the STNR positions.

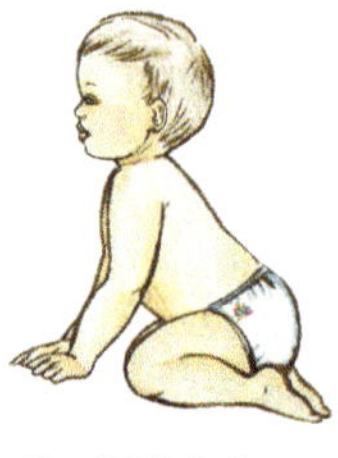

The STNR Reflex.

Results

JC progressed extremely well with his MNRI® program. He exceeded all goals set for him, with the exception of behavior, which was still seen at home on an inconsistent basis, but not at therapy. Highlights of JC's improvements included:

1. Integrated visual and auditory processing.
2. Eye tracking improved on all levels. Previously fragmented and unstable, it became smooth, steady, and differentiated from head movement.
3. Ability to read independently with comprehension.
4. Ability to ask and answer all types of questions, including 'why' questions, immediate present, immediate past, future and past questions.
5. Ability to express himself intelligently and clearly with full sentences.
6. Ability to express feelings, wants, and needs.

7. Ability to share toys and space with other children.
8. Ability to accept unpredictable events.
9. Ability to accept wait time.
10. Ability to accept changes in routines and schedules (although this remained slightly difficult at home).
11. Integration of many of the primary reflexes related to learning, short and long term memory, pragmatics and overall muscle tone regulation.
12. Ability to attempt to 'self calm' in many difficult situations.
13. Ability to listen to lengthy auditory information and provide answers appropriately.
14. Increased success with potty training – both physical production and emotional reaction.

CASL Evaluation:

JC's remarkable progress is strikingly clear in his pre and post scores on the CASL. In Comprehension of Basic Concepts he rose from the 58th to 92nd percentile, and from Test Age Equivalent 4.3 to 6.6, more than 2 years growth. His understanding of antonyms (no pre-test available) now puts him in the 99th percentile, above his chronological age. His ability to complete sentences is also in the 99th percentile, up from 70th, and at a Test Age Equivalent of 7.7, up from 4.7, showing growth exceeding 3 years. Syntax construction went from the 47th percentile to the 97th percentile with Test age equivalent rising from 4.0 to 7.1. His lowest pre-test score, 27th percentile and 3.0 Test Age Equivalent in Paragraph Completion has risen to 87th percentile and 6.5 Test Age Equivalent. Most importantly, his pragmatic judgment percentile went from 27 to the 93rd percentile – improving 3.6 years Test Age Equivalent (from 3.0 to 6.6). Overall, he demonstrated years of growth and moved to significant percentiles in his spoken language skills.

MNRI® Reflex Development Assessment:

Improvement in reflex motor patterns was especially significant in terms of:

- protective responses and attachment: Fear Paralysis, Moro, Bonding
- tonic reflexes: ATNR, STNR, Babkin, Babinski
- righting reflexes: Hands Supporting, Trunk Extension, Foot Tendon Guard
- spinal control: Spinal Galant, Spinal Perez, Landau
- motor coordination and precision: Hands Grasp, Hands Pulling, and Automatic Gait, Flying and Landing
- cognitive activators: Eye Tracking, Acoustic Reflex.

The above results demonstrate an improvement in flexion/extension bio-dynamics, changes in the auditory system for listening and focusing, decreased reactivity to sounds, improved ocular-kinetic and ocular-vestibular reflexes, and improved convergence/ divergence, including the ease of transition of the visual transition point. JC has also demonstrated improvement in attention and long-term memory which now allows for accommodation and assimilation of new information and the ability to use expressive language fluently and easily. JC was also now reading fluently with comprehension.

Interpretation

The extremely positive results (see table at right) achieved by JC after only 12 months of treatment suggest significant changes in his neurological development. In the earliest stages of development, the main purpose of the reflexes is protection and survival. Only when their circuits have gone through precise stages of emergence, activation, and maturation are they able to support higher levels of the brain that need to be free from functions of protection and survival in order to promote learning and development. MNRI® procedures are oriented toward neurological integration of reflex circuits in order to facilitate just this kind of growth.

According to the MNRI® model, the reflex integration exercises JC performed have activated lower motor neurons (extrapyramidal nervous

The Comprehensive Assessment of Spoken Language (CASL)

Sub-test	Raw score		Standard Score		Percentile		Stanine		Test Age Equivalent	
	Pre	Post	Pre	Post	Pre	Post	Pre	Post	Pre	Post
Comprehension of Basic Concepts	21	**36**	103	**121**	58	**92**	5	**8**	4.3	**6.6**
Antonyms		**20**		**146**		**99.9**		**9**		**7.4**
Sentence Completion	13	**25**	108	**138**	70	**99**	6	**9**	4.7	**7.11**
Syntax Construction	12	**25**	99	**129**	47	**97**	5	**9**	4.0	**7.1**
Paragraph Comprehension	3	**27**	91	**117**	27	**87**	4	**7**	3.0	**6.5**
Pragmatic Judgment	4	**22**	88	**122**	21	**93**	3	**8**	3.0	**6.6**

tract) in his brain stem and created the necessary neural connections for maturation, enabling him to gain freedom from excessive protection and survival responses, and to use the higher levels of his brain for learning and development. In MNRI® terms, JC's progress in developing age appropriate inner self-regulation would also indicate improvement in his peripheral nervous system.

Conclusion

The author has used MNRI® systematically during the last 3 years in her work as a speech pathologist with children presenting with a variety of communication disorders such as ASD, apraxia, phonological impairments, auditory processing, Down syndrome, cochlear implants, Angelman syndrome, Fetal Alcohol Syndrome, receptive and expressive language impairments, and various other genetic disorders. Clinical tests and observations demonstrate that the MNRI® method has provided JC with expressive language skills, internal self regulation in behavior, which in turn, have allowed for substantial improvements in his overall learning and communication.

References

Masgutova, S. (2011). *Infant dynamic and postural reflexes. Neuro-sensory-motor reflex integration.* Revised and edited. Scientific-practical manual. Orlando, FL, USA: SMEI. 190 p.

Masgutova, S., Akhmatova, N. (2007). *MNRI® Re-patterning and integration of dysfunctional and pathological reflexes - Svetlana Masgutova® Method of neuro-reflex integration.* Revised and changed. Orlando, FL, USA: SMEI. 82 p.

Thank you to JC and his family for participating in this study and for being true partners in therapy. – Lori Burgess

Success for a Typically Developing Child at Home and School

Diane Whiteside, PT, MNRI® Core Therapist, San Ramon, CA

Eric was a 5½ year old boy struggling in kindergarten. He was a typically developing child without a medical diagnosis. His mother brought him in for an evaluation because of his low self-esteem, although she wasn't sure MNRI® would help him. She stated he had difficulty with attention, fine motor skills, and some gross motor activities. He recently participated in a soccer clinic but got frustrated because he could not kick the ball into the goal. His mother reported Eric verbally expressed that he was no good, he hated his life, and that he was "nothing".

Diane Whiteside

Developmental History

Eric was born full term via C-section weighing 7 pounds and healthy. Mother reported he reached his developmental milestones within typical ranges. He sat at 7 months, never crawled, but walked at 14 months. He was toilet trained at 3½ years. His health has been generally good with the exception of a visit to the ER once. He received nebulizer treatment for a cold/allergy.

Assessment Tools

Eric was evaluated using standardized tests, such as BOT-2, Beery-Buktenica, as well as MNRI® reflex testing and therapist observation. The findings revealed:

Bruininks-Oseretsky Test for Motor Proficiency (BOT-2) Results: The BOT-2 placed Eric as average for strength. He scored an average age of 5 years. He scored below average in all other areas of the test including Fine Motor Precision and Integration as well as Fine Manual Control of Dexterity; Upper Limb Coordination; Manual Bilateral and Balance Coordination; and Body Coordination for Running Speed and Agility with an average age of 4.0 years

Beery-Buktenica Developmental Test of Visual–Motor Integration Results: Eric was average in all areas of the test except Motor Development where he scored significantly lower, with an average score below 3 years of age.

Sensory Processing Profile: The results demonstrated that Eric did not perform in auditory processing at the level of other children his age. He frequently appeared to not hear what was being said to him and didn't respond to his name being called. He had trouble functioning with a lot of background noise, however he enjoyed creating his own vocalizations of strange noises.

Visual Processing: Eric appeared to be generally good with his visual processing except he had difficulty putting puzzle pieces together as compared to children his own age. He also had a hard time finding an object in a cluttered background. His horizontal tracking was noted to be to only 50 degrees.

Eric as he is today.

Vestibular processing: Eric sought out excessive movements including jumping, whirling, climbing, crashing, which often interfered with daily routines like sitting still and he had difficulty standing in line.

Touch processing: Eric strongly disliked having a messy face and hands. He complained a lot about his shoes being uncomfortable as well as haircuts, nail cutting, teeth brushing, and certain fabrics. He occasionally avoided going barefoot and reacted emotionally to touch. He also had a decreased awareness of pain and temperature.

Multisensory processing: Eric had difficulty paying attention, got distracted with other activity going on in a room and frequently appeared oblivious within the active environment. He was also unaware when clothes were twisted on his body.

Sensory processing related to endurance and tone: Eric tired easily when standing or holding a particular position. The muscles in his hands appeared weak. He didn't sleep well, and in the morning appeared lethargic and had no energy.

Modulation related to body position and movement: Eric frequently sought opportunities to fall without regard to personal safety, as it was an activity he enjoyed. He occasionally took excessive risks during play, climbing up high and jumping off a tall structure. As a result, he was accident-prone. Eric became overly excited during movement activities, was constantly 'on the go,' and avoided quiet play activities. He got his feelings hurt easily so appeared to be somewhat defenseless with other children. He frequently was reserved or withdrawn from activities, watching everyone as they moved around a room and occasionally would avoid eye contact. He appeared not to notice when people came into the room.

Behavior and Emotional Responses: Eric was sensitive to criticism and had a low tolerance for frustration. He had emotional outbursts when he was unsuccessful at a task. He had fears, temper tantrums, and nightmares. He expressed that he didn't want to grow up, seemed anxious, could be uncooperative, cried easily, and was overly serious. He also had difficulty making friends and expressed feelings of failure saying, "I am nothing."

MNRI® Therapy Begins

During the three months described in this article, Eric received MNRI® treatment once a week from me, a Core Specialist, and ½ hour nightly from his parents.

October, 2012 Before Eric first began his sessions, his mother said he would jump on his bed and swing his stuffed animals around while his parents read his nighttime stories.

After one week of doing daily reflexes, his mother reported that he began to sit beside her and listen to the nighttime story. He even asked questions about the book. This was such a change that she was truly impressed.

After two weeks of MNRI® therapy, Eric's teacher told his mother he had had a great day. He made two soccer goals and was visibly proud of himself. Since then he began to express more positive attributes.

November, 2012 During week three, Eric was able to sit on the sofa with his father as they watched TV instead of jumping on the sofa around and on top of Dad.

Eric's Thanksgiving class party took place during week four and he was initially unable to participate. The sounds and activities were over stimulating for his system. His mother took him out of the classroom, gave him some quiet time and performed some reflex techniques. He returned to the party after 20 minutes and participated with the other children and even enjoyed himself.

December, 2012 By the seventh week, Eric's parents noticed he unpacked the Christmas ornaments without breaking one. He also decorated the tree and house in one hour. His mother said that in the past, this activity would take the whole evening. He was attentive during the activity and worked quickly.

During Thanksgiving Eric sat on the sofa with a smartphone and played games without interacting with everyone but, at Christmas, the whole family noticed a change in Eric. He sat on his aunt's lap and had a conversation with her as well as played with his cousins.

Parent's report, "Calmer house, calmer child": Eric's parents report their house feels calmer. They are able to finish a conversation without being interrupted. He is able to play quietly by himself and his conversations follow a connected thought, not jumping from subject to subject. He now draws and enjoys sharing his pictures. Most importantly, he is proud of his accomplishments.

Eric enjoying water sports.

Since he started MNRI® reflex therapy, Eric sleeps better and wakes up happier. This is such a change since it use to take him 1 to 2 hours to fall asleep. Now it only takes him 15 minutes to fall asleep and he sleeps through the night instead of waking up several times.

He now asks for his daily reflex therapy. His parents have made it a nightly bedtime routine.

Besides the behavioral, emotional, and motor improvements observed by his parents and therapists, Eric was re-tested after three months using BOT-2 subtests and the results were significant. He doubled his performance in some areas. He improved stationary hopping from 2 hops to 6 hops. He improved side hopping from 0 hops to 3 hops. He improved catching the ball from 1 catch out of 10 tries to 3 out of 10 tries. His handwriting improved to the point of enjoying writing, drawing, and coloring.

Despite her original doubts, today Eric's mother credits MNRI® for giving her son confidence. MNRI® has not only changed his life but has had a positive effect on the whole family, the parent's relationship, and extended family in ways they didn't expect.

Eric and Family, I have seen great progress in Eric's development. He has blossomed in his self esteem, confidence, and excitement for learning! Thank you for participating in the MNRI® program. It has been such a pleasure working with your family and sharing in your successes! – Diane Whiteside

Benefits of MNRI® for a Child with Dyslexia

Patricia Shackleford, Ph.D. in Psychology, MNRI® Core Specialist & Instructor, Melrose, FL, USA

Introduction

Patty Shackleford, Ph.D

Life is so interesting when you work with children. With 43 years in education and psychology I have seen many children and adults with different developmental challenges: physical, emotional, and academic. Over the last fifteen years I have seen a significant increase in children diagnosed with both autism and dyslexia. Autism has been given a lot of press and has generated many research articles on methods to alleviate symptoms and specific strategies for use at home and in school. Less attention has been given to the learning disorder known as dyslexia.

This case study is about a young boy who displayed all the classic signs of dyslexia two years ago. A total non-reader, at age 7½ he was unable to tell the difference between a letter and a number. Today, following 24 months of weekly sessions of MNRI® (Masgutova Neurosensorimotor Reflex Integration) complemented by five days per week of home exercises, he enjoys reading and is able to function at beginning second grade level.

What is Dyslexia?

Unfortunately educators and parents are all too familiar with this term. Defined by the Mayo Clinic, dyslexia is: *a learning disorder characterized by difficulty reading. Also called specific reading disability, dyslexia is a common learning disability in children. Dyslexia occurs in children with normal vision and intelligence. Sometimes, dyslexia goes undiagnosed for years and isn't recognized until adulthood.*

There's no cure for dyslexia. It's a lifelong condition caused by inherited traits that affect how your brain works. However, most children with dyslexia can succeed in school with tutoring or a specialized education program. Emotional support also plays an important role.[1]

Although often undiagnosed before a child begins school, earlier symptoms can include late talking, slow progress in language acquisition, and difficulty rhyming. At school the following symptoms would suggest a diagnosis of dyslexia:

- *reading at a level well below the expected level for the age*
- *problems processing and understanding what they hear*
- *difficulty comprehending rapid instructions*
- *trouble following more than one command at a time and multitasking*

[1]http://www.mayoclinic.com/health/dyslexia/DS00224/DSECTION=symptoms (September 1, 2014)

- *problems remembering the sequence of things*
- *difficulty seeing (and occasionally hearing) similarities and differences in letters and words*
- *an inability to sound out the pronunciation of an unfamiliar word*
- *seeing letters or words in reverse ('b' for 'd' or 'saw' for 'was,' for example) – this is common in young children, but may be more pronounced in children with dyslexia*
- *difficulty spelling*
- *trouble learning a foreign language.*[2]

The National Center for Learning Disabilities describes dyslexia as a lifelong challenge:

This language-based processing disorder can hinder reading, writing, spelling, and sometimes even speaking. Dyslexia is not a sign of poor intelligence or laziness or the result of impaired hearing or vision. Children and adults with dyslexia have a neurological disorder that causes their brains to process and interpret information differently.[3]

History and Background

Mike (not his real name) presents as a beautiful child with no outward signs of a disability. Under no pressure to sit still or to perform academically, he responds as any typical child his age would. Mike is outgoing, loves to run and have fun, and is very creative in play and crafts. Yet when he was required to sit still, follow directions or apply academic concepts, he presented as a student much younger than his age or grade placement. Mike was having a very difficult time in the academic setting. He displayed extreme delays in reading, spelling, math, and phonics, along with frequent letter reversals and number confusion. When our work began he had significant difficulty understanding and following a sequence of directions. It was very hard for his parents to understand why Mike did not respond to requests either at home or at school.

Due to a December birthday he was in first grade at age 7½, when this report begins. He presented most of the above mentioned characteristics of dyslexia. Already in Kindergarten, the local public school had identified him as a child to be watched to see if he would possibly need more individual attention. In December 2009 he was started on the path to evaluation for a possible learning disability and a Response to Intervention (RTI) plan was implemented in his regular classroom. The RTI plan included a revision to a more basic Kindergarten curriculum and tutoring sessions three times a week in a very small group.

Mike's Kindergarten teacher described him as a happy, friendly boy who had a 'tough' exterior and was very concerned about other's perception. Mike had difficulty with blending and segmenting words, recognizing high frequency words, following directions, and completing tasks independently. He often guessed without any thought to his answer, hurried through tasks and needed help to complete assignments.

This RTI plan was carried over to the first grade. No progress was noted with the implementation of the RTI programs even though teachers seemed very committed to helping Mike and his family and worked with him nightly to help him learn the very basic reading and academic skills. By January 2011, Mike could still not tell you if the flash card had a number or letter on it and he was unable to count to 15 without leaving out at least two numbers or forgetting what number came next. Mike tried very hard to please his parents and teachers, but when it came to written or oral language skills, he did not seem to have a clue how to process this information successfully or how to respond either verbally or in writing. Mike also had experienced several significant emotional stressors in his early life. At two his father died in his sleep during the night.

In August 2010, at the beginning of Mike's first grade year, his mother remarried and the family moved into a new home, a short distance away from grandparents who had been involved in his daily life since he was a toddler. There were some limited signs of possible PTSD symptoms due to his father's death and moving from the constant care of his grandparents.

Mike's parents were very concerned and frustrated by his lack of academic progress. They were looking for answers as to why this beautiful, winsome, charming child could not learn. In January 2011 they consulted me, as a psychologist and MNRI® Core Specialist, and we began to develop a plan to help Mike integrate the learning that he had already been exposed to so that he could meet classroom academic expectations in an adequate and appropriate manner. Besides not responding to the modified classroom curriculum and delivery system, Mike was beginning to dislike school. He had one reported behavioral incident during first grade

[2] http://www.mayoclinic.com/health/dyslexia/DS00224/DSECTION=symptoms
[3] http://www.ncld.org/

where he became aggressive toward his RTI teacher. All the evidence gathered from both the school and the family indicated that Mike was exhibiting the characteristics of dyslexia.

Assessments in 2011

In early 2011 Mike had several assessments completed to help assist his family, the school, and me in fine-tuning his academic program.

January 2011: The Developmental Test of Visual Motor Integration-Fifth Edition (VMI)

This test was administered to Mike during a Neuro-Visual evaluation. This activity required him to look at and copy geometric designs of increasing difficulty. He achieved a standard score of only 76, demonstrating a level of visual integration significantly below average, as compared to same-age peers, and below the range expected given his measured intelligence. Mike's performance indicated the following deficiencies:

Oculomotor Dysfunction (eye tracking)

Binocular Dysfunction

March 2011: MNRI® Reflex Parameters Assessment

This was the first of three MNRI® assessments done by Dr. Svetlana Masgutova in 2011, 2012, and 2013. Mike demonstrated non-integrated reflexes for Robinson Grasp, Hands Pulling, Hands Supporting Babkin Palmomental, Foot Tendon Guard, Leg Cross Flexion, ATNR, Fear Paralysis, Trunk Extension, STNR, Spinal Galant and Spinal Perez, among others. His performance indicated a range of integration in his reflex system from mild dysfunction to functional at a very low level of development. The lowest scores were in the spinal reflexes, which are particularly important for postural control, attention, and focus. A more detailed discussion of scores and changes in Mike's level of reflex integration follows in the section on MNRI® Assessment Results.

June 2011: In June 2011 Mike had a Psycho-Educational Evaluation by a nationally certified school psychologist in private practice. A significant strength noted by the examiner was Mike's ability to identify essential details in a picture. Throughout all visual tasks he performed better when presented with picture stimuli than with geometric information. Mike's knowledge/ vocabulary skills, his verbal reasoning ability, and his verbal and nonverbal short-term memory were all within normal limits for his age. The psychologist administered the following tests.

Reynolds Intellectual Assessment Scales

Standard index scores were as follows (mean =100, standard deviation = 15, Asterisks indicate more than one standard deviation from the mean):

Verbal Intelligence Index	101	Composite Intelligence Index	109
Nonverbal Intelligence Index	118*	Composite Memory Index	102

Woodcock-Johnson Tests of Cognitive Abilities-III-NU (WJ-III)

Standard cluster scores calculated from Mike's performance in four areas of information processing ability were as follows (mean = 100; standard deviation = 15, Asterisks indicate more than one standard deviation from the mean):

Long-Term Retrieval	69*	Processing Speed	84*
Visual-Spatial Thinking	98	Phonemic Awareness	120*

The Woodcock-Johnson- Tests of Achievement-III-NU (WJ-III)

Using the standard subtest scores, standard scores in seven academic cluster areas as well as four broad academic areas were calculated as follows (mean = 100; standard deviations = 15, Asterisks indicate more than one standard deviation from the mean):

Basic Reading Skills	86*	Written Expression	84*
Reading Comprehension	72*	Phoneme/Grapheme Knowledge	82*
Oral Expression	99	Broad Reading	71*
Listening Comprehension	100	Broad Mathematics	87*
Math Calculation	91	Broad Written Language	77*
Math Reasoning	86*	Oral Language	100

The results from this evaluation met the diagnostic criteria for Reading Disorder and Disorder of Written Language. Mike's assessment scores were reviewed at his local public school and he was found to be eligible for Special Education Services under the category of Specific Learning Disabilities.

In August of 2011 Mike was enrolled in a private school where he attended their Enhanced Learning Program. He was also retained in the First Grade. Mike was in a classroom with no more than 6-10 children and he received one-on-one instruction for some of his academic needs. He also had time in the regular education classroom during science, Spanish, and P. E.

MNRI® Assessment and Therapy

The Masgutova Method® was chosen as the therapy of choice because Mike had not responded to any of the usual and modified academic interventions given him. Also the possibility of PTSD following his father's death and other life/environmental events led this psychologist and MNRI® Core Specialist to feel that Mike's academic improvements must come from the maturation and improved integration of his overall brain-body system. Beginning in January 2011 Mike would have weekly 1 to 1½ hour sessions with the author as a MNRI® Core Specialist. A Home Program enabled his mother to follow through with a prescribed sequence of reflex repatterning exercises 5 days a week. Both the therapy and home interventions used neurosensorimotor reflex integration techniques offered through various MNRI® programs to remediate the debilitating effects of dysfunctional and immature reflex motor pattern development.

Dr. Masgutova completed yearly MNRI® Reflex Parameters Assessments during the three years that Mike received MNRI® therapy sessions. These Assessments were given on March 2011, March 2012, and April 2013. The following Tables 1, 2, 3 and Graphs 1, 2, 3 show the consistent progress he has made each year.

Level of Development for a Reflex Pattern

Functional level, integration		Level of disintegration/dysfunction/pathology	
Points	**Definition**	**Points**	**Definition**
20.00 and over	High level of integration of a basic r and variants of a reflex pattern.	10.00-11.75	The pattern is on the boundary of normal function and dysfunction. Elements of the correct pattern.
18.00-19.75	Matured level of a reflex functioning . Integration is higher than average.	8.00-9.75	Incorrect pattern with some correct features. Mild dysfunction of a basic reflex pattern.
16.00-17.75	Development of a basic reflex pattern is average. Basic pattern is beginning to support development of variant patterns.	6.00-7.75	Incorrect/ dysfunctional basic reflex pattern
14.00-15.75	Pattern is functional, but at a low level of development.	4.00-5.75	Incorrect pattern. Deep dysfunction of a basic reflex pattern.
12.00-13.75	Basic reflex pattern is functional at a very low level of development.	2.00-3.75	Incorrect pattern. Pathology of a basic reflex pattern.
10.00-11.75	The pattern is on the boundary of normal function and dysfunction. Elements of right pattern.	0.00-1.75	Incorrect pattern. Severe pathology of a basic reflex pattern.

Level of Reflex Pattern Development for Mike

Mike	Table 1. Reflex Patterns A. Sagittal (Left-Right) Motor Coordination System																			
	1		2		3		4		5		6		7		8		9		10	
	Core Tendon Guard		Hands Grasp		Hands Pulling		Babkin Palmomental		Babinski		LegCross Flexion Extension		ATNR		Abdominal		Bonding		Foot Grasp	
1) 2011 Before MNRI	12.5	13	12.5	12	12	12	13.5	13	12.5	12	13.5	14	12	13.5	9.5	9	12.5	12	9.5	9
Average		12.75		12.25		12		13.25		12.25		13.75		12.75		9.25		12.25		9.25
2) 2012 After MNRI	13.5	13	13.5	13	13.5	13	14.5	14	13.5	13	14.5	14.5	13.5	13	13.5	14	14.5	14	12.5	12
Average		13.25		13.25		13.25		14.25		13.25		14.5		13.25		13.75		14.25		12.25
3) 2013 After MNRI	16	15.5	16	15.5	15.5	16	15.5	15	14.5	15	16.5	16	15.5	16	14.5	15	15.5	15	14.5	15
Average		15.75		15.75		15.75		15.25		14.75		16.25		15.75		14.75		15.25		14.75

Above, Table 1. The Reflex patterns for Mike (per Assessments) on the sagittal plane for 1) before MNRI® in March 30, 2011, 2) after MNRI® in March 31, 2012, and 3) after MNRI® program in April 2, 2013.

Below, Graph 1: Same information from Table 1 above, but in graph form.

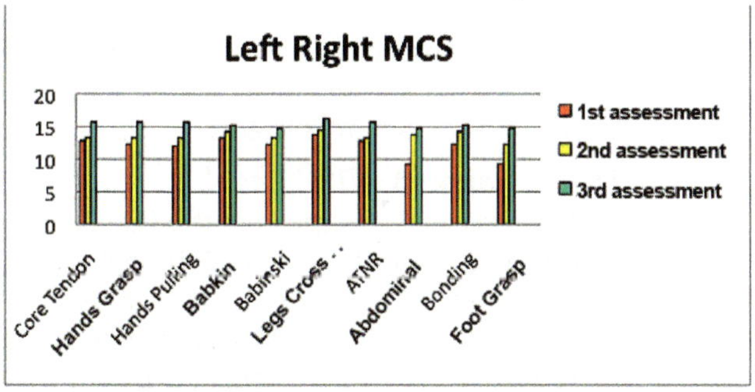

The scores for each parameter are combined to yield a score from 0 to 20, indicating the level of reflex development. Table 1 presents the criteria for scoring. (More information on this is available in the *Assessments* article in this book.)

The Tables and Charts at left present the results of Mike's Reflex Assessment in groups of reflex patterns relating to movements and motor skills within three different body planes: A) Sagittal (Left-Right), B) Superior-inferior (Upper-Lower), and C) Anterior-posterior (Front-back).

Mike's initial MNRI® Reflex Parameters Assessment showed that most of his reflexes in the Left-Right Motor Coordination System (MCS) were functional at a very low level of development. This MCS affects rational and cause/effect thinking, ability to sequence, and to use intuition as the results of past experience. Overall, the MCS affects learning and the development of cognition and action. Abdominal (the protective response of this reflex is help the body to go into the state of sleep) and Foot Grasp (the protective response of this reflex is to escape unsafe ground and maintain balance and stability) fell in the mild dysfunctional range: incorrect patterns with some correct features. Two years later, the last MNRI® Reflex Parameters Assessment indicated all patterns as functional, though still at a low level of development.

Mike's initial MNRI® Reflex Assessment for the Upper-Lower Motor System, which have to do with memory and emotional responses, indicated that the basic reflex patterns were functioning at a very low level of development, with incorrect pattern for Segmental Rolling falling in the mild dysfunctional range (the protective response of this reflex is to coordinate posture and kinesthetic development with functioning of the auditory and visual systems).

According to his last MNRI® Reflex Parameters Assessment Segmental Rolling improved the most, moving from dysfunctional to an average level of development. Average functioning was noted as well for Automatic Gait, Bauer Crawling, Moro, and Hands Supporting. Landau, Flying and Landing, Grounding, and Head Righting had progressed to a functional level of development.

Mike's scores for the Front-Back Motor System, which activates the protection and survival functions of the brainstem and involve reactivity, demonstrate that from 2011 to 2013 the basic reflex pattern for Pavlov Orientation has improved from functional to average. All other basic reflex patterns in this Motor Coordination System progressed to the functional level. It is particularly noteworthy that three spinal reflexes, STNR (the protective response of this reflex is to develop the inhibition mechanism for quieting the body and development of binocular vision and binaural hearing), Spinal Galant the protective response of this reflex is to influence auditory and visual convergence, divergence focusing and concentration) and Spinal Perez (the protective response of this reflex is to organize postural control and circulation of the cerebrospinal fluid) moved from dysfunctional to functional. Integration of spinal reflexes is crucial for the development of a level of postural control necessary for attention, focus, and efficiency in the visual and auditory systems.

Mike	Table 2. Reflex Patterns Serving for Development of Upper-Lower (Superior-Inferior) Motor Coordination System																			
	11		12		13		14		15		16		17		18		19		20	
	Automatic Gait		Bauer Crawling		Moro		Fear Paralysis		Hands Support		Segmental Rolling		Landau		Flying/ Landing		Grounding		Head Righting	
1) 2011 Before MNRI	14.5	14	13.5	14	12	13	12.5	13	13	12.5	9.5	9	13.5	13	14.5	15	13	13.5	12.5	12
Average		14.25		13.75		12.5		12.75		12.75		9.25		13.25		14.75		13.25		12.25
2) 2012 After MNRI	15.5	15	14.5	14	13.5	13	12.5	13	14	13.6	14.5	13.5	13.5	13	14.5	15	14	13.5	13.5	13
Average		15.25		14.25		13.25		12.75		13.8		14		13.25		14.75		13.75		13.25
3) 2013 After MNRI	17	16.5	16	16	14	15	13.5	14	15.5	15	16.5	16	14.5	15	15.5	16	15.5	16	15.5	16
Average		16.75		16		14.5		13.75		15.25		16.25		14.75		15.75		15.75		15.75

Above, Table 2. The Reflex patterns for Mike (per Assessments) on the upper-lower plane for 1) before MNRI® in March 30, 2011, 2) after MNRI® in March 31, 2012, and 3) after MNRI® program in April 2, 2013.

Below, Graph 2: Same information from Table 2 above, but in graph form.

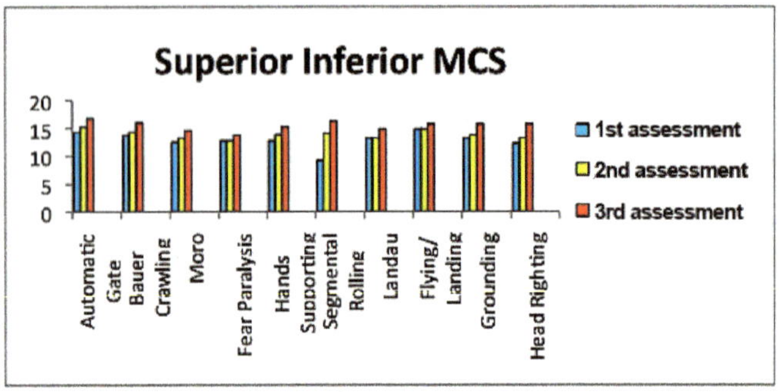

Mike	Table 3. Reflex Patterns Serving for Development of Front-Back (Anterior-Posterior) Motor Coordination System																			
	21		22		23		24		25		26		27		28		29		30	
	Trunk Extension		STNR		Spinal Galant		Spinal Perez		TLR		Foot Tendon		Spinning		Locomotion		Balancing		Pavlov Orientation	
1) 2011 Before MNRI	14.5	14	9.5	9	8.5	8	8.5	8	12	11.5	12.5	12	11	11.5	11	11.5	11.5	11	12	12.5
Average		14.25		9.25		8.25		8.25		11.75		12.25		11.25		11.25		11.25		12.25
2) 2012 After MNRI	15	14.5	11.5	11	11.5	12.5	12	14	13.5	13.5	13.5	14.5	14	13.5	14	14.5	14	13.5	14.5	7.5
Average		14.75		11.25		12		13		13.5		14		13.75		14.25		13.75		11
3) 2013 After MNRI	16	15.5	14	14.5	14.5	15	15.5	15	15.5	15.5	15.5	15	15.5	15.5	15.5	15	15.5	16	16.5	16
Average		15.75		14.25		14.75		15.25		15.5		15.25		15.5		15.25		15.75		16.25

Above, Table 3. The Reflex patterns for Mike (per Assessments) on the front-back plane for 1) before MNRI® in March 30, 2011, 2) after MNRI® in March 31, 2012, and 3) after MNRI® program in April 2, 2013.

Below, Graph 3: Same information from Table 3, above but in graph form.

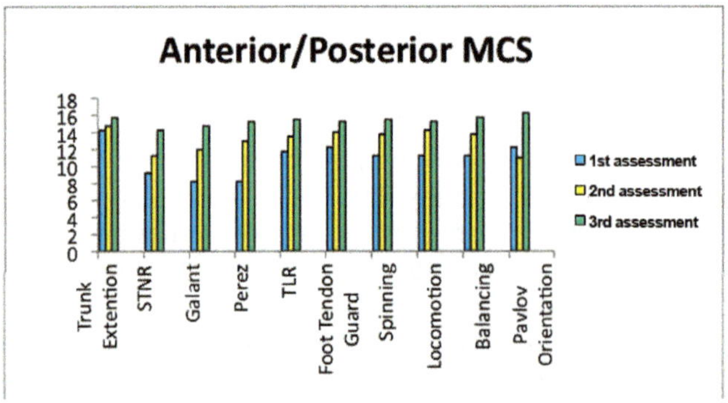

Discussion of MNRI® Results

Mike received therapy from this therapist on at least a bi-weekly basis from February to May and September to November of each year. Already by March of 2011 Mike was able to count to 100 with just a few errors and by June he could correctly count well past 100. In July while playing at the lake, Mike told his grandmother that his Dad had caught 8 minnows and he had caught 6 so all together they had 14 minnows! Sight word recognition was improving and Mike was able to complete his spelling words with 80% to 90% accuracy. By the end of 2011, Mike was able to count to 1,000, sound out CVC (consonant vowel consonant) words and was developing a basic sight word list that he could easily recall, instead of guessing the word from the first letter of the word. Still reversals in some letters and numbers were present and his overall ability to read was very delayed. Eye tracking was still unstable in both the horizontal and vertical fields. Mike had great difficulty in completing the STNR integrating exercises with an easy flow, ATNR, and Hands Pulling. What had improved was Mike's Foot Tendon Guard and overall control of emotional outbursts. In fact, Mike would tell others how much better Foot Tendon Guard made him feel and offer to show them how to do this reflex.

The MNRI® Reflex Assessment from 2011 showed the majority of Mike's reflex patterns functioning in a range from dysfunctional to the boundary between dysfunctional and functional. In contrast, the last reflex Assessment in 2013 found most of his scores squarely in the functional, low level of development to normal range. The move from dysfunctional to functional is extremely significant because only once a reflex motor pattern has become functional can it serve as a foundation for the development of higher level skills. In October of 2012, Mike was finally able to complete the active participation stage of the Spinal Galant reflex. Within days after this session, both his parents and teachers expressed amazement at his improvement in both reading and handwriting.

At this writing (May 2013) Mike has been in the program for 24 months and is on a mid second grade level in math and beginning second grade level in reading, except for fluency. His steady improvement in reflex integration correlates with progress in academics as well as behavioral, social, emotional, and other school related areas documented in the charts and the parent report below.

Achenbach System of Empirically Based Assessment – September 2012 & April 2013

The Achenbach System of Empirically Based Assessment (ASEBA) offers a comprehensive approach to assessing adaptive and maladaptive functioning. It uses an integrated set of forms for assessing and generating standardized data on a broad spectrum of competencies, adaptive functioning and problems. Using T-Scores, the mean (average) is 50, and the standard deviation is 10. A T-Score above 50 is above average. Lower than 50 is below average. For example, if you have a T-Score of 60, you scored higher than 84% of all responses. In the DSM Oriented Scales below, a higher than average score denotes a more significant than average problem. A letter B by the score indicates that this score is falling in the Borderline Clinical Range (more significant than the scores obtained by 93% of the national sample) and a C indicates in the Clinical Range (more significant than the scores obtained by 98% of the national sample).

Child Behavior Checklist

As part of evaluation and measurement of Mike's progress in MNRI® therapy both teachers and parents were asked to complete an ASEBA Child Behavior Checklist. The following scores were taken from these checklist reports, first in September 2012, and again in April 2013 (see Table 4):

The data from the ASEBA indicate that Mike has made significant progress in the DSM IV categories of Anxiety, Attention Deficit/Hyperactive Problems and Post-Traumatic Stress Problems. Mike's regular education teacher reports a positive change in his nervous traits in the classroom, with still concern about his ability to concentrate. The Special Education teacher reports a decrease in Mike's argumentative behaviors, better concentration in the smaller class, and fewer nervous traits noted. Parents note positive change in Mike's abilities to be more independent, diminished fear of doing things wrong, fewer thoughts that others are out to get him, less nervousness, fewer nightmares, and diminished overall fearfulness.

DSM Oriented Scales														
Respondent/ T Score	Affective Problems		Anxiety		Somatic Problems		ADHD		Oppositional Defiant		Conduct		PTSD	
Dates	9/12	4/13	9/12	4/13	9/12	4/13	9/12	4/13	9/12	4/13	9/12	4/13	9/12	4/13
Stepfather	50	50	75C	54	50	56	80C	75C	62	67B	57	63	76C	64
Mother	50	50	51	51	57	50	69B	66B	52	58	60	54	59	59
Classroom Teacher	50	50	56	50	50	50	59	62	50	50	50	50	55	52
ELP Teacher	50	50	56	56	50	50	65B	59	54	50	50	50	61	52

Above, Table 4. Results of the ASEBA Child Behavior Checklist.

Woodcock-Johnson Tests of Achievement-III Form B – 2011, 2012 & 2013

The private school that Mike attends has administered the Woodcock Johnson at regular levels to measure progress in the basic academic areas. The results provided by the school have been given in grade scores. When the last evaluation was completed, Mike was academically completing his instruction at the 2.8 level.

Test	Grade Score 6-2-11	Grade Score 4-30-2012	Grade Score 4-30-2013
Letter Word ID	1.2	1.7	1.9
Calculation	1.7	2.5	2.5
Spelling	1.0	1.4	1.6
Passage Comprehension	K.7	1.1	1.9
Word Attack Skills	1.5	2.2	2.1

Above, Table 5. Results of the Woodcock-Johnson Test of Achievement

The results of the Woodcock-Johnson Test of Achievement (see Table 5) indicate slow steady progress being made in the assessed academic areas. What these results do not show is Mike's improvement in his eagerness and love of reading. He has become very independent in completing homework assignments and in reading on his own each night.

Parent's report

"Since your starting the MNRI® work with my son Mike, he has grown in so many ways. He has gone from a boy who couldn't read a word, to a young man who begs us to sit down and read books with him. He has also gone from a boy who could not perform even the most basic math problems, to a young man who loves math and is excelling. Mike is still immature and impulsive, but since starting the exercises, the changes in both are amazing. He is confident and looks forward to his sessions every week."

Conclusion

As a licensed Mental Health Counselor and School Psychologist, I began using MNRI® in my practice over six years ago. My clients have demonstrated significant progress from the implementation of the MNRI® exercises. Positive changes have been noted in their physical posture, gait, fine and gross motor coordination, academic functioning in and out of the classroom, emotional self-regulation, social interactions, and overall social, emotional and mental well-being. All of these were areas in which Mike needed to improve.

MNRI® practitioners would not be surprised to note that standardized testing showed both social/emotional (ASEBA Child Behavior Checklist 4/13) and academic (Woodcock 4/13) progress concurrent with the maturation and integration of Mike's reflex system. The underlying premise of MNRI® is that when the basic motor patterns of primary reflexes can serve reliably for protection and survival, then they can also serve as the foundation for further motor, emotional, and cognitive growth. This is exactly what Mike's experience suggests. Once his level of reflex integration rose from dysfunctional to functional and in some cases from a low to average level of development, he began to acquire skills. Before MNRI® he had not responded to the high quality educational and professional help that had been available to him since Kindergarden. By the end of first grade he could not read, had difficulty distinguishing a letter from a number and could not count to 15 without omitting numbers. Only when MNRI® gave him access to the neurosensorimotor support of his own internal reflex system, could he begin to benefit from the other resources available at home and school.

There are no words that can express the love and gratitude that I send out to this family for giving me the opportunity to introduce them to the Masgutova Method®. It is delightful to see the positive, significant changes that comes to a child's life when the words on a page come to life and have meaning. The changes came in such a manner that it took us a month or so to realize that we could not spell words in conversations any more. This young boy was understanding all that we said and spelled! I have always known that the MNRI® processes work, but I am still always taken back each and every time I see the evidence again in front of me and see the change it has in the lives of our children and families. – Patty Shackleford

Effect of MNRI® on the Pragmatic Language Skills of a 12-Year-Old Boy

Susan Whitehead, MS SLP, Mancos, CO, USA

Introduction

Susan Whitehead

This study documents changes, both subjective and objective, in the pragmatic language (social skills) of TM-N, a 12-year-old boy, utilizing only the Masgutova Method® as an intervention. The school was concerned because TM-N had very high standardized test scores but did not demonstrate commensurate abilities in the classroom. Parents were concerned because TM-N demonstrated poor organizational skills, immature and inappropriate social skills and interactions, and significant in difficulty creating routines and adapting to change. All these factors affected TM-N's ability to be successful both at school and at home.

In this case study, TM-N received MNRI® therapy for 60-90 minutes per week over a period of eight weeks (approximately 2 months). Data from previous testing done nine months ago was used as a baseline for pragmatic language. His pragmatic language was re-evaluated three days after the eighth session. Three to four therapy sessions with a counselor were held between the pragmatics pre-testing and the initiation of therapy. Parents reported imperceptible changes within that time period. Pre-testing of reflex patterns was done during TM-N's first session. Post testing for reflex integration was done during TM-N's eighth therapy session.

Personal History

Report Date: February 25, 2013, Client: TM-N, 12-year-old male, born on March 22, 2000

Diagnoses:

- 2005 – (evaluation by the Southwest Board of Cooperative Services): educational identification of autism
- 2007 - Adjustment Disorder with Mixed Disturbance of Emotion and Conduct, visual problems, and Celiac Disease
- 2010 - Anxiety and Depression
- A Family Health Nurse Practitioner suggested undiagnosed Celiac Disease and brain inflammation prior to age four as a contributing factor in failure to thrive, all of these manifesting as sensory sensitivities, delayed processing, and difficulty with concentration
- 2012 – (re-evaluation by the Southwest Board of Cooperative Services): no diagnosis, educational identification of autism lifted, qualified for Gifted and Talented program
- Other diagnoses considered - ADD, Autism Spectrum Disorder

Background: TM-N was a 12-year-old boy with a history of health/digestive issues since age 8 to 10 months of age. Results of an evaluation by the school district performed in 2005 when he was in second grade indicated that TM-N exhibited behaviors consistent with an educational identification of autism. However, he did not qualify to receive services through special education because his academics were at an adequate level and his overall language score was in the 98th percentile.

In a re-evaluation by the same agency in the spring of 2012 TM-N no longer met the criteria for an educational identification of autism. He was placed on a 504 Plan stipulating modifications in the classroom and with homework in order to accommodate his challenges with focusing and attention. TM-N additionally qualified to be in the Gifted and Talented Program, a program set up through special education to meet the needs of advanced students.

Parents report that TM-N has continued to struggle throughout the years with social/emotional issues and a lack of organizational skills in school and at home. This has led to significant stress around the home on school mornings when TM-N needed constant prompts and reminders to follow what typically would be a regular routine for most children his age. TM-N needed reminders to take showers, brush his teeth, put on his shoes, take his lunch, etc. His lack of organizational skills made middle school even more challenging than earlier school years as there were higher expectations. Mom reported that by the middle of the school week, TM-N was 'finished'/overloaded. She was concerned that he may flunk out of school.

Birth/Milestones History: Childbirth – his maternal grandmother passed away from Alzheimers Disease in February of 2000 while Mom was pregnant with TM-N. Mom's water broke March 21, three weeks before her due date. Mom feels this is most likely due to stress around the loss of her mother. TM-N was delivered through C-section, as mom did not go into labor after her water broke. Even though TM-N was born three weeks premature, his APGAR scores were high, and he had no trouble with breathing. He stayed in the hospital 2-3 days until Mom was ready to go home. Neither parent smoked or used alcohol, so these were not risk factors for TM-N.

TM-N's milestones were advanced – his first words came very early, TM-N started crawling at 7 months, and walked at 13 months. TM-N knew the alphabet before he was 2 years old and was reading Wright Group books at the age of 4. He is still very advanced academically according to standardized testing.

Health History: TM-N was breast fed for 18 months. Up until 2011, he received all childhood immunizations. TM-N started showing noticeable signs of constipation at 18 to 20 months. This problem got increasingly worse until age 3½ when his medical doctor put him on 4 laxatives daily. This did not help the constipation which became a family issue as the parents had to give him regular enemas.

TM-N was diagnosed with Celiac Disease right before his fourth birthday. This was about the time when he began seeing an acupuncturist. Concomitantly, he started having projectile vomiting. A soy and dairy free diet (except for goat's milk) was initiated in addition to his already established gluten free diet, and he stopped throwing up immediately. At age five, TM-N started to occasionally lose bowel control. Mom explained that this was likely due to a loss of nerve sensations caused by his chronic constipation. The constipation and issues with bowel control gradually resolved themselves although, on rare occasions bowel issues do re-occur. Parents hypothesize that this could possibly be in relation to inadvertent ingestion of gluten.

At age 4, TM-N was prescribed with glasses to improve visual acuity. He also did regular vision exercises to improve visual processing.

At this writing parents described TM-N's health as generally good. He has been on antibiotics for approximately once a year since birth. TM-N suffered from spring and fall seasonal allergies. Sinuses were affected, he got a runny nose and red eyes, and had some breathing difficulties. He had not been diagnosed with asthma. Mom reported that the middle school was full of mold to which TM-N may have had an allergy. When exposed to mold, TM-N's immune system became depressed, and he became more susceptible to various illnesses. His lungs seemed to be most affected. He recently got over a bout with pneumonia.

Diet & Supplements: TM-N was on a strict gluten free diet because of his Celiac Disease. He did now consume dairy products as his medical doctor determined that this was not a problem for him. TM-N ate a lot of sugar when given the opportunity to do so. Parents tried to keep it out of the house. He was not a picky eater. TM-N inconsistently took Emergen-C and a Nature's Life probiotic product.

Emotional Sphere: TM-N was very depressed at the end of 2011-2012 school year. He came home one day and punched the ponderosa post in his living room. (According to parents, he seems to be less depressed now.) TM-N has intermittently seen two different counselors throughout the years. He generally didn't communicate how he was feeling. Hanging out with animals calmed him down. He did take some responsibility for the family dog, but this did not extend to feeding him.

TM-N stopped wearing his glasses in the sixth grade because he said they made him look ugly. He was self-conscious about how he looked, but he did not try to make himself look nice.

Social Area: TM-N's social skills were not age appropriate. He had a few friends, but Mom reported that it was hard to be friends with TM-N as he alienated others. Play/interactions tended to be parallel in nature. TM-N did not see the other person's perspective. Mother reported that kids tended to relate to him about specific interests, but the social part of it eventually messed up the relationship. In school, TM-N had a reputation for being a 'goofball'. TM-N looked forward to and spent time with one particular classmate. Outside of school, they liked to shoot BB guns, play with Legos, and play video games.

Behavior Regulation: According to a parent report, TM-N appeared "pretty put together" at school but was generally lethargic. By Wednesday or Thursday of the school week, he 'checked out' at all levels. School staff members assumed that he was enabled at home. Mom stated that they didn't know how hard it was for him and how much energy it took for him to get through each day. He didn't get in trouble at school.

TM-N lacked organizational skills and the ability to follow routines. Transitions were hard for him. He liked a controlled environment. TM-N resisted change, but at the same time, it was hard to get him into a groove. He did not seem to have habits, did not automatically follow daily routines, and needed constant reminders to do daily tasks. It was hard to tell if TM-N was resisting or just didn't know or think what the next step might be. In middle school, he had to follow routines and a schedule and this was very difficult for TM-N. Because he did not get into a pattern of behavior, he did not automatically transition from one activity to another.

Mom reported that, with eight hours of sleep each night during the week, TM-N appeared zombie-like at school. On the weekends, when TM-N got at least 10 hours of sleep, he was much more lively.

Evaluations

As reported by parents, expressive and receptive language have never been a problem for TM-N, but pragmatic language was delayed.

Speech/Language Evaluation – May 2012: TM-N was given the Clinical Evaluation of Language Fundamentals- 4th edition (CELF-4). This is a standardized assessment that evaluates expressive and receptive language skills and reports results in standard scores, also known as SS. A standard score of 100 is considered to be average with 15 point increments representing a standard deviation. Hence, a score of 115 is one standard deviation above the mean. A score of 70 is two standard deviations below the mean. For the purposes of special education, a student may qualify for services with a standard score one and a half deviations below the mean. TM-N received the following scores:

Overall Core Language SS - 120; Receptive Language SS - 113; Expressive Language SS - 122

These scores are all in the average or above average range.

A Pragmatics Profile was an additional part of the CELF-4. During the language evaluation, this profile was completed by TM-N's parents. Three areas were evaluated as follows: Rituals and Conversational Skills – 22 questions; Asking For, Giving, and Responding to Information – 17 questions; and Nonverbal Communication Skills – 13 questions.

Skills were rated with point values of 1 – 4: #1 = never, #2 = sometimes, #3 = often and #4 = always. "Not observed" or "not appropriate" were additional choices. TM-N received the following scores:

CELF-4 Pragmatics Profile Scores – Pre-test – May 2012

1 pt - never	2 pts - sometimes	3 pts - often	4 pts - always	Not Observed	Not Appropriate
8 X 1 = 8	31 X 2 = 62	13 X 3 - 39	0	0	0
Total Score - 109		Criterion Score = 142		Does not meet criteria	

TM-N's total score was 109. The age-leveled criterion score for passing (indicating appropriate pragmatic language skills) was 142, so TM-N did not meet this cut-off level.

According to the speech/language report, TM-N struggled most with asking for clarification during conversations, responding to introductions and introducing others, and with using appropriate strategies for responding to interruptions and avoiding interrupting others.

Psychological Testing: On November 6, 2007, a psychiatric evaluation determined that there was not sufficient evidence to place TM-N on the autism spectrum.

On April 25, 2012, the Autism Diagnostic Observation Scale (ADOS) was conducted with TM-N. Results indicated that it would be inappropriate to consider an Autism Spectrum Disorder for TM-N. According to the school psychologist's evaluation written on May 11, 2012, TM-N's visual processing speed was in the "Extremely Low" range at the 1st percentile. Sensory integration issues were also of concern. The examiner stated that TM-N struggled with the big picture and an ability to track his environment, had difficulty applying concepts to life situations, and was internally distractible.

TM-N's general adaptive behavior was found to be in the extremely low range of functioning for a student his age. Conceptual skills indicated functioning at or below the 1st percentile, and social skills were scored at or below the 10th percentile.

School Evaluation: Concomitantly, while in second grade, TM-N was referred for a special education evaluation due to focusing issues. The Special Education report, dated November 15, 2007, stated that the school autism team found that TM-N exhibited behaviors consistent with an educational identification of autism. He was not classified as such due to the results of the aforementioned psychiatric evaluation where TM-N was found to not be on the autism spectrum. Other findings were concerns about fine and gross motor skills including an awkward, clumsy gait and writing challenges – both in composition and fine motor control. Language skills were found to be in the 98th percentile both expressively and receptively. TM-N did not qualify for special education services because there were no language, academic, or cognitive concerns at the time.

Miscellaneous Possible Contributing Factors: Parents reported that in 1984, many years prior to TM-N's birth, his parents both went through a series of rabies shots that contained mercury. This toxin may have been passed on to TM-N and his two older sisters. TM-N has not been tested for the presence of heavy metals. He has no fillings in his teeth. Mom believes his unique genetic background may have made him more susceptible to environmental toxins.

Previous interventions: TM-N was given play therapy at 3 years old, regular doctor visits – mostly for constipation, acupuncture, Body Talk for emotional issues (that may have kept him below crisis level), Chinese Herbs for allergies and viruses, counseling, Klong Listening Therapy, was on Prozac for about 6 months (no other meds have ever been prescribed), vision therapy for a few months, the computer program – *Facing Your Fears*, and was prescribed glasses when 4 years old. According to Mom, none of these interventions seemed to have had a significant impact. Any responses at all to interventions have been extremely incremental. Mom described any improvements as occurring in geologic time.

Parental Goals: Parents wanted to see improvement in TM-N's organizational skills, social skills/interactions, and ability to adapt to change. Specifically, they wanted him to be successful in school both academically and socially, and to develop routines at home.

Intervention with MNRI®

Masgutova Method® of Neurosensorimotor Reflex Integration, developed by Dr. Svetlana Masgutova, is a series of techniques and programs that identifies non-integrated reflex patterns and then brings these back into integration. Static and dynamic reflexes and their patterns are the basis for development of postural control and motor coordination; later on they affect the consciously learned skills and abilities for drawing, reading, writing, calculation and more. Reaching the proper reflex neurological maturity establishes the basis for emotional and cognitive development as well as school success, and life achievements. These techniques are individualized accordingly to level of a reflex functioning, age of patient and level of neurological dysfunction/pathology. Additional MNRI® programs support the basic reflex work.

MNRI® therapy began with TM-N on December 13, 2012. Between December 13, 2012 and February 21, 2013, TM-N received eight therapy sessions each lasting 60 to 90 minutes.

During the first session, a reflex assessment identified priorities for reflexes needing integration. The assessment reflected the functional strengths and weaknesses of reflex patterns, and provided the practitioner with information necessary to design a reflex-based treatment plan. The first priorities identified was a hyper response to stimuli for Foot Tendon Guard, Babinski, Fear Paralysis, Spinal Galant, and Spinal Perez. Secondary priorities included Hands Grasp, Hands Supporting, Babkin Palmomental, Leg Cross Flexion-Extension, Landau, STNR, ATNR, and vision.

TM-N's protocol was as follows:

• integrate specific reflexes according to priorities
• provide neuro-structural techniques to release tension in his body
• provide tactile techniques to relieve tactile defensiveness
• provide facial techniques to reduce sensitivity, release facial tension, and improve facial expression.

Results

Observations of therapist: TM-N's negative response to touch has shown marked improvement. He allowed stretching of his limbs without protection at the shoulders. He appeared to be more outgoing and more relaxed in general during sessions.

The following parental observations were sent by TM-N's parents on February 25, 2013 regarding the changes they have seen since MNRI® therapy began:

"TM-N's response to his MNRI® treatments has been significant in relation to all of the interventions we have tried. They seem to have given the most marked results. In the first two treatments with Spinal Perez, we noticed that TM-N's spine was straighter and seemed to stay that way through further treatments. We noticed an improvement in TM-N's posture and how he carried himself. Prior to the treatments, TM-N had a tendency to round his shoulders.

During the treatments, TM-N gradually became more comfortable and more tolerant. He stopped jumping up during the treatments and actually seemed to start to relax. This is significant because TM-N's sensory issues usually make him uncomfortable when being touched or massaged.

With the eight treatments and home follow through, we noticed several things. First, TM-N's posture has improved and he seems to be demonstrating improved physical coordination. On the date of these observations, TM-N stated, "Exercise is awesome!" More subjectively, we have seen more willingness to tolerate school. TM-N still says he does not like it but seems to try a little harder. His overall 'flow' is better. Foot Tendon Guard integration techniques seem to have provided relief for TM-N's sore legs.

TM-N seems to be widening his circle of friends and has definitely improved socially. Teachers have noted that TM-N has been more willing to work with a variety of students and recognizes that working with certain students allows him to complete his work. The teachers report that the work TM-N is producing with his new choice of partners is of higher quality and more complete than what he was producing previously.

In regard to TM-N's performance, when he is asked to do an assignment, a chore or other non-preferential activity, he will complete the tasks with considerably less prompting than prior to treatment. He is less resistant, defiant, and evasive. We are now able to negotiate with TM-N, and he has a reasonable response. Prior to treatment, no amount of persuasion, coercion, or withdrawal of privileges would elicit the desired response. Now positive praise and incentives are highly effective.

TM-N has a lower level of opposition when a request for compliance is made. He is now willing to engage in conversation at the dinner table and sits for longer periods of time than prior to treatment. Finally, TM-N initiates physical contact such as hugs, kisses, saying "Goodnight", etc. This has not been the case for a long, long time."

Post MNRI® Testing: TM-N's first priority reflexes were re-evaluated during his eighth therapy session on February 21, 2013. All of these first priority reflexes showed a significant reduction in hyper response to stimuli.

Pragmatics Profile of CELF-4: On February 26, 2013, TM-N's parents completed the Pragmatics Profile from the Clinical Evaluation of Language Fundamentals – 4th edition as they had previously done in May of 2012.

TM-N received the following scores pre- and post- scores:

CELF-4 Pragmatics Profile Scores– Comparison Pre and Post Testing

	1 pt - never	2 pts - sometimes	3 pts - often	4 pts - always	Not Observed	Not Appropriate
May 2012	8 X 1 = 8	31 X 2 = 62	13 X 3 = 39	0	0	0
Feb 2013	1 X 1 = 1	35 X 2 = 70	16 X 3 = 48	0	0	0
Change in points	-7	+8	+9	-	-	-
Total Change	Pre-test = 109 Post-test = 119 Percentage change = 9% Percentage improvement toward 142 criterion cut-off = 30%					

The Pre and Post Pragmatics Profiles from the CELF-4 indicate the following:

Although TM-N did not meet the criterion score for passing on the post testing, his scores did show an improvement of 9% in skill levels and a 30% shift toward meeting the passing criterion score.

Parent Report: Mom indicated noticeable changes in TM-N in the following areas of the Pragmatics Profile:

- begins/ends conversations appropriately
- makes relevant contributions during conversation
- uses language of peer group appropriately
- shows appropriate sense of humor during communication situation
- asks others for permission when required
- offers/responds to expressions of affection and appreciation appropriately
- presents matching verbal and nonverbal messages
- understands posted and implied group/school rules.

Conclusions

At this writing, TM-N is a 12-year-old boy who experienced just two and a half months of MNRI® therapy between December 13, 2013 and February 21, 2014 to facilitate progress toward identified goals. This was the only intervention applied. TM-N's parents' goals were for him to develop routines at home and to improve in organizational skills, social skills/interactions, ability to maintain routines, and ability to adapt to change.

As of TM-N's final session, the hyper response to all priority one reflexes reduced significantly, and the following presenting issues diminished in frequency or became resolved:

- stress around morning routine getting ready for school (needing reminders for everyday tasks)
- poor organization at school
- difficulty with transitions
- fatigue and overload during the school week
- poor social interactions at school.

TM-N's rapid progress in the integration of his reflexes correlates with his improvement in pragmatic language as measured on the CELF-4 Pragmatics Profile and noticeable progress toward his goals as identified by the parent report. Parents noted they have never seen anything but incremental changes with other therapies – and now they can see a real difference. They attribute this to MNRI®.

I congratulate TM-N on his successes and hope his future is full of easy learning and great success at school. – Susan Whitehead

Masgutova Neurosensorimotor Reflex Integration® Used for Rehabilitation of Brain Stroke Patients

Ewa Kulig, PT, and Aleksandra Szalaj, PT, Dept. of Neurological Rehab., Assoc. of Pulmonological and Rheumatological Hospitals Opolskie, Poland, and NEUROREHA Opole, Center for Specialized Neurological Rehabilitation

Ewa Kulig

Aleksandra Szalaj

Introduction

A new therapeutic method MNRI® (Masgutova Neurosensorimotor Reflex Integration) is being used to successfully treat patients with central nervous system dysfunction. This program, not commonly known yet, is beginning to generate a great deal of professional interest. MNRI® stands along side other known methods such as PNF, Bobath (NDT), and the Vojta method. MNRI® is directed primarily at patients with psychomotor developmental problems as well as both neurophysiological and structural dysfunction related to central nervous system injury or dysfunction. This program is based on multi-year international clinical research and therapeutic application during rehabilitation camps. MNRI® promotes a holistic view of the patient, his development, experience as well as function. This program utilizes techniques designed to support nervous system function, activate and support mechanisms of neuroplasticity in the brain, as well as improving links and whole kinetic chains of movement, neuromuscular coordination, mechanism of self-regulation of muscle tone, and functioning of neural pathways. All of these improvements lead to greater structural and sensory-motor integration within the body.

The main goal of the MNRI® program is to integrate reflex patterns and to promote development and to optimization of gross and fine motor skills. Additionally, it facilitates cognitive and emotional development, especially in children, and improvement of function as they relate to the coordination of sensory systems and systems responsible for motor control with brain function.

The MNRI® program is designed to integrate primary, unconditional motor activity, namely, reflexes. Reflexes are the efferent reactions of the body in response to sensory or proprioceptive stimuli through participation and interpretation of the nervous system. Reflexes serve as the base for motor and cognitive development. A person is born with a set of reflexes that allow the body to adapt to the environment as well as facilitate cognitive development. Each reflex plays an important, underlying role by preparing the organism for the next stage of development. Reflex activation, development, and maturation leads to the integration of reflex patterns. Integration of the reflex, not their inhibition, are always correlated with the acquisition of new skills. Thanks to such integration, the lower levels of the central nervous systems (the brain stem, midbrain, and cerebellum) 'open up' a pathway to the cortex that simultaneously allows for the formation of new neuronal connections, development, learning, and cognition. Retained or inadequate primary reflexes can retard the development of both postural and dynamic responses that are necessary to allow the person to effectively interact with the environment. Such dysfunction may cause compensatory patterns and immature behaviors to be adopted that dominate even after new skills are acquired. As a result, instead of mature and well-integrated automatic function, constant conscious effort will be required since an organism responds to stimuli primarily through reflexes, but falling back to the most developed reactions in a particular situation that it is able to call forth.

The techniques provided within the MNRI® program may be used as part of a single therapeutic regimen or applied as individual techniques in support of other rehabilitative methods used with neurological patients. The choice of approach is based on the requirements of the patient's individualized therapeutic program.

MNRI® Therapy Types

Among the various techniques developed in MNRI®, two types are mentioned most often: Tactile Integration (Masgutova, et al, 2005) and Neurostructural Reflex Integration (Masgutova, 2011) programs.

Tactile Integration Therapy This program is used for its valuable benefit in facilitating the functioning of a patient's sense of touch. The application of tactile therapy serves to stimulate and normalize the sensation of touch (both hypersensitivity and hyposensitivity), proprioception, and kinesthetic awareness as well to regulate muscle tone. These techniques stimulate the mechanisms of interaction between the different skin layers as well as its integration with fascia, muscles, tendons, and skeletal as well as nervous systems (Masgutova, et al., 2005). It is important to note that touch is an important factor influencing a person's emotional state, feelings, mood and emotions. Tactile therapy is based primarily on specific massage techniques and the stimulation of the skin and touch receptors.

Neuro-Structural Reflex Integration Therapy This program is used to regulate the functioning of muscular as well as the proprioceptive and tendoneous systems as they closely correlate to protection mechanisms and protective reflexes. This therapy is directed towards the body's structure as well as symmetry mechanisms and the integration of reflex patterns. Each structural deficiency forces muscles and tendons to work harder and evokes negative protection reactions in the mind-body system. Neuro-Structural Reflex integration therapy program influences the self-regulation mechanism of muscle tone as the reaction of tendons, fascia, muscles and their reaction to external stimuli. This therapy also influences the regulation of nerve impulse transmission. In this manner, the neurostructural integration program serves to normalize the individual links and chains in the kinematic movement system. It influences neuromuscular coordination and helps to restore structural integration based on the principle that structure regulates function. Neuromuscular techniques are used to provide powerful stimulation of the body's natural auto-correction mechanisms, while the reaction is directed into areas where improper tension, pain, or dysfunction presents itself (Masgutova, 2011).

As our experience also shows, this therapy also activates the lymphatic and vascular systems, regulating the hydrostatic pressure within the organism. As a result, the therapy serves to prevent swelling, helps to deliver nutrients to nerves and muscles, and activates gas exchange.

Neuro-Structural Reflex therapy is based primarily on specific manual techniques that work with soft tissues and the proprioceptive system as well as stimulate neurosensorimotor points correlating to reflex patterns. This therapy has broad application not just with patients suffering from neurological issues, but also in orthopedic patients with skeletal dysfunction and pathology.

Application of MNRI® Therapies

Both Tactile and Neurostructural Reflex integration protocols can be applied independently and, based on diagnosis, as individual techniques in order to support the patient's individualized treatment plan. The main goal is to activate and optimize both physical and cognitive function in patients. This is meant to help the patient return to full and proper function and, in children, this often also means proper development.

Initial Patient Evaluation: Adult patient (male, age 65) seen after experiencing four ischemic strokes (over the course of 3 years):

- a stroke in the right hemisphere resulting in paresis of the left side of the body
- two strokes in the left hemisphere resulting in paresis of the right side of the body
- a stroke in brainstem resulting in quadriplegic paresis and increase in muscle tone.

Additionally, the patient had the following conditions:

- Type 2 diabetes
- vascular disease (atherosclerosis)
- high blood pressure.

Initially, the patient was treated at the Department of Neurology and subsequently admitted to the Department of Neurological Rehabilitation for a period of 4 weeks for further improvement.

After admission to the department of Neurological Rehabilitation, the patient's condition was described as:

- quadriplegic muscle paresis, more significant on the right
- general increase in muscle tone, more significant on the right
- disruption of balance, body righting reactions, and ambulation
- contracture of the Achilles tendon
- head positioned in slight reclination due to increased tone in the muscles of the head and neck
- dysfunctional Swallowing Reflex
- disruption of mechanisms of speech phonation and motor function
- positive Babinski response bilaterally
- dysfunctional Foot Tendon Guard Reflex
- hyperactive Foot Grasp Reflex
- dysfunctional Leg Cross Flexion-Extension Reflex
- hyperactive Robinson Hands Grasp Reflex presenting with overflow activation of flexion-adduction patterns of right upper extremity during ambulation
- dysfunctional Hands Supporting Reflex
- increased muscle tone and stiffness in the trunk
- disrupted breathing rhythms with shallowness of breath (Coughing Reflex present with intermediate level of effectiveness)

The patient reported difficulty in transitioning from a seated position to standing (Berg Scale 1-2) – able to stand after 3-5 attempts. Initiation of movement was executed with a great deal of effort and with poor stability while standing. The patient was observed having great deal of difficulty in transferring their center of balance forward as well as over both lower extremities. A hyperactive Foot Tendon Guard response was observed, triggering strong dorsiflexion of the feet and increased tone of the quadriceps muscles. Increased tension of the trunk muscles with a hyperactive Trunk Extension response along with overflow activation of flexion-adduction pattern in the upper extremities was observed. These issues made it difficult to lean the trunk forward while performing the sit to stand transition. This transition was assigned a score of 1-2 on the Berg Balance Scale.

The patient's gait was slow, exhibiting a hip hiking and circumduction pattern. It was also unstable and accompanied by great difficulty in shifting weight onto the right foot. Often, 2-3 attempts were required in order to complete the weight transfer. During the swing phase, the patient presented with right hip hiking, lack of right knee flexion correlated with considerable tension of the quadriceps muscle (this muscle tension visibly blocked movement), as well as plantarflexion of the sole of the foot. The heel strike phase triggered hyperactivity of the Babinski, Foot Grasp, and Foot Tendon Guard Reflexes which hindered appropriate weight bearing and loading response phase of the gait. Counter rotation of the trunk, due to disruption of neuro-muscular coordination, was also absent (disruption visible during activity and at rest). Additionally, the patient reported episodes of vertigo (loss of balance described as slight dizziness and "losing track" of the floor) resulting in falls. Within 10 days of admission, 5 such events were noted. The patient felt unstable, required physical assistance of an attendant and used an elbow crutch held in the left hand. Absent aid, the patient was transported in a wheelchair. The patient required assistance in performing daily tasks. The patient's function was assigned a score of 6 on the Hauser Ambulation Index (AI) scale (Note: 6 pts = "Requires bilateral support and more than 20 seconds to walk 25 feet; may use wheelchair on occasion").

Through palpation, increased muscle tone was observed in the following muscles:

- pectoralis major muscle
- trapezius muscle
- erector spinae muscle
- muscles of the head and neck: facial muscles, supra and infrahyoid muscles, temporal muscle and mas-

seter muscles

• quadriceps muscles: bilaterally, greater tone on the right side

• triceps surae muscles in the calves: bilaterally, greater tone on the right side

The muscles were rigid and stiff. The patient presented with limited range of motion of the knee joints. His range of motion was tested in the supine position to be:

• 45° active flexion of the right knee

• 90° active flexion of the left knee.

Rehabilitation

The patient underwent an individualized and complex physical rehabilitation program consisting of:

• kinesiotherapy based mainly on Tactile, Neuro-Structural, and Dynamic and Postural Reflex Integration

• neuro-speech therapy using MNRI® Program of Oral-Facial Reflex Integration

• psychological therapy

• group therapy

Physical rehabilitation sessions took place daily for a total of approximately 1½ to 2 hours.

Techniques Used During Individualized Kinesiotherapy Sessions

The MNRI® method serves to facilitate integration of two types of reflexes:

• *Dynamic* – programming of automatic sequencing of involuntary movements and reflexes to stimuli

• *Postural* – reflexes in the context of reactions preparing the body to attain particular static postures reflexively.

The goal of the reflex pattern integration techniques is the creation of correct automatic motor responses to correct stimuli. Mature and automatic reflex patterns are essential for development of mature behaviors and help to assure safe and effective interaction of the organism with its environment. All corrective exercises are designed to create an appropriate reflex circuit in response to a specific stimulus (i.e.: a specific stimulus leads to a specific motor response). Three kinds of exercises maybe applied: passive correction, active-assistive, and isometric techniques.

During treatment sessions, exercises to integrate the following reflexes were used: Foot Tendon Guard, Babinski, Leg Cross Flexion-Extension, and Trunk Extension.

Integration of the reflexes noted above positively affects control over posture, postural tone, neuromuscular coordination, coordination of movement, and muscle tone regulation. Ultimately, these exercises allowed for improvement and development of motor control and skills such as sitting, standing, and walking.

Focusing on the upper body, exercises to integrate the following reflexes were used: Robinson's Hands Grasp, Babkin Palmomental, Hands Supporting, and Hands Pulling Reflexes.

Integration of the above reflexes positively affects neuromuscular coordination of consciously controlled movements, muscle tone regulation, and functioning of the upper extremities as well as the activation of mechanisms maintaining symmetry within the mind-body system. These exercises ultimately allowed for improvement in head righting, swallowing, and chewing as well as phonation and motor control of speech.

Results Achieved Through MNRI® Tactile Therapy:

• improvement in perception of light touch (touching and stroking of the skin)

• stimulation of perception of deep touch (squeezing and stretching)

• kinesthetic stimulation of movement and balance (joint and trunk rotation)

Results Achieved Through MNRI® Neuro-Structural Reflex Integration Therapy:

• relaxation of hyperactive Core Tendon Guard response

• regulation of muscle tone and tension of tendons in the entire body

• activation of proprioception and the connections between various body systems in order to enable the optimal functioning self-regulation mechanisms

• activation of receptors of deep touch and pressure

• stimulation of structural body symmetry

• stimulation of symmetrical efferent responses of the central nervous system to stimuli

• activation of movement in the horizontal plane, or in other words, the dynamics involved in moving the upper and lower parts of the body

• improvement in the coordination between alternating (reciprocal), homologous, and homolateral movement of the limbs
• stimulation of nerve transmission
• improved blood circulation
• improvement in the flow of lymph within the lymphatic system
• stimulation of the Breathing Reflex and improved breathing function
• improved gas exchange in breathing
• improved swallowing, phonation, and speech

Final Patient Evaluation

The following results were obtained after applying the therapeutic regimen:
• general improvement in neuromuscular coordination and control over muscle tone
• reduction of the generally elevated muscle tone on both the right and left sides of the body
• clear improvement in balance and righting reactions
• significant improvement in locomotive function (walking, standing up, sitting, etc)
• reduced trunk stiffness and improved regulation of muscle tone
• reduced shortening of the Achilles tendon
• improved functioning of improperly integrated reflexes:
  o Trunk Extension Reflex
  o Foot Tendon Guard Reflex in both lower extremities
  o Babinski Reflex in both lower extremities
  o Leg Cross Flexion-Extension in the lower extremities
  o Foot Grasp Reflex
  o Robinson Hands Grasp Reflex
  o Hand Supporting Reflex
• reduction of elevated muscle tone in the head and neck
• improved swallowing and breathing function
• improved phonation and motor control of speech

After the applied therapy, the patient generally felt distinct functional improvement in activities of daily living (ADL), ambulation, locomotion, and communication. The patient became more independent, improved mood and wellbeing, became more active and more eager to verbally communicate.

The patient's gait became more stable and reciprocal but with some trace of hip hiking and circumduction during the swing phase. He attained a more appropriate gait pattern with better bilateral flexion in the knee joints as well as contra-rotational movement of the trunk. Patient still presented partial weight bearing on the right lower extremity. However, multiple attempts were no longer required. Each of the following reflex patterns was improved: Foot Tendon Guard, Babinski, Foot Grasp, and significant improvement in Leg Cross Flexion-Extension. These improvements allowed for better muscle tone regulation and improved coordination of the gait pattern (initial contact of the heel, loading response, stance phase, and swing phase). The patient began ambulating without assistance while using a single elbow crutch. He improved, reaching a score of 4-5 points on the AI scale. After 10 days of therapy no incidence of loss of balance of falls were recorded.

Statistically significant improvement was seen in both trunk muscle tone regulation and control as well as Trunk Extension and Foot Tendon Guard Reflex patterns. This enabled improved dorsal flexion and better extension of the lower extremities and easier sit-to-stand transitions with fairly good forward shifting of the center of gravity. However, there is still visible asymmetry in weight bearing between the lower extremities. Each time, the patient was able to stand successfully on the first attempt. The patient felt more confident and stable, with less swaying and loss of balance, attaining 3 points on the Berg balance scale. In the stand-to-sit transition, the patient received 4 points on the same scale.

While palpitating individual muscles it was clear that there was a reduction in elevated muscle tone. The muscles became more flexible, pliable, and available for contraction. Through normalized muscle tone of the quadriceps muscle in both lower extremities, active flexion of the right knee was improved to 70^{0}.

The muscle tone in the face and neck became more normalized. As a result, improvement was seen in head righting, swallowing, as well as phonation and motor control of speech. Reduction of hypertonicity of the trunk muscles and improved control over their contraction allowed for improvements in breathing including the lengthening of inhalation and exhalation.

Improvement in the functioning of the Hands Supporting, Hand Grasp, and Babkin Palmomental Reflexes was observed. Due to these reflex improvements, as well as muscle relaxation, normalization of muscle tone, and coordination of movement, hand function was also improved.

Summary

Modern medicine and rehabilitation serves the twin goals of treating patient's suffering from central nervous system dysfunction and improving their quality of life. With the application of a tailored rehabilitation program, cooperation between the therapeutic team and the patient's family, and the use of modern methods such as the MNRI® program, the patient stands a much greater chance of improving various functions. These include improvement in both physical and mental functions, improved sense of confidence, independence, and self worth. This, in turn, translates into a sense of dignity and improved quality of life.

This study has indicated therapeutic benefit of MNRI® therapy as used for stroke patients. An interdisciplinary approach with MNRI® used for patients with neurodeficits demonstrates progressive results.

We wish all our post-stroke Winners good health and great success in their rehabilitation and life.
– Eva Kulig & Aleksandra Szalaj

Case Study of Janek: MNRI® for a Child with Severe Cerebellum Damage

Malgorzata Koczyk, MA, MNRI® Instructor, Parent, Gdansk, Poland

Malgorzata Koczyk

Introduction

This article presents a case of amazing improvement in the case of a boy with severe cerebellar damage through the use Masgutova Neurosensorimotor Reflex Integration (MNRI®).

Key words: MNRI® method, cerebellum, cerebellum damage, MNRI® intervention program.

Background

All prenatal signs promised a normal birth and successful future neurodevelopment for Janek. The routine ultrasound showed no damage in the cerebellar structure or in any other brain parts; brain circulation was considered normal. He was a first child born from a first pregnancy.

Natural delivery with the help of external manipulation took place on October 17, 1999 in the forty-third week of the pregnancy, approximately 3 weeks overdue (pre-natal examination and interview had indicated an expected birth date of September 25). The newborn's weight at birth was 8.8 pounds, with 8 points on the APGAR scale. As a newborn and infant, Janek was passive; he had problems with Sucking, Swallowing, and Breathing Reflexes. He cried a lot, did not like to lie on his belly, and was unable to turn over. The doctors claimed that all these problems were consequences of his delayed birth and long, difficult delivery.

When Janek was 17 months old with many noticeable delays in his development his doctors gave a diagnosis for the first time: cerebellar hypoplasia and severe atrophy of the cerebellar vermis. Magnetic resonance imaging showed major atrophy of the cerebellum and its cortex. This diagnosis meant the most severe structural brain damage and neurological dysfunction. It explained Janek's failure to creep, crawl, sit, and achieve other developmental milestones. It helped us to understand his extremely poor head righting control, poor muscle tone regulation, body clonus, problems controlling hand movements, and the absence of reflex activity in his lower limbs.

A more precise analysis of the MRI results revealed that the brain cortex was profoundly damaged as well. Janek's condition was classified as a disease: cerebral palsy in the cerebellum, with the label "ataxic CP". Although the damage was categorized as very severe, earlier examinations done in the Doppler Ultrasonic Laboratory and at the Medical Academy of Gdansk during the 17th and 24th weeks of the pregnancy had not shown any issues in this area. We can only speculate that the cause must have been one of three things: the delayed delivery in the 43rd week of the pregnancy, too long and difficult delivery, or most reasonable – congeni-

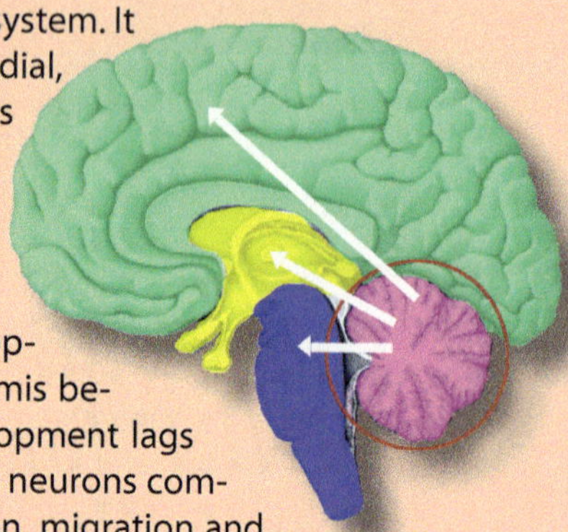

The ***cerebellum*** is a very complex structure of the central nervous system. It consists of two hemispheres connected by a vermis located in the medial, cortico-nuclear zone in the posterior fossa of the cranium. The vermis functions within the spino-cerebellar ascending tract receiving somatic sensory input from the entire body and head. Its basic function is coordination and control of posture and locomotion. Damage to this and other areas of the cerebellum causes deep pathologies in coordination of balance, posture, movement, and overall neurodevelopment (Parisia, Dobynsb, 2003). During neonatal development the vermis becomes fully foliated by 4 months gestation while hemispheric development lags behind by 30 to 60 days. (Loeser, Lemire, Alvord, 1973). The cerebellar neurons complete their foliation by 7 months after birth and their final proliferation, migration and arborization is only complete by 20 months of life (Loeser, Lemire, Alvord, 1973). Disorders and pathologies in neural development of the cerebellum associated with the vermis are usually inherent (Coffman, Dum, Strick, 2011).

The vermis is associated with all regions of the cerebellar cortex and affects functioning of the brainstem, diencephalon, cerebral cortex, and spinal cord. Pathologies in the vermis cause mental retardation influencing the control of eye movements, precise body and limb movement, and motor planning, initiation, and timing. Severe disorders in tissues of the vermis, as well as cerebellar hemispheres are considered non-treatable.

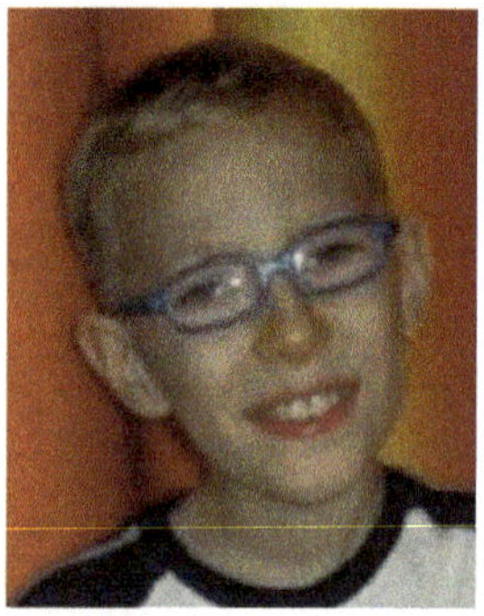

Janek Koczyk

tal malformation and damage that may have occurred during the first trimester of pregnancy.

Other chronic medical issues further compounded Janek's serious neurological condition. In November 2001 at age 2 following a high fever to which he responded with an epileptic crisis, he began pharmacological treatment for this additional ongoing disorder. He also suffered from chronic upper respiratory tract inflammations and spent a lot of time at hospitals. After long periods of hospitalization he regressed noticeably. His ophthalmic examinations showed hypermetrophia, astigmatism, squint, and myopia. Janek was also allergic to dairy products and presented seasonal allergies in spring and fall.

Consequences of Ataxic Cerebral Palsy

Physical Development: Authors in neuroscience (Coffman, Dum, Strick, 2011; Borkowska, Domanska, 2006; Loeser, Lemire, Alvord, 1973; Maryniak, 2004) claim that damage to the cerebellar hemispheres and vermis causes major deficits in motor function: dysmetria (poor ability to estimate and control range of movement of the limbs), hypermetria (tendency of movements to exceed their intended range); and ataxia (overall poor control, coordination, and alternation of agonist and antagonist muscles). Damage to the cerebellar hemispheres and dentate nucleus may lead to tremor in controlled movements such as pointing; visual nystagmus, atony (loss of muscular strength), and asthenia.

It is difficult to detect cerebellar disorders in newborns and infants. Delays in the development of head righting, standing, and sitting or absence of these motor skills (Siwek, 2006) can indicate poorly functioning extrapyramidal nerve pathways in the brain stem and cerebellum due to deficient neural development during gestation. Still, an infant is more likely to be diagnosed with low muscle tone and vestibular problems. At ten months Janek's neurologists claimed that ,although his development was quite different from other children, it was still within the norm for his age and our request for rehabilitation was denied. Only when more significant retardation emerged did it become obvious that he needed rehabilitation.

Walking is an important developmental milestone governed by the cortex. Stable walking, using sequential movements of two legs organizes vestibular, proprioceptive, visual, and auditory systems, helps a child to gain new physical and cognitive skills, and opens new opportunities for living and learning.

At ten months, when his palsy was extremely noticeable and he was still unable to crawl or to sit, Janek began physical therapy. Because of his health problems and frequent hospitalizations, regular rehabilitation became possible only when he reached two years of age. By age three his motor development was described as follows: "he sits in a unsteady way, he has a hunched back. His rolling is poor though he manipulates properly during the turning phase, muscular tension in limbs is low" (Extract from the Specialized Hospital in Koscierzyna, 16.11.2002). Until he was five, he moved by 'sitting-crawling' on the floor. From 2002 to 2006 Janek went with his mother to the best rehabilitation centers in Poland – Zlotow, Stobno, Wagrowiec, and Bydgoszcz – six times each year. He experienced rehabilitation methods such as NDT (Bobath), PNF, Peto, Knilla, and W. Sherborne. None of them resulted in any progress. Moreover, whenever Janek fell ill with infections, he regressed and lost previously gained and mastered skills.

The introduction of MNRI® in 2003 when Janek was 4 years old marked the beginning of a welcome change. Month by month his physical development improved amazingly – his head righting, sitting, trunk extension, standing, crawling, walking with support with increasing stability and gravity: all these achievements in physical-motor area defied medical predictions. The MNRI® programs most effective in improving his motor development were: Dynamic and Postural Reflex Integration, Dysfunctional and Pathological Reflex Re-Patterning, Neuro-Structural Reflex Integration, Tactile Integration, and Cognitive-Proprioceptive Integration.

Cognitive Development: A widely held opinion in neuroscience is that ataxic cerebral palsy, as present in Janek, is a disease "limiting intellectual skills in children" (Brearley, 1999). Scientists (Maryniak, 2004) attribute the features and difficulties of children with this type of neurological disorder to specific work of the nervous system: sensory-kinesthetic, vestibular-proprioceptive, and visual-auditory exploration.

Janek's ataxic cerebral palsy affected multiple areas of his brain: extrapyramidal nerves in his brain stem (lower motor neurons in the medulla, pons and midbrain), cranial nerves, the diencephalon (thalamus-limbic system, hippocampus), and the pyramidal nerve system of the cortex, all playing an essential role in development of cognitive processes and abstract thinking.

A psychological report from April 2002 (age 2½) describes Janek's speech development: "Understanding of the speech is possible, as he matches activities to verbal instructions. His active speech is at a low level and he can communicate only with people familiar to him. The child says some single words (mainly nouns, some verbs, some two-syllable phrases). He repeats a lot."

Janek's cognitive development was slow and uneven, especially at the beginning. He is still slow at motor tasks, tires more quickly than the average child when reading and writing, processes instruction slowly, is slow answering questions, and he needs extra time for memorizing when presenting information orally. Following school rules is difficult for him. Because Janek is still limited in these skills, he still has minor challenges with reading and writing (due to his visual and ocular-motor problems) and his math skills are still on a very low level.

Nevertheless, his comprehension and creativity have developed to a very high level for boy with such a diagnosis. Today he is doing very well in many areas. In his work with MNRI®, the fairytales that he creates and retells (though slowly) are amazing in their logic, structure, and cleverness, surprising many specialists.

We consider that the MNRI® programs that affected his cognitive sphere most effectively were: MNRI® Dynamic and Postural Reflex Integration, MNRI® Visual and Auditory Reflex Integration, MNRI® Cognitive-Proprioceptive Integration, MNRI® Oral-Facial, and MNRI® and Art-Therapy.

Emotional Development: Children with cerebellar type cerebral palsy (CCP) may have problems with emotional stability and can be passive, immature, fearful, and shy. New experiences can cause anxiety; that is why they tend to build long-term dependent relationships with others. Demanding behavior sometimes helps them to make "the environment friendlier" (Siwek, 2006). Unpredictable behavior is not unusual. Sometimes they withdraw and show a limited spectrum of emotions. They can also burst out in anger, aggression, or auto-aggression. It is often hard to find the cause of such emotional responses and behaviors and attempts to calm a child down may result in a bigger explosion of aggression.

Janek had all these challenges in emotional development including severe anxiety. His poor speech, repetitive manner and inability to express his feelings and thoughts around busy people with no patience to listen to him were always challenging. Emotional isolation due to these symptoms, along with immaturity and anxiety, were long-term features, though his motivation remained strong. Janek is now able to control his emotions. Even when he might be provoked at school he is rarely angry and never violent. His special education teacher

The MNRI® program allowed Janek to overcome all the dire medical diagnostic predictions. Month by month his physical developments improved amazingly. His head righting, sitting, trunk extension, standing up, crawling, walking with support with increasing stability – all unpredictably improved in the physical-motor area. The MNRI® programs that mostly impacted his motor development were: Dynamic and Postural Reflex Integration, Dysfunctional and Pathological Reflex Re-Patterning, Neuro-Structural Reflex Integration, Tactile Integration, and Cognitive-Proprioceptive Integration.

describes him as a friend in the class for others.

The MNRI® and Art-Therapy and MNRI® Tactile Integration sessions impacted his emotional sphere and behavior profoundly, and helped change many of his challenges into developmental successes and creativity. Other MNRI® programs also were helpful.

MNRI® Assessment and Intervention

At present Janek's developmental process is facilitated by the Svetlana Masgutova Educational Institute, and since June 2003 he has attended therapeutic and rehabilitation camps organized by this institute twice yearly. He has also had therapy sessions with a 'local' MNRI® Core Specialist – his mother, for a Home Program between the camps based on consultations led by Dr. Masgutova.

Janek has been assessed for reflex pattern integration at MNRI® camps in Poland regularly from 2003 to 2014. His scores for reflex patterns year by year demonstrate his neurodevelopmental journey, his on-going success and his stable results. Reflex scores that were on pathological and severely dysfunctional levels at the beginning rose to the functional level, some close to the norm or even within the norm. See below data from 2003 to the present 2014. Progress was steady from the beginning, seemed to accelerate after four or five years (2008, 2009), and has continued steadily. For a key to the meaning of these scores, (see the Assessments article in this book). See Tables 1-4.

The synthesized diagnostic functions Z_S, Z_H and Z_D (A. Krefft, 2007), measuring the level of reflex integration/development with verification by statistical tests before the MNRI® Program at Camp-1 (May 26 – June 6, 2003) and after MNRI® camps every other year (2004 – 2013) show p values of <0.001 for every year

Table 1. Reflex Patterns Serving for Development of Sagittal/ Left-Right Motor Coordination System (average per a year)

Reflex	X_1	X_2	X_3	X_4	X_5	X_6	X_7	X_8	X_9	X_{10}	Z_s
X_{1-10}	Core Tendon Guard	Hands Grasp	Hands Pulling	Babkin Palmomental	Babinski	Leg Cross Flexion Extension	ATNR	Abdominal	Bonding	Foot Grasp	
Before MNRI Program 2003	4	3,5	3,5	3,5	1	1,5	4	3	6	0,5	0,0855
After MNRI 2004	6	5	5	4,5	2,5	2	5	5	7	1,5	0,1362
2005	7	5,5	6	5	3,5	3,5	6,5	6,5	7,5	2	0,1865
2006	7,5	6	6,5	5,75	4	4	7,5	7	8,5	3,5	0,2322
2007	10	6,75	8,5	6	4,5	5,5	8	8,5	9,5	4,5	0,3272
2008	11	8,5	11	6,5	6	7	11	11	11	6	0,4912
2009	11,5	9,5	12,5	7,5	8,5	9,5	13	13	13	8	0,6494
2010	12,5	11	14	14,5	12	11,75	14	14	14	9,5	0,7783
2011	12,75	13,75	16	15,25	14,25	12,5	15,5	15,5	15,25	12,5	0,8657
2012	13	14	16	15,25	14,5	13	15,5	15,5	15,75	13,5	0,8778
2013	13,25	14,5	16,25	15,5	14,75	13,25	15,75	15,5	15,75	14	0,8868

Table 1. Sagittal reflex patterns in group $X_1 – X_{10}$ presented in the scale of 0-20 points.

and average, indicating statistically significant changes in reflex growth every year. See Table 4.

Positive results from MNRI®

The data in Janek's initial MNRI® evaluation (2003) indicate dysfunction (below 10 points) in 100% of the reflex patterns tested. Ninety percent were within the pathological range (below 3.5 points). His most recent evaluation of reflex integration after the 2014 MNRI® camp shows the following results:

- improvement in 100% of tested reflex patterns – all of them moved several points
- 90% of tested reflexes moved to a positive level (above 10 points) – no longer dysfunctional (see Graph 1)
- only 3 reflexes (10%) remain with scores lower than 10 points: Flying & Landing (6.75), Grounding (9.5) and Balancing (9.75)
- 0% of reflexes remain in the pathological range.

Significant changes in the level of integration reflex growth every year was observed on the trend function of Z_S (see Graph 1) which shows a slope which is statistically significant different from a zero value. For a detailed discussion of how this kind of improvement in two particular reflexes (ATNR and Abdominal Sleep Posture) can influence development and skill acquisition, see the section below on BAEP testing.

Table 2. Reflex Patterns Serving for Development of Horizontal Plane/ Superior-Inferior MCS (average per a year)

Reflex X_{11-20}	11 Automatic Gait	12 Bauer Crawling	13 Moro	14 Fear Paralysis	15 Hands Support	16 Segmental Rolling	17 Landau	18 Flying and Landing	19 Grounding	20 Head Righting
Before MNRI Program 2003	1.5	1.5	3	3	3.5	0	0	0	0.5	1.5
After MNRI 2004	3	3	5	5.75	5.75	3	2.5	0	0.5	3.5
2005	3.5	3.5	6.5	6.5	7	5.5	3	1.5	2.5	3.5
2006	5.75	6.75	7.75	7.75	8.75	7.25	5.5	1.5	3.75	5.75
2007	6.5	7.5	8.5	8.5	9.5	9.5	6.25	1.5	5.5	6.5
2008	8.5	9.5	9.5	8.45	10.5	10.25	7.5	2.5	6.5	8.5
2009	8.5	10.5	11.5	10.5	12	12.5	8.25	3	7.5	9.5
2010	11	12.5	13.75	12.75	13.5	13.5	9.5	4.5	7.75	11.75
2011	13.25	14.5	15.25	14.25	14.5	14.75	10.5	5.75	8.25	11.75
2012	13.25	14.75	14.75	14.5	15.25	15.5	12.5	6.75	9	13
2013	13.5	15	15,5	14,25	16,25	15,5	12,5	6.75	9.5	14.5

Table 2. Horizontal plane reflex patterns in group $X_{11} – X_{20}$ presented in the scale of 0-20 points.

Table 3. Reflex Patterns Serving for Development of Dorsal Body Plane/ Anterior-Posterior MCS (average per a year)

Reflex X_{21-30}	21 Trunk Extension	22 STNR	23 Spinal Galant	24 Spinal Perez	25 Tonic Labyrinthine	26 Foot Tendon Guard	27 Spinning	28 Locomotion	29 Balancing	30 Pavlov Orientation
Before MNRI Program 2003	3.5	2.5	2	2	2	2	0	0	1.5	5
After MNRI 2004	4.5	4	4.5	4	4.5	3.5	2.5	2	3.25	7.5
2005	6.75	5.75	6	6.5	5.5	4.5	4.75	4.5	3.5	9
2006	8.5	7.5	7.5	8.25	6.25	6	6	6	4	10
2007	9.5	8.5	9.5	10.5	7.25	6.5	8.75	8	4.25	10.5
2008	10.25	9.25	10.75	11.75	8.75	8	11.5	11	5.75	11.75
2009	11.75	10.75	12.75	12.75	975	10.5	12.5	12.5	7	12.5
2010	12.5	12.5	13	14	10.5	11	13.75	13	7.5	13
2011	13.5	14.75	14.5	15.25	11.5	13	14.5	14.25	8	13.25
2012	13.5	15	14.5	15.5	12.5	13.5	14.5	14.25	8.5	13.75
2013	13,75	15	14,75	15,5	13,75	13,5	14,5	14,25	9,75	13,75

Table 3. Dorsal plane reflex patterns in group $X_{21} – X_{30}$ presented in the scale of 0-20 points.

Table 4

Coefficient	Features	Before Program		After Program (average for all years)	
	Level of integration of reflexes	Average	SD	Average	SD
Z_S	X_1 to X_{10}	0.3246	0.1668	0.4798	0.1744
Z_H	X_{11} to X_{20}	0.3776	0.1869	0.4922	0.1832
Z_D	X_{21} to X_{30}	0.3654	0.1867	0.4724	0.1846
ANOVA Tests		$p < 0.001$			

Table 4. Statistical analysis of Janek's results.

Other Assessments

EEG: In June 2008, when Janek was 9 years old, medical investigators working at the Masgutova Institute studied his brain. Baseline readings done with a new generation of EEG equipment demonstrated extreme neurophysiological immaturity as shown in Picture 1.

Notable pathological features of the preceding EEG recordings were:

- dramatically disorganized spectral activity
- average alpha activity of 57 µv
- irregular average index of 53% ranging from 8-9 Hz
- absent amplitude modulation

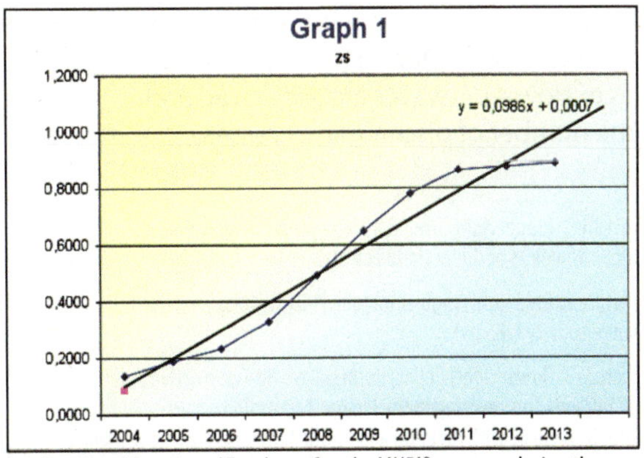

Graph 1. Trend function of Z_S values after the MNRI® program during the period of 2004 - 2013 used for Janek (analysis was prepared by Dr. H. Filipowski, Medical Academy of Poland based on NewKrefft Statistic Mathematic Analysis, 2007).

- irregular β- activity
- β-activity of 34 μv at the highest amplitude; a high index of 74% and low frequency with dominance of sinusoidal-type waves in the temporal (T3, T4) and frontal (F-7, F8) lobes)
- excessive θ activity indicated by a group of waves with an amplitude of 50 μv, an index of 46%, and a frequency range of 4-5.8 Hz (This slow wave activity dysfunctionally occurs primarily in the central, occipital, and frontal lobes.)

The abnormal profile displayed in Janek's brainwave spectrum is exacerbated by δ2 activity of 56 μv with an index of 46% and a dominant frequency of 2.8 Hz. The presence of low frequency δ and θ waves frontally indicates inefficient function, while δ in other brain areas denotes damage.

The extensive abnormality and damage evident in these EEG recordings are totally consistent with the severe life challenges Janek experienced in cognitive, emotional, and behavioral domains.

Description of the examination: The background activity with low spatial variance, weakly reactive to visual stimuli, formed with few alpha 8 c/s waves, many theta 5-7 c/s waves with an amplitude up to 30 mV and many segments of fast beta activity in all leads. On this background, generalized discharges of sharp waves complexes with theta 5-7 c/s waves were registered. Sharp waves have the highest amplitude from the frontal regions. During the Hv, complexes of slow 4 c/s waves are registered and the number of discharges increases.

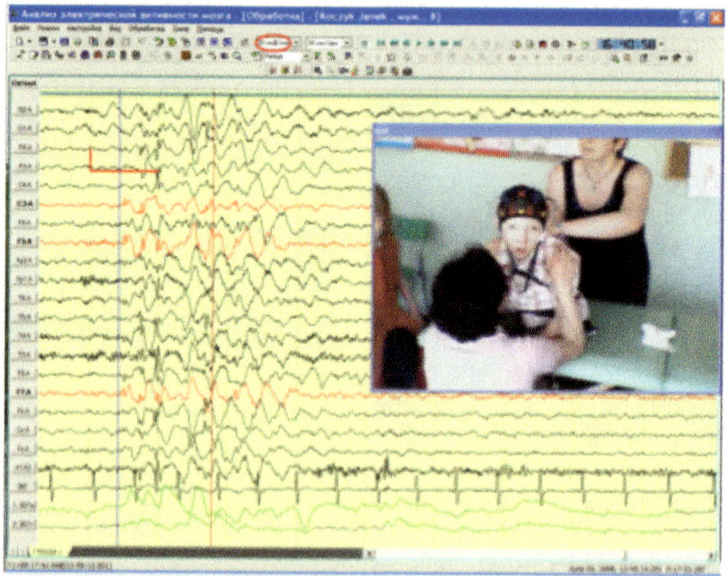

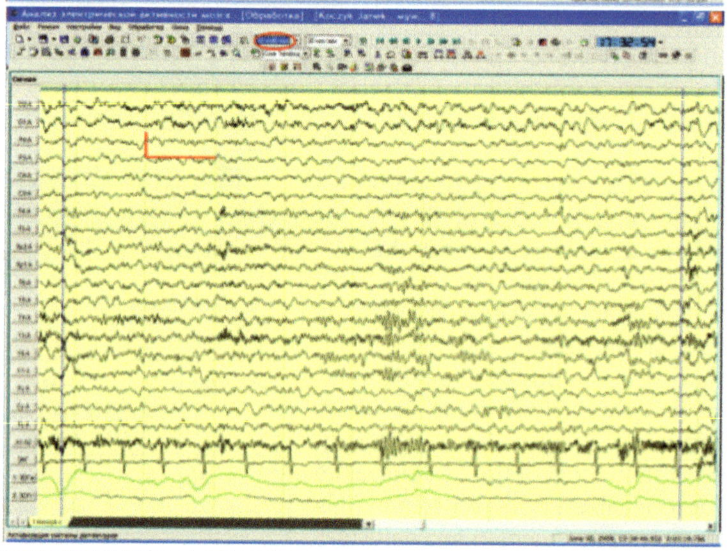

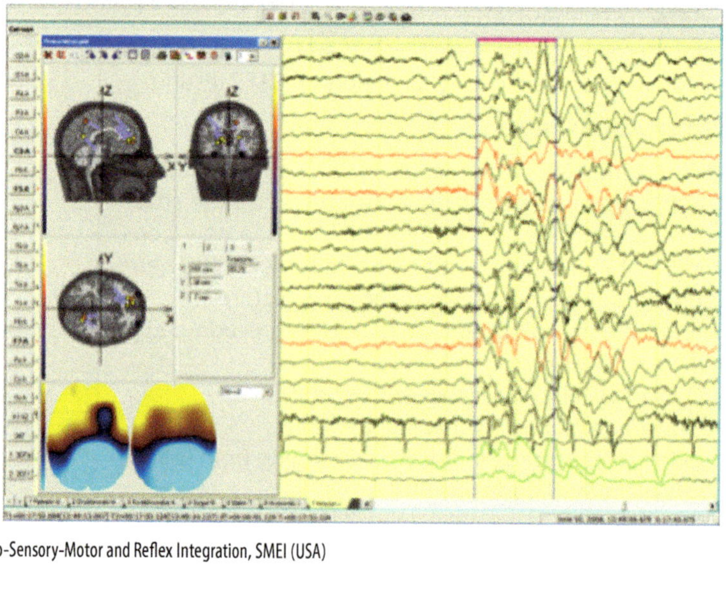

Picture 1. (upper right) Irregular alpha waves in various brain areas as well as occipital theta activity.

Picture 2. (mid right) High amplitude (140 μv/48 ms) – slow wave activity (285 μv).

Picture 3. (lower right) Epileptiform spike-and-wave activity of 140 μv/48ms revealed in the Picture 3 recording occurs simultaneously with occipital and fronto-temporal slow wave activity. The extensive abnormality and damage evident in these EEG recordings are totally consistent with the severe life challenges Janek experienced in cognitive, emotional, and behavioral domains.

The record of dysfunctional background activity with a presence of generalized changes of paroxysmal complexes of sharp waves with theta waves is observed. The record is sensitive to Hv.

Brainstem Auditory Evoked Potentials (BAEP)

Another research project organized by the Masgutova Institute and conducted over a three-year period by Prof. W. Pilecki of the Department of Pathological Physiology of the Medical Academy of Poland used Brainstem Auditory Evoked Potentials (BAEP) to measure excitation waves evoked by auditory stimuli to the brainstem. Two parameters of nerve excitation are measured in BAEP testing: 1) Peak I Latency and 2) Interpeak Latency I-V (IPL).

The Interpeak Latency I-V parameter registers in milliseconds (ms) the amount of time between registration of electric signals in two parts of the auditory pathway: the I segment (auditory nerve, in the ear) and the V segment (inferior colliculi of the mesencephalon, in the brainstem). (Pilecki, Masgutova, et al, 2012).

The IPL I-V value normally decreases as transmission becomes more rapid during maturation of the auditory system. What makes this particularly relevant for Janek is Dr. Pilecki's discovery of a correlation between changes in the auditory pathway and change in a nearby motor neuron. Prof. Pilecki uses BAEP and Visual Evoked Potentials (VEP) to provide an "indirect diagnosis" for children at risk for CNS deficits, based on the connection he sees between the work of the auditory pathway and the motor neuron pathway (Pilecki, Masgutova, et al., 2012).

The Auditory Evoked Potentials (BAEP) Test (70 dB given at a frequency of 10 Hz) was administered to Janek before (10/23/2011) and after 12 days of MNRI® exercises (11/6/2011).

BAEP Results:

Peak I Latency: The norm for Peak I Latency is any value at or below 1.75 ms. Higher than 1.75 ms indicates hyposensitivity of the auditory nerve and weak auditory acuity. The slightly slower speed (1.84 ms on the left side) was due to poorly fitting headphones in that part of the registration and in any case is without clinical significance. Janek's auditory acuity was normal and did not change significantly after MNRI®. (See Table 5.)

BAEP Test Results

Peak I and Interpeak I-V Latency test results before and after MNRI®

	10/23/2011, pre-test		11/6/2011, post-test		Norm
	Right	Left	Right	Left	
Peak I	1.60 ms	1.70 ms	1.61 ms	1.84 ms	≤1.75 ms
Change			+.01 ms	+.14 ms	
Diff. R - L	Right side	.14 ms faster	Right side .20 ms faster		
IPL I-V	3.93 ms	4.13 ms	3.73 ms	4.02 ms	3.85 – 4.25 ms
Change			-.2 ms	-.11 ms	
Diff. R - L	Right side	.2 ms faster	Right side .29 ms faster		

Table 5. The Auditory Evoked Potentials (BAEP) Test results.

IPL I-V Latency: The norm for IPL I-V varies between 3.85 – 4.25 ms. This parameter registers transmission in the vertical section of the auditory pathway, indirectly evaluating nearby motor tracts as well. Elongation of the IPL I-V in children manifests in delayed development and insufficient myelination of the auditory pathway and, as discussed above, frequently correlates with motor skill deficits. Many inborn, as well as acquired damage of the CNS, are associated with high values of the IPL I-V. Accordingly, clinical improvement is often connected with shortening of the previously elongated value. Janek's IPL I-V values improved, that is his auditory processing got faster, by .2 ms on the right side and .11 on the left. Although statistically Janek's improvement seems not so significant, still his auditory processing was improving: by age 10 his auditory system and expressive language were better coordinated and speech became clearer. These changes were stable and very significant for his overall neurodevelopment.

The Masgutova Institute has been interested in using BAEP testing to provide quantifiable, replicable, and objective research-based validation of the efficacy of MNRI® interventions totally independent of the examiner's subjective observations. As is widely recognized, primary motor reflex patterns are involved not only in inherent brainstem survival mechanisms, but also in the development of motor, emotional, and cognitive functions. Two reflexes in particular relate very closely to the parameters under observation in BAEP testing: ATNR and Abdominal Sleep Posture.

In MNRI® testing of the ATNR, turning Janek's head to the left triggered the Abdominal Sleep Posture Reflex while turning to the right elicited the ATNR. Language is processed primarily in Broca's and Wernike's areas in the left hemisphere with its strong connection to the right ear. In the MNRI® model, turning to the right is understood to activate this right ear/left hemisphere channel, so is crucial for language development. Conversely, the left ear/right hemisphere link is involved with synthesizing executive functions and processing the emotional content of language and non-verbal auditory information.

In Janek's case, his hyperactive Abdominal Sleep Posture Reflex on the left side caused him to inhibit auditory processing of non-verbal stimuli (musical or emotional content) in the right hemisphere. At the same time his hyperactive ATNR kept his system hard at work processing stimuli with logical or verbal content in his left hemisphere, but slowly and inefficiently, because the two reflex patterns were in conflict with each other and both active. This influenced Janek's behavior as well. Although his intelligence helped him to decode speech with logical content, he was under stress, had difficulty following instructions, and often ignored them. He liked stories and creative poems only when they were logically linked. Hypersensitive/hyperactive responses to loud and unexpected sounds (complaining, holding his ears, running away) clearly indicated the presence of stress in his auditory system.

MNRI® research (Masgutova, Regner, 2007; also unpublished research by Skhomorokchov, Zakharov, et al., 2008, material of the International Svetlana Masgutova Institute), on the brain wave spectrum indicates that high frequency brain waves (alpha and beta) and the Auditory Stapedius Reflex are activated by the ATNR, while low frequency brain waves (theta and delta), along with inhibition of the Stapedius Reflex, are associated with the Abdominal Sleep Posture. Thus the ATNR creates a state of readiness for auditory processing and the Abdominal Sleep Posture does the opposite, leading the auditory system to ignore stimuli and rest.

With simultaneously active ATNR and Abdominal Sleep Posture Reflexes, Janek was unable to develop efficient binaural hearing. Instead, he tended to use monaural auditory perception with his right ear (hyperactive ATNR) while at the same compensating with his hyperactive Abdominal Sleep Posture pattern to calm down his auditory hypersensitivity. All compensatory strategies exact a cost in terms of inefficiency, fatigue, and stress. In Janek's case, these dysfunctional reflexes may have led him to block auditory stimuli, avoid auditory decoding, and in turn undermine the development of articulation and rhythm in his speech.

Janek's BAEP scores show an increase in processing speed (.20 on the right side and .11 on the left) following 12 days of MNRI® intervention in 2011. This short-term improvement in transmission speed observed on his BAEP corresponds to even greater improvements in MNRI® reflex Assessments over the eleven years of his treatment. From 2003 to 2014 his score for ATNR went from 4 (deep dysfunction) to 15.75 (almost at the norm of 16). His scores for the Abdominal Sleep Posture Reflex improved from 3 (pathological) to 15.5 (functional but at a low level of development). Janek's work towards integrating his ATNR, Abdominal Sleep Posture, Stapedius, and other reflex patterns allowed him to begin to resolve the issue of the improper links among those reflexes.

Resolving the conflict between the ATNR and Abdominal Sleep Posture naturally led to more efficient binaural hearing and processing. Already by 2011, the year of his BAEP testing, Janek's MNRI® scores had reached 15.5 for both reflexes. It was just at this time when his speech began to improve rapidly. Our hypothesis is that BAEP testing over the eleven year period of Janek's MNRI® treatment would have showed a similar pattern of early dysfunction/pathology followed by remarkable progress, with improvements in processing speed as measured by BAEP, correlating closely with improvements in reflex integration as measured by MNRI® Assessments. This result is abundantly clear in his improved language decoding and expressive speech. His story telling and logical thinking amaze both his parents and specialists.

Parent's Observations and Conclusion

When our beloved son was diagnosed with rare severe damage of the cerebellum and its vermis at 17 months we were told that his cerebellum was in a morbid state and that our expectations for his physical, motor, social, emotional, and cognitive development were unrealistic. Still we chose to struggle with the disease and refused to give up.

As his deficits increased and were complicated by chronic health issues, our disappointment grew; still, as parents, we continued to hope that one day our Janek would walk and talk. The many unanswered questions that remained, the lack of explanation for the anomalies in his CNS structure and the fact that we saw no progress in spite of multiple interventions for nearly three years following his diagnosis made us desperate.

After going through various other therapeutic modalities with no visible affects, Janek started the MNRI® program in 2003 when he was 4 years old. This program led almost immediately to many very amazing positive effects in his physical, emotional, and cognitive development as well as remarkable growth in communication and socialization. In the beginning of 2006 we decided to work exclusively with MNRI®. It was important to us to choose a method based on a holistic approach and that could address neurodevelopment in all possible aspects of his life. As parents, we can say that MNRI® has met all our needs.

Today (2014) working with MNRI®, Janek is doing very well in many areas. Despite his cerebellar CP diagnosis and problems with vision, he continues to develop and he has disproved many of the claims in scientific literature regarding limited developmental possibilities for children with cerebellar hypoplasia and severe atrophy of the cerebellar vermis.

Janek's motor skills have improved steadily. He eats by himself using a spoon or a fork. He gets dressed and undressed by himself. He moves using a wheelchair or by walking on all fours on the floor. He is able to stand up with support from objects or furniture. Considering the progress he is making using MNRI®, we assume that his motor development will continue and in the future he may be able to walk independently with no support.

When Janek began elementary school, his cognitive deficits were far more visible than now. Neuropsychologists had diagnosed him with some features of visual agnosia. Problems with memorizing limited his achievement and he used confabulation to compensate for lack of memory. Now, Janek does well at school and does not need to resort to fantasy. He retells facts and stories that actually happened and is able to use his memory as a resource.

Janek still has problems with verbal planning, grammar, and word retrieval. His speech lacks fluency due to pauses during articulation and his slow, irregular tempo. However, the quality of his speech is still changing. Each year brings new improvements. His tempo has become faster and people are better able to understand him. Today we know him as a deep thinker and philosopher. Improvements in his communication skills have made socializing easier and he now enjoys friendships with many of his schoolmates. Traditional specialists admitted his high intelligence only when he started using expressive language. His parents and MNRI® team knew he was smart and worked hard and effectively for his speech to develop. Now Janek is even willing to study new languages. He understands and can speak Kashubian. He also studied Chinese for some time but he had too many other responsibilities. At present he attends English classes.

Janek does not write easily, but he can use a computer. He still has problems with mathematics and reading at school but he does well in biology and history. Janek also takes part in classes in which he creates stories with his friends and he acts in theatrical performances. He also enjoys debates and photographic competitions. He has a keen interest in photography. In September 2013 his photographs were on exhibition in St. Mary's Gallery in Gdansk. He is a member of a school sports team and participates in a class for future athletes in wheelchairs. And this is a boy whose family had to struggle with the diagnosis of mental retardation!

Along with his great progress in other spheres, Janek's health has also improved remarkably. He has not been ill with anything for a long time. If he catches an infection, he gets through it in two or three days. He has not taken medicine for epilepsy since 2008 and he does not have epileptic fits any longer.

Today Janek is a child of high motivation, enthusiasm, and strong personality who knows what he wants and is as persistent and as goal-oriented as all members of his family. We are determined to continue to work hard and are hopeful for even more progress in the future.

Having witnessed my own son's success as a mother, I decided to study MNRI® and I have reached the level of MNRI® Core Specialist and Instructor. My child led me to this choice, for which I am happy and grateful. The diagnostic insights that emerge from MNRI® Assessments and Dr. Masgutova's conceptual understanding of development helped us to understand developmental, behavioral, cognitive, and emotional deficits based on physiological and neurological features of natural reflex patterns. The great success achieved through MNRI® therapeutic programs inspires hope that this method will become widely used by specialists engaged in remedial therapies for children with educational and developmental needs.

References

Borkowska A. R. (2006). *Neuropsychologiczne podłoże trudności w czytaniu i pisaniu:* G. Krasowicz- Kupis (ed.), Dysleksja rozwojowa. Perspektywa psychologiczna, HARMONIA, Gdansk.

Borkowska, A. R., Domańska Ł. (ed.) (2006). *Neuropsychologia kliniczna dziecka,* Wydawnictwo Naukowe PWN, Warszawa.

Brearley, G. (1999). *Psychotherapy for children with motor deficits.* Warsaw, PL: WSiP. p. 19-20.

Coffman, K., Dum R., Strick P. (2011). *Cerebellar vermis is a target of projections from the motor areas in the cerebral cortex.* J. Proceedings of the National Academy of Sciences of the United States of America 108(38): 16068. doi:10.1073/pnas.1107904108.

Brearley, G. (1999). *Psychoterapia dzieci niepełnosprawnych ruchowo,* Warsaw, PL: WSiP

Borkowska, A. R., Domanska, L. (Ed.) (2006). *The child clinical neuropsychology*. Warsaw, PL: Naukowe PWN. p. 98.

Borkowska, A. R., (2006). *Neuropsychologiczne podłoże trudności w czytaniu i pisaniu:* G. Krasowicz- Kupis (ed.), Dysleksja rozwojowa. Perspektywa psychologiczna. Gdansk, PL: HARMONIA.

Borkowska, A. R. , Domańska, Ł. (ed.), (2006). *Neuropsychologia kliniczna dziecka*. Warszawa, PL: Wydawnictwo Naukowe PWN .

Krefft, A. (2007). *Diagnosis of Functions of Invisible Phenomena* (Statistic Mathematic Analysis). Oficyna Wydawnicza Politechniki Wrocławskiej. Wrocław, Poland.

Loeser, J., Lemire, R., Alvord, J. (1973). The development of the folia in the human cerebellar vermis. *Anat. Rec.* 173: p. 109–114.

Maryniak, A. (2004). *Dzieci z guzami móżdżku, Wydawnictwo* EMU, Warszawa.

Masgutova, S., Akhmatova, N. (2004). *Integration of dynamic and postural reflexes into the whole body movement system*. Scientific edition: Prof. N. Akhmatova. Warsaw, Poland: MINK.

Masgutova, S., Masgutov, D. (2005). *Dr. Svetlana Masgutova method of facial reflexes integration. Techniques for facilitation of motor and speech development*. Orlando, FL, USA: SMEI 85 p.

Masgutova, S., Kowal, J., Mazur, G., Masgutov, D. (2005). *NeuroKinesiology tactile therapy™* by Dr S. Masgutova. Scientific Edition: Dr. A. Regner, J. Szymczak. Warszawa, Poland: MINK. – Polish version 133 p.

Masgutova, S. (1998). *The basis of the psychological work with a child using the picture*. Bydgoszczy, Poland: Psychological Inst. Publications.

Masgutova, S., with Kowal, J., Mazur, G., Masgutov, D. (2005). *Neurokinesiology tactile therapy™ by Dr S. Masgutova,* Scientific Edition: Dr. A. Regner, J. Szymczak. Warszawa, PL: MINK, 133 p.

Masgutova, S. (2006). *Neuro-sensory-motor development: visual and auditory reflexes integration. Facilitation program of development and learning for children and adults*. Warsaw, Poland: ISMI. 86 p.

Masgutova, S., Akhmatova, N. (2010). *MNRI® Re-patterning and integration of dysfunctional and pathological reflexes - Svetlana Masgutova® Method of neuro-reflex integration*. 2nd Edition. Orlando, FL, USA: SMEI. 82 p.

Masgutov, D., Masgutova, S. (2010). *Integration of oral-facial reflexes integration. Neuro-structural concept of improvement of motor aspect of speech*. Warsaw, Poland: ISMI. 80 p.

Masgutova, S., Masgutov, D. (2011). *MNRI®: proprioceptive-cognitive integration. Optimize innate auditory-ocular processing and learning skills*. Manual. Orlando, FL, USA: SMEI. 60 p.

Masgutova, S., Masgutov, D. (2011). *MNRI® oral-facial reflex integration – II. Neurostructural concept for facilitation of motor-speech control.* Orlando, FL, USA: SMEI. 122 p.

Masgutova, S., Masgutov, D. (2011). *MNRI® neuro-structural reflex integration*. Revised. Warsaw, Poland: ISMI. 80 p.

Michałowicz, R. (ed.) (2001). *Mózgowe porażenie dziecięce,* Wydawnictwo Lekarskie PZWL, Warszawa, PL.

Parisia, M., Dobynsb, W. (2003). Human malformations of the midbrain and hindbrain: review and proposed classification scheme. *Molecular Genetics and Metabolism* 80 (1–2): p. 36–53.

Pilecki, W., Masgutova, S., Kowalewska, J., Masgutov, D., Akhmatova, N., Sobieszczanska, M., Kalka, D. (2012). The impact of rehabilitation carried out using the Masgutova neurosensorimotor reflex integration method in children with cerebral palsy on the results of brain stem auditory potential examinations. *Advances in Clinical and Experimental Medicine*. Official Organ Wroclaw Medical University. 2012 (3). 2012, 21, 3, 363–371.

Siwek S. (2006). Mózgowe porażenie dziecięce: A. R. Borkowska , Ł. Domańska (ed.), *Neuropsychologia kliniczna dziecka,* Wydawnictwo Naukowe PWN, Warszawa.

Siwek, S. (2006). *Child cerebral palsy*. In book: *Klinical Child Neuropsychology*. Borkowska A.R., Domanska L. (ed.). Warsaw, Poland: PWN. 2006. p. 98.

Encephalan 131-03; Medicom, Russia; Electroencephalographs-recorders computerized portable "Encephalan-EEGR-19/26" http://www.medicom-mtd.com/eng/Products/set_eegr_mini mod.html

Janek is growing to be a strong and interesting personality with clear awareness, goals, and an ability to create, joke, and play. It is such a pleasure to be in his company. Dear Janek, many Congratulations to you! Thank you, also to his parents, Malgorzata and Krszysztof, for having made amazing changes in his postural control, language development, overall health, and for making him happy and joyful! – Dr. Svetlana Masgutova

MNRI® Is for Everything!

Suzanne Amanor-Wilks, MNRI® Core Specialist and Instructor, Minneapolis, MN, USA

Introduction

Masgutova Neurosensorimotor Reflex Integration (MNRI®) is a comprehensive, universal application that mimics nature's original physiological blueprint for human development.

Suzanne Amanor-Wilks

Parents in my office often asked the question, "What exactly is MNRI® for?" Thinking back to the hundreds of clients I've seen over the ten year period in which I've been practicing MNRI®, I realize that the list would be too long to enumerate, so with a twinkle in my eye, I smile and respond, "MNRI® is for everything!" As a colleague of mine once put it, "It's for handwriting and it's for digestion problems caused by stress - who would have thought those two could be the same thing!"

Nature has designed a specific path that is universal to each neurotypical baby. The purpose of this path is to develop sensory-motor, language, and socio-emotional achievement stages. This is done, initially, during the early months of life through a physiological process that is, for the most part, reflexive and therefore automatic, stemming from unconscious processes. Because physiology deals with the functions and activities of life, this reflex pathway is an essential component of early development. It is an essential component of life activity.

Reflex integration is central to the child's journey through his or her own life because it is the physiological mechanism that facilitates the achievement of neurological and behavioral milestones. It is an innate process to which we all can relate because we all went through it. That is why reflex integration is universal.

As a parent, do you remember all the reflexive activities that were going on in your newborn? Do you remember your baby's mouthing reflexes in the first month of life? Your baby typically demonstrated mouthing reflexes that included suck, swallow, and root in order to latch on to the nipple and feed. Every parent remembers the baby's Moro and Fear Paralysis Reflexes when the baby startles in response to a loud sound or to having his position unexpectedly moved. There is a wonderful design behind such reflexes that is intelligent and purposeful, leading to a desired end. This desired end is a combination of sensory, motor, speech, language, emotional, social and behavioral outcomes. Hence my response, albeit a little tongue in cheek, that "MNRI® is for everything!"

By the end of the first month, babies typically can raise their heads while lying prone. This development continues from month one to three, as babies slowly emerge from fully dependent infants into individual personalities who follow an internal program that teaches them to become more aware of their surroundings.

They innately learn to use their eyes to recognize familiar people, objects, and sounds and their mouths to coo and make musical vowel sounds. Also, in this developmental phase, their tightly clenched fists begin to open to grab and bat at objects and they learn to smile and play with other people.

From four to seven months, the journey of coordinating vision, hearing, touch, and movement for exploration continues. All the while, babies are developing strong core muscles as their bodies arch and their chests raise and they rock while on their tummies, kicking legs, and swinging with their arms. Babies learn to roll over, to sit up and even to crawl, exploring the environment with avid curiosity, especially through the mouth. They learn to communicate with their surroundings beyond the cry of hunger or the wail of fatigue. Strong emotional bonds are developing where babies begin to demonstrate powerful attachments to parents, caregivers, and objects. The complex interweaving of human development that occurs during this first year points to an innate, inner genius within every newborn child. This genius is its life. It is intelligent, it is purposeful, and it is wired to survive and to protect itself. If the survival needs of this life inside the child are met, then it is also wired to grow and to develop itself.

An analogy I like to use with parents is that MNRI® uploads and updates protection software into the brain's survival programs in a way that meets the brain's survival needs and, thereby, lays the foundation for future growth and development.

At its core, the Masgutova reflex integration intervention combines the knowledge of infant reflexive patternings with specific techniques that stimulate skin, muscles, tendons, and ligaments to jump start or to enhance development. These techniques are applied in a context of care, comfort, stress reduction, and inner empowerment that create deep connections in the child at the level of physiology, psychology, structure, socio-emotion, cognition, and spirit. We are currently researching how these techniques interface with the child's DNA code in order to fire up dormant pathways. The first result we often see, as is seen during the first year of life, is amazingly fast growth in height and weight where needed.

What sets Dr. Masgutova's work apart from any other that I have experienced, is how this work mimics the genius of nature at an active, motor level. It deconstructs the original reflexive process into its detailed, component parts. It gives each component its due place and required period of maturation. Then it reconstructs these components into an entire process and methodology that honors the life inside each child, providing for its survival and self-protection at a physiological level. This foundational physiological process can be reapplied to any human being, at any age and stage of development in order to guide the internal process back to the original pathway, transforming non-ability into ability: coming to the desired end. What previously didn't work now works. The desired end is always growth and development of some skill, behavior, or purpose.

So if the reader has picked up this book with the question, "Can MNRI® help my child?", the answer is usually unequivocally, "Yes." Over the years, I have seen it help anxiety, stress, sensory-motor disorders, attention priorities, digestive disorders, headaches, insomnia, pain, muscle tone regulation, challenging and oppositional behaviors, sibling rivalry, adoption issues such as bonding and attachment, post-traumatic stress, spasticity, academic concerns, psychological conditions, feeding issues, perceptual distortions, speech and language difficulties, and the list goes on.

MNRI® is not a fix all. It is not a cure. It is not usually quick, though it can be. However, what MNRI® is, is a journey and a gift. It is a journey of transforming life into something better. It is a gift of an intelligent, precise, gracious tool with which anyone, parent or practitioner, can empower themselves to meet their child at the level of his or her need. If your child has no major challenges, then it is simply the gift of greater development and more inner peace.

A key component of the MNRI® approach to therapy is the Home Program. The Home Program is a Monday through Friday list of exercises in which the parent is trained by the MNRI® Core Specialist. Occasionally a parent is too overwhelmed to follow the Home Program at which point the Core Specialist works with the child independently of the parent. Often good results are achieved, even when this is the case, so long as the child sees the Core Specialist regularly. However, best results are achieved when the parent and Core Specialist team together, providing the child with a hearty reflex integration menu for optimal growth and development. Because the MNRI® methodology is premised upon stimulating myelination of neurological pathways, and because this myelination requires daily and consistent repetition, an MNRI® program achieves the best results when the parent follows through with the daily Home Program.

While every child I have seen deserves a whole book to honor their journey and their hard work, I will limit myself to a few examples of how MNRI® has worked in a variety of conditions, "from handwriting to constipation".

Writing Block or Dysgraphia

Presenting Problem: JM was a nine-year old boy in third grade who was referred to me by his Montessori teacher in the school where I was working as Learning Specialist. His mother was a neuropsychologist and she wanted JM to see me due to his dysgraphia. He had a handsome, rakish air about him and was popular with his peers in the classroom. The period of MNRI® treatment was from February 28, 2006 to May 31, 2007. The first time he came to see me, as part of his Assessment, I asked him to write a short narrative. I gave him 5 minutes to do this and I gave him some writing aids.

JM sat there staring at the sheet of paper and holding his pencil. Though he had received an excellent Montessori education and had a good pencil grip and beautiful cursive penmanship, he looked uneasy and uncomfortable. Because I was assessing, I could not give him any further assistance but I felt for this intelligent boy, who tilted his chair forwards, balancing it on the front legs and shifted his body from side to side. As he stared at the paper and tapped his pencil, JM looked disconnected and vague as he attempted to come up with a piece of writing. It felt like the longest five minutes ever and I guiltily wondered if I should change my job description to Public Torturer.

Dysgraphia is a neurological processing disorder that is characterized by written language difficulties. Dysgraphia is an inefficiency in the ability to automatically coordinate the muscle motor sequences involved in forming letters with the ideas that need to be expressed through those muscle movements. Children with dysgraphia often experience writing fatigue, an unusual grip, difficulty thinking and writing simultaneously, and slow and labored writing.

At the end of a very long five minutes, JM came up with his story. It consisted of two words, "Bob died." I wondered if this might prove to be some unconscious foreshadowing of the dysgraphia that was about to die.

History and Background: JM came from a healthy, caring home where child-centered education was highly valued and where he and his siblings were encouraged to interact with the outdoors and to develop their innate gifts and talents through multi-sensory activities.

There had been no significant perinatal concerns though he had colic and had to be held often as an infant. At six weeks, JM had frequent ear infections that subsided after 2½ years when dairy was removed from his diet. He had had some initial delays in language but had caught up without interventions. His mother thought that the language delay had been due to ear infections and, by age 26 months, his language had "taken off."

JM's mother had had her son assessed with a neuropsychological evaluation that suggested "strong signs of dysgraphia." She observed sensory-seeking difficulties with modulation problems and reported that JM was a messy eater and that he was tactile defensive in relation to clothing tags. ADHD had been ruled out.

JM's mother reported that she found her son to be incredibly resilient, facing challenges yet soldiering on. According to her, JM's modulation issues led to either over-arousal, when he would be in everyone's space, or under-arousal, when he would be hypersensitive. She felt he also had a mood disorder and an underlying anxiety which he masked by the need to be in control. He was very thorough and orderly, taking a linear, sequential approach to tasks and demonstrating an ability to organize. He had a high pain tolerance with a hypo response to pain that tended to block sensation and, at the same time, he had a low frustration tolerance. JM had a high average reading ability while his writing was below average. Previous treatments included occupational therapy.

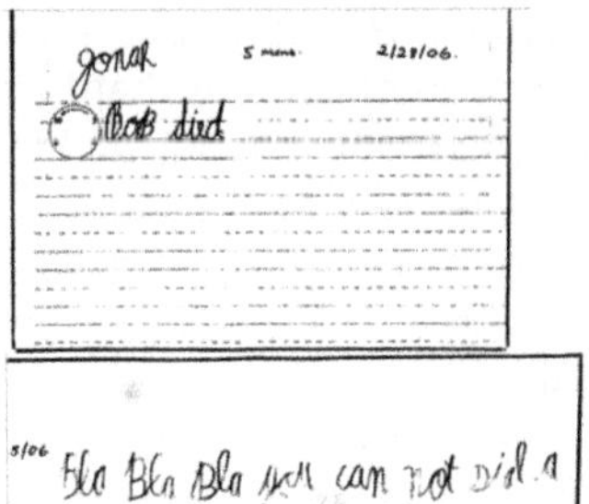

JM's writing progress from the earliest (top) to after MNRI® treatment (below).

MNRI® Intervention: The key reflexes that appeared to be non-integrated for JM were Robinson Grasp, Babinski, Babkin Palmomental, Pavlov, ATNR, Tonic Labrynthine, Spinal Galant, Archetype, Visual, and Moro Reflexes. At that time in the evolution of MNRI®, which is always progressive, we did not have the full understanding of the divisions between Moro and Fear Paralysis Reflexes, but in hindsight, I suspect that his Fear Paralysis Reflex was also active.

On February 28, 2006, JM had his first evaluation session with me. I set a goal for him that "my writing is easy, creative, and fluid." I used his 'Bob died' debut as a pretest of his skill level at this time. In the next two sessions, we did some Archetype Movements to coordinate front, back, and side motor coordination systems and also worked with the Pavlov Reflex which demonstrated auditory and kinesthetic inefficiencies. JM demonstrated a fear of the unknown and a need to feel prepared for it. The Pavlov Reflex, which is a curiosity reflex, appeared to address this fear. In the MNRI® system, a reflex can be integrated directly or it can be integrated by addressing its facilitating or opposing reflex. Pavlov was the facilitating reflex to Fear Paralysis and I used it with him to open him up to new neurological learning. In his next session, I worked with him to integrate his Robinson Grasp Reflex and we did more Archetype Movements. After this, JM wrote another piece of writing on March 15, 2006. This time it was quick, smooth, and flowing. He wrote four whole lines ending with a smiley face: "Bla bla bla," he wrote, "you can not sail a boat across the ocean when so many others have failed yes I can ☺". "Mmmmmm…" I thought pensively as I looked over his writing, "Could this be another subliminal message from his unconscious?"

Sessions with JM continued through to the end of the school year, comprising another eight sessions of reflex integration. After the long summer, we resumed in the fall. On October 12, 2006, JM wrote a whole page involving a fictional tale about a boy called Mark who, though good and obedient to his mother's limits, had overcome his fear of the unknown to venture across the woods. It had a rather hapless ending which was developmentally appropriate for boys of his age.

The interventions that I did with JM to address his writing difficulties were balance board activities (for vestibular stimulation), Archetype Movements (for the coordination of his muscle coordination systems), visual exercises (to integrate his eye tracking with his balance, posture, hearing, and midline awareness), Pavlov, Robinson Grasp (was very difficult for him), Babkin Palmomental, Asymmetric Tonic Neck, Tonic Labrynthine (was a huge integration for him), Babinski (was wired in with his Grasp Reflex), Headrighting, Moro, and Spinal Galant Reflexes.

JM did basic vestibular exercises for his Home Program to support the reflex integration process at school. These involved rocking motions to stimulate his inner ear because of his history of ear infections interfering with his language development.

Results & Conclusion: I saw JM from February 28, 2006 to May 31, 2007. JM's dysgraphia appeared to resolve within 6 months of his initial evaluation with me. This period included the three-month summer break as well as the ten day spring break. JM was able to develop from a blocked writer in the spring of 2006 into someone who could write with fluidity and flow in a variety of genres from fiction to factual reports by the fall of 2006.

Had we continued with JM's therapies, we would have addressed the other aspects of his presenting problems including his mood and anxiety disorders. At that time, MNRI® Family Conferences were not available in America and I did not yet have the complete Core Specialist training. JM continued with his academic career and I left the school to develop my career as a Core Specialist. If I were seeing JM today, I would also address the emotional regulation issues, the mood disorders, and the hand-mouth coordination issues with Neurostructural Reflex integration, Foot Tendon Guard, Tactile, Facial, and Fear Paralysis Reflexes.

Hyperactivity

Presenting Problem: Ezekiel came to see me at the age of 2 years, 8 months. His mother was concerned about his hyperactivity and biting other children at the daycare center. Zeke seemed unable to differentiate his responses to people, objects, or sounds in his environment, and he was very rough with pets and other children. His mother had been taking him for occupational therapy, but felt that she wanted to see more improvement and wondered if reflexes might be a missing piece for him.

She was also concerned about his toe walking and emotional dysregulation with his constant meltdowns. Zeke would frequently get hysterical about toilet flushing and loud noises in general, including sirens. He had potty issues with uncomfortable bouts of constipation or diarrhea which triggered holding patterns where he was afraid to have a bowel movement. Zeke was sensitive to gluten which caused skin rashes and had an aversion to different textures such as shaving cream and playdough. He had a very poor attention span and sleep had always been an issue with frequent bouts of waking up in the middle of the night followed by rising very early in the morning. Zeke had a hard time with transitions leading to constant meltdowns.

As a therapist, I was fairly comfortable with this age group as I was a trained Montessori teacher for ages 0 to 6. Nevertheless, I found Zeke to be extremely hyperactive and very wild. During our initial evaluation, he screamed all the way through and then, at the end, as he unceremoniously took himself off the massage table and out of the room, he stated matter-of-factly, "That was fun." It appeared that extreme dysregulation had somehow been anchored as a comfortable state for his nervous system. He was very undifferentiated at the level of his sensory-motor system and was hyperreactive, tactily defensive, and neurodevelopmentally disorganized with a lack of coordination in the links between his motor system, emotional expression, and cognitive functioning. None of this was unusual developmentally for a two year old but it was the degree and intensity of his responses that removed him from the average. Zeke cried when he was touched and the whole evaluation was a rather hair-raising process for all involved.

History and Background: Zeke's mother had high blood pressure during his pregnancy and he was breach until 3 days before the birth when he spontaneously turned. His was a natural, drug-free birth. Breastfeeding did not go well as he did not have a strong suck and he could not latch on. As soon as he was introduced to formula at 6 weeks due to poor weight gain, he refused to go back to breastfeeding. Zeke had excessive drooling such that he would have a puddle on his shirt due to his poor muscle control in the mouth. Zeke's mother noted that his father has an older son from a previous marriage with diagnosed autism.

Zeke saw an occupational therapist through the school and she reported that he had a Sensory Processing Disorder and questioned if autism or Pervasive Developmental Delay Not Otherwise Specified might be a possible diagnosis due to his poor eye contact and sensory processing issues. He had received some cranial-sacral therapy.

MNRI® Intervention: The period of MNRI® treatment was from April 20, 2012 to June 28, 2013. At the assessment, Zeke was found to have non-integrated reflexes in Core Tendon Guard (this was very hyperactive), Foot Tendon Guard, Fear Paralysis, Abdominal, Bonding, Spinal Galant, and Sucking Reflexes.

The key priorities for Zeke's Masgutova reflex integration intervention were to facilitate the release of the fight or flight response in his sympathetic nervous system. This was driving the physiological anxiety that was contributing to his uncontrollable behaviors. He needed external assistance in organizing his nervous system and calming his hyperactivity.

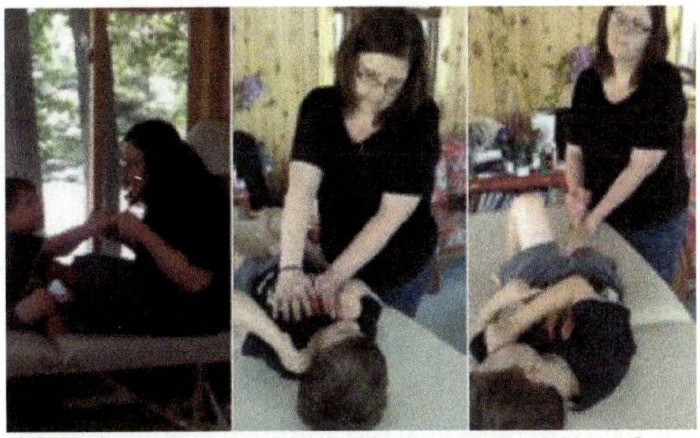

Above, Mom working with Zeke on Bonding Reflex and other reflexes in his Home Program.

Interventions Zeke received over his treatment period were for Fear Paralysis, Spinal Perez, and Core Tendon Guard (to release the fight or flight patterns that were driving his behaviors), Embracing Squeeze (a big key for his tactile sensitivity and his need for bonding with his mother), Tactile, Neurostructural, Fear Paralysis, Bonding, Hands Supporting, Spinal Galant, Spinal Perez, Sciatic Nerve Activation (a big key for his potty issues), Abdominal, Babkin Palmomental, Breathing, Leg Cross Flexion-Extension (was very active and a trigger for his emotional upset and refusal), Foot Tendon Guard, Babinski, Shock Absorption, Trunk Extension, Automatic Gait, Spinning, Archetype, Facial (initially triggered screaming and extreme emotional dysregu-

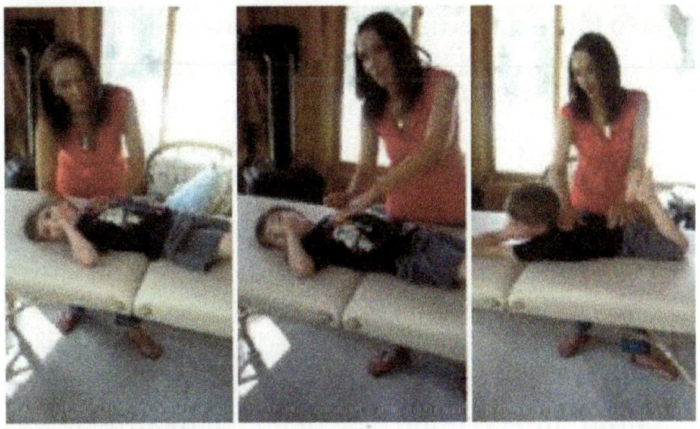

Author working with Zeke on (from left) 1) Core Tendon Guard, 2) Fear Paralysis, and 3) Spinal Perez.

lation), Symmetric Tonic Neck, Asymmetric Tonic Neck, Moro Embrace, Robinson Hand Grasp, and Birthing Reflexes.

His Home Program included Embracing Squeeze, Hands Supporting, Fear Paralysis, Foot Tendon Guard, Core Tendon Guard, Abdominal, Tactile, Babkin Palmomental, Robinson Grasp, Trunk Extension, and Facial Reflexes.

Zeke on balancing ball.

Results & Conclusion: From the initial reflex screening, it took two weeks to see a much calmer Zeke who, for the most part, could be cooperative on the massage table until something would trigger him, such as a non-integrated reflex. Within a few months, Zeke remained on the massage table for the duration of the entire session and cooperated with all movements requested of him. Zeke still demonstrated oppositional behaviors but they were greatly improved.

His sound sensitivity greatly improved whereas Zeke still noticed noises but they no longer triggered emotional meltdowns. Transitions became better – Zeke could go from one activity to another without having a major emotional upset.

Zeke's biting stopped after we started the Oral-Facial work. The clarity of his speech greatly improved – articulation became less slurred. Zeke's extreme drooling completely stopped.

Zeke's toe walking was resolved and his mother only saw it once in a while when he was under extreme stress. He became completely potty trained and his potty issues were mostly resolved. There was an occasional rare reemergence of the holding problem but nothing akin to the previous issues where he required daily coaching sessions on the need to go to the bathroom.

Sleep was an area that needed further improvement. When Zeke was tired or the hour late, he would fight and resist sleep. This area was to become the focus of his continuing therapy with due attention being given to his Abdominal Reflex.

Zeke made excellent progress in key areas of his challenging behaviors. His mother was fully committed to the reflex integration and she took exceptional responsibility for his Home Program. Zeke's mother's exemplary follow-through has provided the conditions for Zeke's excellent results over a fourteen month period. Reflex integration therapy is a key intervention in cases of pervasive developmental delays and sensory processing disorders. When begun early, as in Zeke's case, it promises to have far reaching future consequences for behavioral changes. I see no reason why Zeke should not normalize into a healthy, well integrated first grader in a couple of years when his school career begins.

PTSD (Post-Traumatic Stress Disorder)

A typical MNRI® program takes 2-4 years to resolve some issues depending on the level of severity of the challenge. In some cases, however, as in this one, it took two sessions.

Presenting Problem: FH was an 8 year old girl whose mother attended one of my community lectures. She had a diagnosis of PTSD which was caused by repeated strangulation when she was an 8-month old infant at the hands of her 3 year old adoptive brother with reactive attachment disorder. Her parents had been unaware of these occurrences which had left FH with some Post-Traumatic Stress. Her adoptive brother had been transferred to the care of FH's maternal uncle in order to remove FH from harm's way. FH's sleep was interrupted two to three times each night and she lived in a state of fearfulness, always wanting to know where her parents were. She refused food and had a nervous stomach which interfered with her eating.

After my lecture, her mother came up to me to tell me about her daughter's history. FH presented as sickly, pale, and frightened. Her body posture was hunched forwards and she had no eye contact or social interaction. She stared frozen and clung onto her mother, shy and withdrawn in the face of any public attention. FH was receiving play therapy with a play psychologist and occupational and speech therapy through the school system.

I showed her mother the Embracing Squeeze, Foot Tendon Guard, and Fear Paralysis Reflexes. This took all of 5 minutes. Her mother phoned me the next day saying that her daughter had slept through the night for the first time in her life and booked two sessions with me.

MNRI® Intervention: The period of treatment was from January 2 – 12, 2012. I did not give FH a reflex evaluation as her time was so limited and, instead, I went straight into the reflex integration intervention.

During the first session, I taught FH's mother the Tactile strokes including the Embracing Squeeze for bonding, as well as Fear Paralysis, Foot Tendon Guard, Abdominal Oval technique from Tactile therapy, and I gave FH Neurostructural Reflex integration on the front of her body.

The mother brought FH back three days later, saying: "I don't know what you're doing, but she is eating. She's been eating and eating. Last night as I was doing the techniques on her, she was just sitting there laughing. She was just laughing and laughing and thought everything was so funny. And she's eating. It's a miracle!" She went on to report that normally, her 20 month old baby ate more in one meal than her 8 year old ate all day. Now her daughter was wolfing down her food – bananas for the first time, macaroni and cheese with hotdogs. Her mother reported that her daughter was happy for the first time in her life. She commented that this was only after one session and that they had been working on this since she was 10 months old to the present time. She also reported that her daughter had slept like a rock and that, when she put her to bed, she did not ask to be held as she normally did but fell asleep straight away. The other change she saw was that her daughter was beginning to express herself and had told her mother that, when she talks to people, she felt embarrassed. This was the first time she had ever expressed her feelings.

When I saw FH on her second session, she looked like a completely different child. Her skin was no longer sallow, she was expressive, and her eyes looked alive. She appeared happy and normal. Her mother reported that this fearful child, who previously would not leave her side from one room to another in the house, went to the garage completely unassisted to look for a ball. She also went upstairs alone without noticing that she was unaccompanied. At this time, I did more Neurostructural Reflex integration and taught the mother the Core Tendon Guard, Abdominal, and Spinal Perez Reflexes for the Home Program. I did some Facial Reflex integration therapy on FH as well as the Moro Embrace on a therapy ball. This was the last time I saw them as the mother was going through her own personal struggles and this was as far as we were able to go with the reflex integration therapy.

Her Home Program included Tactile Strokes with Abdominal Ovals, Embracing Squeeze for Bonding, Core Tendon Guard, Foot Tendon Guard, Fear Paralysis, and Abdominal Reflexes.

Results & Conclusion: MNRI® released FH's appetite suppression and returned her hunger mechanism. She began sleeping, expressing herself, and her fear mechanism was released. She saw increased confidence and became happier and developed a more relaxed disposition with a better sense of humor and more inner peace.

FH's ability to express herself and her nourishment needs had been suppressed as a baby and sublimated to a survival program of fear that was triggered by the extreme duress of the strangulations endured under the innocent tyranny of her adoptive brother. Her developing patterns had been frozen into a classic Fear Paralysis response which presented as Post-Traumatic Stress Disorder and an inability to advocate for herself through the expression of bodily needs such as nourishment, sleep, and emotional fulfillment. One small intervention with Fear Paralysis Reflex, Bonding, and some fight or flight releases was all that it took to unlock her innate patterns of development. Once her frozen patterns of fear were unlocked, FH was free of the interference which had stunted the development of her self-expression and ability to nourish herself.

Mitochondrial Disorder

Presenting Problem: Dermot was a 5½ year boy whose parents brought him in to see if I thought MNRI® therapy could address his mitochondrial disorder which had had a global impact on his development. Dermot's body gave the appearance of paralysis. He had extremely low tone with severely restricted range of motion in his limbs giving him a board like appearance. His legs gave repetitive, reflexive flexion motions and he had cortical visual impairment with some intermittent peripheral visual awareness.

History and Background: Dermot had had a normal delivery with traces of myconium present. His hands and arms were almost blue postpartum and he was very cold. He was not immediately transferred to his mother at birth but was heated first. Dermot had a pathological bite reflex whereby his jaw would clamp down and be unable to release. This made breastfeeding impossible for his mother, especially once teeth were growing in.

In the first 3 months of life, Dermot had digestive issues with frequent bouts of colic and vomiting. He had pyloric stenosis, a condition in the lower part of the stomach where food is obstructed from passing into the small intestine. He had surgery at 7 weeks old to address this issue. Dermot had a feeding tube as he could not swallow and would aspirate his food. At 4 months of age, Dermot suffered a grand mal seizure. One month later, he started having facial tic seizures. A 24 hour EEG and MRI came out significantly abnormal, revealing a demyelination of the white matter in his brain.

Dermot had respiratory issues and was prescribed Pulmicort, a mild steroid which helped to prevent his frequent pneumonia. After he started coming to see me, Dermot was placed on oxygen support after a sleep study revealed severe sleep apnea which was waking him up on average over 37 times per hour through the night. Dermot had scoliosis, with a 50° curvature of the spine and was fitted with a body brace. He had dysplasia of the left hip which was 25% misaligned from the socket. Dermot had to wear leg immobilizers at night to keep his legs straight. Because of his craving for oral stimulation after the placement of his feeding tube, Dermot began to chew on both hands to the point of breaking the skin and causing dangerous infections.

Dermot had been receiving services from a physical therapist and an occupational therapist since he was 9 months of age. He started speech therapy involving augmentative communication three months prior to beginning treatment with me. Dermot had neuromuscular electrical stimulation to improve swallowing, involving mouth massages with electrodes applied to pharynx, larynx, and mouth. However, this intervention had no impact on his swallow and Dermot continued to aspirate on his own saliva. Feeding therapy was also attempted. Dermot had Hyperbaric Oxygen treatments after which there was a degeneration in his condition.

MNRI® Intervention: The period of MNRI® treatment was from April 12, 2012 to May 28, 2013. Dermot initially demonstrated the following non-integrated reflexes – Core Tendon Guard, Hands Grasp, Babkin Palmomental, Hands Pulling, Hands Supporting, Babinski, Leg Cross Flexion-Extension, Head Righting, Asymmetric Tonic Neck Reflex, Foot Grasp, Bonding, Fear Paralysis, Moro, Foot Tendon Guard, Spinal Perez, Spinal Galant, and Trunk Extension.

For the first year of therapy, Dermot came to me once a week for a 90 minute session. After a few weeks of treatment, he also started seeing a local speech therapist who had taken the Oral-Facial training and was able to augment Dermot's work with me with extra Oral-Facial integration.

Dermot's treatment followed these priorities: Neurostructural Reflex integration (for the release of his physiological fight or flight patterns), Sequential Arms Opening, Leg Cross Flexion-Extension, Foot Tendon Guard, Babinski, Trunk Extension, Automatic Gait, Tactile Integration (greatly calmed arm/hand hyperactivity), Babkin Palmomental, Asymmetric Tonic Neck Reflexes, Fear Paralysis Reflex, Moro neurosensorimotor point stimulation, Spinal Galant, Spinal Perez, Trunk Extension, Archetype Movements (intermittently), Breathing Reflexes, scapula releases, Facial Reflexes, Abdominal Reflex, and Sciatic Nerve Activation.

When Dermot's mother initially began the Home Program, I felt it was important not to overwhelm her and we took our time building up her level of skill. Once she started seeing results from the therapy, Dermot's mother became very enthusiastic and extremely skilled, gaining expertise rapidly. She came up with her own spreadsheet for tracking the daily exercises that she was consistently doing with Dermot. She has been an exemplary MNRI® parent and has developed Dermot's Home Program into a special bonding routine that she greatly enjoys doing with Dermot, who is one of three boys in the family.

Where previously his mother felt helpless to intervene in Dermot's rapidly deteriorating condition, she now feels happy and empowered and enjoys the MNRI® journey with her son. Dermot's father also does some of the exercises with his son and both parents team together to make sure that Dermot receives an optimal Home Program which has been able to reverse the degeneration spiral that he was in when he first came for therapy. MNRI® is at its most brilliant and satisfying when it takes a team approach. Specialists teaming with other specialists, each taking on an area of expertise within the six core programs and also teaming with the parents for the greater good of the child.

The chart shown is Dermot's mother's spreadsheet of her son's Home Program exercises. She takes satisfaction in checking off each exercise as she accomplishes it and keeps his massage table in the midst of family activity in the home, so she can work on Dermot as she interacts with the rest of the family in the kitchen area.

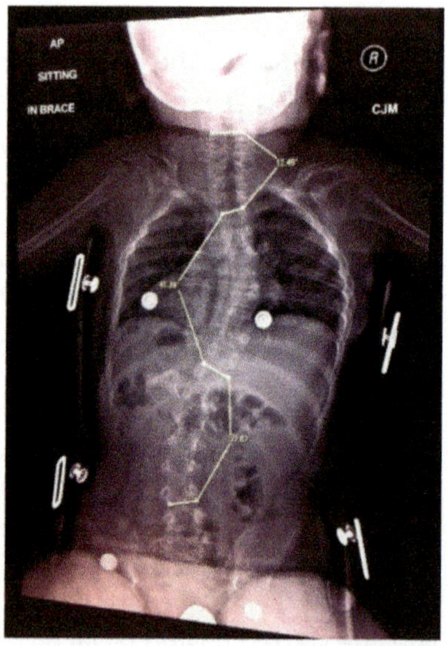

Dermont's X-ray showing curvature of his spine.

Results & Conclusion: Dermont has an increased range of motion and flexibility in his arms with movement development organized into homologous arm movements. Originally, Dermot was unable to lift his arms away from or above his rib cage. He can now do so with ease and can stretch his arms far above his head, reaching outward towards a toy or person close by. Previously, Dermot's hands were clenched in tight fists most of the time but they are now open most of the time. This allows for the exploration of new types of play. Dermot now tolerates the prone position which was previously too uncomfortable for him. He is even able to sleep in the prone position now while supervised. He has begun spontaneous opening and stretching of his arms in an organized pattern. After previous Botox injection, prior to beginning the MNRI® therapy, Dermot's legs appeared limp but now he is kicking and moving his legs quite a bit more.

Dermot's Home MNRI Program				
For the week of - -				
Brief program	frequency-times a week	check when completed		
Tactile integration	3			
Greeting				
stretch limbs				
abdomen ovals				
trunk extentions 21 x's hand below knees (cha chas)				
Stroking				
Embracing squeeze				
infinity 8's				
segmental stretching				
rotations				
Core Tendon Guard	3			
Fear paralysis	1			
Rainbow Arms	3			
Hand Supporting	3			
Babinski	3			
Foot Tendon Guard	5			
Leg Cross Flexion-Extension	3			
Spinal Galant	4			
Spinal Perez	3			
Spinal Dominoes	3			
Babkin	3			
Grasp	3			
Head rightng	2			
Facial Reflex	5			
Spinning				

The Home Program schedule made by Dermot's mother.

He has had a release of the fixation and rigidity of his core and greater relaxation of the muscle tone. Where Dermot's core used to be hard as a rock, it's now more pliable and beginning to soften up.

Dermot had been headed for spinal surgery for scoliosis prior to the MNRI® interventions but the scoliosis is no longer progressing. His hip dysplasia is also not worsening.

Dermot is drooling much less with an improvement in his swallowing and we are ongoing work with his reflexive bite. Dermot has gone from 6-8 bibs to collect drool daily down to 2-4. He has started to suck again and stopped self-injuring his hands and is learning how to use a pacifier again. He swallows faster when eating food purees and has less pooling and aspiration of food. Dermot allows tooth brushing without reflexively biting down on the brush. Dermot has never puckered before but now puckers his lips when expressing distaste or applying lip balm. He can now move his tongue more in different directions.

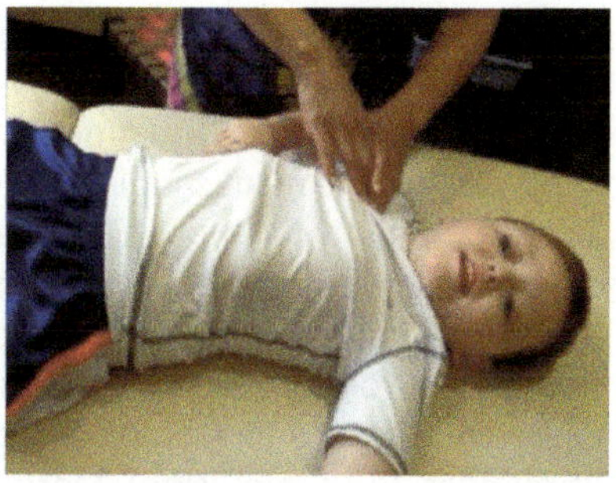

Working with Dermot, a 5½ yr. boy with mitochondrial challenges.

Dermot has begun communication sounds with his grandmother on the phone whom he recognized as they vocalized back and forth while on the speaker phone.

To the eye of most observers, Dermot is still extremely challenged, still non-ambulatory, non-verbal and non-visual. However, to the practiced eye of his parent and specialist, Dermot has made huge strides in one year of MNRI® therapy. In April of 2013, Dermot's mother was able to qualify him for additional services with two more qualified MNRI® Core Specialists which gave us the team approach we had been dreaming about. Working as a team will expand Dermot's possibilities for future development even further and I am looking forward to another year of even greater gains as Dermot rebuilds his neurodevelopmental system from the foundation up.

Autism Spectrum Disorder

When I think of Morgan, a smile spreads across my face. Morgan seems to have that effect on everyone who has the pleasure of meeting him. Morgan looks like a Minnesota version of Harry Potter and he has some indefinable 'magic' effect on people who meet him. When he began therapy with me, he was eleven years old and in fifth grade. Morgan is now thirteen and will be going into eighth grade during the coming school year. Morgan and his mother have a wonderful relationship that is peppered with love, humor, commitment, wisdom, and iPad negotiation. Like many children on the Autism Spectrum, Morgan would spend his entire existence lost inside his electronic gadgetry, if permitted. When Morgan's mother limits his time on the iPad, Morgan turns to his inner gift of being able to run any movie in his mind in full, glorious, technicolor detail. He pops in a rerun into the internal software system inside his brain and laughs uproariously as he watches some of his favorite movies as I work with him on my massage table.

Presenting Problem: Morgan was diagnosed with high functioning autism at the end of kindergarten. He was reassessed at the end of fifth grade when they added a diagnosis of Asperger's. His mother has a quote by Dr. Vincent Carbone on her Facebook page: "Children with autism are not learning disabled, they are teaching challenges. The question is not whether they can learn . . . the question is, can you teach?" Morgan found his elementary school stressful and his mother was a constant advocate for him with various teachers who did not always understand the full implications of what it meant to be a child with autism interacting in a regular classroom full of other children.

History and Background: Morgan's birth was fourteen days late. He was induced and was born with a C-section. His mother had a tumor in her uterus which was pushed up against his shoulder which has led to a lack of pigment in that shoulder. For the first eleven weeks of life, Morgan turned blue once a week and would spend an overnight in the hospital under breathing apparatus. This was not reflux but simply a nervous system overwhelm where his system would just shut down. He was simply too tired to breathe.

Morgan was a very picky eater up to fourth grade but following his participation in a food therapy group with a food therapist, he started trying new foods though he still had problems with food textures.

Morgan was a toe-walker and would often walk on the very tip of his toes like a ballet dancer.

Sleep has always been an issue for Morgan and when he first started with me he would sleep for three hour intervals before moving himself into his parents bedroom where he had a sleeping bag on the floor. He sleep-walked in second grade for almost a year, ending up in the garage on two occasions, at which point, the parents put a chain on the door.

Morgan would get overwhelmed by the demands of the school environment which had a tendency to trigger fight or flight behaviors in him involving running away or getting into conflicts with specific children over the use of playground equipment or to certain behaviors. Morgan was particularly sensitive to another boy in his class with an autism diagnosis who had some vocalizations and self-stimming behaviors. Morgan had a great antipathy towards this boy and desired to distance himself from being associated as having a similar condition. Some of the stress of school life was caused by his sound sensitivity to the teacher's voice level and the background noise of the students. Morgan had a bout of bronchitis and pneumonia after which he was suspended from school at the beginning of December 2010. Morgan had high anxiety and frustration levels with low tolerance thresholds for sensory overloads. He would worry a lot over things and get stuck on little obsessions and compulsions. Morgan had been suffering from constipation all his life which nothing seemed able to resolve. Morgan presented with low tone and had a strong urge for deep proprioceptive squeezes from his mother.

Morgan had information processing lags which required at least four days of integration time before he could output the information he had learned. Spelling tests required that he learn the words a week before the test. If he started two days before the test, he could not remember the words. Morgan was not able to logically extrapolate that if x + y was xy, then y + x was also the same.

During therapy sessions, Morgan was very obsessed with what time it was, asking the same question repeatedly, "What time is it?" He was also very attached to his mother, needing frequent hugs and squeezes from her in order to self-regulate as well as extremely frequent, "I love you's". He chewed his fingernails to the point of infection and demonstrated some self-stimulating hand flapping movements. Other stimming behaviors were head banging with closed fists and reflexive fingers closing motions. More repetitive actions included finger tapping and foot swinging.

Morgan was irritated by too much expectation of conversation and preferred to take flights of imagination inside his head or, even better, to play on his iPad. He could become extremely obsessive about limits set on his time for computer games and, if these were not available to him during a given period, he would become very frustrated and single-mindedly focused on regaining access.

Morgan was extremely sensitive to perceived criticism and was unable to differentiate between people being amused and entertained by his engaging personality and criticism. He seemed to feel inordinately guilty about things which were not even asked of him, saying, "I'm sorry," repeatedly for things which he perceived as being disappointing to his mother, including his autism.

Morgan found it hard to re-regulate if he was upset when things did not go the way he wanted them to and would persevere on the denied outcome repeatedly and irrationally. He found it impossible to let go of negative emotional upsets or cognitive drives towards getting his computer game time.

Morgan did not have any solid friendships except one very highly bonded relationship with a neighbor child which was reinforced through very strong community ties. Unfortunately, this child left the neighborhood in a rather abrupt fashion cutting all ties, which traumatized Morgan and the entire neighborhood. Some previous treatments were behavior therapy, occupational therapy, physical therapy, cognitive thinking therapy, and social skills group therapy.

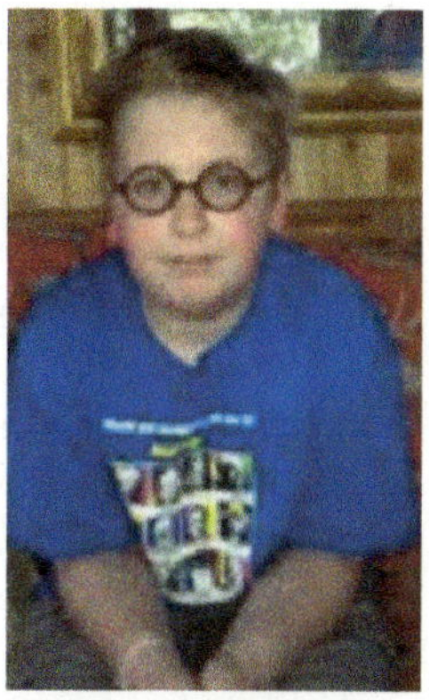

Morgan, now 13 yrs, improving from an autism diagnosis

MNRI® Intervention: The period of MNRI® treatment was from September 17, 2010 to October 29, 2012. Morgan presented with low tone, high sensitivity, poor proprioception and tactility, an active startle reflex and many other retained reflexes. The reflexes were found to be non-integrated in Morgan were Core Tendon Guard, Robinson Grasp, Hands Pulling, Babkin Palmomental, Babinski, Leg Cross Flexion-Extension, Asymmetric Tonic Neck, Abdominal, Bonding, Fear Paralysis, Moro, Foot Tendon Guard, Spinal Galant, Spinal Pereze, Tonic Labrynthine, Symmetric Tonic Neck, and Trunk Extension Reflexes.

Initial priorities for Morgan were to reduce his hypersensitivity and release the fight or flight mechanism. Morgan came to me initially twice a week for 150 minutes of treatment and then once a week for a 50 minute treatment session. We began with Neurostructural Reflex integration, Foot Tendon Guard, Spinal Perez, and some Tactile integration including the Embracing Squeeze which was very important for Morgan and its proprioceptive input greatly helped in his regulation and bonding issues.

We continued his reflex integration intervention with Abdominal, Babinski, Leg Cross Flexion-Extension, Birthing Reflex, Spinal Galant, Trunk Extension, Automatic Gait, Shock Absorption, Hands Supporting (which triggered a big anger response), Moro, Asymmetric Tonic Neck Reflex (which triggered a complete meltdown with uncontrollable sobs moving towards hyperventilation over the loss of his little neighborhood friend who had been taken into foster care). We continued his therapy adding Babkin Palmomental, Robinson Grasp, Auditory-Visual Reflex integration, Facial Reflex Integration, Acoustic Reflex, and Sequential Arms Opening.

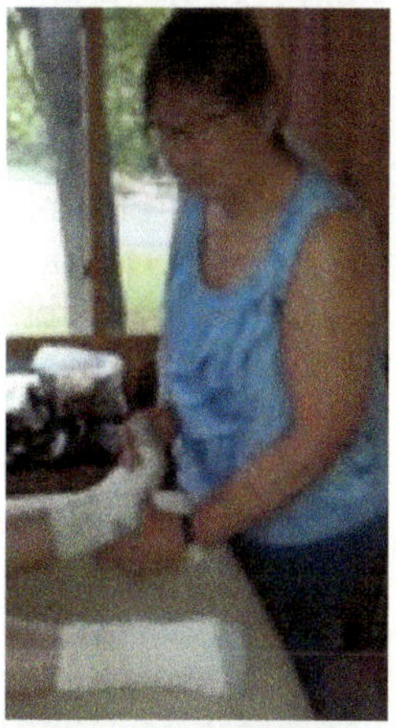

Morgan's mother learning Foot Tendon Guard.

At this point in his reflex integration therapy, we began working with Morgan's fears which were triggered by an away camp organized by the school. His two main fears were 1) the fear that his mother did not want him because of his autism for which he felt a deep shame, and 2) a fear, even terror, of being away from his mother. We continued working with Morgan's Fear Paralysis, Acoustic, Abdominal, and Moro Embrace Reflexes. We added the Archetype Movements and Pavlov Reflex to open up his curiosity mechanisms so camp would become a source of enjoyment and pleasure rather than fear and terror. We later added Hands Pulling Reflex to his program.

Morgan's therapy continued with occasional bouts of sickness or emotional releases which included weeping over the loss of his grandpa or friend. Morgan responded well during such bouts to Neurostructural and Spinal Perez which helped to detoxify his emotions. In April 2011, he was diagnosed with asthma which continues up to the present time.

We worked with Breathing Reflexes, Moro with Bonding Reflex, Babkin, and Birthing Reflex. During the Bonding therapy, Morgan had a huge catharsis over his fears that his mother did not love him. He had a meltdown, repeating, "I just wish I wasn't autistic! I wish I wasn't this way." After this experience, his science teacher reported at school that he was very impressed with how much Morgan had changed and that he was more engaged, with greater concentration.

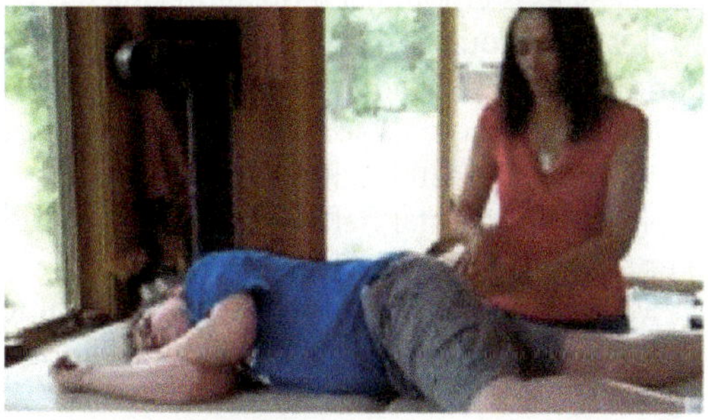

Sciatic Nerve Activation for Morgan

Work with Morgan was interrupted due to family emergencies and resumed after an eight month break. At this point the direction of the therapy was towards the Tonic Reflexes and focused on his nail biting and re-emergent toe walking.

His Home Program was Embracing Squeeze, Tactile Therapy integration, Fear Paralysis Reflex, Abdominal Reflex, Spinal Perez, Spinal Galant, Foot Tendon Guard, Babinski, and Archetype Movements.

Results & Conclusions: Morgan's imitation mechanism switched on as demonstrated in excessive amounts of yawning. He never was able to yawn before and his mother reported that he has yawned more in the past 6 months than in his entire life.

His constipation is now resolved though somewhat dependent on nutritional support.

Morgan's sleep is now fully resolved and he slept for the first time through the night for nine hours in March 2011, six months after the start of the MNRI® reflex intervention and he no longer rises at 5:30 am. His sleeping from 9 pm. to 7 am. is, for his mother, the miracle she never expected to see.

Morgan's toe walking resolved over a six month period but returned when therapy was interrupted for 8 months He no longer flaps his hand for self-stimulation.

Morgan has become much more relaxed about what time it is than previously. He has a much higher tolerance for conversation and is not so easily irritated by interruptions to his internal thought processes. He handles transitions with greater ease, has more inner peace, and has learned to handle losses and deaths as a part of life. Morgan changed schools and is now attending a school for children with autism. He no longer has daily stress about his school life, enjoys the classroom, and has made new friends. He lets go of emotional upsets and cognitive drives much more easily and gracefully and has more self-esteem and confidence about his diagnosis. He has released his deep shame about his autism.

Morgan is less dependent on external regulation with hugs and "I love you's" from his mother. Where he used to say "I love you" and "I need a squeeze" at least twenty times during a 60 minute session, he now says it only once, if at all. He is now even able to go to camp away from home and tolerate being away from his mother.

Morgan was never a team player or interested in sports but he recently won two gold medals in the state championship Special Olympics for the 25 meter dash and the shot put as well as fourth place in the 50 meter dash.

Morgan's breathing and asthma issues are still not resolved and were noticeable during the long 8 month break in therapy. His nail biting also has not been resolved.

The work with a child with autism is multi-layered and has many branches and routes. Health challenges often emerge associated with gut, respiratory, immunological, and nutritional aspects. Spinal Perez is a big key for detoxification and needs to be maintained throughout the therapeutic process. Releasing fight or flight mechanisms in the core with neurostructural interventions is also a big key as is emotional regulation which is often addressed through Fear Paralysis, Moro, Foot Tendon Guard, Tactile and Bonding Reflexes.

The next stage of Morgan's therapy will be focusing on his Breathing Reflexes, the release of his Babkin Palmomental Reflex manifested in the nail biting, and the Tonic Reflexes.

Morgan has gained a greater sense of peace about his autism and wrote the poem about it, at right.

ME

When I was smaller and people were taller
I realized that I was different
I had a power that set me apart
I learned to take it, to use it, to make it
It's not so bad to be different
To do your own thing and do it with heart
Different is hard
Different is lonely
Different is trouble for you only
Different is heartache
Different is pain
But I'd rather be different than be the same
At first I'd wonder what hex I was under
What did I do to be so different
Then I discovered some others like me
Wonder no longer, Together we're stronger
It's not so bad to be different
Be true to yourself
That's what you must be.
– Morgan

Constipation

Presenting Problem: Bella, the sweet, blonde ten-year old, fourth grade girl sitting in the rattan chair in my sun room looked healthy. She had the typical midwestern coloring of the blonde hair and blue eyes with healthy, milky skin, and fresh cheeks.

She stayed close to her mother and her lack of eye contact made her appear somewhat reticent and compliant. She smiled a lot but didn't say much. Her mother was a social worker and had heard about reflex therapy through word of mouth. She hoped this might be able to help her daughter in three areas of difficulty which were 1) school-related anxiety, 2) generalized emotional regulation issues at home, and 3) constipation. Her daughter had a medical diagnosis of Ehlers-Danlos syndrome which affects the connective tissue and is seen in low collagen production, hypermobility of the joints which are loose and can easily hyperextend, muscle pain, and a tendency for dislocations and sprains. Functional bowel disorders are under recognized in Ehler's. Bella had a mild form of the Ehlers-Danlos and had sprained her ankle in first grade which resulted in continued pain. Bella reported ongoing pain in her ankles when she was doing a lot of running or jump roping. Bella was an A student in school and had no academic issues. She also performed well in sports activities such as soccer.

History and Background: Bella's mother had a healthy pregnancy and normal delivery with no issues. Bella had early feeding challenges which her mother felt were at the root of her systemic emotional regulation issues. Bella had a very difficult time nursing which had possibly, according to her mother, established some pre-verbal fear patterns. Bella was her first child and, as a new mother, she was very anxious about the fact that she could not feed her baby. She had been sent home from the hospital without knowing how to nurse. She described how this had left her feeling tense, sweaty, and in frequent tears at not being able to nurse her newborn. For the first month she pumped her breast milk into a tiny tube that she would tape onto her little finger which she gave to Bella to suck. This process took an inordinately long time and Bella was always hungry. The situation resolved itself by 6 weeks and then Bella was able to nurse normally up to the age of 15 months when she was weaned. Bella's mother felt that these early patterns of nourishment were at the core of Bella's emotional regulation challenges.

Bella had experienced constipation since the age of six months which had become chronic and long-term, particularly in moments of stress such as school transitions and testing periods. She would typically miss having a bowel movement for up to four or five days at these times. She had seen the family pediatrician on a number of occasions and had been referred to a pediatric gynecologist and a pediatric gastroenterologist. She had been prescribed Miralax and ate an exceptionally healthy diet with a wide variety of fruits and vegetables including several healthy servings per day of broccoli, asparagus, avocados, carrots, and mangoes yet had been unable to resolve the constipation issues.

Bella had a fever at 15 months which developed into a febrile seizure that lasted over an hour. Emergency Room professionals could not stop the seizure because they had to place a ceiling on the medication dosages they could give to such a tiny body. They kept Bella on a breathing tube and airlifted her to a major pediatric hospital in the metropolitan area where Bella was medically induced into coma via a spinal tap. The hospital said that they believed the seizure was Bella's body's reaction to the MMR vaccination which had been administered 15 days prior to the crisis. They felt that having three shots at one time had been too much for the baby's little body. After the seizure, Bella was very weak and fussy for some time. She had lost muscle mass and regressed from walking to falling over and not being able to bear weight. It took some time for her to regain stability and walking skills.

The emotional regulation behaviors had begun at the developmentally appropriate age of around 2½ but had continued up unto the time when her mother brought her in to see me.

These behaviors presented as explosive reactions over themes of displacement, not getting enough, not getting her turn, having to pick up her bedroom, and such. She had a low frustration threshold and would have long, extended meltdowns where she could not re-regulate. She would have a lot of fear if her mother needed to step away to do something during these long drawn out bouts and would panic, needing constant soothing from her mother for extended periods.

Bella's anxiety was highly triggered by her classroom situation involving peer group dynamics among a lot of behaviorally challenging children whom the teacher had a hard time managing. Bella felt frustrated, overwhelmed, and unsafe in the classroom environment. She was also particularly anxious about having her

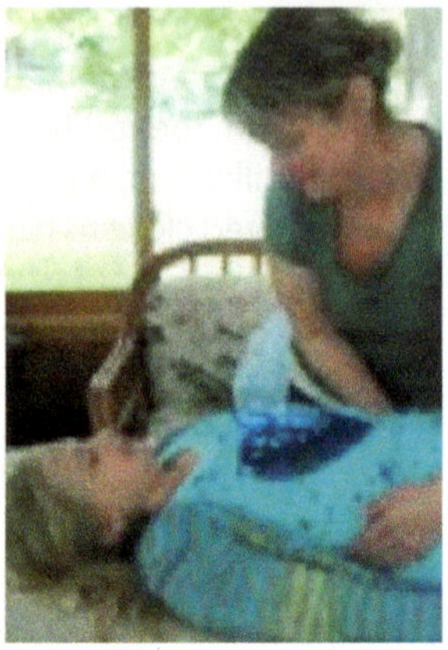
Bella's mother is learning the Core Tendon Guard Reflex.

blood drawn which may have been triggered by the medical interventions during the febrile seizure and reinforced by watching her brother having his blood drawn three years previously which had frightened her. Bella had developed an acute fear of having blood work and throat cultures.

In April 2013, her gastroenterologist had recommended that she be checked for thyroid, gluten allergy, and blood cell counts. Bella had gone for the blood work but was unsuccessful and had to leave as she was too scared. They tried again at the family doctor's office but again Bella was too frightened and the nursing staff's recommendation of holding Bella down was declined by Bella's mother. They decided to defer the blood draw until such time as Bella was able to participate.

Previous treatments were with a pediatrician, gastroenterologist, pediatric gynecologist, podiatrist, and physical therapist.

MNRI® Intervention: The period of MNRI® treatment was from February 28 to June 28, 2013 (present). The initial evaluation of her reflexes demonstrated an overall pattern of low tone, tactile defensiveness, hypersensitivity in Bella's muscle tone response and in her tactility. She demonstrated some non-integration in Core Tendon Guard, Foot Tendon Guard, Babkin Palmomental, Babinski, Leg Cross Flexion-Extension, Fear Paralysis, Trunk Extension, Shock Absorption, Moro (hyporesponsive), Bauer Crawling, and Head Righting. According to her history and intake, I determined that Bella would also need Breathing Reflexes, Spinal Galant, Tactile Integration, and Facial Reflex integration to address the early nourishment issues.

At this time, Bella has had four months of MNRI® intervention. Her mother has taken the Home Program at a slower pace, needing to take the time to feel confident in what she is doing. End of school also interrupted the Home Program for some time so it has been somewhat slow and intermittent but this has not prevented Bella making excellent progress in her therapy. Bella comes to see me once a week for a 50 minute session.

Sessions with Bella have included Tactile Integration (weekly and a key component of her Home Program), Neurostructural (including Core Tendon Guard Reflex which was in extreme hypervigilance), Fear Paralysis, Moro Reflex (which demonstrated a misdirection of the neck movement), Spinal Perez, Hands Supporting, Breathing, Sciatic Nerve Activation (a special technique that appears to interrupt any misfiring at the level of the reticular activating formation at the top of the brainstem), Archetype, Leg Cross Flexion-Extension, Babkin Palmomental, and special Flexion-Extension techniques.

Bella's Home Program included Tactile therapy, Foot Tendon Guard, Fear Paralysis, Hands Supporting, and Spinal Perez Reflexes. Bella's mother is beginning to learn the Core Tendon Guard Reflex.

Results & Conclusion: On May 17, 2013, we focused Bella's therapy session on the goal of having her blood drawn and Bella was able to have her blood drawn within a few days, remaining calm and self-empowered throughout the entire process. She was very proud of herself.

Bella had a small crisis on June 11, 2013 following her therapy session in which she spent a lot of time yawning. Yawning is often a good indication in this therapy that certain mechanisms may be switching on. When Bella left, she was pale, with a fever of 101.8. She threw up, she had diarrhea, headache, chill, loss of appetite, insomnia, and low energy. This kind of response can be a sign of detoxification and is often welcomed by the specialist as a transitional point in the therapy marking a step up to a higher level of function. This was indeed the case for Bella. Following this transitional point, her mother reported improvement in many areas. The slight fever and accompanying symptoms were similar in kind, if not in degree, to the early infant fever which marked the febrile seizure. Following this, all three presenting issues of anxiety, emotional dysregulation, and constipation began to resolve. At the next session, I gave Bella Archetype Movement Integration in order to reset her nervous system and the Abdominal Reflex to calm things down for her.

Bella's constipation resolved itself and, for the first time in her life, this is no longer an issue. It took two months of MNRI® therapy, with weekly visits, to resolve this issue.

Bella's mother reported that her emotional regulation is much better. Bella is much happier, she smiles a lot more, and she has a better sense of humor. She is no longer experiencing the extreme emotional dysregulation and when issues do arise she is able to calm herself down through breathing and positive self talk. She is better able to tolerate limits and has more inner peace.

Bella has better eye contact and is more affectionate. Her muscles feel less reactive and her tactile defensiveness is releasing whereby she can respond to touch in a more pliable manner whereas before she was very guarded.

Bella's therapy will continue as she still has a long way to go in order to fully address all aspects of her presenting concerns. She has made an excellent start and her mother is committed to the journey of reflex integration. The next aspects of her therapy will involve Facial Reflex integration in order to address her early nourishment issues. We will continue with all neurostructural aspects to address the connective tissue disorder, joint mobility, and her muscle tone regulation.

Even though we have not focused on the reflexes supporting early feeding, Bella has had a generalized and global response to the therapy which has impacted the emotional regulation issues that were triggered, according to her mother, by those early nourishment issues.

The genetic connective tissue condition will require longer term therapy and there are still many layers to Bella's anxiety issues that would benefit from continuing attention to her reflex integration process.

My sincerest thanks to all the children who have trusted me with their development and to their families. May you bloom and grow in Peace. – Suzanne Amanor-Wilks

Living and Breathing Easier

Connie Jordan, RRT-NPS (Registered Respiratory Therapist-Neonatal/Pediatric Specialist), MNRI® Core Specialist and Instructor, Denver, CO, USA

This case study is about a 10 year old girl with a medical history of birth at 28 weeks gestation via emergency C-section. She presented with decreased oxygenation and fetal distress. Her APGAR scores were 2 and 4.

Connie Jordan

APGAR is a quick test performed on a baby at one and five minutes after birth. The one minute score determines how well the baby tolerated the birthing process. The five minute score tells the doctor how well the baby is doing outside the mother's womb.

An APGAR score ranges from 0-10 and is comprised of evaluations of five categories which are: breathing effort, heart rate, muscle tone, reflexes, and skin color. Each category receives a rating in the range of 0 to 2 depending on the observed condition of the infant. Scores for each category are added together to produce the overall APGAR score. Normal APGAR scores range from 7-9 for a healthy baby while scores lower than 7 indicate the infant requires some additional assistance to survive outside the mother's womb.

After delivery this infant received assisted ventilation, supplemental oxygen, respiratory stimulation, and airway management. She was transported to the neonatal intensive care unit where she spent 4 weeks.

While in the NICU she received assisted ventilation, suctioning, bronchodilator breathing treatments, CPT (chest physio-therapy), chest x-ray, and arterial blood gas monitoring. Chest x-ray results showed she had respiratory stress syndrome at birth. While still in the NICU she contracted bacterial pneumonia requiring IV drug intervention.

Once released from the hospital she continued to present indications of respiratory distress including chronic cough, shortness of breath, low SaO_2 level (oxygen level in hemoglobin) which was supplemented with oxygen and breathing treatments.

She also experienced chronic ear infections, frequent pneumonia or bronchitis infections, sinus infections, and stomach issues. At age two she was diagnosed with asthma. She was prescribed steroids, bronchodilator (MDI), CPT, and antibiotics for her bacterial infections.

At age ten she became my client. Her respiratory assessment can be viewed on the following chart under week 1. She currently was on a steroid inhaler and albuterol. Her FEV (Forced Expiratory Volume) was 46% of her predicted without bronchodilation.

This child received the MNRI® Respiratory Reflex integration weekly for 3 hours for a period of 5 weeks. Her clinical results are presented in chart 1. (Please see the Abbreviation Key at the end of this story.)

Chart of Clinical Results

		Week 1	Week 2	Week 3	Week 4	Week 5
Height	*Pre:*	44.0″	44.0″	44.1″	44.1″	44.3″
	Post:	44.4″	44.5″	44.6″	44.5″	45.0″
Blood Pressure	*Pre:*	107/60	123/72	112/68	118/70	120/69
	Post:	98/62	102/64	103/60	110/64	96/60
Heart Rate	*Pre:*	109	98	104	100	96
	Post:	90	90	92	88	86
Respiration Rate	*Pre:*	24	26	22	20	20
	Post:	20	20	18	16	18
FEV*	*Pre:*	75 LPM	78 LPM	85 LPM	95 LPM	110 LPM
	Post:	80 LPM	90 LPM	100 LPM	110 LPM	120 LPM

**FEV (Forced Expiratory Volume) predicted value=160 LPM (liters per minute–0_2 flow rate) at weight 50 pounds*

Breath Sound (B.S.) Evaluations: Weeks 1 – 5

Week 1 Pre: B.S. were decreased in both bases bilaterally. Inspiratory wheezing was present RUL>LUL and scattered throughout RML. Expiratory wheezing was scattered throughout both lung fields.

Week 1 Post: Wheezing throughout both lung fields was now only scattered. Increased lung exchange in both lungs noted. Easy breathing vs. decreased shortness of breath was also noted.

Week 2 Pre: Inspiratory and expiratory wheezing was throughout both lung fields. Rt>L. Decreased B.S. noted in RLL only.

Week 2 Post: Air exchange showed noticeable improvement from first session.

Week 3 Pre: Minimal inspiratory wheezing in R>L lung field. Expiratory wheezing in the bases could be heard R>L.

Week 3 Post: Minimal inspiratory and expiratory wheezing bilaterally.

Week 4 Pre: Slight inspiratory wheezing in RUL and RML. Left lung without inspiratory wheezing and clear. Rales (crackles in or near the base of the lung) were noted in Rt. LL. L lung base is slightly diminished in B.S.

Week 5 Pre: Scattered inspiratory and expiratory wheezing Rt. > L.

Week 5 Post: Slight wheezing in both lungs Rt.>L. Increased aeration to both lung bases were now audible.

Breath sounds have been affected by the MNRI® Respiratory Reflex Integration therapy. Increased aeration to both lung bases was now audible. Client has gone from being a mouth breather to a nose/mouth breather with ease. Breathing prior to therapy was only present in the upper lung fields using accessory muscles. At post therapy, the client is now an abdominal breather with corrected abdominal movement IE: inspiratory - abdomen expands out, expiratory – abdomen contracts inward.

Forced Expiratory Volume (FEV) Results

FEV predicted normal value for this client's height, weight, age and sex is 160 LPM. To monitor results of therapy, FEV was used. Pre values obtained were prior to bronchodilation medication use and prior to Respiratory Reflex Integration therapy. Post values were obtained without bronchodilation medication and after 3 hours of Respiratory Reflex Integration therapy. (See graphs on next page.)

FEV Evaluations: Weeks 1 and 5

Week 1 Pre: FEV = 75 LPM (47% of predicted)
Week 1 Post: FEV = 80 LPM (50% of predicted) 3% Increase over pre-therapy
Week 5 Pre: FEV = 110 LPM (69% of predicted)
Week 5 Post: FEV = 120 LPM (75% of predicted) 6 % increase over pre-therapy

Other changes noted after 5 weeks of therapy are:

- height measurement increased 1 inch from 44 inches to 45 inches
- blood pressure dropped from 107/60 to 96/60
- heart rate dropped 28 BPM (beats per minute) from 109 BPM to 86 BPM

- respiration rate dropped 6 breaths per minute from 24 to 18. This indicated improvement in the ventilation to perfusion ratio.

Notable structural changes include:
- improved rib spacing bilaterally
- lung symmetry improved
- released diaphragm
- trachea and hyoid shifted to center midline
- rib cage lengthened and lowered.

Further assessments of this client which support the value of MNRI® Respiratory Reflex integration therapy include:
- increased aeration to both lung fields
- ease of breathing without distress or noticeable shortness of breath
- ability to walk and talk at the same time without needing to stop
- fluent speech without needing to pause to catch her breath
- dry non-productive cough without tightness
- B.S. clear to bases bilaterally
- O_2 use reduced 50% from 2.0 L to 1.0 L and only while sleeping or as necessary upon exertion
- apnea monitor no longer required
- CPT no longer required
- improved activity level in daily activities
- increased school attendance (Prior to therapy absenteeism rate was as high as 23%. Currently, it is 0 % resulting in improved grades.)
- absence of ear, chest, or sinus infections for 4 months
- ability to participate in physical education without fear.

The client states that she can now breathe easier and live life with less stress.

FEV (LPM) Improvement with Therapy

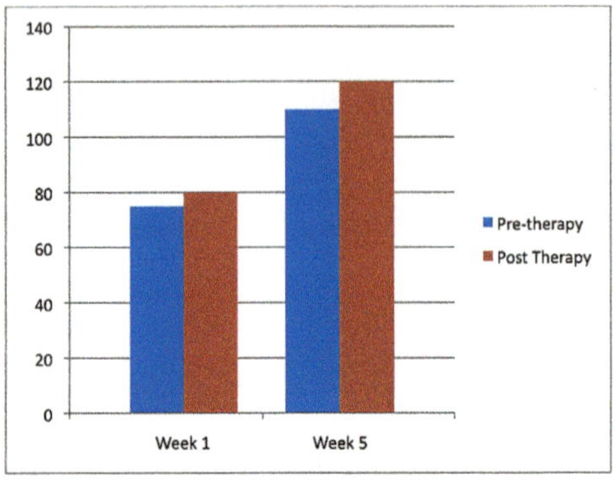

Figure 2

FEV Percent of Predicted Improvement with Therapy

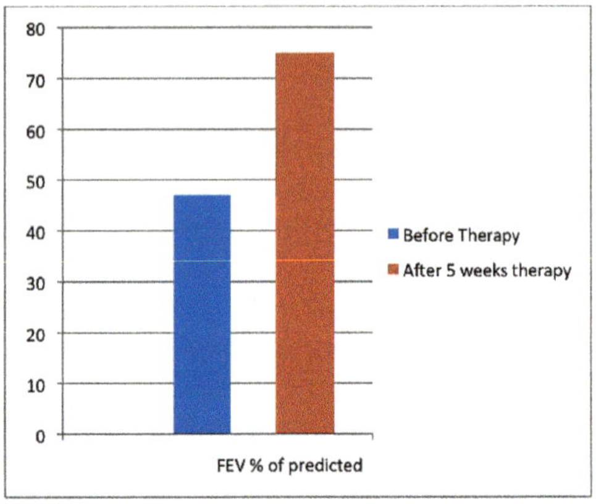

Figure 3

Abbreviation Key:

RUL= Right upper lobe	LUL= Left upper lobe
RML= Right middle lobe	RLL= Right lower lobe

(Rt.>L meaning the wheezing is greater on the right lung than the left lung):

R= Right	L=Left
LMP= liters per minute	CPT= chest physical therapy B.S.= breath sounds
MDI= meter dose inhaler	NICU=neonatal intensive care unit
I= inspiratory	E=expiratory

It is such a joy to see great improvement in the breathing and strength in our children. Breathing should be an ease of living; thank you to everyone who improved breathing reflex functions, showing that it is possible to come through even severe health challenges and to live to our full extent!
– Connie Jordan

The Effects of MNRI® on a Child with Autism

Diane Whiteside, PT, MNRI® Core Specialist and Instructor, San Ramon, CA, USA

Diane Whiteside

My goal in this article is to evaluate the effect of MNRI® therapy on a child with autism and epilepsy. I hope to raise awareness with other specialists in the helping professions of the importance of MNRI® when working with children's reflexes. I will share my success in improving a child's neurological development with MNRI® which demonstrated fast, stable, and optimal motor development and behavioral changes. The changes seen below are shown first, after six months of initial treatment, then again, after two years of extensive treatment.

Case Study

Ian was a sweet 38 month old boy when I first saw him on Sept. 5, 2011 for a comprehensive evaluation. His medical history was significant for autism and epilepsy. Ian had been on the Ketogenic diet (3.5:1 ratio) for the past year which had been very successful in controlling his seizures. He continued on the diet for 2 years. Ian wore glasses for near sightedness and did not have verbal language. He received individual physical therapy, speech therapy, and ABA in his home. He had a sister that very much wanted to play and interact with her brother.

Muscle Tone: Ian had generalized hypotonia in his trunk and extremities and had difficulty holding his head up in prone, on all fours, and when pulled to sit.

Postural Control: Ian demonstrated poor postural control due to generalized hypotonia. When sitting in a chair, he typically slouched causing him to slide forward. This position made it difficult for him to use his upper extremities for good coordination and motor control. He did not readily engage in activities requiring him to raise his arms above shoulder level for more than a few seconds.

Muscle strength: Ian had weak muscle strength throughout his body. His head lagged when pulled to a sitting position so he had difficulty with movement and transitioning from floor to sit to stand. He did not have the strength to climb a step. His facial muscles were very weak producing a flat effect with very little facial expression.

Gross motor: While lying on his stomach, he was only able to prop himself on his forearms and raise his head to 30 degrees for less than a minute before fatiguing. When he reached out for a toy, he often collapsed on his other arm. He also would only reach for a short duration, about 2 seconds. He was able to roll from side to side and to his back but it was very slow and required quite a lot of work. He would roll to his sides but rarely roll all the way over to his stomach or back. He was not able to raise his arms and legs together or push his trunk off the floor. He needed assistance to get into an all fours position. He needed moderate assistance to get to a sitting position, and could still lose his balance backwards. He could hold it with assistance but was not very interested in moving forward in this position. He was not able to get to stand from the prone position without maximal assistance and was not able to balance in a kneeling or half kneeling position. He would bear weight on his legs when assisted to stand. He walked unassisted on flat surfaces around the living room and kitchen but was unable to climb a small step or go up and down stairs without moderate assistance.

While lying on his back he seldom reached for a toy held above his head. He looked at it and followed it for only about 25 degrees side to side, then turned his head slightly to follow the object; his eyes had difficulty working together. He was not able to raise himself off the floor in this position. He attempted to lift his head but not consistently.

Visual-Motor Integration: Based on the Peabody Developmental Motor Scales (PDMS), Ian demonstrated significant delays in his visual-motor integration skills with an age equivalent at the 9 month level and a percentile ranking of 1%. He was able to track an object horizontally and vertically past midline but had difficulty differentiating his eye from head movements when visually tracking. During desk activities, he could not sustain visual eye contact in his lower visual field when engaging with objects but he typically felt around the table for objects rather than look directly at the object. Visual attention improved with musical toys and when objects were held higher up in his visual field. He did not engage in midline play with both his hands together, instead, he used one or the other hand. He would, however, transfer objects from one hand to the other. He was able to clap his hands but imitative clapping is inconsistent. Ian was able to place 7 cubes in a cup with initial demonstration and tactile cueing. He was not currently inserting basic shapes into a form board, building a 2-3 cube tower, or imitating vertical or horizontal crayon strokes.

Peabody Developmental Motor Scales

Assessments:
Peabody Developmental Motor Scales
(The first score was at 3.2 years)

MOTOR SKILLS	Raw Score	Standard Score	Percentile	Age Equivalent
Reflexes	6	-	-	-
Stationary	32	4	2%	9 months
Locomotion	28	1	<1%	6 months
Object manipulation	0	1	<1%	6 months
Grasping	29	1	<1%	7 months
Visual Motor Integration	42	1	<1%	9 months

	Gross motor	Fine Motor	Total
Sum of Standard Scores	6	2	8
Developmental Quotient	48	46	43
Percentile Rank	<1%	<1%	<1%

Raw scores are the total points accumulated by the child on a subtest but it is the standard score that provides a clearer indication of a child's subtest performance, and these scores can be related to one another. Standard scores are based on the distribution of a mean of 10 and standard deviation of 3. Standard scores of 6-7 are considered below average, 4-5 poor, and 1-3 very poor performance. The percentile rank represents the percentage of the distribution that is equal to or below a particular score. For example, Ian scored in the <1% rank for stationary gross motor skills, meaning that 1% of the standardization sample scored at or below his score. The age equivalent also referred to as 'motor ages' are used only to communicate a child's competence in a universal language, and are not always the best measure of a child's performance. Quotients ranging from 90-110 are average, 80-89 are below average, 70-79 are poor, and 35-69 are considered very poor performance. Ian scores in the very poor performance range.

Ian's Individual Treatment Plan

Masgutova Neurosensorimotor Reflex Integration (MNRI®) was chosen to treat Ian. Based on the results of Ian's MNRI® assessment, an individualized program was established for him to strengthen and support physical, emotional, and cognitive growth through the work with non-integrated, retained, poorly developed, dysfunctional, and pathological reflex patterns that were identified.

The home work for Ian included exercises from MNRI® re-patterning, several exercises from Tactile Integration (Embracing Squeeze, stroking of extremities, and joint distraction), Visual-Auditory Reflexes (ear tapping, horizontal eye tracking – slow and fast, digital movement, start and stop), and Oral-Facial Reflex integration (the whole program). These techniques were used 2 to 3 times a week.

His MNRI® Home Program included reflexes performed 5 times per week including Spinal Perez, Spinal Gallant, Embracing Squeeze, Leg Cross Flexion-Extension, Robinson Grasp, Hands Supporting, Trunk Extension, Foot Tendon Guard, and Balancing. Reflexes performed 3 times per week included Babinski, ATNR, Grounding,

Hands Pulling, Segmental Rolling, Spinning, STNR with eye tracking, Bauer Crawling, and Fear Paralysis once per week.

A physical therapist/Core Specialist provided the program one time per week and progressed through the program accordingly. His Mother gave the Home Program 5 times a week with excellent follow through.

Below are four examples of reflex techniques used:

The techniques used to stimulate the Robinson Grasp Reflex were for the hypo-active hypo-sensitive reflex. Ian was lying on his back. I put my thumb into the upper one third of the palm of Ian's hand. With my other hand I closed his hand into a fist with his thumb over the ends of his fingers. I rotated my thumb in the direction of the tips of his fingers. I held it closed for 7 seconds and repeated it 7 times. This created sensory-proprioceptive activation in his palm which stimulated a stronger grasp.

The technique used to stimulate Hands Supporting was the same for hyper or hypo active response. With Ian lying on his back I worked with both hands at the same time in three different positions. Ian's hands were at the sides of his body, palms slightly extended and elbows straight. I placed my thumbs at the bottom third of his palms to give the sensory-proprioceptive stimuli with a pulsating cha-cha rhythm. I pressed pulsing against his lower palms for up to 60 seconds. The second position was with his elbows bent at 90 degrees and the pulsing lasted 60 seconds. The third position was with his arms extended with his shoulders at 90 degrees and repeating the pulsing on the lower third of his palm. I repeated this sequence a couple of times.

The technique used for Spinal Galant included four steps with Ian lying on his side. First, I stimulated the sensory–proprioceptive stimuli, stroking with my fingers down along the side of his spine from the shoulder to the sacrum. I repeated this 7 times. Then I pressed firmly but gently his shoulders and hips and held for 7 seconds. Then I pressed his shoulders and hips together and held for 7 seconds and then apart and held for 7 seconds. I repeated this 7 times on one side and turned Ian to his other side and repeated.

The technique used for the Leg Cross Flexion-Extension included 4 steps with Ian lying on his back with both legs straight. I gave pressure to the middle of the bottom of one foot and held it for 7 seconds. Next I flexed the opposite leg bringing the knee towards his chest, holding it for 7 seconds. Then I brought the leg straight again in 7 seconds and stretched the whole leg when it was straight for 7 seconds. (See additional instructions with the article about Leg Cross Flexion-Extension in this book.)

Results

The initial observations and post–test data showed significant positive changes. Within two weeks of starting MNRI® Ian was able to hold his head up for 10 seconds to look at a toy in a prone position. By four weeks some of his Tonic Labyrinthine and Head and Trunk Righting Reflex patterns changed, which was evidence of the development of his cerebellum and lower motor neurons. He was able to crawl about 5 feet but his head righting continued to be poor and he was not able to hold his head up while crawling. By six months he had developed better facial expression, which was possible evidence of increased development of the extrapyramidal tracks, cranial nerves, and cerebellum functions. He was laughing, smiling, and was much more engaged with people, his therapist, and family. This demonstrated functional improvement of the limbic system.

He started developing his ability for expressive speech, verbalizing his first word, "Mama." This indicated development of the Broca cortex area. He was able to move his tongue out of his mouth and around his bottom lip, left and right.

He began following 2-step directions without prompts which may have demonstrated growth in the cerebral cortex and functional cognitive development.

In the area of gross motor, he began to roll in all directions and demonstrate more freedom of motion of the horizontal planes of the body. He was more stable and was no longer losing his balance and falling backwards while sitting. We did bridging with his hips and he could hold the wheelbarrow position with extended arms.

He could step down two steps, holding the rail without assistance. He needed only moderate assistance for climbing up 2 steps with his hand on the rail (an improvement from maximal assistance). His movements were quicker, more fluid, and precise than before. He was even able to bend forward to walk into an igloo. This was evidence of improvement of his postural control, motor coordination, and motor planning – all demonstrating development of the sensory cortex and the pre and post-motor cortex.

His improvements per age development according to the PDMS over a period of six months were:

• stationary improved from 9 months to 11 months of age

- locomotion improved from 6 months to 14 months of age
- object manipulation improved from 6 months to 12 months of age
- grasp improved from 7 months to 10 months of age
- visual motor improved from 9 months to 19 months.

Results after additional treatment: He increased his visual tracking from 30 degrees to 180 degrees with head movement and he was able to isolate his eye movement from his head movement 30 degrees to both sides from which was previously 0 degrees.

In the area of locomotion, he progressed from a 6 month to 14 month level. He was getting up and down from the floor by himself with some effort. He could walk up stairs with minimal assistance and down a few steps while holding onto the rail.

With object manipulation, he had been at a 6 month level and progressed to a 12 month level of performance. He interacted with rolling and catching a ball. (Previously he did not engage with these activities.)

With stationary skills, he progressed from 9 months to 11 months. He could balance while sitting and standing. (Six months before he had difficulty balancing while sitting and would fall backwards unexpectedly.) His grasp improved from 7 months to 10 months. He was now also able to be interactive and play with his sister.

After two years of MNRI® and four MNRI® Family Conferences, and other programs, Ian was now running, playing with minimal to no assistance, independently rolling both directions, and Bauer Crawling 50 feet. He was also demonstrating an improvement in his emotional and cognitive processing. He could walk up and down stairs holding the rail and a few steps without using the rail. After one MNRI® Family Conference his Mother stated, "Ian has jumped 3 levels in his HPI program," and "We were not even doing the [other] programs while at the conference except MNRI® so I know this is what made the difference." Ian's biggest gain was his development in visual motor skills from 9

Peabody Developmental Motor Scales

Scores at 3.8 years just 6 months after starting the MNRI program

MOTOR SKILLS	Raw Score	Standard Score	Percentile	Age Equivalent
Reflexes	-	-	-	-
Stationary	36	4	2%	11 months
Locomotion	72	2	<1%	14 months
Object manipulation	2	1	<1%	12 months
Grasping	36	2	<1%	10 months
Visual Motor Integration	82	4	<2%	19 months

	Gross motor	Fine Motor	Total
Sum of Standard Scores	7	6	13
Developmental Quotient	51	58	49
Percentile Rank	<1%	<1%	<1%

Ian's Developmental progress pre/post MNRI® after 6 months

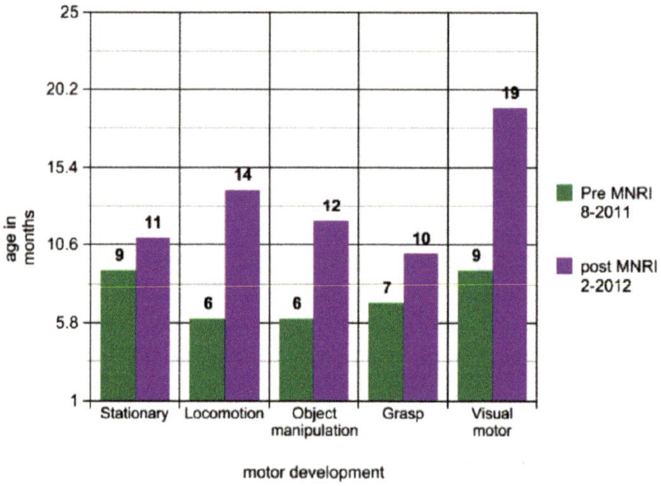

Results of Reflex Patterns Evaluation (after 6 months of MNRI® Therapy)

Table 2. Reflex Patterns Serving for Development of Left-Right Motor Coordination Systems.

	1		2		3		4		5		6		7		8		9		10	
	Core Tendon Guard		Hands Grasp		Hands Pulling		Babkin Palmo-mental		Babinski		Leg Cross Flexion Extension		ATNR		Abdominal		Bonding		Foot Grasp	
	R	L	R	L	R	L	R	L	R	L	R	L	R	L	R	L	1	2	R	L
Before MNRI Program	9.5	9	5.5	5	4.5	4	4	3.5	4	4.5	8	8.5	5	6	4.5	4	5.5	4.5	5.5	5
Average		9.25		5.25		4.25		3.75		4.25		9.25		5.5		4.25		5		5.25
After MNRI Program	12.5	12	8.5	7	5	4.5	4.5	4	6.5	6	12.5	12	6.5	7	6.5	6	9.5	8	7	6
Average		12.25		7.75		4.75		4.25		6.25		12.25		6.75		6.25		8.75		6.5

Table 3. Reflex Patterns Serving for Development of Superior-Inferior MCS

	11		12		13		14		15		16		17		18		19		20	
	Automatic Gait		Bauer Crawling		Moro		Fear Paralysis		Hands Support		Segmental Rolling		Landau		Flying/ Landing		Grounding		Head Righting	
	R	L	R	L	1	2	Up	D	R	L	Up	D	Up	D	Up	D	Up	D	R	L
Before MNRI Program	6.5	6	6.5	6	6	6	5.5	6	3.5	4	4.5	4	3.5	2.5	2	3	3.5	3	4	4.5
Average		6.25		6.25		6		5.75		3.75		4.25		3		2.5		3.25		4.25
After MNRI Program	12.5	12	7.5	7	7	8	7	9	5.5	5	5.5	6	6.5	6	7.5	7	7	6	8.5	9
Average		12.25		7.25		7.5		8		5.25		5.75		6.25		7.25		6.5		8.75

Table 4. Reflex Patterns Serving for Development of Anterior-Posterior MCS

	21		22		23		24		25		26		27		28		29		30	
	Trunk Extension		STNR		Spinal Galant		Spinal Perez		TLR		Foot Tendon		Spinning		Locomotion		Balancing		Pavlov Orientation	
	R	L	R	L	R	L	R	L	Up	D	R	L	R	L	Up	D	R	L	1,2, 3	4
Before MNRI Program	8.5	8	5.5	6	5	4.5	4	4	4.5	5	4	3.5	3.5	4	3.5	4	3.5	4	7	5
Average		8.25		5.75		4.75		4		4.75		3.75		3.75		3.75		3.75		6
After MNRI Program	12.5	12	9.5	9	6.5	7	7	8	8.5	7	6.5	6	6	6.5	5.5	5	8.5	7	12.5	
Average		12.25		9.25		6.75		7.5		7.75		6.25		6.25		5.25		7.75		6.25

months to 19 months.

The graphs at right demonstrate the improvement in the level of development of the reflex patterns which also reflect Ian's cognitive gains.

Summary

The information we gathered showed how impressive the MNRI® program was even for the first 6 months as it was the only program used.

His progress in the next year and a half, when other programs were introduced, showed that the MNRI® program was also very complementary with them. Ian's development with MNRI® was very impressive especially because he had had traditional therapy intervention for the first 3 years of his life and had progressed only to an average age of 7.4 months. Then, at 3.2 years old, the Masgutova Method® was added to his therapy and, in 6 months, Ian had a significant jump in development to an average of 13.2 months. His motor, emotional, and cognitive development increased by an average of 6 months.

With the MNRI® program, and then by attending MNRI® Family Conferences for intensive MNRI® therapy, Ian continued to progress towards age appropriate motor, emotional, and cognitive development.

Finally, after two years of MNRI®, four MNRI® Family Conferences, and other programs added, Ian can now run a mile daily, hang and cross the monkey bars without assistance, roll both directions without assistance, crawl 50 feet, etc. He is now expressing his joys and dislikes, making eye contact, and greeting people and expressing his wants. His parents continue to take him to multiple camps a year and perform the therapy on a weekly basis because of the great results they see.

Graphs below show Ian's Reflex Pattern Assessments: May 2011 (Before MNRI®), November 2011 (Test-1), May 2012 (Test-2), and November 2012 (Test-3) on the three body planes.

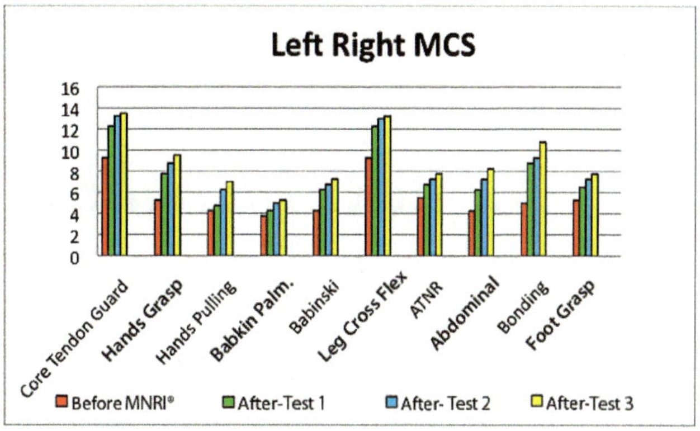

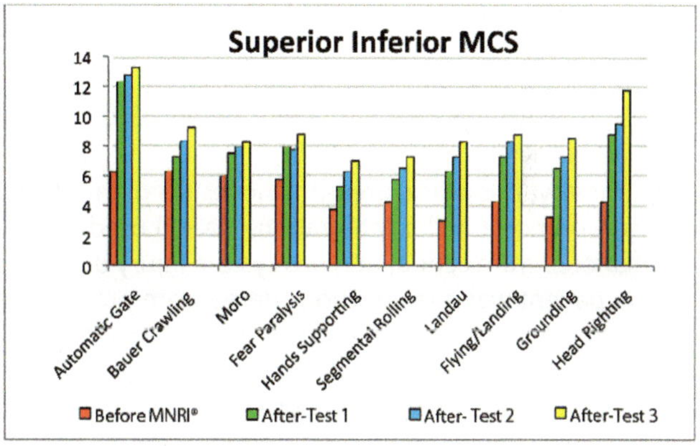

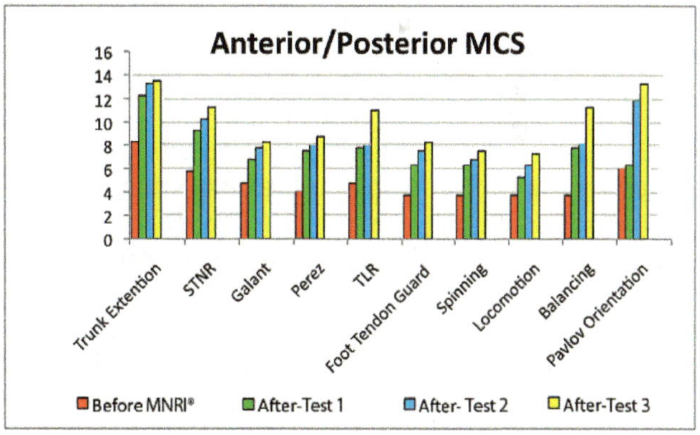

It has been exciting to be a part of Ian's life and watch his growth and development. His awareness of his surroundings and the people in his life stimulate his desire to socialize as his ability to express continues to emerge. Thank you for being a part of the MNRI® family! You are an inspiration to us all. – Diane Whiteside

Wojtek's Incredible Story

Teresa-Ewa Busz, MA, and Beata Oginska-Dutkiewicz, MA, MNRI® Instructors and Core Specialists, Zielona Gora, Poland

Teresa-Ewa Busz

Beata Oginska-Dutkiewicz

Summary: This article presents the results of the Masgutova Neurosensorimotor Reflex Integration (MNRI®) used with a patient who suffered from central nervous system injury during a car accident. The patient was a male who suffered from brain damage resulting from two hemorrhagic strokes which caused a car accident. The MNRI® therapy helped to return sensorimotor function as well as significantly improve both physical and emotional abilities. Ongoing work with this method continues to impact his development of perception and expression and began to rebuild communication, speech development, and cognition.

Description of the Accident

On November 26, 2009 at the age of 30, Wojtek was involved in a car accident. Doctors did not expect him to survive. He suffered injury to the brain stem, had several broken ribs, and other internal injuries. He was unconscious, suffering from both circulatory and respiratory failure. Doctors had to induce a coma. The cause of the accident was likely due to a stroke that caused him to lose consciousness and subsequently swerve into an oncoming truck. For the first couple weeks in the hospital, he was completely unresponsive.

On December 2, 2009 the patient underwent a second operation to remove a hematoma. Afterwards, he showed steady progress. After returning home, the patient underwent daily rehabilitation yielding positive results. On his birthday three months after the accident, Wojtek began to walk. Two weeks after that, on his daughter's first birthday, he began to speak.

Unfortunately, on March 2010 shortly after beginning to speak, he suffered another hemorrhagic stroke that extended to the lateral ventricle. He was once again fighting for his life. Blood was not properly flowing to his brain. An MRI was performed. However, it was too late. The stroke he suffered returned him to the same cognitive state he was in after the accident. The resulting lack of oxygen caused a relapse in his capabilities; he could no longer walk or speak. Constant and intensive therapy had to begin anew.

An MRI of Wojtek's brain stem was performed at Doctor Antoni Jurasz University Hospital in Bydgoszcz, Poland on August 8, 2010. Numerous scans were taken of several damage sites, among them large blood clots and possible seepage of cerebrospinal fluid.

From the time of the accident in 2009 through May, 2011 the following therapies were used for Wojtek: PNF,

Bobath, neurological rehabilitation, reflexology, neuroreflexology, speech therapy, psychological and neuro-psychological therapies, aqua therapy, Adeli suit, and biofeedback.

The MNRI® program was introduced into the therapeutic regimen in May 2011 during a Family Conference organized by the Dr. S. Masgutova Institute in Poland and Bobath, Adeli suit, neuroreflexology, psychological therapy, and biofeedback therapies were discontinued.

Description of MNRI® as Applied

Before the conference Wojtek had a busy month. Shortly before coming to Dr. Masgutova's conference he participated in an intensive program for patients with aphasia. His family was very concerned as to whether he would be able to handle participating and worried about his prognosis after Dr. S. Masgutova's diagnosis. Up to this point, Wojtek's wife Kamila had not seen great strides in his rehabilitation. She was very worried that, with his level of brain injury, there just wouldn't be much improvement in his psychomotor and cognitive development.

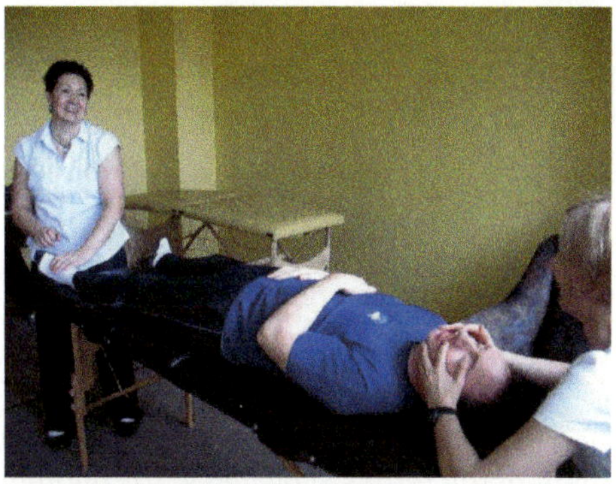

Dr. Masgutova working with Wojtek at an Intensive Conference in Poland.

Wojtek and Kamila arrived in Mielno filled with doubt, but also with a hope that MNRI® might just be able to open the door to faster improvement and a higher quality of life for Wojtek.

The patient was wheelchair-bound. He could sometimes transfer with assistance. He was very anxious, yelled a great deal, hyperventilated often, sweated profusely, and did not tolerate lying on his stomach. His arms were in flexion close to his chest with his hands in a fist. He presented strong tactile defensiveness and strongly resisted being touched. He had very hyperactive protection and survival mechanisms. Due to the stroke in the left hemisphere, his right eye would not close completely. He was extremely sensitive to noise, especially with large numbers of people. He was persistently agitated, irritated, and tense. With his wife's help he was able to participate in rehabilitation. Once on the treatment table he would constantly pick up his head, flex his arms, and yell. He had pathological muscle tone, making it difficult to lie in any position. He did not want to lie on his side at all, tensing his body even more and communicated his displeasure with cries and yells. While lying on his stomach he would always support himself with one arm and would not permit his head to be placed on the table. Always fearful, he tried to control all movement around him. Although he would occasionally tire, he soon started fighting again. His tactile defensiveness and discomfort in combination with physical disability kept him from settling down. The motor control of his extremities was very disorganized. His right leg was very stiff while the left showed better movement responses. His grasp reflex was hyperactive. Wojtek would often dissociate and close his eyes. He tended to roll his eyes up towards the top of the skull. It was difficult to establish eye contact with him. He could recognize close family relations, but was not interested in other people.

Wojtek and I began working together during the intensive MNRI® rehabilitation conference in Mielno, Poland. Each day of the conference consisted of 6 one-hour sessions of MNRI® for 11 days including: the MNRI® reflex integration program, Oral-Facial and Auditory-Visual therapy, Proprioceptive and Cognitive Reflex integration, Neurostructural reflex integration, Tactile, Archetype Movement, and Art Therapy. The main goal of the reflex pattern integration program was to optimize the functioning of his brain and the whole nervous system. Importantly, we worked on lowering tactile hypersensitivity, reducing negative defense mechanisms, as well as regaining the abilities he had lost. Through the reflex pattern integration exercises, we were able to establish a 'language' through which we could reach his nervous system. Each day we worked with his sensory, tendon, and muscular systems. Through stimulation, his brain would begin to recognize and relearn correct reflex patterns. These patterns were built on top of newly organized relationships between sensory system pathways and motor neurons of the brain in order to rebuild proper reflex circuits.

To the surprise of both his wife and mother, Wojtek responded quickly and positively to the applied techniques. Pathological muscle tension was progressively reduced. He exhibited less tactile defensiveness, no longer yelled, hyperventilated less often, and opened his spastic fists. Towards the end of the conference he stood up from his wheelchair and was able to take a few steps with contact guard assistance. There was also improvement in the movement of his eyes and the closing of his right eye.

At the end, Wojtek's family was happy and very proud of him. As MNRI® therapists, our group was glad we could help. After seeing his progress, Wojtek's wife decided to continue the MNRI® program at home as well as periodically bring him to our offices in Zielona Gora for additional therapy.

Over the course of 5 months of systematic therapy, incredible changes were observed. Wojtek started walking even up stairs with assistance. He showed a greater range of emotions. He began to recognize and show the way to various destinations. He was able to concentrate and to remain present for longer periods of time. He stopped yelling, perspiring excessively, and hyperventilating. The motor control of his hands became more stable. He began to recognize people and objects and to communicate with a thumbs up gesture. Although he didn't always comprehend, he began to respond to tonal inflection when asked questions.

Wojtek is now in his second year of MNRI® therapy. He participates in Dr. S. Masgutova's intensive conferences twice a year. All this work has been yielding ever-greater results. Lately, positive changes have been observed in bowel and bladder control, walking and body posture, and speech development.

We still have a lot of work in front of us, but Wojtek's optimism, motivation, and the incredible support of his family all allow him to fully benefit from the possibilities opened up by MNRI®.

MNRI® Developing Integrated Reflexes

Interest in the MNRI® reflex integration program continues to increase in the medical community. More and more doctors of various specialties have become interested in the concepts and underpinnings of MNRI®. There is growing interest among patients and the families of patients suffering from neurodeficits as well as those with neurophysiological and structural dysfunction caused by damage to the central nervous system. Many people with challenges are looking for more effective therapeutic approaches.

The MNRI® programs are based on many years of clinical experience and research as well as ongoing assessment of the results achieved with children and adults participating in MNRI® Family Conferences.

MNRI® offers a holistic view of neurosensory motor development and human cognitive function. It is based on the application of very specific hands-on techniques designed to facilitate optimal functioning of the central nervous system and activation of brain neuroplasticity mechanisms. It significantly impacts reflexive and automatic responses of the body. It influences coordination between the muscular and nervous systems, activating: sensory-motor memory, homeostasis and self-regulation of the nervous system, and muscle and fascia systems for reestablishing structural integration of the body.

The main goals of MNRI® are supporting and facilitating reflex integration, activation and optimization of gross and fine motor control, support and facilitation of psychological and psychomotor development, improvement in sensory-motor aspects of the function of reflex circuits which as we know allows for creation of new neuro pathways.

Reflexes are an automatic response to specific stimuli that have a very specific response described by several parameters: specific motor pattern, direction of response, latency and timing, intensity, and symmetry.

Retained, but not integrated, reflexes have a negative impact on the development of postural responses, balancing and righting responses, dynamic motor control, and overall levels of functioning. They lead to the development of compensatory mechanisms and not-matured and underdeveloped behaviors that prevent the person from achieving their highest potentials and skills. Non-integrated reflex patterns negatively impact brain development at many levels.

MNRI® offers both a diagnostic aspect as well as a therapeutic one (in the form of exercises in the reflex integration program). Integration exercises are movement patterns that are used to remind the body of proper sensory-motor connections. This allows for correction of reflex patterns and reorganization of the nervous system to facilitate optimal functioning.

All innate automatic responses, when integrated and matured allow for proper function of the cortex and serve as a base for the development of higher motor and cognitive functions, including walking, speech, cognition, inner control, emotional stability, writing, and abstract thinking.

Analysis of Wojtek's Progress with MNRI® Therapy

The following represents the Assessment performed by Dr. Masgutova of individual reflex patterns. (See the chapter on *Assessments* for more information.) This data was collected over the course of two conferences attended by Wojtek. The first conference took place in May, 2012 and the second in September, 2013.

The NewKrefft statistical model was used in analyzing the data (Kreft, 2004). The data show that significant and positive improvements occurred in the level of reflex pattern integration.

The following table presents the areas of focus in using MNRI® to improve Wojtek's functional capabilities and motor skills. For each area, we also indicate the level of visible, beneficial, and stable changes after the first and second conference.

Wojtek	Left- Right (Lateral- Median/ Sagital) MCS									
Reflex pattern	1/ Core Tendon Guard	2/ Hands Grasp	3/ Hands Pulling	4/ Babkin	5/ Babinski	6/ Leg Cross Flexion/Extention	7/ ATNR	8/ Abdominal	9/ Bonding	10/ Foot Grasp
Average1	10	5.25	5.25	4.25	6.25	8.25	6.25	4.25	7.25	6.25
Average2	10.75	5.5	6.25	4.75	6.75	10.75	7.25	6.5	8.25	7.25
Average3	11.5	5.75	7	4.5	7.25	11.5	8.5	8.5	8.75	7.5

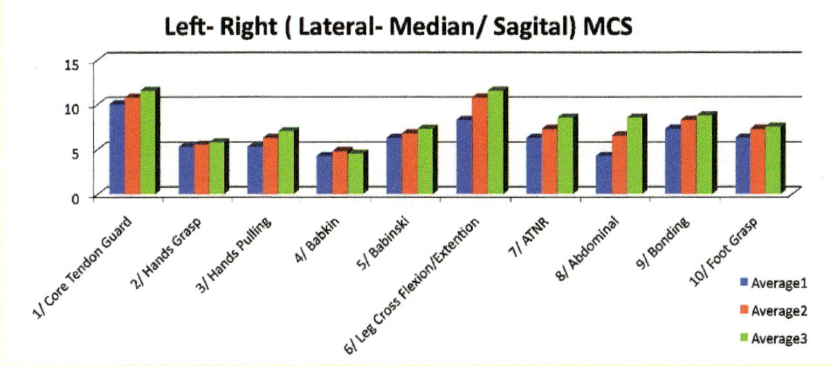

Wojtek	Up-Down (Superior-Inferior/Horizontal) MCS									
Reflex pattern	1/ Automatic Gate	2/ Bauer Crawling	3/ Moro	4/ Fear Paralysis	5/ Hands Supporting	6/ Segmental Rolling	7/ Landau	8/ Flying/Landing	9/ Grounding	10/ Head Righting
Average1	9	7.25	7.75	8	3.25	4.25	2.75	1.25	7.25	6.5
Average2	10.75	8.25	8.75	9	4	5.25	3.75	1.25	8.25	7
Average3	11.25	8.25	9.25	9.75	4.5	5.75	4.25	2.5	8.25	9.5

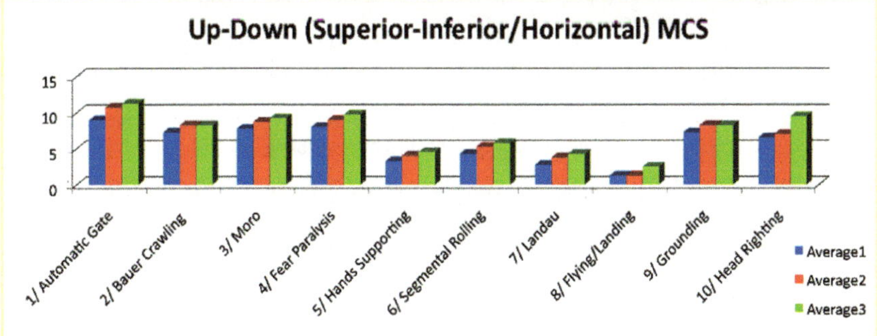

Wojtek Reflex pattern	Front-Back (Anterioir/Posterior) MCS									
	1/ Trunk Extention	2/ STNR	3/ Galant	4/ Perez	5/ TLR	6/ Foot Tendon Guard	7/ Spinning	8/ Locomotion	9/Balancing	10/ Pavlov Orientation
Average 1	8.25	6.5	5.25	5.5	4.5	6.25	5.25	3.25	4.5	5.75
Average 2	10.25	6.75	7.75	6.25	6.5	7.25	5.75	4	6.5	7.25
Average 3	11.25	8	9	7.75	8.5	7.5	7.5	4.5	8.5	7.5

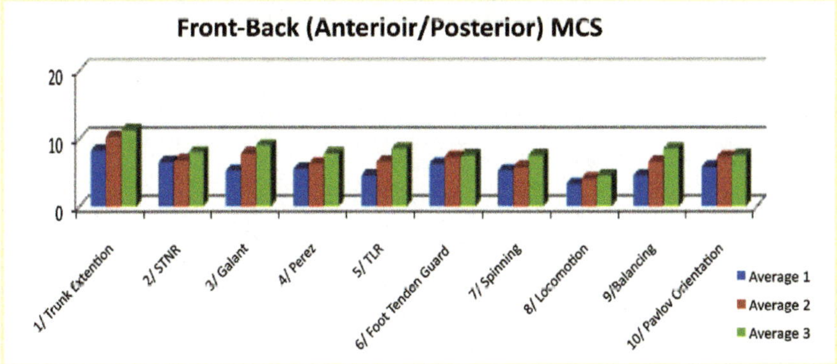

Wojtek's Progress in MNRI® Therapy:

Skill	Before MNRI® Program	After MNRI® Program
1. Gross motor coordination control	Couldn't stand independently or with support. Couldn't walk, perform homological or reciprocal movements. Asymmetric and weak sitting position. Pathological coordination of movement. Very fragmented trunk extension.	Post-conference 1: Improvement in sitting with better spine extension. Attempted to stand independently with prolonged periods of maintaining this position. Post-conference 2: Better symmetry observed in standing posture. Was able to walk independently and better able to move around the house. Could climb stairs with assistance.
2. Muscle tone	Pathological and spastic muscle tone - particularly in the upper extremities, more significant on the right side. Clutched fists, with the thumb inside the palm - impossible to open.	Post-conference 1: Improvement in regulation of muscle tone. Palms open. Incremental, constant, and persistent increase in range of motion of the upper extremities. Post-conference 2: It was possible to more easily and quickly open his arms (bringing arms above his head and then stretching them to the sides), including opening the palms. Beginning stages of improved hand coordination. Able to perform tasks with left hand more precisely and more quickly.
3. Postural control	Significant structural asymmetry of the body and extremities - both in posture and movement. Could not balance statically or dynamically. Tired easily. Avoided uncomfortable positions: lying on the stomach, standing, walking. Absence of coordinated complex movements. Did not stand. Did not walk independently or with assistance. Clenched fists spastically enclosing the thumb.	Post-conference 1: Improvement in structural symmetry. Maintained upright position in sitting. Could independently walk ever-greater distances. He leaned forward and to the right while walking. Post-conference 2: Fully upright body position. Tolerated lying on his stomach and on his side. Tolerated standing and walking. Increased levels of happiness. Demonstrated greater endurance and coordination. Was proud of his postural control.
4. Fine motor control	Right eye did not close fully and exhibited bulging. Uneven alignment of the eyes. Very narrow visual field. Lack of eye movement with hyperactive nystagmus. Pathological oculovestibular and oculokinetic reactions.	Post-conference 1: Open palms. Significant reduction in hyperactivity of hands Grasp Reflex. Post-conference 2:Able to independently perform the following actions: eat, place shoes onto his daughter's feet, and turn pages in a newspaper. Better able to hold a cup in his left hand and drink independently. His movements were more confident, purposeful, and smooth. Able to use independently developed signals for communication - thumbs up for 'yes' and thumbs down for 'no'.
5. Visual responses	Auditory hypersensitivity. ATNR reflex pattern and Acoustic Reflex were not coordinated.	Post-conference 1: Right eye closed more completely. Reduced bulging of the eye. Recognized more people and places. Was interested in looking at objects (e.g. newspapers). Improvement in eye movement. However, oculomotor and oculokinetic reactions still pathological. Post-conference 2: Could focus due to improvement in visual convergence. Concentrated for longer periods of time. Improved 3-D vision. Better eye tracking triggered by objects, sounds, and motions. Increasing visual field. Stabilized nystagmus. Improved oculovestibular and oculokinetic reactions.
6. Auditory responses	Presented with an Abdominal response instead of ATNR which interfered with his ability to regulate reaction and decoding of acoustic stimuli.	Post-conference 1: Began to show animated and positive reactions to music. Had greater tolerance to movement and sounds. He vocalized. Post-conference 2: Significantly reduced auditory hypersensitivity and irritation from noise. Improved function of the ATNR Reflex pattern and coordination with the acoustic reflex. Abdominal Reflex still interfered with ATNR Reflex, but there was improved decoding of auditory stimuli. Improvements in speech.
7. Emotional development, appropriate reactions and emotional stability	Extremely frightened and tense. Strong tactile hypersensitivity. Agitated when confronted by new people. Yelled almost constantly. Perspired profusely. Hyperventilated often.	Post-conference 1: Was less afraid. Defense mechanisms triggered less often. Showed greater interest in surroundings. Smiled more. Began to cry. Rarely yelled. Post-conference 2: Attempted to establish contact, communicate, express happiness. Showed pride in wife and daughter.

Skill	Before MNRI® Program	After MNRI® Program
8. Cognition and perception	Only recognized immediate relatives. Disinterested in surroundings. Poor ability to process stimuli. Poor ability to: concentrate, remember, analyze and synthesize. Poor spatial orientation. Unable to fully comprehend sentence structure and statements directed at him.	Post-conference 1: Began to recognize other people, places, and the way to places. Pointed with his finger to indicate the target of his desire. Said, "No." Liked to look out the window. Liked to watch comedies (e.g.: *Laurel and Hardy*). Post-conference 2: Often laughed in manner appropriate to situation. Increased duration of presence and concentration. Was more careful.
9. Oral-facial reflexes, speech	Weak functions related to eating, including biting, chewing, and swallowing. Poor coordination of these functions with breathing.	Post-conference 1: Improved swallowing and breathing, sucking, and chewing. Difficulty in controlling organs of speech production persists. Post-conference 2: Improved facial expressions and head control. Reduction in excessive facial muscles tension.

The above analysis speaks to Wojtek's success. Every day, his family saw positive changes in his behavior and abilities.

Woven into this story is the love shown to Wojtek by his wife, daughter, and mother and their battle for the life of their husband, father, and son. They showed incredible hope, strength, and determination in order to bring Wojtek back to health. In all of this let us not forget Wojtek's willpower and positive spirit that, through his optimistic nature, was able to lend energy to his family and his therapists to persevere in working with him. He projected a zest for life and a desire to help in his own rehabilitation. This testified to the development of his awareness and control.

Summary & Conclusion

Wotjek riding a bike.

After suffering injury to his central nervous system, Wotjek made incredible developmental, physical, emotional and cognitive strides after MNRI® was added to his rehabilitation regimen. Improvements in many areas have given him a greater degree of independent functioning including: gross motor control, muscle tone, postural control, fine motor control, and ocular, auditory, cognitive responses as well as emotional development.

The therapeutic regimen leveraged the neurosensorimotor resources of reflex patterns. These patterns are units of the nervous system (Sechenov, 1960; Pavlov, 1927/1960) that can be used to positively influence the self-regulation of many processes and areas of development and function. The analysis of ongoing improvements experienced by the patient provides testimony to the efficacy of MNRI®.

Working with reflexes, MNRI® stimulates the proper creation of reflex circuits, stimulating and, using the neuroplasticity of the brain (at all ages and various levels of injury), promotes the integration of the basic sensorimotor reflexes necessary for proper development. Wojtek's case confirms the importance of working with primary motor patterns that never disappear, but integrate into higher brain functions. By stimulating the brain stem in an effective manner we can rebuild lost abilities at any age.

MNRI® can be used effectively in treating patients with serious brain injury. In Wotjek's case, not only did it provide release from emotional stress, it also activated and supported various functions. The therapeutic process underscores the important and non-trivial nature of working with protection and survival mechanisms that form the basis of further development.

References

Krefft, A. (2007). *Diagnosis of functions of invisible phenomena* (Statistic Mathematic Analysis). Wrocław. Poland: Oficyna Wydawnicza Politechniki Wrocławskiej.

Masgutova, S. K., Akhmatova, N. K. (2004, 2007). *Integration of dynamic and postural reflexes into the whole body movement system.* Scientific edition: Prof. N. Akhmatova. Warsaw, Poland: MINK.

Pavlov, I. P. (1960). *Conditioned reflexes: An investigation of the physiological activity of the cerebral cortex.* (Anrep G.V., D. Sc. Trans., 1960). New York, NY, USA: Dover Publications Inc. (First published in 1927).

Sechenov, I. M. (1965). Reflexes of the brain. (Russ. tr. S. Belsky). Cambridge, MA: The M.I.T. Press. (First published in 1863.)

We congratulate Wojtek and his amazing achievements in recovery and his benefits of integrated reflexes which included biking, speech, and happy singing! Congratulations to all his family and thank you for your best support, love, and care! – Teresa-Ewa Busz and Beata Oginska-Dutkiewicz

Case Study: Development of a Child's Fine Motor Skills and Speech

Joanna Kowalewska, MA, Neuro-Pediatric PT, MNRI® Core Specialist and Instructor, Wroclaw, Poland

Introduction

Joanna Kowalewska

The main challenge in therapeutic professions is finding research methods to resolve our client's problems with maximum precision and efficiency (T. Pilch, 1995). This case study describes a girl, Maja, now 5 years of age, who has a dysfunction of the central nervous system. The main hypothesis was that MNRI® therapy, which influences reflex circuits, also will improve her neurodevelopment for growth in fine motor coordination, specifically, manual motor skills and speech. (Consent of the Bioethics Committee of Wroclaw Medical University No. KB – 160/2012 on the use of these tests in the work, *The impact on the formation of precise motor speech development in terms of sensorimotor integration*. Kowalewska J. (2012). Unpublished Dissertation, Wroclaw Medical University Department of Health Sciences, Wroclaw). This research was conducted at the Wroclaw subsidiary of the International Svetlana Masgutova Institute.

A clinical observation and description of the results were conducted at the beginning and end of the research project, allowing for the creation of an individual therapeutic program for the child.

Maja in the sandbox.

Research Techniques and Tools

The research included the following steps and procedures:

Observation, interview, research of documents, both medical and others.

To see the specific level of Maja's development, we observed her: 1) motor coordination – the quality of her movements in time and space, 2) facial expressions, and 3) communication and interaction competence.

These abilities were observed in a free play situation using the following developmental criteria: a) sensorimotor abilities (how she moves and uses her senses), b) oral-facial motor abilities (oral motor control, ie. mouth open/closed or salivating/not), c) communication (who was the initiator, her vocabulary, oral-motor and oral-facial abilities and habits), d) eating and drinking (dependent/independent, the quality of the oral-motor abilities, how much and how long she chewed and swallowed, whether she choked), e) play (who initiated the play, what she played with, her ability to play and relate to peers).

The interview was focused on aspects of development such as motor, sensory, emotional, cognitive, communication, and social abilities, including information about family structure.

Examination of documents from various specialists: medical, hospital abstracts, the results of the medical examinations and opinions: psychological, pediatrician, oral specialists, and the judgment of the disability were studied.

Tests used for research were:

Kielin's Profile of Student Achievements. This profile, based on clinical observation, allowed us to score specific skills of a child in relation to her age. This profile gives a glimpse of 'student achievements' in various aspects: visual and auditory perception, upper and lower motor skills, hand-eye coordination, social skills, receptive and expressive speech, imitation, cognitive function, and self-reliance. This scale can indicate matured/strong and underdeveloped functions, and allows us to track the child's progress and changes (Kielin, 2009).

MNRI® Reflex Assessment was used for: 1) diagnosis of primary motor and reflex patterns, 2) creating an individual MNRI® program aimed at improving the neurosensorimotor system through activation of primary reflex patterns. Diagnosis of reflex integration allows a specialist to see specific disorders in the reflex circuit and establish whether the reflex pattern is appropriate to the age of the individual and neurologically normal.

This Assessment presents a detailed analysis of work of sensory-motor coordination in a circuit, activity of motor reaction; intensity (muscle tone regulation), latency (duration and dynamics of reactions), and symmetry in response to stimuli, and characteristic features of the reflexive motor reaction (Masgutova, Regner, 2009; Mass, 2007; Pilch, 1995; Sadowska, 2011).

Clinical Observation and Research

Developmental Characteristics of the Child: Maja is a child born from her mother's first pregnancy, born after 39 weeks of pregnancy by C-section. There was an indication of oligohydramnios (low amniotic fluid levels) and intrauterine growth restriction of the fetus. The physical examination at birth showed no abnormalities except for a slight decrease in muscle tension. An ultrasound in the second day of life found a second stage hemorrhage but during follow-up visits after the first and fifth months, the brain was normal in an ultrasound.

After 3 months of age the following were noted: some postural asymmetry, low tone for head righting (couldn't lift her head while lying prone), asymmetry in supporting herself in the upper extremities, and tightly clenched fists. At the fifth month the Vojta Method was recommended for rehabilitation; at the eighth month the Bobath Method was also recommended; at the 11 month the Vojta Method was discontinued for Maja because she was crying in pain from the procedure and because her muscle tension had increased greatly. Since her tenth month of life, due to microcephaly and delay in psychomotor development, she was under the permanent care of developmental specialists, including ophthalmic care due to convergent alternating strabismus and myopia, resulting in permanent prescription of eyeglasses. At 18 months of age she had her first seizure. The MRI of the brain showed hypoxic-ischemic changes reaching deeply into the subcortical area of the parietal and frontal lobes of both hemispheres. She also had minor asymmetry visible in her ventricles. An EEG and sleeping study showed pathological changes localized in the right hemisphere as well as the frontal-temporal-center of the right side and the temporal-parietal-occipital of the left side. The sleeping test additionally showed that Maja's physiological sleep was accompanied by multiple and continuous seizures. Despite taking medication (currently Lamictal and Keppra) epileptic seizures occurred periodically and so deep in pathology that she was hospitalized several times. All the dysfunctions described influenced Maja's neuro- and cognitive development significantly. Currently Maja is under the care of many specialists.

Maja creating art.

Functional Diagnosis: An observation of Maja, then 3 years old, showed many dysfunctions before MNRI® was begun, particularly in the following areas of development:

Sensory-motor activity. Maja could physically move: sit, stand, and walk, she tried to move on her own though still she needed support when walking as her movements were unstable, ungrounded, and poorly balanced. She had poor control of lower limbs – her knees were bent; she had difficulty carrying her body weight; she had a lack of proper postural control; and appeared asymmetrical in a standing position because her gravity line was offset in relation to her body midline. She had increased muscle tension and her balance disorder caused difficulties for motor planning and control. Manual fine motor skills (clay modeling, drawing) were a real challenge for her and she was reluctant to use her hands. Muscle tone in the areas of her shoulder, neck, head, and face were hyper and her tactile sensation in these areas was hypersensitive. Rotational movements of the spine were very limited and rigid; so when looking behind, her entire body was involved. She had hyper muscle tone in her trunk and upper limbs and low motor velocity which made her unable to properly fulfill daily visual-motor hands-on tasks.

Oral-facial motor abilities. Maja had impaired muscle tone control and tactile sensitivity. Her mouth was open, she had problems swallowing, salivated a lot, breathed mainly through her mouth, and her tongue was very tight and not mobile enough to swallow or develop articulation. Her facial expression, muscle tone control, and movements of the jaw were also very poor.

Maja using tools before MNRI®.

Language and communication. Maja's first syllables appeared at about twenty months. At that time she also learned the pointing gesture by using her index finger. She said her first words around 27 months. At the time of our study Maja could say simple sentences and her words were pronounced with great difficulty and poor coordination in her breathing. Maja's comprehension was much higher than she could express in her speech and behavior; she followed 30% of instructions. She loved to deal with words – on a visual-oral level and also by recognizing objects by pictures–her receptive vocabulary was much larger. She liked to be both a listener and an active communicator, initiating play and interaction.

Eating and drinking functions. Maja was able to drink from a cup. She could independently put food into her mouth and bite off soft foods. She required assistance when using a spoon and fork. Maja had difficulty controlling closing her mouth which resulted in poor chewing ability and difficulty keeping food in her mouth. Overall, she showed significant difficulties in coordination, control, and precision in her oral-motor abilities.

Play, comprehension. Maja was interested in objects and toys but only briefly. At the same time, she was able to initiate play and give commands. She showed a very high cognitive curiosity and enjoyed listening to stories with specific toys or personalities. Her expressive speech was delayed significantly and she was reluctant to speak. She liked attention and during play she displayed initiative but refused to do the same when requested to repeat the activity with a hands-on task.

Results of Test Procedures

Profile of student achievements according to Jack Kielin

The Jack Kielin's Student Achievement Profile was conducted for Maja in specific areas of abilities with scores converted into percentiles. Her development was evaluated as follows: the average level of completed tasks for a 3 year old was 37.9% for Maja.

Fine motor abilities showed a very low level (4) - 13%, and self-reliance (11) - 18%. She obtained a high score in receptive language (7) - 83 %, but lower for expressive language (8) 54%, which meant her development levels were disharmonious.

Maja's speech development analysis showed that her speech difficulties were caused by problems on a sensorimotor but not cognitive level. Learning to control and regulate her oral muscle tone and tongue movements were most important. The most critical task was to start repatterning her oral reflexes to increase her ability to speak.

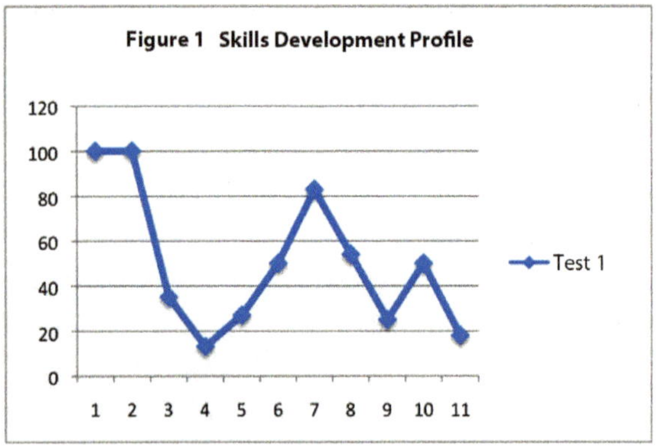

Figure 1. Maja's Profile of Achievements from Jack Kielin's Student Achievement Profile:
1) Visual perception 2) Auditory perception 3) Gross motor activity 4) Fine motor coordination 5) Hand-eye coordination 6) Social development 7) Receptive language 8) Expressive Language 9) Imitation 10) Cognitive functions 11) Autonomy

MNRI® Reflex Assessment according to Dr. S. Masgutova

Results of Maja'a Reflex Assessment showed dysfunctions in a majority of her reflex patterns, namely:

Core Tendon Guard. Maja demonstrated strong overprotection and asymmetry on the left side. This test provided information about her response to stress. The overprotective response particularly in the lower left side were blocking her ease in movements and postural control.

Hands Grasp Reflex. Maja demonstrated a tendency for delay in grasp and a preference for Phase 4 grasp as the result of poor strength in her upper limb muscles (though she had hypertonicity). She was not able to reproduce other phases of the Hands Grasp Reflex pattern. During testing she also showed compensation with Hands Pulling and Babkin Reflexes which was reflected in Maja's poor tool manipulation abilities.

Hands Pulling Reflex. The sequence of response (norm: flexion of elbows, next, head, and lastly core) was not displayed; her head righting in this test was delayed, and her core strength was very poor for flexion. Thus ,she was poor with body midline control in the whole, and her muscle tone regulation in the upper limbs was limited also for supination and pronation. This test also explained the deficits in Maja's ability for proper hand-eye coordination.

Hands Supporting Reflex (Parachute) demonstrated a hypersensitivity to sensory stimulus. The range of extension movement was limited (she attempted to use flexion instead), so the pattern couldn't fulfill its protective function (proper extension). Poor functioning of this pattern affected the gross motor skills of her upper limbs and rotation ability in her shoulder joints causing lack of freedom and rigidity in her head movements and a lack of stability.

Babkin Reflex. This reflex in Maja was pathological. When making a fist, Maja opened her mouth and at closing her mouth she demonstrated gritting of her teeth. The muscle tone in her cheek area was hypo and around temporomandibular joints (TMJ) was hyper, creating confusion in her TMJ movements and mouth closing. Due to a strong link between the oral-facial and hand-motor coordination, every attempt to speak elicited movements in her hands, and, vice versa. The dysfunctional Babkin Reflex negatively affected her hand-mouth coordination and articulation.

Asymmetrical Tonic Neck reflex (ATNR). Turning of the head to both the left and the right was causing significant shortening of Maya's lower limb on the same side of her body, which indicated her ATNR was not functioning correctly and showed a retained reactive automatism inhibiting its maturation and integration. Turning the head to the side was also causing greater muscle tone in her oral-facial and neck area. This reflex pattern was evidently affecting Maja's auditory-visual coordination leading to dysfunctions in auditory and visual processing. Her ATNR was affected also by a poor rooting reflex – in the test (of the rooting reflex) she showed an incorrect tendency to move her eyes in the opposite direction compared to her head movement, which was causing additional spatial disorientation for visual and auditory systems.

Spinal Galant and Spinal Perez Reflexes. These reflex patterns are the basis for development of antigravity, and also for dynamics, proper stability, and balancing. They affect the motor coordination within all frontal, sagittal, and transverse planes. These patterns in Maja were dysfunctional, resulting in poor head control (her neck was sunken in her shoulders) and her gravity line was shifted forward-left from the body's axis midline. Her gait was rigid with a tendency for homolateral postural control and her pelvis was tilted too far forward (the influence of a poor functioning Perez pattern). Her rigidity in spine rotation movements was limiting her torso from turning to the side and limiting limbs movements, involving non-productive movement for her entire body. On a functional level, dysfunctional spinal patterns were affecting her vision by limiting her perceptual field and restricting her focusing.

Plan for conducting therapy with the child

After a detailed analysis of the diagnoses of reflex patterns in Maja, a comprehensive therapeutic intervention program was created for her, which was based on a complex, multi-level approach aimed at specific areas of dysfunction. MNRI® was chosen as the main method for her. The therapy plan targeted the following areas of development: gross motor skills, tactile and proprioception, fine motor abilities, hand-eye coordination, oral-facial and respiratory system. The goals were improvement of 1) the gravity/antigravity mechanism, 2) vestibular system, 3) muscle tone regulation, 4) gross motor skills, 5) reduction of excessive sensitivity to tactile stimuli, 6) the perception of her own body parts, 7) hand grasp and manual skills, 8) coordination and movements of the upper limbs, 9) visual perception and hand-eye coordination, 10) tolerance for touch in areas of the head and face, 11) control of mouth movements (proper closing/opening), 12) oral-facial dynamics – oral-motor control, swallowing and salivation, eating, and chewing, 13) breathing (nasal instead of mouth). MNRI® programs used for realizing these tasks were: Neuro-Structural Reflex Integration, Archetype Movement Integration, Tactile Therapy, Oral-Facial and Visual-Auditory Reflex Integration, Reflex Re-patterning, and Proprioceptive-Vestibular Reflex Integration.

Maja attended MNRI® therapy sessions once a week, lasting about 60 minutes and received a Home Program of 5 days a week from 20-40 minutes to 1½ hours a day. The work with her was organized in a two-fold way: to support of her sensory-motor abilities though reflex integration and activation of her cognitive processes.

At the beginning Maja was not cooperative. At our therapy sessions, her focus and participation was very fragmented with low motivation, and she required constant attention and reminders of instructions. Telling fairytale stories was of great help – she loved to listen to made up stories. Now she can initiate them herself choosing the objects and toys for spontaneous stories.

3.5. Results of the Therapeutic Work with Maja

An evaluation to determine changes in Maja after MNRI® was performed after 6 months was done using the same tools - Student Achievement Profile and a second MNRI® Assessment.

Maja's development was still impaired but improved significantly in all areas and on many levels. Her sensitivity to touch had changed in many parts of her body although still she had tactile defensiveness for touch on the head and upper front of the body. Her motor coordination and postural control improved especially well. She started speaking much more distinctly, building complex sentences. She also was able to control closing her mouth, salivating, and eating functions – swallowing, biting, chewing, though still not at a level of complete automatic response. She started drawings dots and circular lines. Her ability to serve and entertain herself was still problematic. Maja was still reluctant to cooperate, at the same time, she was more interested in socializing with people and initiating more interaction with therapists and others.

Maja's grasp after MNRI®.

Profile of Student Achievements Results

In the second assessment, the results were: Maja's average level of completed tasks increased to 65.4%. The best results were achieved in receptive language (7) - 100%, and expressive language (8) - 86% meaning she showed a dramatic increase in her semantic level of speech. Her speech still needed further sensory-motor support but it improved remarkably. She is more understandable and her speech was better coordinated with breathing which was longer and smoother. The weakest result was recorded in self-reliance (11) - 36%, well below the average level. Detailed results for comparison with the first assessment are presented in Figure 2.

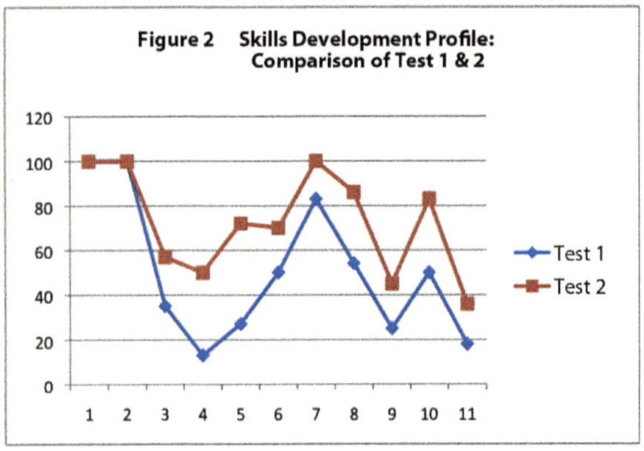

Figure 2. Profile of the Child's Achievements (Assessment 1 – blue; Assessment 2 – red) by Jack Kielin's Student Achievement Profile:
1) Visual perception 2) Auditory perception 3) Gross motor coordination and mobility 4) Fine motor coordination 5) Hand-eye coordination 6) Social development 7) Receptive language 8) Expressive Language 9) Imitation 10) Cognitive function 11) Autonomy

Comparative analysis of Assessment 1. (blue) with Assessment 2. (red); the results *yielded the following changes* for the certain scales:

Visual and auditory perception (scale 1. and 2.) evaluation shows that Maja has achieved the maximum score. The *difference* in the changes of percentages are: gross motor skills (3) - 22%, fine motor coordination (4) - 37%, hand-eye coordination (5) - 45%, social development (6) - 20%, receptive language (7) - 17%, expressive language (8) - 32%, imitation (9) - 20%, cognitive functions (10) - 33%, and autonomy (11) - 18%. The average level of performed tasks changed by 27.5%.

The biggest increase of points were for hand-eye coordination and fine manual skills. Less change was noted in autonomy.

Results of MNRI® Reflex Diagnosis

After a 6 month period of MNRI® treatment, the most significant changes in Maja's reflex patterns were:

Tendon Guard Reflex. Over-protective response improved, she was showing much better stress resilience, which was seen in better muscle tone regulation, symmetry in body, and postural control; her tendency for inclining to left side was resolved.

Hands Grasp pattern. Maja showed a significant improvement in her strength and speed of grasp. There was no longer a negative compensation from the Babkin Reflex pattern, which impacted positively her coordination between opening and closing the hands, and more efficient manipulation with objects. Maja also had begun to imitate certain phases of the Grasp Reflex pattern and manual operations.

Hands Pulling Reflex. An improvement in the sequence of the performed movements (elbow flexion-head flexion-core flexion) has happened, although the movement of the head remains slightly delayed relative to the torso. Her new ability to flex elbows for a stimuli allowed her to facilitate the coordination of the upper limbs and hand-eye links as well as the rotation of the wrist.

Hand Supporting Reflex. There was now no oversensitivity to sensory stimulus. The direction and range of her movement had become correct. There was significant improvement in muscle strength, still, timing was slightly delayed. Propping up was performed automatically, the increase of range of motion in the shoulder joints is now good.

Babkin Reflex. In this pattern Maja moved from pathology to a higher level - dysfunctional, which was a very fast progress in such a short length time (6 months). Still she demonstrated hypersensitivity and hyperactivity in the area of the TMJ (temporomandibular joints) with no previous pathology. Her pathological link between hands activation and eliciting the mouth responses was resolved, which allowed her to regulate better her mouth-motor abilities, eating, and speech.

Asymmetric Tonic Neck Reflex (ATNR). A needed decrease in hypersensitivity and hyperactivity of this pattern has happened, resulting in better ear-hand-eye coordination, and her ability to listen more attentively and comprehend increased.

Spinal Perez and Galant Reflexes. Maja's line of gravity returned to overlap with her midline of the body, resulting in her posture becoming more upright. Her gait has become more grounded, stable, coordinated; the swaying movements of her pelvis are now correct, more flexible, and dynamic. There was improvement in head righting and control.

Conclusion

To effectively support children with developmental disorders, it is important to look at a child as a whole and to be aware that disorders in one area often result in developmental problems in a different area (Luria, 1966), sometimes seemingly areas not connected with each other. When planning therapy, one needs to keep in mind that the sensory and motor systems are closely linked and yet are interdependent of each other (Masgutova, 2005; Mass, 2007; Sadowska, 2011). The MNRI® method gives therapeutic techniques that take into account these facts.

Based on the case study results and its analyses, the achieved results allow us to conclude that:

• an improvement in processing of sensory information caused an improvement in motor functions

• improvement in sensorimotor processing achieved through MNRI® positively affected the manual abilities and articulation, as well as the overall neurodevelopment of a child.

References

Kielin, J. (2009). *Profile of student achievements*. Gdansk, Poland: Gdansk Psychological Publishers.

Luria, A. R. (1966). *The human brain and cognitive processes*. University of Moscow. New York & London: Transl. Harper & Row Publishers.

Masgutova, S. (2005). *Neurokinesiology tactile therapy*. Warsaw, Poland: MINK.

Masgutova, S., Akhmatova, N. (2005). *Integration of dynamic and postural reflexes with the movement system of the entire body*. Warsaw Poland: MINK.

Masgutova, S., Regner, A. (2009). *The development of a child's speech in the light of sensory integration*. Wroclaw, Poland: Continulo.

Mass, V. S. (2007). *Sensory integration and neuroscience - from birth to old age*. Warsaw, Poland: The Foundation for Innovation.

Pilch, T. (1995). *Principles of educational research*. Warsaw, Poland: Zak.

Sadowska, L. (2011). *Neurokinesiological diagnosis and therapy of children with impaired psychomotor development*. Wroclaw, Poland: AWF.

I congratulate Maja and her parents for their step-by-step positive changes, for their enthusiasm, amazing care, and for the impact of all that was done to secure Maja's bright future! – Joanna Kowalewska

ATNR Integration as a Key for Successful Learning at School: Case History

Kim Willkom, OT, MNRI® Core Specialist, Stevens Point, WI, USA

Background Information

Kim Willkom

A Masgutova Neuro-sensory-motor Reflex Integration® (MNRI®) program aimed at correction of the Asymmetrical Tonic Neck Reflex (ATNR) was successful in our occupational therapy (OT) in pre-school, elementary, and secondary schools during the 2013/2014 school year. Forty-six students, including 33 males and 13 females received OT services during that school year. The students were ages 3 to 14 years old, with grade levels from Early Childhood, 4 year old kindergarten, and kindergarten to grade 9. Their challenges ranged from neurotypical children with learning issues to Autism Spectrum Disorder (ASD) and Asperger's syndrome.

Learning in school requires the ability to listen, follow directions, organize thoughts, and communicate in written or spoken form. There are many reasons why learning becomes a challenge for students: difficulties in cognitive abilities, learning skills, focusing, memorizing, or being tired, hungry, angry, irritable, or unhappy. Poor reflex integration is one major cause of both learning difficulties, emotions, and behavior.

Presenting Problem

Many of the students presented emotions and behaviors such as crying, laughing, yelling, screaming, hitting, spitting, kicking, swearing, pushing, behavioral outbursts, and overall defiance of rules. The students who qualified for special education and were identified with cognitive delays, speech delays, and/or behavioral challenges demonstrated direct evidence of immaturity and lack of integration in their reflex systems. As an Occupational Therapist for 21 years, including 17 years in a school setting, and 7 years of MNRI® training and coursework, I can provide evidence that lack of reflex integration, especially a delayed or dysfunctional ATNR, can cause coping and compensation challenges for children. However, a therapist can impact their reflex deficiencies with special knowledge and tools. When exploring many treatment methods to help children, I found that the most effective intervention is integration of reflex motor patterns as practiced in MNRI® programs, rather than reflex inhibition, support of the sensory system, or behavioral tools aimed at increasing conscious control. The lack of consistently effective professional tools in my field motivates me to share these positive results of using MNRI® to integrate the ATNR in our students.

The Asymmetrical Tonic Neck Reflex (ATNR)

Because an integrated ATNR is so essential for support of the neurostructural and physiological aspects of perception in the learning process, it became the focus of our intervention.

The Asymmetrical Tonic Neck Reflex (ATNR) in an infant is an automatic postural response to passive or active head turning: extension of the limbs on the side to which the head is turned and flexion of the limbs on the opposite side. It is designed genetically for several functions: protection of the auditory canal (by 'resetting' binaural hearing to monaural in stress and danger), muscle tone regulation, development of body rotation mechanisms, orientation in space (proprioceptive and vestibular systems), and most importantly for academic achievement, support of auditory perception and language development (the cognitive system).

The influence of ATNR maturation on cognitive function is due in part to the fact that it serves as the neuro-structural basis for activation of alpha and beta brain waves. These fast frequency brain waves allow access to executive functions such as auditory perception (monaural and later binaural hearing), focusing and processing of auditory information (coding and decoding), auditory memory, and language/speech (links between the Wernike and Brocka brain centers).

In head turning to the right, the ATNR 'teaches' our system to activate the executive functions of the left hemisphere: auditory discrimination, classification of language sounds and decoding, comparison, analysis, conscious control of the auditory-speech and auditory-visual coordination systems and expressive language, as well as fine motor-coordination. In head turning to the left, the ATNR supports the executive functions of the right hemisphere: auditory coding, comprehension, automatic oral-motor control, subcortical connections, receptive language, emotional memory, and other gestalt functions (Masgutova, 2010). The matured ATNR protects the auditory system (resetting from binaural to monaural hearing in stress, discriminating auditory 'figure and ground', prioritizing sound), and, in safe situations, supports binaural hearing, conscious auditory perception, language development, and the development of fine and gross motor coordination.

When the ATNR is delayed and immature, it acts as a hypo- or hyperactive response blocking the executive functions described above, and often causing difficulties in neurodevelopment such as: lack of sensory-motor integration, poor 'hand-eye' and 'hearing-seeing' coordination, language deficiencies, inability to listen, memorize and comprehend, and poor use of thinking processes (analysis, comparison, and abstract thinking).

Any incorrect development of this reflex can lead to negative compensation, specifically by the Abdominal Sleep Posture Reflex, an opposite pattern.

Abdominal Sleep Posture Reflex (ASPR) is a life-long postural reflex. It presents an automatic response in infants lying on their stomachs: when they turn their head to the side and flex it downward, then the core flexes laterally, limbs also flex on the side to which the head is turned, and limbs on the opposite side extend. The ASPR slows down perception and processing, leading to a passive or restful state as it triggers slow frequency brain waves (theta and delta), serving generally for inhibiting perception of auditory and other stimuli and promoting asleep (Masgutova, 2010).

When head turning in the ATNR activates the ASPR response, an inappropriate flexion of limbs on the same side of the body, this signals a non-integrated ATNR and confusion between these two patterns. Conflict between the ATNR ('be aware', 'listen') and ASPR ('calm down', 'rest', 'ignore') results in an inability of the brain to differentiate these patterns and process efficiently. In this conflict the child's brain typically 'is pushed' to ignore the auditory stimulus. For children who are highly motivated to learn, the result can be extra effort, over-focusing, muscle hypertension, and thus, suppression of the vestibular system, cutting off binaural hearing. Other results can be reactivity, impulsive behavior, and repetitive or obsessive-compulsive behaviors (Masgutova, 2010). This can also make the auditory system (Stapedius Reflex) hypo- or hyper-sensitive and immature. Children find themselves in a vicious circuit: immature ATNR—>auditory sensitivity or inability to regulate perception—>discomfort and stress —>use of the ASPR (to calm down this sensitivity and stress)—> pushing more on ANTR to fulfill its task (immature, it cannot fulfill its functions). This tendency was seen in our children in their passivity, sleepiness, and inattentiveness when receiving information during instruction or lectures, also during reading and writing, or in the opposite—>hyperactivity and irritability. When hearing and listening trigger the ASPR instead of the ATNR, auditory processing is significantly inhibited.

Levels of integration of the ATNR can vary both in degree and from side to side, making many combinations possible: immature on one side only, immature on both sides, immature on one side and dysfunctional on the other, etc.

Reflex Diagnosis

The 46 students below were tested by the school Occupational Therapist to determine their level of development of this pattern. We rated their responses as follows, A representing the highest level of development and D the lowest:

A. Integrated – normal; the reflex has developed and matured sufficiently to serve its protective function and to support efficient processing of auditory stimuli

B. Immature – inefficient regulation of perception for protection or processing

C. Dysfunctional – perception and processing under significant stress in the presence of auditory stimulation, including speech - poor processing

D. Severely dysfunctional – system is overwhelmed in the presence of auditory information and tends to ignore the stimuli or shut down

All but 2 of the students assessed demonstrated significantly immature responses to the MNRI® ATNR test. The assessment showed that, in 44 of 46 students (September 2013), head turning triggered the incorrect response, ASPR, on at least one side.

Intervention

MNRI® repatterning exercises were used to work with the ATNR reflex pattern and its sensory-motor components. They use natural, genetic motor resources to facilitate neurodevelopment of auditory perception, processing, decoding, analysis, and memory.

All 46 students participated in correction of the ATNR pattern: 31 neurotypical with overall challenges in learning (LD) in Group 1, and 15 diagnosed with ASD in Group 2. The students were initially assessed in September 2013 before treatment and then again in April 2014 after treatment (see Table 1). MNRI® exercises were done in their weekly 20 to 60 minute occupational therapy sessions.

Results

Both their teachers and the Occupational Therapist observed that positive changes in the students' ATNR correlated with an increase their ability to focus, listen, hear, and comprehend school material on a higher level.

Table 1: Pre and Post Assessment Results

At right, the results of the ATNR tests before and after the MNRI® therapy are presented. The children are divided into two groups, as follows: Group 1 – LD and Group 2 – ASD. The ATNR pattern displayed by students before and after was differentiated into four levels: A) Integrated, B) Immature, C) Dysfunctional, D) Severely dysfunctional.

Table 1. ATNR Level of Development

Group		#	Test	A. Integrated # (%)	B. Immature # (%)	C. Dysfunctional # (%)	D. Severely Dysfunctional # (%)
1.	LD	31	Before	**1** (3.23%)	**7** (22.58%)	**14** (45.16%)	**9** (29.03%)
		31	After	**3** (9.68%)	**14** (45.16%)	**11** (35.48%)	**3** (6.45%)
2.	ASD	15	Before	**1** (6.67%)	**1** (6.67%)	**7** (46.66%)	**6** (40%)
		15	After	**1** (6.67%)	**1** (6.67%)	**10** (60%)	**3** (20%)

Group 1 – LD - neurotypical development with learning challenges (31 students):

Integrated ATNR: Only one student presented at the level of integration in the pre-test. After the intervention 2 students moved from immature to integrated (normal).

Immature ATNR: Seven students showed immature ATNR in the pre-test (6 with ATNR and 1 mixed ATNR and ASPR). After the intervention the number of students reaching this level doubled, which means that 7 students moved up from dysfunctional to immature.

Dysfunctional ATNR: In the pre-test 14 students presented with a dysfunctional ATNR (1 with hyperactive ATNR, and 13 with mixed ATNR and ASPR). Following the intervention 11 students remained on this level, and 3 moved up to the immature level.

Severely Dysfunctional ATNR: Twelve students presented severely dysfunctional levels of the ATNR in the pre-test (7 with ASPR on both sides and 5 mixed). After the intervention 6 students moved up to the next level and 3 remained with no change.

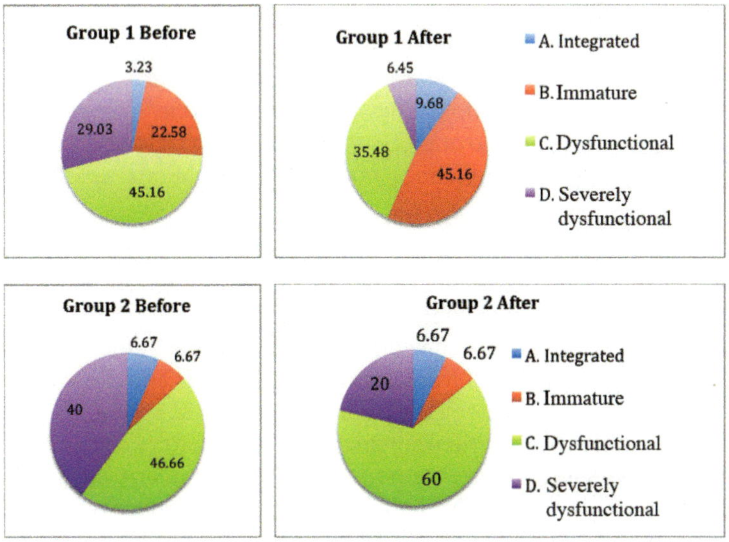

Group 2 – ASD – children with an autism diagnosis (15 students):

Integrated ATNR: one student presented with a normal ATNR before and after the intervention. No other students achieved this level.

Immature ATNR: 1 student presented with an immature ATNR both before and after the intervention. No other students reached this level, which means that positive change in ANTR expression was more challenging in the ASD group of students.

Dysfunctional ATNR: 7 students presented a dysfunctional level of the ATNR (1 with immature ATNR and 6 with mixed ATNR and ASPR). After the intervention the number of students in this group grew to 10. This means that 3 students from the lowest level moved up.

Severely Dysfunctional ATNR: 6 students had a severely dysfunctional level of the ATNR in the pre-test, (4 students with ASPR only, and 2 with mixed ASPR and ATNR). After the intervention exercises 3 students moved up to the next level, so only 3 remained with no sign of significant change.

Clinical observations revealed more detailed changes in our students. Those who had improved their mixed patterns for 'right' ATNR also showed improvement in:

• sensory integration: auditory tolerance for louder noises, selective and less fragmented perception, and better faster processing

• fine motor coordination: better auditory-visual orientation, improved hand-eye coordination, and manual skills

• gross motor coordination: planning of homolateral and cross lateral movements, conscious balance, stability and postural control

• cognitive processes: better focusing, memorizing, sequential and analytical thinking

• speech: improved articulation, phonetic decoding, and expressive language.

Students whose 'left' ATNR had improved also improved in:

• sensory integration: better gestalt perception and processing, better eye/tongue/auditory coordination, better auditory orientation

• fine motor coordination: more logical steps for manual skills, better auditory-visual orientation, less of use of retained Babkin Palmomental Reflex pattern (mouthing and putting the fingers and pen into mouth), improved eye tracking

• gross motor coordination: automatic sequencing, spontaneous control and performance in running, skipping, stretching, rotational and other everyday movements, long distance action, better anti-gravity and grounding mechanisms

• cognitive processes: improved focus on the big picture, long term memory, memorizing songs and poems, creative thinking

• speech: better rhythm and emotional expression in speech.

Informal Observation

The following comments from school staff attest to the real life changes our students experienced:

"Greatly improved in comprehension" – teacher

"Improved in his eating at lunch time" – educational assistant

"Decrease in chewing on pencils, objects, clothes" – teacher

"More focused and able to attend regular education classes" (versus spending all of the school day in special education classroom, under the table, or in a seclusion room) – teacher

"Improved in speech with greater vocalizations" – teacher
"Greater ability to transition from one classroom to the next" – teacher
"Decrease in aggressive behaviors at school" – teacher
"Much improved fine motor skills for writing his name" – parent
"Now he can focus and attend in the kindergarten classroom, whereas, before he could not stay long and needed frequent breaks throughout the school day." – educational assistant.

Discussion and Conclusion

Once a week from September 2013 to April 2014 the students in this study experienced a brief (one 20 to 60 minute session) intervention aimed at correcting aberrant ATNR reflex pattern expression. This MNRI® corrective ATNR program provided significant positive changes for these students in learning, communication, fine and gross motor skills, behavior; validated by improvement in MNRI® post-testing of the ATNR reflex and in informal observations of their behavior by their OTs and the school staff.

The ATNR supports hemispheric dominance for speech and language as well as the formation of specialized executive left and right hemispheric functions: auditory discrimination, classification of language sounds, auditory memory and comprehension, vocabulary, language programming, and output to speech muscles. Thus, knowledge about the ATNR and the corrective work necessary for its integration is essential for improving the neurophysiological aspects of auditory processing and comprehension in learning.

The ATNR pattern is often immature even in neurotypical children with learning challenges as well as in children with neurodeficits such as ASD. Only 2 children in this study demonstrated the matured ATNR pattern.

Delay in maturation and integration of the ATNR is often due to its being mixed with the opposing ASPR. Neurological differentiation of these two opposing patterns must occur. A majority of the children in this study presented mixed ATNR/ASPR, which makes auditory processing even more challenging because of their conflicting tasks: ATNR ('get ready to listen') and ASPR ('ignore auditory stimuli').

Improvement in the ATNR occurs within specific dynamics: when it is mixed with the ASPR, repatterning of the ATNR circuit must be the first priority in order to connect the ATNR trigger with the ATNR response instead of the ASPR. Only then can the positive dynamic for development and integration begin. Those working with delays in the ATNR must know about this frequent compensation in order to support the neurophysiological aspect of the reflex pattern with an appropriate intervention, especially in children with learning challenges and ASD. Specialists need specific tools. MNRI® Assessment and repatterning are particularly useful.

Integration of the ATNR is a key to improving students' selective auditory perception, focusing, listening, and comprehension at an optimal level for learning. The MNRI® program has proved to be an excellent support for both neurotypical children with learning challenges and those with neurodeficits.

References

Masgutova, S. (2010). *Integration of infant dynamic and postural reflex patterns – MNRI®: neurosensorimotor reflex integration.* Orlando, FL, USA: SMEI.

Masgutova, S., Masgutov, D. (2011). *MNRI® proprioceptive and cognitive integration: Optimize Innate, auditory-ocular processing and learning skills.* Orlando, FL, USA: SMEI.

I wish to thank the children, parents, and staff of my school for their support and understanding during the assessment of this new treatment method. – Kim Willkom

Rehabilitation Progress with a Child with Cerebral Palsy

Karolina Niezgodzka, PT, Core Specialist, SMEI, Warsaw, Poland

Karolina Niezgodzka

This case study follows a child born prematurely and diagnosed with various dysfunctions of the central nervous system. In this article, we hope to assess the efficacy of the MNRI® on gross and fine motor development.

Patient History

Ewa Janicka was born on July 12, 2007 in the 36th week of gestation through a vaginal birth. She was diagnosed with hypoxic ischemic encephalopathy and cerebral palsy. A genetic karyotype test (assessment of her chromosomal characteristics) was performed and came back normal. Developmental delays with hypotrophy and increased muscle tone were observed. At the age of 5 months, she underwent a percutaneous endoscopic gastrostomy (PEG) surgical procedure for treatment for severe reflux. She had been diagnosed with difficulties swallowing and secretions in the upper respiratory tract. An EEG test indicated past seizure activity. A test of evoked potentials of the visual nerves indicated demyelination of the left ocular nerve and normal results on the right.

Since then, Ewa has been treated with several therapeutic approaches: Vojta method, NDT (neuro-developmental therapy), and traditional physical therapy, however, no significant change or progress has been observed. Due to severe spasticity, doctors recommended botox injections and surgical tendon releases. Such procedures were postponed to allow for a longer period of physical rehabilitation with the hope that Ewa's condition would improve.

Ewa started receiving MNRI® intervention in October, 2010, at 3.3 years of age, at the Svetlana Masgutova Institute in Warsaw. Her level of development in primary reflex patterns was assessed at that time. Ewa's height and weight were within normal limits, muscle tone was decreased with tendency to hypertonicity. No head or trunk control was present and she tolerated supine position only. A prone position was not tolerated due to a G-tube and protective responses. Her ankle joints were positioned in increased plantar flexion and internal rotation due to increased muscle tone and pathological Tendon Guard responses.

Ewa had very poor control over the function of her upper extremities. Sensory stimulus used to trigger the Hands Pulling Reflex pattern demonstrated a pathological response. Lack of control of the head and trunk was demonstrated with tendencies for hyperextension of the arms.

When positioned in the STNR position (Symmetrical Tonic Neck Reflex – sitting on the heels with hands propped on palms), Ewa immediately responded with extension and a strong backward thrust of her legs. Her only skill was the ability to use her legs to firmly kick the floor. She was unable to crawl or roll over. She was only able to use her legs to spin herself around the axis of her shoulders.

The assessment of the tactile system showed a severe dysfunction of the sensory system with hyposensitive responses of the upper extremities, but hypersensitive reactions of the lower extremities and face. Ewa engaged with visual stimulus presented within her visual field but she demonstrated delays in visual perception, very narrow visual field, poor convergence, poor distal and 3-dimensional vision. Tests of the auditory system did not show any abnormalities in function of the auditory nerve. No auditory hypersensitivity was noticed.

Ewa was non-verbal but was able to make sounds with qualities of emotional communication, like showing discomfort, etc. She had significant difficulties with sucking and swallowing responses and demonstrated strong hypersensitivity in both external and internal oral-facial areas.

A moderate cognitive disability had been diagnosed related to dysfunction and delay of neurophysiological development. Ewa was responsive, recognized relatives, and her understanding of speech (decoding) was rather good. Her emotional development was not concomitant with age.

Applied MNRI® Programs

Based on the results of Ewa's initial Assessment in 2010, an individual therapy plan was created. Since then Ewa has been participating in intensive two-week therapeutic camps twice a year. Over a period of 3 years, Ewa J. participated in 5 two-week MNRI® camps in Mielno, Poland and one two-week series of intensive sessions at the International Dr. Svetlana Masgutova Institute in Warsaw, Poland.

Based on results of the MNRI® intervention, Ewa's parents decided that MNRI® will continue to be the only therapeutic tool/modality used in their daughter's rehabilitation and has remained this way for the past three years. Due to the parent's own physical handicap, Ewa has not received therapy as part of a daily home program. All results to date have been achieved during sessions with therapists applying MNRI® techniques.

The main goal of Ewa's therapeutic program has been the integration of primary reflex patterns and primary movement for development of correct and optimally functioning motor coordination and postural control, resulting in the development of cognition, communication, and learning skills. The main mechanism of the MNRI® approach is the stimulation of components of brain plasticity through facilitation of the neural connections of the reflex circuits.

Ewa's MNRI® Assessment and diagnosis are as follows:

The MNRI® Assessment tests for 30 reflex patterns. Reflexes are scored on a scale of 0 to 20 points (five parameters with four features each). 0-1.75 is severe pathology; 2-3.75 pathology of a basic reflex pattern; 4-5.75 deep dysfunction; 6-7.75 average dysfunction; 8-9.75 light dysfunction; 10-11.75 pattern on the boundary of dysfunction and normal development; 12-13.75 basic reflex pattern is functional (very low level of development); 14-15.75 low level of development; 16-17.75 norm in development, and 18+ as very good reflex pattern. (See the article on *Assessments* in this book.)

Ewa J. has had MNRI® reflex integration Assessments 3 times over a period of the last 3 years. All resulting data are presented in the form of graphs. The results have been organized into three bar charts based on the body planes in which reflexes are presented.

Graph 1 Assessments (right) presents 7 reflex patterns diagnosed in the sagittal plane. It shows that the lowest level of development during the first diagnosis was presented with Leg Cross Flexion-Extension and Hands Pulling responses, while the highest level was the

Graph 1 Assessments for Ewa. J: Sagittal plane reflexes

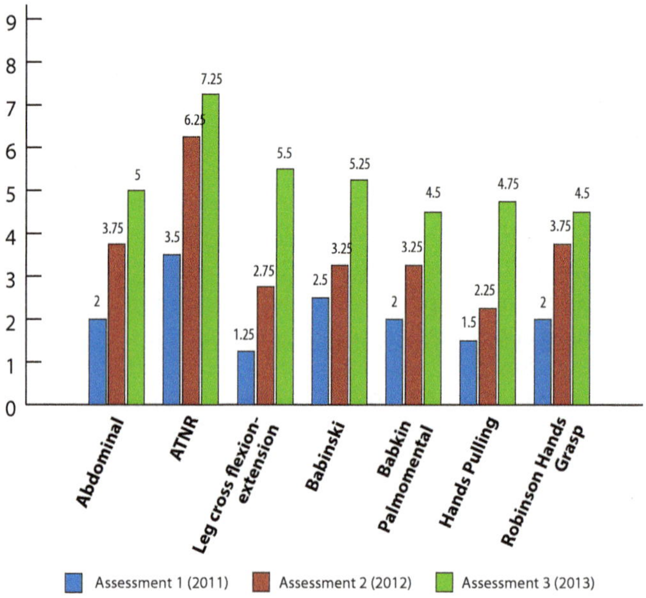

Asymmetrical Tonic Neck Reflex (ATNR). The most significant improvement during the therapeutic process over the period of 3 years is demonstrated with Leg Cross Flexion Extension and ATNR.

Graph 2 Assessments presents 8 reflex patterns diagnosed in the horizontal plane. It shows that the lowest level of development during the first diagnosis was Flying and Landing, Thomas Automatic Gait, and Landau responses. The highest level was with Moro. The most significant improvement during the therapeutic process over the period of 3 years is demonstrated with Segmental Rolling, Fear Paralysis, and Moro responses.

Graph 3 Assessments presents 7 reflex patterns diagnosed and classified along the dorsal plane. The lowest level of development during the first diagnosis was with Trunk Extension, Tendon Guard response, and Spinning. The highest level was with Spinal Perez. The most significant improvement during the therapeutic process over the period of 3 years is demonstrated with Spinal Galant, Spinal Perez, and Trunk Extension.

Graph 2 Assessments for Ewa. J: Horizontal plane reflexes

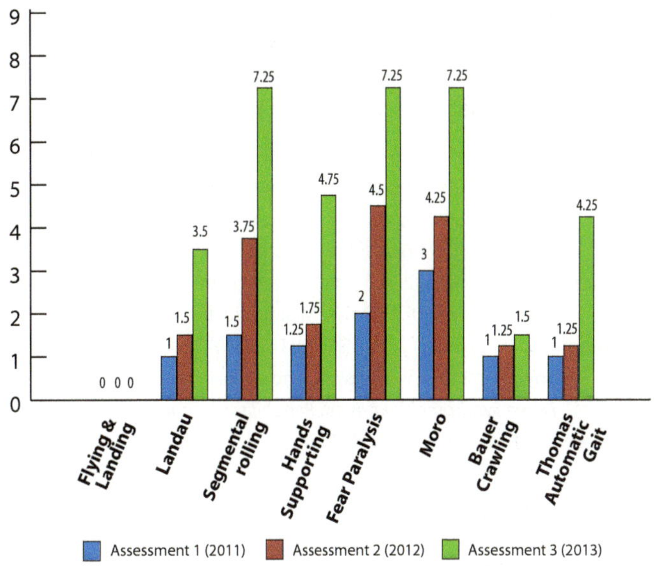

Reevaluation of Ewa's Level of Development after MNRI® Intervention

Significant progress has been made with respect to coordination of movement. Control over the neck and head is improved. Improved breathing function has been observed and she is better able to clear her breathing pathways. Ewa can now roll from her back onto her stomach, and back again. She can also transition from lying to a stable sitting position. Additionally, she now attempts to stand by using furniture for stability.

Ewa shows a greater acceptance for touch and therapy in the facial area. She also demonstrates greater tolerance for lying on her stomach. Expansion of Ewa's visual field has been observed.

Although Ewa still has the Percutaneous Endoscopic Gastrostomy tube, the Swallowing Reflex and tongue action have improved to the point that attempts are being made to feed her naturally.

Ewa has begun to utter new sounds and her communication has improved. She shows improved perception and concentration, has great interest in new objects, and eagerly participates in play. Ewa shows greater overall tolerance in the activities designed to increase her capabilities and improve quality of life. It seems that she has greater understanding and conscious acceptance of her surroundings. Control over her emotions has also improved.

Graph 3 Assessments for Ewa. J: Dorsal plane reflexes

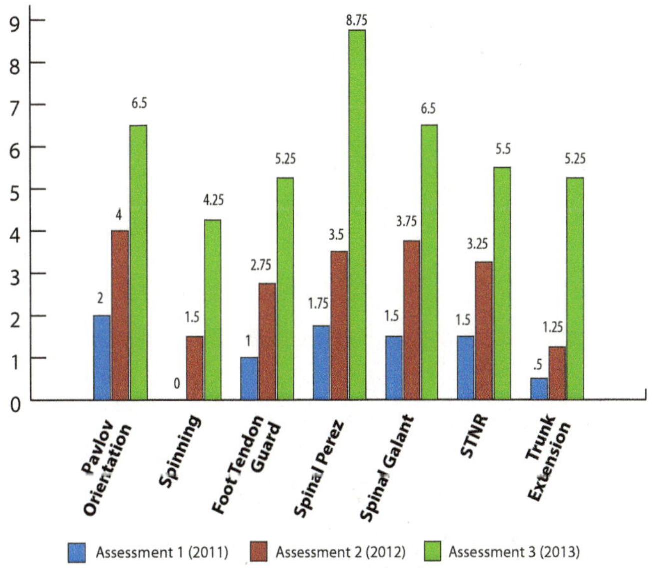

Conclusions

The combined use of MNRI® diagnostic assessments and therapy programs can be used to support the development of the nervous and sensory-motor systems and the systems of perception. The method has been able to improve both the functioning of Ewa's reflexes as well as gross and fine motor development. It has allowed the improvement of the quality of life of an individual with serious injury to the central nervous system.

Through the use of MNRI®, significant improvements have been observed in oral-facial functioning and ability to communicate in this particular patient.

The MNRI® therapy positively influenced Ewa's perception and concentration. She is now able to focus longer on a particular object and learn more quickly.

References

Masgutova, S. (2008). Neuro-sensory-motor reflex integration in children with developmental challenges as the successful facilitation for their every day activities. Diagnosis and correction program for children with cerebral palsy, autism and FAS. Conference Materials. *How to Assist Handicapped Individuals Successfully?* Warsaw-Sandomirż. 26 p.

Masgutova, S., Regner, A. (2008). *Language development using sensory-motor integration approach.* Scientific edition: Prof. T. Galkowski, Prof. J. Mezwa, Dr. B. Dolyk. MISM, Wroclaw, Poland: Continuo.

Masgutova, S., Sadowska, L. et al. (2008). The MNRI® - Masgutova neurosensorimotor reflex integration program of diagnosis and therapy for children with down syndrome. *International Scientific-Practical Conference - XXXII National Day of Child Rehabilitation: Down Syndrome. Treatment, Rehabilitation and Education.*

Masgutova, S. K., Akhmatova, N. K., Shubina, Z. H., Kiselevsky, M. V. (2008). Immunological assessment of effectiveness of reflex integration therapeutic program for chronic inflammatory respiratory diseases. *Russian Journal of Immunology.* 2 (11), 4. (Russian) p. 454-463.

Masgutova, S. (2008). *Masgutova method of reflex integration for children with cerebral palsy.* Editor: S. Wenberg. Orlando, FL, USA: SMEI Published in Internet.

Masgutova, S., Akhmatova, N. (2010). *MNRI® Re-patterning and integration of dysfunctional and pathological reflexes - Svetlana Masgutova® method of neuro-reflex integration.* 2nd Edition. Orlando, FL, USA: SMEI 82 p.

Masgutova, S. (2011). *Infant dynamic and postural reflexes. Neuro-sensory-motor reflex integration.* Revised and edited. Scientific-practical manual. Orlando, FL, USA: SMEI 190 p. English version

Masgutova, S., Masgutov, D. (2013). *Parents' guide to MNRI®: neurosensorimotor reflex integration.* Orlando, FL, USA: SMEI 60 p.

Pilecki, W., Masgutova, S., Kowalewska, J., Masgutov, D., Akhmatova, N., Sobieszczanska, M., Kalka, D. (2012). The impact of rehabilitation carried out using the masgutova neurosensorimotor reflex integration method in children with cerebral palsy on the results of brain stem auditory potential examinations. *Advances in Clinical and Experimental Medicine.* Official Organ Wroclaw Medical University, Poland. 21, 3, 363–371. http://www.advances.am.wroc.pl/pdf/2012/21/3/363.pdf

Congratulations and thanks to Ewa and her parents for your support in our research. Ewa, you were so good and so motivated. I wish you great health as you move forward! – Karolina Niezgodzka

Visual, Behavioral, and Physical Changes in a Child with Autism Spectrum Disorder

Mary T. Wirth, MS, OTR/L, C/NDT, MNRI® Core Specialist, Antioch, IL, USA

As an occupational therapist I have pursued ongoing continuing education in Neurodevelopmental Treatment (NDT); the eight week basic and the advanced three week baby course, as well as other intensive NDT courses. I was also a clinical assistant for some of those workshops. Additional training was received in Craniosacral (CST), MyoFascial Release (MFR), sensory processing study/training, and Neurolink Training. Despite the extensive tool box I held, I believed that something was missing in supporting the treatment of others. I revisited both Jean Ayers' focus on inhibiting the presence of primary reflexes as well as the Bobath approach to inhibiting reflex movement, while facilitating more functional patterns of movement, without satisfaction.

Mary Wirth

I further explored these primary reflex patterns and others to address them in treatment. Their presence was obvious when clinically observing children but what was missing in our collective treatment approach to looking at children with challenges? I was not about to develop another sensory diet for a family or staff to follow. Sensory diets have their benefits at the time of implementation but were unpredictable in their effectiveness over time. With my knowledge gained in further exploration of primary reflex pattern, it became apparent that sensory diets did not reach the genetic blueprint for the primary movement pattern and it's associated sensory links in order to impact the lower motor neurons and create permanent change. Only by reaching this level will permanent change occur which creates the foundation for functional skills and learning in all paradigms: motor, cognitive, language, social, and emotional. In the search for more effective strategies to promote lasting change, I found Masgutova Neurosensorimotor Reflex Integration (MNRI®). This approach has provided me with the missing component that establishes the foundational framework for all other treatment approaches studied. This article is just one example of the positive results I have observed utilizing this approach as the major provision of my occupational therapy treatment.

Case Study: Noah

Noah was a 9 year old boy who was diagnosed at the age of 6 with profound Autism Spectrum Disorder through Alexian Brothers. His mother sought various therapeutic services to assist her son in attaining greater ability to function in the world with comfort and engagement. Traditional occupational, speech, and physical therapy were initially utilized along with integrated manual therapy. Nutritional assessment and supplementation to support and balance his cellular health was primary for the past five years along with elimination of gluten, casein, and artificial dyes. He had seen an optometrist yearly. My period of MNRI® treatment was from November of 2011 to March of 2013. I saw him for 2 hours monthly, and his mother provided a Home Program.

In November of 2011, Noah was referred to me by his optometrist, Dr. Zelinsky, to assess his underlying reflex movement patterns because Dr. Zelinsky observed that Noah's visual processing difficulties seemed to have an underlying cause.

Noah initially presented with significant attention deficit issues in school and at home, a poor ability to coordinate and organize movement for sequential actions, and inability to coordinate visual and auditory information (he could not coordinate visual with auditory, thus social interactions were poor). He had poor auditory comprehension, auditory processing delay, and had sensory seeking behaviors. There was a significant challenge with self regulation, engaging in peer interactions and group settings at school, and emotional over-reactions to daily experiences, anxiety, and OCD (obsessive compulsive disorder) tendencies. He also had sensitivity around the ears.

He was able to see clearly but not efficiently process incoming visual sensory signals because the incoming signals need to be appropriately filtered before accurate motor responses could be made. (When the signals are not filtered, confusion results and often the person relies on protective mechanisms.)

For Noah's Assessment, I utilized MNRI® to look at his reflex patterns and their underlying parameters of physical organization as well as the neuro-sensory links for sensory processing.

Some patterns should be present in their functional form when given the sensory input such as Hand Grasp. The Assessment looks to determine if the links between stimulus and motor response are congruent. Other patterns, when tested with the sensory input, should not be visibly present BUT should be available when the individual is under stress and require their action for safety such as when the Babinski is needed for balance. (See the article on *Assessments* in this book for further information.)

Noah demonstrated the following highlighted issues in his initial assessment:

Body Patterns. Noah had hypersensitivity to touch in general with both startle and flight reactions. He had significant tension patterns through the left arm and leg that were evident when the extremities were stretched to trigger protection. His upper body patterns had significant asymmetry in their presentation. Hands Grasp was hypoactive (with more evidence unilaterally), and not coordinated bilaterally. Hands Pulling had an incorrect sequence and increased intensity. Hands Supporting had the basic parameters but functional control was not present. Babkin was indicative of significant associated oral reactions with hand use, clinical observation of stabilization with lip tightening/pursing for postural control, atypical control for eye movements, and jaw deviation to the right.

Noah's lower body patterns were significant for hyperactivity and mixed directional qualities of the motor response. Leg Cross Flexion-Extension was unable to be generated from the hip and pelvis with compensation of plantar flexion and toe grasp for power.

His spinal patterns presented as hypersensitive on the right side and poor ability to generate the movement patterns with dissociation and control. Asymmetric Tonic Neck Reflex presented with Abdominal response to both sides. The Symmetric Tonic Neck Reflex was unable to be functionally sustained in a hands and knees position, with his arms collapsing, and an inability to sustain head against mild resistance.

Treatment Program

Noah was able to attend therapy sessions on a monthly basis for two hour sessions. His mother was educated on Home Program strategies to carry over between sessions.

Strategies utilized included Tactile integration and Neurostructural to decrease the holding tension through his body and to integrate his responses to touch. Noah's mother was sent home with some basic Tactile strokes including Embracing Squeeze.

Repatterning using MNRI® protocols for Spinal Reflex patterns and upper and lower body reflex patterns. His parent worked on various reflex patterns between treatment sessions.

Oral-Facial and Auditory Visual Reflex repatterning. His mother was given some of these strategies to work on between treatment sessions.

In the year since he has received this therapy modality, the following are documented highlights of the positive changes that were not evident prior to introduction of the MNRI®:

Visual Changes:

Noah's central eye sight remained 20/20 without glasses. In November 2008 Noah had developed a habit of avoiding using peripheral eyesight which persisted until January 2013, but he did not exhibit this in March

2013 after MNRI® treatment. As of March 2013, Noah's overall oculo-motor control had improved at near point. He was now able to be prescribed a pair of specialized eye glasses that were designed for stimulating proprioceptive input from his eye muscles. (In April 2011, Noah had difficulty in accurately judging object locations due to a lack of binocularity.)

Behavioral Changes:

Noah had increased reasoning skills, was thinking more flexibly (more open to initiating negotiating skills), and more aware of trying to follow the rules. He developed of a sense of humor, began to read social cues (which resulted in big shifts at school and the development of friendships), and became calmer in general although he could still get wound up and in other's space. Noah began to extend beyond himself with interest in giving to others, and began to be interpreted as defiant but, when examined closer, was actually doing the same behaviors he had been doing but now was doing them with visual regard to others and thus was being perceived differently.

Physical Changes:

Noah had improvement in his structural body symmetry, spinal patterns (grading and sustainability against gravity), and his response to touch (he began relaxing). The timing, alignment, and efficiency of Hands Supporting improved. He had an improvement in his grasp while applying traction in different planes of movement, a decrease in hyperactive responses of the feet, and improved symmetry of his upper body patterns for Grasp, Hands Pulling, and Hands Supporting. He was able to engage in the visual component with Hands Pulling. Noah has even been motivated to work on the movement patterns himself.

Summary

Noah presented with multiple areas of non-integrated primary movement patterns/reflexes at his initial occupational therapy evaluation. In order to obtain more functional motor control and higher cognitive processing, these underlying reflex/movement patterns were addressed. It was because of how hard his system was working to organize information that he was not able to attend, respond to requests, or interpret environmental information efficiently. The documentation of changes in his visual system, social interactions, and physical gains are believed to be directly related to the introduction of the MNRI® approach. The protocols utilized addressed Oral-Facial, Auditory Visual, Tactile, Postural, and Dynamic Reflex integration. These protocols focused at the underlying neurological links that provide the foundation for the skills that were required in Noah's home, community, and school.

I thank Noah and his family for their participation and support of his therapy with MNRI®. It has been my pleasure to work with you and watch your skills and life unfold. – Mary Wirth

Developmental Gains for a Child With Dyslexia and Allergies

Beata Oginska-Dutkiewicz, MA, and Teresa-Ewa Busz, MA, MNRI® Core Specialists and Instructors, Zielona Gora, Poland

Summary: The authors of this article share their experience, clinical observations, and remarks about the case study of a boy with severe dyslexia and allergies. They describe the significant impact of allergies on the functioning of the central nervous system and positive changes in his health and neuro-sensory-motor development as a participant of the MNRI® therapeutic programs from the age of 6.

Szymon is the son of author and MNRI® Core Specialist, Beata Ogińska-Dutkiewicz.

Beata Oginska-Dutkiewicz

Teresa Ewa Busz

Problems of Allergy

Szymon was born on March 3, 2000 via C-section. He weighed 4kg (8.8 lb). Szymon received 10 points on the APGAR scale which was supposed to indicate that he is in a good condition and health. For the first 24 hours he was separated from me due to my recovery after epidural anesthesia. We stayed in the hospital for 8 days, the standard recovery time after a C-section. Szymon was doing well, with a good Sucking Reflex and was able to sleep easily.

After returning home, my son developed intestinal difficulties. I only breast fed him but he exhibited a negative reaction to my milk. This was a big surprise. This started the battle of figuring out what I should eat and what I should avoid eating to avoid triggering his sensitivity. Carefully following my doctor's orders, I followed a progressively more restricted diet, but this still didn't have a significant impact on my child's health. For many months he had great digestive difficulties, diarrhea and long and painful bouts of colic. He slept very little for a newborn and was restless. He developed a rash on his entire body causing severe itching, unease, hyper vigilance, and exhaustion. Disoriented, confused, and worried, we traveled from one specialist to another trying to find answers to help our son. Frequently doctors would give contradictory treatment plans for Szymon's symptoms. At last at the age of 8 months, Szymon was diagnosed with atopic dermatitis (eczema), food and airborne allergies.

Atopic dermatitis is a hypersensitivity of the body mast cells that are found in the skin, mucous membranes of the nose, throat, ear, larynx, lungs and intestines. Mast cells are best knows for their role in allergy and releasing histamines. However, they also play an important protective role in wound healing and defense against pathogens. Szymon's immunoglobulin E (IgE) blood test results showed strong allergies. Other tests were also positive showing reactions to various foods, pollen, and fungus.

We went through countless visits with allergy specialists. Medications recommended by one doctor where questioned by the next. This chaos of information seemed to have no end.

Szymon had a list of restricted food and other items that he was not even allowed to touch: wheat products, fruits, flowers, leaves, and basically anything that is connected to nature. All the restrictions and ongoing challenges made us all feel exhausted and hopeless.

Over the period of two years our house became almost empty. We got rid of all the plants, window covers, couches, and anything that could collect dust. The washing of floors, dusting of furniture, scrubbing bathrooms (sources of dangerous fungus and mold) became my daily routine. After 2 years, Szymon's allergist recommended he stop all contact with my sister, Szymon's aunt, who has a cat and a dog at home. That was more than I could bear and pushed me to start my search for alternative solutions to help my son.

We started homeopathy. After a few big episodes of rashes we noticed positive changes in the appearance of the skin. There was no more itching and old sores caused by scratching began to heal. My son felt better and was also calmer. Unfortunately, soon after that, Szymon started developing problems with his breathing. He developed bronchitis and multiple bouts of pneumonia. He was eventually diagnosed with asthma. Inhalers and steroid medications stopped ongoing inflammation and prevented future asthma attacks.

For 6 years Szymon was on a special diet: no gluten, milk or dairy, eggs, seafood, nuts, or citrus. Food allergies would be activated immediately after his contact with the allergen. He would have difficulty swallowing, would start choking, produced excessive amounts of saliva, and suffered from abdominal pain. Even today Szymon's diet is very limited and the introduction of any new product is always undertaken with great caution.

As an MNRI® Core Specialist I was aware and noticed that besides poor immune system and severe allergies, Szymon's motor development has also been impacted. His attention span was short, his emotions and mood were unstable, he had fears and phobias. In terms of motor function, Szymon presented with poor static and dynamic balance, non-uniform muscle tone, and unwillingness to perform any fine motor skills including writing, drawing, etc.

After studying and researching huge amounts of material, I realized that everything that he had gone through (struggles and challenges with his health, medications used to control his allergies) were aggravated by the constant fear and stress of his parents who tried to keep him safe. All of those components impacted his development and especially the development of his brain. The brain is not an isolated organ. All physical experiences of the body will impact the brain and central nervous system.

Dr. Danuta Myłek, specializing in allergies, author of the book *The Allergy* (2001) wrote, "The brain doesn't sneeze, or itch, or cough. But it does manifest symptoms typical to this organ: behavioral changes, difficulties with attention and concentration, poor memory, difficulties with sleep or nervous ticks."

From the beginning, in trying to protect Szymon and following doctors' recommendations, I was teaching him how to be very careful and how to avoid the risks of potential allergens. In doing that, I limited his natural curiosity and desire to explore and discover. That, of course, had an impact not only on his motor development but also his emotional development. Szymon started to be uneasy and fearful of unexpected allergic reactions of his body.

Children and parents coming to my office are usually searching for answers. They want to find the reasons for their children's difficulties with concentration, attention, memory, elevated anxiousness, uncertainty, poor psychomotor endurance or delayed motor and psycho-emotional development. Based on the 'parent interview' we gather, the data indicates that 70% of children presenting with the above symptoms have some kind of allergies and 80% of these children had complications during delivery.

I am convinced that allergies, as chronic disease, have a very strong impact on a child's psycho-motor development. I witnessed that with my own son and continue seeing it with the children I work with.

Impact of MNRI® on Treatment of Allergies and Strengthening of Immune System Function

Our clinical observations have been validated by scientific experiments conducted in Russia. Based on the results of the research, MNRI® has been shown to have a significant positive impact on hormonal regulation and the immune system of patients (children and adults) treated for recurring pulmonary infections. Chronic inflammatory reactions of the pulmonary system are closely correlated with poor regulation and activity of the immune system.

It has been demonstrated that using MNRI® allows for positive changes in functioning of the immune system. It allows for normalization of levels of T lymphocytes (CD3, CD4, CD8) and the NK natural killer cells as well as metabolic functions of leukocytes, and levels of regulatory and anti-inflammatory cytokines.

A statistically valid increase was observed in the level of cells of various antigens and NK natural killer cells CD16. The above cells are key actors in the innate immune system and are known to have significant biological function in control against cancer cells, viruses, and parasites. (Masgutova, Akhmatova, Kisielewski 2008).

Besides the immuno-regulatory impact, MNRI® also increases the level of cortisol to within physiological norms. Scientific studies show that MNRI® can affect the level of the Adrenocorticotropic hormone (ACTH) of the pituitary gland, and through that, regulation of the production of IL-1. (Masgutova, Akhmatova, Kisielewski, 2008).

Dyslexia

During the early school years my son appeared to have a predisposition for dyslexia. Several authors studying this subject claim that dyslexia is a symptom presented by children with a normal intellectual development but having difficulty with processing of information. Frequently accompanied by difficulties with: perception, planning, memory, speed of information analysis, differentiation and decoding of time-space orientation, sensory and cognitive modeling.

Dyslexia and the accompany symptoms impact the following abilities: writing, reading, counting as well as motor coordination (fine and gross motor control), programming, and control (Dubrovina 1991; Gross, 1987; Bogdanowicz, 2004)

Possible causes leading to dyslexia are related to dysfunctions in integration of cognitive and motor functions, neuropsychological deficits in visual, auditory and motor functions of language as well as memory (visual, auditory, and motor), atypical lateralization of higher cognitive functions, poor awareness of body and space orientation, or poor orientation in static motor control.

The etiology of dyslexia has multiple levels. There are elements of genetics, anatomical dysfunction, and neurophysiological dysfunction of the central nervous system during pregnancy and delivery, delayed neurodevelopment, environmental factors, family structure and function, not to mention a lack of proper early diagnosis and intervention (Bogdanowicz, 2004; Goldberg, 2009)

Szymon's Diagnosis from the Psychological and Educational Center

Based on the diagnosis provided by the Psychological and Educational Center, Szymon presented with perceptive-motor deficits, and dysfunctional cognitive processing (poor visual, auditory and kinesthetic-motor processing), diagnosed as severe developmental dyslexia.

Szymon's diagnosis was done relatively early, his physical development as a baby was within normal limits. He started to meet all milestones on time including sitting up, crawling, and walking at 12 months. Only at the age of 3 did I notice small tremors of the hand whenever Szymon tried to do something more precise or complex that required fine motor control. Building a tower of blocks would be a very difficult task. Szymon couldn't assess distance, level of strength, or pressure to use. His body was expending a great deal of energy into ineffective movements. Negative experiences led to avoidance of fine motor functions like playing with puzzles, drawing, playing with clay, or playdough. Even as he tried to draw, Szymon was only able to do it for a short time and would usually draw chaotic lines using non-uniform levels of strength and with poor precision. His first attempts at writing during preschool revealed Szymon's grapho-motor problems.

Szymon liked to move, he liked running, climbing, and playing on the playground. As a young boy he liked to place his body with the head positioned upside down.

When falling asleep he would lie with his legs curled under his belly (TLR position) or on his back with an extended neck and open mouth. His speech started developing at the normal time, but when it came time to start using longer sentences, Szymon was slow to find the right words. Many children and adults did not have enough patience to wait for him to finish his sentences. Szymon's stuttering became worse while his resistance to speaking altogether increased. Up until the age of 10, Szymon wasn't able to pronounce the rolling "r" sound.

Therapy

After noticing more negative symptoms with Szymon's development I continued searching for ways to help him. I started taking classes, workshops, and became a Sensory Integration Therapist, a Special Education teacher, and a Brain Gym® specialist.

In 2005, I participated in the *Dynamic and Postural Reflex Integration* class where I met Dr. Svetlana Masgutova. From that day on I continued my studies with her to become a Core Specialist in MNRI®.

Szymon at age 10 on balance board.

At the age of 6, Szymon had an MNRI® Assessment performed by Dr. Masgutova that confirmed his diagnosis of sensory and sensory motor integration dysfunctions. Szymon most certainly needed help to improve the functioning of his nervous system. Szymon's life was filled with constant challenges: the situation of his delivery (being born via C-section), genetic predispositions, side effects of long term allergies and medications used to control it, as well as a diagnosis of severe dyslexia and other learning disabilities.

His early school years were difficult. He exhibited difficulties with writing, reading, and ability to focus his attention. His nervous system was prone to quick overload. Emotionally, Szymon also had a difficult time. He would hold on to stressful responses for a long time after the fact. After receiving a poor mark, his feelings of upset and failure would persist.

With all of these challenges, Szymon didn't like school. He was stressed, fearful, uncertain, and would withdraw easily. The diagnosis of perceptive-motor delays, dysfunctional cognitive processing and severe developmental dyslexia made it very difficult for him to learn and effectively participate in school activities. The recommendations we received from the Psychological and Educational Center was for Szymon to read and write more in order to practice. Such advice sounded to me like a cruel joke!

Instead of following the advice that didn't make any sense to me I started using Brain Gym® exercises, art therapy, and from 2006 on, MNRI® to facilitate the development of my son's nervous system.

MNRI® reflex integration diagnosis demonstrated clearly the areas of need within Szymon's sensory motor processing. We started working with the reflex integration program looking for improvements in motor development, cognition and perception. To our surprise, MNRI® positively impacted many other areas of Szymon's development.

In working with Dr. Masgutova, it took us time to understand fully the meaning of such statements as, "there needs to be a strong focus of working with body stress/protective responses; working with Spinal Perez will help with his allergies" or "we need to pay attention to the development of muscle bulk, bones, and joints; we need to work with structure using the neurostructural integration protocol", "stimulation of skin receptors and proprioception is also very important." Deeper understanding of all these statements, their scientific and neuro-physiological meaning and the power of the method came to us step by step.

With deep faith and respect for Dr. Masgutova's knowledge and experience, we continued to use MNRI® with Szymon. We could see changes even at times when life would not allow us to follow the recommended intensity and frequency. The journey brought challenges and successes. Still, with all of these ups and downs, Szymon continued to make progress. Using MNRI® we were impacting not only motor development but also the physiology of Szymon's whole body. His immune system got much stronger. At the age of 9 after pulmonary function tests at the Center of Pulmonary Diseases in Karpacz (metacholine challenge test, spirometry test) *Szymon's asthma diagnosis was overturned!* Other positive changes manifested themselves in weight gain and finally the growth of muscle bulk. For years prior to starting the MNRI® program, Szymon looked very skinny, had poor strength and underdeveloped musculature.

Another big achievement came at the age of 10. After participating in the MNRI® Family Conference where Szymon received 6 hours of therapy daily, for 10 consecutive days, he started pronouncing rolling "r"! As we know altering the pronunciation of this specific consonant at the age of 10 is very difficult to achieve. This new ability immediately translated to emotional wellbeing and increased his self-confidence. The clarity of his pronunciation also improved significantly and there was a decrease in stuttering. All those changes positively impacted Szymon's willingness to speak/converse and dramatically improved his confidence in communication.

At present Szymon participates in school theatre performances, which he loves. He can speak clearly, projecting his voice loud and clear. His speech is fluent and pronunciation flawless.

Currently we are working on 'school stress' which still has a negative impact on Szymon's development. The other priority is strengthening his pulmonary system. Based on researched conducted by Dr. Masgutova we know that children with dyslexia have significant challenges with developing and using higher executive functions of the brain. The harmony of psychomotor development is also more challenging for them.

Nonintegrated, not matured dynamic and postural reflex patterns will not support the development of higher levels and functions of the brain. They will not be able to optimally support development of auditory or visual perception. Additionally non-integrated reflexes disorganize the neurophysiological processes of activation and inhibition, thus negatively impacting self-regulation mechanisms and proper functioning of the whole brain. As a result they will impact behavior, motor and cognitive development as well as emotional regulation.

A reflex is a natural, genetically coded developmental program realized by the brain-body system from the moment of conception. It is a genetically given response for a specific stimulus. The basic function of a reflex is as a protection and survival response of the body. As the body matures, reflexes provide a basis for further development. Receiving sensory and proprioceptive stimuli through afferent pathways, the central nervous system and its network process the information and sends it out through efferent pathways back to muscles and organs/glands to organize and perform a particular motor response. (Masgutova, Akhmatova 2004, 2007). Each reflex pattern has specific time of development and maturation. If for whatever reason it does not go through that process it will not be able to properly support development of higher functions and skills.

Dr. Masgutova performed Assessments of dynamic and postural reflex patterns within a group of 580 children aged 6-19 with a diagnosis of developmental dyslexia. The results of the Assessments demonstrated that the level of reflex pattern integration has significant and in direct correlation with the children's abilities to process and decode information that serve as the basis for the processes of learning and memory function.

Table 1 presents the degree to which various nonintegrated reflex patterns appear in a sample of 580 children with dyslexia (age 6-19 performed between 1989-2002).

Data presented in Table 1 demonstrates the results showing that children with dyslexia often do not possess matured and integrated reflex patterns, especially ones related to development of auditory-visual functions and fine motor control.

• ATNR Asymmetrical Tonic Neck Reflex was not integrated in 83.3% of assessed children. This reflex is responsible for experiencing and developing one's own proprioceptive system, structural coordination of the body and auditory reflexes: mono and binaural hearing, language development, auditory processing, auditory memory, space orientation.

Table 1. below presents the degree to which various nonintegrated reflex patterns appear in a sample of 580 children with dyslexia (age 6-19 performed between 1989-2002).

	Table 1. Reflexes	% of children w/ dyslexia w/o integrated reflex	Reflex Patter	Number of children
1	ASYMETRICAL TONIC NECK REFLEX	83.3%	ATNR	483
2	BAUER CRAWLING	73.0%	BAUER CRAWLING	423
3	ROBINSON HAND GRASP	71.4 %	ROBINSON HAND GRASP	414
4	MORO EMBRACE	71.4%	MORO	414
5	LEG CROSS FLEXION EXTENSION	71.6%	LCF	412
6	SYMMETRICAL TONIC NECK REFLEX	70.1%	STNR	409
7	TRUNK EXTENSION	61.8 %	TRUNK EXT.	389
8	HANDS SUPPORTING	64.3%	HANDS SUPPORTING	373
9	BONDING	64.1%	BONDING	372
10	PAVLOV ORIENTATION	63.8%	PAVLOV	370
11	BABKIN PALMOMENTAL	56.9%	BABKIN	339
12	SPINAL GALANT	52.8%	GALANT	320
13	FLYING AND LANDING	52.2%	FLYING and LANDING	301
14	SPINAL PEREZ	44.8%	Perez	260
15	HANDS PULLING	43.1%	HANDS PULLING	250
16	BABINSKI	41.6%	BABINSKI	241
17	LANDAU	40.0%	LANDAU	232
18	THOMAS AUTOMATIC GAIT	35.7%	AUTOMATIC GAIT	207
19	TONIC LABYRINTHINE REFLEX	67.2%	TLR	121

• Bauer Crawling was not integrated in 73% of assessed children. Reflex responsible for development of: gross motor control and activity of the whole body, reciprocal/cross lateral movements, head righting, binocular vision, binaural hearing.

• Robinson Hands Grasp was not integrated in 71.4%, Babkin Palmomental in 41.6%, Hands Pulling 43.1% of assessed children. These reflexes are responsible for development of manual skills like writing, drawing, eye coordination, speech production. Additionally, phases of sequential fingers opening and closing underpin differentiation, calculations, and other math skills.

• TLR Tonic Labyrinthine Reflex was not integrated in 67.24% of assessed children. This reflex is responsible for postural control, protective mechanisms of freezing and fight or flight, selective visual-kinesthetic perception and information processing.

• STNR Symmetrical Tonic Neck Reflex was not integrated in 70.1%, Trunk Extension in 61.8% of assessed children. Both reflexes are responsible for head and trunk position, binocular vision, binaural hearing, horizontal and vertical eye tracking, convergence and divergence, vestibulo-ocular and ocular-kinetic reflexes. They also participate in development of visual skills, including comparison, visual analysis, visual prioritization, reading, and writing.

• Spinal Galant was not integrated in 52.8% of assessed children.This reflex is responsible for postural control of the spine, motor coordination, reciprocal-lateral coordination of posture and movement, development of attention, consciousness, fast brain waves activation serving to stimulate awareness, alertness, ability to notice and memorize.

• Spinal Perez was not integrated in 44.8% of assessed children. This reflex is responsible for development of spinal postural control, and detoxification of the brain through activation of peripheral nervous system.

Each column in the graph below corresponds to the rows of Table 1 presented above.

Szymon's assessment results were representative of the data of the group of children with dyslexia.

Our therapeutic work with Szymon included several programs:

• Dynamic and Postural Reflex integration
• Neurostructural integration
• Archetypes
• Facial and Auditory Reflex integration
• Tactile integration

All programs facilitate and support each other and allow for a holistic approach to treating the entire musculo-skeletal structure of the body, proprioceptive and tactile systems, motor coordination and, above all, organizing the nervous system.

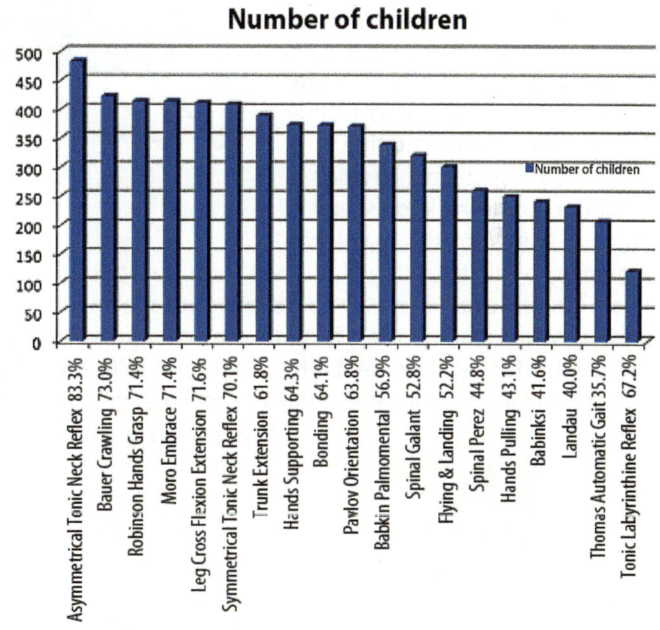

Neuroplasticity of the human nervous system and sensory integration allowed for positive changes in the functioning of Szymon's brain. The genetically coded resources embodied by motor reflex patterns that are naturally available to us can be recognized by the brain and allow for proper foundations for further development to be created.

Month after month Szymons's body was changing. Among these changes were improvements in his level of motivation and his ability to concentrate and learn. After each diagnostic round by Dr. Masgutova (February 2006, March 2011, and April 2012), ongoing positive changes in the level of development and maturation of reflex patterns was noted. As a mother and MNRI® Core Specialist, the correlation between the level of integration of reflex patterns and Szymon's development and maturation was unquestionable and served to motivate our continuing work.

The tables and charts (on the next page) present the results of three consecutive Assessments of reflex patterns integration level performed in February 2006, March 2011, and April 2012. Analysis of the results of the Assessments demonstrates that development and maturation of the reflex patterns improved from being functional at a very low level (results from 12-13.75) of development of basic reflex patterns to the next level, functional but at a low level (14-15.75). Mathematically these results are significant as analyzed with the statistical model developed by Prof. A. Krefft.

An important factor having significant impact on the quality and speed of development and maturation of Szymon's reflex patterns was his very fast growth. Over the period of 3 years Szymon's body structure and mass had changed dramatically. He grew from 138 cm (4' 6.3") and 34 kg (75 lbs.) in 2010, to 164 cm (5' 4.5") and 48 kg (105 lbs.) in 2012. Such fast growth of a young man in such a short time had a strong negative impact on his coordination, proprioception, musculo-skeletal system. Our therapeutic approach was to support his development and body structure with all MNRI® tools. At the present time in December of 2013, Szymon is 173 cm (5' 8") tall and it seems that his body will continue to grow.

Szymon Dutkiewicz	Left-Right (Lateral-Median/Sagital) MCS										
			1	2	3	4	5	6	7	8	9	10
Reflex pattern			Core Tendon Guard	Hands Grasp	Hands Pulling	Babkin	Babinski	Leg Cross Flexion/ Extention	ATNR	Abdominal	Bonding	Foot Grasp
Date:	//2006	Average 1	13.75	15.00	15.00	14.75	13.75	14.25	13.75	12.75	15.50	13.50
	//2011	Average 2	14.75	15.00	15.25	13.75	13.75	14.75	14,25	13.25	15.50	14.75
	//2012	Average 3	15,25	15,25	15,50	15,25	14,25	15,25	14,75	14,00	15,75	15,25

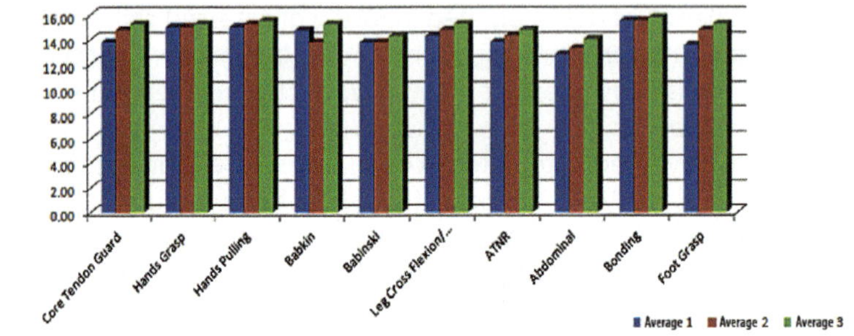

Szymon Dutkiewicz	Up-Down (Superior-Inferior/Horizontal) MCS										
			1	2	3	4	5	6	7	8	9	10
Reflex pattern			Automatic Gate	Bauer Crawling	Moro	Fear Paralysis	Hands Supporting	Segmental Rolling	Landau	Flying/ Landing	Grounding	Head Righting
Date:	//2006	Average 1	15.25	14.75	13.25	12,75	15,25	14.25	14.25	14.75	14.75	13.75
	//2011	Average 2	15,25	14,75	13.75	13.25	15,50	14.25	14.75	14.75	14.75	14.00
	//2012	Average 3	15.50	14.75	13.75	12.75	15.25	14.50	15.25	15.25	14.75	14.00

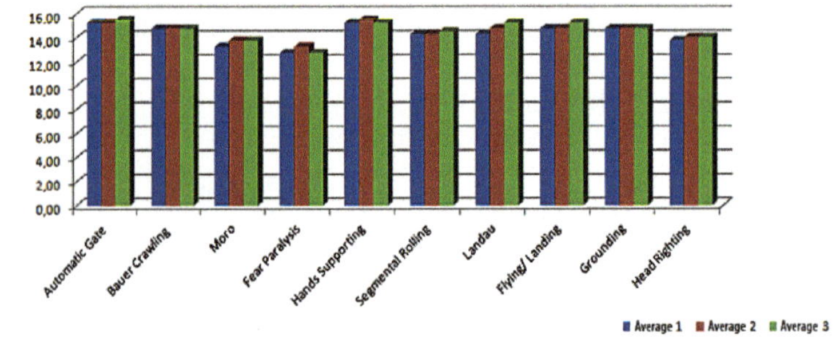

During such intense structural changes and ongoing growth, we will have many more therapeutic and developmental challenges to overcome.

We have a deep trust in the approach that we started a few years ago and will continue working with MNRI® to support Szymon's further development and maturation.

At the present time Szymon is in middle school, he knows how to read and write (abilities that at the beginning of his educational journey were questioned). His writing is clear but his graphomotor skills are below his age level. He still has problems with spelling and his reading, although relatively slow, continues to improve. When reading 'in his head' he can do it much faster. His attention span has grown to about 1-1½hours; a very important achievement is his progress in self-regulation in stressful situations. Szymon is more resilient in the face of stress, but when it does happen he will ask for MNRI® sessions.

Szymon today at the age of 13 years.

Szymon has great communication with his peers, he is very understanding, empathic, and easily forgives. As a sensitive, cheerful, and optimistic young man, also accepting of others and their differences. His personality is very likable and he has many friends who love and respect him for his qualities. Szymon continues to participate in school theater and enjoys acting. He also loves soccer.

In November of 2013 Szymon again visited the Psychological and Educational Center and was reassessed. Results demonstrated very positive changes in many tests. Szymon reached the norm in reading comprehension and showed significant improvement in: fine motor control, visual and spatial orientation and analysis, audio-visual skills, and visual-motor skills (visual-motor coordination, visual memory).

I am very happy and proud of my beloved son! I trust that he will become a wonderful, curious, and responsible adult.

I would like to thank Dr. Svetlana Masgutova for her teaching and years of support. I also send my thanks to all my colleagues from Poland who continue to support our therapeutic work with Szymon and my international colleagues in the USA, Canada, Belgium, Estonia, and Sweden for ongoing sharing of information, kindness, and friendship. To all of you, I send my warmest regards.

References

Bogdanowicz, M. (2004). *Dyslexia risk: Problems and diagnosis.* ('Ryzyko dysleksji. Problem i diagnozowanie'), Wydawnictwo. Gdańsk, Pl.

Dubrovina, I. V. (1991). *School psychology service: theoretical and practical aspects.* (Russ.). Moscow, Russia: Scientific Research Institute of General & Educational Psychology.

Goldberg, E. (2009). *The new executive brain: Frontal lobes in a complex world.* New York: NY, USA: Oxford University Press.

Gross, D. R. (1987). *Psychology: The science of mind and behavior.* London, UK: Hodder and Stoughton.

Krefft, A. (2007). *Diagnosis of functions of invisible phenomena* (Statistical Mathematical Analysis). Oficyna Wydawnicza Politechniki Wrocławskiej. Wrocław, Poland.

Masgutova, S. K., Akhmatova, N. K. (2004, 2007). *Integration of dynamic and postural reflexes into the whole body movement system.* Scientific edition: Prof. N. Akhmatova. (Pl.). Warsaw, Poland: MINK.

Masgutova, S. K., Akhmatova, N. K. (2004). *Children with challenges – Movement development and reflexes integration.* (Pl.). Warsaw, Poland: ISMI.

Masgutova, S. K., Akhmatova, N. K., Kiselevsky, M. V. (2008). Clinic-immunological assessment of effect of the therapy program of neuro-sensory-motor integration of reflex patterns at chronic inflammatory diseases of respiratory system. *Russian Immunology Journal.* 2(11), #4. (Russ.). p. 454–463.

Masgutova, S. K., Kowal, J., Mazur, G., Masgutov, D. P. (2005). *NeuroKinesiology Tactile Therapy™* by Dr. S. Masgutova, Scientific Edition: Dr. A. Regner, J. Szymczak. (Pl.). Warsaw, Poland.

Masgutova, S. K., Regner, A. (2007). *Language Development Using Senso-Motor Integration Approach.* Scientific edition: Prof. T. Galkowski, Dr. B. Dolyk. MISM, Continulo. (Pl.), Warsaw, Poland: MINK. p. 196.

Masgutova, S. K., Sadowska, L. (2005, 2006). The use of Edu-K for children with learning difficulties: early developmental dynamic. In materials of Polish National Conference: *New Methods of Movement Development and Learning Facilitation: Children with Learning Difficulties, Hyperactivity and Autism.* Scientific edition: Dr. S. Masgutova, Dr. Anna Regner, Prof. M.D. E. Ronin-Walknowska. Maternity Department, Pomorska Medical Academy in Szczecin MINK, (Pl.). Warsaw – Szczecin, Poland. p. 164-176.

Merzenich, M. (2013). *Soft-wired: How the new science of brain plasticity can change your life.* San-Francisco, CA, USA: Parnassus Publishing.

Pavlov, I. P. (1960). *Conditioned reflexes: An investigation of the physiological activity of the cerebral cortex.* (Anrep G.V., D. Sc. Trans., 1960). New York, NY, USA: Dover Publications.

Piaget, J. (1976). *The grasp of consciousness: Action and concept in the young child.* Cambridge, MA, USA: Harvard.

Sechenov, I. M. (1961). *Reflexes of the brain.* (Russ.). Moscow, Russia. p. 28-54. (Original work was published in 1863)

Venger, L. A. (1988). The origin and development of cognitive abilities in preschool children. *International Journal of Behavioral Development,* 11(2). (Russ.). Moscow, Russia. p. 147 – 153.

Vygotsky, L. S. (1986). *The child psychology: The problems of child development.* In 6 Books. Book – 4. (Russ.). Moscow, Russian: Pedagogika.

Szymon, our dear son, you grow showing how beautiful development can be and how it is possible to move into different conditions! Congratulations to you! – Your mother, your friends, MNRI® Specialists, and Dr. Masgutova, with love and respect.

Effective Treatment for Selective Mutism

Jo Anne Tierney, OTR/L and MNRI® Core Specialist, Mendota Heights, MN, USA

Jo Anne Tierney

Taylor is a 7½ year old girl with a diagnosis of Selective Mutism. She has attended occupational therapy sessions twice a week since the age of 4½, with the major emphasis of therapy based on the Masgutova Method of Neurosensorimotor Reflex Integration® (MNRI®). This paper will describe Taylor's history, presenting problem, the therapy approach utilized, and results of therapy up to this point.

History and Background

Taylor was born at 37 weeks following an uncomplicated labor and natural delivery, weighing 6 lb. 7 oz. Initial difficulties with feeding included some tongue thrusting, but she had no difficulty with sucking and was bottle fed. She was a colicky baby and would scream, arch, and cry much of the day; she was unable to lie flat and would wake if laid down, so she slept upright in a bouncy chair at night. She did not tolerate tummy time and was not relieved or soothed by swinging, swaying, or white noise. At three months of age she was diagnosed with silent reflux. Zantac gave her some relief, but at seven months of age it stopped working altogether. She was put on Prevacid, and within three days she was "a new child." She rolled over for the first time, and became a happy baby who enjoyed wiggling and exploring. She began sleeping in her own room in her crib at eight months of age, and was weaned off Prevacid at 11 months, when she was eating more solids and sitting upright on her own.

Due to the effects of the reflux, Taylor's developmental milestones were initially met at a later age than would be considered typical: she did not roll over until the age of 7 months, at the point when the Prevacid succeeded in controlling her reflux. After that time, however, she began to make nice gains. She sat independently at eight months, crawled at 10-11 months, cruised around furniture at 11-12 months, and walked independently at 12-13 months. At that time it was noted that Taylor seemed to trip over her toes frequently, especially with the right foot, which turned in. As she began preschool at age three, she exhibited severe anxiety, and began to toe walk. Her leg muscles were very tense and tight, and she complained of pain behind her knees. Her teacher reported that Taylor was unable to cross her legs in order to 'tailor sit,' and when sitting with legs straight out in front of her, one leg tended to roll in.

Per parent report, Taylor had a hard time slowing her body down, would crash into her parents when hugging them, and played roughly with her little brother. She had disliked swings since infancy. Her parents started holding her on their laps when swinging, and she began to swing independently for a short period of time. However, she continued to be very tentative on swings and had difficulty getting in and out of the swing, even if close to the ground.

Taylor began babbling at four months of age, and was very "talkative" by five-six months. She spoke her first word at nine months, but by age two she spoke just a few words and a few two-word phrases. At that time she suddenly surprised her family by saying, "My name is Taylor…T…A…Y… L…O…R." It was also discovered at that time that she knew about 70% of the alphabet in random order. She has always been exceptionally observant and aware of her surroundings.

Severe separation anxiety was noted at age 18 months; she screamed constantly when placed in the church nursery, and, other than her mother, she would only accept being left with her father or grandmother. At age 2½ she began to shut down or freeze upon arrival at her play group. She would eventually run around with the other children, but she did not speak. At age three she started preschool where she also did not speak and often remained curled up in a fetal position, never looking up. Her affect was blank and expressionless. She did not play or take part in any group activities. She would barely walk, and instead shuffled very slowly. She often covered her head, and her mother described it as seeming as if she were in a silent panic attack. The moment she was removed from these settings, she snapped out of it and started to talk. At that point, her pediatrician gave her a diagnosis of Selective Mutism. She began play therapy sessions with a psychologist. Taylor was lively and animated in the sessions, but these improvements did not generalize to other areas of her life, so this intervention was discontinued.

As a preschooler, Taylor was placed in a small classroom of five children, all of whom had delayed development. The teacher was experienced in dealing with severely anxious children and Taylor began speaking to her. She did well and was placed in a more advanced class. She was then mainstreamed in a preschool class of 20 children with a special education teacher assisting in the classroom. She began to speak in class, raised her hand to answer questions, and showed interest in joining group activities. Anxiety continued to be a significant issue, however. Per parent report, if she wanted to join an activity or game at school, she would often stand very close to the activity, but wait for someone to include her; she did not ask to join in. She continued to be unable to answer questions from people she did not see on a regular basis, and she did not speak or make eye contact with strangers. She would close her eyes and freeze nearly every morning when the school bus arrived, and needed to be carried onto the bus. According to her mother, she was unable to control this response, but knew it was about to happen, and would sometimes cry out or try to warn her parents before it happened. She was able to verbalize that she wanted to do or say things but could not. Her parents found that movement would sometimes slow or stop the shutdown.

Taylor had severe temper tantrums between the ages of three and four; these tantrums could last for hours, and she would often scream, "I can't stop! I can't calm down! I need help!" At the time, her emotional maturity was described as being closer to that of a 2 year old than a 4 year old. At age 4½, Taylor was not yet dressing herself, and made no attempt to do so. She was still in a diaper due to anxiety and fear of the toilet.

At the time of her initial assessment at age 4½, Taylor's strengths were described as: seemed academically advanced, very observant, excellent memory, and very left-brained, with mechanical knowledge and interests.

The Presenting Problem

At the start of therapy in February of 2010, Taylor presented with very severe anxiety, which affected her ability to function in social situations such as day care, preschool, the church nursery, and the community. She presented all of the behaviors noted above. Motor skills were delayed and ranged from the second to the fifth percentile when compared with same age peers. Taylor also presented with very significant sensory integration dysfunction, especially in the areas of vestibular, tactile, multi-sensory, and auditory processing.

Taylor has a medical diagnosis of Selective Mutism (ICD9 code: 313.23). Her treatment diagnosis, determined by the treating occupational therapist, is Specified Developmental Delay (315.8), Lack of Normal Physiological Development (783.40); and Abnormal Reflexes (796.1).

Selective Mutism (SM) is a childhood anxiety disorder characterized by an inability to speak in one or more social settings (at school, in public places, with adults) despite being able to speak comfortably in other settings (at home with family). Those with SM experience anxiety related to speaking, including being unable to use other non-verbal forms of communication such as nodding the head or pointing. More than 90 percent of children with SM also meet the diagnostic criteria for social anxiety disorder, now termed social phobia (www.selectivemutism.org). *The Diagnostic and Statistical Manual of Mental Disorders*, 4th Edition, Text Revision, or

DSM-IV-TR (APA, 1994) states that the criteria for SM includes "an inability to speak in at least one specific social situation where speaking is expected (e.g. at school) despite speaking in other situations (e.g. at home)", and "the disturbance has interfered with educational or occupational achievement or with social communication."

Assessment, Intervention, and Results

Taylor has participated in MNRI® Reflex Integration Therapy for the past 3 years, since the age of 4½. Therapy has been provided 1-2 times per week, with parents providing strong Home Program carryover.

Reflex Assessment: Taylor has participated in MNRI® Assessments by her occupational therapist (February 1, 2010 and ongoing), and by Svetlana Masgutova, Ph.D, (February 25, 2011 and November 15, 2012), for evaluation of the motor reflex patterns, in order to define the level of their development and the sufficiency of their functioning (see Table 1). The results of the reflex Assessments are as follows:

February 1, 2010: Initial reflex Assessment at age 4½ indicated retention, or non-integration, of a number of primary reflexes that typically should have been integrated into her motor system by that time. It was hypothesized that the delay in development and integration of these reflexes was contributing to, and was a reflection of, many of the challenges that Taylor was dealing with. This Assessment provides a description of Taylor's neurosensorimotor functioning at the start of therapy.

Table 1. Taylor's Pre and Post- Assessment results by Dr. Masgutova

Reflex Evaluation Results

Taylor	Reflex Patterns Serving for Development of Left-Right Motor Coordination System																			
	1		2		3		4		5		6		7		8		9		10	
	Core Tendon Guard		Hands Grasp		Hands Pulling		Babkin Palmomental		Babinski		Leg Cross Flexion Extension		ATNR		Abdominal		Bonding		Foot Grasp	
	R	L	R	L	R	L	R	L	R	L	R	L	R	L	R	L	1	2	R	L
MNRI Assessmt 2/25/11	12.5	13	8.5	9	12.5	12	8.5	7	8.5	9	12.5	12	8	7.5	7.5	7	13.5	12	7.5	6
Average		12.75		8.75		12.25		7.75		8.75		12.25		7.75		7.25		12.75		6.75
MNRI Assessmt 11/15/12	14.5	15	14	13.5	14.5	14	13	12.5	12.5	12.5	15.5	15.5	13.5	14	12.5	12	15.5	15	11.5	11
Average		14.75		13.75		14.25		12.75		12.5		15.5		13.75		12.25		15.25		11.25

Taylor	Table 3. Reflex Patterns Serving for Development of Superior-Inferior MCS																			
	11		12		13		14		15		16		17		18		19		20	
	Automatic Gait		Bauer Crawling		Moro		Fear Paralysis		Hands Support		Segmental rolling		Landau		Flying and Landing		Grounding		Head Righting	
	R	L	R	L	1	2	D	D	R	L	Up	D	Up	D	Up	D	Up	D	R	L
MNRI Assessmt 2/25/11	12.5	12	8.5	8	7.5	9	7	9	8.5	8	7.5	8	8.5	8	12.5	12	7.5	9	8.5	9
Average		12.25		8.25		8.25		8		8.25		7.75		8.25		12.25		8.25		8.75
MNRI Assessmt 11/15/12	15	14.5	12.5	13	11.5	12.5	11	12.5	13.5	14	12.5	12	12.5	11.5	13.5	14.5	12.5	13	12.5	12
Average		14.75		12.75		12		11.75		13.75		12.25		12		14		12.75		12.25

Taylor	Table 4. Reflex Patterns Serving for Development of Anterior-Posterior MCS																			
	21		22		23		24		25		26		27		28		29		30	
	Trunk Extension		STNR		Spinal Galant		Spinal Perez		Tonic Labyrinthine		Foot tend		Spinning		Pavlov Orientation		Locomotion		Convergence/ Divergence	
	R	L	R	L	R	L	R	L	1	2	R	L	Up	D	R	L	R	L	R	L
MNRI Assessmt 2/25/11	12.5	13	7.5	8	7.5	7	8	7	9.5	9	6.5	6	11	9	7.5	7	8.5	8	12.50	10
Average		12.75		7.75		7.25		7.5		9.25		6.25		10		7.25		8.25		11.25
MNRI Assessmt 11/15/12	14	14.5	12.5	13	12.5	13	14	13.5	13.5	14	12.5	12	14.5	14	12.5	12	13.5	13.5	15.00	14
Average		14.25		12.75		12.75		13.75		13.75		12.25		14.25		12.25		13.5		14.50

Figure 1. A graph of Taylor's Pre and Post- Assessment results by Dr. Masgutova for the Left-Right Motor Coordination System.

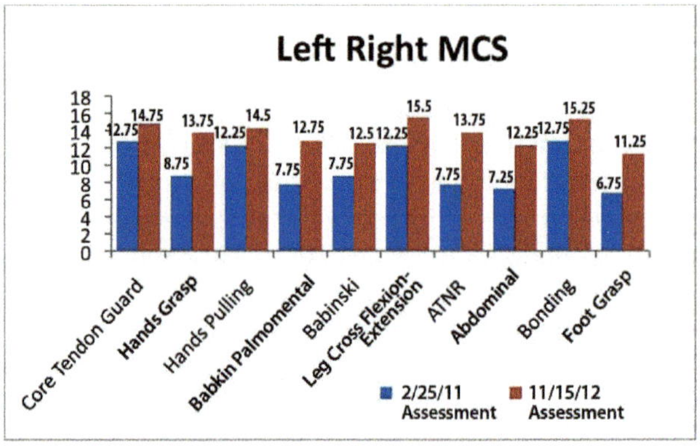

- *Tendon Guard Reflex:* a protective reaction of the whole body to physical or emotional stress; if not integrated, the entire postural system is tense. It can be assessed by looking at the body's symmetry. The Core Tendon Guard was very strongly hyperactive at Taylor's initial assessment, with ¾" of asymmetry for lower tendon guard. Testing the upper Tendon Guard was not possible at that time, as she cringed, pulled away, and sat up. This appeared to be very scary for her. Under chronic stress, the Core Tendon Guard remains active beneath our conscious awareness, and in Taylor's case, it caused her to freeze or become immobilized in an internal posture of withdrawal and search for safety.
- *Fear Paralysis Reflex:* a protective response to sensory stimulation, including tactile, visual, and auditory. Taylor's FPR was strongly hypersensitive and hyperactive, with the greatest sensitivity noted in response

Figure 2. A graph of Taylor's Pre and Post- Assessment results by Dr. Masgutova for the Superior-Inferior Motor Coordination System.

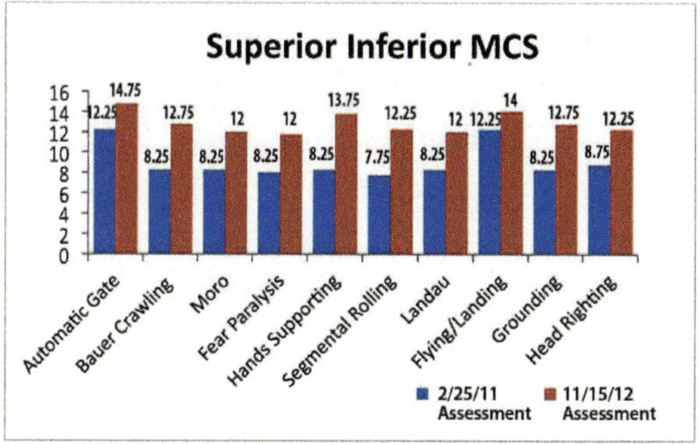

Figure 3. A graph of Taylor's Pre and Post- Assessment results by Dr. Masgutova for the Anterior/Posterior Motor Coordination System.

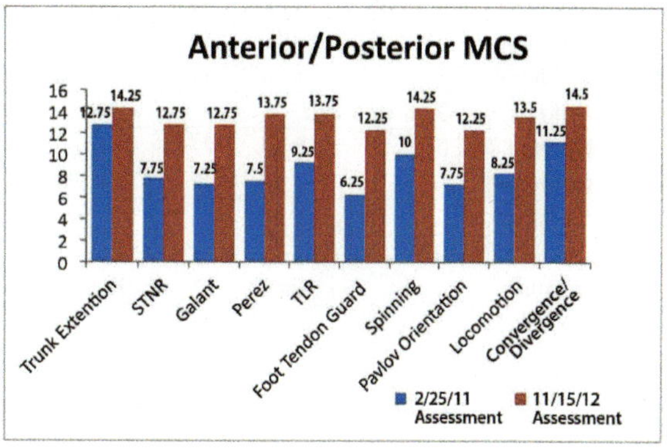

to visual stimuli.

- *Foot Tendon Guard Reflex:* hyperactive, with elements of an incorrect pattern.
- *Babinski Reflex:* hyperactive, with incorrect (opposite) pattern of response.
- *Hands Supporting Reflex:* hyposensitive, hypoactive, asymmetrical, with incorrect pattern and direction of response. Taylor remarked that "this is scary."
- *Hands Pulling Reflex:* she responded with either no engagement of the arms, or an incorrect, opposite pattern of arm extension, and significant straining of her neck muscles.
- *Robinson Hands Grasp Reflex*: hyposensitive, with use of an immature grasp pattern. She grasped objects with her thumb along the side of her closed fist, rather than the more mature, and secure, pattern of thumb across the fingers of the fist. She remarked that it was "too hard." At the same time, her functional grasp was a primitive, gross grasp on a writing implement, and grasping skills were at the 20 month level (her chronological age: 54 months).

Table 1 shows the results of Dr. Masgutova's Assessment of Taylor's subsequent reflex development and level of integration. The chart presents data from her Assessment on February 25, 2011 (one year after the start of MNRI® therapy) and compares it with data from her Assessment on November 15, 2012 (1 year, 9 months after her first Assessment with Dr. Masgutova, and 2 years, 9 months after starting therapy). See also Figures 1, 2, and 3 which show this information in graph form.

During this period of time, very significant changes are seen in the development and integration of primary reflex patterns. On the pre-test, most of Taylor's scores placed her reflex pattern development at a level of dysfunctional to mild dysfunction. On her post-test, 1 year 9 months later, Taylor's reflex development had reached a level ranging from mild dysfunction to the boundary between dysfunctional and normal function.

Standardized Assessments:

Standardized assessments have been administered on an annual basis, in February of 2010 (initial assessment), with reassessment in February of 2011, 2012, and 2013. These assessments include the following:

The Peabody Developmental Motor Scales-second edition (PDMS-2) is a standardized assessment providing information in both Gross Motor and Fine Motor domains, for children from birth to 5 years of age. The therapy provid-

Table 2, below. Peabody Developmental Motor Scale Assessment, February 2010, at age 4½ (in parentheses); re-assessment in February 2011 at age 5½ (**in bold**).

Subtest	*Raw score*	*Age equivalent*	*Percentile*	*Standard score*
1. Stationary	(40) **51**	(28 mo) **51 mo.**	(2%) **16%**	(4–Poor) **7-Bel. Avg.**
2. Locomotion	(149) **165**	(43 mo) **53 mo.**	(9%) **16%**	(6-Bel.Avg) **7-Bel. Avg**
3. Object manipulation	(31) **34**	(37 mo) **40 mo.**	(9%) **9%**	(6-Bel.Avg) **6-Bel. Avg**
4. Grasping	(42) **45**	(20 mo) **37 mo.**	(1%) **2%**	(3-Very Poor) **4-Poor**
5. Visual-Motor Integration	(131) **140**	(51 mo) **71 mo.**	(37%) **63%**	(9–Avg) **11-Avg.**
Gross Motor Composite			(2nd) **8th**	Poor / **Poor, Bord. Bel. Avg**
Fine Motor Composite			(5th) **16th**	Poor **Bel. Ave.**
Total Motor			(2nd) **8th**	Poor / **Poor, Bord. Bel. Avg**

Subtest	*Total Score*		*Scale Score*		*Standard Score*		*% Rank*		*Age Equivalent*		*Description Category*	
1. Fine Motor Precision	(26)	**30**	(11)	**12**					(5:6-5:7)	**6:3-6:5**	Average	**Average**
2. Fine Motor Integration	(31)	**33**	(14)	**14**					(6:3-6:5)	**6:9-6:11**	Average	**Average**
Fine Manual Control			Sum: (25)	**26**	(44)	**45**	(27%)	**31%**			Average	**Average**
3. Manual Dexterity	(13)	**18**	(5)	**8**					(4:4-4:5)	**5:6-5:7**	Well bel avg.	**Below Avg**
7. Upper-Limb Coordination	(1)	**13**	(2)	**7**					(4:0-4:1)	**5:6-5:7**	Well bel avg.	**Below Avg**
Manual Coordination			Sum: (7)	**15**	(26)	**33**	(1%)	**5%**			Well bel avg.	**Below Avg**
4. Bilateral Coordination	(7)	**13**	(5)	**8**					(4:0-4:1)	**5:0-5:1**	Well bel avg.	**Below Avg**
5. Balance	(21)	**33**	(8)	**15**					(4:4-4:5)	**7:3-7:5**	Below avg.	**Average**
Body Coordination			Sum: (13)	**23**	(32)	**40**	(4%)	**16%**			Below avg.	**Below Avg**
6. Running Speed & Agility	(24)	**26**	(13)	**12**					(5:10-5:11)	**6:3-6:5**	Average	**Average**
8. Strength	(8)	**11**	(8)	**7**					(4:6-4:7)	**5:0-5:1**	Below avg.	**Below Avg**
Strength & Agility			Sum: (21)	**19**	(38)	**36**	(12%)	**8%**			Below avg.	**Below Avg**
Total Motor Composite					(32)	**34**	(4%)	**6%**			Below avg.	**Below Avg**

Table 3, above. Bruininks-Oseretsky Test of Motor Proficiency – February 2012, at age 6½ (in parentheses); Taylor's scores in February 2013 at age 7½ (changes noted in bold).

ed during this time period centered almost exclusively on MNRI® with some added activity (through play) with her vestibular system. Scores for her initial assessment (shown in Table 2) in February 2010, at age 4½ (54 months), are in parentheses. Scores for her re-assessment in February 2011, at age 5½ (66 months), are in bold type.

The Bruininks-Oseretsky Test of Motor Proficiency, second edition (BOT-2) is a standardized measure of motor skills in individuals aged 4 through 21 (see Table 3). More specifically, it highlights motor performance in the broad functional areas of stability, mobility, strength, coordination, and object manipulation. There are four motor area composites, each comprising two of the eight BOT-2 subtests, and a Total Motor Composite, which comprises all four composites and provides a measure of overall motor proficiency. Scores from February 2012, at age 6½ years, are in parentheses; scores from February 2013, at age 7½ years, are in bold type.

The Sensory Profile is a questionnaire filled out by caregivers identifying behaviors related to how the child processes sensory information and connects these behaviors to the child's functioning in daily living skills. This questionnaire compares the child's behaviors to those of a sample of over 1,000 typical children. Results are classified as falling into three categories: Typical Performance (at or above 1 SD below the mean), Probable Difference

Table 4, below. Taylor's Sensory Profile Questionnaire; initial assessment, February 2010; re-assessment in February 2013 (changes noted in bold).

2010 *Typical Performance*	*Probable Difference*	*Definite Difference*
Visual Processing Oral-Sensory Processing Items Indicating Thresholds for Response	Auditory Processing Modulation Related to Body Position and Movement Modulation of Movement Affecting Activity Level Modulation of Visual Input Affecting Emotional Responses and Activity	Vestibular Processing Touch Processing Multi-sensory Processing Sensory Processing Related To Endurance/Tone Modulation of Sensory Input Affecting Emotional Responses Emotional/Social Responses Behavioral Outcomes of Sensory Processing
2013		
Auditory Processing Visual Processing Oral-Sensory Processing **Modulation Related to Body Position/Movement** Items Indicating Thresholds For Response	**Touch Processing** Modulation of Movement Affecting Activity Level Modulation of Visual Input Affecting Emotional Responses and Activity **Behavioral Outcomes of Sensory Processing**	Vestibular Processing Multi-Sensory Processing Sensory Processing Related To Endurance/Tone Modulation of Sensory Input Affecting Emotional Responses Emotional/Social Responses

Table 5, below. Developmental Test of Visual Motor Integration, February 2011, 2012 (in parentheses); re-assessment in February 2013 (changes noted in bold).

	VMI		**Visual**		**Motor**	
Raw Score:	(13,15)	**17**	(16, 23)	**23**	(12, 14)	**20**
Standard Score:	(89, 85)	**87**	(95, 123)	**110**	(77, 80)	**95**
Percentile:	(23,16%)	**19%**	(37, 94%)	**75%**	(6, 9%)	**37%**
Age Equivalence:	(4-10, 5-6)	**6-3**	(5-0, 8-6)	**8-6**	(3-9, 4-6)	**6-11**

Table 6, below. Standardized testing by NWEA (Northwest Evaluation Association) provided by Taylor's school district.

Mathematics			***Reading***		
	Taylor's score	*Percentile*		*Taylor's score*	*Percentile*
Fall 2012	166	59%	Fall 2012	158	43%
Spring 2013	204	98%	Spring 2013	187	75%

(at or above 2 SD below the mean), and Definite Difference (more than 2 SD below the mean). (See Table 4).

The Developmental Test of Visual Motor Integration (VMI) is a standardized test using a developmental sequence of 24 geometric forms which the child copies with paper and pencil. Supplemental Developmental Tests of Visual Perception and Motor Coordination were also administered. The scores from February 2011 and 2012 are in parentheses. The scores from February 2013 are in bold type: (2011, 2012) **2013** (see Table 5).

Standardized testing by NWEA (Northwest Evaluation Association) provided by Taylor's school district also shows solid gains (see Table 6).

Discussion

Motor Skill Development after one year of therapy:

Within one year of the start of MNRI® therapy, Taylor showed significant improvement in all areas of motor skill development, as assessed by the PDMS-2. Balance skills improved by nearly 2 years, from the 2nd to the 16th percentile, and from poor to below average. Locomotion improved by 10 months, from the 9th to the 16th percentile. Grasping improved by 17 months, from the 1st to the 2nd percentile, and from the very poor to the poor range. Visual-motor integration skills improved by 20 months, from the 37th to the 63rd percentile. A 3 month gain was seen in the development of object manipulation (ball) skills. Overall gross motor skills improved from the 2nd to the 8th percentile, as did her 'total motor' skills. Although still in the 'poor' range, they were at the border of 'below average'. Fine motor skills improved from the 5th to the 16th percentile, and from 'poor' to 'below average' when compared with same age peers.

After two years of therapy:

Due to Taylor's age, a new assessment that was more appropriate for older children was used. Because of this, direct comparison could not be easily made between 2011 and 2012.

After 3 years of therapy:

Very nice improvement was seen in Taylor's motor skill development. As assessed by the BOT-2, her skills were now in the average to below average range. Over the course of one year, between February 2012 and February 2013, she improved on all subtests, with gains ranging from 3 months to 3 years of age equivalence. She was now functioning in the average range for Fine Manual Control, Balance, and Running Speed and Agility. This demonstrates significant progress from the start of therapy 3 years earlier, when all of her gross motor and fine motor scores (other than Visual Motor Integration) were below average, poor, or very poor. The improvement in Taylor's balance has been phenomenal, with more than 2 years of growth during her first year of MNRI® therapy, followed by 3 years growth in the past year. This represents 5 years of growth over a period of 3 years. This corresponds with the very significant improvement seen in grounding, stability, vestibular processing, and coordination between her core and limbs.

Taylor's VMI scores indicate gains of 3.5 years, and from the 37th to the 75th percentile, in the area of Visual Perception between February 2011 and February 2013. Over the same period, she gained over 3 years (from the 6th to the 37th percentile) in the area of motor coordination and 17 months in the area of Visual Motor Integration. These gains coincide with the maturation of upper limb reflexes and the development of a more mature grasp pattern.

Sensory Integration: Very nice gains were seen in the area of sensory processing. Since starting MNRI® therapy 3 years ago, Taylor achieved measurable improvement in her ability to process tactile input: from definite difference to probable difference. At the same time, her auditory processing moved from probable difference to typical performance. This indicates an improved ability to use these sensory inputs adequately. Although still in the definite difference range, very significant improvement is seen in her processing of vestibular input. By the end of the first year of therapy, she had become much more comfortable with vestibular input, and loved to swing prone in the hammock swing, asking for "higher" and "faster." She demonstrated improved registration of sensory input, and inattention/distractibility was now in the range of typical performance.

Social & Emotional Skills: Taylor demonstrates very strong improvement in this area, to the point where school staff no longer believe that she has a diagnosis of Selective Mutism. She no longer freezes in social situations or in response to transitions. She is able to talk, answer questions, and participate at school. Despite these improvements, she continues to need work on issues of anxiety, the art of conversation, and development of matured social skills. Although she continues to become rather easily overwhelmed, she no longer has frequent or sustained temper tantrums. Anxiety no longer causes a protective freezing response, but instead is manifested in frequent repetition of questions such as "What's for lunch?" and "Are my boots in my bag?" Instead of freezing to the point of becoming immobile and shutting down, she is now dealing with her anxiety verbally and by preparing herself physically and emotionally to deal with it. This is a more positive level of protection, one that allows her to grow, learn, and adapt.

Self Care Skills: As sensory processing improved and anxiety decreased, Taylor began to develop greater self-care independence. By the end of her first year of MNRI® therapy, and without addressing these skills directly, she went from almost total dependence for dressing skills, to being able to dress and undress herself almost independently (other than fasteners), was toilet trained, and managed all toileting tasks other than flushing the toilet. These skills continue to strengthen. She is now able to dress herself completely except for snaps and some buttons, and is fully independent with toileting skills.

Cognitive Skills: A standardized assessment provided by the school district at the beginning and the end of 1st grade indicated very significant improvement, with reading improving from the 43rd to the 75th percentile, and mathematics improving from the 59th to the 98th percentile. Perhaps some of these gains can be attributed to Taylor's decreased anxiety and improved neurophysiological maturation, allowing her to more easily access her innate skills.

Summary

Taylor is a 7½ year old little girl with a diagnosis of Selective Mutism, who has not only overcome her diagnosis, but has also made significant and wonderful gains in motor coordination, self-care independence, release of anxiety, and development of a sense of inner peace. Play therapy, which was provided prior to the start of the MNRI® therapy, did not bring about changes that were stable or that generalized to other situations in her life. The reflex development and integration provided by MNRI® therapy has produced steady and stable results which have generalized to other areas of her life, including home, school, and community. During this time, MNRI® techniques have been used almost exclusively, other than some directed vestibular play. Motor skills, self-care skills, anxiety, cognition, and behavior have not been directly addressed in therapy through traditional means, and yet significant gains are seen in all of these areas.

Taylor's strongly hyperactive Core Tendon Guard response at the start of therapy indicated a strong, protective reaction to informational stress. This information was interpreted by her Reticular Activating System (RAS) in the brain stem as 'not safe', and elicited a negative protective response – in her case, it was a freezing response, both of movement and speech. Reflex responses that should bring positive protection, such as Hands Supporting, Hands Grasp, and Hands Pulling, were very hypoactive, and were actually scary for her. Fear Paralysis, Moro, and lower limb reflexes were hyperactive. She demonstrated a very narrow visual field, as her tendency to freeze when under stress resulted in strong visual convergence and difficulty accessing her peripheral vision. As these reflexes were addressed and appropriate nerve pathways developed and myelinated, her responses moved and continue to move in the direction of positive protection, giving her a greater feeling of safety and inner peace. At the same time, functional changes are evident in Taylor's emotional development and control, speech, and social interaction. Physical changes are also apparent, as relaxation of tendons and muscles now allow her to sit cross-legged, her grasp has improved from an immature gross grasp (use of fist to grasp a pencil or marker) to a grasp that varies between cross-thumb and tripod – and she has become a bit of an artist, with a love of drawing blue macaws! The developmental progress that Taylor has experienced over the past 3 years is both wonderful and significant. Despite this, Taylor continues to have some challenges, both physically and emotionally. Work with reflex development and integration will continue, in order to give her the neurophysiological support that will provide safety and inner peace, and that will support her in her quest to reach her full potential.

One of Taylor's blue macaws.

Taylor's progress using the Masgutova Method® indicates that this can be an effective tool in the remediation of Selective Mutism. MNRI® techniques are meant to be interpreted by the RAS as safe. These techniques bring safety, and with that, the opportunity to change old patterns. When inner peace is reached on a physiological level bringing feelings of security and safety, then the door is opened for physical development, cognition, emotional maturation, and curiosity. This does not happen at the cortical level. A child like Taylor cannot just be told to relax, or to talk. This happens through work with subcortical organization, developing and maturing the nerve net system and nerve pathways for greater ease of learning and interacting with the world around her.

References

Masgutova, S.: information drawn from courses and manuals.

Selective Mutism Group/Child Anxiety Network: SMG, part of the Child Anxiety Network, is a non-profit organization dedicated to providing information, resources, and support to those impacted by a child with the anxiety disorder known as Selective Mutism (SM). (www.selectivemutism.org/)

Diagnostic and statistical manual of mental disorders, 4th Edition, Text Revision (DSM-IV-TR).

I would like to thank Taylor and her amazing family for the opportunity to be a part of their lives and their journey. I have had so much fun working with Taylor, and watching her personality blossom as her anxiety melted away. She is a bright and beautiful girl with a world of potential. (And I will always have a soft spot in my heart for blue macaws!). – Jo Anne Tierney

We Proudly Present Mia Corinne Terrell: Life Start with MNRI®

Abi Terrell, Parent; Rachel Shimer-Lal, MOT, OTR/L; and
Diane Whiteside, PT, MNRI® Core Specialists, San-Ramon, CA, USA

Background History

We had just reached 20 weeks in our pregnancy when we learned that our baby wasn't growing at a normal rate. Previously, I had an ultrasound at 9 weeks which read as completely normal. We had wanted to have a minimally invasive pregnancy and deliver Mia at a birthing center in a bathtub. The news we received that fateful November day changed not only the birthing plan, but our lives, too! We didn't want to consider terminating the pregnancy, even though doctors informed us of the high probability for a very sick baby if she survived at all. But this was our love child. We were high school sweethearts who had just been reunited after over 20 years of being apart. We were in a hurry to start our lives together and this was the baby of our dreams. We knew that Mia would need to be delivered early and finish developing in an incubator but we had no idea how scary the journey would be.

Abi Terrell

Rachel Shimer-Lal

Diane Whiteside

Each week, the ultrasounds showed grim results. She was hardly growing. The umbilical artery, which carries nutrients to her, was showing signs of slowing and reversing the flow. This evidence caused our doctors significant stress, which, in turn, caused us stress. We didn't know what any of it meant really, and so we looked to the doctors with complete trust. This pregnancy was no longer in our hands. All I could do to help my baby was to sing and chant to her, "I'm in love, love is in me, we are well," (a mantra I had been given) and imagined a bubble of safety surrounding us. On January 18, 2011, I was on the beach playing with my dog when I got the call from our doctor to check in to the hospital. The rest of my pregnancy would be spent hooked up to monitors in the hospital. The nurses constantly adjusted two monitors, one for Mia's heartbeat, and another measuring contractions. Every so often, her heartbeat would drop off the monitor and they would run in to my room to readjust the monitor and find it quickly. The daily ultrasounds continued showing signs of the umbilical artery stopping and reversing flow.

After a week, the decision was made to schedule Mia's delivery. They pumped me with steroids to speed up her lung development and gave me two magnesium treatments (recently found to preserve the brain). The treatment, fully worth the agony, made me feel like I was dying of the worst flu, and burned as it passed through the IV. I was more miserable than I had ever felt but didn't care how much pain I had to go through to help my baby. You really don't know how strong you are until you have to be and you just are! There's no effort, it's pure survival mode – survival of your baby. A couple of hours before my scheduled C-section, I went into preterm labor. Mia's heartbeat was lost again, and the doctor switched the plan to an emergency C-section. As they rushed me into surgery, a more seasoned doctor entered the operating room and calmed everyone down. I was so relieved when he calmly found Mia's heartbeat and switched the plan back to a non-emergency C-section.

Mia was blue when they delivered her and as tiny as a Barbie doll. I couldn't believe she was real. She breathed on her own from the start, but was placed on a respirator anyway, due to protocol. Less than 24 hours later, Mia was placed on a C-Pap. The doctors were amazed. Later, when the doctor conducted the pathology on the placenta, I was told that it was the smallest placenta he had ever seen in all his years, and that the reasons for our troubled pregnancy were unknown. Mia's survival is a true miracle. Many children that are born under these circumstances and survive, have lasting difficulties, such as poor eyesight, cognitive and other developmental delays, asthma and breathing problems, and sometimes cerebral palsy.

Our Miracle Mia is healthy, happy and developing like a normal child. How was this possible, I asked myself. How is this possible that she is so perfect? I know the answer to this question. I know that it's a combination of the love and the intervention she has received. From the start of her life, Mia attracted admirers from all over the world. We uploaded her picture on Facebook and it quickly spread. Over the months, many people watched her progress, sending encouraging words and support.

When Mia was 12 days old, Rachel, my best friend, and Dr. Svetlana Masgutova, came to see her in the NICU. During this visit we witnessed the most profound and amazing miracle. I could see that Dr. Masgutova's subtle touch was having an instant effect on Mia. She explained that the techniques would help Mia's underdeveloped nervous system to become activated and set up to develop normally. We truly believe that Mia's perfect development started that magical day in the NICU. As Dr. Masgutova gently worked on Mia, whose body was smaller than the doctor's hand, Mia softened into her touch. Mia's face softened too, and I was amazed to see my feisty little baby look relaxed for the first time. Her tight little muscles let go for a moment, letting go of the fight or flight survival mode and trusting completely what was being done to her. Up until that day, I had felt so helpless and at the mercy of the doctors. While we were grateful for the medical interventions that kept Mia alive, we also knew that some of those interventions came with lasting side effects that would later cause trouble, such as the C-Pap causing Retinopathy of Prematurity (ROP). Mia was surviving but not thriving at this point, and we wondered how healthy she would ever be.

The MNRI® intervention addressed Mia's long-term development. Dr. Masgutova gave us private training on the MNRI® techniques that she used with Mia on that day at the hospital. We did not understand how these techniques worked and it took us some time to learn them but we saw instant and consistent results with the MNRI® techniques. When Mia came home from the hospital, the MNRI® specialists came to visit again. This time, Mia received additional intervention to help her develop a deeper suck. She had learned to breastfeed while in the NICU, but her suck was shallow and weak. She fatigued quickly and couldn't receive the nutrients she needed before becoming too exhausted. Andrea, one of the MNRI® specialists, taught me how to do the MNRI® Suck Reflex technique and within 4 months, we were breastfeeding almost exclusively. Over the next 2 years, Mia saw the specialist team periodically. Every time Mia saw Dr. Masgutova, we noted a significant growth and/or developmental spurt in Mia over the next few days. My husband and I even joked that

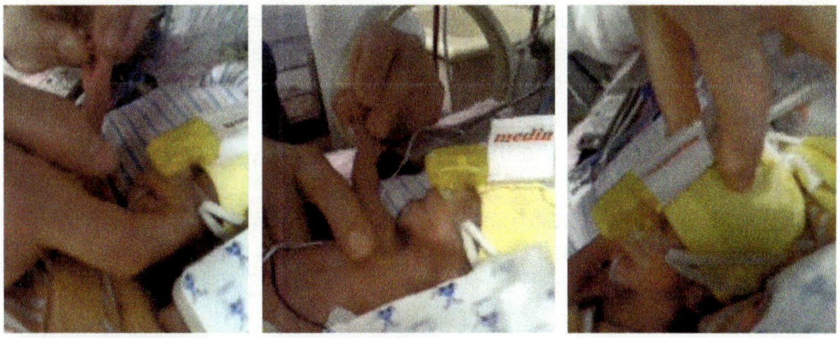

Mia at 12 days old in the Neonatal Intensive Care Unit receiving MNRI® from Dr. Masgutova: (left:) Oral-Facial Reflex Integration, repatterning of the Babkin Palmomental Reflex pattern; (center:) Archetype Movement Integration: Sequential Arm Opening for stress release; (right:) Oral-Facial Reflex integration: work with the Crowning Reflex pattern for release of birth stress.

we wanted Svetlana to keep Mia for a couple days afterward so that she could deal with the difficult behavior the intervention would bring on! After the difficult days passed, though, we saw a huge and positive change in Mia. We observed that the MNRI® techniques helped her muscles to loosen, helping her to move easier and rotate her body. We also saw her better able to regulate her moods and relax. Over the past three years, we have continued to do MNRI® techniques at home, and have follow-ups with Rachel and Dr. Svetlana. Everywhere we go, Mia amazes and charms. Although she remains tiny, she is a picture of perfect health and the gap between her and her peers is narrowing. She is smart, focused, inquisitive, emotionally intelligent, empathetic, a problem solver, a performer, can count to 20, speaks clearly and well in two languages, and even reads her name! She has perfect eyesight, no lung or breathing issues, no significant delays in her development, and can keep up with her full-term peers in every category. We know that the intervention done by the MNRI® specialist team has helped Mia to avoid those common problems micro-preemies experience, and we are forever grateful.

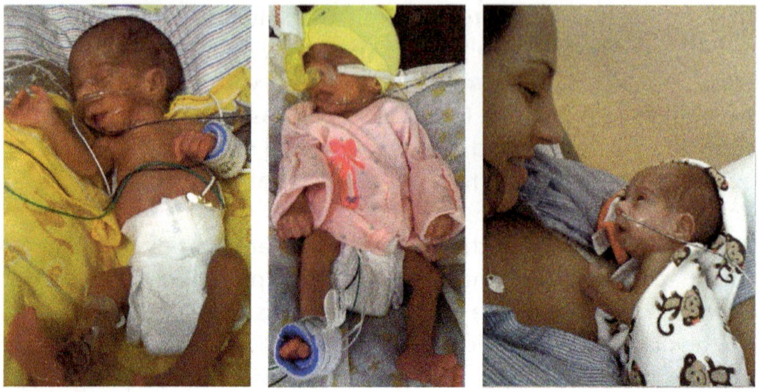

Mia at 12 days old still in the Neonatal Intensive Care Unit.

MNRI® Intervention

Per Diane and Rachel: As mentioned above, MNRI® was started for Mia at the Neonatal Intensive Care Unit at the University of California at San Diego Hospital, 12 days after her birth. Mia's parents invited us to work with Mia and were waiting for us with great hope. Mia was in deep survival, with no promises for it. We all were inspired to help Mia thrive and continue her successful neurodevelopment. She was so small she could fit in the palm of your hand. She did not have an efficient sucking pattern and was being fed via an NG tube down her throat then, later, in her nose. Mia lacked a clear voice, distinct facial features, and eye movements and the pacifier was too big for her very small mouth. Mia demonstrated unstable breathing and heart rate, and protested when nurses conducted medical procedures. However, she responded amazingly well to MNRI® Reflex Repatterning and Neuro-Structural Reflex Integration Programs.

Medical personnel warned us that Mia would only be able to tolerate our intervention for a maximum of 10 minutes. However, Mia's session went on for almost two hours, with short interruptions by doctors and nurses who conducted routine medical procedures. She immediately displayed an amazing desire to be touched and responded with comfort and inner peace to our tactile techniques. She was expressing this desire even with her voice, cooing with a vibrating tone of her voice. The MNRI® Core Specialist team also visited Mia the next day and it was noted that her chances for survival and healthy development were already improving. Mia's strength and determination inspired and touched her nurses and doctors, and they were amazed that she was doing so well despite the fact that she was so tiny, and not growing very quickly. The MNRI® Specialist Team visited Mia at home shortly after discharge, at almost four months old, and was amazed at the improvements they observed in her development. In less than a year, Mia was demonstrating extraordinary skills of gross and fine motor coordination, emotional and social development, and a great inner strength and determination.

Date of Assessment	Mia's Reflex Patterns Group 1 - Sagittal plane of body and movements). Scores (norm 16-17.75 points)									
	Core Tendon Guard	Hands Grasp	Hands Pulling	Babkin	Babinski	Leg Cross Flexion-Extension	ATNR	Abdominal Sleep Posture	Bonding	Foot Grasp
Feb. 6-th, 2011 12 days old	8.5	6.75	5.75	3.75	6.5	6.5	9.25	6.25	8	7
	Group 2- Horizontal plane of body and movements									
	Autom. Gait	Crawling	Moro	Fear Paralysis	Hands Support.	Segment. Rolling	Landau	Flying/ Landing	Grounding	Head Righting
	4.25	4.25	8.5	8.5	7.25	3.25	7.75	3.25	4.25	3.75
	Group 3- Dorsal plane of body and movements									
	Trunk Extent.	STNR	Galant	Perez	TLR	Foot Tendon Guard	Spinning	Locomotion	Balancing	Pavlov
	4.5	6.75	6.75	6.25	8	3.25	3.25	3.25	3.25	7.25

MNRI® Reflex Pre-Assessment

On February 6, 2011, at 12 days old, when Dr. Masgutova visited Mia at the hospital NICU, she administered an MNRI® Reflex Assessment of Mia. The results of this Assessment are presented on the previous page.

Mia's scores for the reflexes Assessed at that time indicate pathology for 8 schemes according to A. Krefft: (criteria: 2-3.75 points), deep dysfunction for 4 reflex patterns (criteria: 4-5.75), average dysfunction for 11 schemes (criteria: 6-7.75), light dysfunction for 6 (criteria: 8-9.75). (See the article about *Assessments* in this book for further information.) There were no reflexes at a functional (low) or normal level. The results from this Assessment show that Mia's sensory-motor system and its neurological circuits were not well developed and matured. These results also show a very high level of stress which actually was seen in her behavior at the hospital NICU. It was evident that Mia was suffering when any even simple medical procedure was administered to her. During this Assessment, several overall features were noted in Mia including a hyperextension reaction and hyperactive extensor tone when her head was turned to the right. Mia also demonstrated significant asymmetry in body flexion to the left. Her reflex Assessment conducted by Dr. Masgutova was recorded on video and presents unique material from which others can learn. The MNRI® Assessment allowed Dr. Masgutova to create an individual program for Mia's parents and her therapists for further work.

Other Pre-Assessments

Peabody Developmental Motor Scales Test at 2 months

MOTOR SKILLS	Raw Score	Standard Score	Percentile	Age Equivalent
Reflexes	4	12	95th	4 months
Stationary	8	8	25th	1 month
Locomotion	4	8	25th	1 month
Object manipulation	0	N/A	N/A	N/A
Grasping	2	6	9th	1 month
Visual Motor Integration	1	5	6th	1 month

	Gross motor	Fine Motor	Total
Sum of Standard Scores	28	11	39
Developmental Quotient	96	73	85
Percentile Rank	39	3	16

Peabody Developmental Motor Scales were assessed on June 3, 2011 when Mia was chronologically 5 months old, with an adjusted age of 2 months old.

Raw scores are the total points accumulated by Mia on a subtest. Each subtest has different possible raw score totals, therefore the raw scores are not comparable to each other, nor do they provide any information regarding motor performance. It is the standard score that provides a clearer indication of Mia's subtest per- formance, and these scores can be related to one another. Standard scores are based on the distribution of a mean of 10 and standard deviation of 3. Standard scores of 6-7 are considered below average, 4-5 poor and 1-3 very poor performance. The percentile rank represents the percentage of the distribution that is equal to or below a particular score. For example, Mia scored in the 25% rank for stationary gross motor skills, meaning that 25% of the standardization sample scored at or below her score. The age equivalent, also referred to as 'motor ages', are used only to communicate a child's competence in a universal language, and are not always the best measure of a child's performance. The developmental quotient provides information regarding performance of overall fine motor, gross motor, and total motor skills. Quotient scores are based on a distribution of a mean of 100 and a standard deviation of 15. Quotients ranging from 90-110 are average. Scores of 80-89 are below average, 70-79 are poor and 35-69 are considered very poor performance. Mia was in the Below Average range.

Performance Observations/General Descriptions

To help understand areas of assessment, each subtest is explained below as described in the Peabody Developmental Motor Scales examiner's manual.

Reflexes measure the child's ability to automatically react to his/her environment. Stationary measures the child's ability to sustain control of his or her body within its center of gravity and retain equilibrium. Locomotion measures the child's ability to transport his or her body from one base of support to another. Object Manipulation measures the child's ability to throw, catch, and kick balls. Grasping measures the child's ability to use his or her hands and fingers. Visual-motor Integration measures the child's ability to integrate and use his or her visual perceptual skills to perform complex eye-hand coordination tasks.

Neuromuscular Status

Mia at 3 months old.

Mia's neuromuscular status was assessed on June 3, 2011 and the results of this evaluation are shown below.

Muscle Tone: Mia had a combination of low tone in her flexor muscles and high tone in her extensor muscles. She demonstrated higher tone in her upper extremities as compared to her lower extremities. She had poor head control with low muscle tone of neck and trunk flexors. While being pulled to sit up, Mia demonstrated neck extension and spine extension with extreme flexion of her upper extremities.

Mia's occiput was extremely tight on the left side. She was in a constant Stage 2 Suck with a forward swallow so her sucking was extremely inefficient. Mia used her jaw and buccinator muscles to swallow. Mia demonstrated positive protection on her upper and lower lips and positive Rooting Reflex bilaterally. Mia's muscles were tight, with limited elasticity.

Strength: She generally had poor neck and trunk and extremity strength.

Endurance: Mia had poor endurance and fatigued with minimal activity such as nursing, interacting with Mom, or in any antigravity position.

ROM/Flexibility: She had the ability for trunk and extremities extension except for her upper extremities where she had excessive tightness of flexors.

Righting reactions: Mia had minimal head righting reactions.

Movement Patterns: She demonstrated fair to poor quality of movement. Her movements were influenced by poorly developed reflexes and tone. Her movements were jerky, with poor muscle grading. Her hands were in flexion 80% of the time. She had cortical thumbing, greater on the left than the right hand. Her grasp was extremely forceful/inefficient in the distal aspect of her fingers, with difficulty getting therapist's fingers in her palm to assess grasp secondary to fisting bilaterally. Mia did have symmetrical movements of her hands and legs.

Visual Motor Skills: While lying supine, Mia was able to horizontally track a rattle to the right approximately 45 degrees and to the left 30 degrees. She was able to vertically track a rattle upward 20 degrees. All eye movements were challenging as she fatigued quickly.

Sensory: Mia was hyper sensitive to tactile stimuli, greater on her feet and face. She did not tolerate the C-pap device on her face. She was hyper sensitive and hyper active on her trunk and abdomen. She was desensitized to the auditory stimuli in the NICU. She was able to tolerate high levels of noise.

Vestibular: Mia enjoyed gentle rocking but hated riding in the car.

Proprioceptive: Mia enjoyed being swaddled.

Motor Skills: While lying prone, Mia lifted her head 25 degrees and would rotate her head to both sides. She had increased flexor tone in legs and neck. She kept her arms extended and abducted with fisted hands. She had minimal movements of her legs in prone. While lying supine, she was able to flex and extend lower extremities. She would kick her legs in that position both at the same time. Her arms were held in extension and abducted with bilateral hands fisted. She was able to keep her head in midline. In side lying Mia would flex her legs and her hands would come together. She did not tolerate side lying well except with support. While supported in a sitting position, she was able to hold her back in a rounded position for 3 seconds.

MNRI® Reflex Post-Assessment Results

Mia's scores for reflexes at the second Assessment showed significant positive changes in reflex scores: no pathological reflexes were evident (versus 8 reflexes at the beginning), deep dysfunction was still present for 10, average dysfunction for 3 schemes, light dysfunction for 4, on the boundary of functional and dysfunction for 3, and functional on a very low and low level of development for 7 and 2 schemes. These results show that Mia's sensory-motor system and neurological circuits were progressing very rapidly and evidence that the level of stress of her prematurity was major. Her behavior, communication, and perception were very close to a neurotypical infant, though her gross motor coordination was still significantly delayed and demanded the special attention of therapists. The reflex Assessment also indicated an increase of tone in her hands (though precision was very high); hypersensitive and hyperactive reaction in her feet bilaterally; and also in Spinal Galant and Perez Reflexes.

Spinal Galant: lower quadrant more hyperactive bilaterally, but more hypersensitive on left than right.

Cortical thumbing: greater on the left than the right, grasp was extremely forceful/inefficient in the distal aspect of her fingers with difficulty getting therapist's fingers in her palm to assess grasp secondary to fisting bilaterally.

Hands Pulling: opposite pattern was present, neck into extension and whole body went into extension with very tight flexion in her upper limbs.

Date of Assessment	Reflex Patterns Group 1 - Sagittal plane of body and movements). Scores (norm 16-17.75 points)									
	Core Tendon Guard	Hands Grasp	Hands Pulling	Babkin	Babinski	Leg Cross Flexion-Extension	ATNR	Abdominal Sleep Posture	Bonding	Foot Grasp
02/06/ 2011	8.5	6.75	5.75	3.75	6.5	6.5	9.25	6.25	8	7
06/03/2011	12.5	12.25	13.25	7.25	8.5	12.25	12.25	14	14	8.5
09/04/2011	14.25	15.25	15.5	11.25	12.25	10.5	14.25	13.25	14.25	12.5
02/25/2012	14.5	15.75	15.5	12.25	12.75	11	14.75	13.5	15.25	12.5
06/03/2012	15	16	16	15	13.75	13.75	15.5	14.5	16	13.75
	Group 2- Horizontal plane of body and movements									
	Autom. Gait	Crawling	Moro	Fear Paralysis	Hands Support.	Segment. Rolling	Landau	Flying/ Landing	Grounding	Head Righting
02/06/ 2011	4.25	4.25	8.5	8.5	7.25	3.25	7.75	3.25	4.25	3.75
06/03/2011	7.25	5.5	10	10	13.25	4.75	5.25	4.25	8.5	4.75
09/04/2011	9.25	12	13.75	13.75	14.25	9.25	12.25	9.25	9.75	9.5
02/25/2012	11.25	13.25	14.25	14.75	15	9.25	12.75	9.25	10.25	12.25
06/03/2012	14	14.5	15	15	15.5	11	13.5	12.75	12.5	14.5
	Group 3- Dorsal plane of body and movements									
	Trunk Extent.	STNR	Galant	Perez	TLR	Foot Tendon Guard	Spinning	Locomotion	Balancing	Pavlov
02/06/ 2011	4.5	6.75	6.75	6.25	8	3.25	3.25	3.25	3.25	7.25
06/03/2011	5.75	12.25	7.75	8.5	11.25	4.75	4.25	4.25	5.25	12.25
09/04/2011	12.5	13.75	12.25	12.25	12.75	12.25	8.75	8.75	6.25	14.25
02/25/2012	13.75	14.25	13	13	13.25	12.75	9	9.75	9.25	14.75
06/03/2012	14.75	15.5	14	14	14.5	14	12.5	12.5	12.75	17.5

ATNR: when head was turned to the right she had appropriate pattern but with a hyper active extension of upper and lower extremities. When head was turned to the left the pattern displayed inconsistent left and right leg flexion present, her left arm was in extension and her right upper went in and out of extension and was not as strong a pattern to the left as it was to the right. Mia demonstrated with a hyperextension reaction and hyperactive extensor tone when head turned to the right.

Babinski: went into a Foot Tendon Guard pattern.

Foot Tendon Guard: correct pattern bilaterally.

Toe Grasp: grasp on the right and more extension on the left in the pathological range.

Leg Cross Flexion Extension: positive reaction, although very hyperactive.

Trunk Extension: positive extension but seemed to block around the hips and did not flow all the way up to her head.

Automatic Gait: appropriate response with alternate stepping.

Bauer Crawling: positive reaction using Babinski with both feet when she was pushing forward.

Flying and Landing: opposite pattern, she pushed down or went for extension when lifted through her upper extremities.

Hands Supporting: opposite pattern, when facilitated into compression she initiated a little extension but her elbows were flexed and they went into abduction more than a horizontal 90 degree direction.

Hands Grasp: she was able to close her fingers in a tight grasp around the examiner's finger. She presented with cortical thumbing bilaterally. She had difficulty opening her hand in order to reach for an object or release an object. However, while breast feeding she demonstrated the ability to open her hand to reach for her Mom's breast. While asleep there was decreased tightness of the fisting.

Based on results of the Assessments, the following MNRI® programs were recommended: Neuro-Structural Reflex Integration, Repatterning, Visual-Auditory and Oral-Facial once a week for every program. Her parents were given a MNRI® Home Program consisting of Repatterning exercises during work days (5 times a week).

Peabody Developmental Motor Scales at 32 months

MOTOR SKILLS	Raw Score	Standard Score	Percentile	Age Equivalent
Reflexes	-	-	-	-
Stationary	40	10	50%	28 months
Locomotion	102	6	9%	22 months
Object manipulation	16	6	9%	22 months
Grasping	44	11	63%	34 months
Visual Motor Integration	98	8	25%	26 months

	Gross motor	Fine Motor	Total
Sum of Standard Scores	22	19	41
Developmental Quotient	83	97	88
Percentile Rank	13	42	21

Post Assessments

Mia at 3 years, 3 months and her mother, Abi.

Peabody Developmental Motor Scales was re-assessed on August 1, 2013, when Mia was 32 months old. Her neuromuscular evaluation is shown on the chart below.

Muscle tone: She had improved in her muscle tone but still generally had increased extensor tone. She had greater tone in her upper extremities than her lower extremities. She demonstrated greater strength in her upper extremities than her lower extremities. She was able to hang from a bar holding her body weight for 8 to 10 seconds.

Reflexes: Mia did not like her head tilted back or lying on her back. This elicited a hyperactive Moro Reflex and Fear Paralysis making it difficult to change her diaper as well as float on her back in the swimming pool. Mia demonstrated a good pattern for Hands Pulling Reflex. She spent some time on her stomach so she demonstrated good prone extension strength and was able to maintain Landau Reflex for 30 seconds.

Motor Development: Mia had good motor planning development. She was able to imitate three different positions. She could walk down stairs holding a hand two feet per step. She could walk backwards and run forward. She could walk sideways both directions. She could walk on a line 6 feet. She had difficulty in gross motor with continuous rolling, jumping with feet, and walking on her tip toes. She demonstrated a static tripod grasp with her right hand while scribbling. She could insert 3 shapes into the proper holes and she could screw on a lid. She had mild difficulty with fine motor skills such as buttons, stringing beads, and building an eight block tower. She had good balance in kneeling and standing, poor balance with single leg stance, and was not able to stand on her toes.

Quality of movement: She had very quick movements and we were unable to grade her movements. She had difficulty performing fine motor skills like unbuttoning large buttons on her shirt.

Visual motor: She continued to be hyper-vigilant with her eye tracking. She was in constant movement searching the horizon. Her system was in constant protection but continued to improve with MNRI®.

MNRI® Post Assessment results (09/04/2011 – third assessment; 02/25/2012 – fourth assessment; 06/03/2012 – fifth) demonstrated steady improvement of every reflex pattern in Mia.

Summary

Mia with Dr. Masgutova.

The use of the MNRI® programs supported Mia's survival and neuro-development and was highly beneficial for Mia as an infant born prematurely. She received reflex repatterning and the neurostructural program in the NICU at 12 days of life. This work has made a significant impact on her motor development as displayed in the assessment of the Peabody Developmental Motor Scales. This is the basis for all of her motor development. Mia made significant gains in the grasping and visual motor subtests, indicating above age level in fine motor skills. She has managed to escape many of the problems that preemies typically face, such as poor eyesight, lung-related illnesses, allergies, and cognitive and developmental delays.

Dr. Masgutova and the MNRI® Specialists are so pleased and happy for Mia. Her parents are full of hope and continue to practice the MNRI® techniques at home. Mia has an annual visit to her specialist team. At four years old and weighing 22 pounds, she shows perfect cognitive development, minimal delays in gross motor skills, (mostly due to size-related challenges), and is in good overall health.

Mia is truly a miracle baby, a small and mighty warrior who is so dear to all of us. She is a reason for celebration for her amazing parents, Abi and John. These parents did all possible for Mia to become a True Winner! – MNRI® Team

I Have a Mattress in My Room for You

Anna Krasowska, Special Education Pathology Specialist, MNRI® Core Specialist, Instructor, Rogow, Poland

Anna Krasowska

In this article I will share a story of my work with a child with cerebral palsy using the MNRI® program.

The parents with their 2 year old daughter, Zuzia, visited me in March, 2012. Zuzia was adorable, with beautiful big eyes and a diagnosis of damage of the central nervous system (CNS) in the form of moderate cerebral palsy (CP) with overall delay. She was not able to walk, speak, or communicate her physiological needs (eating and toilet functions). Her mouth was constantly open and she drooled intensely. During the two years of her life they had been involved in rehabilitation using the therapy programs of Vojta, Bobath, SI, speech therapy, psychological-educational therapy, and other developmental programs. At her earlier visits to doctors and their neurologist, Zuzia's mother was told that her daughter's development was normal, and that her pace was slowed down to some extent but that she shouldn't worry. She was hesitant about this information as Zuzia was their second child, and her daughter's development was very different in comparison with her elder son. Zuzia developed her ability to sit up with some delay at 10 months, for example, while her brother did this at 6 months. Mom observed delays in other areas: eating, manipulating objects, focusing, and other milestones. So she made a decision to start intervention using the Bobath method and then, a month later, the Vojta therapy.

Zuzia today as a 4-year-old.

During the first visit Zuzia's mother asked me what progress and time frame using the MNRI® with her daughter they can expect to see. I shared my long-term experience working with children with similar symptoms and the effects of MNRI® techniques. I wanted to prepare them for possibly years of work with Zuzia because CP is serious damage to the brain. As a therapist working at MNRI® clinics and camps in Poland and abroad, I have witnessed remarkable positive changes in children. But every child is different as 'they build their own history' of improvement or full recovery. It is a challenge to predict the results and to respond to these kinds of frequently asked question by parents, "How many days a week do we need to do the exercises?", "How long will it take?", "How many sessions a week do we meet with the therapist?", "How can we recognize if the method is effective?", "Will our child improve?", "What can we expect for our child's future development?" All these questions are so heart-touching as they are about "MY CHILD", their mostly beloved child in the world. I conversed with them on all these questions and shared examples – some with easy results and other more challenging ones - that depend on age, therapy time length and intensity, severity of the neurodeficits or nerve system damage, on quality of the Home Program by parents or helpers. One very important observation, from my professional point of view, is that MNRI®, under any and all these conditions, brings better and more stable results within a shorter time period than I have seen in any other method.

Zuzia: therapy can be fun!

With Zuzia, we started with weekly visits and I trained the parents for their Home Program. At the beginning of our work, Zuzia was very hypersensitive for touch, with extreme tactile defensiveness and a fear of height, new places, and people and her motor abilities and skills were delayed significantly.

During her first visit, the initial MNRI® evaluation of reflex patterns (April 4, 2012) showed that the development of Zuzia's major number of reflex patterns was on an immature/dysfunctional level with average score of 5 points (indicating incorrect reflex patterns with deep dysfunction) to 8.5 points (incorrect pattern with some correct features; level of average dysfunction of a basic reflex pattern) compared to the norm – 16-17.75 points (see the *Assessments* article in this book). Here are some brief descriptions of her Pre-Assessment results on reflex patterns (see more in Table 1):

• Upper limb reflexes – Hands Grasp (5.25 points), Hands Pulling (6.25), Hands Supporting (6.25), Sequential Fingers Opening and Closing (6.5), Babkin Palmomental (6.25) were causing a lack of manual dexterity; and Oral-Facial – Sucking Reflexes (7), Swallowing (6.5), Chewing (5.5) were resulting in poor oral-motor coordination, articulation, and language. These reflexes were causing Zuzia to have limited development of her nonverbal and verbal communication.

• Lower limb reflexes – Babinski (6.25), Foot Tendon Guard (7.25), and Foot Grasp (4) were not allowing her to coordinate the links between her ankle, knee, and hip joints and for standing with grounding. Her Crawling Reflex (5) and Automatic Gait (6.25) were strongly delayed resulting in her inability to crawl and walk.

• Spinal Galant (6.75) and Perez (6.75), Trunk Extension (6.25), and Landau (5.25) were negatively affecting her vestibular-proprioceptive system and thus her postural coordination and control were very poor. Her Perez was causing improper links with Leg Cross Flexion-Extension additionally inhibiting her walking and crawling.

• Rolling over patterns – Segmental Rolling (4.25), Spinning (5.5), Locomotion (4.25), and Flying and Landing (3.25) were dysfunctional and pathological; they were resulting in a lack of mobility and lack of ability for rational movements, poor work with her vestibular sysem, and a lack spacial orientation development.

• Zuzia's Asymmetric Tonic Neck Reflex (ATNR - 8.25) which affects the Stapedius Reflex was dysfunctional.

• Her visual reflexes – the Eye-Tracking, Ocular-Vestibular (7.75) and Opto-Kinetic (7.75) were at a very low level of development narrowing her visual span, limiting the work of her eyes for horizontal and vertical viewing and tracking, and affecting the vestibular system.

• The low level of development of reflexes responsible for the development of gross motor skills were Crawling (5), Automatic Gait (5.25), Leg Cross Flexion-Extension (6.25), and the Symmetric Tonic Neck Reflex (STMR, 7.75) which resulted in poor coordination and low precision of movement.

• Her Moro (7.75) and Fear Paralysis (7.25) (responsible for protection and survival) were the reasons for her tactile defensiveness and fearfulness. Based upon her reflex pattern pre-Assessment, I determined the MNRI® exercises and techniques that were essential to use for immediate and future benefit. Knowledge of reflex parameters and features is always very helpful in choosing techniques and programs; so, in case of Zuzia, this truly helped to accelerate her delayed neurodevelopment and to achieve success for very specific results, and the results that followed were incredible, almost unbelievable. The MNRI® techniques selected, from Neuro-Structural Reflex Integration, Reflex Re-patterning and Integration, Archetype Movements Integration, Oral-Facial and Auditory Reflexes Integration, Breathing Reflex Integration, and Tactile Integration specifically reflected Zuzia's needs for repatterning and integration of her reflex circuits.

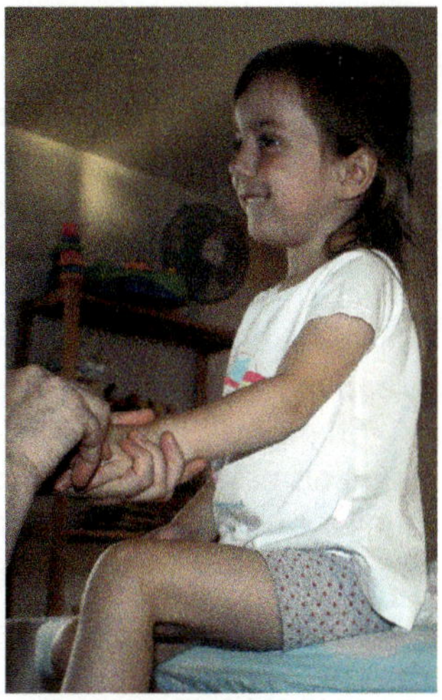

Zuzia receiving MNRI® therapy on her fingers.

During the first sessions, she cried a lot, so I focused on Tactile Therapy to reduce her tactile defensiveness and to create more harmonious relationship for us both. I also used exercises for Abdominal Sleep Posture to help reduce her stress from sleeplessness (which was significant) and to release her emotional overprotection. She needed time to get used to my new therapy space, the situation, and me. At the end of the first session, she already calmed down and looked happier. During our next session, I gradually began introducing more new MNRI® techniques and teaching her mother the exercises to do at home on a regular basis, as this is part of the MNRI® concept, to teach and equip parents with techniques for them to use as support. Zuzia's mother was consistent and followed my suggestions with great enthusiasm. The techniques I taught her were simple and manageable, and with time, the parents grew in their knowledge of profound 'parental' skills to deal with Zuzia's problems.

The following therapy sessions with Zuzia were much calmer, she became cooperative and our work began to run very smoothly and efficiently. She was absolutely happy to have the reflex integration therapy. As the result of the techniques, Zuzia achieved a considerable improvement in the development of reflexes (see Table 1 and Figure 1 below) which influenced her positively in her many abilities and skills (Post Assessment on August 5, 2014):

Upper Limb reflex patterns – Hands Grasp (10.5 points), Hands Pulling (12), Babkin Palmomental (11), Hands Supporting (12.5), Sequential Fingers Opening and Closing (9.5) explained her great improvements in manual dexterity. The first results that we noticed in Zuzia was improvement in the Hands Supporting Reflex pattern. Its improvement together with the Moro Reflex and Fear Paralysis, allowed Zuzia to better cope with falls and get up easier.

Zuzia showing Trunk Extension.

• Lower limb reflexes – we saw improvement in her Babinski (10.5), Foot Tendon Guard (11.5), and Foot Grasp (11) which allowed her to keep her feet straight with no improper adduction inwards and improve rotation and mobility in her ankle joints, coordination with knees and hips, and standing being grounded. Her Crawling Reflex (10.5) and Automatic Gait (12) improved significantly and Zuzia became able to crawl and walk.

• Spinal Galant (11.5) and Perez (9), Trunk Extension (12), and Landau (9.5) started supporting her proprioceptive system well and improved her righting mechanisms and postural control. The problems with negative compensation of Perez and Leg Cross Flexion-Extension dissolved and she became capable of cross-lateral movements, walking and crawling. Subsequently, her grounding and stability (with better distribution of body weight on her feet) also improved. Also, we saw a significant improvement in her balance and mobility – walking, sitting, and moving on knees, etc.

• Rolling over patterns – we saw improvements in the Segmental Rolling Reflex pattern (9), Spinning (9), Locomotion (7.5), and Flying and Landing (5) increased Zuzia's mobility and abilities for rotational movements, spatial orientation of her overall flexibility in joints and movements.

• Zuzia's Asymmetric Tonic Neck Reflex (ATNR, 10), which affected the Stapedius Reflex and her auditory processing, became functional and impacted her language coding/decoding and speech significantly.

• All her visual reflexes improved – the Eye-Tracking was more free from her Head Righting, Ocular-Vestibular (9.5) and Opto-Kinetic (11.5) created wider visual span with more visual stability, fluency, and mobility of her eye movements. Her whole body movements improved with her eyes for horizontal and vertical gaze and tracking, which affected the vestibular system.

• The change of reflexes responsible for the gross motor coordination – Crawling (10.5), Automatic Gait (12), Leg Cross Flexion-Extension (13), and STNR (11.5) resulted in better motor coordination, precision of movement, spatial, and motor-goal orientation. Zuzia became stronger physically and so confident that, after 6 months of our work, she did her first independent steps and started walking. What a joy it was for all members of her family and myself to watch this special moment of her development!

• Changes in her Moro (11) and Fear Paralysis Reflex patterns (10.5) resulted in her ability for self-protection mechanisms, as seen in a decrease in tactile defensiveness and desire to touch things and people, and to communicate with them. She became much more courageous in her home environment and, outside of home, she had less fear and began communicating with new people.

• In the meantime, her mother started toilet training exercises for Zuzia using Babinski, Foot Tendon Guard Reflexes, and also Spinal Perez and Galant. Zuzia very soon started regulating her toilet physiological needs. The changes came very quickly.

We continued working with Oral-Facial Reflex techniques to minimize her intense salivation and support her speech development. It took a lot of time to work with Sucking (11.5), Swallowing (9.5), and Chewing (9) Reflex patterns. After surgery on her tonsils, in between sessions, Zuzia still drooled badly. We used the Oral-Facial Reflex integration to target this problem and it brought quick results – the hypersalivation was resolved. The role of MNRI® was obvious – it caused an increase of oral-motor coordination, better articulation, and a desire to self-train her language development. She began initiating communication herself, and her non-verbal and verbal communication became richer and more interesting. Her abilities for thinking processes improved for comparison, classification, analysis, and comprehension.

When she pronounced the first words, "Mom", "Dad", "do not touch", "release", "give", "go" – it was endless happiness for us. But, with the next step, her speech improved with such speed, it amazed all the people that knew her. All of a sudden, one day I heard not just words or just sentences, but the whole concepts, such as, "You are my favourite person, I love you so very much" and "When are you moving to my house?" "You can sleep in my room. I have a mattress in my room for you." Tears filled my eyes and my heart was full of joy and gratitude... what can be more wonderful than to see such growth in a child and the development of such a warm and loving personality?

Today, Zuzia is a most lovely four year-old girl, full of joy and grace. She goes to a public kindergarten, walking and moving independently. Talking to Zuzia is a real pleasure. Her way of perceiving events, exploring the world, and her way of thinking allows you to see the surprising world of development and its beauty in process. We still need to continue working on her flexibility, stability, and motor coordination control. And we know that we will need to work long and hard to reach more successes as we continue on this road but the rewards have been great and we have tremendous incentive.

Table 1. MNRI® Assessment results: Level of Reflex Patterns in Clusters (average points before and after 2 years and 4 months of the MNRI® Program; (for more information about the Assessments, see the *Assessment* article in this book).

Reflex Groups/Clusters (Points before and after the MNRI® program)	**Before MNRI®** (average points) Date: April 5, 2012	**After MNRI®** (average points) Date August 5, 2014
Cluster 1. Upper Limb Reflex Patterns (Hands Grasp - 5.25/10.5, Hands Pulling – 6.25/12, Babkin Palmomental – 6.5/11, Hands Supporting – 6.5/12.5, Sequential Fingers Opening and Closing – 6.5/9.5)	**6.2** Patterns were incorrect; dysfunction of a basic reflex pattern; sensory system was hyposensitive and motor response hypoactive	**11.1** Patterns are functional, at a low level of development. Sensory and motor system improved significantly.

Reflex Groups/Clusters (Points before and after the MNRI® program)	**Before MNRI®** (average points) Date: April 5, 2012	**After MNRI®** (average points) Date August 5, 2014
Cluster 2. Tonic Reflex Patterns (Core Tendon Guard – 8/13.25, TLR – 6/8.5, ATNR – 8.25/10, STNR – 7.75/11.5, Abdominal -6.25/10.5, Spinal Galant - 6.75/11.5, Spinal Perez - 6.25/9)	**7.04** Patterns were incorrect; dysfunction of a basic reflex pattern; sensory system is hyposensitive and motor response hypoactive.	**10.61** The pattern is on the boundary of normal function and dysfunction. Elements of the correct pattern.
Cluster 3. Righting Reflex Patterns (Trunk Extension- 6.25/12, Head Righting – 8/13, Spinal Perez-6.75/11.5, Landau-5.25/9.5, Foot Tendon Guard- 7.25/11.5, Hands Pulling-6.25/12, Hands Supporting-6.25/12.5, Leg Cross Flexion-Extension-6.25/13, Thomas Automatic Gait-5.25/12, Spinal Galant-6.75/11.5, Flying and Landing-3.25/5, Segmental Rolling-4.25/9, Spinning-5.5/9, Locomotion-4.25/7.5, Balancing – 6.5/11) – 6.5/9.5)	**5.5** Incorrect/ Deep dysfunction of a basic reflex pattern.	**10** The pattern is on the boundary of normal function and dysfunction. Elements of the correct pattern.
Cluster 4. Lower Limb Reflex Patterns (Babinski – 6.25/10.5, Foot Grasp-4/11, Foot Tendon Guard-7.25/11.5, Leg Cross Flexion-Extension-6.25/13, Thomas Automatic Gait-5.25/12, Bauer Crawling-5/10.5, Trunk Extension-6.25/12, Core Tendon Guard-8/13.25, Grounding-3.25/11.5, Flying and Landing-3.25/5)	**5.48** Incorrect pattern. Incorrect pattern. Deep dysfunction of a basic reflex pattern. Sensory system was hyposensitive and motor response hypoactive.	**10.02** The pattern is on the boundary of normal function and dysfunction. Elements of the correct pattern.
Cluster 5. Gross Motor Reflex Patterns (Core Tendon Guard-8/13.25, STNR-7.75/11.5, Trunk Extension-6.25/12, Thomas Automatic Gait-6.25/12, Moro Embrace-7.75/11, Fear Paralysis-7.25/10.5, Spinal Galant-6.75/11.5, Spinal Perez-6.25/9, ATNR-8.25/10, Segmental Rolling-4.25/9, Spinning-5.5/9, Locomotion-4.25/7.5, Flying and Landing-3.25/5, Landau-5.25/9.5, Grounding-3.25/11.5, Hands Pulling-6.25/12, Hands Supporting-6.25/12.5)	**5.72** Incorrect pattern. Deep dysfunction of a basic reflex pattern.	**10.04** The pattern is on the boundary of normal function and dysfunction. Elements of the correct pattern.
Cluster 6. Oral-Facial/Visual and Auditory Reflex Patterns (ATNR-8.25/10, STNR -7.75/11.5, Convergence- Divergence – 7.75/11, Babkin Palmomental-6.25/11, Hands Grasp-5.25/10.5, Hands Pulling-6.25/12)	**6.92** Patterns were incorrect; dysfunction of a basic reflex pattern; sensory system was hyposensitive and motor response hypoactive.	**11** The pattern is on the boundary of normal function and dysfunction. Elements of the correct pattern.
Cluster 7. Protection and Survival Reflex Patterns (Core Tendon Guard-8/13,25, Fear Paralysis-7.25/10.5, Moro -7.75/11, Hands Supporting-6.25/12, Hands Grasp-5.25/10.5, Bonding- 8.25/13.5)	**7.13** Incorrect pattern. Deep dysfunction of a basic reflex pattern.	**11.8** The pattern is on the boundary of normal function and dysfunction. Elements of the correct pattern.
Cluster 8. Curiosity/Cognition Supporting Reflex Patterns Pavlov Orientation-7.13/12, Core Tendon Guard-8/13.25, ATNR-8.25/10, STNR-7.75/11.5, Babkin Palmomental-6.25/11, Hands Grasp-5.25/10.5, Hands Pulling-6.25/12, Bauer Crawling 5/10.5, Leg Cross Flexion- Extension 6.25/13, Landau-5.25/9.5, Flying and Landing-3.25/5, Segmental Rolling-4.25/9, Spinning-5.5/9, Bonding-8.25/13.5)	**6.19** Patterns were incorrect; dysfunction of a basic reflex pattern; sensory system was hyposensitive and motor response hypoactive.	**10.7** The pattern is on the boundary of normal function and dysfunction Elements of the correct pattern.
Cluster 9. Reflex Patterns Supporting Emotional Stability, Maturation (Bonding - 8.25/13.5, Fear Paralysis-7.25/10.5, Moro Embrace-7.75/11, Core Tendon Guard – 8/13.25, Pavlov Orientation-7.13/12, Babkin Palmomental-6.25/11, Hands Grasp-5.25/10.5, Hands Supporting-6.25/12, Landau-5.25/9.5, Flying and Landing-3.25/5, Segmental Rolling-4.25/9, Spinning-5.5/9)	**6.2** Patterns were incorrect; dysfunction of a basic reflex pattern; sensory system was hyposensitive and motor response hypoactive.	**10.52** The pattern is on the boundary of normal function and dysfunction Elements of the correct pattern.

Table 1 (continued).

Summary and Conclusions

The use of the MNRI® concept and techniques brings results, even when it is used for a rather short time frame for a child with CP. In my more than 15 years of experience working with children with developmental challenges, I have learned that it is important to begin the intervention as soon as possible – the sooner treatment is started, the faster are the effects to their development. The choice of strategy is important, thus the Assessment role is big and an essential step for deeper analysis of integration techniques and procedures to build a strong neurophysiological foundation through reflexes. Another important aspect of success is equip-

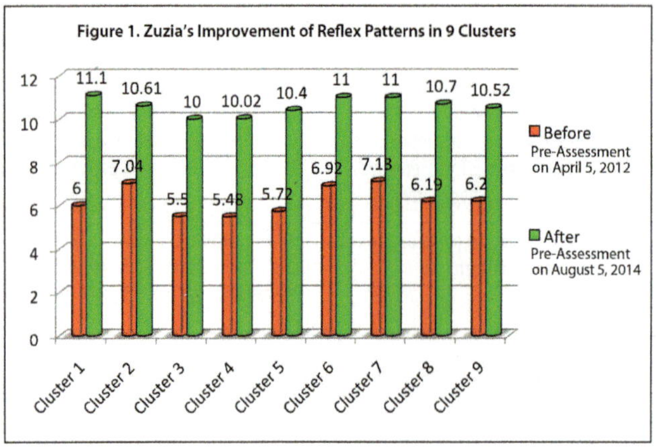

Graph 1. MNRI® Assessment results: Level of Reflex Patterns in Clusters (average points before and after 2 years and 4 months of MNRI® therapy

ping parents with manageable techniques for a continuous Home Program.

Reflex integration must be a first method of work for children, especially with those with developmental challenges. Other methods can then be used to build specific abilities and skills, but without MNRI® first, those therapies are likely to be ineffective. Without working on reflex development, using other therapies are like building a house on muddy ground. MNRI® integration processes allow the children to succeed as it focuses on two levels of integration at the beginning of the program that is not well presented in other methods: a) coordination of the elements of a basic reflex patterns (unconditioned naturally developed from 0 – 4 months of life) taking into account their developmental chronology, and then b) the integration of the biomechanical aspect of a reflex pattern, with its defensive tasks. These two levels of integration offer opportunities for rerouting and strengthening the neurophysiological circuits of reflexes while maximizing the maturation (myelination) of lower motor neurons and extrapyramidal nerve pathways. It guarantees the foundation for all other aspects of our being: physical-motor, behavioral, emotional, and cognition which is what the child, adults – therapists and parents, are all focused on.

Zuzia, a child with CP with whom I used the MNRI® techniques, achieved a great results in many important spheres: motor and postural coordination and control, transition of defense responses into positive protection and courage, improvement in eating and speech functions, and also in her comprehension. Along the path of these gains, she also found her own unique personality and the gift to her and her family was huge. She still needs a lot of work to reach her maximum potential, and most importantly, we are confident she will reach it.

I congratulate Zuzia and her parents on their great success in everything - physical skills, emotional growth, speech, and her cognition. I wish you good health and the realization of all your dreams for Zuzia's development! I am also thankful for her invitation to live in her home and sleep by her on her mattress! – Anna Krasowska

Malia's Impressive Growth and Development!

Donna Sanchez, MNRI® Parent, Belmont, CA;
Magdalena Gorecka, Pediatric PT, MNRI® Core Specialist, San Francisco, CA, USA

I said, "Please God, we will take her home and love her anyway you give her to us… please, please, don't take her." (Quote from the book, *Angels Unexpected,* 2015, written by Donna Sanchez, Malia's Mom.)

Donna Sanchez

Magda Gorecka

Malia wasn't breathing when she was born on March 6, 2006. She was pale and limp. She didn't cry at all – there were no sounds coming from our newborn baby girl…. "Why am I not hearing her cry?" Babies are supposed to cry upon entering the world. Why am I only hearing the pediatrician's voice saying, "Hurry up, give me the baby!"? The answer is, Malia was at death's door–but thank God… she never entered.

She suffered from oxygen starvation during her delivery. It was termed H.I.E. or Hypoxic Ischemic Event. Due to that lack of oxygen, Malia, who is now 8 years old, lives with the diagnosis of cerebral palsy (CP). She doesn't walk, has difficulty using her hands, has a difficult time with speech, and she is totally dependent. However, she is the happiest child we know. Her spirit is amazing and she is very smart.

Malia has a Special Needs Trust due to birth injury and because of that we have been able to research and try the many different methods of therapy that are out there for CP. Starting from the very early age of about 10 months Malia started participating in therapies to help her motor development. At the beginning she was participating in a traditional physical therapy program in a local clinic. In 2007 when Malia was about 18 months old we traveled to a clinic in the Ukraine for 12 days of intensive therapy and once again 6 months later for another 12 days. I have to say we learned a lot from the eastern Europeans. Their therapists, doctors, and parents have a different mindset about how to handle this type of situation and we have been *so fortunate to learn that early on.* Their approach was to be more intensive with therapy from the start and work hard to reach a higher level of development for the child. There is no time to be wasted and the results will not come if we do not work hard for them. American western medicine is great and we appreciate it very much... however.... when it comes to matters of brain-injury, we have a lot to learn because we've lost the ability to work hard at anything. Brain injury takes serious work and diligence, there's no medicine or quick fix for it.

After further research for more help for our daughter we also learned about The Institute for the Advancement of Human Potential in Philadelphia and traveled there with our little Malia when she was two and a half years old. This was a big learning for us as well. Malia has been diagnosed as having severe brain damage, yet

Malia and Dr. Masgutova

during our visit at the institute we got many tools and so much needed hope. We were committed and ready to work hard on her recovery and to follow a Home Program with her after we got back home.

At the end of 2009 Malia started working with the physical therapist and ABM practitioner Magda Gorecka who started her MNRI® studies soon after. Until then we were following the home program we learned at the Institute, doing traditional physical therapy at the local clinic and also participated in a therapeutic horseback riding class once a week.

At the age of 4, Malia's development was assessed by a therapist using objective tools. The results from the Peabody Developmental Motor Scale marked her fine motor skills at the equivalent of 5 months old, and her visual skills at the equivalent of 13 months. She had Gross Motor Function Classification Scale (GMFCS) level of IV-V.

In recent years we have been so blessed to have met Dr. Svetlana Masgutova and learned about her method, Masgutova Neurosensorimotor Reflex Integration (MNRI®). We now have been to six Masgutova Family Conferences (dates: August 2010 – August 2013), and we have also been practicing this method at home for 3½ years. It makes absolute sense that we use this method as a basis for all other methods. It's about integrating the most primitive reflexes that these children often do not have integrated, or the reflexes are dysfunctional or pathological. The way I remember and summarize this method is: reflexes are organized in the lower brain and if you have chaos going on down there, you will never be able to reach higher learning or higher function... in other words, these children cannot reach their fullest potential in life without basal organization. Dr. Masgutova is the most genuine, loving specialist I've ever met... she truly, truly cares about these children and it's her life's mission to help them. Because we trust her, we continue with her method. She looked me in the eyes at the first camp and told me, *"I know I can help your daughter, but it's going to take work and commitment–at least 4 years"* she said. I believed her. I was always optimistic and hopeful, but she gave me much more hope and we have seen our daughter only progress with doing what we're doing; just the absence of regression can already be a huge achievement for many children with severe cerebral palsy!

Malia and Dr. Masgutova

At the beginning of our journey Malia couldn't do anything without help. She was completely dependent on our assistance. At the age of two Malia would try to move small distances around the house by trying to crawl, it wasn't easy but we saw an independent spirit in her early on. Around the age of three we learned that is was important to 'not skip' the stage of belly crawling; we taught her to do this but it was very hard for her and she needed tons of encouragement. In the beginning, it was very difficult for her to lift her head up. She would move her legs and try to use her arms but it seemed that gravity was just trying to glue her body to the floor to make it more difficult. Head control was a big challenge and something we knew needed to be a high priority. Malia's body was governed by low muscle tone with weak muscles demonstrating spastic organization every time she tried to move. Her vision was developing but the visual field was narrow and her eye movements were poorly differentiated from head movement with poor teaming of binocular vision. She would startle easily, her meals took a long time because her chewing abilities were very limited.

First Experience with MNRI®

As we mentioned earlier, Malia's first Family Conference took place in August of 2010. As parents we learned about MNRI® from another parent, Malia's neurologist, and from her therapist, Magda.

Our first experience with the method was overwhelming. The amount of work and learning that took place during those 8 days was huge. Malia did great at participating and staying happy and compliant. At that age (4.3 years) she was still very limited in her motor function. Trunk and head control were very challenging, as well as reciprocal movements of the legs, or the use of the arms to reach to an object or grab a toy were very difficult. Malia's speech was limited to less than 20 words and now she's speaking sentences – labored sentences – but speaking and expressing herself nonetheless. Her movement had dyskinetic qualities demonstrating poor functioning of the cerebellum and cerebrum, and poor differentiation of postural and distal muscles of the body. Malia improved her belly crawling and started to move around the house in a more upright position by scooting and hopping on her bottom with legs positioned in a 'W' for balance. Frequently though, her hands wouldn't move fast enough and she would lose balance and fall forward without being able to block with her arms.

Cognitive development was always progressing faster with Malia. We always knew that she is a very smart girl though her difficulties with expressive language due to motor impairment, poor motor control, and pathology in muscle tone were impacting and slowing down her emotional and cognitive development compared to children in her age group.

Malia's first Assessment of reflex integration levels (August 2010) done by Dr. Masgutova was highly anticipated by all of us. We wanted to find out and learn more about her prognosis and how we could help our Malia.

Listening and witnessing the Assessment for the first time was an incredible experience. There was so much information and all of Dr. Svetlana's words seemed to fit exactly to what Malia's needs and challenges were as if she had known her for years. It was amazing to see the proficiency and knowledge presented by Dr. Masgutova in relation to reflex patterns and their impact on the development and function of a child.

That first Assessment demonstrated a deep dysfunction in a majority of Malia's reflex patterns and, in a few cases, pathology of reflex patterns. Our first Family Conference started us on a long and intense journey of helping Malia to correct the pathology and to start developing more correct reflex patterns and, through that, to facilitate motor, cognitive, and emotional development.

The results of Malia's August 2010 Assessment are presented in the tables below:

On a scale from 0 to 20 points, the data demonstrate that Malia's results were well below the norm (16.00-17.75 points). They were also below the transitional point (10.00-11.75 points) where a reflex pattern is considered to be on the boundary between the development of normal function and dysfunction, or at the point that both correct and incorrect features of a reflex present themselves. A result below this level means that the particular reflex pattern is dysfunctional.

Table 1 A. Reflex Patterns (1-10)

Reflex #	1	2	3	4	5	6	7	8	9	10
	Core Tendon Guard	Hands Grasp	Hands Pulling	Babkin Palmomental	Babinski	Leg Cross Flexion Extension	ATNR	Abdominal	Bonding	Foot Grasp
Average	9.5	4.75	5.5	3.75	3.25	3.25	4.75	5.25	8.5	3.75

Table 1-B. Reflex Patterns (11-20)

Reflex #	11	12	13	14	15	16	17	18	19	20
	Automatic Gait	Bauer Crawling	Moro	Fear Paralysis	Hands Support	Segmental Rolling	Landau	Flying/ Landing	Grounding	Head Righting
Average	3.25	4.25	6.25	4.75	3.75	3.25	0.75	0	1.5	3.25

Table 1-C. Reflex Patterns (21-30)

Reflex #	21	22	23	24	25	26	27	28	29	30
	Trunk Extension	STNR	Spinal Galant	Spinal Perez	TLR	Foot Tendon	Spinning	Locomotion	Balancing	Pavlov Orientation
Average	3.75	4.25	4.25	4.25	2.25	3.25	4.25	3.25	2.25	6.00

Considering the above criteria, most of Malia's reflex patterns were deeply dysfunctional or even pathological. Only two reflexes, Core Tendon Guard and Bonding, could be considered to have only light dysfunction. There were no reflex patterns showing positive functioning when she started the program.

As parents and therapist receiving the results of the Assessment, we needed time to process the information and translate it over to functional limitations and understanding why and how development of reflex patterns and their maturation can impact motor development.

The results of the Assessment and the knowledge that it is possible to correct them was giving us hope but also brought up many questions, "Can we really impact and change dysfunctional automatic reflexes?", "How long do we need to wait to see the results?", "Is MNRI® really the guaranteed method to make good progress in comparison with all that we tried before?", " How are we going to manage with the Home Program?", "How often do we need to come back to the conferences?" These are questions that parents and therapists ask again

The MNRI® Team and Malia held by her mother, Donna.

and again in order to find the best for their child and not to miss the opportunity for them to recover to the maximum of their possible potential.

At the beginning of the conference we didn't know any of the answers but gradually started observing, learning, and understanding. In addition to learning from the Core Specialists at the conference, we also participated in the evening lectures organized for parents to facilitate their learning more about MNRI®. We could ask questions, practice hands-on techniques and also share experiences with other parents. Gradually, one little step at a time, we started understanding the importance of primary reflexes and their role in the organization and development of the brain. We began to understand the power of reflexes as automatic responses given us by nature that explained so many of the difficulties that we saw in Malia's daily life.

The first conference opened up the door to further development. We started seeing the first positive changes in our little girl and we were determined to make every effort to continue the work with both the Home Program and by participating in the next conference in California.

Home Program

The Assessment allowed Dr. Masgutova to create an individualized MNRI® Home Program for Malia. Malia's leading therapist, certified MNRI® Core Specialist, Magdalena Gorecka, helped us with implementing the recommended Home Program.

Magda continues to be Malia's physical therapist. She provides direct hands-on therapy, as well as the support and guidance related to school IEP meetings, equipment needs, etc. ...

In late October 2010, we started working with the intensive Home Program for Malia. It included 5 days a week of MNRI® including, reflex integration exercises, Neurostructural integration protocol, elements of Tactile and Archetype, Oral-Facial program, and balancing board exercises. Continuously, month after month we started seeing progress in Malia, sometimes it seemed slower than we wanted, and sometimes she surprised us with her progress!

Changes in her development that we, and her specialists, have noticed:

• ongoing improvement in gross motor coordination: She has the ability to move around the house independently while on a level surface using modified creeping (for now the movement continues to be homolateral), able to come to high kneeling and balance for up to 15-20 seconds, able to stand up straight with minimal assistance at the pelvis. Malia can sit on a small chair with her feet on the floor and maintain that position, with standby assistance for safety, for several minutes.

• big improvement in the following reflex patterns: Progress is seen with Trunk Extension, Head Righting, Automatic Gait, Leg Cross Flexion-Extension pattern which facilitates her reciprocal movements and control of lower extremities. At the beginning of MNRI®, all the reflexes of Hands Supporting, Hands Pulling, Bauer Crawling, Spinal Galant, Spinal Perez reflexes were exercised passively; right now Malia is able to actively participate and activate the muscles responsible for very specific movements. Her coordination, strength, and balancing responses continue to improve.

• fine motor coordination improvements: Malia started reaching for and holding objects. She still needs several attempts to be successful, but her abilities of purposeful hand movements have changed greatly. She started playing with toys (something that all of

Malia and MNRI® Core Specialist, Connie Jordan.

us have dreamed of). Her determination is inspiring! Because of her love for music she learned how to put a CD into the CD player and start playing the music!

• vision: Malia's ability to move her eyes independently of the movement of the head has improved and continues to develop. Her visual field has extended and coordination of eye movement has improved.

• auditory perception: Malia's understanding of language and auditory perception are at a high level. Her startle response due to sudden sound stimuli has decreased significantly, demonstrating reduced hypersensitivity of auditory Fear Paralysis and less mixture of Moro with Fear Paralysis.

• speech: Malia's speech came late but it is blooming! At the age of 5½ – 6, she started using more words and began communicating more spontaneously. It was a gradual process of gaining more confidence and motor control to be able to start producing words. These days we continue to work on clarity of speech and pronunciation of consonants. We are also working on extension of exhaling and timing of breathing rhythm and speech.

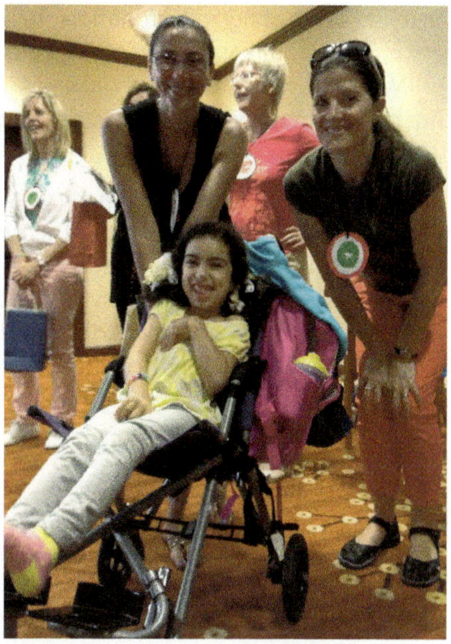

Malia with MNRI® Core Specialists, Magda Gorecka, right, and Boutaina Rosen, at left.

Malia has made great progress over the years and she still continues to improve her skills. She is very smart, has an amazing sense of humor, and is a very determined young girl. We work as a team, in which Malia is the most important member, to support her continued development.

As she becomes older, and body proportions change, her center of gravity shifts further and further away from the ground. It becomes even more important to maintain a regular therapeutic schedule and to continue with the application of MNRI®. We know that by using reflex patterns, we stimulate the innate genetic resources that serve as the keys to communicating with the brain. The richness of MNRI® gives us tools to address all aspects of child development. From reflex integration, Tactile, Neurostructural, Archetypes to Facial, Auditory-Visual, and cognitive.

Malia's last Reflex Assessment in February 2013 shows significant progress in the level of reflex patterns integration. That supports and explains why she could make such good progress in her motor development and motor control. The tables below compare results of this February 2013 Assessment against the first assessment in August 2010 (see table 2 A, B, C):

The results of the last Assessment indicate that all Malia's reflex patterns have improved significantly. There are NO reflexes at the pathology level! Only 6.7% (2 of 30) reflex patterns are at the level of deep dysfunction, 40% of reflexes

Table 2 A. Reflex Patterns (1-10)

Reflex #	1	2	3	4	5	6	7	8	9	10
	Core Tendon Guard	Hands Grasp	Hands Pulling	Babkin Palmomental	Babinski	Leg Cross Flexion Extension	ATNR	Abdominal	Bonding	Foot Grasp
Feb 2013 Average	13.5	8.5	10.25	7.5	11.25	11	14.25	13.75	15.5	8.75
Aug 2010 Average	9.5	4.75	6.75	3.75	3.25	3.25	4.75	5.25	8.5	3.75

Table 2-B. Reflex Patterns (11-20)

Reflex #	11	12	13	14	15	16	17	18	19	20
	Automatic Gait	Bauer Crawling	Moro	Fear Paralysis	Hands Support	Segmental Rolling	Landau	Flying/ Landing	Grounding	Head Righting
Feb 2013 Average	9.25	11.5	12.25	9.25	10	10.25	5.75	5.5	9.75	9.75
Aug 2010 Average	3.25	4.25	6.25	4.75	3.75	3.25	0.75	0	1.5	3.25

Table 2-C. Reflex Patterns (21-30)

Reflex #	21	22	23	24	25	26	27	28	29	30
	Trunk Extension	STNR	Spinal Galant	Spinal Perez	TLR	Foot Tendon	Spinning	Locomotion	Balancing	Pavlov Orientation
Feb 2013 Average	11.75	13	10	10.25	9.75	9	9.5	8.75	9.75	9.50
Aug 2010 Average	3.75	4.25	4.25	4.25	2.25	3.25	4.25	3.25	2.25	6.00

reached level of light dysfunction, 30% are at the boundary of normal function and dysfunction, and 16.7% are functional at a very low level.

Malia has made amazing progress and she continues improving in her physical abilities and efficiency of her communication. We will continue to support Malia in her journey of ongoing learning and development.

Conclusion

From Magda Gorecka: From my very first experience with MNRI®, I knew that it has amazing depth and provides insight to help understand the physiology, development, and pathology of the human brain. Listening to Dr. Masgutova explaining the role of primary reflexes and her approach to facilitating their integration to open possibility for development was a professional, "Aha," moment for me. Participating in Family Conferences and seeing so many children (and adults) change so significantly in such a short time was and continues to be an amazing experience! It is a privilege to be a part of the MNRI® professional team. I want to say 'Thank You' to Dr. Masgutova for her gift of knowledge.

Thank you to Malia's parents, Donna and John, for their deepest dedication, kindness, and love and also for their support and perseverance in the journey of unlocking possibilities and ongoing development of their dearest daughter, Malia. – MNRI® Team

MNRI: Four Letters that Stand for Massive Neurological Repair In Ryan

Kyle Van Dyke, MD, and Annette Van Dyke, RPh, Chippewa Falls, WI, USA and Peg Johnson, MNRI® Core Specialist, Madison, WI USA

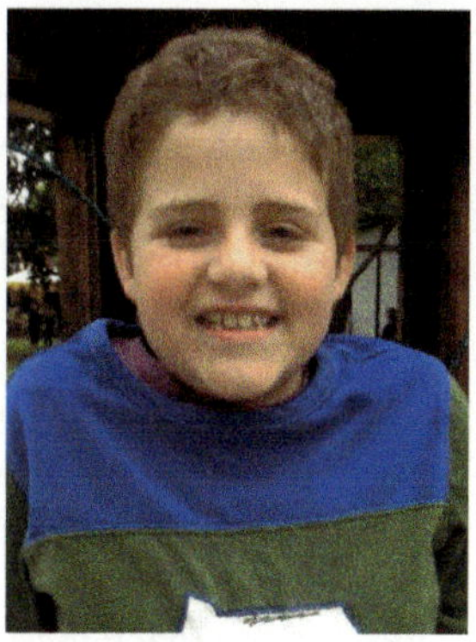
Ryan Van Dyke

Our story began in 2004 when Ryan, at the age of 25 months, was diagnosed with severe autism. This diagnosis hit us hard since we had watched him regress right before our eyes. Our once social, verbal, happy baby became a drooling zombie in just a few months. At 12 months of age, Ryan had been talking, pointing, cruising and was so very excited to explore his surroundings. Within hours of his immunizations at 13 months (MMR, varicella, pneumococcal, influenza) Ryan developed a high fever, began crying nonstop and had massive constant diarrhea. Days turned into weeks and into months, the crying continued and all language ceased, eye contact disappeared, and Ryan was unable to point or stand.

Our quest began by meeting with several developmental pediatricians/neurologists to figure out what happened and how to proceed. After repeatedly being told, "Sorry not much future/success to be expected," we got very angry and determined. My husband and I are both healthcare professionals and could not accept that our once healthy child no longer had a viable future. If a similar course had occurred in an adult, the medical system would have worked up the case looking for an underlying cause and treatment. Instead we were told Ryan would never speak, never be potty trained, and never know who we were. "The brain damage is just too severe and also please get on a waiting list for an institution since he will likely be violent by age 7." Ryan finally began walking (toe) at 25 months.

With that advice we made a decision that would change our lives and Ryan's prognosis. We moved to be near an amazing doctor who was working with an organization called the Autism Research Institute (ARI). Kyle (Ryan's dad) changed his career to work with Dr. Elizabeth Mumper in Lynchburg, Virginia and to focus on treating children with autism. Through this work Ryan was able to receive biomedical treatment, including being a subject in a clinical trial for mild hyperbarics (mHBOT.) Hyperbarics brought Ryan's speech back and also alleviated much of his gastrointestinal issues. Ryan's medical care today continues with a colleague, Dan Rossignol.

Ryan had ups and downs for a couple of years with lots of infections, auditory, visual, and motor planning processing issues. Eventually, Ryan was diagnosed with PANDAS/PANS and that really helped us to understand his extreme obsessive compulsive behaviors. To try and decrease his infections, Ryan had tonsils and adenoids removed by surgery. Unfortunately, Ryan regressed completely following the anesthesia. So we lost Ryan again to the nonverbal zombie state. This was a heartbreaking set back. Around this time we also received test results showing that Ryan was positive for an autoantibody to his own brain.

We could have given up but instead decided to treat his PANDAS and auto-immunity issue and, with this, a move to Wisconsin seemed optimal. We moved to the wonderful state of Wisconsin where behavioral services were not only recognized but covered by waiver/insurance. Equally as important was that now Ryan could have regular access to hyperbarics.

The move to Wisconsin was a miracle in the making because it was in Wisconsin that we met a developmental optometrist, Dr. Valerie Frazer. At this point in 2008, Ryan was 6 years old and he could not hold a crayon or use a fork. Ryan toe walked 100% of the time despite years of vision therapy, occupational, and physical therapy. Ryan spent his spare time doing self stimulatory behaviors (spinning, flapping) and scripted TV programs frequently. In the fall of 2010, desperation led to Ryan having tendon lengthening surgery to try and address the extreme toe walking. Post surgery, the university physical therapists did not give us any hope that he would remain 'flat.' We tried theratogs, hippotherapy, weekly physical therapy, and a home PT program. Ryan's gait and hip rotation were not able to be fixed or trained. Then insurance refused to pay for any further physical therapy due to these clinical assessments.

Luckily, to help with vision reflexes, Dr. Frazer recommended Theresa Fischer, an Illinois physical therapist who is trained in MNRI®. Theresa began seeing Ryan and also arranged for an evaluation by Dr. Masgutova in December 2011. The Assessment by Dr. Masgutova uncovered numerous areas to be addressed. We started a MNRI® Home Program. The improvements were miraculous. Ryan's hip rotation improved, his gait normalized, and his extreme sensory reactions faded. Theresa was amazing at keeping up with Ryan's progress and tweaking his program.

A few months later in June 2012, we were thrilled to participate in the first MNRI® family conference to be held in Madison, Wisconsin. The Family Conference was truly the highlight of our life. Several community professionals attended the evening lectures, embraced MNRI®, and are attending advanced trainings. Ryan made amazing progress in those five days and literally gained years of skills.

Today, Ryan is quite different than in 2004. He can not only walk properly, he can run! He can jump and even stand on one foot. Motor planning and spatial awareness are vastly improved. Much to his OT's surprise, coloring is now a favorite past-time. His language is conversational and lively. Social and cognitive changes abound as well. When Theresa was no longer available, we found Peg Johnson. Ryan loves his MNRI® sessions and often asks when he will see Kim Willkom, our current MNRI® provider in our new location.

With Ryan's improvements, the stress on our lives has eased, and we have begun to pay it forward. Kyle is very involved with the newly formed organization called the Medical Academy of Pediatric Special Needs. MAPS aims to teach medical professionals to implement a sound standard of care that is built on scientifically tested and proven methods. MNRI® is a perfect mate to MAPS.

I, Annette (Ryan's mom), was a volunteer coordinator for Talk About Curing Autism (TACA) Wisconsin Chapter (www.tacanow.org) until we relocated to Chippewa Falls, WI where small community life is well suited to our family.

I often muse the 'what could have been' and wish we had discovered MNRI® earlier. Especially during the regression: the years, money, and tears that could have been saved. However, now that we know what needs to integrate, it is just a matter of continuing his MNRI® program. My dream is for MNRI® to become part of every neurologist's tool kit and be used to evaluate developmental delays. Even better would be for new parents to understand primary reflexes and how to incorporate everyday life activities to facilitate typical development.

Autism rates are soaring but advances in science like MNRI® and biomedical treatments are also progressing and need to be shared widely.

Observations from Peg Johnson, MNRI® Core Specialist, Madison, WI

Ryan is an 11-year-old boy that was diagnosed with severe autism at the age of 2. Approximately three years ago he started his MNRI® journey. I believe his story is the 'missing link' in action.

Ryan and I worked together for ten months, bimonthly for one hour. During that time I continued to see significant improvements in his development. Ryan's parents are committed and diligent in implementing the MNRI® Home Program exercises, which has been paramount in his success. This commitment to the MNRI® Home Program cannot be emphasized enough, aiding in maturation and integration of the reflex patterns, impacting daily functions and development of an individual.

An MNRI®/Van Dyke Reunion! from left, Dr. Kyle Van Dyke, Core Specialist Peg Johnson, Ryan Van Dyke, Core Specialist Stephen Best, Dr. Masgutova, and Kylie Van Dyke.

Based on the in-depth Home Program already in place, I focused primarily on Oral-Facial and Visual/Auditory Reflex integration in our sessions. During this time period, his vision stabilized while demonstrating better teaming of his eyes. His eye tracking had improved significantly, which became smoother and steadier. Ryan's vision showed continual improvement in his convergence and divergence.*

Additionally, he made improvements in his auditory processing, which were evidenced by his enhanced interactive communication skills. Ryan responded to questions freely, with less delay. He has become more involved in conversations with emerging ability to interject with questions, such as "What?" and "Where?"

Over the course of our time together, he demonstrated more appropriate social behavior and manners. His Mom reported remarkable gains in fine motor skills, such as cursive handwriting and a love for drawing and coloring, which had not been an interest of Ryan's before. Ryan now enjoys interacting with other children in his neighborhood. Ryan is able and motivated to learn new skills to help him cope with stress. For example, after spending only one session teaching him deep belly breathing techniques, I observed him practicing the these new skills on his own while waiting for his Mom after the session.

According to the MNRI® model, reflex integration is key in addressing delays in neurophysiological development by directly addressing protection and survival strategies of the brain. In Ryan's case, freeing him from these unproductive protective responses has improved his ability to connect with his world through his visual and auditory systems.

**More documentation of his vision improvements that correspond with Ryan receiving regular MNRI® Visual/Auditory Reflex Integration are noted in the following report submitted by Dr. Valerie Frazer, mentioned above:*

I have been treating Ryan's vision since he was six years old. I'm an optometrist who specializes in vision development and optometric neuro-rehabilitation. Ryan has difficulty with eye tracking and teaming (the ability to use his eyes together). He has an eye that turns out (exotropia) more at distance than near. That eye also shows a vertical deviation that has been very variable. I have utilized several treatment options including vision therapy to help improve these skills over the last few years. One of the difficulties that I encountered with Ryan is that his visual skills and abilities would vary greatly from visit to visit. Since beginning the MNRI® program, Ryan has been much more consistent with his visual skills and shows better eye alignment and response to depth perception targets at distance. One of the tools I use to measure improvement is the Visual Evoked Potential or VEP. Sensors are placed on the forehead and on the back of the head (occipital lobe) where vision is processed. This test offers objective data on what information from the eyes actually gets to the back to the brain for processing (amplitude) and how quickly it gets there (latency). The test is done with large, medium, or small check sizes. Ryan underwent his first VEP testing in August of 2012. He had already been utilizing the MNRI® program but was no longer in a vision therapy program. Ryan's initial VEP when trying to use both eyes was at the 85% level, was 10.45 for amplitude which is within normal, and 119.9ms for latency at the smaller check sizes (64x64 pattern). Normal latencies are 80 to 117ms. A change in latency must be 10% or more to be considered significant. This test was repeated in September and latency was again 119.0ms. In January of 2013, we repeated the VEP and latency had improved to 94.6ms. At his most recent testing in April of 2013, his VEP latency has remained stable at 92.6ms. Ryan still has additional work to do with his vision, but I have seen progress and increased stability in his vision since beginning MNRI®.

It has been amazing to be part of Ryan's journey and a pleasure to work with him and his parents, Dr. Kyle and Annette Van Dyke. Their determination not to give up and recognition and implementation of this powerful work have made all the difference in their son's life. – Peg Johnson

Change Is Possible

Lisa Ortego, Speech Language Pathologist, MNRI® Core Specialist In Training, Carencro, LA, USA

Lisa Ortego

Austin was a fifth grade student who was diagnosed with autism. Due to his diagnosis and the need for special education services, he attended his local public school rather than the Catholic school where his mother taught and siblings attended.

In the spring of his fourth grade year, Austin took the fourth grade LEAP, a required standardized test for grade promotion. Unfortunately for Austin, he failed it. Even though he attended summer school and took the LEAP test at the end of the summer, he failed it again. This second LEAP failure resulted in Austin being retained in fourth grade. Austin was devastated that he had to repeat the fourth grade. His Mom, therefore, switched him to the Catholic school where she taught, where his siblings attended, and where I provided speech therapy services.

When Austin started speech therapy with me, he was a shy child, unable to complete his thoughts, spoke in the tiniest whisper, and appeared to live in his own little world.

The standardized test that I gave Austin at that time revealed the following scores:

Listening	59	Organizing	73
Speaking	59	Grammar	54
Semantics	67	Spoken Language	56

A 77 and below is needed to qualify for speech therapy services in the school system. Austin's scores were significantly below this level.

As I began to work with Austin, it was apparent that he could easily become over-focused on certain things and could not move beyond it. For example, if he made a bad grade on a test, he wanted to stay angry for a few weeks until something else came along to distract him. Once I began the MNRI® protocol from the *Children With Challenges Class* with him, Austin's frustration soon changed to relaxation and peace. His classroom aide saw significant changes after our speech therapy sessions; he was always calmer and his thinking processes seemed clearer. Things began to change quickly as Austin's inability to find words changed, and he could actually participate in conversation. It did not take long before he was able to socialize and blend in with other students because he finally had the ability to communicate.

As his reflexes began to integrate, he giggled all of the time. After 1-2 months of receiving MNRI®, Austin's giggles turned into reflections that upset him. The trauma of failing the LEAP surfaced and Austin's perception of this event led him to believe that his old school was a monster that had kept him from coming to the school where his family (mother and siblings) was. I began to incorporate Dr. Masgutova's 'Good Day,' 'Bad Day' drawings in between our reflex work during therapy session. The following is a brief summary of Austin's drawing:

Austin's 'Bad Day' drawing was his old school (the public school where he failed the LEAP). It was drawn as a monster that kept him from attending his new Catholic school where his mother and siblings attended. Austin drew himself as a stick figure saying that he did not want to go back and leave his family. His 'Good Day' drawing showed his new school saying that he could stay with his family. Austin drew himself as a stick figure saying that he was happy at his new school with his family. It was encouraging to see Austin then change his 'Bad Day' to his 'Good Day,' by turning the old school, that was drawn as a monster, into his new school, which was the Catholic school where his mother and siblings attended. He changed his statement of wanting to be with his family (a frustration of his at the public school) to a statement that he was happy to be with his family at his new school. This allowed him to come into the present and realize that he was safe and happy.

Austin soon came to the realization that he had autism and one day truly mourned at home with his parents. His mother told me that she had never seen him cry like that before. She stated that he was always in his own world. He was never told nor did he realize that he had autism. Autism was never mentioned in our sessions. His mother figured that something 'clicked' allowing him to remember conversations in the past that she had with him about receiving help at his other school because of his autism. As soon as Austin started therapy with me, we began the 'Good Day/Bad Day' drawings again with the MNRI® reflex work. Here is a brief summary of his 'Good Day/Bad Day' drawings:

His 'Bad Day' drawing revealed the word 'autism' and he stated that it was in his brain and he wanted it out of his brain. His 'Good Day' drawing revealed the word 'a-u-t-i-s-m' written with little dashes between each letter because he wanted autism to break up in his brain. With astonishment, I watched him change his 'Bad Day' drawing into his 'Good Day' drawing by drawing a ghost that was eating his autism and breaking it up in the 'Bad Day' drawing. He was now smiling and ready to continue moving forward. Several months later, he states, "Mrs. Lisa, I sure don't get upset about autism anymore or my old school either."

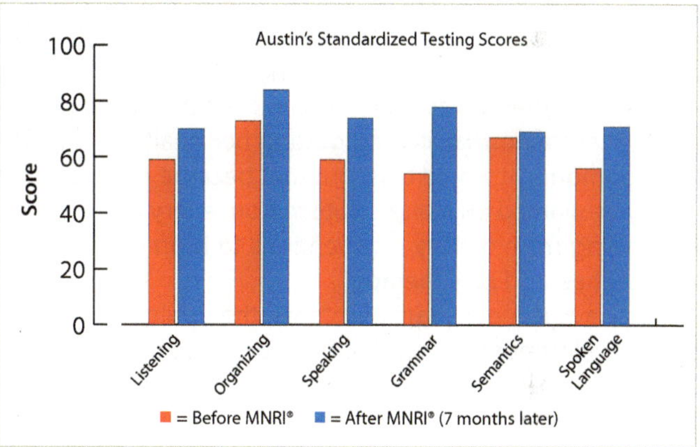

Seven months later, a reassessment revealed the following scores:

Listening	70	Organizing	84
Speaking	74	Grammar	78
Semantics	69	Spoken Language	71

The graph at right shows the progress. In the 13 years that I have been a speech language pathologist, I have never seen changes to this extent with a child with autism, especially within a 7 month period. Scores can change by 15 points in 7 months, especially with MNRI®!

Austin continues to do well as he and his family embrace the MNRI® journey. I cannot wait to see where he is a year from now. This work is such a blessing and a true journey that accomplishes what once was thought to be the impossible. It restores hope again as the saying, "Change in possible" becomes a much appreciated reality!

I would like to thank Austin and his family for joining the MNRI® journey and trusting in the MNRI® process. It has and continues to be an honor and privilege to work with Austin! He is a great teacher and student! Here is to the rest of the journey ... where the impossible becomes possible and dreams come true! – Lisa Ortego

Finding Something Missing: Extraordinary Talents in a Child

Hela Bali, MNRI® Core Specialist-in-Training, Parent, Irvine, CA, USA

Hela Bali

At age two years and three months, my daughter Malak was diagnosed with autism so I decided to quit my job and dedicate my life to her recovery.

Nothing was important in life other than giving my child a second chance to have a normal life and getting her healthy . . . I did not know how much time, energy, or money it would take, but that became my mission in life.

I have worked on Malak's autism from every angle possible. I have read and researched everything from dietary interventions to biomedical treatments, to developmental base approaches, and play therapy.

We have done 4 years of intense therapies from HANDLE, OT, Speech, RDI (Relationship Development Intervention), Floortime, Son-Rise, Yasko protocol, listening therapies, Brain Balance Program, neurofeedback, hyperbaric chamber, osteopathic treatment, Brain Technologies, diets from gluten free to Body Ecology, GAPS, pH Miracle Diet, homeopathy – from classic to sequential ... I am sure I am forgetting other things we dove into trying to find answers for her. I will say we tried everything.

All that we did was helping somewhat but I always felt that my beautiful angel was still struggling. She was still toe walking, having trouble communicating, had sensitivity to certain foods, sensitivity to touch, issues with hand-eye coordination, could not enjoy playing ball with her sister, could not move forward in her swimming lessons, still had trouble coordinating her upper and lower limbs, had high anxiety, very highly sensitive auditory system, could not socialize with friends although she loved being with kids, did not know her right from her left, etc. I really felt that with all the intervention I had done, and all the knowledge I acquired all of these years trying to help my child, that there was still something missing.

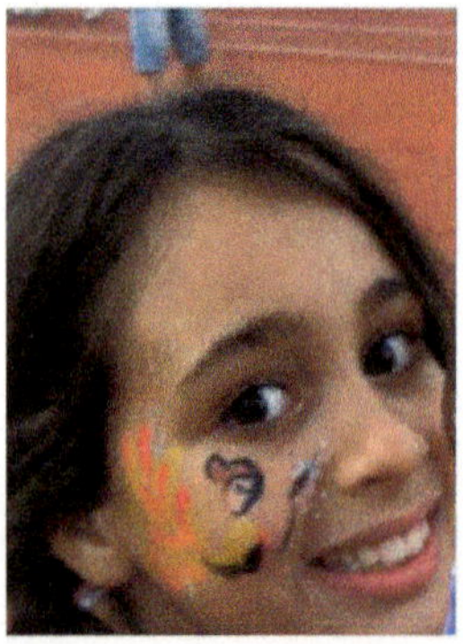
Malak Bali as artist.

I started looking further and was introduced to the word *reflex*. So I did my research, read books about primitive reflexes and reflex inhibition. I went into the path of reflex inhibition and tried it with a couple of programs offered in the Los Angeles area. We finished the program and the only thing that changed for Malak was that she starting using her bike without training wheels, but she still had all her other issues. We heard about reflex integration with Quantum laser. I received training and a home program, bought a laser for $3,400 and did the program that was supposedly dedicated to speed reflex integration for 9 months every day. We saw no major changes. At that point I decided to dig further and was introduced to the Masgutova Method®. Dr. Masgutova was not local, and her last clinic of the year was in Texas. I had an urge to sign up, take my daughter, and fly to see her. I wanted the truth about my child. Every professional I have worked with never gave me the answer. She was still delayed and needed help, and I just needed to find the tools and the keys to help.

Dr. Masgutova had the answers for me, and I found out that my daughter's reflexes were very delayed across the board. MNRI®changed her life and brought her from a place of fear and anxiety to a joyful child that now is 8 and called a "Mozart little version" by her music teacher (see the letter below), a talented artist by her art teacher, and a sweet and loving girl by her family.

Malak at swimming classes.

We have done the MNRI® Home Program for almost 3 years. We diligently do our Home Program 5 days a week. We also incorporate every other protocol of it and we see amazing changes every time we add new techniques. Malak participates in homeschool three times a week, goes to a typical school twice a week, and loves to sing and compose her own music now. She wants to specialize in music when she grows up. She swims in a regular swimming school three times a week, and is moving fast with her level (almost ready for stroke clinic). She participates in a neurotypical art class and makes beautiful drawings full of imagination. She can verbalize her emotions now and talks a bit about it. She is not 100% conversational yet, but we get beautiful fluid communication coming from her that we never had before MNRI®. She definitely knows her right and left sides now, enjoys playing soccer with dad and sister, her anxiety is way manageable, and her mood is always happy. She is a joy to be around at school and at home. Every piece of her development is coming along. I know now that I have the right tools, the correct answers, and I am so grateful for this amazing program.

Parents out there with kids with challenges: you have to rebuild the brain from the ground up, and this is the only therapy that recognizes the importance of the brain stem. You have to work your way developmentally like nature intended. Dr. Masgutova has done all the work and research and still comes up with new techniques and answers every day. Do not underestimate the importance of reflexes and their proper integration. This is the way to recovery if you are looking for one.

I have been working with Malak for about a year. Malak is gifted with a rare talent that only advanced music students can acquire through years and years of training.

Malak possesses a perfect ear and she can immediately repeat any given music performed on piano, which is rare amongst students her age, in addition to, she plays with such expressiveness that the composition requires.

Malak's memorization skills are very well developed, as well, and due to her concentrated work in the classroom, she can remember any given music and play it by memory.

We are currently working on a vast repertoire which is considered to be very advanced for her age of a student and she is perfecting in each and every piece.

Sincerely,

Prof. A. K.

Thank you for your remarkable dedication and perseverance in finding treatment that works for your beautiful daughter. We are very happy that her many talents are now blossoming!

Dr. Svetlana Masgutova & MNRI® Team

Luke: Change and Growth

Shaaron Freeburg, MNRI® Grandmother, Ramsey, MN, USA

Introduction

My grandson Luke, now 18 years old, was born extremely premature at 27 weeks, 15½ ounces, 10½ inches. His early deficiencies were many and we were given no treatments to help him. This is his story.

Shaaron Freeburg

Basic Information

As a preemie, Luke had severe intrauterine growth retardation; pneumonia at six weeks; hernia repair at two months; G-tube placement due to failure–to–thrive; two eye surgeries for strabismus; and six surgeries for hypospadias. His diagnoses were PDD-NOS, autism, mental retardation, Cerebral Folate Deficiency, and seizure disorder. Luke's physical growth was delayed and his motor coordination was limited by rigidity in his joints, lack of postural control and balance, limited eye movements, and myopia. He was easily irritated, interacted little, had eating problems, poor facial muscle control affecting speech and vocalization, and high tactile defensiveness. His eye tracking, convergence/divergence, and ocular-vestibular and optokinetic reflex patterns did not allow him to focus. With such poor visual perception, Luke remained mostly in his own inner world, ignoring the world around him. He could barely tolerate being outdoors: a slight breeze or sunshine was overwhelming. His hands would drip with perspiration.

Luke's expressive speech was also very limited. Questions resulted in a delayed "yes" or "no" but no conversation. He mostly repeated scripted phrases he had memorized from often-watched videos or gave echolalic responses.

At age 11 he began having tonic-clonic seizures two/three times weekly. His eating disorder required a feeding tube, and he had multiple developmental delays: auditory processing disorder; fine motor delay; gross motor delay; receptive and expressive language delays; obsessive-compulsive disorder; sensory integration dysfunction; and a visual processing disorder.

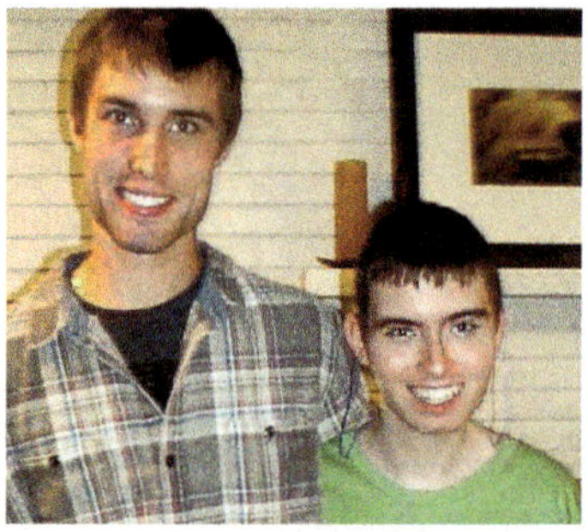

Luke, right, with his cousin, Zachary Brown.

Luke has been on a casein-free, gluten-free diet for 12 years. He eats slowly due to his oral hypotonia, so in order to maintain enough caloric intake we supplement with formula via feeding tube. In 2006, after 40 hyperbaric dives his sense of smell seemed to materialize along with more tolerance to outdoors. Luke has attended a six week Boost-Up Plus program each summer for the past three years along with hippo-therapy and music therapy. His exceptional MD, Dr. Dan Rossignol, has helped him through many issues (hyperbarics, IVIG, cerebral folate deficiency, gastro/gut problems, methyl B12 injections), and, along with forward-thinking local MDs and an amazing osteopath (glutathione IVs, heavy metal chelation), have accommodated his medical needs locally. He receives an hour of occupational therapy a week and has had speech therapy and auditory listening therapy.

Luke, right, with his brother, Spencer.

At age 16, Luke was introduced to MNRI® and received an evaluation from Dr. Masgutova. He started two hour weekly sessions with a MNRI® therapist, then months later added another two hours with another MNRI® therapist. Months after that, I learned to give Luke home exercises for an additional two hours weekly.

Dr. Masgutova set forth a substantial, very workable regimen for Luke. Her MNRI® program has allowed him to organize, awaken, and blossom! There are other therapies available as well but Luke cannot engage in them until some of these basic, fundamental connections, via MNRI®, come together. Luke is definitely 'in process!' His reflex patterns are now more developed and we are hopeful they will mature even more, serving to connect his body, mind, and soul!

Results

Luke's vision skills are much improved: he is more present, aware, and interested in the world around him. He is quick to notice a movement or a sound or a sigh, and even watches facial expressions.

One example of MNRI®'s impact was Luke's response to the Babinski Foot Reflex. Just after the session, he took off his socks off and ran outside to walk barefoot on the ground. This was so abrupt and out of character, it seemed he was connecting with the ground for the first time ever and liking the feeling. Shortly thereafter, he climbed up on the hood of the car and wanted to jump way down to the ground. Feeling natural sensations in his own feet gave him such joy! Luke's physical movement has become more relaxed and coordinated; his shuffling walk is changing to a more mature cross-lateral gait.

Luke can now interact more spontaneously, both physically and socially. He has become less mechanical and comprehends humor more. He imitates characters from stories, sounds, and music. Luke is opening up to others verbally and physically: more touching and non-mechanical chatter. He is not as defensive to sensory/tactile stimulation and no longer vulnerable to emotional reactivity like lashing out with hitting or screaming. He is more relaxed and fluent with words and language and is more able to share information than he ever could in the past. He frequently wants to know what words and phrases mean. As his receptive language and understanding deepens, his cognitive skills continue to develop.

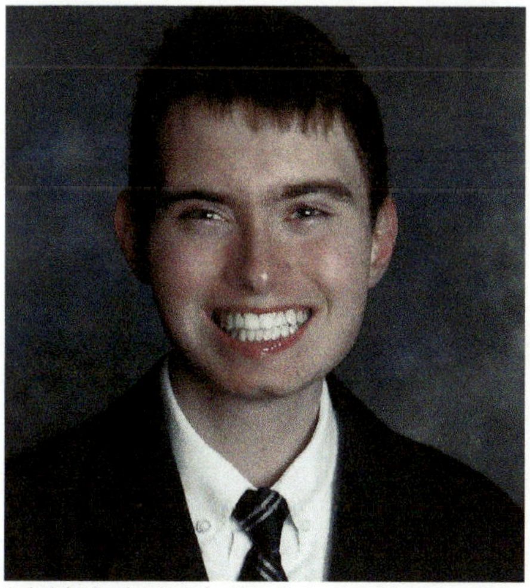
Luke's high school graduation photo.

Oral sensitivity is also no longer a big problem: he can tolerate the hair dryer, lip balm being applied to his dry lips, and three years of braces and the ensuing adjustments.

At a recent MNRI® session, Luke became more aware of his therapist's feelings and was empathetic with her. We feel his sensory system is reorganizing. He responds extremely well to Pavlov Orientation Reflex repatterning. This exercise, in which he is blindfolded and uses his hands to "feel" different qualities, has helped Luke with unlearned elementary concepts: differences between hard and soft, round and square, smooth and rough. Now that he is comfortable with these basics, a cascade of new awareness seems to be coming forth. He reads at grade level 2 or 3; does simple two-column math addition and subtraction, which makes us proud of him. He has had no seizures at since March, 2012.

Recently Luke's dad, Ragnar, brought home a used VCR that he had found at a garage sale on a Friday. Luke's eyes lit up when he saw it (he loves videos and VCRs). Dad said he would hook it up on Sunday when he returned from running in Grandma's Marathon in Duluth, MN. Sunday they would decide if it worked or not. While the parents were gone for the weekend, Luke rummaged out in the garage, found the VCR, brought it into the house, dismantled his current working VCR, and hooked up the 'new' garage sale VCR. The parents were flabbergasted that it was hooked up and running. Luke had it all figured out and wired up prior to their return. He was quite pleased with himself.

Summary

I think how beneficial MNRI® would have been to Luke if he had started it when he was small. I believe his progress would have been much more rapid. For infants born under stressful conditions like premature birth and loss of oxygen, it is critical to their neuro-developmental well-being to receive reflex integration without delay. Although Luke began MNRI® at age 16, he is benefitting greatly. MNRI® can make all the difference in the life of each child.

The changes in Luke – his increased presence, focusing, communication, enhanced sensory-motor integration, and physical strength and skills – are just amazing! We congratulate Luke and all his family, his grandparents and parents, for such great results! – Minnesota MNRI® Team

Finding Treatment that Works for My Boys

Candy Heinrich, MNRI® Parent, Houston, TX, USA

Candy Heinrich

I have two boys. One is almost ten years old and has been diagnosed with autism and ADHD. He had anxiety and sensory processing challenges. The second is eight years old and (unofficially) has ADHD and some sensory issues. While I cannot yet call us an MNRI® success story, we are definitely an on-our-way to a success story. We began the program two months ago and just came home from our second conference.

We have been working on recovery for more than four years. We have spent that time trying out various approaches and protocols to deal with and remove stored toxins and subclinical infections that are inevitable when your immune system is not operating efficiently. We made significant progress with both boys with this approach. Once we started to reduce the toxic burden on their bodies and remove viral and other pathogen infections…their behavior and functioning improved significantly. However, it was just not enough. I realized that no amount of detox or immune system work was going to optimize existing brain function and/or grow new neural connections. I wish I had started addressing neural connectivity dysfunction via the Masgutova Method® four years ago!

I made a false-start of sorts trying out Brain Balance first. I got sucked into their promise of recovery in 12 weeks. I understand now that the idea that my boys could recover in 12 or 24 weeks is just not plausible. How can I expect 3 or 4 years of neural development to be gained in 12 or 24 weeks?

When we did the initial Brain Balance evaluation, the director informed me that the core issue for both my boys was dysfunctional primitive reflexes. Well, I had never heard of that. As we were approaching the end of the 12 week BB session it was clear that I needed to re-evaluate BB for my children. What was surprising is that my older son (autism/ADHD) had made good progress, but my younger son (ADHD only) had made very little progress at all. I dove into research and discovered that MNRI® had a much more foundational approach to addressing primitive reflexes than other programs that were available.

Before I made the time and money investment in MNRI®, I wanted to make sure it was the right choice. So I drove 3 hours to the closest Core Specialist. After a short evaluation, she gave me insights into how my boy's neural connectivity dysfunction was affecting their cognitive function, behavior, and personality. It was absolutely amazing the insights she had into my children…whom she had only met a few minutes earlier. She was able to tell me details about their personality and functioning…all from ten minutes (or less) of pushing and prodding my boy's bodies.

So we took the plunge and signed up for a conference.

The Heinrich boys

The first conference was tough. I didn't know what to expect. Both boys went into what some people refer to as a regression. I understand now that remapping/rebuilding/building new neural connections can be tough on some of the kids. Their brains and bodies are used to working a certain (dysfunctional) way and suddenly they are 'asked' to work in a new/different/more efficient way. Even really good change can be tough. We didn't really notice any improvements while at the first conference. After we were home for a few weeks...that is when the changes began to unfold. I understand now that for some kids (mine) it takes a few weeks for their brains/bodies to adjust to the new neural connections before the improvements will be seen.

For my older son the improvements have been physical and mental. He grew two clothing sizes in less than two months after coming home from the first conference. He has more muscle tone and is more coordinated. Loud noises don't bother him any more. His skin isn't as sensitive. He is sleeping better and his anxiety level is better. Because his sensory system is less sensitive, his brain has a higher capacity to pay attention to his school work. He is not so distracted by what is going on around him and now he can complete his math work with very few prompts. Because his sensory system is less sensitive, his brain is now free to think deeper thoughts. He is starting to plan for his future. He is curious why I have certain opinions. He is wondering about his world. These are all significant improvements for a boy who at five years old used to sit in the corner by himself lining up his cars...only speaking when he wanted to quote his favorite movie or ask for food.

My younger son's improvements are pretty amazing, too. I always knew he had so much potential. I knew he could be really smart in school and could be a really good athlete. However, he seemed satisfied doing nothing. No amount of positive reinforcement or encouragement affected him at all. All of this changed dramatically within three weeks of coming home from the first conference. He now cares about his school work. He asked to join a baseball team. He really wants to do well in life.

We are by no means finished, however, I know this is exactly where my boys need to be. I know that the Masgutova Method will set them up to be the best that they can be.

Presence, increase in logical thinking, better tolerance for tactile stimuli, and inner peace – all these are amazing changes in both boys. We are so proud of you! We also thank your parents for their comprehension of the essence of the MNRI® program and their regular work! – The MNRI® Team

Ethan Catches Up

Laura Carlson, Parent, Mankato, MN, USA

Laura Carlson

The Masgutova Method® has given Ethan an openness for learning that he may have never achieved, or achieved this quickly, had we not started this journey. By age 3, it was clear to me that Ethan was displaying many symptoms that were cause for concern – lack of eye contact, hand flapping, never crawling, delayed speech, foggy awareness of surroundings, melt downs, and an overall general unhappiness. On the contrary, Ethan also had periods where his symptoms were mild and his struggles went unnoticed or were mistaken for discipline and behavior issues, laziness, tiredness, lack of attention or simply 'quirky' hand flapping. We soon discovered Ethan did indeed qualify for a diagnosis of PDD-NOS with Anxiety Disorder and jumped immediately on the path of biomedical intervention.

MNRI® was mentioned to us in passing by Ethan's biomedical doctor and I was fortunate enough to connect with a parent in Minnesota who had traveled to Poland for a few family camps. Because Ethan was so high functioning, there were very few traditional therapies that seemed a good fit for us or affected his body deep enough to enhance social and emotional growth. Despite his diagnosis, it had been difficult to even qualify Ethan for additional services through the school district. Ethan had a year left of preschool and I was very concerned about his social, emotional, and cognitive development heading into kindergarten. He was lagging behind the development of his peers and beginning to show more and more frustration. It was as if there was something blocking his brain from allowing information to absorb fluidly.

Ethan in TaeKwonDo class.

As I researched MNRI®, the concept of reflex integration resonated with me most notably because Ethan had either been delayed, or missed completely, developmental motor milestones such as sucking, rolling, and crawling. Looking back, I now believe Ethan was working extremely hard and expending extensive amounts of energy to compensate for his body's lack of foundational reflexes and doing such a good job that for the first few years of his life even the pediatrician was not concerned! With no prior experience in MNRI®, we decided to attend our first Family Conference when Ethan was 4½ years old.

Following his MNRI® Assessment, we could now see where his body was

Ethan with his family.

not properly processing input for learning and growth. Armed with that knowledge, Ethan and I got to work on a Home Program that included 45 minutes a day, 5 days a week and a monthly check-in with a Core Specialist for the next seven months. These were intense months fitting in therapy with the rest of daily life. Finding time to do the techniques and figure out how to get Ethan to cooperate may have even been harder than learning the techniques. Fortunately, the rewards spoke for themselves. As preschool progressed, Ethan's special education teachers noticed his social and emotional regulation and motor skills growing by leaps and bounds. Ethan was now processing and absorbing information at a faster level than ever before. By the end of the year, Ethan was academically and cognitively functioning at a level equal to that of his peers and more importantly, he was growing confidence in his own body's ability to TRY and be successful at new tasks!

While our journey continues, I feel as though we have discovered a foundational piece of the puzzle in allowing Ethan to function at his best level. There is no question in my mind that we have stumbled upon a program that allows traditional therapies to work better and faster or will eventually even change their future trajectories! We look back at our first year using MNRI® with appreciation and gratitude and look forward to watching the MNRI® program develop to its full potential and our Ethan right along with it. Thank you, thank you!

Ethan, you are such a great and lovely personality! We congratulate you and your parents for all the achievements in learning that you have reached – like memorizing and actively thinking! These achievements show you are strong and talented. Great thanks to your Mom and Dad for their optimism and faith in you, Ethan. And thank you for bringing us this story which can be an example of hope and action for so many kids. We wish your family future success! – MNRI® Team

Bringing Joy and Hope to Vietnamese Families

Ha Nguyen, MNRI® Parent, Hanoi, Vietnam

Ha Nguyen

I'm writing this story with tears in my eyes, tears for the little boy that has gone through severe motor and sensory challenges impacting his daily functioning in almost all areas. Tears of happiness and joy in witnessing his progress and tears of the deep gratitude to the kindest people in the world I have ever met. Only three years ago, MNRI® was just a dream I thought would never come to us. But MNRI® has finally come to Hanoi, a once French colonial city in southeast Asia, thousands of miles away from SMEI based in the US and Poland. The treatment brings stronger hope than ever, it brings us what we had never dreamed of before. We are grateful for the loving and healing hands and spirits of the Core Specialists. We feel truly blessed that we have met them.

I'm mother to a boy who turned 10 last January but he looks like a 5-6 years old. Due to significant motor and sensory challenges, he does not speak a single word and is not going to school with children of his age. He was not toilet trained, he needed another person to feed him and assist him in his daily life, his left eye is tight and over-converged, he sometimes walked on his toes, and his hip and his knees were rigid. His tongue lies outward. We could not touch his face to do OT facial exercises. He was always fidgeting, his fingers were always seeking wet and tough textures. He either went to bed after midnight or woke up at 3-4 am. in the morning. (That made my life as a working mother so hard.)

Three years ago, I was working with a consultant that guided us to restore the relationship with our son through the Relationship Development Intervention (RDI) program. After some time, we got stuck and could not make further progress. We tried another neurodevelopment program from which some miracle improvements happened. But it was not enough and the improvements did not remain long and we felt stuck again. Our consultant and I talked about how we could move forward and that maybe MNRI® may help. As soon as I read about MNRI® – about reflexes – I wanted to try it immediately. I feel a big regret that I did not know about this method in the first year our son was born when I saw abnormalities in the control of his head and neck, low muscle tone, difficulty sucking, and his delay in all movement milestones. And only 3 years ago, I was very upset to hear that I might have to wait for at least another 3-4 years before Vietnamese families would be able to receive MNRI® treatments.

Ha's son with Sharon, Brea, & Shannon.

My dream came true much earlier than that. A Core Specialist-in-Training was permitted to give families under her supervision a few basic activities. I tried to contact a MNRI® Core Specialist with a speech language pathologist background on the MNRI® website and on different forums moderated by parents of children with neuro challenges. I found Brea Maday on the website of the Children Place where she used to work in Minnesota. But the institute told me she no longer worked there. Upset again, a Google search did not find many results. Luck smiled on me in return for my desperate search. I finally found her and she emailed back me saying she would help us!

Ha enjoying a moment with her son recently.

Before Brea could travel to visit us in June of 2012, we started in March and worked over Skype. She gave us more than a dozen simple activities that would help with our son's mood, sleep, and toilet training. When she came to work with us last summer, we were given a tremendous amount of information and work: Tactile, Neurostructural, Repatterning, Archetype, Oral-Facial, Eye-tracking, ATNR, Spinal Galant, Spinal Perez, Abdominal, Balance Board, Hands Supporting, and Moro techniques.

I spent 45-60 minutes every morning before or right after my son woke up and 30-45 minutes at night when he was about to sleep. In the morning, I would play some morning bird songs or Mozart when I did MNRI® for my son.

When we first started MNRI®, occasionally we would see some new disorders. Some other parents that started at the same time panicked and gave up. But we knew we needed to be patient and have a very strong belief: our hard work has paid off and we are excited to witness our child's continued progresses. I have read about and used many other therapies and treatments. My experience tells me MNRI® is the most comprehensive treatment and we need MNRI® before any other treatments.

My husband and I once had a very sad thought that our son might have to wear nappies forever and that he would need another person to assist him for his whole life. But now he lets us know when he wants to go to the toilet. Without being told, he knows to turn on the light in the bathroom, and he's becoming gradually more independent in brushing his teeth and other things.

Results

For the first time in his 8 years, my son went into a normal sleeping pattern – going to bed before 10 pm. and waking up before 8 am. with taking a short hour nap. He no longer coughs at night.

His teacher tells us he's the best boy in his class. She does not have to remind him when the children need to set up tables for lunch, to take chairs back to their place after they have finished lunch, or move to different rooms for different classes. When he went to English class with his helper, he would take a chair for the helper first before he got one for himself. He no longer stims, throws bottles, or tears papers in his class but is calm and cheerful most of the time. He's confident in daily activities in his class and at home. He packs his backpack to go to school.

At home, he smiles at me when I come home from work and will help me carry my bags. He once cried whenever we asked him to do laundry with us but he now happily runs to us helping us with the laundry. He always smiles in his eyes to me, his Daddy, and grandparents. He helps us putting shopping items into bags. He changes his clothes by himself now (he could not do it before due to tight shoulders). He understands almost every thing we tell him. He puts my coat on the hanger (I did not even have to ask!). He wears socks (he never did that before!). He refrains from scratching objects. He plays ball games with our volunteers and can feed himself with minimum support from caregivers. We no longer see the Moro we saw in him 2 years ago; it has changed. He's the super problem solver!

His metal levels have dropped faster than any children that were on biomedical programs. We use a lot less biomedical stuff; we think we may not even need it any more.

The list of benefits goes on.

In 2013, we had the honor to welcome Brea back with her wonderful team: Sharon Heller and Shannon

Desilets. Another dream of our life came true: receiving full 5 days of MNRI® programs with MNRI® Core Specialists! I was very nervous before they came to visit us. How would they be able to work on our son? He had not let other strangers work on him before and it will be five days! But I, my husband, and other families were in awe with their great talent, dedication, and the love they gave us and our children. They not only worked on his body, they even worked on his super sensitive hands that he cannot raise high. They worked on his face when he was awake! They worked on his TMJ! All this was greater than my expectations!

Ha Nguyen (far right) with her new MNRI® friends, Pamela Curlee (top), Amy Cameron (left), and Sharon Heller (center) in Baton Rouge, LA.

From this work, I'm confident we will continue to see encouraging improvements in his movements, senses, and cognition. Neurostructural helped build his muscles stronger and helped him feel his body better. Neurostructural and Spinal probably helped with his toilet training. Tactile helped reduce stimming and gave him mood stability; Babinski, Foot Tendon Guard, and Moro calmed him down, putting him in a ready-to-learn mode.

He now has total bowel and bladder control. He has gained 8 kilos (17 lb. 10 oz.) and is 10 cm. (almost 4 inches) taller after 2 years of no change in his weight and height. I am looking forward to the day when our child will be able to express himself in a functional manner, to be another Carly Fleischmann story to tell the world! And before that, we are looking forward to welcoming Dr. Masgutova to our country, where she can even speak Russian with families here!

Inspired by my child's progress and by the loving spirit of the MNRI® Core Specialists, I decided to take MNRI® core training classes and have been further motivated by the *Archetype Movements Class* with Connie Jordan and *Dynamics and Postural Class* with Diane Whiteside. I plan on continuing other MNRI® classes in the future. I have been so honored to have received the approval from Dr. Masgutova to attend the *Neuro-Structural Reflex Integration Class* later this year.

I send our love to Brea, Sharon, Shannon, Geri, Jessica, Pamela, Susan, Lori, Connie, Diane, and all the Core Specialists for their support in making these long trips and bringing the MNRI® therapy program to families in Asia.

We thank Ha and her son for their devotion and belief in the MNRI® program and for bringing it to other families in Vietnam. We wish you ongoing progress and good health! – MNRI® Team

All We Had Hoped For and Far, Far More

Mary Smith, MD, Cardiologist, MNRI® Parent, USA

Our story started out like many others – we were blessed with a much awaited, beautiful little girl on Sept. 14, 2006 – our little princess, Anika! She was all that we had hoped for, and we could hardly get enough of her. She was bright-eyed, interested in things, loved being taken outside for walks and loved being read to.

We started to have some worries when she showed no signs of walking by a year of age, but we were repeatedly reassured by our pediatricians that all was good. Over time though, our worries continued to grow: she seemed to move slowly, didn't walk till she was 18 months old, started to get horribly constipated after age 1 (when cow's milk was introduced), and then started to get frequent colds. She also did not sleep well, and after age 2, she started to get increasingly upset by live music and loud sounds. At the same time, we were busy welcoming her little sister into the world, and blamed a lot of issues on that. But all our false reassurances came crashing to the ground when she turned 3 and we took her to her first day in pre-school. Her teacher called me the same evening and suggested an evaluation and then everything went into a high stress blur. Evaluations, meetings, appointments and before anyone used the word, we realized that we were dealing with autism.

Through the horror and the grief that the diagnosis brought, came a rapid realization that we were going to have to figure it all out without any help from the medical community. Her pediatrician had no useful suggestions, and the pediatric neurologist was even worse. Her speech was in the first percentile, her gross motor skills tested below the level of a 1 year old child and fine motor skills were nonexistent. She had a decent vocabulary but could not make even a brief sentence. The school district team suggested placing her in a special education setting, as she was not felt to have enough skills to be mainstreamed even with a full time aide.

Fortunately, we ran into a wonderful lady who we had hardly known beforehand, but who had helped her own daughter navigate her way out of the autism spectrum. She was our first guardian angel. She told us about changing her diet and possibly using supplements, concepts which seemed to be contrary to everything the doctors were telling us. We were lucky enough to decide very early on, that we would follow the path suggested by the one person who had actually helped their own child out of autism rather than listening to a hundred experts who had never successfully treated it.

The next year was the busiest year of our lives. We placed Anika on a strict gluten/dairy/soy free diet, started seeing a wonderful DAN! Doctor, started sensory integration therapies, tried several supplements, enrolled in a highly recommended special-ed school, started working with an RDI consultant (Relationship Development Intervention) and also started seeing a homeopath. Very early on, in fact 2 days into changing her diet, her brain fog started to lift. By 6 months into all this, she was sleeping through the night, got fully toilet trained, stopped getting sick all the time and her horrible constipation disappeared. But along with this wonderful progress was the sickening feeling that, while she was making great progress compared to other kids on the spectrum, she was still falling farther and farther behind her neurotypical peers. About a year after her diagnosis, we had another chance conversation with a friend-of-a-friend-of-a-friend, and he told us about Dr. Masgutova's work. He was our second guardian angel.

Our first MNRI® camp was a 5 day camp and we actually found it rather overwhelming and a bit confusing. But we also saw some unmistakable improvements which left us very excited. Before attending our first camp, Anika used to limit her foods to certain textures. Things had to be soft, apple sauce was her favorite. In fact most hard/dry/beany things would simply make her gag. Two days into the camp, she ate an entire gluten free rice tortilla, to my complete surprise. She also exhibited the very first glimpse of resilience. Before going to the camp, she used to get upset very easily and would not re-try things if they seemed difficult at the first attempt. During the camp, she fell while trying to run across a concrete floor (she had just gotten out of the hotel pool), and instead of the usual dissolving-into-tears, she quickly calmed herself and moved on. To us, this was far more impressive than the slow gains she had been making with the usual physical therapy, speech therapy, and occupational therapy, and we decided to dive into Dr. Masgutova's work more.

The second camp we attended really confirmed our initial impression. Anika made such strong gains in her motor skills that her physical therapist and her speech therapist were completely surprised. They had to rewrite their goals for her when she returned from the camp. At the third camp, she was enrolled in a clinical trial and had a thorough pre- and post- assessment, which helped us quantify what we were seeing. She gained 14 months of skills in 10 days, whereas with the usual physical therapy, she had gained 4-5 months of skills in 6 months. This was the last proof I needed. We pulled her out of her speech therapy, physical therapy, as well as occupational therapy, and began using MNRI® as our primary therapeutic modality.

About a year into MNRI®, we found a wonderful Feldenkrais practitioner and started seeing him regularly, while still working closely with MNRI®. We saw stronger and stronger gains, with doing both the modalities together. When Anika was ready to enter kindergarten, we met with the school district to re-write her IEP. This was 3 years after her initial evaluation. The team could hardly believe that this was the same child they had assessed 3 years ago. She no longer qualified for an aide at all, and tested in the completely normal range for speech. To qualify for any help at all, the team had to really work and identify areas of concern. She is now in first grade, has been mainstreamed since kindergarten, and only receives social skills help about 10 minutes a day.

We home school Anika half the day, mainly so we can use her time more efficiently to do her home therapies (we currently do vision therapy, Brain Gym®, and MNRI® therapy at home). She attends regular gymnastics and just won her white belt with a black stripe in Tae Kwan Do. She is an excellent swimmer and reads well above grade level. She is on the Kumon® honor roll for both mathematics and reading. She is able to do all that we had hoped for, and far, far more.

We used to read stories of other children online, in books, and on blogs – and we were always so heartened by happy endings. We feel so grateful that we have been able to share our own happy ending with you, and we can say for sure that Dr. Masgutova is our third guardian angel.

Anika has achieved excellent results with the MNRI® program. We congratulate her and all her family members and wish them all good health and new successes! – The MNRI® Team

Gains and Improvements with MNRI® Are Remarkable

C's MNRI® parents

C.O. is a 39 year old female with a diagnosis of autism and disarthria. She does not take any medications but has been on various supplements throughout the years. She was home-schooled because of a lack of appropriate placement options. She can read, write, and do basic math. From a young age she has demonstrated an ability for music, solving jigsaw puzzles, memorizing names, phone numbers, and addresses, spelling, and other visual skills. Her social skills, logic, reasoning skills, spacial awareness, proprioception, balance, etc. have always been very poor.

She has been exposed to various therapies from the age of 3 years. The list of therapies include speech therapy, occupational therapy, Behavior Analyst services, Doman-Delacato method, Interactive Metronome, Neuronet Learning, Hemispheric Integration, Quadriciser, HANDLE, Rhythmic Movement Training, Anat Baniel Method, Neurofeedback, hippotherapy, dancing, and, most recently as of January 2013, MNRI®. The therapies previous to MNRI® produced some results and improvements in behaviors, intelligence, social skills, language, speech, etc., however, it was always the case that the majority of the gains were lost after reaching a climax. This proved very frustrating for her and for us as parents. The potential was seen, but then we would find a regression to previous behaviors.

In the 5 months since starting MNRI®, however, her gains and improvements have been remarkable. C had a history of violent behaviors and outbursts. When she got angry, she would physically attack property and people, many times requiring physical restraint for her own protection and the protection of others. Since participating in her first MNRI® conference, she has not had any volatile behaviors but has found very acceptable alternatives; removing herself from situations that anger her and being able to verbally express her feelings. This is a huge improvement considering her history.

Her speech articulation improved (about 25% clearer – several friends have commented they understand her on the first try now) and her language has improved greatly. Before MNRI® she would respond to questions with one word, or would usually say, "I don't know" in a very slurred manner. Now she responds in full sentences, many times expanding beyond what is needed. She follows multiple step instructions without any problem, such as "Go take this to the kitchen, put it in the refrigerator, then bring me a bottle of water and a pen". This was impossible for her to follow before MNRI®.

Her understanding of conversations has improved dramatically. An example would be when we were talking and I mentioned to someone that I would have to get something from my purse, C immediately went to get the purse from another room and brought it to me, without being asked.

She has become aware of and concerned about the feelings of others. When her grandfather passed away recently, we were on the way to the funeral home when she said, "Don't worry about me. I left the old C at home. When we get to the funeral home I will be fine. I will be calm, all I will do is say 'Hi' to people and sit quietly." When she arrived she was true to her word. At the funeral home, she went to the coffin and said "I'm sorry Grandpa, I will miss you and see you in paradise," then she touched his forehead and went and sat down. She appropriately greeted people as they arrived and sat through the service without any problem. The level of maturity was outstanding. Even after such a difficult event, she showed no regression.

C also has an increased awareness of what is happening in the moment and a desire to be recognized as being present. She now wishes to participate more often in activities with the family. Her problem solving and planning skills have shown dramatic improvement. She plays a game called Rummikub which requires planning, strategy, and organization skills. She is able to play the game independently and often wins. Her father said that she has played the game while working on a search-word puzzle simultaneously and she made some very good plays that required planning about two or three plays ahead and won the game fairly (no concessions made). I think she is very intelligent.

C's self-confidence and independence have also improved. She has no problem going into an elevator alone (in the past it was an ordeal to get her to go in an elevator, many times impossible). Car trips have become pleasurable, whereas, in the past it was an ordeal requiring many stops and with various outbursts. The list of improvements goes on and on. We look forward to more improvements as we continue with the program.

P.S.: Update after last Therapy Clinic: A frequent problem we had with C was, when my husband and I were eating with C, she would begin screaming and become hysterical. She sometimes refused to be in a car with us, and to go to restaurants with us. This behavior was present when C started MNRI® in January 2013, but for some reason that we do not know, the frequency, intensity, and duration of this behavior appeared to have been increasing to the point that we, the parents, did not know what to do. This, I think, was the only setback we observed since she started MNRI®. Although, she gets angry, she does not try to hurt anyone but just goes running upstairs to her bathroom, slams the door, and takes a long bath in her bathtub. This behavior was repeated up to eight times in a day because she had this same reaction when she saw that a meal was being prepared and also when we got a drink, even water.

So during the clinic, Svetlana and a therapist did a psychotherapy process one day with C. (Her Mom was not present at the time.) Svetlana said, "It was a very strong process of learning to please parents and taking care of them and to be good to them, and so on. We did goal setting."

We saw the results very quickly. We noticed a big change in C's behaviors within a couple days of coming home. There was an immediate reduction in explosive anger episodes after the MNRI® Clinic. My husband also observed positive changes in her behavior in general. Last night C went to sleep without the 'white noise', she seems happier, and even is sharing more time with us. I have not tried to push issues like her eating with us yet but we had to make two trips to the hardware store on Saturday and I observed she was acting much more normal. She still refuses to eat with us but does not seem to get angry, just takes her food to her bedroom. Also she has not been taking baths when we eat downstairs because her bathtub has been disabled due to a present plumbing problem. The out-of-order bathtub issue, although a coincidence, may be working to break her old bad habit. I am not sure if the positive changes in her behaviors are a result of the procedure Svetlana did with C during the Friday morning MNRI® session or if maybe it is a combination of the cumulative effect from all the other sessions as well. Anyway, I thank God, Dr. Masgutova, and other therapists who worked with C that week (and before) for the positive lifesaving changes.

C, we wish you many new achievements in your future; your dance abilities are just amazing! We look forward to seeing your other skills developing as well! Thank you for your parents – so giving, caring, and loving. We wish C her best in taking good care of them! – The MNRI® Team

Impacting Seizures and Cognitive Abilities

Elizabeth Rosalina, MNRI® Parent, Jakarta, Indonesia

Elizabeth Rosalina

I am a mother of two children, and I want to share their story with MNRI®.

My daughter, Jane Gabriella, 13 years old now, was born with a 10 of 10 points on the APGAR score and a normal head circumference. But suddenly, when she was 6 months old, she had a motor delay and then, when she was 7½ months old, macrocephaly (overly large head; a sign of possible complications within the brain). A CT scan showed that subdural fluid had collected in her frontal brain area that might result in a motor delay. Even with motor delay problems, she still had very good social contact with other people. We went to physical therapists to improved her delay. Then at 15-18 months, she suddenly started developing some autistic features and, also, her language (baby babbling) disappeared along with her sociability.

She was able to walk at 18 months but with poor motor coordination and control. She often fell down and hit her lips on the floor. She also had poor oral-motor abilities and drooled, couldn't chew, and could only eat porridge until 3½ years old. She couldn't swallow food with textures. She had poor fine motor skills and she couldn't hold and grasp an object or imitate finger movements. Objective studies showed that at 2 and 3 years, her EEG was normal except around speech centers, but the Magnetic Resonance Imaging (MRI) showed a mild myelination problem at the temporal parietal bilateral area.

At 3, her development began improving. However, she had an accident – she fell into an open deep hole when she was playing in a public garden. Her lips were broken as her jaw hit the metal ring around the hole. She got a vaccination after that to avoid tetanus. After this, she regressed significantly; she couldn't point her finger to an object anymore and her improvement from any therapies thereafter was slow.

We tried so many therapies, techniques, and approaches for Jane including Traditional Chinese Medicine overseas. Then we attended our first seven day MNRI® Family Conference when she was 7 years old and have continued participation in them since then. Our first Conference was not easy for her and she was irritated by the touch and exercises as they were asking her to do something different than her routine behavior and it was interpreted by her as limiting her 'freedom.' She also didn't want to be on the massage table, vocalized a lot, and so on. In spite of all that, I saw good progress in her tactile defensiveness and improvement of her hypersensitivity (tactile and auditory) after the third day of the conference. She started wanting to play with sand(!), and it became easy for me to cut her nails (wow!). One month later, she was able to grasp and have better eye contact. Two months later, she attempted to hold a pencil and try to write letters. I was so happy with the result!

My happiness didn't stay long because at 7½ years of age, she suddenly had her first seizure which increased until she was 8½ years to 9 years old. I was very scared when I saw the electroencephalography (EEG) showing spiky waves in her frontal brain area at that time.

I know kids with autism have a greater chance to get epilepsy when they are 7½ to 19 years old because hormonal changes do not match with their brain development. At that time her body was very sensitive to medications and food supplements, so I used kinesiology as a non-invasive approach to control the epilepsy and continued using MNRI® to avoid any regression in her brain. And surely, the exercises did help her – I didn't see any major regression after the epilepsy attacks as I had in earlier times. Still, she was continuing to have poor muscle tone affecting her fine motor coordination and she became tired easily when walking long distances.

Jane Gabriella (left), Elizabeth Rosalina with her husband and son.

The same MNRI® exercises have helped my son with his speech delay in the past 4 years. He had some tendencies for hyperactivity, didn't make friendships, focused too much on toys (OCD), poor eye-motor coordination, refused to look at my eyes, wet the bed, thumb sucking, allergies, and, yet, he was amazingly creative at the same time. We were visiting an osteopath doctor once in awhile for him to fix 'tension' in his back and neck, before I knew the MNRI® method. (I didn't want to bring him to a doctor who might confirm a diagnosis of Aspergers.) I decided to work on prevention and started applying MNRI® to him also. What a surprise for me when his habit of thumb sucking disappeared when I did the Babkin Palmomental exercise with him! I worked with Visual Reflex patterns and got eye contact from him after just 3 times of Symmetrical Tonic Neck Reflex and Tactile eye area activation. His bed wetting stopped after I applied Spinal Perez. Also, his behavior was and is changing a lot. He has become more patient and polite, he has started to make friends (and now even feels lonely without friends and family members). His school problems until now were low motivation to learn and over-focusing on toys and play. He didn't like to do the MNRI® Home Program or school homework. He had challenges at school, though his mind was sharp and he was SO smart. Subjects for primary school in my country are strange, as in many countries possibly – kids must learn many things that are not important or interesting and thus it does not motivate them. Still, they must learn it and we must teach our children patience. And again MNRI® is helping me to keep him patient and smart. We do MNRI® exercises before tests and exams, and his results are good so far. And his talents are growing – now he loves music and singing, he even started playing piano. Before, he didn't like music (saying it was too noisy), he didn't like to sing and it was difficult to teach him to sing especially in the classroom. His remaining learning challenges (handwriting and composing longer sentences) are now improving. He loves having sessions with the MNRI® team and is eager to improve all of his skills and his personality.

Because of my kids' developmental needs and because I want them to grow and be happy in this life, I keep inviting the MNRI® team and organizing the MNRI® Family Conferences in my country of Indonesia. I want to support MNRI® development for my kids and for other kids with challenges. I am open and happy to share my experiences with other parents. My daughter, having multiple developmental challenges, is now improved very far compared to her previous condition. She is now cooperative in the Conferences, able to follow instructions, and she wants to learn to speak (closing and opening her lips upon instructions). Her oral-motor and also visual-manual imitation are much better. She is able to use a spoon and fork together independently when eating(!). She is able to chew and can eat crackers and nuts. She can easily swallow capsules and tablets. She is able to hold pencils (although still not firm enough). Now she can type in her mini calculator dictionary. She also has better comprehension and eye tracking movements. She must wear glasses but, because they are not comfortable, we need to find other ways. I hope and believe that MNRI® techniques may improve her vision. Clinically, she no longer has seizures although we need to repeat an objective EEG results.

Jane is an amazing abstract painter using her hands and limited painting tools. At her art exhibitions, audiences evaluate her paintings as unique, beautiful, and they want to buy them.

MNRI® Conferences and Mini Clinics in Indonesia have shown significant changes in kids and parents support these results with consistent implementation of MNRI® Home Programs. We know that the parent's patient and enthusiastic work has resulted in kids that have become more intelligent, with better speech development, are less spastic (children with CP), improved motor planning and coordination, and their behavior has become more mature and easily controlled. Their oral-facial skills and face/head structure improves in symmetry and functions, they find voluntary control of articulation to be easier, more relaxed and they smile and joke more; their vision is improving significantly and much faster than with other therapy modalities.

I like Dr. Masgutova; she is honest in assessing the kids. During her MNRI® reflex Assessments, if there is any progress, she will say, "Bravo", while if not or only a little bit progress, she will give parents encouragement to do more work or give them specific exercises for certain developmental areas (also for her specialists at debriefings). I watched the MNRI® Team during their debriefings and they are a team of great integrity, discussing solutions for different challenges in kids; they love and respect the kids and their parents so very much!

I believe MNRI® is the best technique I have ever known. In organizing Family Conferences or Mini Clinics, I always feel that the MNRI® Team is my big family.

I send love and hugs to Dr. Svetlana Masgutova and her MNRI® Team and also all the kids with challenges, who are True Winners!

We are so grateful for Rosalina for her excellent work in organizing the development of the MNRI® Conferences in Indonesia. We wish Jane and her brother good health and more new amazing pictures. Blessings to all her family. – MNRI® Team from Poland

Bringing Inner Peace and Trust to Your Child

Diane Hunter, MNRI® Parent, Life Coach, Los Gatos, CA, USA

Diane Hunter

My son Ian and I first met Dr. Masgutova when he was 5 years old.

He would not sit still for therapy and I was a bundle of stress and emotions suppressed and bottled up in my body. His body was in constant motion and state of protection. Although he wasn't using words to communicate, he had no issue with communicating what he felt.

This is how Ian expressed himself in the world back then. He would glance at you in fleeting moments with his sparkling green eyes. He was in there ready to show me his true self but he didn't feel safe yet in his body or his environment. He would constantly bang on the windows as if to say "get me out of here." He would run up and down the hallways, hit therapists and yell and scream. I chased him around trying to keep him from getting hurt or into trouble. I needed the environment to be very clean and free of allergens for him. This is what I thought would protect him and keep him safe. He would sit for only a few seconds and the only way to do therapy with him was outside in open space.

Ian with MNRI® therapists

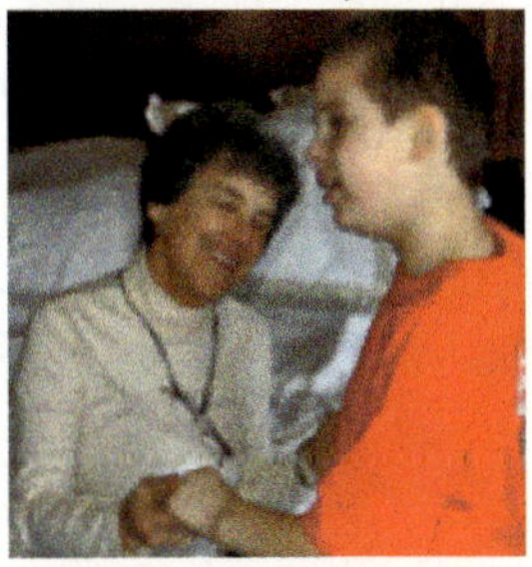

What I began to see was how my level of intense focus and vigilance activated his state of protection and influenced the way Ian expressed himself. One of the most powerful lessons I've learned through working with Dr. Masgutova and MNRI® is how sensitive Ian is to the way I feel. He is far more sensitive than I ever imagined. I had all these emotions trapped inside my body and I wasn't allowing myself to feel. When I am stressed, scared, and confused, it does not feel safe to him and it would activate his protection response. It was as if he was a mirror of my emotional state and he would run around yelling and pulling items off the counter and banging windows.

So I began to shift some of my attention on taking care of myself and finding ways to bring in more acts of self-compassion and self-care into my daily life. Back then, when raising a child with autism, I often put my needs to the side and completely ignored what my body was requesting. Why would I not give myself the same level of love and compassion that I was giving to my son? Why did I think I deserved any less?

As I started to learn to take care of myself and feel my emotions and let them move through me, I began to breathe deeper and I noticed he started sleeping better, would sit for longer periods of time and a relaxed, curious, playful nature emerged.

One evening I was feeling sad but not allowing myself to really feel it; I was trying to hide it. Ian was pacing around the house and kept trying to get out the back door. It was way past his bedtime and he showed no sign of settling down. I stopped trying to hide it and sat on the couch and cried. Something very surprising happened next. Ian came and sat down right next to me, peacefully, while I cried. It lasted for about 5 minutes. When I felt done and complete, he stood up, walked into the living room, lay on the couch and promptly fell asleep. It felt as if he was waiting for me to release the emotion.

This turned everything around for both of us. I really started paying attention to the way I showed up in our relationship. If I was stressed or in any kind of state of fear, I would first be honest with him and let him know it wasn't about him and he didn't need to take it on. Then I started finding ways to take my fears out of the dark and bring them into the light and look at them one by one. As I dissolved my fears, Ian relaxed. We've been on such a powerful learning journey together and I consider him my greatest teacher.

Two years ago Ian's emotional world and comprehension improved in an outstanding way. He was looking for communication with specialists and other children, though non-verbally. He was expressing himself with behavioral and body movements – showing amazing interest, sympathy, and even support. When Dr. Masgutova asked him jokingly, "Ian, help me to help you" or "Dear Ian, tell the new boys that sessions are important for growth, tell them to follow techniques with good motivation to change," he started coming up to new children and vocalizing sounds corresponding to, "Please, listen, it is true! Please, follow the techniques!" The MNRI® team was accompanying his emotional changes with great love. Dr. Masgutova was so proud the first time Ian kissed her after her work with him. Ian then started imitating more and understanding play better.

Ian kissing Dr. Masgutova.

Two conferences ago he started peeing in the toilet. This was HUGE!! Now he will hold his pee through the entire night and wake and immediately go to the toilet and pee. For any parent with a child that has not been grounded in their body enough to feel the sensations and messages, this is a huge shift and evidence of integration. He feels safer and safer in his body and the neural pathways are repatterning. He also now walks to the bathroom to let us know he needs to go.

After the last conference, Ian started sleeping 12 hours a night. Sleep and rest are essential for integration. The more we do MNRI®, the better Ian sleeps, the more present and engaged he is and the more joy he expresses.

It has definitely been a test of patience and commitment, and at times my attention shifted in other directions. But I always came back because I believe the repatterning is the core component that ties everything else together and provides the web of support.

Ian was 12 in Feb. 2014 and after almost 7 years of using MNRI®, I say with great confidence that MNRI® provides the foundation for his integration. It supports and compliments all the other therapies we offer him such as the Son-Rise Program® and diet interventions. Without MNRI®, Ian would not be as integrated as he is today.

This is how Ian expresses himself in the world today. He follows what feels good to him. He eats when he's hungry, he plays when he's feeling playful and curious, he rests and sleeps when he's tired, and he is pure love all the time. He is more and more interested in being present with the family and I truly believe this is because he feels safer in his body and in his environment. He teaches me every day to be present and his level of patience with me blows me away.

I'm so deeply grateful for the commitment, wisdom, brilliance, compassion and love that Dr. Masgutova generously gives to the world. She has given our family such a gift. We have found our way home to love.

Ian's way of exploring and communicating with the world teaches us all about non-verbal communication and is an example for us to understand the inner world of another person. We thank his dedicated Mother and Dad for all they do for Ian to be a Winner. – Diane Whiteside & MNRI® Team

Autism After the Masgutova Method®

Debby Gans, MNRI® Parent, West Palm Beach, FL, USA

Debby & Benji Gans

I watched in awe as Benji spoke in front of a group of therapists, parents and children last night at the start of the Masgutova Method® intensive therapy Conference, "Hello, I'm ten years old and I started coming to these conferences when I was five because I have Autism Spectrum Disorder..." He spoke clearly, had great eye contact, and peppered his chat with appropriate humor. Afterward, several therapists and parents approached me and said, "Autism spectrum? Really? It does not seem that way to me!"

Today, autism no longer defines Benji but it wasn't always that way. We spent many desperate years first wondering if he would ever stop crying, then whether he'd ever learn to speak, and later whether he'd ever be comfortable in his own skin. We tried play therapy, OT, PT, speech therapy, ABA, The Listening Program, Relationship Development Intervention®, and HANDLE, all of which helped him make some progress, although it seemed glacial. I watched other children easily learn to talk and walk, easily transition from bottle to cup and easily try new things and meet new people. My child resisted everything different. He was inconsolable if I didn't push his stroller in the same direction each and every time we went for a walk. When we finally summoned the courage to take away his bottle, for almost a year, he refused to drink from anything but the garden hose. He didn't sleep through the night, no matter what sleep expert we consulted, until he was six years old.

Benji Gans in 2014.

After endless amounts of searching, we found some therapies that really worked. The summer after he finished kindergarten, we went to his first Masgutova conference. To me, the therapy looked like a relaxing massage. My son, on the other hand, viewed it as an attack. Fortunately, my Dad came with me to that conference. It took everything in both of our power to keep Benji on those massage tables so that the therapy could begin to calm his overactive nervous system. We stuck pennies in his toes to distract him, we took turns singing in his ear, we bribed him, promised him a swim in the ocean if he'd just hold it together for 10 more minutes. It was the longest four days of my life.

As we drove home, I already knew it had been worth it...the child who could previously only express his feelings by crying or cowering in fear said, "Whose stupid idea was it to come to this place? I never want to do that

Benji Gans in 2014.

again!" During the next few weeks, the changes were apparent and easily visible. Benji's entire affect relaxed. His ability to self-regulate improved. He began to express his feelings. He became curious about the world and began to fear it less. Suddenly, he could walk without tripping. And, he quickly decided that his "cushions" (orthotics) were uncomfortable and asked that they be removed from his shoes! Within a month the shape of his foot began to change. His curved toes dramatically straightened.

The most dramatic shift was evident one evening about 6 weeks later…

"There were a lot of policeman in the Chinese restaurant tonight. I wonder why they were there."

He was quiet for a minute, then continued "the policeman who came to our table called me 'little buddy.' That's because he didn't know my name." Suddenly, he could both process and analyze interpersonal relationships. For so long, I feared he'd never be able to do that. After five years, we continue to see changes thanks to Dr. Masgutova and her amazing team.

The Masgutova Method was a key, major element in Benji's struggle against autism. But, as all special needs parents come to learn, there is no one magic, silver bullet that can fix all of the issues. So, while doing our daily Masgutova exercises at home, I continued to search for other tools to help Benji move more easily through the world.

We found a developmental pediatrician near us in Florida who understood Benji and offered tons of practical advice. Dr. Udell discovered that Benji could tolerate casein, but not gluten. When we took him off gluten, he could focus more. Later, Dr. Udell suggested we give Benji a small dose of caffeine each morning. It further improved his focus and it lacked the side effects of traditional medications.

Last year, we stumbled upon an autism research study that also made an enormous impact. It evaluated brain changes in mildly autistic children after intensive work using the Lindamood-Bell learning process (LMB). Benji attended both school and four hours of intensive therapy each day, 65 miles from home. It was tedious, but certainly worth it! He had easily learned to read words years before, but until LMB he couldn't understand what he read. He learned to draw conclusions, especially when eaves-dropping on my conversations. And he understood that his grandmother was dying even though no one told him. He insisted on speaking at her funeral and spoke with the eloquence of an adult, not a child on the spectrum.

This past weekend, he returned from an amazing overnight camp for boys with social anxiety and ADHD. A few years ago, I never dreamed he'd be able to function independently ever, let alone at ten. He thrived there and returned more organized, thoughtful, and seemed to have matured two years in three weeks!

Not only did he finally learn to talk, he also leaves an indelible impression on everyone who knows him. He is sweet, loving, and will sometimes still hold my hand in public places. In the beginning, I expected to be the teacher. Now, I realize that he came into the world to teach me many things…the power of hope, the ability to think outside of the box in order to solve problems, the strength of optimism, and above all, to believe that anything is possible.

We thank Debby and Benji for their confidence in the MNRI® program and her tenacious love of Benji and perseverance in finding hope and treatment for him. – Dr. Masgutova and her MNRI® Team

Opening the Door to Potential

Danielle Mixon, MNRI® Parent and Core Specialist-in-Training, Prairieville, LA, USA

MNRI® has changed my son's life. Will was born full term, and met all of his developmental milestones on time as an infant and young toddler. I remember taking him to his doctor for his 2 year check-up and for his vaccinations at 26 months of age. He was happy, healthy, and age-appropriate for all of his milestones.

Danielle Mixon

Within weeks, I started to notice that he wasn't talking as much and seemed to be drifting into his own world. He became extremely anxious and started having frequent severe meltdowns for seemingly no reason. He stopped following commands and responding to his name when we called him. He stopped playing with his toys, started banging objects on windowsills, opening and closing doors and drawers repeatedly, and would run his fingers up and down our wooden blinds repeatedly, to the point where he has broken many of them. He would walk around the house, much like the Energizer bunny, and go from one nonfunctional activity to another. He lost physical skills as well in that he could no longer push the pedals of his tricycle to make it go. He would sit on the tricycle and just look confused as to how to work it. He stopped pointing at things and became completely nonverbal.

Will during sad times.

By 31 months, we went through the process of having him evaluated by Early Steps, an early intervention program for children under 3. Results of the testing showed that his receptive language abilities were greater than 3 standards below the norm, or an age equivalency of 13 months. His overall communication scores were less than the 1st percentile for his age. His overall cognitive scores were in the 6th percentile for his age. As a pediatric occupational therapist, I knew that these scores indicated a poor prognosis for his future. At the time, I worked with children in schools and knew that kids with test scores this low are put in self-contained classes.

Will qualified for a speech therapist and a special instructor to come to our home to help him. I called his sessions "scream therapy" because that's basically what they consisted of. The therapist and teacher tried to work with him and pull him out of his 'world', and he protested by screaming and having tantrums. I also tried to do therapy with him with no luck. He would end up crying and having a fit if I put any demands on him.

As he approached age 3, he was evaluated by the public school sys-

Will enjoying a pony ride.

tem. I remember vividly during the 45 minute evaluation with the speech therapist, Will did not say one intelligible word, but seemed to speak to his hand in a high pitched jibberish. He also opened and closed a drawer repeatedly, did not respond to his name, and didn't seem to notice that other people were in the room. When given a toy horse, he banged it on the table repeatedly. When the educational diagnostician evaluated him, she took him into another room, and came running out within 10 minutes because she couldn't get him to stop screaming. He qualified for speech therapy and special instruction for the school system. When they would work with him, it was more of the same scream therapy. Some of his goals were to respond to his name and to follow a 1 step command.

Soon after he turned 3, a friend suggested that I look into Dynamic Therapy Specialists, a therapy group in Baton Rouge. I made an appointment and Will started MNRI® therapy once a week with Core Specialist, Vicky Roy. She showed me some exercises to do with him on a daily basis and, very soon, things started to change. The first signs of improvement were that his tantrums started to decline, and he stopped talking jibberish in his high pitched voice. After his therapy sessions, he seemed relaxed and would come home and lie on the couch for a while. Within a month, he started to look at me again, and his repetitive behaviors started to decline. I continued his weekly therapy sessions with Vicky and I did MNRI® with him at home for about 20 minutes per day, 5 days each week.

I started taking MNRI® courses, starting with *Lifelong Reflex Integration, Children with Challenges,* and *Dynamic and Postural Reflex Integration,* which were extremely helpful in learning new techniques to help Will overcome his challenges. He started to notice his surroundings, started talking a little bit, and developed a close bond with me. He began verbally labeling objects but his functional speech was slow to develop. He had no concept of words that did not have a visual representation such as pronouns, prepositions, adverbs, or adjectives. The MNRI® took his brain and body out of such a severe state of fight or flight that he was eventually able to focus on learning new information. We started a Home Program consisting of MNRI®, fine motor and gross motor skill training, as well as verbal behavior training to teach him language in a way that he could understand. He also attended traditional speech therapy once a week to work on verbal skills. He was finally able to accept other interventions without getting completely overwhelmed. Other professionals that worked with him, such as his preschool teacher and traditional speech therapist, said that they have never seen a child learn so fast. I can thank MNRI® for that. He started to learn new information and, with a lot of teaching, was able to start bridging the gap between where he was and where he needed to be. As a therapist, I was blown away with how well MNRI® worked for Will, especially when so many other strategies had proven unsuccessful.

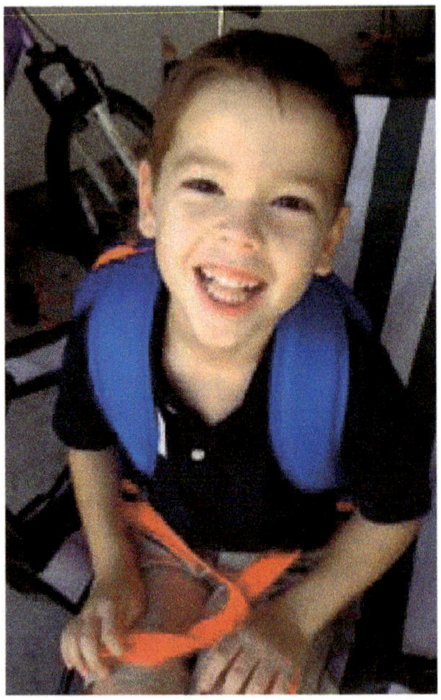

Will engaged and enjoying life.

We continued working at home with Will daily as well as weekly therapy visits with Vicky for 2 years. He would show progress and then would seem to have a decline from time to time, which I later learned was called a transition phase, in preparation for higher levels of growth. During transitional phases, I was told to continue the exercises, and then I soon saw improvements again. For a while, he would stop one nonfunctional behavior and then seem to gravitate to something else equally nonfunctional. We kept working with him and, eventually, he stopped all behaviors such as opening and closing doors, etc. One remarkable memory is how he was fascinated by letters and numbers and would read website addresses on TV, signs in the environment, etc. to the point that this is all he wanted to do. I took the *MNRI® Auditory Visual* course, and did the protocol with him for 2 weeks (4 times total) and he completely stopped this nonfunctional behavior! He started noticing the pictures on TV and in books instead of just the words. As a result, we soon saw a jump in his language and comprehension skills.

He is now 5½ years old and is in a regular education kindergarten class. All of his autistic characteristics have faded and he fits right in with his neurotypical peers. He loves school, has several friends as well as a charming personality, and always gets green lights for good behavior. He is a little chatter box and is excited to tell me all about his day when he gets home from school. He is able to draw pictures, write a paragraph, and he typed 9 sentences with his sight words in a computer lab lesson! His teacher was blown away because she has never seen a 5 year old type so much correctly. His peers even voted him as student of the month! He has mastered most of the kindergarten skills and is performing well above average for his age.

Will waiting for the school bus and what the future will bring.

His latest test scores are as follows: Receptive language – 5 years, 9 months; Expressive language – 6 years, 3 months; Literacy – 7 years, 2 months; and Math – 6 years, 0 months. He was 5 years, 5 months at the time of testing. His communication scores, from age 31 months, to current, have improved by 4 standard deviations! This is type of progress is unheard of! The future looks promising for Will and I cannot thank the Masgutova Method® enough for giving my son his life back.

MNRI® has not only changed the life of my son but it has also changed my professional life and the way that I practice therapy. I am in training to be a MNRI® Core Specialist and I have recently opened a therapy clinic to help other children and their families with similar challenges, and to let others know of the many benefits of MNRI® intervention.

We thank Danielle for her heartfelt story and congratulate Will in his progress out of his 'world' and into ours. Will shows us that he is a Winner and plans to have a bright future of good health and many accomplishments! – MNRI® Team

Child on Autism Spectrum Makes 4-6 Months Gains in Language in 5 Months

Stephen Best, Sp. Ed. Ed. S K-2 CDC Teacher, MNRI® Specialist, Tellico Plains, TN, USA

Stephen Best

A 5 year old boy had received Special Education Speech Services in school and testing from private outside sources for three years. His average gains according to testing by a private outside clinic were 2½ months per year or 5 months gain in 2 years (by testing with Preschool Language Scale 4 English Edition). In August 12, 2010 the child began to receive *an average of five minutes per day* of MNRI®. In 5 months, outside testing revealed gains in age equivalency ranging from 4 - 6 months.

Basic information: The child was diagnosed with autism at age 3. Speech services began at that time. In August 2009, the child began pre-school Special Education. Services including OT, PT, speech, and outside speech therapy. He was tested by outside sources (a hospital rehabilitation center) using Preschool Language Scale 4 English Edition testing on July 21, 2010. In August 2010 we received permission to utilize the Masgutova Method® technique.

Results: In October 2010 the child began to talk using one word utterances and by December 2010 he was communicating using three word utterances, and recognizing numbers to 28. This use of outside source testing provides a transparency because they had no knowledge of any new methods that had been implemented. The family structure had not changed and the school setting had basically stayed the same. The only difference seemed to be the small amount of Masgutova Method® therapy being used.

Summary: In August 2010, prior to MNRI® therapy, this boy spent most of his time on his tip toes, flapping his hands, and humming. Parent comments in October 2010 indicated that he was now able to start telling them what he wanted and everyone that came into contact with him stated a noticed improvement in social skills. It was noted in his IEP in Dec. 2010 that he now only occasionally walked on his toes. Although no further testing was done, the child was dismissed from PT in May 2011 and Masgutova therapy was discontinued in November 2012 by a school system administrative request. As of March 2013 he was now doing simple paper-pencil work in writing and math and spending more time in regular education settings. He was beginning to self regulate toilet skills. The parent reports he is still making progress.

Preschool Languages Scale 4 Results

7/21/10	Standard Score	Percentile Rank	Age Equivalent
Auditory Comprehension	50	1	1.3
Expressive Communication	50	1	1.5
Total Language Score	50	1	1.4
12/29/10	**Standard Score**	**Percentile Rank**	**Age Equivalent**
Auditory Comprehension	50	1	1.9
Expressive Communication	50	1	1.10
Total Language Score	50	1	1.8

I am thankful to each and every child that I have the honor to work with as they are my true teachers! – Stephen Best

My Son Grew into an Absolutely Great Young Man

Nancy Morris, Parent, PT, MNRI® Core Specialist-in-Training, Mine Hill, NJ, USA

My first experience with Dr. Masgutova and her work came when my son, Zachary, attended the first Family Conference in Palm Beach, FL in 2006. Zack was 8 years old at the time and had difficulties with vision and vestibular processing, attention and focus, and hypotonia with motor delays.

Nancy Morris

His birth history was significant for a transverse lie (sideways position) and a difficult C-section delivery.

While working with Pamela Curlee during his reflex repatterning session, on perhaps the fifth or sixth day, he had an amazing experience. Pamela was working with his spinal reflexes and head crowning. She shifted her focus to more work with the head crowning, and Zachary followed spontaneously with what I now know to be the sequence of birthing reflexes. He then got off the table and began exploring the room like he had never been in there before, studying the space and items in it with intensity. We both stared and watched him and Pamela said to me, "He just went through the birth process!" I cried, she cried, as the true significance of what had just happened hit us both.

By the end of the conference, he was present to us, interactive in a way he had never been before, and had more confidence initiating conversations with people he didn't know.

At 15, he attended a second five day Family Conference because he was still experiencing some challenges with his strength and coordination.

Zack is now 16 and a honors junior in high school, active in marching band as well as Boy Scouts, where he is a Life Scout. He is bright and funny and an absolutely great young man!

I am now a Core Specialist-in-Training.

I am forever indebted to Svetlana and her amazing skills!

To say that Zack has become a great young man is a perfect description. Watching him grow and unfold his dynamic potential since that first family conference has been such an inspiration to so many people. And now, Nancy, we watch as you turn around and share this gift of inspiration with others . . . We are so happy that you both are part of the growing MNRI® family. – Pamela Curlee

Excited about Beth Joy's Great Progress

Sarah Davis, B.S. Nursing, Educational Therapist, Licensed Brain Gym® Instructor, Certified Cognitive Enrichment Instructor, Proud Parent, Oak Ridge, TN, USA

Sarah Davis

Looking back gives perspective on our journey. Beth Joy Davis has had the privilege of experiencing MNRI® since she was 23 years old. Now at 34, she continues to progress, in fact, at an increasing (rather than decreasing) rate of change as she ages. How does this happen? Why is this possible? Perhaps the path was paved by many years of well-chosen intervention programs. Perhaps these faster changes are happening now due to less restriction from dental appliances and foot orthotics and her more frequent access to MNRI® sessions.

Dr. Masgutova answers these questions of 'how' and 'why' Beth Joy continues her neurodevelopment so successfully based on her professional knowledge. She explains that Beth Joy's extrapyramidal nerve net system, which carries messages from the body to the brain and back out to the body, has become gradually mature enough for her to change at a quicker rate. Dr. Masgutova's concept is this, "Delayed development that is taken care of through effective interventions can promise on-going changes and we must orient on other measures, not only age." She says, "Let's give her the chance for growth to happen and let's organize the best support based on the rules of neurodevelopment." Her course of action is to work with reflexes because they are the basis for overall neurodevelopment and its plasticity. Dr. Svetlana Masgutova's MNRI® programs are greatly impacting Beth Joy's development and the development of many other children and adults around the world, some of whom I know. Beth Joy is not the only one in our family who has benefited from the MNRI® programs. MNRI® is a strategy for us – parents and other family members – to promote our own well-being and help us cope with the stresses of caregiving and all of life. I found MNRI® just in time. MNRI® was a key player in my 5-year journey recovering from caregiver burnout. Beth Joy's sister also benefits from MNRI® classes and sessions.

Throughout Beth Joy's life journey, she has been living out the promise of her name which means "God's House of Joy". The developmental challenges she has carried on her journey include: cerebral palsy neuromuscular hypotonia (low muscle tone predominant form); intellectual disability; seizure disorder; hypothyroidism; scoliosis; postural, balance, coordination, and gait issues; sensory integration disorder; visual and auditory perception problems; lack of binocularity due to eye muscle imbalances; 2 missing teeth next to front top center teeth; problems with gross, fine, and oral-motor skills; and dysfunctions in articulation, fluency, speech, and language. As an adult, Beth Joy was diagnosed with a genetic condition, Koolen-DeVries Syndrome, which is a microdeletion on chromosome 17 (17q21.31).

As Beth Joy began her journey in utero she was not very active. At birth she was a 'floppy baby' and was unable to suck well which required supplemental bottles in addition to breast feeding. I spent many hours feeding her (10 hours a day lessening to 6 hours a day before she went off the bottle at 1 year of age). She had several hospitalizations for developmental evaluations, the first at 7 weeks of age. From 5 to 12 years of age, each week she had three hourly sessions of the Verbotonal Method of speech therapy. This therapy was movement based and included auditory training and vibro-tactile input. Without the Verbotonal Method of speech therapy and further stimulation to the body from physical and occupational therapy, I believe she would have remained non-verbal. Throughout Beth Joy's life journey, God has led our family to various traditional and alternative medical, therapy, and educational programs, which have been of benefit to her development.

At age 23, Beth Joy completed her special education schooling and began adult services in the community. For 11 years she has lived in a supported living home with one other woman who has disabilities and a family. The mother of the family provides the needed supports for the two women she cares for. Beth Joy also has a supported employment day staff person who is her job coach and helps her in the community on weekdays. She has worked in the dining room of a nursing home nine hours a week for 11 years.

Currently, Beth Joy has weekly physical therapy appointments (including Integrative Manual Therapy™). The Beckman Oral Motor Program is carried out by trained staff and her mother. She also benefits from weekly Feldenkrais movement class, chiropractic services, and therapeutic massage. When she was 20 years old, Beth Joy ventured into Brain Gym® and found it to be fun and helpful. A Brain Gym® instructor referred her to Dr. Masgutova.

Beth Joy's MNRI® opportunities began at 23 years of age. Through the years she has been a student in 6 MNRI® classes, had 37 hours of sessions with Dr. Masgutova, attended MNRI® conferences (13 total conference days spread over four conferences, from 2008 to 2014), and had 3 MNRI® clinic appointments in one day (August 2012). Since May 2013, she has had access to specialists within 2½ hour's drive and has had 1 to 2 times a month MNRI® sessions with Core Specialists Stephen Best and Priscilla Roberts. Twice she has also been able to travel 3 to 5 hours to have sessions with specialists who were teaching courses. Core-Specialists, Pamela Curlee and Sharon Heller, have worked with Beth Joy since the 2008 Conference. Throughout her MNRI® experience, I have taken MNRI® courses and provided implementation of her MNRI® home programs.

With an eclectic mix of interventions and the unknown contribution of her personal developmental readiness, how can the impact of MNRI® (or any intervention) be measured? Beth Joy is the authority on what works for her! She craves and is willing to work on therapies and activities that she needs (like MNRI®). She will even suggest to families that their special needs person "needs to do them, too." Another measure of effectiveness of interventions is observing changes that occur with the newest and/or latest intervention, particularly those changes that might be expected. Some examples follow. The addition of physical therapy, including Integrative Manual Therapy (IMT™), to Beth Joy's life in 2007 resulted in significant changes in her motor skills, posture, gait, and speech articulation skills. Developmental growth spurts have occurred after starting MNRI® and through the years after Beth Joy participated in various MNRI® sessions, classes, clinics, and conferences. In August of 2012, Beth Joy had three clinic appointments and an additional hour with Dr. Masgutova over several days. One month later, she finally learned to tie her shoes (age 33 years). My summary of her progress compiled three weeks after the April 2013 conference also includes: improvements in motor planning, coordination, pace and accuracy of movements (gross, fine, and oral-motor), better balance standing on one foot, progress on physical therapist's pre and post MNRI® conference testing. Her swim coach noted improvements in the form of some of her strokes, greater reach, more ease of movement and her walking gait was more typical. Beth Joy did her best at her adaptive cheerleading and dance show: doing more complex steps, moving faster, doing the turn arounds and higher one-foot kicks, better with Bal-a-Vis-X, more fluent speech and articulation (a friend's mom said, "I usually just understood part of Beth Joy's words, now I understand most of them"), and singing better.

Seven weeks after the April 2013 conference, the progress report included noticing activities and suggesting we do them for Father's Day, suggesting gift ideas, asking questions about how to help a friend be more social, and reading better.

Beth Joy's physical therapist aptly described her post-April 2013 conference experience as changes which have dynamically accelerated in the rate of change in coordination and balance (she has now risen from level 2 to level 3 on a scale of 1 to 5), verbalizing then executing motor plan for tasks, better concentration, evolving self-image (becoming a woman), more self-confident attitude toward life, and a variety of changes in mouth and head as her development is evolving.

At the April 2013 conference, Dr. Masgutova and MNRI® Specialists Renata Chodyko and Elzbieta Cieplinska worked with her oral-motor patterns using MNRI® Oral and Facial Reflex protocols. Since then, I have been encouraged to see the change in Beth Joy's oral-motor skills due to the increased physical freedom of her hard palate and the improved strength of her oral muscles. Beth Joy has a delay and structural deficits of her mouth cavity: narrow palate, improper position of teeth, poor biting and chewing, and limited mobility of the tongue. These problems significantly affect her oral-motor control in eating and speech.

We have worked on these deficiencies for many years. The surprise in Beth Joy's progress is that significant change is happening at 33 and 34 years of age. Dr. Masgutova was concerned about the need to extend her hard palate, "to give a stretch for the central connection of the hard palate to show the brain and bones which way to move…" and she was thinking that this could also change her narrow head for the better. What was the problem? Dr. Masgutova thought that her dental retainer was limiting the needed changes. Traditional dental and orthodontic thinking says that Beth Joy's right and left palate bones would have been permanently fused together since she was about 14 years old (as an orthodontist told me). However, as Dr. Masgutova worked in Beth Joy's mouth, she felt that it became more symmetrical (left-right) and higher, and possibly also affected to some extent, its width.

I had used various oral-motor programs with Beth Joy throughout her whole life. I knew the shape of the bones of her mouth. I felt an amazing change in her palate wideness and mouth space with my own hands! At home, her physical therapist of 6 years perceived the bones in the roof of her mouth had moved as well as the temporal bones expanding laterally. After the April 2013 conference Beth Joy's dentist suggested leaving the retainer and mouth guard out to allow freedom for further changes. He also made an impression of her upper mouth and created a model. New impressions have been taken every 1 to 5 months to observe changes. Comparing models has shown changes in the width and shape of her palate.

In retrospect, I now believe that since she was 16 years old Beth Joy's dental appliances restricted her natural developmental processes. These devices included: braces, failed glue-on bridges, retainers replacing teeth, and mouth guards for teeth grinding. When the braces were taken off, continued muscle imbalances and her inability to use a lower retainer resulted in many of her teeth not maintaining the new positions. I think the limitation of movement related to dental appliances affected not just her mouth but also her general physical and cognitive development. Beth Joy was free from dental appliances from May through mid-August of 2013 and the widening of the palate continued. Also by early August of 2013, a movement upward of the roof of the mouth indicated that there has also been a repositioning of the sphenoid bone. This butterfly shaped bone that is behind the eyes has moved into a position that is better for brain functions related to development and the important thinking skills of cognition. Why are these changes in the position of the bones of the mouth and head happening now, at 33 and 34 years of age? Dr. Masgutova says it is occurring now because of Beth Joy's developmental readiness for the growth and movement of the bones. She explains that Beth Joy's upper palate, mouth cavity, and even lower jaw (that were all delayed in growth) are now growing like happens in a young child. This is unexpected good news!

In August after the next MNRI® conference, Beth Joy got a smaller retainer to replace the missing teeth and hold space for possible future implants. It also assists her in biting and speech. MNRI® techniques help sucking, biting, chewing and swallowing and lessen tongue thrust. Changes in these functions are important to prepare her for the possibility of implants.

For the past six years, Beth Joy has also experienced restriction due to firm orthotic supports in her shoes. She has very high arches. The orthotic supports helped resolve her plantar fasciatis. Dr. Masgutova recommended changing the orthotics saying that the foot reflex patterns and TMJ (jaw joints) are in tight correlation. Beth Joy's overactive Core Tendon Guard and Foot Tendon Guard reflexes keep her system in protection mode and affect her foot deformities. Since August 2013, she has used orthotics with lower arch support to facilitate releasing of these overactive reflexes. Her feet now have more freedom to move and develop muscle strength, flexibility, sensation of balance, and grounding. The plantar fasciatis has not returned.

Another important area of change for Beth Joy has been her visual skills. Her eye muscle imbalances did not allow her eyes to work properly for binocular vision. The Visual and Auditory Reflex Integration Program that was another area of special focus of the MNRI® specialists contributed to Beth Joy's progress in the visual

Beth Joy answers a question in the Miss Shining Star Pageant

skills of eye tracking, convergence/divergence, eye leveling, ocular-vestibular, and optokinetic reflex patterns. She tested binocular for the first time in her life in August of 2011. These last three years, she has increased in both binocularity and acuity of her vision. Each year the strength of her glasses prescription has been decreased. In October 2013 her optometrist added a mild prism to further promote visual efficiency. One, three, and six months re-checks showed continued increases in acuity and binocularity. The change of her reflex patterns could be considered a reason for improvement in her binocularity.

On Beth Joy's life and MNRI® journeys, I have had many "Aha" moments of increased understanding of why techniques are working and Beth Joy is progressing. The traditional statement that we had heard was that after 12 years of age in individuals with cerebral palsy, the brain is not progressing in any significant way. Our Beth Joy is proving this is a false statement – in the case of proper work, the brain continues to develop.

Revisiting Beth Joy's MNRI® journey and current events in her life in December 2013 has left me again awed at the progress she has made through the years. My "Aha" moment increased my understanding of how her lack of stability due to the Trunk Extension and Gravity Reflexes not being integrated keeps her arms and hands always on-call to protect her body. Thus the development of her hand skills is delayed. After the April 2013 conference, she began a progression of more easily moving into the positions of the Robinson Hand Grasp Reflex. Recent work on her Trunk Extension Reflex contributed to her hands and arms developing new freedom. Clearly this is good news for her development of fine motor manipulative skills, but it also opens a new window of opportunity for oral-motor and speech development. The Babkin Palmomental Reflex connects the functions of the hands and the mouth affecting her oral-motor and speech skills.

Beth Joy responds well to the combination of MNRI® programs and the Integrative Manual Therapy (IMT™) provided by her physical therapist. After recent sessions with MNRI® specialists and her physical therapist's treatments, Beth Joy has begun to move her chin side-to-side and open her mouth wider. The long-awaited movements will help her articulation and oral-motor skills. By Christmas 2013, Steve Best found that Beth Joy was able to relax more quickly with MNRI® Neurostructural techniques. Her body was responding a lot faster to releasing survival/protection.

In summary, it is of the utmost importance to address core body developmental issues which are at the base of the pyramid that builds up to the higher motor, cognitive, and communication skills. Beth Joy says it best. She chooses to continue to pursue MNRI® interventions. We are very thankful for the major contribution of Dr. Svetlana Masgutova, her team, and the MNRI® program to her continued development.

Beth Joy's cheerleading and dance adventures provide glimpses through windows revealing her developmental progress. Beth Joy's adaptive cheerleading and dance journey parallels in time her MNRI® journey. Initially Beth Joy said, "I can't dance." Her difficulty with balance and gravity limited her kicks, jumps, and turning around. Even simple steps eluded her due to her motor planning, directionality, and rhythm problems. As noted above, Beth Joy did significantly better after the April 2013 conference. By December 2013, I was stunned to see her do 8 consecutive alternating 90 degree kicks; higher jumps; and walk backwards while circling her hands around each other. Her on-beat rhythmic movements were beautiful to behold. Beth Joy's progress with cheerleading and dance has been made possible by her improvement in various reflexes stimulated by MNRI®. Now Beth Joy says, "I love to dance!"

Beth Joy is also more frequently saying with confidence, "I am smart!" in the midst of life situations. Beth Joy is now sharing of her own thoughts, feelings, beliefs, heart, and spirit. On December 23, 2013 Beth Joy's family received an early Christmas present. Beth Joy, her sister, and I spent 40 minutes in heart-to-heart talks. She conversed on a wide variety of topics. She was making comparisons, evaluations, and connections. Beth Joy now has the emotional maturity to express her hurt in tears, words, and questions rather than holding it inside herself. The deep level of the thoughts, feelings, and beliefs which have been locked inside her is now being revealed.

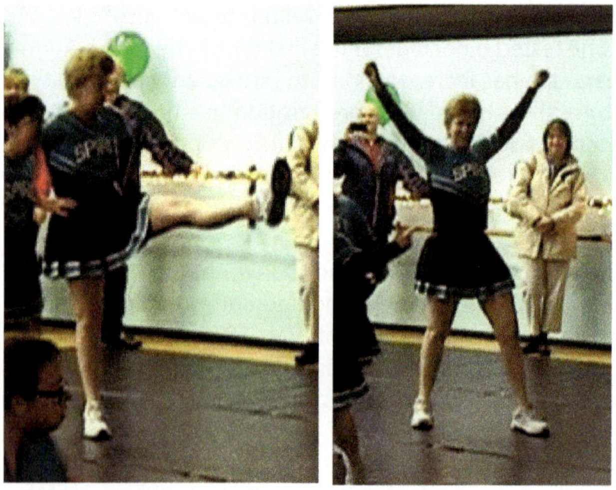

Beth Joy does high kicks with Revolution Spirit Cheerleaders

Beth Joy had a new opportunity in January of 2014. She was in the *Miss Shining Star Pageant*. Beth Joy has not liked dresses, uncomfortable dress shoes, or itchy jewelry. She would not speak in a microphone because it was too hard to get her words out. In the pageant, she walked across the stage comfortable in her party dress, jewelry, and dress shoes. It was inspiring to see Beth Joy on stage with a microphone singing and speaking confidently as she answered her question. The next day she spontaneously asked, "Should I wear my cheerleading uniform again next year?" (in the pageant). Like a butterfly, Beth Joy is spreading her wings and flying into new adventures.

Beth Joy attended her most recent MNRI® conference in January of 2014. During her three days, she showed significant progress in many areas, particularly in cross lateral movements. The Flying and Landing Reflex has been one of her most challenging reflexes. In 24 hours, she improved jumping in all four directions. It was amazing to watch her focus intently then jump higher and further. It was a sacred time as we watched her have fun learning to jump more naturally.

Since the conference, Beth Joy has shown progress in many skills and she is experiencing more success in many of her favorite activities:

• significantly improved focus which positively affects all areas of her life

• speaking with more clarity, fluency and mature speech patterns. She talks a lot more and her slow, deliberate pace of speaking is now more often faster and more natural

• as Beth Joy scrapbooks, her independence is revealed in her increasing creativity, planning, and accuracy in placement of pictures and decorations

• Beth Joy had not practiced Bocce ball for 11 months. At the Special Olympics in April, she played much better. She was focused and coordinated her brain, eyes, and body to roll the balls to the target much more effectively. She was thrilled to earn 1st Place!

• Beth Joy made a big change in her singing voice during the January 2014 conference. Her vocal consultant confirmed that she had increased her vocal range to include higher notes, shifting her from second soprano to first soprano range (without singing at all!). Beth has a varied music ministry. The impact of the words and music is amplified by Beth Joy's presentation which touches the listeners' hearts and spirits in special ways.

As Beth Joy continues her life journey, she is transforming into the person God created her to be. Her family is awed by the amazing grace, peace, and joy of God revealed through His daughter, Beth Joy, God's House of Joy!

Amazing story of the development of an amazing young woman. Congratulations to Beth Joy, you are a real Winner coming a long way, showing the world your inner strength and determination! Thank you for setting such an example for others! Our gratitude also goes to Sarah, her Mom, for her belief in the MNRI® program. – The MNRI® Team

My Son Alfred, Asymmetrical Tonic Neck Reflex, and Dyslexia

Jam Hoa Liong, Parent, MNRI® Core Specialist-in Training, Jakarta, Indonesia

Jam Hoa Liong

My son, Alfred, was born in 2000 in sunny tropical Jakarta, Indonesia. He was as healthy as many other babies. During his growing period though, he missed several milestones in his development, and was late in several reflexes, especially crawling, segmental rolling, and spinning. He had some issues with sensory integration but this was not clear at that time. He grew as a rather overweight baby and seemed lazy to move. When he was two and half years old, Alfred started kindergarten, and we learned that he had challenges to adjust, felt insecure, was too emotional and cried easily. Knowing his challenges, our family made a decision to find a school oriented to an individual approach for kids. At last we found a Montessori school, which he enjoyed. He got good practical life lessons, like pouring water, brushing, sweeping, and fastening buttons and zippers. He also had lessons oriented to sensory activation which helped Alfred to coordinate his senses to some extent. It was amazing that, at the same time his fine motor abilities for drawing were so good, his cross motor coordination was a challenge. We learned that his basic modalities for cognition were visual and kinesthetic.

In his second year of kindergarten, I realized that Alfred could not spell words easily, was very short and fragmented in focusing, panicked easily, was stubborn, had problems with his vestibular system, flat-feet, and at the same time was very curious in asking too many questions. Often I lost my patience when all that was expressed when I was in a hurry ("Oooohhhh, Alfred, can you stop asking or at least keep your questions for later?"). He asked the same question many times, and it could be for a week, two weeks, or a month – it was frustrating to get the same questions again and again. Sometimes he did ask them to different people. Of course, I was confused and didn't understand what was going on, and why he was doing this constantly – now it is clear, he was looking for links while processing information and needed time. His rigidity in thinking and coding processes were not matured for his age, which led to dyslexia. His sister also had dyslexia, but the symptoms between the two were totally different! And it was really tough to understand my child's symptoms!

Based on my experience of everyday solving his sister's dyslexia problem, I made a decision to start to work with Alfred earlier. When he was five years old I found a facilitator using the Davis Dyslexia Program to help Alfred in reading and spelling. It was a 5-day program then we continued the method at home. Alfred's facilitator told me that his brain was not able to process the sounds of alphabet. So he learned spelling using the alphabet instead of the alphabet's sound. And his difficulty was to understand words that have no picture, like a, an, the, some, to, too, am, are, is and many other words that total about 274 words. For these words we had to explain using pictures or clay or a dictionary to show him the meaning of the words and how to use them in sentences. Alfred is talented as a visual thinker, so the facilitator trained him to focus using this talent, and gave him a model of scale to motivate himself. Alfred improved his spelling and reading, and did well when he was in grade 5. In this way, he developed his language in writing, poetry, and reading comprehension. But Alfred had difficulties not only with language, he also had difficulties grasping the concepts of math. Again we used specially oriented material to help him to understand.

This was a car-free day. We were biking as a family activity on Sunday morning. From the left: mother (author), her brother, Alfred's Dad, Alfred, Alfred's sister, and his young cousin.

We had been working intensely with Alfred academically at home and he was busy also with sports, swimming, bicycling, outbound, Tae Kwon Do, and Tai Chi to strengthen his muscles, vestibular system, and balance. All this support helped him develop more confidence. Sometimes I made playdough to help strengthen his fingers. I also suggested he do Brain Gym®. Alfred improved to some extent and he had very good motivation for learning. Still he was missing some information and asking to repeat sentences or he got only the end of the sentence. I was still seeing the problem of poor focusing and tried to teach him to regulate it but it happened again and again.

Things changed for Alfred after I started using the MNRI® program. At the *MNRI® Dynamic and Postural Class* in 2011 while learning about the ATNR Reflex pattern, I realized that many of Alfred's problems were because of the immaturity of his ATNR and its improper influence in auditory processing – here was the reason he missed the last words in conversation and instructions and did not comprehend the content well. And it came to me that all his irritation was from trying hard to make sense of links and decoding. He was in our world, decoding and working hard to understand what we were saying and how to make sense of our daily routine. Of course, it was not easy for him to make friends in this kind of state and to develop social skills. His dyslexia, I became aware, was associated with his poor ATNR. I also found an explanation of why he was hypersensitive to some sounds and even panicked and was easily angered.

Alfred went through the MNRI® Assessment by Dr. S Masgutova to find out the level of development of his reflex patterns and determine a home program. Over 60% of his reflexes were not matured at the age of 10. And sure enough, she pointed out that his ATNR was a problematic one and that he needed the repatterning of this reflex in order to improve his postural control for auditory system support.

Looking at the definition of this complicated reflex: the trigger for the ATNR is reflexive and passive head turning to the side. As a result of the head turning, the limbs extend on the same side of the body to which the head is turned, while the limbs on the opposite side flex, and that impacts organization of the auditory processing. It was clear that we needed to work on the postural aspect of this reflex. In this reflex the head turning activates the levator scapula, scalene, and upper trapezius muscles causing them to contract, and this triggers the extension of the limbs on that side, and the sacrospinalis and quadratus lumborum muscles contract, flexing the limbs on the opposite side. It became more clear to me that there was disorder in the sequence of the work of these muscles, his postural control was in confusion, and Alfred's brain couldn't act properly. This reflex pattern had no chance to mature in Alfred. I was questioning again and again if this is possible; that just because of improper links in ATNR he can face such an auditory processing disorder. Dr. Masgutova told me that according to her testing, he is OK with perception of music – and yes! He was and is!!

The reasons for poor ATNR in Alfred, according to Dr. Masgutova's interpretation are:

• the ATNR was not supporting Alfred's vestibular-proprioceptive system, as it deals with the acceleration-deceleration mechanisms within the axis of the body – turning of the head in ATNR posture was still activating his whole body as in a young child:

• the ATNR was not serving as a well organized structure for the Stapedius Reflex pattern

• the ATNR was hypersensitive and hyperactive in him and not matured still, which meant that ATNR was based more on reactive response and not allowing for the freedom of controlled skills to happen

• the ATNR was mixed with the opposing reflex – Abdominal Sleep Posture (ASPR), which was most of the problem for decoding auditory information, as ATNR activates the auditory system and ASPR calms down it.

Alfred joking with his young cousin while they are biking.

Continuing on the topic on the ATNR, it is worthy to note that chronologically, the duration of active response to trigger the ATNR is the 4th–6th month of the infant's life. This is the time when ATNR is being linked with monaural perception of bits of auditory information and thus integrates with Stapedius (auditory) Reflex allowing next for binaural hearing to take place. The ATNR must be integrated into the whole motor system between the 6th–7th month. Language development on the level of decoding human speech continues in the infant until 17 months in infancy, and if not, the working speech decoding system will be late. It seems that our Alfred was dealing with improper decoding which was taking a long time for him to link, and it was important for him to link and not to ignore the information as his brain really was very smart, it didn't want 'empty spirts in flow of information'. This is why he had so many questions on the same topic, as he needed to cognate it again and again and was trying to memorize on a cognitive level what was done by other kids on an automatic level! In the case of Alfred, he was certainly using the posture of the ASPR when sitting at a desk. He was triggering this reflex while processing the auditory information… actually being passive or 'sleeping' on that level of brain activity organization.

Alfred showed improving his vestibular.

Unfortunately, in my country, we don't have MNRI® specialists yet. So I worked with Alfred regularly with ATNR and five or six other reflex patterns a day according to the priority given at the Reflex Assessment, and rested on the weekend. After months, we all saw the results of our work. It started as less questions, better balance, improvement of posture, better space-time orientation and vestibular responses. In academic skills, he showed surprising progress especially in math.

Alfred had not been able to ride a bicycle, and I often felt sad when the whole family would go out on Sunday morning to ride bicycle and only Alfred stayed alone at home. At the end of the month on Saturday night we made a decision to set the goal: "to ride a bicycle easily, to enjoy, and to have fun", and worked on repatterning of all his reflexes marked as priority one in his Assessment. The repatterning included was ATNR, Leg Cross Flexion-Extension, Foot Tendon Guard, Fear Paralysis, Hands Supporting, Babinski, Spinal Galant and Perez, and Trunk Extension. The result was just amazing… he now rides a bicycle without fear and his balance is perfect. Next, he improved his ability to do out bound, climbing the net, and crossing the shaky bridge.

A year later, we attended a MNRI® Clinic. And we continued at home with Alfred's MNRI® Home Program according to Dr. S. Masgutova's suggestions.

Following the second year, Alfred showed a huge achievement in every part of his skills. He has demonstrated significant progress in sensory, cognitive, and proprioceptive system development. Improvements in his fine motor skills and Hand Grasp were seen through his beautiful and smooth drawing. His auditory processing abilities have developed very well. He is completely adequate in conversation, with no longer any long and repetitive questions that were making him look strange. When he is facing a problem in studying or homework or forgetting something, he is calmer, without panicking, and capable to deal with problems, being good in sequence and timing. His language improved and became typical as with other children. When Alfred started school on the next level with a new teacher and I informed her that Alfred had dyslexia, she was very surprised. Alfred still works on interpreting written language skills of structure and tenses and is making on-going progress.

Among all his achievements, the most I like is that Alfred grew in his motivation for learning in all subjects. He knows what he wants and how to reach the result with confidence, commitment, and in a natural and positive way. In his last session with Dr. Masgutova, he surprised her with his high level of abilities to set very clear and mature goals. And I believe that the MNRI® method has transformed my son to be a human with a good heart and open friendly communication. This is why I have made a commitment to study MNRI® and to bring this knowledge to other specialists and parents. I am on my way to realize my goal!

Our MNRI® Team congratulates Alfred for his positive changes in learning and strength! We wish many more new achievements for you, young man! We thank Liong Jam for being such a great Mom and for her organization work for MNRI® in Indonesia. – MNRI® Team from Poland

My Daughter's Quick Success at a Family Conference

Priscilla Roberts, Parent, Speech Pathologist, MNRI® Core Specialist-in-Training, Chattanooga, TN, USA

Priscilla Roberts

I am writing to share the story of my daughter's success after attending the MNRI® Family Conference in Minnesota in the summer of 2012. Della and I made the decision to go to the conference because Della had been dealing with some issues for quite some time. She struggled with making good grades in school. She would study and study and still not make good grades. She loved to ride her horses but she was afraid many times because she just didn't feel safe on them. If they moved quickly or unexpectedly, she would be easily scared and just couldn't get past that no matter how hard she worked. She was also very emotional when it came to having difficulty with something. She would become frustrated with a task and then her emotions would just take over. We wanted to find some help so that life would be easier for her. So, we chose to attend the full eight day MNRI® Family Conference to see if it would help and to learn the MNRI® techniques to self-manage stress, to increase confidence, and to turn learning challenges into new possibilities.

Dr. Masgutova completed an Assessment of Della's reflex patterns and found that her sensory motor system for her upper and lower limbs lacked coherence. The reflex patterns of her upper limbs were hypo-sensitive and her motor responses were hyper-reactive. In contrast, her lower limbs reflex patterns demonstrated hyper-sensitivity and hyper-reactivity. This explained how it was very difficult for Della to operate with such differences in upper body/limb organization and lower body/limb organization. While Della had received other therapies to help, none addressed these issues.

Della's receptive and expressive abilities in processing of information and language were blocked and this had a significant impact upon her life. Although she demonstrated a high level of intelligence, she was trapped in this non-productive sensory-motor system that limited basic organization and easy learning (coding and decoding of information). She had rather strong challenges in expressing herself to the level of her intelligence. She had wonderful ideas of things that she wanted to do but could not organize herself to complete the tasks to the level that she wanted. This made her life and mine very frustrating. She was challenged socially, physically, emotionally, and academically.

During the eight day program, Della attended six sessions daily: Reflex Repatterning, Archetype Movements, Oral-Facial Reflex Integration, Proprioceptive and Cognitive Integration, Tactile Integration, Visual and Auditory Reflexes Integration, and Neurostructural Reflex Integration. I didn't know what to expect from the work. I didn't know if it would take a few days, weeks, or months to show an Improvement. I didn't know if it would work at all. It just sounded so logical that I thought it had to have some type of impact.

Here is the experience from three perspectives:

Della: On the first day with Shannon, I was able to stand on the balance board and move, but my movement wasn't flowing and I had to stare at one spot. I had to have a focus point in order to balance and move at the same time. On the second day, I was asked to do a twist walk on the balancing board and I was having a lot of trouble figuring out how to move it with a flow in my hips. I was moving it, but the movement was very sticky. Finally, after playing with it, my hips unlocked and I was able to move my hips back and forth with a flow in them. After that, I tried to hula hoop and balance on a big ball among many other tasks. I was able to figure out how to hula hoop and I was so proud of myself! I was even able to sit on the ball, bring my knees up onto the ball without holding on to anything and balance sitting up on my knees while on the ball without looking at one spot! I could talk and look around at the same time while balancing. When my hips found the freedom of movement, my world opened so much in what I was able to accomplish, and my confidence grew as well!

Della standing on her knees on balance ball.

Della's Mom: On the second day of the conference, Della attended the first three sessions of the day and we noticed no changes. After lunch, Della had her session with Shannon Desilets for Proprioception and Cognitive integration. Shannon started her off on a balancing board and asked Della to work on moving up and down the hallway while standing on the board. She had to rotate her hips to move herself. At first, only one hip rotated, the other one was somewhat stuck. (That hip had been stuck for years.) After about 5 minutes, all of a sudden, Della's hips both began to rotate and she could move. Her face lit up and she was flying up and down the hallway on the balance board. Then she asked what else she could do. Shannon gave her the hula hoop to see if she could do it. She immediately figured it out and began walking up and down the hall with the hula hoop just swinging, while moving to music that was playing (I have to say here that Della had tried many times to master the hula hoop and wasn't able to do it). So, it was amazing just watching her do this. Then she wanted to do more. She went into the room and got the big ball and asked if she could get on it. Shannon helped her and the next thing we knew she was 'standing' on the ball on her knees balancing and rocking. It was amazing to watch. The joy on Della's face cannot be described. She had found her center!

I have seen many approaches to helping with motor coordination, but I have never seen anything change a person so much in so little time. This work truly impacts the core issues and opens new doorways of possibilities! We still have the video of her new movements and are amazed every time we watch them…

Shannon Desilets: What I witnessed in Della:

On Day-1 Della showed a low level for control for balance and antigravity organization which was challenging her motor planning, control, and coordination. Her ocular-vestibular and optokinetic reflex patterns were more rigid and were limiting her eye motions and blocking her body.

Della transformed on Day-2 of our sessions. Like a tightly closed flower… into a blossomed rose! The tension in her body released ... visibly at her core, her neck, and her jaw. Her feet found stability. She no longer needed her eyes to be locked on an object in order to maintain her balance on the balance board. Her breathing became more relaxed. Her core awakened and suddenly she found freedom of movement. Her hips moved with fluidity as she reached her goal of dancing. Each day, her skills grew more and more. This young woman who initially had a difficult time maintaining balance and breathing, now was able to control her core and balance while tall kneeling on a therapy ball – unassisted.

Her confidence grew. Her face softened. Her feet found 'stable' ground. Her core awakened and found freedom. Her eyes got freedom from the movements of the head (her Head Righting Reflex started functioning in coherence with eye movements). Della transformed.

We are very thrilled with Della's progress and congratulate her and her mother for their hard work and endurance to greatly improve Della's prognosis! – The MNRI® Minnesota Team

Fred Learns to Crawl

Mary Rentschler, M.Ed., MNRI® Core Specialist and Instructor, Washington, DC, USA

Mary Rentschler

When I opened the door I saw an adorable rosy-cheeked bright-eyed and smiling 14-month-old in the arms of his attractive young mother. She had heard me speak at a local child development organization and came to see if I might help her little boy with crawling. As often happens, the pediatrician had said, "He's really fine, let's just wait and see." In fact, Fred was the picture of good health and there was nothing traumatic in his history to cause concern but the Mom didn't think it was fine for her 14-month-old son not to be crawling, or even attempting to crawl.

The Mom put Fred (not his real name) down on the floor while she took off her coat. He immediately went into a full Landau, smiling happily and bouncing a bit on his tummy with arms and legs lifted like a 5-6 month baby. I put a very nice toy truck on the floor about 4 feet away from him. He was clearly interested and excited by it. The bouncing accelerated a bit, as did his breathing, and he continued to smile, but there was no movement away from his spot on the floor.

Landau Reflex

Landau is one of very few reflexes that have no protective function. We call it the joy reflex. In an infant the motor pattern is symmetrical lifting of head, arms, and legs in response to visual perception of an interesting object and to the proprioceptive stimulus of being prone. Muscle tone in the abdomen is stimulated by contact with support and triggers extension of the spine and limbs. In older children and adults we see a version of this when our team wins a point (Yea, team!!) and when we jump for joy. The reflex is important developmentally for positive emotional communication, visual perception, differentiation of spinal muscles and circulation of cerebral-spinal fluid. Because of the release of endorphins it activates, the Landau can become hyperactive, even addictive.

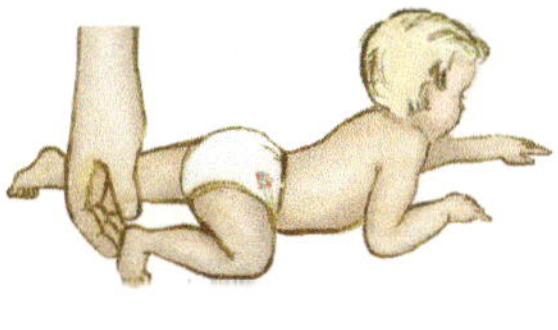
Bauer Crawling

Fred certainly had a hyperactive Landau. That explained his stamina in the position as well as his lack of frustration and his smiling good cheer, even when the desired toy was out of his reach.

I decided to begin right away with work on Bauer Crawling. Up on the table on his tummy, Fred was still smiling, but wanted nothing to do with the asymmetrical Bauer position. If I moved one leg into flexion, the other one came too, no matter which side.

Leg Cross Flexion-Extension

It was impossible to hold him in position long enough to give the stimulus and move him through the pattern. "OK then," I thought, "what about Leg Cross Flexion-Extension?"

The motor pattern for Leg Cross Flexion-Extension is in opposition to that of Landau. It is a key reflex for differentiation and coordination of the legs. Fred stopped smiling when we put him on his back and began to move one leg at a time. The mom and I worked together, one holding a leg down in extension while the other brought the opposite leg into full flexion. Fred's resistance was strong but not enough to stop us. It seemed extremely uncomfortable for him to have one leg bent and the other straight. Nevertheless, we continued. Then suddenly the resistance stopped. The Mom and I looked at each other. "His brain just recognized the pattern," I said. We then worked very thoroughly through all the steps of MNRI® Leg Cross Flexion-Extension Reflex repatterning. Fred was calm and compliant.

Soon it was time for them to go. Mom put Fred down on the floor again to organize her things and again he went right back to a smiling full Landau. It occurred to me that he was stuck in extension, so I moved him into full flexion and had him push out of it into full extension as we do in Archetype Movements. After two or three repetitions, they went home with that and Leg Cross Flexion-Extension repatterning for their Home Program.

The next afternoon I found a message from Fred's Mom on my voice mail. "He's crawling!" she said. When they came for a second appointment the following week, Mom again put Fred on the floor while she took off her coat. This time he shot all around the room exploring everything with perfectly coordinated cross-lateral crawling on hands and knees. He smiled when he was happy or amused, otherwise not. I completed a short assessment and finding everything age appropriate, congratulated them and suggested they return only if they felt in need of help. I've never seen them again.

I wonder how things would have gone differently for Fred if he were never to learn to crawl. Neurophysiologically, the sensory-proprioceptive stimulus for crawling reaches the cortex, so it can definitely be considered a learning reflex. Many years ago as a special education reading teacher, when I visited classrooms I used to ask the children to stand up and do a 'cross crawl,' (a kind of marching in place in which one touches opposite hand to opposite knee). In a group of 15 or more children there would always be two or three who touched the same side arm to the same side knee, failing to cross the body midline. The teacher would confirm that these were the ones having problems learning how to read.

The cross lateral motor coordination used in crawling is involved in so many areas of development that it is no wonder it has for so long been considered one of the most important milestones of early development. Bearing weight on the hands opens palms and strengthens fingers and wrists in ways that will later support the fine motor control necessary for hand-writing. Scanning for a desired destination and avoiding obstacles builds spatial orientation, depth perception, and binocular vision. The eyes focus off in the distance, then back on the hands, just the way later they will use focus recovery to copy from a school blackboard. By activating the vestibular system, crawling helps to integrate the Head Righting and Symmetrical Tonic Neck Reflexes and enhances vestibular, auditory and visual integration. Psychologically, crawling is related to goal setting and motivation: the child crawls for a purpose. Early success in obtaining a desired out-of-reach object builds confidence in the notion that effort brings a good reward!

Most MNRI® practitioners have similar stories. Of course, we love them because they make our work seem magical. Lightning-like progress is far from magic though; it's simply what happens when precisely the right nudge is given to a system that is fully ready. With severe challenges we also get excellent results, albeit with lots more work. Fred had gotten stuck in his motor development somewhere along the way. We don't need to know exactly when or why. He was otherwise very, even extremely, healthy and when the stimulus for Leg Cross Flexion-Extension connected with and awakened the motor pattern already present in his brain, further development happened naturally. Suppose you trip and accidently unplug a lamp. When there is no damage to the lamp, after you plug it back in again, how long does it take for the light to go on?

Although I only saw Fred and his Mom twice, I am grateful to them for this delightful experience and for how useful it has been to me as a teaching story. – Mary Rentschler

The Babinski Journey

Lisa Ortego, Speech Language Pathologist, MNRI® Core Specialist-In-Training, Carencro, LA, USA

Lisa Ortego

As a speech language pathologist who is working on certification to become a MNRI® Core Therapist, I find that God sets different obstacles in one's path to bring learning experiences to a whole new level. With that information, it allows the therapist to have a better understanding of some of the difficulties clients may be experiencing due to non-integrated reflexes. This knowledge provides a deeper compassion and understanding, especially when the body experiences it firsthand. Here is my story….

It all began when I was in college and I noticed how my feet began hurting. This pain intensified during my time working at a hospital and having to climb up and down stairs daily. The end result was feet that hurt constantly and I felt a deep exhaustion even after a good night's sleep. If the tightness was not released in my feet, it progressed to tingling, very tight arches, and pain up my legs into my back. I was told by a podiatrist that my bone structure in my left foot was completely different, causing me to pronate inward with a collapsed arch. He basically diagnosed me with plantar fasciitis, prescribed orthotics, told me never to gain weight, and that I would always have problems with my feet. Unsatisfied with his answer, I began seeking other methods to at least help with my energy levels because, whenever I had flare ups in my feet which happened often, I felt like my energy was zapped. I turned to massages, craniosacral therapy, acupuncture, and even a chiropractor in efforts to break up the tension in my feet and keep the pain from climbing up my back.

It was not until I took my first reflex integration class with MNRI® Instructor Pamela Curlee who explained that my Babinski Reflex was not integrated and showed me exercises to do. Immediately, I felt my feet wake up and the pain was gone. However, each day was war with my feet – as I got relief, the tension would try to return as I walked. Through each MNRI® class, I began to feel more and more relief as I was able to have specific MNRI® Babinski techniques practiced on my feet. The moment that changed everything occurred in the *MNRI® Children with Challenges Class* where we were working with the Babinski Reflex. As I lay on the table, I soon realized that, despite trying my best to move my big toe as I went into Babinski, I could not. However, it was not until I was instructed to look at my hands as therapists rotated my feet and hands at the same time that my body took over and I had no control. There was a specific coordination and my hands literally were moving in the same way as my feet! As my feet were rotated, my hands would feel the restrictions in the same place that my foot had the restrictions. I gasped as I saw my finger get stuck at the same time that my big toe hesitated during the rotation. Once we were back in our chairs, Pamela asked us to flex our feet into Babinski. I was stunned because my hands and feet were in such tune and coordination I never felt before! I kept thinking of the homunculus chart with the hands and feet next to each other on the brain.

Little did I know that this was only the beginning of improving my Babinski. On the way home that night, all of the passive movements that my feet were given during the class basically woke up my brain. I moved my hands and feet into Babinski and out of Babinski the whole way home. After my hands and feet started moving as one, my body began wanting to do Archetype movements. I was thankful that my sister took the class with me and was driving because I could not stop my body's movements!

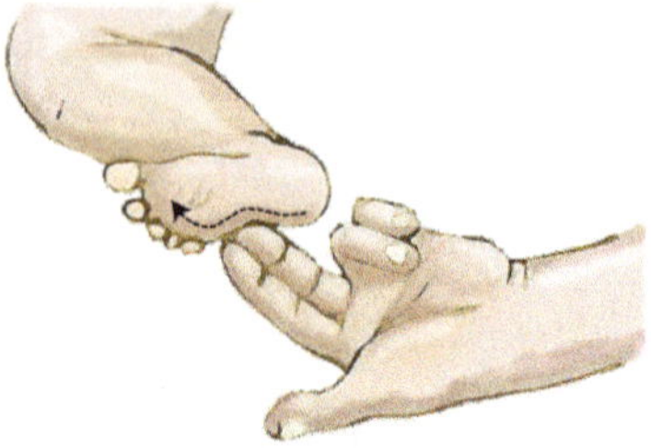

Activating the Babinski Reflex.

The next morning, I wanted to continue the movements but in a homolateral manner. My right hand and foot wanted to move together, then my left hand and foot. I felt like I was in the twilight zone as I tried to process what my body wanted! Once we got to the hotel, a coworker who was also taking the class, continued with the integration process for me. During class, Pamela explained that I was going through homologous movements and the next step would be homolateral movements. When I explained I was already there, she basically stated that it was integrating. The next step, of course, was cross lateral. During that week, my right foot felt as if it actually had shock absorbers for the first time. The interesting aspect was that it didn't matter how hard I tried to walk correctly with my left foot, I could literally feel the difference between the reflex in my right versus my struggle with my left foot. It was as if I had a spring that hooked correctly on my right foot and one did not snap in correctly on my left foot. Strengthening exercises would not even come close to fixing it. I then experienced a tremulous craniosacaral rhythm to the point that I had to give myself Fear Paralysis because I just felt jumpy. I wondered if this was a small taste of what dysregulation of reflexes felt like before they anchor in. I was relieved that it only lasted a couple of days. As I met with my coworker and sister that week, they could see how my walk had changed, along with my pronation. The impossible was changing!

As I reflect on the whole experience, I truly understand on a whole new level the information that Dr. Masgutova presents about reflexes going through homologous, homolateral, and cross lateral movements. I have felt the power of passive movements as the body experiences the correct movement patterns. I agree that "simple is profound!" It definitely was a learning experience that made me think of what so many of the children that we see go through! I hope that my story will help parents as well as professionals understand how it feels to go through an integration process because many of our children cannot tell us what they are feeling when a reflex pattern becomes repatterned and matured.

Of course, I also want to thank Dr. Svetlana Masgutova and Pamela Curlee because they are true angels here on earth!

Thank you, Lisa, for sharing your experience! Is it so helpful to know more about our internal worlds and the changes in these worlds through integration and rerouting of reflex neuropathways! – The MNRI® Instructors

MNRI® Techniques in the Classroom

Tina Marks, Speech Therapist, MNRI® Core Specialist in Training, Opelousas, LA, USA

Tina Marks

I am a speech therapist who works in a variety of settings. I've tried many times unsuccessfully at getting MNRI® into my regular education classes until a month ago. About two months ago, our kindergarten teacher referred a student to me for academic and speech concerns. I did my traditional required screening procedure, then did the MNRI® reflex Assessment. I showed his teacher the results of his reflex Assessment and had next to each reflex the specific issues related to it being non-integrated. Every reflex assessed on this student was non-integrated which explained his academic struggles as well as his expressive and receptive language issues.

Not only was she amazed but the MNRI® seed was planted outside my room. Since he was in the Response to Intervention Process, I was not allowed to provide much time for services for him. Therefore, I gave his teacher my Archetype DVD and asked her to have her class do it daily as part of their morning routine. After the fourth day, she pulled me into her room to show me significant handwriting improvements on another student. She could not get over the progress of her class after only doing Archetype Movements for four days. She also noted improvements in behavior. It was amazing to watch the student I assessed do the movements. He went from not being able to do any Archetype movements independently to currently being able to do cross-lateral movements with about 90% accuracy and was starting to self-monitor his movements. He now knows 6/10 sight words; whereas all year had only known 2/10 sight words. The student whose handwriting improved was unable to gallop or skip at the beginning and is now galloping and skipping. The overall behavior of the class has improved.

Due to the excitement and gains of this class, other teachers wanted to know what was going on and wanted "that DVD" in their classrooms as well. As a result, our principal found out about these nice changes in her students and agreed to pay for her teachers to attend the *MNRI® Reflex Integration and Learning Skills* course so that all the classes can have MNRI® throughout their day.

I would like to not only thank this student for his hard work and success but also his teacher, April Dunnehoo, and our principal, Brenda Green, for being open and allowing MNRI® to be used in our school. This success allowed for numerous teachers and others to be trained in MNRI® techniques which has had a greater reach of students. – Tina Marks

Using MNRI® in a School Setting

Brian Esty, MNRI® Core Specialist, San Francisco, CA, USA

Brian Esty

I am a Manual Therapist on the roster of complementary care providers for one of the largest private elementary and high schools in San Francisco. The schools embody an academic tradition which de-emphasizes fact memorization before the permanent teeth emerge, and integrates movement as an essential quality of child development. Additionally, as expensive private schools in San Francisco, the average student has access to extraordinary resources and is of above average intelligence. The academic tradition of these schools recognizes that restrictions in movement or posture will resonate both emotionally and cognitively in a child, and therefore have established relationships with local therapists that can address these issues.

In my practice I see children with structural, emotional, or cognitive issues. Some of the structural issues I encounter relate to posture and balance, challenges with movements or actions such as catching a ball, car sickness, and visual issues related to tracking and convergence. Common emotional issues are oppositional behavior, shyness, phobias, low tolerance for stress, and aggression. Cognitive issues often relate to attention, either over focus or inability to focus; difficulty with math or language arts, speech issues, fidgeting, and poor personal boundaries. As a MNRI® trained therapist, I always look for an underlying theme that can tie the structural, emotional, and cognitive symptoms together with the level of maturation of a developmental or lifelong reflex (automaticity). The beauty of MNRI® lies in being able to test a hypothesis by checking the functional attributes of the automaticity, and then validate the hypothesis by working to develop the automaticity and checking for structural or functional changes. This process can be quickly adjusted to work with the context of what is presented in a session as a child develops.

My general sense of the children I work with, and the adults that they eventually become, is that they are essentially too smart for the good of the full maturation of their innate automatic sensorimotor functions, which are the foundation of conscious awareness. (Panksepp, 2012) A common example of this is an infant who figures out how to walk before their gait reflexes, which are cultivated by crawling, fully mature. When infants crawl, they not only develop their postural and gait dynamics but also integrate these with vision and balance. If they figure out how to walk, but without crawling, they do so in a more conscious, less automatic manner that is inherently less robust and adaptable, and therefore inherently more stressful. Generally, industrialized societies assume that emphasis on the earliest possible acquisition of skills leads to a more successful (safer) life. Thus, it follows that the automaticities with which we interface with the physical world may be riddled with adaptations and compensations, a consequence of being deployed before maturation. The trillions of dollars spent on pharmaceutical and recreational drugs (including alcohol) to alter the regulation of our emotional and cognitive functions, as well as medication for aches and pains, suggests a widespread issue of incomplete maturation of our innate self regulatory mechanisms. MNRI® views these self regulatory mechanisms as the primary functional expression of the link between our structure and function which we call 'reflexes'.

Coaching the further maturation of children's developmental and lifelong automaticities is inspiring in that, as the function of reflexes are strengthened and their structural, emotional, and cognitive capacities increase, a therapist can easily feel that they have contributed to the wellbeing of that individual over the course of their life, and to the lives of others in that individual's community.

For example, I recently worked with a boy who has been struggling with math. Although he was still at the level of learning mathematical fundamentals, he had already concluded that he "hated math". He had additionally established an oppositional relationship with his math teacher, reinforcing his perception that math was "useless". It has been established that incomplete maturation of the Asymmetrical Tonic Neck Reflex (ATNR) will challenge the type of associative cognition called upon to make sense of the relationships between objects which we call mathematics. It was a simple matter to check how comfortable he was with homolateral movements with the head turned to the side, which employs the ATNR. He disliked being tested in this way almost as much as he disliked math. However, he was simply disinterested when other automaticities such as those related to cross lateral movement were checked. I worked with him in over several sessions to build capacity in sensorimotor dynamics where stress had been identified. Additionally, I gave him homework from the MNRI® Archetype Movements, which his mother became very invested in doing with him, especially before sitting down to doing math homework. Within a few weeks his mother came in beaming to say that he was knocking out his math homework without any struggle. This led to a rather amusing conversation with the boy as to why he would hate something that was easy for him. A current challenge is to rebuild his relationship with his math teacher, where the idea that he could suddenly become teachable, is possibly beyond the scope of her experience. Educating educators to the possibility and potential of changes emerging from these therapies is an ongoing vocation. Another challenge is to integrate the Archetype Movements into the boy's awareness as a resource he can call upon whenever he notices he is challenged in associative cognition. One can easily see how a relatively small amount of therapy and coaching may impact the direction of this boy's life.

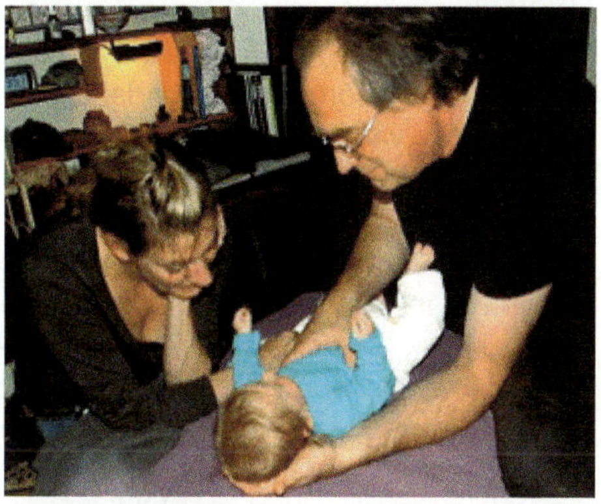

Integrating visceral and cranial reflexes with vision and tactility in an infant.

This type of story is common both in my practice and in MNRI®, which specializes in the needs of developmentally challenged children. My interest is in furthering the reach of these tools into the community of high functioning and gifted children and adults, to enhance their capacity for self regulation and access to their potential. It is interesting to me that a privileged lifestyle, high intelligence, and a sensitive and caring academic and home environment will not on their own guarantee that a child will express their full potential. However, when fuller maturation of a child's sensorimotor automaticities is added to the mix, in my experience, the blossoming of that child's potential appears more likely.

References

Panksepp, J. & Biven, L. (2012) *The archaeology of mind.* New York, NY, USA: W.W. Norton & Company.

I would like to acknowledge and express gratitude to the researchers, parents, children, and others who made, and continue to make possible the evolution of reflex integration modalities. – Brian Esty

Success for Children with Severe Neurodevelopment Disorders

Agata Palasz, PT, Paulina Turzynska, Psychology, Michalina Lelek, Special Education, Therapy Center for Rehabilitation & Education in Pierszczewie & Strzepczu, Poland

Agata Palasz

Paulina Turzynska

Michalina Lelek

We would like to share our experiences working with children having various developmental problems while treating them with MNRI®.

We are therapists of the Kashubian Foundation for the Rehabilitation of the Disabled and provide MNRI® therapy at three Centers for Rehabilitation, Education, and Early Childhood Development. Under the supervision of the Foundation, there are 59 pupils with a variety of developmental disorders including multiple and profound disabilities, Autism Spectrum Disorder, Down syndrome, and cerebral palsy. In addition, the Power-Assisted Early Development Foundation includes another 25 children. Working with these children are 24 therapists, teachers, speech therapists, and psychologists. All have taken at least one training program with MNRI®, with most having completed 3 or 4 training sessions. All the above personnel of the Foundation implement the therapeutic program based on MNRI® and, in each of the children, we see visible and positive developmental improvements in physical, cognitive, emotional, and language development. Through the case history of these four children, Nadia, Paulina, Krzysztof, and Zosia, we would like to show the effects of MNRI® therapy. The basic therapeutic programs used with these children are Tactile Integration Therapy, Oral-Facial and Visual-Auditory Reflex integration, Dynamic and Postural Reflex integration, Re-patterning, and integration of Dysfunctional and Pathological Reflexes.

Nadia

Nadia is a little girl who was born prematurely with low birth weight (24 week pregnancy; birth weight of 650 g. or 1½ lbs.) with impaired vision from retinopathy of prematurity. She stayed in an incubator for a few weeks and was hospitalized for a long time after birth. Nadia underwent two unsuccessful eye operations and was revived several times, and is now blind. During our first meeting with Nadia, we

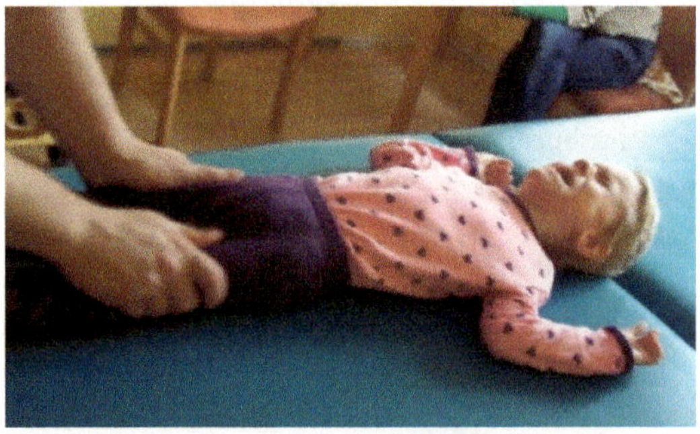

Beginning therapy was difficult for Nadia; she reacted to the sensation of being touched.

observed extreme reactions to touch and in the orientation of her body in space. She seemed to be afraid of everything, over-reacted to stimuli and, during our meetings, clinged to her mother and father. According to the opinion of her parents, Nadia very often responded to many various stimuli by screaming, crying, or scratching and biting people who were close to her. At the age of two years was not able to sit alone, turn from her back to her stomach (and vice versa). She presented with many of the most difficult consequences of extreme prematurity. Prior to 2011, Nadia's parents had taken her to an Early Development Support Rehabilitation at a resort. There, Nadia received therapy from a variety of methods including NDT-Bobath, however, they saw no positive effects.

Nadia: in January, 2014, she practiced walking intensely.

She began to attend our Center for Early Development Support in Pierszczewie when she was 2 years of age in October 2011. When we met Nadia, she did not trust anyone. Our early attempts at communication with her were very difficult. She only wanted to be alone with her Mom. She struggled, screaming, scratching, and biting. The team of therapists (speech therapist, psychologist, and physiotherapist) and parents decided that for awhile Nadia should be limited to only attending classes at the resort run by our Foundation. Her parents resigned from therapy in other centers so she could focus on getting stable here. With this decision, it was possible to start building a sense of security in Nadia. Our therapy was based on the MNRI® method. Nadia regularly came 3 times a week (about 3 hours), at the same time each day, in the same room, and with the same therapist.

At the very beginning of our program with Nadia, Tactile Embracing Squeezing was primarily used. When we started using this, thanks to regular classes, we saw changes in her. She became calmer. We later were able to work in a sitting position, then it was possible to position Nadia on a massage table. Currently, she works with us very well and lies quietly, listening to requests. She has begun to communicate, "yes" answers "aaa" or "er"; when she wants to eat, she says "amm". When she objects and wants to convey the message "no", she just shakes her head. In addition to Tactile Therapy, we are giving her a priority of corrective techniques for Re-patterning Reflexes and Oral-Facial Reflex integration. After regular therapy for 3 years, we have achieved very good results. Nadia sits alone, moves from a simple sitting position on the ground, and, with the help of a walker, moves around. She firmly holds objects in her hands. She is much calmer now, not screaming even when she is in a new place. She is cheerful, smiling, greets everyone, and takes our hand. She no longer shows aggressive or self-harming behavior and conducts herself appropriately to the situation when a stranger wants to take her in their arms. Now she is 5 years old and attends preschool group at the Rehabilitation Education Center for Children in Pierszczewie. She seems to feel safe among her peers and rides the bus to the resort by herself, without Mom. She began to attend classes at the pool for Halliwick method therapy and she is very fond of classes in this group.

Paulina

Paulina is an eight year old with congenital encephalocele with hydrocephalus. She started attending the Rehabilitation Education Center in Pierszczewie when she was 3 years old. We saw that Paulina seemed to be largely developmentally stuck in her upper limbs, which we saw as a base from which to develop to her other spheres of autonomy. But Paulina showed a high sensitivity in her hands. She gave great resistance to various activities such as painting and touching different textures. She could not hold a spoon or cup during meals, sit well alone, or rotate her body.

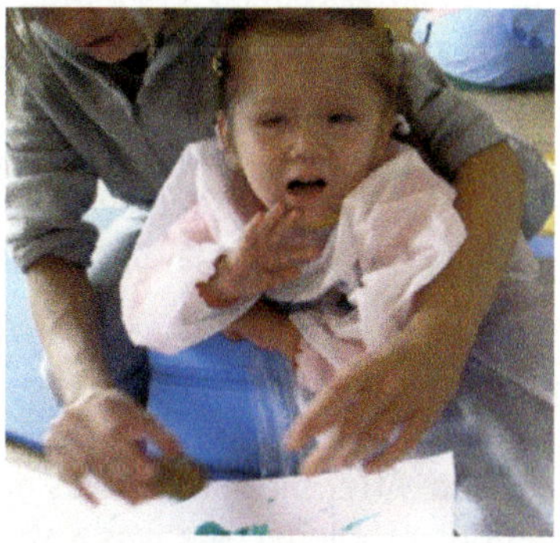
At occupational therapy, Paulina at first protested and required a lot of help.

Paulina was introduced to the MNRI® therapeutic program using re-patterning techniques and integration of MNRI® reflexes. For five days a week before each occupational therapy and meal we worked on the integration of Robinson's Grasp Reflex and developed all phases of this reflex. Other reflexes we worked on were Hands Supporting (parachute), Hands Pulling,

Paulina, after therapeutic effects of MNRI®, began to enjoy tactile and sensory stimulation.

Spinal Galant, Spinal Perez, and Spinning. After several weeks of intense work with Paulina, we saw good progress. She became more independent and began to take different actions. Today, she eats her meals by herself with her own utensils. Food now gives her pleasure. The effects of MNRI® are also visible during occupational therapy; she enjoys painting with her hands and playing with different textures (eg., starch, jelly) – she does not complain and it even gives her satisfaction and joy.

Paulina has also begun to sit alone, turn, and crawl. She has become more active in her wheelchair. She's now smiling, cheerful, and talkative. She is very active and likes group activities such as the activities in *Original Play* by Fred Donaldson.

Krzysztof

Krzysztof was brought to classes in Early Development Support beginning in May, 2012 at the age of 4 years old with a diagnosis of Autism Spectrum Disorder. At this time he had major difficulties in establishing and maintaining social relationships and communication (especially verbal) and did not maintain eye contact. He also had a lot of self stimming behavior. Most of the time Krzysztof did not know how to play with toys – he just screamed and cried. Krzysztof also screamed and cried in new places and unfamiliar situations. He displayed the typical difficulties of children with autism. Until May 2014, Krzysztof was given the therapeutic program based on the MNRI® method. He went to therapy regularly, 2 hours per week in classes with a speech therapist and a psychologist. The intervention included techniques from such therapeutic MNRI® programs such as Repatterning, Visual-Auditory Reflexes, and Oral-Facial Reflexes integration with a special focus on Robinson's Hands Grasp, Babinski, and Babkin Palmomental.

The effects of this therapy surprised his parents, therapists, and teacher. Now, in December 2014, Krzysztof is 6½ years old and does not differ significantly from his peers. He attends a compulsory pre-school in a normal public kindergarten. In September 2015, he will begin the first grade. Krzysztof is successfully communicating with others, asking questions, and talking about current events. He maintains eye contact with the participants of his conversations and has a rich vocabulary. He has established friendships with his schoolmates. He no longer shows fear in new places and situations.

Historically, Krzysztof did not tolerate other people's music but, today, there is no trouble with this. Formerly, he was terribly afraid to go to church (probably difficulty with sound over-stimulation which reverberated in the church building). His problems associated with auditory hypersensitivity have been resolved. Today, he has no problem participating in weekly worship services with his family. Moreover, Krzysztof's ability to read and write simple words surpasses many of his peers. The only areas that require some support are some minor obsessions and atypical eating habits (a less varied diet). It is worth noting that these issues do not impede the free functioning of Krzysztof with his family and school. According to current diagnostic criteria, Krzysztof today shows no symptoms of his original diagnosis of autism.

Zosia holds and lifts the glass from the table, keeps it in her hands, and tilts it to her mouth.

Zosia

Zosia is a 7 year old girl who is very social, likes music, having fun, and the sea. After birth, there was no indication that Zosia would have any serious problems. She was a strong, healthy infant and met her developmental milestones on time. Then, at the age of two years, her parents

But that's not all ... today Zosia can spend time in a way that she likes most – now she can also browse through a book and turn the pages one by one!

saw some worrying signs and began to suspect Rett syndrome. After several medical tests, the diagnosis of Rett Syndrome was confirmed when she was 4. According to literature, a person with Rett syndrome person loses their motor skills with age. Thus, due to the significant loss of purposeful hand functions, Zosia was referred to as a "girl without hands."

When she was 6 years old, she began to attend our Rehabilitation Educational Center in Strzepczu. Despite its serenity, Zosia struggled for a long time with difficulties regarding functioning in everyday life. Due to the low efficiency of her fine motor system—primarily using her hands, she did not cope well with basic functions, such as eating food. She could not drink alone because holding a glass was impossible for her. Large amounts of food would fall out of her hands while eating, so she needed the constant assistance of an adult. The very use of her hands required great effort, but Zosia had a lot of motivation. One of her favorite activities was looking at picture books. Unfortunately, trying to turn pages caused the closing of the entire book or remaining still on the same page and any attempt to grab toys also ended in failure. With her physical awkwardness, she often wiped most items off the table. Low efficiency of her fine motor system prevented many other activities because of all of the time and assistance needed for her. But Zosia was interested in the world and wanted to learn new things so she persevered.

From the beginning of her stay at the Center, Zosia was given therapy techniques aimed at integrating her Robinson's Grasp Reflex with MNRI® methods. For two years, the so-called 'exercise sticks' were her everyday therapeutic practice. It should also be noted that a key program for Zosia was also Tactile Reflex therapy. Zosia is currently undergoing Phase III of Rett syndrome. According to the literature, this period is followed by the largest fine motor regression. However, Zosia's case shows that this is not necessarily the case. In the opinion of therapists and her parents, she has seen significant fine motor improvement. Today, Zosia copes well with food and is able to drink from a glass. She requires only a little help with the removal of glasses from a table.

It is clear that she continually needs to have therapy. However, we also believe that, by using the MNRI® program, Zosia will have further positive improvement and achieve more successes.

Conclusion

Here we have highlighted the progress and improved development of Nadia, Paulina, Krzysztof, and Zosia to illustrate the substantial progress and positive trends which are marked by the introduction of the therapeutic MNRI® method. These children's stories are an example of the success that we have observed with great satisfaction and have also seen with children in our other wards at the Foundation.

References

Bodzioch M., Zespol, R. (2010) [w:] *Medycyna Praktyczna Neurologia*, PL.

Chrzan-Degoś M., (2012) *Wcześniaki – Rozwój psychoruchowy w pierwszych latach życia,* wydawnictwo Harmonia, Gdansk, PL.

Delacato, C. H. (1995) *Dziwne, niepojete. Autystyczne dziecko,* Fundacja Synapsis, Warsaw, PL.

DSM-IV (*Diagnostic and Statistical Manual of Mental Disorders*) – klasyfikacja zaburzen psychicznych Amerykanskiego Towarzystwa Psychologicznego

Pisula E. (2000) *Autyzm u dzieci. Diagnoza, klasyfikacja, etiologia.* Warszawa: Wydawnictwo Naukowe PWN, Warszawa.

Szulca A., Czekalska, M. (2002) *Dziecko z przepukliną oponowo-rdzeniowa: diagnostyka i leczenie zespolow neurodysraficznych,* Osrodek Wydawnictw Naukowych, Poznan.

We thank all our children and their parents! We are enthusiastic and hopeful for improvements for all in various rehabilitation centers that use our training. They are all Winners and work very hard to gain new skills. – Authors

Impacting Sensory-Motor Function in an Adult with Multiple Sclerosis

Laura Paquette, OTR, MNRI® Core Specialist, La Crosse, WI, USA

Through this story, I will show how MNRI® program has helped T.W. regain sensation and motor function that she thought was lost due to multiple sclerosis (MS).

Laura Paquette

Introduction and Presenting Problem

T.W. was a 31 year old woman who was diagnosed 10 years ago with MS. Over the years since her diagnosis she has undergone four medical treatments including subcutaneous injection, an intramuscular injection, a low-dose cladrabine/steroid treatment, and a once a month infusion treatment. She had received physical therapy on and off between the years of 2001-2006, and in 2010 she was prescribed a Walk Aide, which is an electrical stimulation device used for foot drop. At the time of our first MNRI® session, she was still receiving a monthly infusion treatment but is no longer.

T.W. and I had our first session using MNRI® techniques on November 25, 2011. At that time, her main issues were loss of sensation in her left leg with decreased motor control and poor balance. These challenges caused her to have significant difficulties in safely and confidently managing stairs, a necessary daily activity for getting to and from work. She also was experiencing a decline in activities she previously was able to do, i.e., now she frequently required support to maintain her balance while walking, assistance to dress her lower extremities due to inability to lift her left leg, inability to carry out fitness activities and exercise programs, or to run errands independently due to balance and bladder control issues. She has also sustained injuries from several falls where she was unable to catch herself.

Evaluation of reflex pattern responses showed an over-active 'red light' or 'freeze' response in her Core Tendon Guard Reflex, giving her a flexed, protective posture, causing asymmetry especially on her right side. Responses to sensory stimulation of her foot reflexes (Foot Tendon Guard, Babinski, and Foot Grasp) were not symmetrical and presented mixed patterns for all. She did not have sensation on her left leg. Her Trunk Extension was limited, with movement stopping at hip level. Her Leg Cross Flexion-Extension showed an incorrect motor pattern and direction of movement, i.e. when the correct motor response was to come down with her leg to 'catch' and protect herself, instead her leg moved up and off to the side. Her comment at this time was, "I have no idea where down is." During the pre-test for Thomas Automatic Gait, she was extremely unsteady with her forward gait, unable to maintain her balance with backward gait, and described her sideways gait as "very difficult, uneven, and scary." Her spinal reflexes were asymmetrical. She responded to Spinal Galant with hypersensitivity and inability to shorten and lengthen between her shoulder and hip. She responded to Spinal Perez with inability to activate sacral movements. The Hands Supporting Reflex was asymmetrical and she was unable to move her right arm into full extension or in the correct direction.

Treatment & Results

During our first treatment session we facilitated integration of her Core Tendon Guard Reflex pattern to establish symmetry. We then facilitated Foot Tendon Guard and Babinski Reflex patterns to establish the correct sensory-motor response and it was during this process that she regained feeling in her left leg and foot (which she thought was permanently lost). When facilitating Leg Cross Flexion-Extension, we first trained the protective step of the reflex pattern: her leg coming down to catch herself. When she was trying to use a conscious process to direct the movement, she had no control of her leg. We needed to initiate the integration process by using the command to 'just think about the movement' and this eliminated the effect of trying too hard. Once the pattern was established, we were then able to continue the integration process to the point where the motor response was automatic when the sensory stimulus was presented. She no longer relied strongly on cortical thinking. We then addressed her spinal reflexes. We were able to establish the correct motor response, but symmetry and timing were still not accurate. The Hands Supporting Reflex pattern was facilitated so that if she did fall, she would have the upper limb reflex response to catch and protect herself.

Prior to this treatment session, she used both hands to hold the railing and the wall when she walked down stairs, looked down at her feet to direct them, and used a robotic gait, kicking one leg out in front of her and having it land on the step as it came down. This was a terrifying process for her. After this first session, she walked down the steps with one hand on the railing, using a relaxed automatic pattern, and turned to look behind her during mid-step to say something to me without realizing what she had just done! In the days that followed, she was able to dress herself independently, manage the steps going to and from work, walk without needing to hold someone for balance (with the exception of times of fatigue or illness), carry out integration exercises that we established to continue the integration process, and, during episodes of tripping or falling, she was able to catch and protect herself using her arms.

We had a second session one month later. During the pre-test, she showed improvements in her spinal reflex responses (she had been doing a Home Program), maintained the feeling in her left leg, and maintained the results of the other patterns we had addressed the month before. During this session we focused our attention on reflex patterns for balance, grounding, and stability, along with Thomas Automatic Gait. Following this, she was able to step forward, backwards, and side – both directions without support and with significantly improved control and confidence. Following this session, when she went to put on her Walk Aide, she hesitated as she was unsure of what leg had needed it! She has not used this device since. Functional changes were made in her ability to run errands independently, pick up a few things from the store without using a cart for balance and support, walk and turn her head to carry on a conversation or look at something without losing her balance, and maintain her balance on the train (using her legs for weight shift instead of relying on her arms to hold on). She began doing exercises she had received from physical therapy for core strengthening. She developed improved bladder control and enjoyed an overall increase in her energy level.

We have had three more sessions since the initial two over the past year and her progress and gains in motor control and functional skills continue to grow. Her immune system appears to have strengthened as demonstrated by shorter episodes of illness and more rapid recovery with no loss of skills. Our frequency of sessions has been limited by distance but she faithfully carries out exercises and activities to support the continued integration of the reflex patterns (including following the Archetype Movements).

MNRI® techniques enabled T.W. to gain motor functions that were thought to be lost. Her neurosensorimotor system has moved from a state of strong protection and survival to one in which her primary reflex patterns were responding with a more appropriate or adequate sensory motor response. The effects of this have been improved motor coordination, improved functional skill and quality of life, decrease in anxiety, improved immune system function, and a higher energy level. It is our hope that, with continued work using MNRI® techniques, T.W. and others with a diagnosis of MS will be able to regain functions thought to be lost and explore the possibilities of influencing this disease process.

I congratulate T.W. for her positive outlook and sense of humor in her courageous approach to life. I am so grateful for what she has taught me through this journey. – Laura Paquette

Collin's Success and Development

Diana Lilja, Pediatric PT, MNRI® Core Specialist, Roscoe, IL, USA

Patient's history

This is a story about an adorable little boy named Collin, who has learned to interact in this world, despite some major setbacks. Collins' Mother showed me a picture from the MRI of his brain one day. I said, "This can't be true; he could not maintain life without a brain". She assured me that the tiny layer of cells on the very edge of the picture were his brain. Later, I realized that it was only one slide of many. Her doctors shared the worst of the bunch with her. Regardless, Collin did have a major loss of brain tissue. His mother and I agreed that we would just have to build new pathways to make up for the missing tissue, as much as we possibly could.

Diana Lilja

Collin was born at 37 weeks gestation by a scheduled C-section. He had been diagnosed with Aqueduct Stenosis Hydrocephalus in utero (at 24 weeks gestation). He had a VP shunt placed the day after his birth. His diagnoses include: Lobar Holoprosencephaly, absence of the Septum Pellucidum, Chiari 1 Malformation (found at birth), failure to thrive (at 9 months), feeding tube placed (at one year), low growth hormone levels, low muscle tone, Nystagmus (at 18 months), and far sightedness in the left eye.

Progress with Therapy

Collin was evaluated twice by another physical therapist and was on a waiting list before starting weekly physical therapy with me. He also received developmental therapy, vision therapy, occupational therapy, and speech therapy as part of his early intervention program. Mom reported that the biggest shifts that she noted were from the reflex integration therapy which served as building blocks for the other therapies to work off.

When I started treating Collin, he had limited visual tracking. His head was generally turned to the left and he mostly only looked to the left, with the ability to focus visually for only very short durations. He could not hold his head against gravity at all. He had difficulty moving his limbs against gravity, with very limited movement noted secondary to his extremely low muscle tone. He had a strong startle to sounds. His reflexes were mostly hypotonic (with a few hypertonic). He had some limited hand movement (a little more if his shoulder was supported). He was starting to develop tonal patterns in his legs.

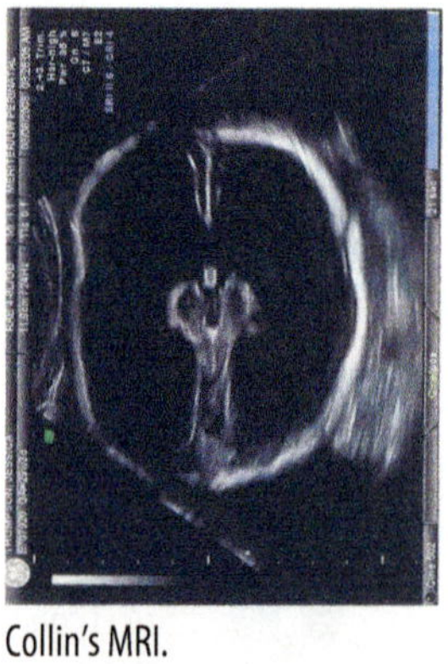

Collin's MRI.

Collin started receiving weekly 1 hour MNRI® treatment on September 22, 2009.

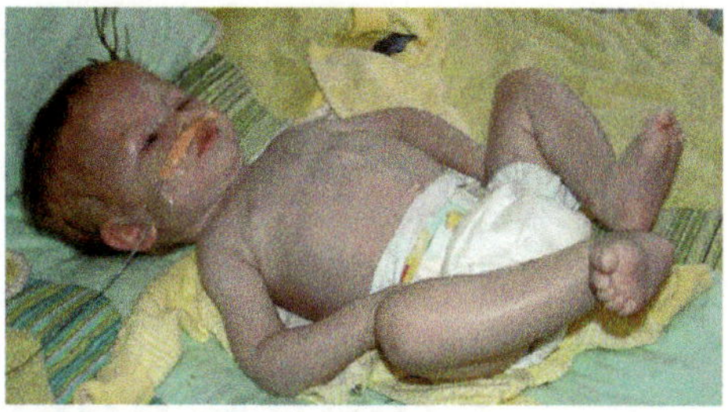

He received weekly physical therapy and a pretty intensive Home Program (about 5 hours per week). We worked on Reflex Integration Exercises, Neurostructural Reflex Integration, Tactile Integration, Oral-Facial Reflex Activation, and Visual and Auditory Reflexes Integration. Mom reports that cha-cha rhythm techniques and their sounds have become very familiar in their house and lots of embracing squeezes were used to increase body awareness.

Above, Collin as a baby.
Below, Collin as a toddler.

Mom noted that Collin came out of his strong 'freeze' stage after a few months of MNRI® therapy. He was calmer and happier overall. He was able to be in a room with more than a few people, without getting overstimulated. She could see that he was changing, from initially having no idea where his body was in space, to having a little control, so he could sit in an externally propped sitting position. His head control and his visual tracking improved significantly. He eventually progressed to being able to independently sit in a propped position, and even lift one arm occasionally. He progressed from being in apparent pain with his tonal patterns that emerged as he started to move, to looking much more comfortable as he gained motor control. He started paying much more attention to people, enjoying their interaction, rather than being stressed by the input. He learned how to roll, which gave him the chance to explore his environment independently for the first time in his life. He could roll anywhere he wanted to go. He went from hating tummy time, to choosing it. He was able to play with more toys and eventually carry them from place to place. He learned to push up to a hands and knees position and hold this position for up to 30 seconds, with some rocking. He started to stand, requiring less assistance as we progressed.

Collin today after treatment.

The oral-motor reflex integration sessions made a huge change for him. He still did not have words, but could make sounds to let people understand his needs more. His oral-motor sphere started changing crucially. Collin progressed from no oral feeding to oral feeding and from G-Tube feeding to tolerating some oral feeding.

By the time of his discharge from Early Intervention physical therapy at 3 years old (10/02/2011), Collin had progressed from having very little ability to interact in his world, being in a constant state of 'freeze' to being a mobile, communicative little boy. This was after approximately two years of weekly MNRI® therapy, with an intensive home program. He still has many limitations, but has made amazing progress for his diagnosis, most importantly learning to enjoy his life!

Currently (March 2013), Collin is able to eat whatever his family is eating. He will probably have longer term tube feeding at night, secondary to his low growth hormone levels.

Working with Collin and his devoted Mother was a labor of love. It was a great pleasure to travel along their path for awhile and watch Collin progress in all areas. Thank you, Collin, Jessica, Ryan, and P.J. for the opportunity! – Diana Lilja

Ethan's Story

Rachel Nice, MNRI® Parent

Ethan arrived five weeks early and stayed in the hospital for three days for observation. After another three days at home, Ethan re-entered the hospital for jaundice. He also experienced early issues with severe reflux, agitation, difficulty calming and sleeping, and sound sensitivity. By six months of age, Ethan's fists remained adducted, thumbs curled into tiny fists. He did not grasp or reach for things. At about eight months of age, we noticed he avoided eye contact at close range during face-to-face interaction, turning his head to avoid looking into someone's face. At that time, a developmental pediatrician evaluated Ethan, and she diagnosed him with left-hemiplegic cerebral palsy and determined that it was too early to tell whether he suffered mental retardation. Ethan could not lie on his stomach without extreme discomfort. He could not roll over, sit up, or crawl. A second doctor's evaluation confirmed that Ethan was globally developmentally delayed, missing all of the typical milestones for infants his age. Physical therapy services did not begin until Ethan was a year old due to delays in the medical referral system. Occupational therapy started later still, and due to hospital policy at the time, Ethan did not have access to speech therapy until eighteen months of age. Ethan's severe anxiety around non-family members, even at that young age, made therapy sessions difficult and less productive.

As he grew older, Ethan continued to experience auditory and tactile sensory issues, problems with fine and gross motor skills, delayed speech, poor balance and coordination, lack of protective responses, allergy sensitivities, mixed muscle tone, and poor body awareness. At three years of age, dyspraxia/apraxia, severe leg cramps, and night terrors added to the list of observations. By five years of age, we noticed odd unexplained physical movements in certain postures. Ethan also received a diagnosis of mild scoliosis. Conventional therapies produced slow though continual results. While still delayed, Ethan possessed greater mobility though of poor quality and remained less stable. Hand splints and occupational therapy helped improve hand function, though still delays persisted. At eight years old, a doctor confirmed Ethan's difficulty focusing with a diagnosis of attention-deficit disorder (ADD).

Despite the many challenges he faced, Ethan proved to be a cheerful and loving young boy, always ready with a quick smile, a curious mind, and demonstrated himself an apt pupil. Ethan's life began to change more dramatically starting at age five. Diet regimens to avoid allergens and biomedical interventions helped improve his health and reduce the number of ear infections. Conventional therapies continued, and auditory intervention training helped Ethan's hyperacusis (hearing sensitivity). During this time, we discovered MNRI®. His first treatment achieved immediate results! Ethan's gait improved immediately upon getting down from the treatment table!

We learned the MNRI®Dynamic and Postural Reflex integration techniques and continued with the exercises at home daily. In the next months that followed, Ethan's overall anxiety levels improved. Goals that he had for adaptive physical education, not reached in the previous two years, became accomplishments! Both Ethan's gross and fine motor skills showed dramatic gains. Ethan could stand on one leg, run, hop, skip, and gallop.

The MNRI® Visual and Auditory Reflex integration techniques improved Ethan's visual tracking skills and helped him to participate in ball play more successfully. His protective responses improved and he fell less frequently, and when he did fall, he did not hit his head or face as he did previously. At the end of Ethan's kindergarten year, he completed speech therapy services and did not need to continue.

Over the following years, we incorporated many other therapy modalities in addition to MNRI®with varying degrees of success. We utilize some therapy modalities off and on, while others we have tried and discarded, but MNRI® has remained the one constant throughout Ethan's life. MNRI® has complemented all of the other therapies in Ethan's life beautifully, helping them to achieve even greater levels of success. Regardless of the other therapies that are tried, the MNRI® education we have earned through MNRI® course attendance helps us to make better sense of other modalities we come across, because MNRI® has provided us a framework with which to understand our son's development. We share our experiences with any families who are interested in listening because we want to help other families help their children and know the same kind of hope and success that we have obtained.

Today, only highly trained specialists might be able to ascertain signs of Ethan's original diagnoses. Ethan successfully participates in any activity he desires. Muscle tightness in some areas persists, but sensory issues are no longer problems. Ethan can run, jump, and climb, and for the very first time in his life, Ethan has scored in the average range on the BOT-2, a well-known physical standardized test! Ethan's scoliosis and ADD also continue to improve. MNRI® has helped our son forge a bright and promising future.

Ethan made amazing progress in all spheres – sensory-motor, cognitive. Being a gifted child, he is an example of how hard work and knowledge can help us to move forward constantly. Thank you to his mom, Rachel, for helping create the amazing results by working with your son with MNRI® and your deep love for him. Our gratitude to you also for organizing MNRI® lectures and getting the word out to other parents. – Dr. Masgutova and MNRI® Team

Miracles Come in All Sizes: Billy's Success

Heather Stewart, MNRI® Core Specialist, Toronto, Ontario, Canada

Introduction

When parents do not accept the limited box their child is labeled with, it opens up the door to many possibilities. When given a very restrained view of the future potential for their son, Billy, this family set out on the road to develop his potential to whatever it could be.

Heather Stewart

This story is being told to encourage those parents in similar situations to step onto the road to other possible outcomes. Miracles come in all sizes – big and small, and this is the story of one family's miracle that is still developing and unfolding.

Patient's History

As reported by his mother, Billy was 10 days late for birth, and labor was induced. After 23 hours, a C-section was performed because his head was slightly twisted to one side and a vacuum extractor was tried with no luck. Initially, it was believed he had a heart valve problem and Billy was transferred to HSC but it normalized without intervention.

Blood pressure was elevated in one side of his body and low in the other but then that normalized also. No one could explain this. Billy was jaundiced for four days and required phototherapy. He had mild torticollis and always slept on one side but this improved at around 9 months. Billy never babbled or cooed as a baby. He could not wave bye-bye or clap.

Billy as a toddler

Billy never crawled but bum shuffled and protested his hair and nails being cut and his face washed. He refused fresh fruit and vegetables unless it was pureed and ate very quickly putting large amounts of food into his mouth creating a choking hazard. Billy started walking at 15 months and was very clumsy and would fall a lot. A thumb sucker with many allergies, Billy had his tonsils and adenoids removed due to sleep apnea. He sought oral stimulation by biting on books and toys and tended to touch people and objects frequently.

At the age of four, six days after his MMR shot, Billy woke up with crossed eyes. A CAT scan determined a muscle problem.

The family travelled through the maze of assessments from a range of professionals including neurologists, occupational therapists, physio-therapists, speech-language pathologists, psychologists, chiropractors, Early Intervention Services, Family & Children Services, genetic testing, developmental pediatricians, and underwent a range of assessments: neurofeedback evaluations, ABA Therapy, school support services, tutors, and audiologists while looking for help for their son, who identified many labels and characteristics. A few were: dyspraxia; global cognitive, intellectual, developmental delays/disability – 1st percentile; significant attention weakness, impulsive; learning disabilities; poor motor control: gross and fine 1-4th percentile; decreased muscle tone, joint instability and posture control; dysgraphia; weak executive and strategy functions; needs social skills, (2nd percentile) shy, mild anxiety, immature play and not good at taking turns; apraxia – significantly delayed receptive and expressive language skills, vocabulary and comprehension (4th percentile); dysarthia; had Strabismus (eye muscle) surgery – (visual processing 4th percentile); non-motor visual problem solving 2nd percentile; symbol copying 1st percentile; visual reasoning 36 percentile; and working memory 0.1 percentile. He test negative for Fragile X Syndrome.

It is unbelievable seeing this boy today that this kind, caring, compassionate human being has ever been labeled with such a dim outlook for his future.

MNRI® Facilitation of Billy's Neurodevelopment

Billy started his MNRI® therapy sessions with me, his local MNRI® specialist when he was 10 years old, coming to me approximately once a week.

Just before Christmas 2010 the family received a special miracle when Billy discovered his arms and realized how to use them. This was from our work with the Hands Supporting Reflex pattern. He then began hugging people! This improvement gave Billy's family enthusiasm to attend a MNRI® Conference in 2012 with Billy.

Billy's MNRI® pre-evaluation of reflex patterns in 2012 showed that a majority were still below norm and dysfunctional (May, 8, 2012 – before MNRI® Program - see Table 1-3, below). Thirteen reflexes out of 30 were below 10 points (on scale 1 to 20). The next two evaluations were done on May, 17, 2012 after Billy attended an 8-day MNRI® Conference, and later on April, 27, 2013 after their MNRI® Home Program. The last two evaluations after MNRI® work demonstrate that Billy's reflex patterns improved significantly.

Reflex Development Evaluation for Billy

Based on the results of the MNRI® Assessment, an individualized program for Billy was established for his MNRI® Conference and Home Program. The main purpose of the individualized program was to strengthen and support physical, emotional, and cognitive growth through the work with non-integrated and dysfunctional reflex patterns. This included: Repatterning of Infant and Lifelong Reflex Patterns, Neuro-Structural Reflex Integration, Tactile Integration, Archetype Movements Integration, Oral-Facial Reflex Integration, Visual and Auditory Reflex Integration, and Proprioceptive-Cognitive Integration.

The main MNRI® exercises for his Home Program were targeted especially at ATNR, Babinski, Babkin Palmomental, Balancing, Bauer Crawling, Core Tendon Guard, Fear Paralysis, Foot Tendon Guard, Grounding, Hands Pulling, Hands Supporting, Segmental Rolling, Spinal Galant, Spinal Perez, Spinning, STNR, and Trunk Extension (in spite of the higher score).

Billy	Table 1. Reflex Patterns (X_{1-10})									
Reflex Patterns	X_1 Core Tendon Guard	X_2 Hands Grasp	X_3 Hands Pulling	X_4 Babkin Palmo-mental	X_5 Babinski	X_6 Leg Cross Flexion Extension	X_7 ATNR	X_8 Abdominal	X_9 Bonding	X_{10} Foot Grasp
May, 8, 2012 Before MNRI Program	13.75	12.25	9.25	8.25	7.25	11.75	8.75	9.25	11.25	7.25
After MNRI Conference May, 17, 2012	14.5	13.25	10.25	8.75	8.25	12.75	10	10.25	11.25	8.25
After MNRI Home Program April, 27, 2013	15	14.25	12.25	10.5	9.25	13.75	11.75	11.25	13	9.75

Billy	Table 2. Reflex Patterns (X_{11-20})									
Reflex Patterns	11 Automatic Gait	12 Bauer Crawling	13 Moro	14 Fear Paralysis	15 Hands Support	16 Segmental Rolling	17 Landau	18 Flying/ Landing	19 Grounding	20 Head Righting
May, 8, 2012 Before MNRI Program	14.25	9	10.25	7.75	7.25	8.75	9.25	8.75	11.25	13.25
After MNRI Conference May, 17, 2012	15	11	11.75	8.75	9.25	10.25	10.5	9.25	12	14.25
After MNRI Home Program April, 27, 2013	15.25	12.25	13	11.75	11.75	11.75	12.25	10.75	12.5	14.75

Billy	Table 3. Reflex Patterns (X_{21-30})									
Reflex Patterns	21 Trunk Extension	22 STNR	23 Spinal Galant	24 Spinal Perez	25 TLR	26 Foot Tendon	27 Spinning	28 Locomotion	29 Balancing	30 Pavlov Orientation
May, 8, 2012 Before MNRI Program	14.25	8.75	9.75	11	12.25	7.75	10.75	7.25	12.25	12.50
After MNRI Conference May, 17, 2012	15	9.25	11.25	11.75	13.25	8.75	11.5	8.25	13.25	13.50
After MNRI Home Program April, 27, 2013	15.5	11.25	13.25	13.25	14.25	11.25	12	10.25	13.75	14

Tables 1, 2, 3 above show the level of various reflexes prior to and after two MNRI® Conferences.

The data on reflex in-

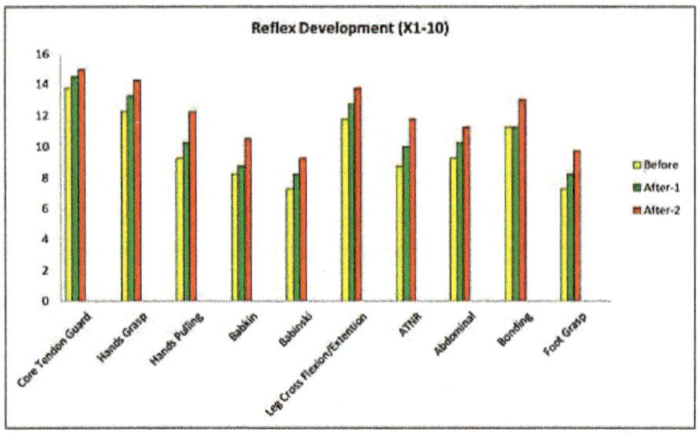

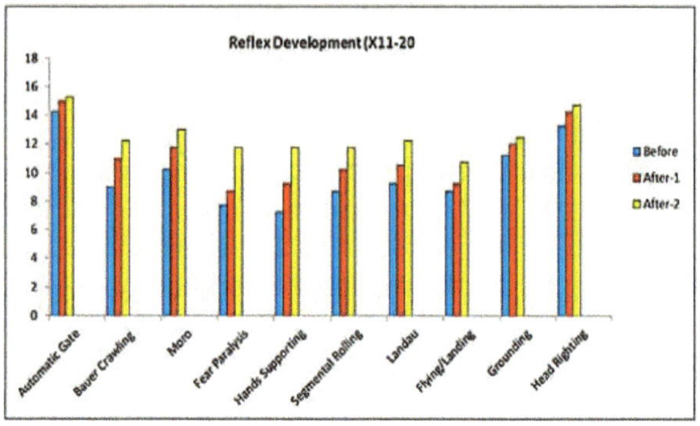

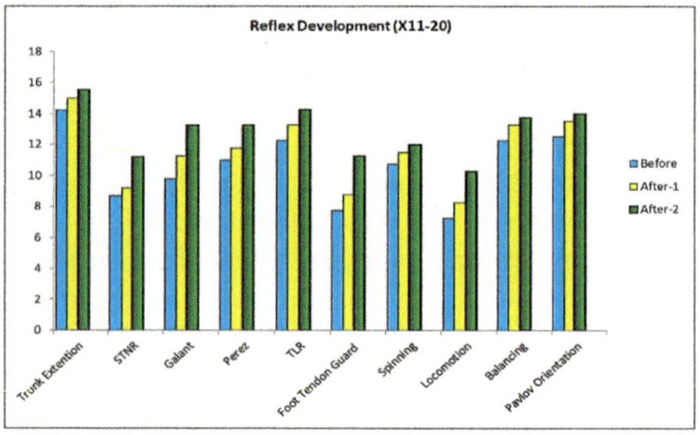

Graphs 1, 2, 3 above show the level of various reflexes prior to and after two MNRI® Conferences in graph form.

tegration after the 2012 Conference demonstrates that 22 reflexes were on transitional and higher level of development (versus 14 reflex patterns in pre-evaluation), and after work at a Home Program (2013), 28 reflex patterns reached the scale for positive development. The data shows that Billy moved from a level of Reflex Integration Dysfunction to a level of positive scale of development.

Other pre- and post-tests on OT and PT and speech development done during the Conference also demonstrated positive changes in various gross and fine motor skills and language abilities. With MNRI® therapy he has never plateaued and continues to improve. Billy exhibits aspects of literacy development that far exceed his predicted intellectual development.

This young lad will require more work for his continued development and his family is up to the challenge. They have seen more changes than they ever dreamed. I truly believe his mother, brother, and possibly, Billy himself, may become MNRI® therapists in the future.

Recently the school called for a parent meeting. His parents were worried until his teachers started the meeting questioning what type of therapy they were doing with Billy as the school has seen huge social gains. Billy did a dance routine in gym with his peers and totally kept up pace with the others. But just the fact that Billy performed in front of teachers and classmates impressed everyone. He is also more assertive and is starting to advocate for himself. For the first time in his life, Billy has three friends his own age!

Miracles come in all sizes and there is still more potential here!

Thank you to Billy and his family for being willing to try the MNRI® approach. It has been an honor to be a part of his healing path. I wish him much success and health in his life! – Heather Stewart

The Inspiring Success Story of Lily

Diane Colonnese, MNRI® Parent, Frisco, TX and Pamela Curlee, MNRI® Trainer, Co-Founder of the Svetlana Masgutova Educational Institute, Denver, CO, USA

Diane Colonnese

Pamela Curlee

Lily Grace Colonnese was born thirteen weeks early on September 15, 2003. That made her a 27-week preemie. She weighed 1.9 lbs. and was 13 inches long. Her APGAR score was 8. She was ventilated immediately and was admitted to the NICU for respiratory distress, sepsis, and prematurity. Her other diagnoses while in the NICU included Stridor, Chronic Lung Disease, Patent Ductus Arterious-Indocin, pneumonia (twice), Anemia of Prematurity, IVH Grade 1, Apnea, Neurologic, Gastroesophageal Reflux, Inguinal Hernia, Hyperbilirubinemia, Retinopathy of Prematurity, and vocal cord paralysis. She was reintubated over a dozen times. Lily had surgery to repair the hernia, a Nissen fundoplication to help stop the reflux, and a g-tube (feeding tube) placement all at one time.

Because of the reintubation, Lily still has a subglottic stenosis scar which makes her airway the size of a coffee straw as compared to a regular sized straw. She also has reactive airway disease because of this scarring under her vocal cords. She suffers with croup anytime her airway is inflamed (from a cold or allergies), screaming too loud, or coughing too much.

Lily had a severe Sensory Processing Disorder after being born so early and being in the NICU for over four months. She would jolt as if an electric current was flowing through her when she was touched. She would scream and cry about anything different in her environment. Sight, sound, and touch were painful to her. She was very developmentally delayed across the board. Her language, fine motor, and gross motor were all very delayed.

Lily at age 7.

We had Lily in all therapies the second week home from the hospital. She received Early Childhood Intervention at home and then we took her to private therapists. We tried everything and she was still so little (12 pounds at 1 year), so sick (her immune system was non-existent), so delayed, and had such severe sensory problems. We didn't know where to turn.

Our friend, Chuck H., knew Pamela Curlee and he was excited to share with us that Pamela was working with a gifted woman from Poland. We learned that she had just recently started to come to the United States to begin her work here. He told us that if anybody could help Lily, it would be Dr. Svetlana Masgutova, because her work was unique and was based on years of experience with children all over the world. We later learned that her work was actually based on her experience with working with children and adults who had survived traumatic catastrophes. Our Lily was a survivor of birth trauma and needed this special work offered by the woman who had pioneered her work using the concept of Neurosensorimotor Reflex Integration.

March 2005, Little Rock, AR, age 18 months

We immediately began to establish the link to get Lily to be seen by Dr. Masgutova. Lily was 18 months old. She was not able to sit with stability, crawl properly, walk, talk, eat, or leave the house, except for therapies and doctor appointments. She was so sick all the time.

Our friend Rick M. was able to get funding together to fly Svetlana and Pamela to our house in Arkansas, and they worked on Lily for 2 days.

My huband, Gary, and I are normally hesitant to let a stranger into our house, let alone to work on our precious daughter, but the minute we met them, we were at peace. They came in the middle of the night and were going to work on Lily while she was asleep. Well, Lily woke up, but instead of being frightened, she looked like she was having a party. She clapped and smiled and watched Elmo while they worked on her. The first thing they did was to work with her breathing reflex pattern which was under stress. Dr. Masgutova worked with precision with the various aspects of this reflex and Lily responded very well. The next day they showed us how to repattern and activate reflexes like the Babinski and the Foot Tendon Guard, Trunk Extension, and Spinal Perez Reflex. It seemed as if this work with the Masgutova Method® was able to awaken so many things that were missing inside of her body. Lily's body responded with incredible speed within two days.

Lily learning to crawl, above, and, below, learning to stand.

Before this work, when people would pick her up and then try to put her down, she would pick up her legs and never tried to stand on them. Following the short time with Dr. Masgutova, she was able to extend herself in accordance with the Trunk Extension Reflex which was finally emerging with over a year's delay. Prior to MNRI®, Lily had only been crawling with a half crawl and she could not stand alone. She was significantly developmentally delayed and had begun her modified crawl when she was around a year in age. After this work, a more solid cross-lateral crawl began to emerge, serving as the foundation for walking.

Lily climbing rocks with Dr. Svetlana, Pamela, and Mom.

Lily had never stood before. After 2 days, she stood on her own for only a few seconds, but it seemed like a miracle. Dr. Masgutova and Pamela took her outside and activated numerous reflexes by carrying Lily in different postures, letting her play in the leaves and the dirt at the park. They even had her climbing a rock 'mountain'.

Lily could not sit until she was over one year of age. She did not use her left hand at all, only her right hand. So if she fell over while sitting, she had no protection to prevent herself from falling down. Following the use of the Masgutova Method, she started learning how to stack blocks with her left hand.

We continued to work on Lily, mainly while she was asleep, and Lily had so many significant changes that continued. Working with the sounds of animals, I was able to also make a connection of many of the sounds of phonemes with her as well. Within three months of Dr. Masgutova's visit, she had a 500-word vocabulary. Lily had no speech or language prior to this work.

Lily was feeling better too. Her sensory system began to improve and she was then able to walk in the grass barefoot without screaming.

Age 22 months

The second time we saw Dr. Masgutova was in Austin, TX in July 2005. Following this session, Lily grew 2 inches very quickly. She was also talking, singing, and laughing with deep belly laughs! Lily's physical skills

Lily learning to walk.

continued to blossom following this second session. She ran barefoot across the back yard (this is the child who couldn't even pull herself up or walk 4 months prior!). The backyard even had a small hill that she ran up and down. She loved to talk and sing. She knew all the kid's songs and could spell her name. We still have the video of her doing these things!

Age 2 ½ years and beyond

In Florida, Dr. Masgutova did a great deal of work integrating reflexes. She worked with Lily in the water using her Aqua Reflex Integration Therapy. After her work in Florida, Lily's intellect started to blossom. Her reading was incredible, as was her vocabulary. Her sensory integration was excelling and we used Embracing Squeeze a lot. As her speech emerged she would not say things like "look at the doggie," instead, she would say "look at the Dalmatian." (She knew all the breeds of dogs and has always called them by their correct title....she never used the phrase "doggie"). At the age 2, she would count out loud to 25 as we were shopping or checking out of a store. Also at that age, she would hold a pencil and try to write.

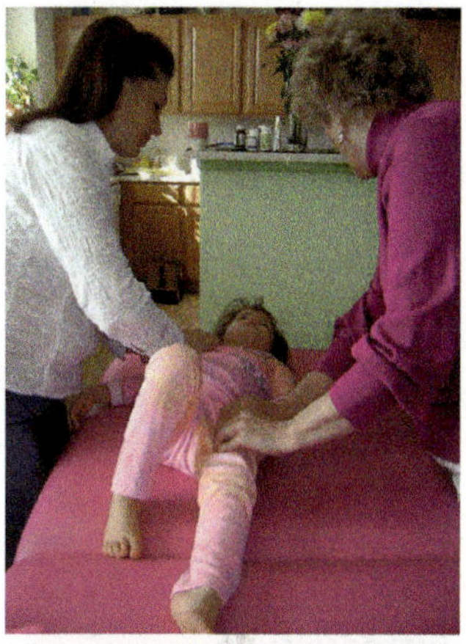
Lily receiving MNRI® session. Lily said, "Is this what you did to wake up my body when you came to my house when I was little?"

At the age of 3 years, she started taking ballet lessons. A home video of her performing a dance at the age of 4 shows her off her toes and moving solidly on her feet.

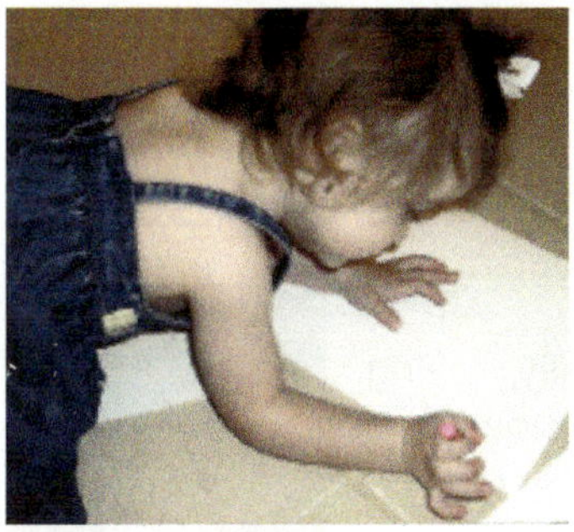
Lily busy as a toddler.

We also had the honor of having Dr. Masgutova work with Lily when she was 3½ (after which Lily stopped walking on her toes), 5½, and age 7. We continued working with over a dozen reflexes, especially the Babinski and Tendon Guard for many years. We did Embracing Squeeze daily. We also worked a lot around her jaw, neck, ears, shoulders and ribs – all as demonstrated by Dr. Masgutova. In pre-school her handwriting and focus greatly improved. We did not tell the therapist that we had been to see Dr. Masgutova and the therapist commented on the great changes she was noticing. I, as the mother, had watched Lily in pre-school one week and when I went back to help in class 3 weeks later. I could also see visible improvements!

Lily is now 10 years old and just took a test to get into a prestigious school in Dallas. She excelled on the entrance exam and ended in the 98% of aptitude. Her social and academic skills are excellent. We are currently working to begin piano lessons, based on the insistence of a piano teacher who declares that Lily "is a natural". When in dance lessons now, the other children watch her and learn the movements from her. She has a great attitude, is social, and sweet as well as beautiful inside and out. And, clearly, she loves Dr. Svetlana!

As a follow up, we just saw a developmental doctor in the developmental clinic at Scottish Rite Hospital in Dallas. The doctor could not believe how well Lily is doing based on her history. She asked me how her language and development was so good. I told her that the Masgutova Method® was what we used. She had heard of it and was very interested in learning more after seeing Lily.

Lily, we are so grateful that we have had the honor of knowing you and your extraordinary Mom and Dad! You inspire us all with the blossoming of your innate wisdom and intelligence and we wish you the greatest fullness and enrichment of life! We know you will continue to be a Winner, again and again. – InJoy, Pamela Curlee

A New Year's Resolution

Terri Zartman, Occupational Therapist, MNRI® Core Specialist, Grand Rapids, MI, USA

Introduction

This is the story of Cole and his amazing success to show the strength of MNRI® and how achievements can be reached in a rather short time. The MNRI® program supports our professional motivation and values.

Terri Zartman

Cole was an eleven year old boy, who experienced a stroke prenatally, leaving him presenting with right sided hemiplegia, which affected his gross motor coordination, walking, and other abilities affected by strong asymmetry, poor postural, motor control, and coordination.

It was his own 2013 New Year's Resolution to be able to walk flat footed. It happened, for the first time in his life, following eight sessions of MNRI®, in January of 2013!

Patient History

Cole had a diagnosis of porencephaly. He had a stroke during the third trimester in utero in his left left hemisphere. He presented with cerebral palsy and a right sided hemiplegia. Cole had been in physical therapy and occupational therapy since he was two months old. He had infantile seizures but has been seizure free since the age of two. As an occupational therapist, I had the pleasure of working with Cole when he was two years old for approximately one year following a more traditional therapeutic approach. Cole had received PT and OT therapy through the school and on-and-off through private clinics his entire life. He has had serial casting done on his right lower extremity. He wore an ankle-foot orthotic but has not in the past two years. When Cole was approximately nine years old, he began physical therapy services with a local, private physical therapy office. Although he was making small gains, therapy was painful and he seemed to have hit a plateau.

Cole's parents received my contact information from a local naturopathic practitioner, suggesting that the approach I was using (MNRI®) may have something new to offer. I saw Cole for the first time in November of 2012 for an initial evaluation. Both Cole and his mother were intrigued by what they saw, in terms of how Cole's body responded from the perspective of a reflex response to sensory stimuli. The evaluation suggested work was indicated with a number of reflex patterns, especially with Babinski, Foot Tendon Guard, and Hands Supporting. After some work, I saw Cole one more time in December of 2012, in which he demonstrated that his Babinski Reflex had calmed immensely and Hands Supporting helped him gain elbow extension. Seeing this unexpected progress, his mother requested intensive therapy over the school's Christmas vacation (so he didn't have to miss school), using the MNRI® method.

Cole was seen January 2, 3, and 4 of 2013, for a total of 8 hours of therapy. Four of the hours were provided by an occupational therapist, three by a speech therapist, and the last hour was a co-treat with both the speech and occupational therapist. Therapy included a combination of MNRI® programs, such as Neuro-Structural Reflex Integration Dynamic and Postural Reflex Integrations, Archetype Movements Integration (both active and passive), as well as, individual techniques including Core Tendon Guard, Foot Tendon Guard, Babinski, and Sciatic Nerve Activation. He also had two sessions including the Oral-Facial and Visual and Auditory Reflex integration therapy. During Cole's last session on January 4, his mother and best friend were present during the co-treating session. The last techniques used during this session included the work with gravity, balancing, and centering from the Lifelong Reflex Integration Program, as well as active lateral spine flexion-extension. Following this, Cole stood on his own, for the first time, with his right heel touching the floor! He then took a few steps, and was able to have his heel touch the floor with each step. His New Year's Resolution came true, only four days into the New Year! Everybody was touched deeply and were happy!

Cole has continued occupational therapy services, emphasizing the MNRI® approach, approximately two hours a month. It had been his request to work on his affected arm/hand, instead of his leg, for three visits after his amazing results with his leg. He was able to achieve active arm extension, wrist extension, and the very beginning of active grasp! There was an unfortunate moment – the family did not manage with follow through on his Home Program. This lack of the Home Program, combined with a spring break vacation, which included a great deal of walking, has led to a slight regression in his lower extremity functioning, including slightly more toe walking and decreased range of motion in the hips and ankles. However, following one session addressing the lower extremities, he was able to again rest his heel flat on the ground. The videos and pictures of transformation of Cole's walking show very clearly his great success! It was amazing how his body still held the memory of the achieved ability, and it was a reminder of how important it is to support the results by regular work while the ability is still in a state of development or transformation. Cole's body responded wonderfully to repetition of the exercises, and with continued work, both within the clinic and at home, the possibilities seem endless! Even his outside physical therapist discharged him, recommending that he use the MNRI® approach exclusively. We are happy for Cole and wish him new achievements – the same way – surprising and great!

I would like to thank Cole and his family for being open to the MNRI® approach and all of its benefits. It has been an honor to be a part of his journey to healing. I wish him as much success throughout his life as he has had with the MNRI® work! – Terri Zartman

Journey of a Lifetime: Destination Unknown

Heather Taylor, MS, OTR/L, Occupational Therapist, Bella Vista, AR, USA

This article will, I hope, give other professionals interested in MNRI® an idea of why I prefer to use MNRI® techniques in my work. I also want to share some of the amazing changes that can be seen – even when kids are not able to participate in a daily program due to limited time and resources.

Heather Taylor

My journey with MNRI® began in July 2011 with the four day *Dynamic and Postural Reflex Integration Class*. Never in my wildest imagination could I have predicted the path this class would lead me professionally and personally, not to mention geographically.

As an occupational therapist who worked in pediatric settings the great majority of my continuing education experience had been with fine motor, handwriting, and sensory integration classes. In 2010 I had finished all the necessary steps to become certified in the Sensory Integration and Praxis Test and, while a sensory framework helped me to identify areas of concern for the children I worked with, I still struggled to develop effective intervention programs for many of them. Standard desensitization techniques for sensory sensitivities appeared to help but were often slow going and difficult to get children with significant signs of defensiveness to willingly participate in the process. Likewise many of the techniques and products designed to help the sensory system become calm worked for some of the children I saw but often they either lost their effectiveness over time and required a more intense intervention or became something the child became dependent upon. I was looking for an intervention that would address these areas of concern and allow mature patterns and behaviors to emerge.

I knew many of the children I worked with had non-integrated reflexes, however most of the programs I had researched were based on the child completing motions that they could not do, which was a little counterintuitive to me. It seemed like a case of, "You know that thing you can't do? Do it until you can!" This was an experience that I felt was doomed from the start and was not something I was willing to put my children through.

I had been searching for continuing education courses in reflex integration but I continually found classes that I felt would be too short, too superficial, too emotional or not scientifically based. When I first read many of the success stories from therapists and families who completed MNRI® (www.MasgutovaMethod.com) and the intensity of the course descriptions, all of my hesitations were gone. I thought, if only half of the stories were true, the program was still worth looking in to.

While the *Dynamic and Postural Class* I took in Wisconsin had its fair share of emotional events, it definitely surpassed my expectation of intensity – both in the amount of content it included in those first four days and in the science behind it. I arrived home in a state of pure giddiness – which is not typical for me. I am not sure if it was because the course met my intellectual needs, if I happened to integrate a few of my own reflexes which were long overdue for some attention, or if I was just excited about the possibility of what all this new knowledge meant for my children but, most likely it was a combination of all three.

When I returned to school I fumbled my way through the exercises, choosing to complete a few particular reflexes to work on with all of the kids, which helped me to refine my technique. I was amazed at the different reactions I received from each child! I was shocked to see that kids who had tremendous physical limitations actually had quite a few skills under their belt while other kids who seemed to be pretty coordinated struggled with some fairly simple exercises. Working with a variety of children on the same exercises also helped highlight the importance of the pattern, direction, strength, timing, and symmetry of each reflex.

I also started to pay much more attention to the movement patterns of everyone I came in contact with. I was amazed at the number of 'signs' that had been sitting right in front of me that I had not noticed before. If I had learned just a tiny fraction of the MNRI® program and was starting to see so many areas I could help people with, I could only imagine what Dr. Masgutova, Pamela Curlee, and the Core Specialists were able to see with their highly trained eyes!

At the time of this writing I have been working using MNRI® techniques for just under two years and have taken seven MNRI® classes. I feel much more confident in my understanding of the reflexes and the proper way to complete the exercises, however, I continue to struggle with getting the work done as often as it should be completed. As a therapist working in the school system, the amount of time I have with students is limited as is the time the teachers and para-professionals have that work with the students. When the para-professionals have been able to follow through with the exercises daily, we have all seen amazing results!

One of those amazing examples was a little girl named Sierra who I started working with when she was two. At one year of age she had undergone open heart surgery and died on the table numerous times before finally being revived. Because of this tragic event she had a great deal of tone throughout her body that made voluntary movements difficult. When I first began working with her, she was basically 'trapped' within her body. She had some control over the flexion and extension of her trunk but little to no control of her neck, arms, hands, legs, or feet. I am embarrassed to report that her situation had not changed much in the five years that I had seen her – regardless of the amount of time she spent in occupational, physical, and speech therapy – that is, until I began using MNRI® techniques.

I had always felt as though Sierra movements were heavily defined by her reflexes and, after attending the *Dynamic and Postural Class*, my suspicions were confirmed. Sierra's body existed in such a state of protection that she even stopped breathing for a few seconds when she was picked up. During her OT sessions we worked mainly with her spinal reflexes and those on her feet. Within a month or two of weekly 30-minute sessions, I began to see changes in not only her tone but also in her interest in completing the exercises. She began to make eye contact more often and seemed to laugh when I started to work on her feet.

Once the para-professionals began working with her for 15 minutes daily, her progress quickened immensely! Within one month she began to reach across midline when completing exercises for Foot Tendon Guard. Within another month she was able to keep her head up while lying prone for five minutes. This far exceeded the goal we had established (that she would hold her head up for 30 non-consecutive seconds) – a goal we had worked on for years. Within a month she began to play with true *intentional* gross motor reaching movements. While she still struggled with grasping items, Sierra was able to use her arms in a way that let us all know that not only did she know what was going on around her but also that she had opinions! A whole new child had emerged!

One day as I walked in to her classroom, I heard laughter coming from the table that her wheelchair had been pulled up to. I heard the teacher say, "Sierra, if you don't start behaving we're going to have to write your name on the board!" which earned an even bigger fit of laughter from the kids. When I asked what was going on, the kids informed me that while they were attempting to finish some second grade math Sierra was distracting them by pulling on one of the braids of a little girl at the table. Sure enough, she had one arm reaching far to the side with her fingers slightly extended with the full intent of pulling another student's braid and laughing. I had never heard her laugh like that before! She still could not speak, feed herself, or crawl but she had managed to finally become 'just one of the kids' if only for a few minutes. Over the summer Sierra lost some ground but her skills began to rebound when the para-professionals began completing her exercises again.

Heather Taylor with Austin

Sierra's progress was slow and steady but sometimes the changes I have seen have been almost automatic. Shortly after taking the Visual and Auditory Reflex Integration Class, I chose one student in particular to work with to practice my new skills with. Austin and I specifically were working to improve his ocular motor skills in particular so I thought the Visual and Auditory Reflex protocol would be a great fit for him. We weren't able to finish the entire routine in the time that we had, however, at the end of his session I told him to pick a prize out of my prize box. While he was rooting around in there he shouted, "Hey wait! That doesn't bother me!" I turned to see what he was talking about and found him holding a toy with a lenticular surface and he was scratching the surface repeatedly. It was a texture like a phonograph record and gives the image a 3D effect. These puzzles drove Austin crazy! He could be working on one side of a fairly active classroom and when someone on the other side of the room would rub one of those puzzle pieces against each other, the noise would cause him to yell and cover his ears. The sound didn't register as disturbing to anyone else. I had chosen to use the Visual Auditory protocol to improve Austin's ocular motor skills yet I had completely forgot about his particular auditory sensitivity, which disappeared after just one session!

A few months after this amazing session, Austin had a visit with his behavioral optometrist. In preparation for the visit we completed the Visual and Auditory Protocol on a weekly (sometimes twice weekly) basis, hoping she would observe a drastic improvement in his skills. She noted that his prescription had not changed however his visual tracking skills had greatly improved.

More incredible news was shared with me two weeks after this visit! Austin's teacher reported that his ability to read had greatly improved! She said that at the beginning of the school year he had read 94% of the first grade word list and 78% of the second grade word list. Eight months later, he was now reading 96% of the words from the fourth grade list! She shared an even more remarkable observation that when he was tested in March he could read only 10% of this list before he would become frustrated and quit but by May his reading improved to 96% of the list. This improvement 'coincidentally' occurred during the two months Austin and I were completing the Visual and Auditory Protocol every chance we got to impress his behavioral optometrist!

I have also seen automatic changes in myself as well! When I took the *Dynamic and Postural Class* the first time, I had quite a few difficulties maintaining the proper position for the Grasp Reflex. I took the class again almost exactly a year later and had no troubles with hand placement while completing the exercises. When we returned to our tables to take notes on the next reflex, I noticed that something felt a little odd and when I looked down at my hand I realized that I had a perfect tripod grasp! Ironically, the OT teaching pencil grasp did not have a proper grasp herself until she had some therapy!

Not all of the changes are automatic but can be profound, nonetheless. One parent I work with firmly believes her child would not be potty trained if it was not for the MNRI® therapy. Their family made frequent trips to specialists who told them (when he was in kindergarten) that there was a chance he may never be fully potty trained as they were not sure he had enough sensation to know when he needed to use the rest room. Now in second grade, he has no accidents during the day and is working towards having no accidents at night.

I also work with a home health nurse that insists the child she cares for never learned to crawl during the two years she gave weekly physical therapy but now she has over the past five months. This is astounding considering I only see her one hour twice a month and she no longer has PT services. While we continue to work on her ability to crawl with good neck extension, she does get where she wants to go without assistance. I attribute her significant progress to her nurse who is very diligent about completing the reflex exercises that have been recommended. She reports that this child has come to love the exercises and will often repeat her feet exercises after the nurse is finished!

MNRI® has been important for us at home as well! My son had a very rough first year or two that started with colic and an inability to sleep unless he was sitting up. In fact, Brendan was only content if he was held which meant I slept sitting up as well for the first year and a half of his life. While we struggled with his nutrition, trying practically every formula on the market, even those held behind the counter at the pharmacy – his motor skills seemed advanced. We were always astounded at what wonderful head control he displayed at such an early age. From a prone position Brendan held his head up while looking around for 30 seconds at only 2 weeks of age. He was able to bear weight on both feet at a month and a half (which is a 5 month skill) and even rolled from back to front two months early.

At around one year of age we discovered that Brendan's adenoids were enlarged and were blocking 60% of his airway. We had them removed and things seemed to improve for a short time. Because Brendan was crawling and walking, he didn't need to be held as often during the day but he still preferred to be held upright when sleeping. Some nights were rather calm but on others he was rolling back and forth from one of my shoulders to the other. Some nights he also continually climbed higher and higher up my shoulder. When I spoke to Brendan's pediatricians about this, I was told that sometimes children could feel our nervous energy and act upon it. I bit my tongue at that point although I wanted to tell him that because of our sleeping situation over the past year, I didn't have any energy at all – more or less nervous energy.

Heather Taylor with husband and Brendan at 3 months.

It was about this time that I took my first MNRI® class and had a most enlightening lunch with Peg Johnson who was starting to co-teach the *Dynamic and Postural Class.* She listened to my story and told me to pay special attention when we covered a few particular reflexes during class. When we covered the Trunk Extension Reflex, so many pieces of the puzzle came together. The reason Brendan had such great head control and was able to bear weight on his feet early was all because his Trunk Extension Reflex had kicked into high gear to protect his body from having a blocked airway! Once I made it home, I immediately began working with Brendan on the reflexes Peg had suggested and started seeing his overall tone decrease.

Shortly thereafter Brendan started having tummy troubles. Our pediatricians took stool samples and recommended a few different diets, however, we saw little relief. We switched to a new pediatrician who referred Brendan on to a gastroenterologist at a Children's Hospital for further testing. After numerous rounds of stool and blood samples came back within the normal range, the specialist informed us that Brendan had "infantile diarrhea" that would most likely just "clear up" by 3-4 years of age. I questioned him about what happens developmentally at 3 or 4 that would change his bowel patterns but was told it was a mystery.

Brendan Taylor at 3 years.

A few weeks later I took the *Tactile Integration Class* with Pamela Curlee and asked her if there were reflexes that would help with our situation and she told me to look at the status of five reflexes in particular. I did and all five needed a great deal of work. The more often I did the exercises the more normal his tummy seemed to get. With the help of a probiotic and these exercises, Brendan even began potty training a few months down the line! In the past he had used the restroom when taken but never outright told us he needed to go, so he still had accidents during the day. One weekend we did quite a bit of work on those reflexes and the following Monday was his first day staying dry at preschool and his first night staying dry throughout the night!

Coincidentally, he also started being able to ride his bike a few days later, which I thought was interesting. Brendan never had much luck riding a tricycle and often 'scooted' on it with both feet. When my husband and I bought him a bike with training wheels Brendan could push the pedal forward slightly but would then hit the break by pushing the pedal backwards. After working on the reflexes really intensely on one weekend, Brendan jumped on his bike and pedaled away! We had focused on reflexes specifically addressing hip movements, which I believe allowed him to experience the flexibility he needed to move more freely. Watching this process unravel before my eyes seemed just like magic!

As with all programs, consistency is the key and I have not always been as consistent with Brendan's exercises as I need to be. In trying to juggle motherhood, work, and being a MNRI® Core Specialist-In-Training, the exercises have unfortunately taken a backseat from time to time. When we have worked diligently, I have seen Brendan flourish and it is simply wonderful!

Brendan and I recently had the privilege of visiting Dr. Masgutova for an Assessment which was a truly amazing experience! There are no words to describe the incredible insights she has or the improvements that can be made with her hands! I cannot wait to attend a Family Conference to soak up more of her knowledge and see firsthand the results that can be achieved when her intuitive hands and knowledge direct such intensive treatments!

As the title of this article implies, my destination with MNRI® is unknown. Before taking the first class I had no idea that I would be taking so many more classes – especially so soon! However, each class I have attended has answered so many of the questions I had while simultaneously producing an insatiable thirst to know more! Becoming a Core Specialist is the direction I am travelling in with my professional life. However, I am not fully convinced it is my true 'destination.' Every time a new class is listed on the schedule, my heart can't help but skip a beat!

Geographically MNRI® has also lead me to La Crosse, WI; Dallas, TX; Ann Arbor, MI; Lafayette, LA; Overland Park, KS; Baton Rouge, LA; and Orlando, FL in the past two years… none of which I had planned to visit but now a few of those cities are almost becoming a second home to me. I have learned so much from Dr. Masgutova, Pamela Curlee, MNRI® Instructors and Core Specialists, as well as my fellow students – many of whom have quickly become some of my closest friends!

The work Dr. Masgutova has developed is truly a gift not only to those who receive MNRI® but to those of us learning to provide these techniques as well!

I would like to thank Dr. Masgutova for sharing not only her brilliant knowledge of the inner workings of the body but also her passion and for this work with those of us lucky enough to have studied with her. I would also like to thank the children that I have the overwhelming privilege of working with, as they have always been my greatest teachers! – Heather Taylor

So Here We Go . . .

Tricia Borsch, MNRI® Parent & Founder of Bridge to Healing Foundation, West Palm Beach, FL, USA

As a little girl playing with my baby dolls, I always knew that being a mother was at the top of my list of priorities. When I hit 35, I realized my clock was ticking and I needed to make my wish come true.

I became pregnant and was on top of the world. I never felt better or happier. Because of my age, I had an amniocentesis done at 18 weeks and everything came out perfect. I also found out that I was going to have a baby girl.

At 27 weeks I asked my doctor if I could have another ultrasound so that I could see what she looked like. The room became very quiet and the doctor was called. He looked at it and asked me to get dressed and come into his office. He then told me that it looked like my dear baby girl had hydrocephalus – water on the brain – as her ventricles were enlarged. I was referred to a specialist who performed a Level 2 ultrasound every two weeks until her scheduled C-section. Some weeks were good with no enlargement and some weeks were bad with more enlargement. It was a total roller coaster ride for all of us.

Because of a possible pending surgery for a shunt, we were referred to a pediatric neurosurgeon. He basically told us that we didn't need to worry about the hydrocephalus as he could put a shunt in but we needed to think about all of the other issues she would be facing. He said, "You will basically be taking a vegetable home that will need lots of attention." WOW! As you can imagine, the next 12 weeks were unbelievably hard.

Holly was born via C-section on January 8, 1997 and was immediately whisked away to the NICU Unit as she needed to have a cat scan to determine if she had hydrocephalus. A long day was ahead of me as I waited for the results. Around 4 pm., I finally asked my nurse if the doctor had come in to read the results and she told me that he been in that morning. As she did not have hydrocephalus and didn't need surgery, he didn't feel the need to stop by my room and give me the news. I was very relieved when I read his obituary as no other parent would ever have to suffer from his true lack of bedside manner as I did.

Through an MRI the next day, it was determined that Holly had an underdeveloped corpus callosum which would lead to developmental delays. Because she was in the NICU unit, she was put through every test by every doctor you could imagine – cardiologist, geneticists, neurologists, etc.

I brought my bundle of love home and just loved her and watched her develop the best she could. Holly was late sitting, crawling, walking, and talking. As I really didn't have anyone to compare it to, I wasn't that concerned . . . I was too busy just loving her and loving being a Mom!

At her 2 year check up – she was 2.4 years – I proudly told the doctor that she had about 18 words even though that was an exaggeration on my part. OOPPSS . . . she should have had about 250 at the time . . . I couldn't even exaggerate that! We were referred to Child Find and Holly was finally diagnosed as Developmentally Delayed at age 3. The therapies – speech, OT, and PT each three times a week began. We eventually added sensory integration and social group therapy as well. I was doing the best I could to help my Holly to be the most she could possibly be. My belief was that I would push her as far as she could go without breaking her spirit! This roller coaster continued for the next ten years. Yes, we did see improvements but not the real changes we were looking for.

Holly Borsch today.

By age 13, I was done with testing and therapies that were not making any improvements. I just wanted her to be happy and well adjusted and we would deal with the quirks in the best way we could. In spite of it all, Holly was very happy and loving as long as she was in her 'comfort zone.'

Because of her difficulty in school and problems with reading comprehension, it was suggested that I have her eyes checked for tracking issues. She went through the testing and the doctor said that there may be some tracking issues but she felt there were other issues that needed to be addressed and referred us to Dr. Leah Light. We did the additional testing with Dr. Light's staff.

I will never forget the day we went for our results with Dr. Leah Light – December 22, 1999. She went over all of the results and started talking about lack of integration of primary reflexes and that there was some wonderful therapy available to help with these issues. I politely listened while inside saying,"Yeah, right, and where do we go for this and how do they do it and how much is this going to cost me and what in the heck is it going to do for her?" But I was also somewhat intrigued and went on the internet that night and did some research and was very excited about what I was reading. When I found a checklist of signs of neuro-developmental delay I was shocked – 33 of the 35 things on the list applied to Holly! I thought, "I have got to check this out!"

After all these years of therapy, I didn't want to do another long drawn out therapy program that we may or may not see results . . . I wanted to see immediate results so that I would be inspired to continue on. Dr. Light told me that we could do a Home Program with her or we could go to an intensive Masgutova Family Conference and be worked on by Core Specialists for eight days. The eight day intensive was the choice we made. We were on a plane to California on January 7, 2000 and started our first day of camp on January 8, 2000 – Holly's 13th birthday. What a fabulous birthday present it was and continues to be for us!

At this point, she still had downcast eyes, rolled shoulders, foot pronation, poor balance, poor coordination, poor handwriting skills, learning difficulties, awkward gait, couldn't cope with change, anxiety, poor social skills, hypersensitivity to light and sound and touch, and was slow in processing information.

We started the Family Conference and within three days I was seeing amazing changes. Holly had spent hours in PT trying to learn to jump flat-footed off a foot high block but could never master it. On day three of the Conference, she was standing tall on a bench and said, "Watch me jump!" It scared me as it was a cement floor and she had never successfully jumped from anything that high. WOW, she landed a perfect 10 with both feet flat! She also was starting to come out of her shell and make conversation and eye contact with her therapists and other children. She was smiling and interacting in a way I had never seen before. She could now run like a little gazelle instead of the unsteady awkward gait she had before. By the time we got home she was doing jumping jacks, jump roping, and somersaults. These were all huge gains for her in a very short period of time.

The changes that people saw when we came back from Family Camp were unbelievable. Our very special guardian angel, Donna Wexler, who was Holly's Speech Therapist/Social Instructor/Lifeline and dear friend since age four, gives her observation:

"I was amazed when Holly walked into our clinic after the first MNRI® Family Conference experience with a smile on her face, her head up, and her voice audible. Watching in amazement, my mind wandered back to the previous problems I saw in Holly at four years of age. As a very young child, Holly presented with severe apraxia of speech. She had great difficulty learning to form words and her oral motor skills were weak. She drooled profusely and appeared

unaware of this problem. In addition to her difficulty with communication, she had extreme difficulty in regulating her body, often jumping about the room, moving from place to place with great disorganization. As she improved, she was able to communicate; however, she continued to have difficulty regulating her volume, speaking in complete sentences, and thinking about others. She did not look at her communication partners and found interacting with other children very difficult.

In contrast, as she entered the clinic after her first MNRI® Family Conference, she began greeting everyone she knew. Previously, Holly had walked in with her head down not looking at or talking to anyone. In addition, I noticed a change in her gait, her facial affect, and the clarity of her speech. As I watched in amazement over the next few years, I began commenting on the changes in Holly's self-awareness and social interaction.

Many of the skills I had been teaching Holly were suddenly falling into place. Holly was no longer afraid to speak to others. She started to express a desire to be with her peers and she began demonstrating creative imagination through both written and verbal storytelling. Her creativity had been released!

After watching on the sidelines for a year or so, I decided I needed to take some MNRI® courses. As I began to use this work with my patients, I started to see astounding results. The work not only helped to integrate reflexes but also made it blatantly clear to me that teaching 'speech/language' skills were futile without involving the rest of the body. When I was in graduate school, the teaching was only about higher cognitive functioning. Sadly, I believe this continues to this day.

I now have a partner in reaching out to other professionals to let them know that children and adults can be helped by this approach to healing. I am seeing remarkable progress with my students who are receiving 'whole body' intervention and improvement in my speech therapy students is moving at a remarkably faster rate. I am grateful to Holly and her Mom for introducing me and many of my colleagues to this effective approach."

In the past three years we have spent much time on our own massage table and also have had the opportunity to attend some other Family Camps. At each one, we see reflexes that have been integrated and stayed integrated and huge progress in so many other areas.

I wish I had known about Dr. Magustova's work when Holly was young as it would have saved lots of time and tears. However, it has also showed me that it is never too late for your body to make wonderful changes.

Because of Holly's successes, I wanted each and every parent with a special needs child to know about this wonderful work. I also wanted to help therapists get training in these methods to add to their tool boxes. I took the paperwork that the Institute had drawn up to start a non profit organization and turned it into Bridge to Healing Foundation, Inc.

Bridge to Healing Foundation (BTH) is a (c)(3) non profit organization that was formed to help people do their own fundraising using BTH as their vehicle for tax deductible donations. All checks are made payable to Bridge to Healing thus making the donation tax deductible to the donor. BTH keeps 10% as a way to raise our own funds to provide scholarship opportunities for families and therapists. A check is then issued to you and/or the place providing the therapy.

Amazing story of a growth of an amazing girl… Congratulations to Holly – she is a real Winner coming from a long way and showing the world her inner strength and desire! Thank you for setting such an example for others! Our gratitude also goes to Tricia, her Mom for support of the MNRI® program. – The MNRI® Team

The Gift

Lisa Ortego & Tina Marks, Speech Language Pathologists, MNRI® Core Specialists-in-Training, Louisiana, USA

Lisa Ortego

Tina Marks

Our 90 year old great aunt had come to Louisiana from Texas for a visit with her family. We had been worried about her because we could see her shuffling gait getting progressively worse as her lack of ability to rotate her body crept in her like a thief in the night. It seemed as if her memory was being stolen from her slowly as confusion began to set in and find a permanent home in her body. To make matters worse, she was legally blind due to glaucoma and a cataract transplant that did not go quite as planned. She still had her delightful sense of humor and loved to travel to Louisiana to visit her Godchild (our mother) and her nieces (us).

Since she stayed the week with us, I could not help myself as I asked her if I could work on her. She giggled and said of course so I began the facial protocol on her. I was amazed how I could feel the rigidness in her body. Her cranial bones felt tired and stuck – it felt as if they lost their flexibility and movement. No matter how hard they tried, it felt as if they could not do it on their own. After working with her sphenoid for awhile, I finally felt it begin to melt in my hands as it stretched and opened at last. As I moved on to her eye reflexes, it felt as if they woke up in my hand as well. Before I knew it, it was if the facial bones were reviving from a long sleep and rigidness just slipped away as movement and freedom returned.

It was now time to address the rotation in the rest of her body. My sister, who is also training to become an MNRI® Core Specialist, joined in to help me continue the work. As we walked her to the bed so she could get comfortable, she said the one thing that she really wanted to do was to be able to lift her feet when she walked. Again, as we began the Archetype protocol on her, we could feel the rigidness in her body. It was as if rigidness seized her bones and muscles and captured movement, causing her to forget how to walk and turn. This rigidness felt as if it exhausted her body because it had to work so hard to move the little it could. However, her body was a fast learner, for it did not forget how it once moved. All it needed was a little help! Her body melted in our hands as her rotation came in on her right side. By the time we got to her left side, my sister and I were shocked because her left side literally awoke before we even started the protocol. She hardly needed our assistance as she was able to rotate her head, shoulders, hip, and leg. She giggled and giggled as movement returned to her body and she experienced freedom again. The more she moved, the more she said, "This feels so good!" When we stopped an hour and fifteen minutes later, we all were stunned as we watched the movement that returned in her body, freeing her body of rigidness and fatigue. As she walked, she was able to lift her feet! The dragging of her feet was no longer there!

The most amazing thing happened next…. my sister asked her to look at a magazine. My aunt began to read a recipe out of a *Southern Living* magazine! The very aunt that was legally blind and could not even see a picture clearly was now reading the small print in a magazine! In an hour and forty-five minutes, our aunt was given the gift of movement, allowing her feet to walk instead of drag. She was also blind and could now see! This work is truly a gift from God and Dr. Masgutova is sharing it with the world. Thank you Dr. Masgutova for your sharing your gift!

This is an amazing story of help for our geriatric population. Thank you, Lisa, Tina, and your aunt, for reminding us about such resources living in all of us, including adults and the aging. How good it is to refresh the young energy in our bodies!

– Dr. Svetlana Masgutova

Reflex Integration with a Child Diagnosed with Hypoplastic Left Heart Syndrome

Laura Paquette, OTR, MNRI® Core Specialist, La Crosse, WI, USA

Laura Paquette

Through this story I will describe the progress of a child who has a diagnosis of Hypoplastic Left Heart Syndrome and how he responded with significant changes by the use of MNRI® techniques.

Wesley was born on April 6, 2009 at 34 weeks gestation with Hypoplastic Left Heart Syndrome, paralysis of his left vocal cord, and feeding difficulties requiring placement of a G-tube. He underwent Norwood/Sano open heart surgery soon after birth and had his second surgery of right superior vena cava to right pulmonary artery anastomosis in August of 2009.

I first met Wesley when he was two years old in September, 2011. At that time I was replacing the occupational therapist who had been seeing him since he was 8 months old. I was warned that he would most likely vomit upon seeing me, as he was very sensitive to change and easily upset. He was significantly delayed in his motor skills, had orthotics for support in his shoes, he inconsistently and selectively accepted tastes of foods often gagging upon swallow, and frequently vomited when touching, looking at, or smelling foods. Leaving the safety and familiarity of his home was stressful and a challenge in itself due to his tube feeding schedule, etc. The family had also just lost their in-home nursing care, which they had been receiving 8 hours per day.

On the day of our first meeting, Wesley looked me over and allowed me to sit on the floor next to him. Given his developmental history, I anticipated that most of his primary reflex patterns would not be adequately integrated. After playing for several minutes I asked if I could touch his feet. He agreed and I began to provide the sensory stimulation and motor response for Babinski and Foot Tendon Guard. He looked at me with curiosity and proceeded to move himself into various reflex positions. We had been told during MNRI® trainings that children will often lead you to what they need by assuming the reflex posture or motions so I followed his lead. This is how our relationship and therapy sessions proceeded. He never vomited during our visits and always took an active role in telling me what he needed. It was not unusual for me to begin to work with one reflex pattern and he would tell me to work on another such as, "Not feet, back."

Within approximately six months (two visits per week) Wesley no longer needed his foot orthotics, his motor skills progressed rapidly, his language skills improved from single words to multiple word sentences, his vomiting decreased to the point that it was usually only triggered by coughing and illness. They were able to increase the size of his bolus feedings, and he began to slowly take more foods by mouth and eventually took enough to be considered significant calories allowing the gradual decrease in tube feedings. (His nutritionist commented that she was amazed, as she had doubted that he would ever eat orally). His sensory defensive responses significantly reduced and he was generally less fearful, allowing the family to go out of the home and into public without distress. He became stronger and more physically active. Upon turning three, he did not have delays significant enough to qualify him for Early Childhood services through the public schools.

We continued monthly visits from the time Wesley aged out of the Birth to Three program until Wesley had his third open-heart surgery to finish the reconstruction of his right ventricle at Mayo Clinic in Rochester, MN in September of 2012. To the surgeon's surprise, Wesley's left ventricle had grown and they were able to reconnect the flow of blood to this side of his heart. His parents were told that over the past 15 years of performing this surgery, Wesley was only the second child where they had seen this occur. While in the hospital, Wesley began therapy to aid in his recovery process. Mom called and said he wanted to come home for therapy and asked where I was.

Once he was home I began seeing him on a weekly to every other week basis. He was in strong protection, showed defensive responses to touch, movement, refused to go outside, and did not take foods orally, etc. We slowly began the process of integrating his protective reflex patterns again. Over the next few months he made rapid progress. He was now taking the majority of his calories by mouth (although not enough to remove his g-tube yet), engaging in activities with other children his age, became potty trained, enjoyed impressing people with his physical abilities, began sharing a bedroom with his brother, and sleeping in his own bed. His recent echo cardiogram revealed that his left heart function improved from 52% following surgery to 62%. His medical team refer to him as their miracle child.

It appears that the use of MNRI® techniques enabled Wesley's nervous system to move from a constant state of strong protection and survival and a lot of tension to one of greater balance allowing him to mature and develop higher brain functions. I propose that this influenced his ability to respond more appropriately to sensory information, allowed his body and heart to grow, and supported his motor, emotional, and language development. Our goal of Wesley's feeding tube being removed came true a few months later. He is now moving toward in his next goal to no longer need therapy services and live his life more freely, joyfully, and to realize his full potential.

I want to thank Wesley and his family for allowing me to be a part of their lives and share their incredible journey with them. This would not have been possible without the immense love and devotion from his parents and brother, along with Wesley's courage and persistence. – Laura Paquette

My Improvements with Multiple Sclerosis

Theresa Wojtalewicz, Palatine, IL, USA

Theresa Wojtalewicz

I will give the details here of how the MNRI® method has changed my overall abilities and improved the motor skills that I lost due to multiple sclerosis (MS).

Throughout nearly five sessions working with an MNRI®Core Specialist who uses the Masgutova Method®, I've regained feeling in my left foot and gained balance and alignment that I had lost previously. My experience with this method reveals the benefits of this therapy for individuals suffering with motor loss as a result of MS. My story is unique as I've tried several different medical treatments as well as physical therapy sessions, but I've not retained any of the benefits from those treatments in the way that I've kept the results of the MNRI®method.

I am 31 years old and I've lived with a diagnosis of MS for nearly 10 years. I have received four different treatments for this disease including a subcutaneous injection, an intramuscular injection, a low-dose cladrabine/steroid treatment, and a once a month infusion treatment. I've had symptoms of MS for nearly 15 years. My symptoms have included numbness and loss of motor function from the waist down – primarily on the left side, early on issues with the optic nerve when I'd get overheated, consistent issues with balance causing some falls, bladder issues that have caused incontinence and urgency, and overall fatigue.

I started working with MNRI® Specialist Laura Paquette a few years ago (fairly sporadically) and, after the first treatment, I immediately felt the results. Since that time, I've had nearly five more sessions that have changed quite a few things for me including better control of my bladder, continued feeling in my left foot (which I couldn't feel for more than three years prior), improved balance, and better coordination. I previously could not stand with my feet together and my eyes closed but can now. I've also learned techniques to help in dealing with the anxiety that I've felt as a result of my symptoms which has made a world of difference in my daily life and my overall attitude.

My suggestion for future growth is to ensure that more individuals with MS get access to this type of treatment. The benefits far surpass anything that I've ever done prior. I understand the importance of stopping this disease's progression through FDA-approved MS therapies but I also know that, in conjunction with this method, my symptoms can be managed efficiently for a more functional life with MS.

Thank you Theresa for your courage, bravery, and your exceptional ability to bring humor into your approach of addressing your MS. I am honored to be a part of this journey with you. You are truly an inspiration for so many people. – Laura Paquette

Conor's Story

Tamlyn Hedemann, MNRI® Core Specialist, Albuquerque, NM, USA

Tamlyn Hedemann

I met with Conor, a 16 year old boy, at my office for treatment. Conor was born with learning differences and had severe brain inflammation in 2001. He lost the use of his left side and underwent brain surgery to rule out a tumor. He had fairly good motion in his left leg but limited movement in his left arm.

The first week we worked with the Grasp Reflex. He had a difficult time doing the Sequential Fingers Opening and Closing postures, especially with his left hand. After working with the Grasp Reflex, Conor immediately noticed a difference in his left hand. In the post-activity, he could do all of the finger postures in half the time – and accurately. As we walked out of the office, his mother and I noticed that his speech was clearer – no stuttering at all. This was amazing.

The next week we reviewed the Hand Grasp and he was excellent at all postures. He had noticed increased mobility and strength in his left hand.

At his request to work with lower limbs, we worked on his feet with the Babinski Reflex. As we were working, he told me that his brain was "getting smarter." Just after we completed the exercise, he told of a dialog that he was processing in his mind about the five elements, and said, "Wood, Tam, what's wood about?" I told him that 'wood' or 'tree' in ancient philosophies is about transformation, changes, and choices. He replied, "Oh yeah, that's it, Transformation. We need to transform the left side of my body." I looked at his mother and asked how he knew to ask about 'wood'. She said he didn't know about any interpretation of this to her knowledge.

After the session, he asked if he could draw me a picture of what he had learned for the day. He drew the massage table and his body on the table. Next he drew his feet. (See drawing.)

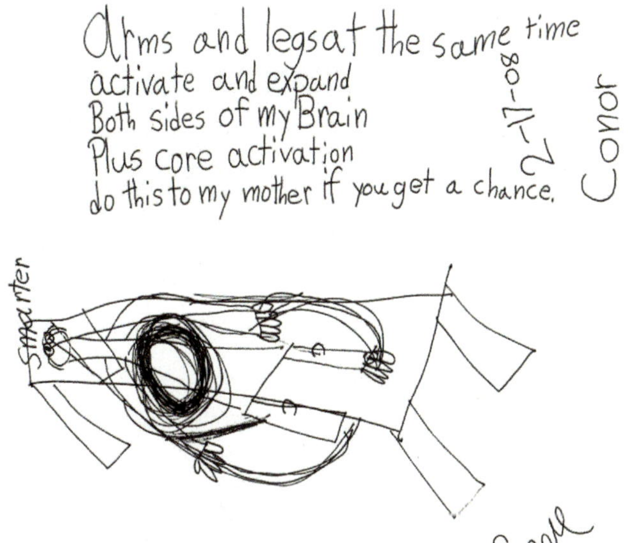

"When you worked on my feet, it made my brain smarter. Then, when you expanded my feet, it activated my hand on the same side, oh, and woke up my knees."

Then he drew two large pea-like objects at the top of his head.

"My brain split into two halves. I could feel the right side of my brain when you expanded my left foot. And I felt the left side when you worked on my right foot."

Then he drew a large circle around the center of his body. He was searching for a word to name it. He finally pointed to the center of his body and asked me what this was called. I told him that I often refer to this part as the core, to which he replied, "Oh, the core. Yes, core activation."

I kept checking with his mother to see if 'expansion' and 'activation' were words that he typically used. She had used the word 'activate' with him occasionally, but never 'expansion' or 'activation' or talk about brain hemispheres. It's amazing how work with the body and its functions awaken the language and expressive possibilities of our own feelings and sensations!

Conor's mother's response to the changes in him is,

"...people can find it far-fetched. Still, yes, it really did happen that way, but it's almost too much to be believed. The week after the first session of reflex integration, Conor was able to alphabetize and shelve in half the time at his volunteer job at the Public Library. During his swim lesson, he swam across the width of the pool for the first time. (He's been in swim lessons for 4½ years.)

The week after the second session, Conor spent the entire swim lesson in the deep end of the pool and swam the width again. He played basketball last night better than he has ever performed. He even used his left hand a few times during the basketball scrimmage. As we drove to basketball he told me he wanted to work on his pronunciation and began reading the words on signs, enunciating the sounds. He has moved to new levels on the video games he plays."

I did the Hands Grasp and foot Babinski exercises with him yesterday in one of our sessions and his sensory-motor training was amazingly successful.

Conor is turning 16 and big successes in neurodevelopment are opening for him. Kids often go through emotional/physical growth around the time of their birthday but Conor's improvements, up to this point, have surpassed anything I've seen before as a professional in my area.

It is with great pride and humble heart that I thank you for the experience we had together, opening up new wisdom in Conor's body. I continue to marvel at the power of this work and the gifts that it brings to recipients and practitioners alike. Thank you for trusting me in this marvelous process, and for reflecting the insights that will have great impact on others. You, Conor, are a brave young man and I cherish the time we have spent together. – Fondly, Tamlyn Hedemann

More Than Bad Body Image

Kim Masters, Minneapolis, MN, USA

I'm presenting this story as an example of how MNRI® can work with a specific population of adults – those with eating disorders – and to suggest that MNRI® should be incorporated into existing treatment programs.

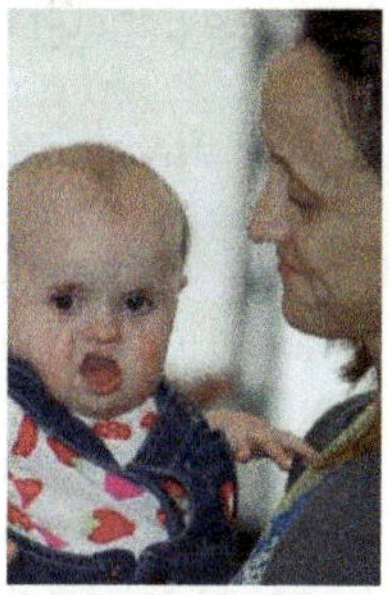

Kim Masters with her niece

I am quite different than the typical person who seeks MNRI® and I expect that my story is also quite different. I began working with an MNRI® professional at the age of 27, halfway through a long stay in a residential treatment program for an eating disorder. The Art Therapist at the treatment program, someone with whom I had been working for many years prior to that time on an outpatient level, had a nagging sense that they were 'missing something' with me. By that time, my journey had included 12 years of eating disorder symptoms, deep depressions, paralyzing anxiety, countless intensive treatment programs including an out-of-state residential program, psychiatric and medical hospitalizations, behavioral and dietary interventions, medications, and a myriad of individual and group therapies.

As I see it now, the presenting symptoms were all the by-product of feeling like a 'freak' in my own skin and not knowing how to relate to the world around me in a consistent way. Because I never had any difficulties learning as a child and because getting the input I craved was fairly easy when I was young, no one noticed anything was wrong. However, I often drove my parents crazy (would put on twenty outfits a day because nothing felt right, insisted on somersaulting everywhere in my house for months on end, would cry at the drop of a hat, absorbed the thoughts and feelings of everyone around me, craved big bear hugs and physical touch, ran circles around the house while my sister read a book, bit my nails voraciously, and bounced my legs constantly, causing the furniture to shake).

I believe now that the eating disorder developed as a protective shield, both from the physical chaos raging in my body and the mental chaos raging in my mind. I was oversensitive to the world around me and, at the same time, was constantly craving physical touch and movement. This was an incredible incongruence, which only magnified as I grew older and the things I desperately needed were no longer socially acceptable as a teenager. I hated myself. I hated that my experience of normal life situations seemed so different, so exaggerated, so much more exhausting than it seemed to be for other people. I quickly came to believe that I was insatiable, oversensitive, dramatic, immature, selfish, bad, wrong, a liar, and never satisfied despite others' best efforts to accommodate me. I collected a wealth of evidence that I was vile. This is who I believed I was at the core. And this was the person I wanted to destroy.

At right is a painting I did while in the residential treatment program, before starting MNRI®. This was an image I would go back to frequently as it portrayed how I felt when I would get 'stuck.' I attempted to describe it to therapists and dieticians but, like much of what I felt on a daily basis, there simply were not words to explain it. In this image, my body has merged with a large, almost chair-like, blob of flesh. I felt a very heavy weight all around me and could not move up from the flesh. I used the word 'stuck' quite often. When I would feel this way, my body image was an incredible challenge for me and it resulted in self-hatred and urges to self-destruct or disappear – both of which fed and were fed by the eating disorder.

Eating disorder symptoms served, at least in the short-term, to keep me in my brainstem where I was safe. Restricting my intake allowed me to disconnect from my body, keep my mind distracted by food and weight and isolate myself from the chaos happening within and around me. My anxiety spun out-of-control. As I lost more and more outlets for my sensory-seeking needs due to restrictions on exercise by therapists and dieticians and as I worked harder and harder to pass as a 'normal' teenager within the sedentary confines of junior high and high school, the more my body image plummeted and the more the eating disorder raged. The more severe the eating disorder got, the more afraid I became of everything. To the insanity of everyone around me, I could not sit still and could not stop working and yet I could not find the energy to get out of bed and could not concentrate on anything.

The eating disorder and the sensory processing issues became a chicken and egg cycle. After so many years of living in a chaotic relationship with my body and with a very broken image of myself so well-established, it became difficult to distinguish myself (spirit, mind, soul) from my body. It makes sense now, looking back, how things played out and how the transition from 'normal' to living in a dysfunctional hell happened so quickly. My sensitivities to textures and fabrics translated quickly into an obsession with body image (nothing feeling right against my skin translated into, "I'm too fat"). Me bouncing my legs and wanting to move became an obsession with exercise and burning calories. On some level, I think I wanted to stay small because I felt safer as a child – if I was a child, running and playing and big hugs would still be ok. But I also realized that as I lost weight, my sensitivity to the world around me quieted a bit. When you're starving, basic functions (breathing, getting through the day) are the only things your body and mind are concerned with. I had no energy. I was depressed. I spun into the starvation mindset that fuels the most irrational of eating disordered thoughts. On several occasions, I seriously contemplated ending it all. Dying was not really what I wanted, but the other option – living – often felt unbearable. I couldn't live inside my own skin if I was 'fat' and I didn't have to live in my own skin if I was very, very thin. I lived out this paradox for years, teetering between life and death. I lived out a mess of anorexic/bulimic behaviors for a very long time. As I would begin eating and gaining weight, my mind began to feel clouded, my thoughts and emotions would 'spin' so fast I couldn't catch them and I was easily overwhelmed by everything. So I would relapse, return to eating disorder symptoms, and suddenly things would become simpler again. I tried so hard, so many times, to get better. Everyone in my life was understandably frustrated with me which only added to my self-hatred. Just like them, I didn't understand why I couldn't just stop . . . or, worse yet, why it seemed I didn't want to stop.

Despite having gained incredible life skills and being in a much better place because of all of the hard work I had done over the years

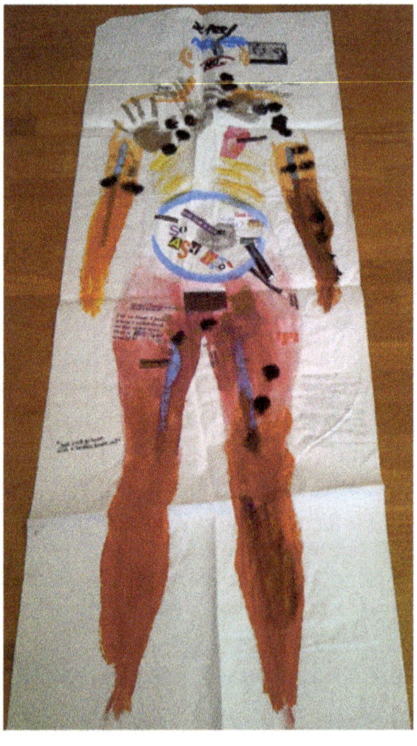

At right is a body tracing I did while in residential treatment, shortly after starting MNRI®. This is a graphic depiction of how it felt to be in my body at times, when it felt so uncomfortable to be inside my skin that I would say over and over, "It hurts; it hurts!" Because it came on so suddenly and so intensely, and because there was nothing I knew of that I or anyone else could do to make it feel better, I feared these episodes. I noticed later, after beginning MNRI®, that I did not include hands or feet in this painting! The footwork and Babkin had a profound impact!

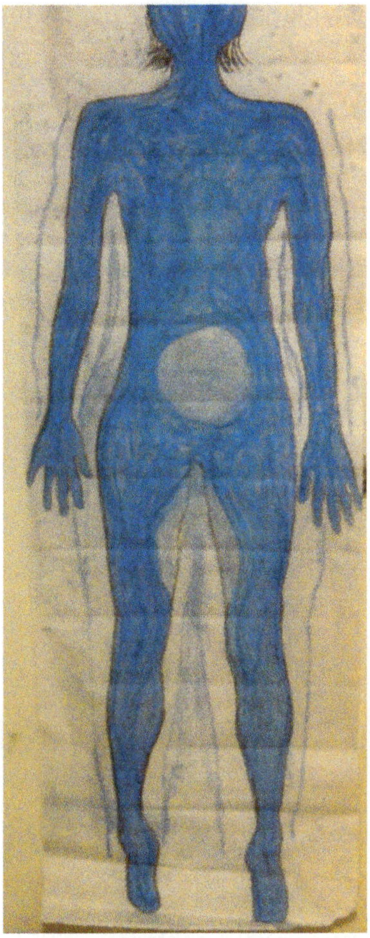

At left is another body tracing I did prior to MNRI®. I had an image of water flowing through my entire body but I was very aware of the space where the water would not go. Long before I knew anything about the physiological components, I knew innately that I was disconnected from my core. MNRI® gave me the ability to hear this wisdom from my body and also to enable me to become a more integrated whole. The outer, faint blue lines are what I saw when I looked at my body, or in the mirror. My understanding of my physical boundaries, how I moved in space and the space I took up was largely skewed.

in my various treatment programs, it seemed that no matter how hard I would try, I just couldn't get past a certain point. It was just too hard and I knew on some level that I would only 'fail' once again. I hated the way life was with the eating disorder but, without the numbing effect of symptoms, life seemed much too loud and overwhelming. Better to fail at recovery than to fail at life, it seemed. Needless to say, I also battled deep depression and debilitating anxiety at times, never feeling like anyone understood me and like the things I desperately wanted to do were always out of my reach. As I had grown in my faith and come to understand the truth of who I was created to be, the incongruence only intensified but my resolve to find freedom intensified as well. I, like the Art Therapist, felt deep down that something was missing, that something unseen was blocking me.

Halfway through my time at the treatment program, the Art Therapist recommended I go for an occupational therapy evaluation. The entire team at the center agreed, supporting the idea that perhaps there was a missing link for me in my recovery. I went to a clinic that primarily worked with children; bright colors were on the walls, toys flooded the waiting room, swings and bubbles were everywhere. Thinking back on that first visit now, I think it was funny that my reaction to the whole thing wasn't one of embarrassment or confusion but one of excitement – I wanted to climb into the swings, jump on the trampolines, and get messy with all the kids! I soon learned that I was a sensory-seeker. And my world began to make more sense. Luckily for me, the occupational therapist I was working with and the owner of the therapy clinic recommended that I see a Speech-Language Pathologist at the clinic who was doing some work with reflex integration. I began twice weekly sessions and continued at that pace for almost two years. Integrating my reflexes through work on my Core Tendon Guard, embracing squeezes, foot work, and the Archetype Movements have been incredible agents of freedom in my recovery. Once I began to feel safe in my body and more confident in my ability to self-regulate in novel environments, the reflexes allowed me the sense of safety, presence of mind, and the courage I needed to begin processing things that had seemingly been stuck as far down as the cellular level. As I was able to listen to my body, I finally heard what it had likely been trying to tell me for so many years. In its own healing and gentle way, my body gave me both what I needed to know and what I needed to do with that newfound knowledge in order to be free. The sudden and incredible pain I would experience before starting this work has dissipated greatly. The pain still comes at times but I am able to hear from my body why it is hurting so much and I now have the words and the courage to share it with others that are safe.

Above is a painting I did shortly before leaving the residential treatment program. I titled it: "When I came…" Sadness, anger, pain; Hurt, emptiness, shame. Seeking to fill my heart basket with polluted, unsatisfying water, Stuffing and emptying my physical basket, futility in the desperation. Empty eyes, tainted by my own shame and on guard for any threat of danger. Layers of sadness, fear, filth unhidden beneath the surface of what I saw. Mind spinning, disconnect from the flesh that seemed to threaten eternal infinite hatred. I was exhausted. All that was good and beautiful smeared with all the confusion.

In contrast to what I now know things can feel like, I lived for years completely outside of reality – outside the reality of my body and outside the reality of the present moment. Reality had become too scary. Over the years, several things happened that exacerbated both the physical and emotional issues playing into my experience of the world and my strategies for coping. When I began MNRI® work, I was a mess. I was anxious, depressed, out-of-control, spinning, and feeling as if I was crawling out of my skin. This experience of life remained even after countless intensive therapeutic interventions to address the eating disorder and related symptoms. At that last residential treatment center, I had done fairly well following the program, taking some big steps toward healing some past hurts, and was in a much better place than when I started. The staff there listened to me and though they did not necessarily understand it, they did their best to respect my voice as I learned to use it again. However, after starting MNRI®, I was able to hear my body, to notice my environment and how it was impacting me and, most importantly, to identify what I needed in that moment in order to stay present and function despite being uncomfortable. Beginning MNRI® at that time allowed me to finish the recommended treatment course and to gradually enter the world with new understanding, new tools, and new hope for real and lasting recovery.

The MNRI® professionals that have worked with me have been absolutely incredible and a huge part of my healing process. Though they had never worked with anyone with an eating disorder – rarely even working with adults at that time – they sought to listen to and learn from me and my body, which allowed me to do the same. There is incredible wisdom in my body; I am fearfully and wonderfully made! I am grateful that my body protected me the best it could with the tools it had at the time, grateful that I have been forgiven for all the abuse I have caused my body, and grateful that my body is now strong and healthy and able to embrace everything that comes my way.

I'm also grateful to have found MNRI® and the clinicians who have walked with me on this journey. In fact, I have taken my first MNRI® course and brought my precious niece who was born with Down syndrome and a serious congenital heart defect to see Dr. Svetlana Masgutova, having experienced the power in the work that the folks at MNRI® are doing first hand. I will be forever indebted to this work and all those that have encouraged me along the way!

Above is a painting I did after a year or so of MNRI®. It was a slow process, but I had lots of layers of things to deal with. This is what I wrote about it:

"Being little should be pink, full of hearts and energy and joy. This is how it should look. I think it looked that way for me, at least in part. But I don't remember.

The black boxes are how I feel – or felt – when things go blank, when I land after spinning or when things are just too much. It's cold and empty and untouchable. It's not 'dangerous' but it's unapproachable. It begs to be left alone. But the healing is here, I know it. The cement that weighed on top is gone. I want it back sometimes; the safety of it. But it won't seal again, I know. I have tried.

My body is contained. Held. And generally 'mine.' And the heart - MY heart - it feels in pieces still sometimes but the pieces are together.

The water is a healing image and it is there as a reminder. There's a path and movement and life. I am still broken. But someday it will be different. I've already decided."

Kim, thank you for sharing your story. It is so detailed and comprehensive – a great example of the growth and discovery of internal tools for survival and action. Your experience is unique and, for sure, will be helpful to SO many people. – The MNRI® Specialists

Marta Awakes!

Sophia and Jerzy Jaszczyk, dedicated MNRI® Parents, Poland

On May 8, 2002 our daughter Marta, a graduate student of the Department of Dentistry, Medical University of Silesia in Zabrze, Poland, was badly hurt in a serious car accident, while leaving home to get her medical diploma. (After years of investigation which, in our opinion, lacked the zeal to determine the cause of the accident, the prosecutor decided to discontinue the case.)

After the accident, our daughter Marta was admitted to the ICU at Medical University Hospital in Katowice - Ligota, where it was found that she had suffered brain trauma, without any other significant damage. After two months (July 8, 2002) Marta was discharged from the ICU with a diagnosis of "conscious, without logical contact with quadriplegic paralysis". Later, on the basis of specialist research (mainly QEEG), doctors gave another diagnosis that she suffered a "cerebral coma" or "sleeping brain".

First, Marta was flown to the Department of Rehabilitation at Bydgoszcz Medical University Hospital, where the first stage of our efforts to get her back to conscious life began. The Rehabilitation Clinic in Bydgoszcz was the first place where we were given hope that Marta would get consciousness back and would be able to start independent life again. Professor Jan Talar of the Bydgoszcz Rehabilitation Clinic is a pioneer and expert in the field of brain and brain stem injuries and rehabilitation of people suffering from such conditions. In our opinion, Prof. Talar's efforts did a lot for Marta's rehabilitation, but most of all, gave us hope that nothing is impossible.

One of the first steps Professor Talar did was ask the question, "Are you ready to disconnect the feeding probe and the catheter, because Marta must take responsibility for her nourishment and must settle her own physiological needs!" He added that "it would involve our purchasing diapers for Marta at our own expense and feeding Marta to give her sufficient amount of calories a day without using a probe!?" Of course, we agreed immediately. Thus, all the tubes were disconnected and the next step in Marta's rehabilitation began.

It was very difficult at the beginning, but over time it became clear that Prof. Talar's recommendation was right. Marta's case is described in his book on physiotherapy of individuals with brain stem damage (Talar, 2002).

After discharging Marta from the Bydgoszcz Rehabilitation Clinic, we tried many different treatment programs. What used to be our living-room, the largest room in our house, was converted to a therapeutic room full of rehabilitation equipment: mats and balls, a standing frame, to name but a few, and specialists in physiotherapy, speech therapy, and others constantly roaming about our place.

Another specialist, who strengthened our belief in our darling daughter's improvement, was Dr. Kim from Korea, a specialist in natural medicine, who came to the Zabajka Rehabilitation Centre in Stawnica near Złotów,

Marta Jaszczyk

Poland. He said, "Everything would be fine. It is just a matter of working hard with Marta and time."

He worked with Marta for two weeks doing professional acupuncture every day, which was followed by two weeks of acupressure. Over the time of his involvement in Marta's therapy, her condition improved significantly, in our opinion. Unfortunately, Dr. Kim left Poland and his method of rehabilitation was discontinued.

After four and a half years in a coma, Marta was examined by Dr. Svetlana Masgutova. She examined Marta's reflexes and stated that they were all of pathological level, "below zero". This was evident as Marta's brain had been silent for a few years. However, the question remained how to open it to start working again.

We expressed interest in Dr. Masgutova's involvement in Marta's therapy. She agreed but said firmly that she would use her MNRI® program (Masgutova Neurosensory Reflex Integration) to open the reflexes that were inactive in Marta, but were ready to be opened. However, she said she could not make any promises as to the success of the therapy! She said that the goal at that time was to improve Marta's health and make every effort to bring her to consciousness, or at least to make her eyes look eloquently and lucidly.

We trusted her opinion and the next chapter of our fight for Marta's health improvement began. Year after year, we worked both at rehabilitation centers and home following the principles of MNRI®.

Marta also went through a lot of MNRI® examinations and research in Dr. Masgutova's Institute. One of the tests was an EEGR test done by Dr. Masgutova and Russian specialists. The EEGR results were interpreted by Professor Alexander Zenkov, a renowned Russian medical specialist in the field, to answer the question if Marta's brain was reacting to MNRI® stimulation.

Professor Zenkov's opinion was as follows, "The assessment of scans, fragments of clinical description, and the given EEG recording show that Marta is in the second phase of coma, the clinical patterns correspond to stupor (9 points on Glasgow scale). The dynamics and topography of electrical activity show a possibility of favorable dynamics of the state (of which nothing is known for sure)." The EEGR test showed a positive effect of MNRI® treatment on Marta's brain.

During MNRI® intervention he stated, "Polymorphous low-amplitude EEG (25-30 mV) activity periodically replaced diffuse theta and delta activity to 50-70 mV. Short bursts of alpha activity of 8-9 Hz to 40 mV amplitude in the frontal parts of the fusiform and flash beta activity of 18-20 Hz, 30-50 mV amplitude in the frontal regions of the spread on the anterior temporal lobes are observed occasionally. Sometimes these spindles include a series of waves up to 130 mV amplitude. The EEG diffusely altered by type of non-specific disorders rhythmogenic reticular structures of the brainstem and thalamus, including mechanisms for the regulation of sleep and wakefulness. The predominance of disturbances in the frontal divisions corresponds to defeat front regulatory and executive functions, mechanisms for focused attention, general control, and socialization. Dy-

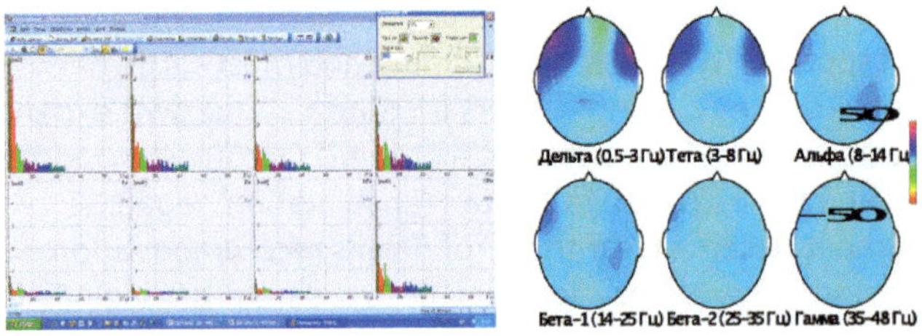

Left and below, showing Marta's brain scans response after application of passive MNRI® Hands Grasp Reflex (left and right hands individually).

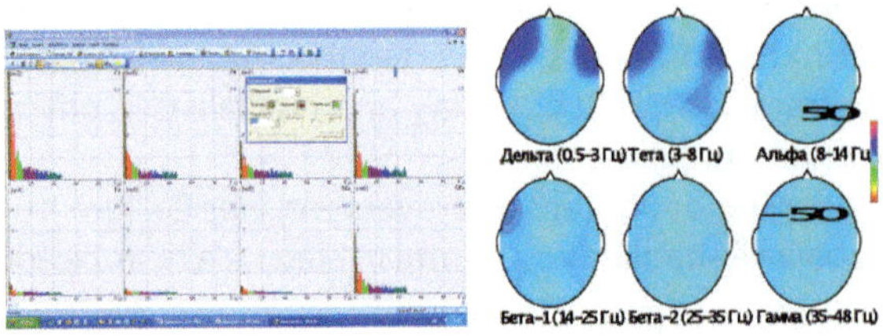

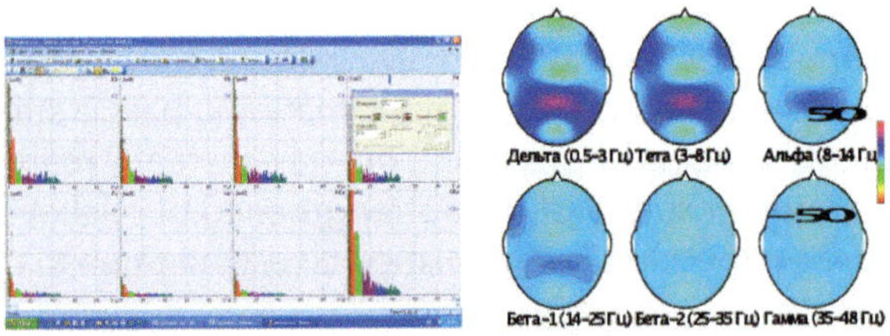

namic pattern of electrical activity and its topographic diversity suggest the possibility of favorable dynamics."

In this way Marta went through many reflex repatterning exercises and for some she responded more strongly, like the Spinal Galant and Perez. During the research of Marta with the EEGR, she demonstrated a possibility for her brain to respond to the MNRI® exercises.

In 2008 Marta underwent QEEG tests in Warsaw where Dr. Rusek evaluated her condition. Her diagnosis was consistent with Professor Zenkov's. Dr. Rusek did the QEEG test before EEG-biofeedback sessions as tests necessary for my plans to further Marta's rehabilitation.

In 2009 I went to a first stage course in EEG-biofeedback administration program in Milanówek where Dr. Rusek was the teacher and facilitator. We bought the equipment necessary for EEG-biofeedback trainings and I did the trainings with Marta under Dr. Rusek's supervision. Dr. Rusek had agreed to instruct me on how to proceed with Marta's rehabilitation. I would mail Dr. Rusek the results obtained in the course of Marta's therapy and she would send me her feedback on how to set the equipment to be able to continue the therapeutic procedures.

Dr. Rusek is the person whom we owe all our successes with Marta's therapy using EEG-biofeedback method. Thanks to her involvement and Professor Zenkov's descriptions we now have loads of data on how Marta's brain functions. Having both Professor Zenkov's opinions and positive results of Dr. Rusek's therapy in our hands we have hope and faith that it does make sense to continue with Marta's rehabilitation. The results of tests show that Dr. Masgutova's MNRI® therapy was having a positive effect on the working of Marta's brain.

Another question that we posed was whether Marta could hear. The first place we sought help was the Institute of Physiology and Pathology of Hearing in Kajetany where, after giving Marta special tests, we were told, "The auditory system is normal." At the same time they could not estimate how the brain perceived sound.

Dr. Masgutova invited Prof. Pilecki from Wroclaw Medical University to participate in one of her MNRI® sessions. He was doing research on influence of the MNRI® therapy of auditory reflexes. After the first EEG study, he said, "The response of her brain to sound is negligible and the readability of waves is questionable." However, he confirmed the positive impact of MNRI® therapy.

In 2012, during another of Marta's MNRI® sessions, Dr. Pilecki made some new tests following the pattern of tests given to Marta earlier. After those new tests, he said that Marta's auditory results were comparable to an average person!

That indicated that Marta could hear like anyone of us. This meant that the three years of therapy, including MNRI® therapy, had positive effects on Marta's hearing, or maybe even let Marta get her hearing back! Unfortunately, still Marta's brain continued sleeping. How long could the parents wait for their child to wake up and have hope for her to wake up? Every parent wants a miracle for their child to recover from an illness fast, but pathology is a difficult issue to cope with. Yet, we always had hope and were determined to wait for the miracle to come.

There was something in Dr. Masgutova's words that gave us faith both in her and in her MNRI® therapy. She was always determined and encouraged us intelligently to make more effort to help Marta in a smart way. She very often said, "Listen to every reaction of Marta's brain while doing reflex repatterning exercises, even if it doesn't seem to respond. Remember, we go with the pace of Marta's brain and the reflexes are the steps that Marta's brain makes. The door to her brain opens with her reflexes and Marta's reflexes are her brain's activation."

In 2012, our dearest daughter woke up, after 4½ years of work on the reflexes! It all happened on one of the many rehabilitation camps when Dr. Masgutova was with Marta, doing a session on Oral-Facial Reflexes integration. She was testing a new technique just after she had created it. This technique was assumed to work on sucking reflexes and it actually did work! Suddenly, our daughter responded to stimulation showing her sucking reflexes, then her eyes started looking around chaotically and finally converged to focus on Dr. Masgutova's face, as if she was looking at her and wanted to see her.

Dr. Masgutova greeted her, "Celebration, Marta, you're back with us! Dear Marta, Hello!". We were both moved, very happy and started crying. It was an incredible experience. It was wonderful that Dr. Masgutova was at that time with us and taught us some new exercises! Because of that, this moment of Marta's awakening was recorded on video tape!

This was the moment from which our Marta started a new phase of her way towards returning to a normal life. It was the most amazing moment when Marta woke up. Dr. Masgutova says that time after time she watches the tape, because it is a "historic thing" in rehabilitation of people with heavy brain damage. The MNRI® specialists kept asking if it was true that Marta had woken up. Sometimes they seemed to be in deep disbelief as if wondering if Marta had actually woken up or perhaps it might not last long. Despite all this, the whole team felt so proud and optimistic that they worked with even greater determination and enthusiasm.

Four months after Marta woke up, Dr. Masgutova checked her understanding and was happy to find that Marta understood commands in both Polish and English. Videos show how Marta moves her arm down voluntarily during a briefing for specialists present at the rehabilitation session as Dr. Masgutova asked Marta to do so. This was incredible. Marta remembers both the Polish – her native language and English that she had mastered before the accident.

Today our Marta shows slow step by step progress. We know it will take a long time before she gets full control of her body again.

Sometimes there are only small signs of movement, but we can see how hard she works and how much she is focused on the ultimate goal of her life. We, her parents, are very proud of her. Marta has the personality of a true hero. She teaches us how to survive and how to love.

We still have a very long way to go to get Marta back and we are well aware of that. However, it has been our conscious decision to do everything in human power to restore our beloved daughter Marta to normal life!

We would like to say a big "Thank you" to all the professionals and others who have helped us and are still helping us on this difficult path of Marta's recovery after the accident. Their dedication to first, saving Marta's life and then, improving its quality, has been the most valuable.

We have found that we live in a world of wonderful and dedicated people. And Marta's case is also a reproach to companies and individuals who promote euthanasia and treat sick people only as potential organ donors. The ICU team in Katowice expected us to give up fighting for Marta's life and to agree to make her an organ donor, too. (This is true! It is unofficial information coming from the sources close to Katowice Medical University). We refused to euthanize Marta because we knew that one day she would wake up and live again!

And she has done that.

More information about Marta can be found on: www.martajaszczyk.home.pl

References

Talar J. (2002). *Strokes of the brainstem: Complex diagnosis and therapy.* Bydgoszcz, PL: Rehabilitation Department and Clinic of the Medical Academy by L. Rydygier.

Congratulations to Marta for coming back to our world! We wish Marta more achievements and health! Thank you to her amazing parents – so caring and loving; many blessings. – MNRI® Team from Poland

My Experiences with MNRI®

Jocelyn M. Pare, M.Ed, Health Care Practitioner, MNRI® Core Specialist, Alameda, CA, USA

Jocelyn M. Pare

I would like to share how MNRI® has impacted my private practice. I have been studying with Dr. Svetlana Masgutova since 2004 and her contributions in helping children with autism, cerebral palsy, and other challenges are respected in our communities for bringing dignity and support to all. I became a Core Specialist in 2011 and continue to study in her program and work at Family Conferences and Clinics. My background is in nursing, special education, teaching at the college level, consulting, health education, and various forms of body work. I travel internationally with high school students and I'm aware of what is available in other countries. So, I can say with confidence, there isn't anything out there that comes close to Dr. Masgutova's reflex integration and therapy methods.

My Mother received an Assessment by Dr. Masgutova when she was 83; it was a great experience for her. She had suffered from PTSD and trauma since age 13 when she had a ruptured appendix, was in a coma for 1½ months, and in the hospital for 6 months. Her doctor said her heart stopped on the operating table and they were able to restart it. She still feared death and was on prescription medication.

She has received various kinds of alternative therapies such as cranial sacral, Brain Gym®, biokinesiology, touch for health, acupuncture, QI Gong, meditation, etc. Although these therapies have helped, MNRI® has helped her more by allowing her more control of her feelings and body sensations, be less anxious, and have less fear of dying. She is now living with more hope, love for life, and less fear of being alone. The MNRI® techniques for PTSD gave her more freedom within her body, mind, and spirit. Now at 92, she still loves to learn from Dr. Masgutova and watch her work on children and adults in class or Conferences. Just being around her ignites a passion for learning, maturing, and connecting.

I also use MNRI® in my private practice. I began to study MNRI® because individuals were coming to me in need of something more than what I had learned in college and so I was seeking new ideas to help them. I often suggest that the child, if possible, receive an Assessment with Dr. Masgutova to give them a sense of direction. I teach my parents the techniques for their Home Programs and assist them or their children in how to incorporate it into their daily lives. Often, children receiving MNRI® sense how good they feel and, in their own way, ask for more therapy by directing their sessions. I have seen improvements in social and emotional behavior, self regulation, sensory systems to improve balance, matured fine and gross motor skills, and academics. Parents often want to learn what they can do to help their children and this offers them more support. They can feel some hope instead of being told your child will never . . . They can now see possibilities for their children, especially when they see their children becoming more responsible and being included in society.

My adult clients require a different approach. An elderly client came to see me for cranial therapy as a referral from another client. She had tried various kinds of natural medicine – holistic, chiropractic, cranial sacral, reiki, acupuncture, etc. She was walking with a cane and I could see that her core was not in alignment with the rest of her body. She looked exhausted by just walking. Following our greeting, I described MNRI® and asked if she would be interested in me using this method for her. She was very open to it. I helped her to notice how her body felt and if she could sense where her body midline was located. She was amazed that she had no sense of this. I began checking her responses for some reflexes: Hands Support, Hands Pulling, Hands Grasp, Trunk Extension, ATNR, and Foot Tendon Guard then began to work with her reflexes, especially Trunk Extension. I completed the session with brushing homologous, both arms, legs, then homolateral movements, same-side arm and leg, then cross lateral, opposite arm and leg, to give her the sense of her core and limbs as we do in the Archetype Movements. I then asked her to sense how she was feeling and her body midline. She was amazed that she could feel her energy lifting, more strength in herself, and her center point. I was surprised at how good she looked – so aligned lying on the table. She seemed to be energetically filled with light all around her body. I assisted her to sit up on the side of the table and drink some water as her body processed the work. We then discussed the session a bit and I reviewed a Home Program for her. She was so delighted as she walked out without her cane. I still have such good memories of this session. This client continues to enjoy receiving MNRI® and has integrated the Archetype Movements into her daily routine and still walks without a cane (except for emergencies).

Another individual came to me with acute back pain which he had had for sometime. He had already sought treatment from doctors, chiropractors, and acupuncture with no substantial improvement. He walked in slow movements, showing pain and discomfort. I described MNRI® as I felt it could help clear some lymph toxins throughout his back and whole body. We diligently worked together, he giving me input throughout our session. As I was working on the front of his body, we both could feel his tensions. He began feeling some hope as we continued onto his back and more and more of the tensions were released. When we finished he easily got up from the table and felt more relief than he had felt for some weeks. I gave him some recommendations on back care such as lifting with knees bent, using the core in front of the power zone, using deep exhaling breaths, and doing Archetype Movements. I asked him to let me know how he was doing in a couple of days. He later told me that he continued to have relief, was able to work, and was grateful for our session. He continues to come to me for MNRI® therapy once a month and has been pain free and more flexible since the first session a year ago.

I have been transitioning my clients toward MNRI® therapy. As a result, they have generally noticed such a change in their freedom of movements, flexibility, balance and coordination, strength, inner peace, and ability to have more comfort in their daily activities. I believe once we tap into our inner reflex state of being, this allows our structure to guide us through life mentally, emotionally, and physically.

We are very fortunate to be able to study with Dr. Svetlana Masgutova and to share her integrative methods with our professions here in the US.

I am very fortunate to work with the children and parents at the MNRI® Family Conferences. My teachers firstly are the children as they direct the inspiration and momentum of the therapies. I have great respect for all of the hard work and energy that it takes to pursue this beautiful work together, and the results are so magnificent. I give a big Thank You from my heart to all of the children and parents for allowing us to be a part of their difficult yet wonderful journeys. – Jocelyn Pare

Multiple Sclerosis and Neurosensorimotor Integration

Laura Paquette, OTR, MNRI® Core Specialist, La Crosse, WI, USA

Laura Paquette

It is my intention in this story, to provide a summary of the results of two, 2½ hour sessions using MNRI® techniques with an adult who has had a diagnosis of multiple sclerosis (MS) for 19 years.

Hans is a musician who is in his 50's and was diagnosed in April of 1994 with MS after experiencing paralysis of his left arm and hand during a performance. He was diagnosed with MS after ruling out Lyme disease, diabetes, and a brain tumor. He was treated with IV steroids and regained movement but continued to experience decreased dexterity of his left arm and hand. At that time he was advised not to run as it could advance the disease. After considering that advice, decided to start running regularly for fear he would someday not be able to. He continued running up until five years ago at which time he experienced a relapse and his walking ability deteriorated significantly. At this time he began taking Copazone and was told he probably would not regain his motor function. He has had acupuncture treatments, massage therapy, and continues to take Copazone.

A mutual friend of ours suggested seeing if MNRI® techniques could help him as she knew I had a niece with MS that had responded well. He decided to give it a try. During our initial visit and assessment on Oct. 21, 2011, he presented with a shuffling, unstable gait that required a great deal of energy, effort, concentration, and caused a lack of confidence regarding his balance and safety. Upon assessment he presented with asymmetry with his left side being more rigid and less mobile and his right side over-compensating. He had decreased sensation in his feet (especially his left). He lacked the ability to move his hips freely. When asked to demonstrate various movements and to give feedback on his sensation and movement awareness, it was clear he had difficulties with his overall body scheme. We initiated our treatment with the intention of establishing greater symmetry by facilitating the integration of his Core Tendon Guard Reflex pattern. During this process we both were acutely aware of his Tendon Guard release. We then facilitated the integration of his Babinski and Foot Tendon Guard reflex patterns resulting in a greater awareness of his movements, mobility, and feeling through his feet and legs. The assessment of his Spinal Galant Reflex pattern indicated he was hyper-sensitive and hypo-active with little active hip motion. By working with this pattern and Leg Cross Flexion-Extension, we were able to achieve active hip motions, the ability to shift his weight from one leg to the other, and establish a better sense of his feet contacting the ground. We ended our session agreeing to be in touch. Several days later he called and described feeling a bit overwhelmed and pleasantly surprised and optimistic with the changes he felt in his walking. He stated that his friends and family had noticed the changes as well. We met again the following week.

During our second session we used several of the Archetype Movement Reflex integration techniques to establish his foundational movement patterns to support the integration of his Dynamic and Postural Reflex patterns. It was during this process that his body awareness improved significantly and he established improved midline orientation and control. We further integrated the automatic responses of many of the reflex patterns addressed on our first visit and then put them into intentional movements by marching, dancing using hip movements, rotation, cross crawl pattern, and lateral flexion and extension of his spine. When he walked to the kitchen to get a glass of water, he realized he had walked there without thinking about it and said with great surprise, "Did you see what I just did? That's amazing."

Several days later Hans went to Arizona for the winter and emailed me to say that he had just gone for an hour and a half walk in the desert! Prior to this walking was not something he looked forward to and did not engage in walking activities that required distance or length of time more than a few minutes. He bought a used bike and made a goal of walking or riding his bike verses taking his car whenever possible. Several months later upon his return to his home to the midwest, he confided with excitement in his eyes and voice that he had started running to and from the mailbox (which is approximately 300 feet) and running to and from the garage (which is at least 75 feet each way).

During an outdoor musical performance (in extreme heat which can be quite challenging for people with MS), I watched him standing for the entire performance, dipping his guitar, shifting his weight, moving around the stage over wires and equipment with relative ease, smiling his infectious smile, and radiating confidence. I sat back with a feeling of immense gratitude for the efficiency and complexity of our brains and for Dr. Masgutova and MNRI®, which provides us with techniques that can influence the organization of our primary reflex patterns through our sensory motor responses in a safe, natural way that enables us to recover functions thought to be lost.

We have had a couple of additional sessions but these have been limited by distance and weather. Hans continued to gain motor skills and maintain the skills he acquired on previous visits. We look forward to future therapy sessions, anticipating continued advancements in his skill, and hope to provide support and information that may help others experiencing the effects of multiple sclerosis.

I would like to congratulate Hans and his openness to try MNRI® techniques, and I am grateful for the outcome that brought him movement, joy, and happiness that he so freely gives to others.
– Laura Paquette

Our Journey to the Right and Appropriate Treatment for Our Son!

Maria & Peter Wallin, MNRI® Parents, Falun, Sweden

This is a story about our son Jake who was born in 2007. We live in the small city of Falun situated in the county of Dalarna in Sweden.

Already at the age of one year, we knew that something was not right with our son's speech. He was quiet and did not use his infant talk. When other children said their first words such as "Mom", "Dad", he said nothing. We turned to the Care Centre Britsarvet in Falun to get a referral to the ear clinic to do a hearing test. The ear specialist thought that we should implant tubes in Jake's ears to facilitate hearing and thus see if his speech would start. After several ear infections, Jake had tubes implanted at the age of 1½ years. After the implant, he began to talk more, but he was still talking less than his friends at the same age. We contacted the Child Welfare Centre and asked for an examination of his speech and language development. It was found that he was very late with his speech and a referral to a speech and language pathologist was sent immediately.

Jake was a happy and social 2½ year old who was late with his speech. He was diagnosed with a moderate speech and language impairment. He understood exactly what was being said but had difficulties conveying and communicating with others so that they could understand.

In April of that year we got an appointment with a SLP at the clinic of speech and language in Falun, who gave Jake the diagnosis of moderate phonological language impairment. Jake's primary difficulty was with pronunciation and he had a specific disorder of articulation ability. We had a schedule and programs using different sounds and words that we practiced with him a little every day.

It was difficult and hard on us that Jake got this diagnosis. We have always known that there was something not quite right with his speech. Jake is the youngest of three siblings. When we compared him with his siblings, he deviated in speech.

He started kindergarten in 2010 to stimulate his speech and language development. The induction itself took a lot of time. He sometimes felt insecure which often manifests itself in that he did not want to eat in kindergarten. The higher pace in kindergarten made it more difficult for Jake to make himself understood. We were in constant communication (meetings) and dialogue with the kindergarten staff and the head teacher to mentor and provide information about Jake's difficulties. My husband and I devoted a lot of time to that. Unfortunately, there is little knowledge about children with speech and language impairment in kindergarten and they needed guidance in this. We believe that Jake developed but this is mainly due to the work we put into practice at home.

We read on the internet about the speech clinic at Danderyd Hospital In Stockholm that they had been successful with exercises and new therapies. We called and consulted both the clinic manager and SLP Helena

Björelius at the Oral Motor specialist Center in Danderyds Hospital and described Jake and his difficulties. SLP HB had started to use the Masgutova Neurosensorimotor Reflex Integration (MNRI®) at the speech clinic since fall 2010. We thought that the treatment method sounded very interesting.

We visited SLP Helena Björelius at the speech clinic for the first time in November 2011 and she assessed Jake to make recommendations. Jake did not produce any fricative sounds, or made them with a 'snorting' sound and he had slurred speech. She said he was hypotonic in the muscles used for speech and his Oral-Facial Reflexes were not integrated. He was also generally hypotonic especially in his joints. Jake was diagnosed with developmental dysarthria by SLP Björelius. We learned both about Jake's disabilities and also got answers to many other questions. In summer of 2012 we had our first session with the MNRI® program. I had an instructive tutorial on how we could practice at home with Jake. We used reflex integration with the exercises from Oral-Facial Reflex integration, Bonding, Babkin, Babinski, Legs Cross Flexion-Extension, and Spinal Perez. We also used Visual Reflex integration and some Neurostructural Reflex exercises for the diaphragm. It was really a relief for us, being his parents, as we felt that we had finally found the correct and appropriate treatment for Jake. All the body parts must work and be synchronized in order to function in a complete way. The speech is part of this interaction. SLP Björelius at the Oral Motor Center came to us as an angel, and confirmed what we always felt; it was not only the speech but the relationship between speech and his generally hypotonic muscles which was the key to success. This knowledge and experience can not be found in Falun. In spite of this important fact, the county council of Dalarna found that we had to pay for this treatment for ourselves. And we did. We also shared our new experiences and knowledge with the speech clinic in Falun, so they would know what they had missed.

We feel that Dr. Svetlana Masgutova's treatment has made our son develop in a positive direction very fast. By spring 2013 he is generally less hypotonic and has now developed all speech sounds and does not slur his speech any more. He also has developed more self esteem.

We are eternally grateful that we have been treated by the MNRI® program at the Oral Motor Specialist Center in Stockholm and we will continue to recommend all parents who have a child with speech difficulties to turn there. Many thanks to Helena Björelius for her expertise, skill, and knowledge! You are the best!

We wish Jake additional achievements in learning, growth, and personality development! We congratulate and thank his parents for all the small and big achievements and wish him great health and continued success on his journey for more positive changes. – The MNRI® Team

The Miracle of Noah with the Oral-Facial Program

Priscilla Roberts, Speech Pathologist, MNRI® Core Specialist-in-Training, Chattanooga, TN, USA

Priscilla Roberts

Makala brought Noah to me 6 months ago (October, 2012). Noah was 3 years old at the time. Another parent had told her about me and she came with a lot of skepticism. I started doing the Masgutova protocols on him. I started with the Neurostructural protocol with some reflex work. I showed his mother how to do Squeeze Embrace, Babinski, and Foot Tendon Guard. Makala brought Noah to me once a week, missing a session here and there through the holidays.

When he first came to me, Noah said no words that were distinguishable. He would say "u" at times but this was the extent of his speech. He may have intended a word but it was difficult to tell. He came with a diagnosis of apraxia. It was difficult to determine the extent of his understanding as he had no expressive language and he didn't always follow what was said to him or follow directions with any consistency.

He also demonstrated poor ability to imitate movements. As I worked with him, I would ask him to perform a movement or task and he might or might not do it. By the third session, Noah was showing improved awareness and attention to what I was asking him to do. I have a clock that chimes every 15 minutes and on the third session, he imitated the chiming of the clock. That was a huge step as his mirror neurons for the auditory system began to turn on. He began to imitate animal sounds and other sounds that he heard around him. It was small attempts but the mirror neurons were beginning to turn on both for auditory and vision. I added more work with the reflexes, the six pointed star of Archetype movements, and Tactile integration. At this point he began to imitate movements as well. I encouraged his mother to work with Noah on these imitative skills for both visual/motor activities and vocalizations.

His progress was incredible. He began to speak in single word productions that were difficult to understand as he was not producing consonants. His vowels were accurate. Then slowly, some consonants began to appear in his speech and Noah's words began to take form. Two word utterances began to appear.

All of this time, I had continued seeing Noah once a week and doing a variation of the work listed above. In March (5 months later) I worked with Noah using the Oral-Facial II protocol to stimulate his Oral-Facial organization and suck, swallow, breathe synchrony. A couple of weeks later (in April), his family took a video around the time of Noah's birthday. It showed Noah singing, "Happy Birthday" to himself. He sang the whole song by himself without any help from anyone! This was a far cry from the little boy that I met the previous October who had no speech at all and a diagnosis that he would be unlikely to ever speak. The little boy that came to me in October has developed so much, and so quickly. His life and his family's life has changed significantly with the Masgutova work. The power of this work is enormous. He only received one session per week for 6 months to stimulate this type of change!

Noah's mother put the following post on Facebook about her son. I think that it tells Noah's story well:

"Apr 10, 2013 10:15 am.

FOR ALL THE PEOPLE WHO HAVE BEEN PRAYING FOR MY YOUNGEST SON, NOAH.....I watched my baby miss all of his milestones and at the age of three still was not speaking (At all). I, like most mothers, knew something was not right. After NUMEROUS doctor visits, tests, and puzzled faces I finally had a diagnosis! Because isn't that what we all want when we know something is wrong? Then at least we can start fixing it,,,,,RIGHT??? Noah was diagnosed with Periventricular Leukomalacia. That meant he had white matter brain loss. He was born that way AND there was nothing we could do about it. Shortly after we received another diagnosis. Speech Apraxia, YAY! Something I can fix,,,RIGHT? Speech Apraxia is the brain's lack of ability to get the mouth to do and say what it's suppose to. We were told less than a year ago that our little boy may never be able to speak. PRAISE the LORD, a woman named Priscilla Roberts was placed in my lap. I was quite skeptical about going to ANOTHER therapist, especially one who was a little "different", because I had put Noah and myself (selfishly) through so much already. But God made it clear to me, this is where Noah needed to be. Priscilla let me know that not everything has a diagnosis bc we still know so little about the body and brain. "He is neurologically disorganized" she said....That was the first time anyone had made sense to me in my gut. "We just have to start him from square one and help him map out his brain, he'll be fine," she said with a smile (as I wept).

Keep in mind when you watch this video that, less than a year ago, despite lots (and lots) of speech therapy, Noah was still not speaking. Thank you, thank you for all your prayers. Thank you GOD for your guidance and for holding my sweet boy in your arms! I have to go, this kid won't shut up!!! ;)

Makala"

I want to thank Makala and Noah and their family for their belief in me and the Masguatova work. At first, it was just something else to try, then it quickly caught their attention and Makala and Noah began to work in earnest. It has meant a lot to me to be a part of their journey. Watching Noah change and grow and develop has been such a special opportunity for me. I thank you with all my heart. – Priscilla Roberts

Seeing the World Through a Different Lens

Susan Terris, BCR, MNRI® Core Specialist, Edmonton, Alberta, Canada

Susan Terris and grandson

It has been a lengthy search for answers that brought me to find MNRI®. Dr. Masgutova was a great contributor to the field of Educational Kinesiology, to which I was a practitioner and instructor. It was in one of her written contributions to an Edu-K journal, and then learning that Dr. Masgutova was coming to a community near me, that motivated me to pursue her knowledge and understanding of human development, growth, and learning. This personal search for answers led me into the fields that I pursued, and I truly feel I now have the answers that nobody else was able to provide. For children with developmental challenges, more homework, stimulant drugs, and behavioral therapies were not the answers to my burning questions, yet this is what traditional protocols offered. I am so grateful to now have the MNRI® knowledge – for personal reasons – and to pass on to others who are in search of something more. Below are three examples of results I observed when using MNRI® with children:

A mother and her 5 year old son who was diagnosed with autism and struggled with a variety of behaviors came to my office. My office is in my home and I have two large fluffy cats. Upon entry into the house, the boy appeared to not know how to react to the cats. He flapped his hands, screamed, and yet, I got a sense he wanted to reach out to touch and engage with the cats but this ability seemed to be blocked from him. I quickly escorted him and his Mom into my office where he spent the first part of the session on her lap in great distress. I talked quietly to him and worked on his feet – which didn't particularly thrill him, but he tolerated it, all the while, fussing and crying about the intrusion. I worked on his hand, and soon he pushed his other hand towards me to continue, implying that he was beginning to be okay with the session. I next moved to manipulations on his back where I used some Neurostructural protocol in the lower lumbar area. At this point, I noticed that, within a few moments, his body was relaxing, his head bobbing, and his eyes started to close. I finished that area and moved to the next area up the back, but this woke him up, so I returned to the former posture. Again, his eyes closed and soon he was so deeply asleep that we moved him from his Mom's lap onto the massage table to continue. I went through a variety of protocols, using Tactile Therapy and some reflex repatterning, as much as I knew at the time. I fit the program into a 1½ hour time frame as this is what the parents allotted for. At the end of the session, the parents were provided with a small Home Program, although I am not sure what they were able to accomplish at home – they claimed it wasn't much – but, within a month of seeing him once a week, we noticed that this young man was becoming increasingly more comfortable coming into my house, seeing the cats and interacting with them, and getting the therapy. He soon began articulating more clearly, and the parents said that play school was more comfortable, he was engaging more with others, he was happy, and seemed to fit in so much better with his peers.

Another little boy came to see me with a diagnosis of cerebral palsy. He was 4 years old. He had contraction in his arm and hand and on the right side of the body. The arm and hand had not grown, therefore, there was a lack of symmetry and use of the smaller hand. We worked with the Spinal Reflexes, Sequential Arms Opening, Hand Mapping, Babkin Palmomental, and other reflexes that were deemed important for him. His Mom provided a great Home Program, as well, and some nice things began happening. The right arm showed signs of lengthening, and his hand began opening and growing! This little guy began crawling using both hands which had not happened before; he began reaching to retrieve objects with his right hand, and the right side of his body was releasing its hold. It was very exciting to see the changes happen over time!

My third story is of a boy, Simon, who I have worked with since September 2013. Simon, at this time, was 3 years old, and diagnosed with autism. When Simon was tiny, he was challenged with many food intolerances and his world was very limited. When I first met him, he wouldn't engage, didn't like touch, and struggled to just lay on the table to receive any therapy. We continued diligently to do what we could for him, and, eventually, his nervous system settled down enough to allow more therapy.

His Mom happily shared with me in December 2013 a 'short list' of the changes they had observed:

1. He has gone from screaming before getting to school, through school at the top of his lungs (for a relatively calm little boy) for about 3 hours in total, to calmly going to preschool, and participating. This was a sudden change.

2. He went from a palm grasp to an index finger extended grasp while drawing/holding pens. He can now roll over. No more toe walking.

3. His speech is exploding. He is learning and retaining vocabulary, the rate at which this is happening is dramatically better. His is also using words appropriately. He has gone from mostly one word at a time, to regularly using two word phrases and even coming up with three word phrases from time to time. Another big change is his willingness to try and say a new word. I can say, "Say cheese" and he will say something like, "Eeeze" – this is very exciting for our family.

4. Comprehension – his receptive communication has grown. For example, Mom can call his name and he turns to look, whereas before, was his responding to a command but never his name. He now understands that calling, "Mommy" is a way to get Mom to come to him; this is new. He is also able to understand unfamiliar requests and two step commands now. Before he would respond to familiar requests that were given only in the exact same way. For example, he could understand, "Boots and coat", but not, "Let's get dressed to go out" – now he reacts appropriately to both.

5. More affectionate: he is receptive to snuggles and wants to be held when he is hurt. He calls for Mom and wants to be comforted. Before he would just scream out of pain or frustration and would get angrier if Mom tried to snuggle him. He now wants Mom around. He is now enjoying her company. He wants Mom in the room with him while he is playing. He's checking over at her to see if she is watching and smiling at him.

6. Development in play skills – finally playing with toys, and enjoying it. He's playing pretend and feeding his stuffies pretend things. He understands representative objects a little better. He is doing things he never did before but that his parents had been working on for months.

7. Drawing knees up when jumping – has only come on when we have done MNRI®. When stopped, this skill goes away despite constant access to various trampolines.

8. He understands concepts like first, then (next), and bribery – when you do this, you can have.... he never got this concept before.

9. The speech is huge!

Now it is December of 2014. I have given Simon MNRI® once every two weeks in four day, two hour infusion interventions every couple of months. As Simon's nervous system is continuing to settle and clarity of input is traversing appropriately between his peripheral and central nervous systems. He can focus on information which he pays attention to and he is flourishing. Many of Simon's behaviors are becoming typical of a 3 year old. He is now independently using the toilet, asking for what he wants, and his interactions are becoming increasingly more advanced. His parents are very happy with his progress.

Simon and other childrens' progress, came about from the diligence of their Moms all learning from many programs, yet, risking learning a new program, one that utilizes a different system: the reflex system. The presentations that brought them to this work have made much sense, and their commitment to a Home Program has greatly enhanced their outcomes.

I have spent many years in schools working with children with a variety of challenging behaviors, postures, learning and ability issues, sensory processing concerns, and much more. I worked tirelessly to support the use of the pincer grip to grasp small objects, helping children to find their centers by supporting them while sitting on bolster seats, having children lay on balance boards to enhance the strength of their core, and utilizing behavior modifications to change behaviors that were not serving the child's cause. The Individual Educational Programs at the beginning of the school year present with a handful of parent/therapist/teacher agreed upon objectives to work on within a certain amount of time. The school's focus was working in the parameters of their objectives, but imagine what could happen if all these objectives and more could be accomplished within a shorter period of time and with greater depth, as the brain and body get clarity – not just with the fingers working together to pick up small objects, but for a much deeper fulfillment of desired mental attributes. This is what I have witnessed in working with MNRI®.

It is with much gratitude that I write this. The more I learn, the more I realize there is so much more to learn. As everything learned and understood is linked to the body, mind, and spirit, the reflex work is proving to be a mighty resource to understanding what has been missing in our ultimate endeavours. Thank you, Team Masgutova! – Susan Terris

Hanna: Longer Sentences and Increased Vocabulary

Asa and Janne Pettersson, MNRI® Parents, Sweden

This is a story of our daughter Hanna who turned 12 years in August, 2013. In August 2010 we received a referral to the Oral Motor Center at the Speech Clinic, Danderyd Hospital in Stockholm, Sweden. It is located about 150 km from our home, and we came into contact with speech and language pathologist, Helena Björelius. Hanna was given a motor speech disorder diagnosis; childhood apraxia of speech and developmental dysarthria.

Asa & Janne Pettersson

During her life she has undergone a fair number of tests assessed by physiotherapists, psychologists, speech and language pathologists, and teachers. She has regularly practiced oral motor and various speech and language exercises as well as memory and physical activities. Skills like walking and cycling (demanding balance) developed slightly later than average. She has been considered to be late in her gross and fine motor development. At 4-5 years of age she was diagnosed with a developmental delay (biological age minus 1½-2 years) and speech and language disorder. Hanna, however, has always been very communicative and found it easy to connect with people. She has compensated by observing and reading situations. She has a good order of things and generally has a good memory. Since Hanna started school, it was clear that Hanna's speech, language, and also several other areas had a general developmental delay. She then was diagnosed with mild intellectual disability and attends a small class. She still has difficulty with reading, writing, and mathematics.

At our first meeting, SLP Helena Björelius observed, in addition to Hanna's speech dysfunction, that she suffered from dysfunctions in primary motor schemes/reflex patterns associated with oral motor functions, speech, body movements during walking, running etc. Her specialist suggested working with MNRI® treatment.

Hanna

Before we began to regularly practice MNRI®, Hanna seemed to be hypertonic in her body muscles, she could not sit still for long periods. Orally, she was hypo in tone, unstable in the jaw and used gross mouth and lip movements during speech. She had in periods difficulties with swallowing and also drooled a lot, especially when she had a cold coming up.

Hanna has improved considerably after nearly three years of regular training in various areas, in accordance with recommendations from Helena Björelius and the

Left, Hanna at a MNRI® Conference in Poland; right, Hanna sailing with her brother and little sister.

Above, Hanna outside her school

custom designed training programs offered at the meetings Hanna had with Svetlana Magutova, Ph.D (May 2011, November 2012, and March 2013). Hanna has been regularly visiting the Oral Motor Centre since the beginning of October 2010 ranging from two times a week to once a month and more – based on the guidance given to us by our specialist, Helena. We have been practicing with various basic programs at home. Hanna, also participated in two Camps in Mielno, Poland (November 2012 and March 2012) as Hanna worked very intensively with different types of exercises and got a new 'base program' to work on at home. We have experienced these events as very positive for Hanna, for us, and for Hanna's development.

What we clearly see is that Hanna, since she began working with MNRI®, now has a much more mature body language, she has more coordinated movements, and her whole way of moving, walking, running, jumping, swimming, cycling etc. are made with a completely different body control than before. She can sit for longer periods without bagging up or get 'ants in the pants.' Also, her fine motor skills have improved, as evidenced for example, when Hanna writes, plays computer games, or uses their mobile. When she speaks, her face muscles move more adequately, which means she has a more relaxed facial expression and a softer, more restrained appearance. Hanna is now treated more age appropriately, which was a problem earlier. Drooling is significantly reduced and the sensitivity around the mouth has increased slightly. She still has trouble articulating but can often produce sound combinations that were previously impossible or difficult to do. Regarding word production and language structure, Hanna has still a big challenge ahead. She has clearly made great development with expressive language since October 2010. The sentences are getting longer and vocabulary has increased. Receptive language and comprehension has improved. She has great social skills and several playmates of varied ages. The school has also remarked about Hanna's performance since she started her training program. She can hold attention longer and focus better.

Hanna has developed in many ways in a relatively short time, although with a lot of training, support, and encouragement. We believe that a continued effort on this track can give Hanna a great foundation to challenge her disabilities!

Dear Hanna, Congratulations on your many successes with MNRI®! And how many more you will have because you are goal oriented and so motivated! Thank you to your parents for being your best friends! Smiles and joy from us! – The MNRI® Team from Poland

Initiating Speech in a Delayed Toddler

Lou Bosch-Wilson, MNRI® Core Specialist, Mill Valley, CA, USA

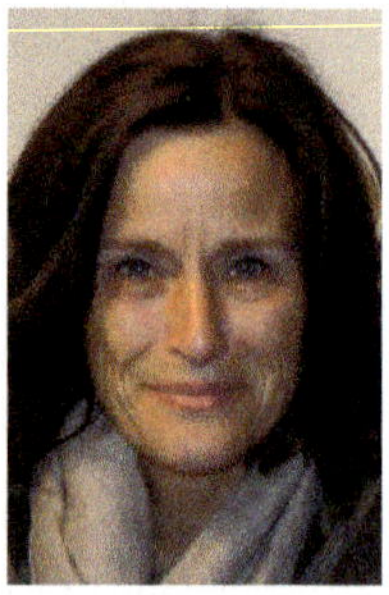

Lou Bosch-Wilson

Jace's mother's reports, "We started seeing Lou when my son, Jace, was about 23 months old. I was concerned about his language development, because he was not making very many sounds, certainly not verbalizing or babbling, all possible signs of speech aphasia. He was born via C-section with IUGR (Intrauterine Growth Restriction) and low birth-weight, and he had been slower developing than other kids, even with some movement therapy, so we were particularly sensitive to this. He was also very shy around other people, taking a long time to trust and accept them. After hearing an interview with Lou and learning about her work with reflexes and movement, I sought her out."

My initial assessment of Jace was that he was fully mobile and mellow in nature, content to play with toys, though he made sure to keep close to his parents. His mother communicated with him in Greek, which he seemed to understand fully. He still nursed and any time he was uncomfortable with me getting closer, he would immediately indicate that he wanted to nurse.

Jace made very few sounds and would make a simple grunt to indicate he wanted something. I suggested that we start to focus his attention more on everything to do with making sounds – using tongue, lips, teeth, cheeks, hands, etc. as fun ways to make different types of sounds. The key was to focus on fun and imitation often through song and games, without expectation or judgment. We did that and sneaked in some reflex integration work as I could.

We began by working on the floor, as he was much happier and cooperative when playing with toys/trucks, and I was unable to get him to sit still for any period of time prior to this. During the first two sessions, I was able to do most of my work while he nursed, so I focused on his back, neck, and head.

Jace receiving MNRI® from Lou.

Within a couple of weeks, he was able to tolerate most touch – as long as he was sitting, so I began to add facial therapies. Soon he was able to relax enough to lay on his back on my table.

I wanted to see that the muscular apparatus both outside and inside of his mouth was intact and operational. For example, I wanted to see if his tongue and jaw were able to move in different directions on cue.

I also checked his Tendon Guard throughout his body, to determine his body's response of stress. His neck and his Core Tendon Guard Re-

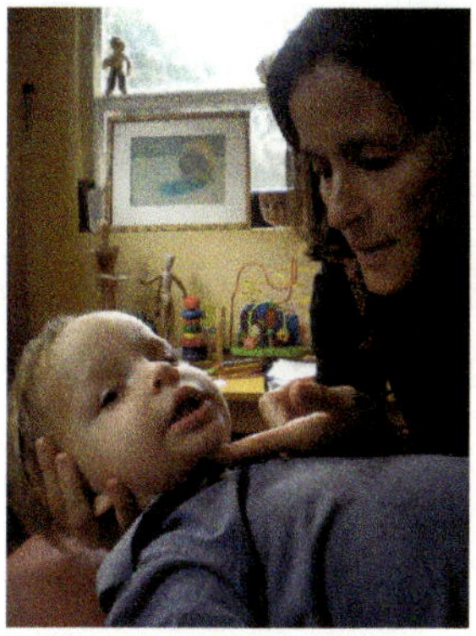

Jace receiving Oral-Facial MNRI®.

flexes were tight to the touch and he gradually allowed me to work in these areas. His ankles also reflected a tight Foot Tendon Guard Reflex.

By the third week of working together, I was able to begin to work extensively in his mouth with no resistance – though he did like it more when I had some coconut oil on my fingers! In fact, he even opened his mouth wider when asked. I had also showed his mother the many infinity 8 strokes, embracing squeezes, the spinal reflexes in addition to facial stroking and parietal tapping.

Soon he began to put his fingers in his mouth himself to indicate that he wanted work and stimulation in his mouth – and would even ask his mother to work on him at home, in his mouth and around his head and his ears.

Also, by three weeks, his mother said that Jace was loving the "massage" touch at home, more than she ever thought likely. He would lay on his side so mom could work on his Spinal Galant Reflex.

I reminded his mother that it was important to manipulate his hands and fingers – with stroking, traction and the important Babkin Palmomental Reflex to enhance his hand-mouth connections. I also began to build in eye tracking into our sessions.

At the one month point, I introduced the Landau Reflex into his home program. This is the reflex of joy and happiness and influences coordination of upper and lower parts of the body; as well as front and back. It especially develops movement control of the back of the body and influences the development of the nervous system. I thought it was important as he expressed an anxiety going into an arch position and it was difficult to do the Spinal Perez Reflex patterning with him. In addressing his Tendon Guard, Landau supports the body posture of orienting backwards to widen one's perspective. I also included Fear Paralysis and the Abdominal Reflex into our sessions to help him with his fears and tendency to withdraw.

I worked with Jace regularly one to two times a week as our schedules permitted. His mother was very motivated and diligent by following a Home Program I gradually built on. After six months of working together, Jace was fully conversational in Greek and able to communicate in short phrases in English.

His mother reported, "In the first few months we worked together, I was impressed, amazed, 'bowled over' by Lou's interaction with both me and my son. I had never seen Jace respond so quickly and trustingly to any practitioner. It was only a matter of several weeks before I was able to leave the room without Jace freaking out and only a few more visits before he complied with more and more of Lou's requests for cooperation and trust – JOYFULLY.

In terms of results, Jace has made progress in more ways than we can count. Not only have his language skills exploded, he's become a much more loving, affectionate child: confident, outgoing, adventurous, playful."

We continued to work together for another nine months, during which time we focused more on other areas of development, particularly his vestibular system, rotational work, crossing the midline and more. He became more and more self-assured and outgoing. Some of his final accomplishments included that he jumped into the air, clearing the ground; blew out his first candle – something he had not had the prior exhalation capacity for; and he was able to accurately and consistently throw and catch balls of different sizes, from varying distances and angles.

Jace after receiving MNRI®.

I would like to thank Jace and his wonderfully supportive family for the opportunity to be a part of their lives and their journey. I had so much fun working with Jace, and watching his curiosity and personality blossom as his anxiety melted away. Six months after our work was completed, I met with Jace and his Mom at a nearby park. Jace chatted up a storm with me while he explored on his own, confidently weaving his way up and down the highest jungle gym. – Lou Bosch-Wilson

How MNRI® Found Me

Marie-France Renaud, Sexologist BA, Victimology MI, Crimonology MI, Pol. Tech, Social & Public & Health Communication Science MA, MNRI® Core Specialist, Eastern Canada

My personal journey with MNRI® started about five years ago. After self diagnosing my 4 year old daughter, C, with Selective Mutism (SM) and a possible sensory processing disorder, I spent five months searching high and low for a professional that could understand SM and would treat her sensory issues as well. I finally found Héloise Allard, a wonderful occupational therapist located just an hour and a half away from home. She was not a specialist in SM, but her style of work and personality was perfect for my little C. After an evaluation session with Héloise, I heard a few laughs from my daughter in her presence. C was jumping, running around, and having fun during her evaluation! Clearly, my daughter felt comfortable at that point and even the evaluation had already changed her from the 'frozen' child I knew (not being able to move a muscle in social settings, staring at the floor, and rolling her tongue in every direction) to this.

Marie-France Renaud

The following week, we went back to visit our OT, Héloise, to get the evaluation results. When we entered the small waiting room crowded with other parents and kids with challenges, the secretary greeted C and tried to engage her in conversation but it was too much for her. She was already velcrowed to my leg as I attempted to manage an office crisis through a cell phone conversation.

When Héloise, with her radiant smile and gentle enthusiasm, came in to greet C, she was reluctant and yet eager to go 'play' again. Héloise addressed me right there, saying, "I am just back from a class this weekend, is it OK for me to try some techniques with C?" She took her by hand to a room on the other side of the corridor to work with her. I was still talking on my phone, busy with my office crisis, and didn't realize that this was the first time I ever left my daughter out of my sight with an adult she hardly knew.

Ten minutes later, my previously little frozen daughter opened the door all bubbly, speaking, and bouncy – skipping in front of me, the other kids, parents, and staff. She was jumping, she was loud, she was smiling and she was SPEAKING in front of other people. My surprise was huge. "What is this?" I said. I went into the room where C and her specialist had spent the last ten minutes and asked her to explain what she did for C. She explained that she did a reflex integration exercise and showed it on me, pressing inside my hand on extended arm. She handled me the book on MNRI® and I saw that this was an exercise for Hands Supporting. This was my first contact with MNRI®.

I was in shock! My eyes filled with tears as my daughter was now completely alive for the first time in front of other kids and adults. WOW! I had so many emotions running through all my heart at that very moment. My mind was trying to grasp, intellectually, the miracle that had just happened. My soul was jumping to the clouds with absolute joy. At the same time my stomach squeezed as I felt so guilty that, I as a therapist, could not achieve the same result that was achieved in ten minutes by another therapist.

I held 'the book' and I sat in silence in the therapy room for next two hours while C had a fantastic regular OT treatment and I read this book, trying to very quickly absorb the text. The book was called, *MNRI®: Masgutova Neurosensorimotor Reflex Integration, Dynamic and Postural Reflexes.*

As an ex-Police Instructor and Communication Consultant for the Quebec Police Academy, I was fascinated with how the 'fight or flight' response influenced communication strategies, specifically the choice of using force by experienced officers. They often seemed to repeat the same poor methods in intervention. I always tried to help them to understand what was really motivating them to act or react beyond the symptom in another more strategic way. I often thought they did not have the appropriate response to a specific situation and tried my best to train them to change to a more analytical response. When I read about reflexes, I understood then, that this reality transcended their culture, their excellent education, and repetitive training (with professional actors), their personality, and especially their own INTENTION. It was something the vast majority of them could not control; this invisible reality was controlling them.

Even by just reading the title of this book, I was sold. That's it! The reflexes! The neuro-physiological part of the communication phenomenon IS the bottom of the proverbial iceberg. As I was reading, I realized right there that what I was holding in my hand was 'the missing link' I had been seeking for, as a police trainer. It's all about survival. I realized the potential of MNRI® for victims of trauma because, not only was I holding the key of knowledge and understanding, I was holding the key to repattern the reflexes. To INTEGRATE them? How is this possible?

As my eyes were mesmerized and travelling between the book and my then speaking daughter, I still could not figure out how and why Héloise's MNRI® intervention had resulted in her speaking in just 10 minutes. I left that session with a new mission, a new vision, and the beginning of a new comprehension. My whole world just shifted right then and there. It was the birth of my MNRI® journey. My new life mission became to understand the scientific miracle I had just witnessed and share it with whom ever would listen.

How MNRI® Found My Body, Mind, Heart, and Soul

A week or two later, I entered my first MNRI® course in Montreal and encountered one of the most amazing beings I have ever met in my life – Christine Poulin. She was the MNRI® Instructor and welcomed me with strength and grace. I tried my best during these following four days (sitting in the first row right in front of her) to sponge up the information my Attention Deficit Disorder (ADD) brain could retain so that I could understand why my daughter expressed herself with such joy in front of strangers after just ten short minutes of MNRI®.

The longer I sat there, the more questions that came to my mind. A lot of information, new concepts, and the reflexes were such a novelty! New terms were 'sensory stimulation', 'neurological pathway', 'circuit of response', the 'TLR', 'STNR', 'Babinski', 'Babkin Palmomental', 'The Foot Tendon Guard', 'Moro,' and 'Fear Paralysis'.

The Fear Paralysis Reflex attracted my interest, especially as this was an explanation for the freezing and dissociation response in children with SM. Being the last resort for survival and protection when the fight or flight response becomes impossible, it brings survival in making my child freeze, even when there is clearly nothing to fear. This reflex, programmed in our phylogenetic memory, is surviving actually in two specific situations: when we drown and when we are surviving in extreme cold weather. When we are drowning, the respiratory system shuts down, the vocal cords stretches for a last attempt to scream and seek attention, then the mouth closes, the throat closes, and the sinuses, too. Our blood travels to our brain, lungs, and heart to deliver its last oxygen reserve until being rescued or die. That is the FREEZING response. Freezing was the response my little C had around people other then her immediate family and she was showing this in a completely safe surrounding, which meant that the perception in her subconscious mind of the situation was extremely stressful, though she communicated with joy and happiness when interpreting the social context or people

Communication with joy and happiness starts with an integrated Babkin Palmomental: ☺=sensory stimulus for Babkin.

as secure and safe. It's her freezing response that has been rescuing her all this time, and the moment it was addressed and worked out in play and with a new technique, its overprotective state changed!

At that first MNRI® introductory course, I learned more about my daughter than I thought possible. I learned how her sensory hypersensitivity could, in itself, trigger the freezing response causing her body to paralyze in social settings. It wasn't the social setting itself, it was her sensory system interpreting the environment as too loud, too bright, too crowded, too tactile, too strong smelling, or just too much sensory information which caused her body to crash under this overload of stimuli.

I remember listening to Christine, our MNRI® Instructor, with my sunglasses on, sitting in front to avoid distractions, being annoyed by the computer fan and the person breathing five rows behind me. My brain hurt from the daylight and that was perfectly normal for me, thus the sunglasses. I remember realizing right there that even after reading the book, *The Out of Sync Child* by Carol Stock Kranowitz, it never occurred to me that I could have this sensory hypersensitivity as well. I thought everybody perceived the world like this.

I then realized I was far from being 'normal' or 'typical'. I had suffered from ADD all my life but refused to take medication for it. I managed in school, though I don't know how. My thirst for learning and curiosity always overshadowed my learning challenges. As I raised my hand for the twentieth time, I gathered my courage and asked, "Is there any link between ADD and the freezing response?" The instructor explained that when the reflexes are not integrated, the dominant survival response will take over, it can be fight, it can be flight and freezing and added that, "ADD is often a combination of the flight and freezing responses acting at the same time". Here I was completely lost, "Oh my God, how can I understand this?" Fortunately, it was time for her to demonstrate Babinski patterning and she offered to demonstrate it with me on the table. She said this would help my daughter a lot. After getting the sensory stimulation on my right foot, I felt that my vision and my auditory system were affected. It felt like the eternal cloud around my head was slightly clearing up. I was awakening right in front of my own internal 'eyes.' Meanwhile, the other people there just stared at my feet while all the changes were happening inside my body and brain.

This was the moment when I really started understanding that a sensory stimulus was an awakening of the reflex pattern, its sensory circuit, and how powerful MNRI® and reflex integration was. I left this class in absolute awe! While my body felt physical chaos from trying to respond to the stimulus, my mind was trying to think of which exercises to choose for myself to support my own developmental needs. I had a million more questions. I was determined to work with my daughter with all I learned at the course. I called Héloise's superior at that time, Kathleen Sirard, and asked her if she would work with me, an adult who needed MNRI®, also. After some time and discussion, she agreed.

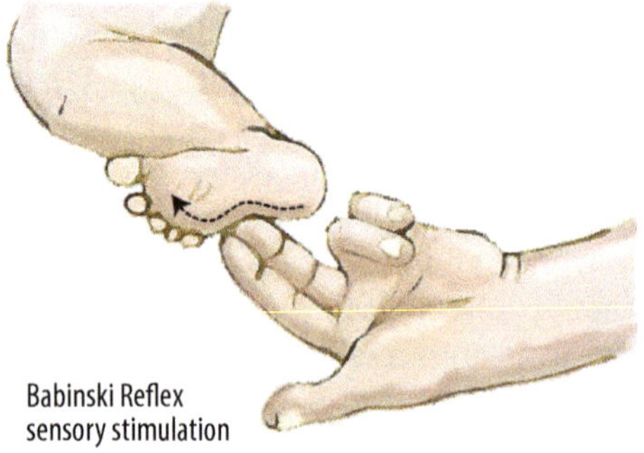

Babinski Reflex sensory stimulation

That was September 10, 2009. In December of that same year, my daughter was able to express herself with her teacher, some of her peers, to eat and go to bathroom at school. She participated for the first time at the school Christmas presentation in front of several hundred people. At the end of presentation instead of usual kids' question, "Did you see me?" she asked, "Did you HEAR me, Mom?"

This is how I found MNRI® and MNRI® found me. Kathleen worked with me, an adult, and this is how my journey of healing with MNRI® continued.

On Our Way to Inner Peace

While other SM kids and families were struggling, we are doing better. My daughter still demonstrates freezing responses from time to time in social situations when people speak a different language – it's not perfect but she is usually communicating normally at school and many other places. We continue to work on her focus, concentration, fine motor coordination, and memory organization. She is growing into a joyful child, full of curiosity and talents, maturing emotionally, and is loving and lovable.

C already knows that when she grows up, she wants to be an MNRI® Core Specialist. She is determined to share the gift that gave her the freedom to express herself and contribute to her various social spheres and say, "I AM". She looks forward to every MNRI® conference as they give her quantum leaps to facilitate inner peace, academic performance, kinesthetic freedom for movement, and postural control.

On my end, I have been pursuing that mission ever since I was struck by the MNRI® lightening bolt – the first time I laid eyes on the MNRI® book. My on-going personal healing story and plans to open MNRI® to the world of first responders, military, police, and firefighters will be available via my website echomotion.ca. Please check this website and enlighten me with your ideas.

MNRI® is one of the most enlightening gifts I have received. It's now time to share it infinitely.

Thank you Dr. Masgutova for contributing to the butterfly evolution of my angel, for showing me my pathway to freedom from PTSD, sensory processing issues and ADD; a syndrome you call "reflex integration disorder" via your deep knowledge, love, dedicated life mission, and wisdom! – Marie-France Renaud

It's Not Magic, It's Science

Dawn Burnell-Powers, Neurodevelopmental and MNRI® Instructor;
Core Specialist, Vancouver, British Columbia, Canada

Dawn Burnell-Powers

This is the title I have given to my two hour talks about MNRI®. I arrived at this title because so often people tell me, "If I was not watching with my own eyes, I would not believe that change could happen so quickly. It looks like magic." We need to be careful when we are in the realm of new scientific information, especially information that may not be new, but is new to us. Our first inclination may be to dismiss it. Because some aspects of this work can bring change about quickly, people may not believe it is possible or believe it is real. Doctors sometimes use the word miracle in place of magic. Regardless, of the word used, we often dismiss a scientific concept that contradicts our previous understanding of the body and, in the case of MNRI®, the nervous system in particular.

I have chosen to tell the stories of three clients that left their doctors scratching their heads and dismissing the results as a 'miracle' because the rate of change was very quick (with respect to the condition). But even though the doctors describe these instances as miracles, they continue to refer patients to me with comments like, "I'm not sure exactly what this specialist does, but she gets results no one else gets." My hope is that one day, they will attend one of my *It's not Magic, It's Science* talks. Then, hopefully, they we will come to a shared understanding of how the nervous system works.

I will limit myself to three case studies, although in the years since adding MNRI® and other modalities to my private practice I think I have accumulated enough stories for a separate book of testimonials. I am choosing these three stories because, in two of the three cases, Dr. Masgutova never saw the clients described. In one of the cases, she saw the client for a total of three hours. I am emphasizing this, not so you will conclude that I am an extraordinary therapist, but because I believe the strength of a method can only be evaluated if students can replicate the results of the originator of the program.

A Story that is Very Hard to Swallow

A client came to me very early in my MNRI® training in an unusual manner. He was referred to me by my daughter's 12 year old friend. This man was her grandfather, a seventy-two year old gentleman, who had six months previously suffered a brainstem stroke.

He had experienced a rather good recovery considering the severity of the stroke. Six months post stroke, he was walking with canes, dragging his left leg slightly. Apart from the dropped foot, there was no visible paralysis, although he still struggled with some fine-motor issues. His speech was clear. He was happy with his recovery except that he still could not swallow.

He could swallow nothing, not even his saliva. His epiglottis appeared to be paralyzed. He had no feeling on the back of his tongue. He said the experience was like dental freezing that never wore off. As a result, he was fed by g-tube. According to him, the most severe consequence of his not being able to swallow was he had to sleep facing down so his saliva could run out of his mouth at night. This caused him incredible pain in his back. Instead he had chosen to sleep on his side, but this required him to wake up and spit every five minutes. He said he felt the lack of deep sleep was making him psychotic.

His doctors told him he might never regain his swallow. They referred him to SLP's who told him that if he visualized swallowing, perhaps one day he would remember how, despite the lack of sensation in his mouth. But they had little else to offer him, except to help him psychologically prepare for the fact that he may never swallow again. He traveled a very, long distance to see me because he was desperate. I decided to try MNRI® Oral-Facial Reflex repatterning. At the time, I had been fortunate enough to watch some of the best MNRI® Oral-Facial specialists do this although I had not done it myself. I had seen them bring about phenomenal results in minutes so I knew this was possible and wished desperately that one of them were here to guide me.

But I could see the desperation in this man's face. I began by doing every technique I could think of to his head and face to wake up and organize the facial nerves. Then I used several techniques in his oral cavity. I was very careful to avoid him accidently aspirating saliva into his lungs. It was a real concern. Finally, I did the techniques along the throat for swallowing that I saw more experienced MNRI® specialists do and, in less than a minute, this man began to swallow. He was ecstatic, "How could this have happened so quickly?"

He reported he could now feel the entire inside of his mouth and he wanted to eat food immediately. I made him promise not to do so, explaining that although the epiglottis was definitely moving, we did not know if the movement was sufficient to prevent him aspirating food into his lungs. I showed his wife what to do to stimulate the swallow and showed him how to systematically stimulate the inside of his mouth to ensure that complete feeling would return. They practiced on their long journey home.

During the trip home he became a bit angry at how the doctors had been complacent, leaving him in such an awful place, when the solution had appeared relatively simple to him. He was tested the following day and the doctor was amazed that the results indicated that he could safely swallow and eat. I saw this gentleman only one more time to help him with his foot drag and that, too, was resolved with a single intense session of Leg Cross Flexion-Extension Reflex repatterning and hooking up all the pieces of his gait. His neurologist still sends clients several hundreds of miles to see me when he has a 'tricky' case.

This experience taught me to trust MNRI® results and to strive for deeper comprehension of the neuro-sensorimotor modality as a new piece missing in other methods. Even in the hands of someone who has not mastered all the techniques, a client will experience change, because the nervous system is waiting for the information provided by this modality. This was certainly true in this grandfather's case. His body was waiting for the sensory-motor stimulus that would evoke the expected complex proprioceptive-motor response of a swallow. After all, swallow as a genetically originated sensory-motor response, was a motor pattern that he had been practicing months before he was even born, as well as all his life before the stroke. All he needed was for that pattern to be reawakened.

It is Good to Crawl Before We Walk: A Story of the Extraordinary

Many children in my practice have cerebral palsy (CP) as the result of birth trauma. This is a story of 'Sam,' a little boy who was brought to me at 18 months of age. The parents had recently received confirmation that he had CP. They were told that their child would probably never walk and they needed to begin preparing for this eventuality. Sam's spasticity, affected both sides of his body to different degrees, in both upper and lower limbs. His legs already showed mild deformity due to it. He had health complications, exacerbated by a serious GERD condition, that prevented him from gaining weight. To his advantage, he did not have seizures.

I am lucky to collaborate with a wonderful group of professionals, who cross-refer. I must point out that I have not been the only one offering this little one therapeutic support and I was not only using MNRI® with him. He received regular cranial sacral therapy treatments from me and two other practitioners. He also received regular chiropractic support from me and a naturopath. And his mother used other programs.

In CP, It is extremely important to use whatever we know to help the development of the symmetry of the skull, increase its capacity, and create space for the brain to grow physically and to develop. There are many reasons for constricted sutures that go beyond the scope of this story, but creating more possibility for brain development is always one of my priorities and one to which MNRI® pays particular attention.

I usually begin with MNRI® Neurostructural Reflex integration therapy to integrate protection and survival mechanisms but in this case, I used it first and foremost to boost Sam's immunity and support necessary muscle bulk and weight gain. I did several sternum techniques to address his GERD. Core Tendon Guard helped to stabilize his system and he began to gain weight and build muscle. Over the next 24 months, he had periods of illnesses when he would lose weight; however, he has now gained both weight and muscle mass and has gone extended periods (weeks on end) without reflux or vomiting.

Sam's family was very motivated to work and did all the exercises I offered very regularly. They made all the necessary dietary changes and were faithful with supplements. Initially, I visited him very often at night, usually once per week. He did not like being touched and would cry if I touched him while awake. As with other children, this quickly faded, he began to enjoy touch and I was able to work on him awake or asleep.

I began with repatterning the spinal reflexes. The family did feet and hand reflexes and Embracing Squeeze daily at the beginning. We quickly moved to repatterning crawling. At 22 months, 4 months after the doctors said he probably would never walk, Sam was walking without support. By 24 months he was walking in bare feet and able to adjust easily when stepping from the sidewalk to grass without falling. By 28 months, he was running.

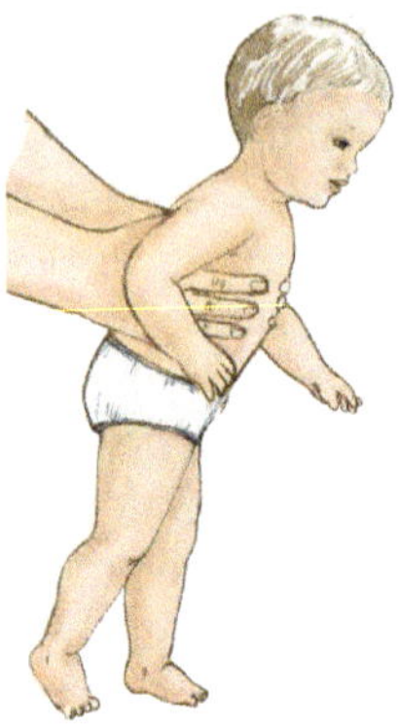
Automatic Gait shown in an infant.

What sets this therapy apart from others, is how we teach children with CP to walk. I recommend that parents only use orthotics if it means the difference between walking or not walking. Sam never used orthotics in the process of learning to walk, much to his physiotherapist's dismay. Immobilization caused by orthotics prevents the information from reaching the limb and inhibits the reflex pattern.

Refusing orthotics or minimizing them to 2-3 hours a day takes great courage on the part of the parent, but it always seems to pay off. If we inhibit, for example, Crawling or Automatic Gait Reflex, the nervous system is unable to learn from feedback. As result, the reflex cannot mature and the nervous system cannot create pathways that would lead to further development. In this particular case, the little boy's gait and foot alignment changed monthly. Any customized orthotic would have been outdated by the time it was manufactured, not to mention the expense of changing the orthotic monthly to reflect the current relationship between the foot and leg.

Body righting and postural control was challenging. We would get it turned on for a day, sometimes a week, and then his posture would collapse again. Sometime after 27 months, when (following a session with Dr. Masgutova) we started activating sensory-motor points for Trunk Extension, Sam became able to maintain whole body extension. His Trunk Extension continued to improve with work on these points and I was grateful to have learned more techniques.

Neurological conditions manifest differently in each child. With Sam, the key appears different from month to month. One month, joint rotations are the key. The following month, proprioceptive compressions, vibration or neurosensorimotor points activation may be the way to unlock the system. Although this complicates the task for the specialist and the Home Program for the parents, it is crucial to work with dynamic changes in reflex patterns as they occur.

Just previous to our first visit, the doctors told Sam's parents he would never walk. A simple assessment of his reflexes told me immediately that he still had a chance. Although his development has out stripped any possibilities that his doctors saw for him, understandably his parents want even more for him and hope that some day his gait will show no trace of CP.

At age three, there is no visible evidence of spasticity left in Sam's limbs though his joints in one wrist, knee, and ankle can be very stiff. His one foot still rotates internally and on some days it is worse than others. However, he can and does march well, drawing his knees up with great trunk extension.

He often forgets and points with his middle finger rather than his pointer. He holds a pencil with a fully developed pincer grasp and developmentally his fine motor skills are on par. His head size and shape is within normal range, but we will continue to pay attention to the sutures as he grows. He still has periods of digestive upset and needs more oral work to improve his tolerance of different food textures and to help his chew become more circular.

In the meantime, this family has to struggle with the side effects of their child having received MNRI® for two years. Sam is fluent in two languages and has perfect articulation. He is an absolutely brilliant child who understands complex mathematical concepts well beyond his years. It is difficult to see how he will be satisfied in a typical school setting. This is consistent with my experience using MNRI® with CP. The child's cortex develops at an increased rate leaving parents with a brilliant child. We experience this when we begin using MNRI® in infancy with any child, but it appears particularly pronounced in many children with CP.

A Fall into a Journey of Endless Discovery

Just over two years ago I received a distress call from a mother whose son was in the spinal cord unit at our local hospital. Ryan was 23 years old and had fallen eight stories out of a downtown high-rise. (This was not an attempted suicide.) She had heard about my work with spinal cord injuries and was hoping I could help her son. I entered the hospital under the guise of alternative pain management but the family hoped that I would be able to offer more.

This young man's injuries were substantial and complex. He managed to right himself while falling, allowing him to land on his feet. Unfortunately he landed on concrete. Ryan seriously damaged his left foot and ankle joint. He drove his tibia through the femur on the right leg. He had a complete equine caudal sever and his spine became detached from his pelvis. He also fractured the scaphoid and some of the smaller bones in his right wrist.

My first question on the phone to Ryan's mother was, "Did he hit his head on the way down or upon landing?" The answer was no. I wanted to work with this young man immediately. His brainstem was not injured. Even though he was 4 weeks and 9 surgeries post accident, we were still in a window of tremendous possibilities.

Pain management was a struggle for Ryan and was the priority of every visit for the first two months. Fortunately MNRI® and other modalities were very efficient at organizing the nervous system. Using these methods, I was usually able to relieve his pain within 20 minutes (unless it was a day immediately following a surgery, then it would take up to an hour). He went on to have 6 more operations in these next two years. He was very proud that he was fully free of pain medication at 7 months post accident and 2 months after his last major surgery.

During this first visit, after his pain was under control, I asked Ryan if there was anything else he hoped I could do for him. He asked if I thought he would ever walk. He did not beat around the bush about what his main concern was. In MNRI® we are trained to always work with the client's primary motivation because this is where we find strength in the system. I knew at this point the doctors had not directly given him a definitive prognosis, but it was written in his charts and the nurses had spoken to his mother about preparing him for his life in a wheelchair. His left foot was in an air cast with an entire piece missing from his heel. He was in an isolated unit for MRSA (flesh-eating disease). They were delaying fusing his ankle due to the MRSA and trying to stabilize his pain management. Understandably, the staff were managing his condition on a day-to-day priority basis and were not dwelling on what the prognosis one year out was going to be.

Ryan was well over six feet tall and largely immobile, but I did my best to test his reflexes. All his reflexes were active and within the range of functional. It was clear he was very athletic before the accident. This was a very encouraging base with which to work. I said that I did not know what his 'walking' would look like, but it was clear from how his reflexes were performing, there was a great deal of possibility. I believed he would be able to walk in some fashion, even if they ultimately amputated his left foot.

Initially, I visited him 3 times per week for two months with breaks for surgeries and my working out of town. I structured each session around showing him that he could now do something that he was unable to do at the beginning of the visit that day. This meant at every visit he and his mother could see some change. This was extremely important because they were swimming in a sea of pessimism. He was in a hospital ward where there was only one outcome for these patients.

Either he was very unlucky, or the hospital perceived his condition in a very grim light, because he was assigned a physiotherapist who usually cancelled on him or did not attempt to do more than stretch his limbs. I can only think she was unsure about how to work around his pain, for she did almost nothing with him for two months. When he gained more mobility, she spent 30 minutes once a week with him for the next 2 months. After two months of working with Ryan, the doctor gave permission for him to put some weight on his right foot. But his physiotherapist refused to allow this during therapy. She would say that they would eventually work with him at the rehabilitation hospital where he was going next.

This thinking is entirely opposite to MNRI®. We want to work as close to immediately after the accident as is reasonable. This gives us the biggest window of possibilities. The very first day, I showed his mother reflex exercises for the feet. He could not feel his feet (except for neuropathy). Even though the left foot was in and out of casts, she did what she could. If the toes were exposed, she rotated them. With permission, I was able to remove the air cast and do embracing squeeze on the foot and so was she. She gave her son tireless attention doing any exercise I asked of her for the first five months she was there.

Initially, I did MNRI® Neurostructural Reflex integration therapy only on the front of his body – Ryan could not turn over as they had just reattached his spine to his pelvis. Within a week, he was able to turn on his side and I was able to do Spinal Galant. At two weeks, we were able to get him prone for a few minutes and I did Spinal Perez. When I could not get him into ideal positions, I would put my MNRI® knowledge together with my chiropractic knowledge and do the best I could by reaching underneath him. Therapy was very imperfect at first.

Although the family was encouraged by the small changes they saw each day, we turned a corner at three weeks when Ryan was able to control his own urination. At four weeks he was able to manage digital stimulated bowel movements on his own. I now had his attention. At five weeks, the first signs of sexual function returned. You can imagine how these gains in function impacted his psyche. The notes in his file indicated that the medical staff assumed that bowel, bladder, and sexual function would never return. It took a bit of convincing the doctor to leave the catheter out permanently. Understandably, they were concerned that without feeling, he would not evacuate entirely. Eventually he demonstrated through tests, that although he had next to no sensation, he could fully void.

Four months post accident, the flesh in his extremely damaged foot was responding to a skin graft and the doctors wanted to go ahead and fuse his ankle. I asked the family to delay this surgery knowing that if the foot was fused this would limit what function we could return to the foot. Understandably, the family felt tremendous pressure from the doctors to proceed with conventional treatment. The medical staff still had no hopes that he would ever walk. The goal of the surgery was that he would be able to place enough weight on the foot to aid his balance in transferring in and out of the wheel chair. My biggest challenge was working with a vision of Ryan's recovery that differed so greatly from that of his medical team.

After this surgery, he was sent to our local rehabilitation hospital where he would receive more therapy. At this point, I began seeing him only two times each week as the rest of my work schedule had increased for the summer months. Again there were weeks I did not see him at all due to my working in MNRI® clinics.

At the rehab hospital, I was not invited to work with Ryan in the rehab room so I had to work with him entirely in his hospital bed. At this point, I began working Trunk Extension and Leg Cross Flexion-Extension in earnest. At six months post accident, he stood up on both his feet. Even though he could not feel his feet, he could put weight on them if he held onto something. I remember being so surprised at how tall he was. The doctors came by weekly to see this feat. They were amazed, but did not want to hear about the special therapy he was receiving. They decided he was a miracle…one of the lucky ones. They decided this because they knew that the rehabilitation he had received through the hospital could not account for his current level of functioning.

When he could stand, the rehab hospital wanted to give him a walker. This time I was more forceful with the family. I explained that although the walker would give him increased mobility sooner, it would take him out of Trunk Extension. I believed it was much more beneficial to delay this 'false mobility' where he would be forced to drag his feet because he could not feel them, in trade for 'real walking.' Thankfully, this time they did agree to try to follow this philosophy. Three weeks later, he walked with assisted crutches while maintaining Trunk Extension.

With the assisted crutches he relearned homolateral walking and Automatic Gait again. Feeling began to awaken in his left glute and some sensation returned to his right foot for the first time. And more importantly, Trunk Extension had be-

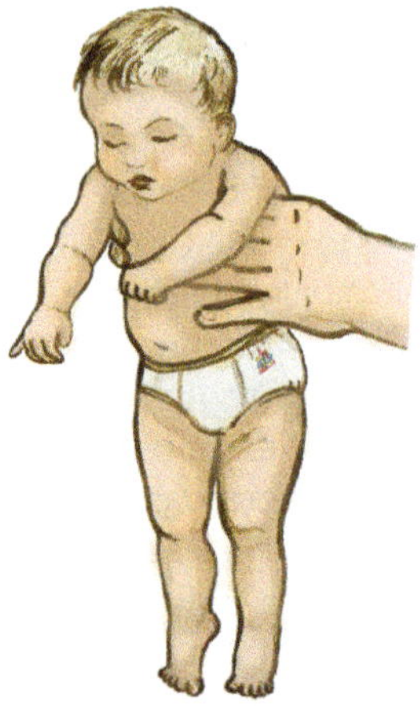

Trunk Extension shown in an infant.

come integrated and now was a resource we could build upon to recover matured gait.

A week before he was discharged, a doctor interrupted our therapy session and demanded that I explain 'scientifically' how it was possible that this patient could stand when his most recent x-ray still showed a complete sever of his pelvic nerves. I explained how reflexes educate and build the peripheral nervous system and that the plasticity was extremely dramatic in this case. Furthermore, I intended to use this strategy to teach Ryan to walk. Days later I heard this doctor repeating this same explanation to a group of doctors who came into the room to see the 'miracle' for themselves. I smiled silently to myself; maybe there is hope.

Seven months post accident, Ryan was discharged from the rehabilitation hospital. He was supposed to return two days each week for ongoing physiotherapy. However, this was not possible due to the waiting list. He could not continue his therapy for another three months. In the meantime, all he had for support was me and a few 30 minute visits to an ad hoc physiotherapist who did not have any specialized training around spinal cord injury. Ryan could stand and balance briefly, but he was wheelchair dependent. He had very little stamina and his muscles fatigued quickly when using the assisted crutches.

We worked together twice a week. (At this point, his mother returned home 1,000 miles away ending her seven-month extended stay working with her son.) In these two months he went from walking short distances with assisted crutches, to getting around entirely on his feet using ski poles for support. Understandably, because he was a young man, he preferred to use canes as they drew less attention than ski poles. Again, I asked him to be patient and use the ski poles. With the poles, he was able to maintain Trunk Extension while he walked. Leg Cross Flexion-Extension was not adequate on the left leg due to the injured foot and I was afraid he would wear down his hip unnecessarily if he persisted with the canes. At one year post accident, Ryan was solidly using walking for getting around with only the poles for support. Enough dorsi flexion and sensation had returned to his right foot that enabled him to start driving again. He was glad he took my advice and did not retrofit his car with hand controls.

Throughout this rehabilitative process Ryan was very confused by how greatly the MNRI® approach conflicted with the approach of traditional physiotherapy. He was a male athlete who thought in terms of muscles. He understood when therapists emphasized core strength. All their exercises were to support his core or to strengthen his legs using a classic weight training approach. It was true; he was very weak. In contrast, I explained that building muscle in isolation to other muscles without understanding how the muscles communicate with one another was not efficient and did not ultimately guarantee increased function.

For example, at one point, he could not fully extend his right leg. The physiotherapists said if he spent 3 weeks strengthening his quads, he would be able to straighten his leg. I worked with him and he straightened his leg in a single session. This was not a problem of strength. It was a problem of communication. My approach was to track the information as it traveled to the hamstring and to see where it was blocked. There was, after all still a complete sever of the equine caudal nerves. His body needed to create and learn an efficient pathway from his lumbar nerves to deliver the information to his tendons and ligaments of his legs. We used reflex repatterning and other modalities to build up the peripheral nerve net to carry this information more efficiently.

Ryan has had several victories and many setbacks along the way. Two years and two months and 15 surgeries post accident, he walks unassisted. No supports! He has an orthotic on his fused foot that supports him up the back of his calf. On his right foot he has an elastic band that attaches to his shoe that gives him additional support. He has about 60% feeling in his right foot and about 30% feeling in his left foot. He can stand for a minute in his bare feet, unsupported on a balance board. It is not surprising that his goal is no longer to walk, but to run.

His only regret was allowing them to fuse his left foot. He understands now, that they did so thinking he would never walk so they did not take the care that was necessary to ensure that full function would return to the foot. They did not try to reattach ligaments, etc.

Today, feeling has not fully returned to Ryan's lower extremities, but every day it continues to change. This again goes against what has been drilled into him. Every professional in the rehabilitative process has told him that possibilities of improved function end at two years post accident. He is now two years and two months post accident and he has not completed the trajectory of his full recovery. For the last year, on average I have seen him two times per month; yet, he continues to make considerable gains.

Ryan takes great pleasure when people who do not know his story ask him, "What happened? Did you sprain your ankle?" For this is how he presents. With orthotics, he walks with a slight limp. But neither he nor I believe his journey is at an end. Every day more function develops. He is now getting the first sign of dorsi flexion in his fused foot. School never came easily for Ryan. He is now attending university and admits that learning is much easier than it used to be – an inadvertent side effect of MNRI® therapy. The numerous videos done by us show all the truth of restoration of mobility and coordination of the gait and other motor functions.

I have learned a lot with Ryan. MNRI® and other techniques that educate the nervous system will continue to recover function if one is patient and perseveres. It is hard for families not to listen to the experience of traditional medicine when it places limits on our expectations. And it is understandable for traditional medicine to place these limits, when they have one particular understanding of the nervous system. However, I believe we have produced enough consistent results, that it is time to stop dismissing these results as miracles.

It may look like magic, but it is science. It is clear, this is science we have yet to fully understand and appreciate. But it is science that has the potential to quickly and dramatically transform peoples' lives.

I am not fortunate to work with any other MNRI® Specialists in my area. Every day I wish there were colleagues to bounce ideas off. But I am so grateful for the parents of my clients. They work so faithfully with their children and have given me so much of their trust. In the absence of colleagues, they are my inspiration each day that I go to work. I thank them for allowing me to share their stories of incredible courage, perseverance, and hope. – Dawn Burnell-Powers

Celebrating Parker: Our Son's Success

Kevin and Samantha Olenick, MNRI® Parents, San Diego, CA, USA

This is a story of our amazing boy's journey. We were told by his pediatrician and orthopedic surgeon that he would probably never walk independently but thanks to MNRI®, he has proved them wrong.

Parker with his parents, Kevin and Samantha Olenick, and Dr. Masgutova (center).

Parker was determined to enter this world and arrived 10 weeks early. This was after Mom was put on hospital bed-rest for an incompetent cervix starting at week 22. During this time, she rested and needed little medical care except for two steroid injections to help the lungs develop. At week 28 she was surprised to be leaking fluids due to a ruptured membrane and was kept under continuous watch until contractions began at week 30. Since Parker was breech and wrapped in his umbilical cord, he was born after an emergency C-section. Despite this, he was breathing well and received a healthy APGAR Score. He was incubated for 12 hours to lessen stress on his lungs. Parker was fed his mother's milk and extra nutrients by tube for three weeks at which time he was strong enough to breast-feed for half of his feedings. After almost six weeks in the NICU he was sent home with only a diagnosis of prematurity. During the next few years, Parker was delayed in his mobility, but on track for premature babies. His verbal skills came early. At age two he was still not walking and pulled himself to standing by rolling over his toes. At the age of 2½ he was diagnosed with Cerebral Palsy – Spastic Diplegia.

During the first year of diagnosis, Parker followed the traditional route for Cerebral Palsy – physical therapy, Botox, assistive devices, and the likelihood for future surgeries. After a year of following this traditional protocol, his spasticity increased and his range of motion decreased. This is when we began to research brain damage and trauma and look for alternative treatments. At age 4½, Parker was seeing a therapist that was trained in many modalities. During one of Parker's sessions she asked us if we noticed if Parker startled at noises. She then explained primitive reflexes and their importance. She told us about her training in the Masgutova Method® and that she thought it was the missing piece to Parker's program. This therapist then taught us a few exercises and asked us to do them every day before her sessions for the week. At the end of the week, Parker had made much more progress than he typically had been making. We were filled with hope.

For the next year, Parker received monthly sessions and we were taught a few more exercises to do at home. We were seeing results, but knew we needed to learn more. We researched our options and decided that the best for Parker would be to go to an MNRI® Family Conference. The intensive therapy for over a week, receiving a complete personalized Home Program and personal training on what Parker needed was the deciding factor. At that time, Parker was walking with a walker and crawling to get around. We fundraised for a year and attended our first MNRI® Family Conference when Parker was 5½. By the end of the Family Conference he was able to bear crawl on his hands and feet without his knees touching the ground. During this first Family Conference we also noticed that his cognitive processing had increased, he was structuring his sentences differently and putting clearer order to his thoughts. He was now able to use all of his fingers, instead of just his pointer and thumb. His digestion improved, he potty trained himself, and skin color changed. When we arrived home we made a schedule to fit in as much of his Home Program as we could. We averaged about two hours of therapy a day plus visited a Core Specialist a few times to confirm the techniques. Four and a half months after our first Family Conference, Parker took his first unaided steps. We were sold and started saving for our next camp.

Now at age 10½, Parker walks everywhere unassisted.

Parker continues to receive his Home Program from his Mom for an average of ten hours per week. He has attended ten intensive MNRI® Family Conferences in the United States and five in Poland. He has a lot of personal motivation to continue the work since he notices the great results. He is now constantly making new goals for himself. He is an honor student, plays many sports, and is extremely personable. We believe the reason for Parker's steady improvements are due to the Masgutova Method®. It formed the foundation that he needed to be able to use his body and mind with ease.

Parker running to his base in baseball.

Parker has gained many successes all due to MNRI®. He has had a decrease in spasticity, an increase in gross and fine motor skills, and improved balance and coordination. He started walking and then increased his distance, went from sitting to standing unaided, began climbing stairs unaided, and learned to kick a ball while walking. Parker learned to catch and throw a ball while walking, walk up and down curbs without a rail or hand, and his hand and leg tremors stopped. With these gains, he also had improved dexterity, more legible and then easier handwriting, his eye tracking improved, and his reading level increased. Soon his auditory processing increased, he developed better sequencing of thoughts, his sensitivity to noises and lights decreased, and he stopped involving his mouth when doing other skills. His hair sensitivity has decreased and his startle sensitivity has decreased.

With MNRI®, anything is possible for Parker.

Dear Parker, Kevin, and Samantha: We have all watched the inspirational journey of Parker and celebrate his extraordinary spirit, enthusiasm, and achievements. We love watching him run races, participate in water sports, hitting a baseball and making that amazing basketball throw. We also deeply appreciate your reaching out to other parents and sharing your enthusiasm for this work. – MNRI® Team

Sidney's Story

Shannon Desilets, PT, MNRI® Instructor and Core Specialist, Pembroke, NH, and Sara Christensen, MNRI® Parent, Hilsborough, NH, USA

This is Sidney's story, as shared by her mother, Sara Christensen, in early 2012:

"Sidney was born with a large cancerous tumor on her spine, compressing all the nerves to her lower body, paralyzing her. She underwent major back surgery at 9 days old and then chemotherapy to remove the mass. She was left with no signs of feeling or movement in her legs, and she was basically going to be wheelchair-bound. We set out to find the best therapist for her, someone that was going to be motivated and wasn't going to give up on her. We met Shannon, an MNRI® Core Specialist, a year and a half ago. When she started working with her, Sidney was almost one, and she couldn't crawl, she couldn't feel her legs, she couldn't bare any weight, and she could barely kick at her knees. Sidney is already doing so much more than the doctors expected, and I truly believe that this therapy has played a huge role in this. It's amazing to see her become aware of parts of her body that she hasn't been able to feel or move, and it is amazing to see them awaken right before our eyes. She now shows signs of some sensation in her feet. She has learned to crawl, stand, and, with braces, she can now walk. She leaves her walker behind her most of the time, and she loves to run to Daddy."

Shannon Desilets

Sidney dancing in the You Tube video.

Sidney update, as shared by Shannon, April 2014:

Running, maneuvering over/around obstacles, and even dancing are now parts of Sidney's daily everyday life. A video of Sidney (www.youtube.com/watch?v=9x_ZQ8zeWos, photo at left) dancing along to Amy Purdy of *Dancing with the Stars* recently went viral. Amy Purdy, who has prosthetic legs, also saw this and spoke of Sidney in an *Access Hollywood* airing. Sidney was invited as her guest on *Access Hollywood!*

To Sidney: you are one of my heroes on this earth. You teach me all is possible, to never give up, and you are a shining star. To Sara and Scott Christensen, thank you for entrusting me with the care of Sidney, and thank you for your strength, hearts, and dedication for Sidney. Your courage and perspective are of such integrity and genuine heart. You are remarkable parents of a remarkable little girl! – Shannon Desilets

Tom Kretsch: On the Winning Way

Brian Kretsch, Brother, Minneapolis, MN, USA

I want to introduce my brother, Tom Kretsch.

Tom played on Benilde St. Margaret's hockey and tennis teams in St. Louis Park. In college he was on the University of St. Thomas Varsity tennis team, both in singles and doubles. He played school club sports including racquet ball, squash, table tennis, and badminton. He graduated with a double major in Business and History. Out of school, he managed the Williston Fitness Club in Minnetonka and was also a tennis instructor. At the age of twenty-four, he suffered a brain injury after a car accident on November 12, 1975.

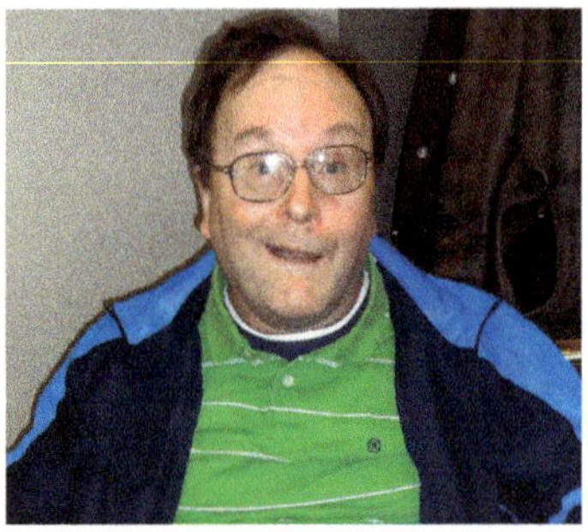
Tom Kretsch in 2013.

He was completely paralyzed. He could not walk, talk, or remember anything. He retained some very long-term memory but nothing short-term. After time in a rehabilitation hospital, he was diagnosed as having no hope and sent to a nursing home to vegetate for the rest of his life. The PT in the nursing home refused to give him therapy because she said there was no hope for him. We took him out of the nursing home in order to pursue our own therapeutic journey because we felt he was better off with the family than in the nursing home. At least we could offer him hope. After many years of conventional physical, occupational, and speech therapies with mixed results, we felt like Tom had the potential to recover further.

Tom's rehabilitation journey took a major step forward when Tanya Hallman recommended MNRI®, the neurological reflex integration therapy originated by Svetlana Masgutova, Ph.D., and made available by a dedicated team of MNRI® professional therapists. This therapy has produced amazing results for Tom.

He has improved in the following ways: more fluid walking, his memory was activated well enough for him to start his speech again, and now he has much clearer articulation which is beginning to be understood by others. He has dramatically improved facial symmetry, better swallowing, and is calmer instead of defaulting to a hyper-reactive yell response. He has improving social skills and amazingly good comprehension which is increasing in complexity to include the beginnings of decoding, analysis, and visualization, and singing along with the radio as well as a returning sense of humor.

We are seeing many postural improvements such as more self-supportive balance in standing, smoother and more precise movements in gait, greater stability in his joints and feet, and better head righting which has taken away his need for high pillows to support his neck. His visual responses have changed such that his eyes are no longer locked and have become more mobile. His eye tracking on the horizontal plane is emerging as well as he has better convergence and divergence indicating improvements in his binocular and distance vision. His new visual skills have helped him with balance and stability. These good changes have become permanent. We take a six month break every year due to seasonal business responsibilities and he does not regress after these breaks in the therapy process. Dr. Masgutova's work has direct, concrete results that can be applied immediately.

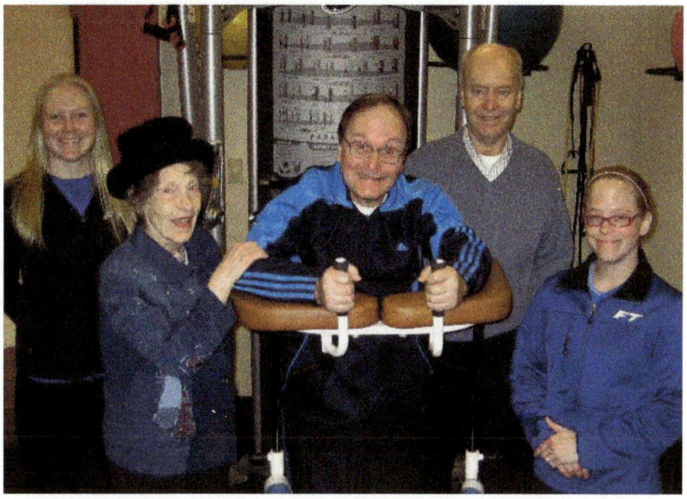

Tom Kretsch, center, with brother, John, mother, Josephine, and health staff in 2014.

Every time we take Tom to a Masgutova Family Conference or a Mini-Clinic, we see new improvements. Most currently, we are seeing that Tom has increasing motivation to progress and his speech is developing into whole ideas expressed in two to four sentences. He is attempting to sing along with old Beatles songs and he is marching when supported in his walking. He has a healthy color and glow to his skin and is smiling a lot. He feels proud of himself and happy about his achievements.

He and all our family love our MNRI® specialists for their devotion, knowledgeable work, and optimism!

We congratulate Tom and his entire family for their endless hope and hard work. It has paid off! Tom's progress is an inspiration to so many. – MNRI® Team

Therapeutic Horseback Riding: What Are Equine-Assisted Activities and Therapies?

Stephanie Keene, Hearts & Horses Therapeutic Riding Center, MNRI® Core Specialist, Buxton, ME, USA

Stephanie Keene

Throughout the world, hundreds of thousands of individuals with and without special needs experience the rewarding benefits of equine-assisted activities and therapies (EAAT). A physical, cognitive, or emotional special need does not limit a person from interacting with horses. In fact, such interactions can prove highly rewarding as well as therapeutic. For instance, experiencing the rhythmic motion of a horse can be very beneficial. A horse moves its rider's body in a manner similar to the human gait. Riders with physical needs often show improvement in flexibility, balance, and muscle strength.

Whether it's a five-year-old with Down syndrome, a 45 year-old recovering from a spinal cord injury, an elderly person recovering from a stroke, or a teenager struggling with depression, research shows that individuals of all ages who participate in EAAT can experience physical and emotional rewards. For individuals with emotional challenges, the unique relationship formed with the horse can lead to increased confidence, patience, and self-esteem. For corporate teams or an individual seeking better leadership, team building or communication skills, working with horses provides a powerful new paradigm.

Stephanie Keene on a favorite horse, Magic.

Hippotherapy is a physical, occupational, and speech-language therapy treatment strategy that utilizes equine movement as part of an integrated intervention program to achieve functional outcomes. Equine movement provides multidimensional movement, which is variable, rhythmic, and repetitive. The horse provides a dynamic base of support, making it an excellent tool for increasing trunk strength and control, balance, building overall postural strength and endurance, addressing weight bearing, and motor planning.

Equine movement offers well-modulated sensory input to vestibular, proprioceptive, tactile, and visual channels. An example would be during gait transitions, the patient must perform subtle adjustments in the trunk to maintain a stable position. When a patient is sitting forward astride the horse, the horse's walking gait imparts movement responses remarkably similar to normal human gait. The effects of equine movement on postural control, sensory systems, and motor planning can be used to facilitate coordination and timing, grading of responses, respiratory control, sensory integration skills and attention skills. Equine movement can be used to facilitate the neurophysiologic systems that support all of our functional daily living skills.

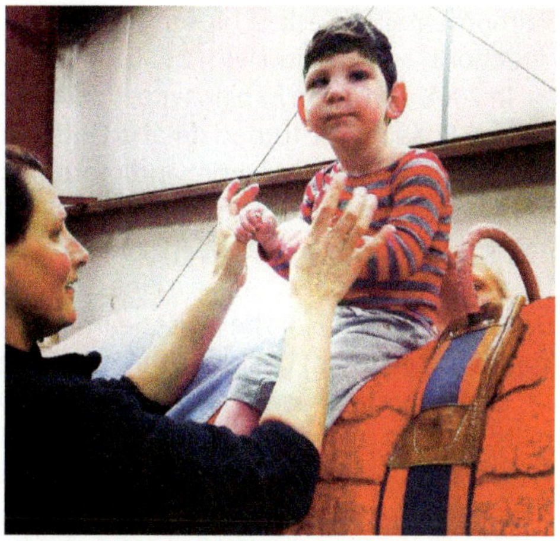

Working on the Hands Grasp Reflex is more fun on a horse.

In our therapeutic equine program at Hearts & Horses we combine MNRI® and hippotherapy therapy for maximum success. Through 20 years of using these therapy techniques, it has become my belief that experiencing and riding horses significantly improves the lives of individuals with physical, emotional, and developmental disabilities. For our physically challenged riders we have seen drastic improvements in balance, muscle tone, range of motion, and strength through therapeutic riding. Our riders who are working through emotional difficulties quickly learn to understand their horses, a first step in understanding themselves. Bonding and empathy happen naturally with the horse, staff, and volunteer helpers. Nurturing is mirrored in every session. Our riders with learning and cognitive challenges benefit from the sensory integration provided through the horse.

Our programs incorporate MNRI® into every session. This is based on the individual's assessment and needs at that time. No two sessions are ever the same. We use the horse as our 'co-therapist' and massage table. These horses are specifically selected to meet the needs and goals in each session. For example a 6 foot tall 70 year old man with balance issues will be on a horse with a large barrel and smooth steady stride. An 8 year old child with low muscle tone and attention issues will be on a small horse with a choppy more forward gait. Horse selection is extremely important to the success of each ride. Horses do not recognize human limitations. Physically, during a 10 minute ride, the rhythmic stride of the horse moves the rider's body over 700 times in a motion similar to that of walking. Deep muscles are allowed to contract and relax. This motion cannot be duplicated in a clinical, OT, or PT setting. This makes our program very unique.

Example 1: Taylor was a 22 month old girl with undiagnosed developmental delays who was non verbal, reaching no bench marks, not sitting, and not crawling. Taylor started MNRI® and therapeutic riding twice a week. She worked on Babkin Palmomental Reflex, Crawling pattern, Spinal Gallant, Trunk Extension, reaching while riding backwards in a prone position, and Flying and Landing at dismount. Within 3 weeks of this therapy, the child began crawling. After 6 months she was pulling herself up on furniture. She began babbling and humming many new sounds. She has been experiencing new changes every day. While still riding only 1 time weekly, Taylor is starting to walk using only one hand for support.

Example 2: Sammy was a 4 year old boy with Lyme Disease who presented like a spastic cerebral palsy patient with right side hemiplegia. He started hippotherapy with MNRI® riding twice a week for 1 hour. At the time he utilized two-side walkers and a leader for assistance. I started him with Tactile Integration, Babkin, Fear Paralysis, Foot Tendon Guard, and Neurostructural exercises. The immediate relaxation in his body was amazing.

Currently he is working on STNR doing visual eye tracking exercises in kneeling position on the horse at a walk. His speech is becoming very audible while the actual physical mechanics of speaking have become much easier. He can now hold this pattern for up to 3-5 minutes. We have introduced riding backwards prone, working on Grasp Reflex pattern while having Sammy roll up paper in each hand. This is done while the horse is walking and also standing still. Then Sammy flicks the paper balls off the horse's rump! It helps to have it aimed at someone!

Example 3: Griffin was a 3 year old boy in 2007 with a diagnosis of Selective Mutism, low tone, sensory processing issues, self-stimming behavior 90% of the time, and who used two-side walkers and a leader for assistance. He started working with hippotherapy and MNRI®, riding 1 hour a week. We began with Tactile Integration, Babkin Palmomental, Archetype Movements, Landau, and STNR with eye tracking. I explained how the horse has 2 sides to their body and how the horse moves and moves him rhythmically, while we walked around. Griffin switched positions all over the horse and started to put his hands out to ask for Babkin

Palmomental. We asked him to take the STNR position for playing basketball, throwing at the hoop from all directions, and passing with us.

In 2008 we began Babkin with him riding backwards. Then he drew Lazy 8's on the horse's rump as I told him stories. Speech just 'turned on' and Griffin began adding to the stories. Soon we began trading stories one line at a time. Griffin added more and more stories as he rode around the arena. Currently his speech has become fully functional. Griffin is now riding a very challenging pony independently at the walk and trot, including trail riding. It is clear there has been a 100% change in his auditory processing. He can now hear and locate a sound from any direction and make a kinesthetic movement to direct the pony. Griffin recently made a huge connection with empathy for the horse, learning to squeeze rather than kick to get the horse to walk forward. Now he communicates with the horse through light pressure.

Griffin's vision has also improved greatly with tracking exercises while riding, reaching for balls, following balls, looking for letters in the arena, and coming up with a word that starts with a letter he recognizes. Griffin is still riding 1 hour weekly. He continues to make amazing new discoveries about himself and his horses every time and every ride! A quote from Griffin: "Horses is FUN therapy; you don't just have to sit somewhere. We can't skip horseback riding. It's my body's medicine!" A quote from Griffin's Mom: " MNRI® and hippotherapy have been the primary part of my child's success! We can't thank you enough!"

We have been thrilled with results of the combination of MNRI® and hippotherapy – it has produced proven results. I want to thank my parents and families for their support, trust, and for sharing their beautiful children and family members with this new therapy. I also want to thank Dr. Masgutova for sharing her wealth of knowledge and caring. – With love and light, Stephanie Keene

Amazing Grace's Story: My Path from Curious Skeptic to Committed Believer...

Cindy Sprenger, OTR/L, MNRI® Core Specialist, Sarasota, FL, USA

Cindy Sprenger

It's April 2015. I have taken 18 MNRI® courses and have 5,000 hours of experience using it in my private practice. It has completely revolutionized my career. I believe it is the most profound approach that exists for treating children and adults with many types of challenges. However, I was rather skeptical in the beginning. I was a seasoned occupational therapist in March 2010 with 20 years of experience in the healthcare field. I happened to take my very first MNRI® class from Svetlana Masgutova herself. During class she frequently spoke about the profound improvements in her clients with many of them making life changing gains in just a few sessions. I, being a seasoned professional and knowing the normal path for improvement, felt she might be stretching the truth but wanted to keep an open mind.

Despite my skepticism, I left Dr. Masgutova's class hopeful that the handful of MNRI® exercises I learned would add something positive for my clients. Perhaps it would give me another tool in my already large 'bag of tricks'. I frequently used popular and widely accepted therapeutic modalities such as Sensory Integration (SI) methods, Interactive Metronome®, Therapeutic Listening® program and Neuro-Developmental Theory (NDT) as the basis for my treatments. So, I went home and practiced three or four of the MNRI® exercises on my husband for a few days until I felt proficient with them.

I decided to try MNRI® for the first time on Grace, a six month old baby girl who had already been on my case load for two months. This baby was drug-exposed in utero and was currently in a medical foster home. She had originally presented with relentless crying along with significant motor delays. Over the two months I had been seeing this baby, although I was able to bring her to a happy disposition, her motor skills were that of only a two month old. She was unable to bear weight through her legs when placed in standing position. She was unable to lift her head up to 45 degrees or push up onto her hands while lying on her stomach. She was unable to roll onto either of her sides from her back. She could not yet lift her legs and grab her feet when lying on her back.

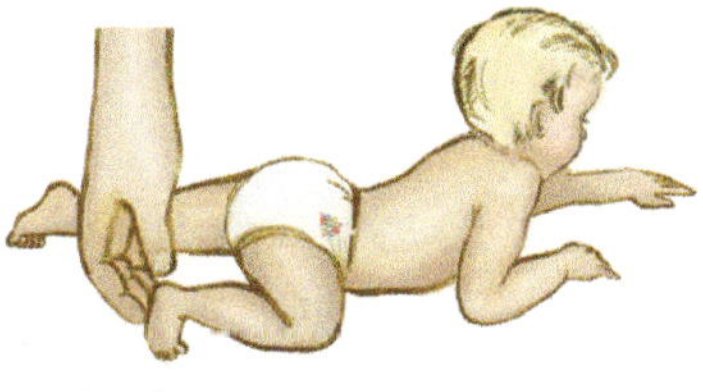

Bauer Crawling

March 22, 2010 was the first day to incorporate any MNRI® into our sessions. I had practiced and felt confident using the Bauer Crawling, Spinal Galant, and Hands Supporting Reflex patterns and exercises. Toward the end of that first MNRI® session this baby made the first lateral trunk flexion (side bending) motion I had ever seen her perform. I sent the family away without making any changes to their Home Program which consisted of traditional techniques. When they returned on March 31, they were so excited to tell me that she had learned to roll onto both of her sides. Her skills for this were perfect! Encouraged by this, I did my initial three exercises again and added the Trunk Extension Reflex pattern and exercise. Following this I stood her up onto her feet and surprisingly she straightened her knees and stood! She was able to stand this way for over two minutes. Again, I sent the family away without changing their Home Program. When she returned on April 7, I was amazed. She could hold your hands and bounce up and down and tilt her pelvis forward and backward. Overall, she had made four months of gain in just two weeks!

Looking back, the astonishingly rapid changes this baby made were due to the ability of the Masgutova Method® to unlock the body's hidden potential. She had the genetic code for all of these movements. Her brain just needed a little reminder about using them.

This was just the beginning of my journey. My appreciation for the Masgutova Method® quickly grew to the point that it has become the predominant technique I use with ALL clients in my private practice. Dr. Masgutova's approach to reflex integration is truly unique. In my opinion, MNRI® represents a paradigm shift in the way reflexes are viewed and treated. My ongoing success stories continue to inspire me about its potential. Over five years later, I remain as committed as ever to learning all that I can about MNRI® from Svetlana. It excites me to think about all the people, both children and adults, who I can reach to truly help improve their lives. The Masgutova Method® is a true Godsend for me and my families!

I feel very blessed to have discovered the Masgutova Neurosensorimotor Reflex Integration Method®. The positive changes I see in the lives of the many people I treat through my Occupational Therapy practice are beyond anything I would have imagined. Thank you baby Grace for opening my eyes! – Cindy Sprenger, OTR/L

Silas Shows a Skill He's Never Shown Before!

Melissa Brown, MNRI® Parent, Buxton, ME, USA

Matthew, Melissa, and Silas Brown

We are just back from an amazing and life changing clinic and are ready to walk this new and exciting road on Silas' Journey! There were so many feelings going into this new experience: curious, excited, nervous, and ready for information overload!

As we had already started down this road earlier in the year doing the Brain Gym® work with Silas' therapist Jean, I never really grasped how life changing it would and could be. It really took this week to 'sell' me on this Masgutova Method® and really seal the deal! And boy, am I glad we went because I am 100% a fan of this work! And I will URGE other special needs families to look into this. Before we went into our first session I didn't know what to expect. I had asked other parents who have their children involved to see what may come out of this clinic and they said it could be physical, behavioral, and emotional but it really depends on the child and how they react to it. I knew going in that sleep was a big struggle for us lately and I was hoping to see change in this area. I was also hoping to see Silas show a skill he has never shown before. As we walked into our first day of clinic, I became so excited to see what this new therapy was all about. We worked with an amazing Core Specialist, Mary, who Silas responded so well to. She has a very calm soul and you can just tell how dedicated and interested she is in this work. I won't go into using all of the terms that they use because they are very confusing and hard to understand, but I will tell you the gist of what happened.

During Silas' evaluation by Dr. Svetlana Masgutova he showed many things and revealed secrets! Dr. Masgutova was able to tell that Silas is a very intelligent child with a huge potential for speech. We told Silas that his secret was out and he better start piping up with some of those words he's hiding! Silas showed potential in many areas and, although a lot of his body is still so affected by his brain injury (especially his right side) and traumatizing birth, he still has the ability to build those skills and reach goals. We will have to work very hard with him but the more we work, the faster we will see the change. Sometimes it is within minutes, sometimes overnight, and sometimes change will take longer. Dr. Svetlana stressed the benefit of tummy time for Silas and how to achieve the best results when doing tummy time. She showed us the big areas to work on with Silas, a lot of which we have already been working on, and some new things. When she finished her evaluation, Mary finished her session and we were on our way. I was overwhelmed but so excited!

We continued therapy on Silas that day totaling about 5-6 hours of therapy! Jean and Stephanie worked very hard with Silas and he barely cried. His body loves this work! Silas was so exhausted from all the therapy he fell asleep around 8 that night (normally he stays up until 10 pm.). He slept so well that night, we even had to wake him in the morning! Already a positive change! The next day we went in for our second session. Mary could already see changes in Silas' body. His muscles and reflexes were responding quicker with less delay and he was more flexible and open to stretching.

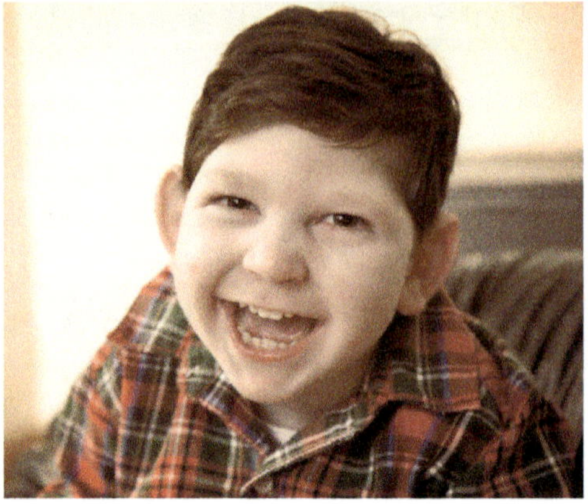

Silas today.

Then Dr. Svetlana came in to work with Silas. So many wonderful things happened during this session including getting Silas into Hands Supporting position where he was bearing weight on his hands and knees! His arms were straight and he was bearing weight mostly on his left arm all by himself! He held it for about ten seconds! This is huge! Silas' therapists have been trying for almost TWO YEARS to get him to do this and Dr. Svetlana had him in this position within minutes! How do you explain this? This work is incredible! And with happy news came homework for us, of course. We were told that Silas' addiction to his iPad was not good for him. That his addiction was to the screen with flashing lights and that wasn't good for his brain. I mean. . .we knew this would be an issue. I had made a monster out of my child with this iPad! We needed to wean him off of this. SO we went back downstairs and continued therapy with him. This time without his iPad. He was to only have it if he worked hard and was cooperating. We were no longer able to use this to break his temper tantrums. He could only have it if he understood that when he stopped crying then he could see it. But even so he could only have a story read to him on it, no more TV shows. Matt and I were very skeptical since we used his iPad all day every day to entertain him. Silas whined for a while and eventually forgot about it.

We took the kids out on a nature walk and Silas, even four hours without his iPad, was so much more alert and open to exploring his world around him. No longer stuck with his face in a screen. Of course, I scolded myself as his parent, thinking I was a bad Mom for letting him get so addicted to this possession. But knowing that I was really trying to break him from this habit made me feel that much better about it. For the rest of the day he had no iPad and was happy doing his therapy with Steph and Jean. He was rewarded once at the end of the day but didn't show interest in it. That night he slept better than ever again. We had to wake him the next morning as well!

He had his final session that day and did more amazing things! Showing better head control and no delay in some of his muscles and reflexes! He had the iPad once the entire session but other than that, he did not use it and did amazing! Matt and I were dreading the ride home seeing as we absolutely depended on his iPad to entertain him during the car rides, but he did amazing! He slept and, when awake, he interacted and cooed at us in pure happiness! His iPad addiction was cured! Thank the Lord!

We left the clinic with so many new feelings. Although we were so incredibly happy and amazed with our experience we were also very confused. We needed to make some very important decisions for Silas now. Will he continue with his OT and PT? Do we still put him in his stander? After parenting Silas one way for 2½ years, we have been shown a new way of healing, one that didn't involve

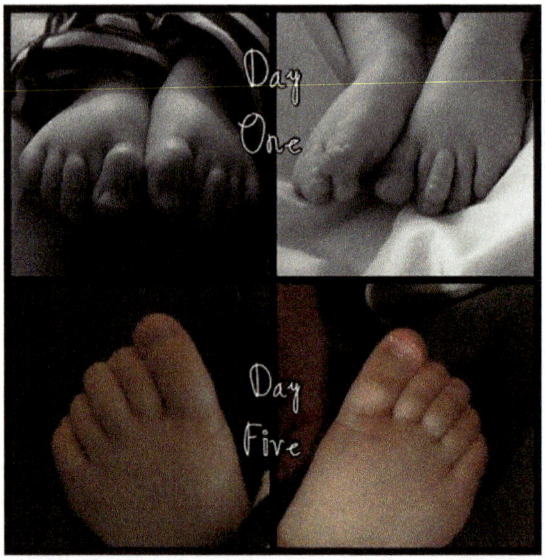

photos by Melissa Brown

Progress for Silas during a Family Conference: truly amazing ... a Core Specialist was working with Silas and focused on 1) Foot Tendon Guard for all toes plus big toe/foot with inhibition, 2) pronation of 1-3 metatarsals, 3) pronation of metatarsus for the Foot Tendon Guard, and 4) Babinski Reflex repatterning. Repetition of all above noted reflexes. . .and it worked!

bracing and equipment to help Silas learn to reach his goals, but instead a very alternative method which actually makes much more sense! So here we are. Completely changing Silas' weekly schedule. Having to let therapists go sadly, and welcome this new beginning into his world. But at the end of the day I know I am doing what is best for him. He has already shown so much growth, his body absolutely loves this therapy and we are so thankful and feel so lucky to have found this for Silas at such a young age where is brain and body is still so open to healing and change.

Silas in late 2014.

I cannot thank everyone enough for helping us to get here. We will continue to fundraise for Silas, and unfortunately, will probably always have to fundraise for him. This therapy is not covered by insurance and is very pricey but we will not let the obstacle of money get in the way of Silas' healing because he deserves this!

I am also happy to say that, as of this morning, Silas is bolus feeding for the first time since he got his feeding tube! We have never been able to successfully bolus feed Silas. This is huge and means he will no longer have to be hooked to his feeding bag 24/7. Silas will also be able to feel the sensation of hunger and food filling his stomach. Big things are happening out here! Thank you so much to the Svetlana Masgutova Educational Institute for changing our lives. Here is the start to an amazing new chapter in Silas' Journey! We are looking to go with Silas in the spring to an 8-day or 6-day intensive Family Conference where we would have 6 hours of therapy every day with Core Specialists and Svetlana. Because, as Dr. Svetlana said within the first ten minutes of meeting Silas, "I want to see this kid in conference!" She also said that the progress Silas showed during clinic is equal to that gained in 7-8 months time frame! SO here's to getting to a Conference this spring!

The positive changes in Silas are growing with the highest speed we can imagine… He is an example of how keys for reflex integration can change so many aspects of our being! There are still many challenges in Silas' life but we know that he will reverse them into successes! Thank you to his Mom and Dad for being amazing personalities and great 'doers.' – MNRI® Team

Reaching for the Greatest Miracle

Katarzyna, Tadeusz, Karolina, and Michal Kubicki Family, Poland

Karolina was born in the most beautiful time in our lives. Wonderful, expected, love was born on August 12, 1995. The miraculous creature got 10 points on the APGAR scale and was 59 cm (23 in.) in length, and weighted 3,300 gms (7.3 lbs.). August 12 seemed to us as the happiest day of our lives and would continue indefinitely. Karolina's development was exemplary, she sat, walked, started her speech on time, and was a curious baby. She was a small girl, with a storm of curls around her face, and large, ever smiling, deep blue eyes. Happiness, joy . . .

Karolina, Katarzyna, Tadeusz, and Michal Kubicki

However, fate made itself known shortly on September 29, 2006 (when Karolina was 11) and beat us in one minute. Karolina was a passenger in a car which got into a terrible accident. After two hours of struggle by the rescue team and firefighters, she was transported by helicopter to the Upper Silesian Centre for Maternal and Child Health in Katowice-Ligota. The first evaluation revealed the appalling news of a "girl with polytrauma who was admitted to the Intensive Care Unit of surgery . . . the child was unconscious, GCS 4 points, incubated, ventilated mechanically, pupils narrow, symmetric, ICP values were > 30 mmHg. She had traumatic brain contusion bleeding around the brain frontal and temporal brain lobes, frontal bone fracture, cerebral coma, right lung contusion, and fracture of the right femur." The doctor explained that my daughter had no chance to survive. If not for machines that support vital functions Karolina would not be alive. Doctors were bringing us new frightening information every day. We were to say goodbye to her any moment because all the time there was a real threat that she would not survive the next hour. So the first two weeks passed. Finally, the situation stabilized somewhat and there was no immediate threat to her life. However, further support from breathing machines was required. The prognosis was dire. Doctors told us often, that after such severe brain injury, hypoxia occurred and was irreversible, she would not talk and be paralyzed. Nobody gave us hope ...

We did not allow ourselves to believe this terrible scenario. After a month, they released Karolina from the ICU, unfortunately, in a coma. Coming out of the hospital, I can still hear the doctor's 'consolation' that we have to look for a hospice situation because she would not survive. Sometimes friends ask me what a mother feels at that moment . . . ANGER, ANGER, ANGER! ! It woke me up to a huge challenge, to gather my strength and determination because NO ONE has the right to talk about my child like that when they have no idea of the future. My husband swore to prove to all that our CHILD would return to the living.

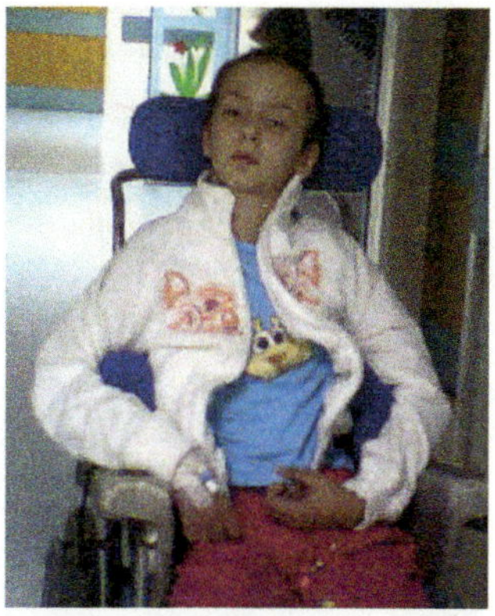

Karolina – paralyzed in the past, and with no chance for future, according to doctors.

In the following months, nothing changed in Karolina's neurological state and she continued being in a coma. I, her mother, was with her all the time and I read as much information on comas, brain injuries, and rehabilitation as I could. One day I came across the website of Dr. Carla Hannaford who wrote, *Smart Moves: Why Learning Is Not All In Your Head.* One of the chapters that I read with bated breath, was about Svetlana Masgutova, a doctor of psychology, who saved a lot of kids in the train accident in Russia in 1989 and her new method of treatment of brain injuries. I was enthralled and became convinced that I had to bring my daughter to this specialist. But how do I do that? I'm in Poland and she was in Russia. One thing I knew for sure, I would find her and SHE will help my child. From this belief, I developed an outright certainty but do not know how I did it. Maybe it's intuition, maybe it's God who directs our fate and goes behind us for obvious help ... I do not know how this happens, even today.

On August 12, 2006 the GREATEST MIRACLE happened in our lives. On her 11th birthday, after 2 months in coma, Karolina AWOKE. In spite of her diagnoses, she said, "Mom, do not cry, today is my birthday." I do not write here about all our feelings, thoughts, considerations as they are only and exclusively reserved for our family. For us they are sacred and it is not that I want to get the sympathy of a reader. But this day to us became the day of her rebirth (because of that day we celebrate her birthday twice: this year on August 12 Karolina turned 18 and 7 years old) it was our first but not the last miracle. Although we thought that from that day everything would only get better and happier, unfortunately, it was only the beginning. Admittedly, Karolina regained consciousness but was paralyzed, could not control her physiological functions, and was fed with a tube. And on top of this, everyone told us that such damage would not change, that there was no cure, and there were no such methods to help Karolina. Then, began our struggle for her health and for Karolina to return to everyday life.

In early 2007, I learned that Dr. Masgutova's Institute was in Warsaw. We took our paralyzed Karolina on a 700 km trip to visit Dr. Masgutova. It was an ordeal (our child was crying in pain for the whole 8 hour journey) but we finally arrived at the Hotel Panorama. Finally, in front of us was a woman who was the first specialist who did not want to look at Karolina's medical records but wanted to examine the child and the function of her sensory-motor sphere first. The reflex Assessment test took approximately 1½ to 2 hours. We were looking forward with great anticipation and nervousness, although she impressed us with her sense of peace and confidence. At the end of the Assessment .... we stood and waited for her 'judgment'. Dr. Masgutova smiled

Karolina in a play at the Masgutova Clinic.

On Aug. 12, 2013, her eighteenth birthday, Karolina presented a dress from the LaMani collection as a model in the Sheraton Hotel in Krakow City, Poland

Karolina (in white) with her mother and brother. She traveled (as other healthy participants of the tour) to Israel, to Jerusalem and Bethlehem, walked on the way that Jesus Christ went to Golgotha.

Karolina (above) won the Three Crowns - the highest level of Pienin's Crown. She also traveled (with other healthy participants of the tour) to Israel, Jerusalem, and Bethlehem, walking on the same paths that Jesus Christ traveled.

softly, her dark eyes stared deeply into ours and I heard the words which to this day and every day I hear in my head, "If you undertake certain therapies, she will recover. She has sensory-motor developmental resources, and her reasoning and thinking abilities are very logical." God, once again thank you for the wonders that we have been given. God saved my child, who woke up in August, and now He helped me find the person that can cure my child with her vast knowledge, wisdom, life, kindness, and commitment. I have no words which could describe the joy in finding the knowledge of Dr. Masgutova, her son Denis, and other therapists.

Without hesitation Dr. Masgutova and her team started helping us in April, 2007 at a subsidiary of the Masgutova Institute in Wroclaw. We traveled 200 km every other day from Bogatyni to Wroclaw. Karolina started controlling her physiological functions and feeling more sensations in her body. Next, she started attending the rehabilitation Camps in Szklarska Poreba, Tleń, Ustrori Marine, Mielno. At these camps our child slowly but steadily and permanently improved her sensory-motor responses and reflex patterns; she started her motor exercises. Like a young child, first Karolina re-learned to creep, crawl, and walk and even now she continues these exercises, though she does all this very well. Next, in conjunction with MNRI® we also undertook other rehabilitation methods: functional rehabilitation, Sensory Integration, Biofeedback, hippotherapy, Pet Therapy, and psychological and neurological consultations which led us to this day.

Karolina achieved her goal – she walks, speaks very well, and logically. She is a student at the Technical College today and works on the aesthetics of her gait being a fashion model! Karolina cannot jump yet and we do not know when she will, but today I know one thing, that God sends a savior. In our case, it was the whole multitude team of good specialists, beginning with Dr. Svetlana Masgutova – which our family calls a savior because of her knowledge, intuition, sensitivity, and talent to devote to all the needy. She sees not only the human being as an individual but looks into their eyes and touches their hands, she sees their soul, their worries, fears, and helps them skillfully.

Doctor Masgutova, thank you - it is a simple, common word but our whole family gives our immense gratitude for your work and dedication. Our child is a living example that the MNRI® therapy is effective, sufficient, and unique. This is the method that presents a holistic approach, which is not limited to the compensation for the individual's 'elements' but also treats a person as an integral whole. It helped our daughter to not only regain her movements, but also to think, feel her emotions, and to live her life using all of her senses.

P.S. Here are Karolina's pictures for everyone to see. Our Karolina paralyzed in the past, and with no chance for future, according to doctors… And today she has climbed one of the mountains in the summit of Three Crowns – Pieniny (Poland-Slovakia), the highest peak; also toured Israel, retracing the path of Jesus as he was crucified; came to Golgotha with us, including touring Jerusalem and Bethlehem as other healthy participants in the tour. Then on July 12, 2013, her eighteenth birthday, she presented a dress from LaMani collection as a model in the Sheraton Hotel in Krakow, Poland.

Congratulations to Karolina, a girl who is a True Winner. She overcame her fears, crying, tears – and then her disability. I thank her family for all the support and effort that they put into her recovery to keep her motivated and enthusiastic. I congratulate this family from deep in my heart! – Svetlana Masgutova

I Took a Blind Leap and It Worked

Barb Gregory, MNRI® Parent, Sherwood Park, Alberta, Canada

I am a mother to 4 beautiful girls. In March of 2011, my youngest daughter Teagan, who was 4 years old at the time, suffered severe brain damage from a prolonged seizure. The seizure was subclinical after 2.5 hours, meaning her body had stopped convulsing but her brain continued to seize. The doctors did not realize this until the next morning because of the enormous amount of drugs she was given to stop the seizure. She was unconscious. The seizure lasted over 24 hours. The damage left her blind, unable to communicate, and Teagan had lost complete control of her entire body although she was still aware of what was going on around her.

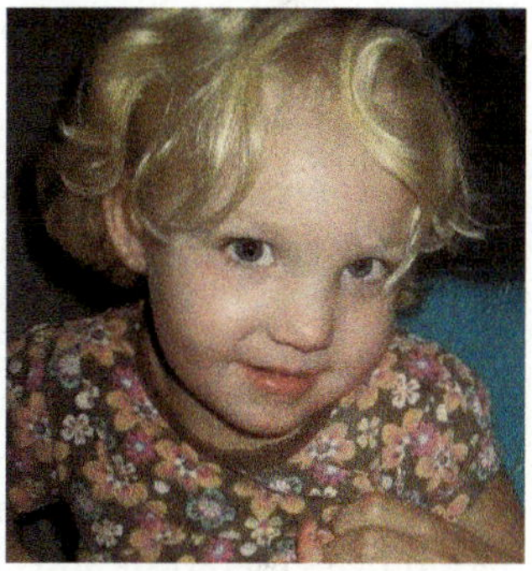

Teagan in late 2010, before her seizure.

The doctors and therapists at the hospital did not give me much hope. They offered very little to help Teagan other than heavy amounts of seizure drugs and basic living aids. Yes, she was provided an occupational therapist, a speech therapist, and a physical therapist but as I watched and participated in Teagan's therapy five days a week, I could tell even they did not hold much hope for her. Her therapy sessions consisted of playing with switch toys for hours. Teagan often fell asleep during her therapy sessions (I almost did, too).

Teagan in June of 2011.

I was desperate to find someone to help my daughter. I was not even close to giving up on her and I knew there had to be someone or something that could help her regain her normal body functions. I searched and searched daily, tried several unconventional therapies for which I was chastised and cautioned against by Teagan's doctors. Teagan did start to show signs of improvements because of these therapies. Her eyesight was the first to come back through light therapy. She was now at least seeing people and toys in front of her. Several rides in the Gyro Stem helped her to slowly regain the use of her right arm and hand. Her muscles started to strengthen but she was barely able to sit up for more than an hour at a time.

After almost a year of intense therapy, Teagan's homeschool teacher told me about MNRI® therapy and how she was able to rehabilitate her father who had suffered from a stroke. I was a little skeptical because everyone I spoke to had an opinion about what I should try. I did not look into MNRI® therapy immediately. I waited and tried to find information about it online. It was difficult to understand the material online, but out of desperation and the fact that two other people I know recommended I try MNRI®, I took a blind leap.

I registered Teagan for an eight day Family Conference in San Diego, California. I honestly had absolutely no idea what to expect. My family and friends feverishly helped us fundraise to get there even though they, too, were skeptical about this therapy.

The first day I arrived in San Diego I met several families who had been participating in MNRI® with Dr. Masgutova and I could not believe the stories I was hearing! A man with a brain injury was recovering from his injuries after being bedridden for almost 40 years, a child with cerebral palsy was running around, another boy came in with a walker and left without it. For the first time since Teagan's injury, I had hope. Teagan underwent 6 hours of therapy each day and I waited. I was given eight days of hands-on training myself. Dr. Masgutova's therapists were wonderful. I was extremely impressed with their knowledge, far out weighing that of the therapists at home. They were working on Teagan like it meant something. They knew this would work even if I didn't, yet. She was being pushed, twisted, lifted on her knees and encouraged to participate. I was impressed because this was what I was looking for. Let's get her up and show her what she can do! I didn't see one switch toy! There was even a respiratory therapist there who amazed me! She straightened out Teagan's twisted rib cage and had her breathing correctly in a couple hours. Her color returned to normal! She no longer had purple fingers and toes. This first conference was very overwhelming; I had to absorb a lot of information and was exhausted from the trip.

I returned home with a wealth of knowledge still not really knowing what to expect but I was determined. I did my best to stick with Dr. Masgutova's Home Program schedule. As people visited they were impressed, too, saying things like, "She is so much more aware!" Teagan was now starting to make eye contact with people, something she did not like to do at all before. Her posture was improving, people started to notice how much straighter she was sitting in her wheelchair. Months went on and I continued to do MNRI® that I had been taught. Now the home therapists were noticing, too. One said to me, "You are an inspiration Barb, I don't know what you are doing but keep it up!" I knew it was the MNRI®.

We were fortunate enough to return to see Dr. Masgutova in California, six months later for another eight day Family Conference and this time Dr. Masgutova was impressed! She did Teagan's Assessment on day two and she said Teagan's brain was learning at a very fast pace. That was music to my ears. One of the fathers who was returning for another conference for his son said, "She looks like a completely different child!" Teagan underwent another eight days of therapy with Dr. Masgutova's therapists and I just can't say enough.

Teagan in 2014.

This time I started seeing improvements immediately. Teagan was 100% tube fed when we arrived. Yes, her oral feeding had slowly been improving but not enough to reduce her tube feeds. She was eating chips, french fries, and little bits of things like chicken and pizza, but never enough to sustain her caloric intake. A couple days into the therapy we went out for dinner and Teagan ate two whole chicken strips and wanted more. I was surprised. Next time out I ordered her pasta and she ate a bunch of that, too. This was a huge improvement in the volume she was eating. When we got home I met with Teagan's Home Nutrition team and Teagan ate a whole bowl of pasta with steak in front of them. Yes, I think it was safe to say they were surprised, too. Teagan is no longer tube fed but instead is eating breakfast, lunch, and supper by mouth. We saw other improvements, too. Her eye tracking improved quite a bit and so did the strength in her back muscles which helped stabilize her core. She is able to reach and twist about while sitting, grabbing, and reaching now, something she was very hesitant to do. Her balance has improved quite a bit too; she is much more stable on her feet.

I am working hard at fundraising so I can bring Teagan back to Dr. Masgutova and her team. I am convinced with time and work my little girl will walk and talk again, thanks to God and Team Masgutova!

We congratulate and thank Teagan for her success and motivation. We congratulate the whole family for their care, support, and great enthusiasm, and trust in Teagan's development. We wish you good health and many achievements. – The USA & Canadian MNRI® Specialist Team

Hard Work Pays Off

Dave & JoAnn McKenzie, MNRI® Parents, St. Paul, MN, USA

My daughter Diana was born normal in 1983. At the age of 10 months she had a status seizure (which is a prolonged seizure) and from that she was induced into a coma. She was in a coma for 3 days. When she awoke from the coma she had brain damage throughout her whole brain and severe liver damage.

Dave & JoAnn McKenzie

This left her completely floppy with no muscle tone, and no reflexes. She was diagnosed with viral encephalitis, yet the cause was still unknown. Since she was under the age of 2 years her final diagnosis is athetoid cerebral palsy. The cause is still unknown. Diana is quadriplegic and fully dependent on others for all of her care and needs. She has lived at home her entire life and I, her mother, am her primary care giver.

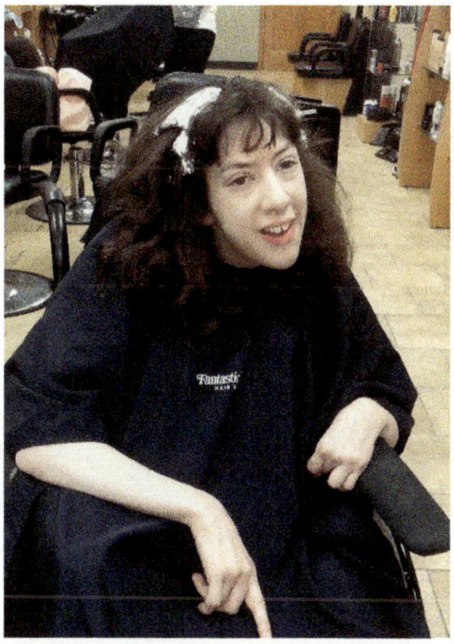

Diana McKenzie out at the hair salon.

I have been providing many types of therapy in our home for Diana for years. Shortly after the time of Diana's diagnosis we began working with Gillette Children's Hospital with outpatient therapy. This meant I had to bring Diana to Gillette several times a week. Bringing Diana back and forth just about every day became too much. It was at this time outpatient therapy was stopped and we started working with A Chance to Grow. While working with ACTG we had speech and occupational therapists who came to our home for several years, eventually the state made cuts and we had to start to bring Diana to ACTG for speech and OT. We also began a neuro-physiological therapy program as a family which was run by myself and volunteers in our home six days a week for six hours a day. Diana did this program for 18 years which helped to keep her body well maintained.

In 2005 at the age of 22 Diana was first evaluated by Dr. Masgutova. Dr. Masgutova set Diana up with a program, that we did at home daily for a year. Diana attended Dr. Masgutova's Children's and Adult Conference in 2006 in Florida. We came home and did 5 programs and worked daily on reflexes that we had learned at the Conference. Since 2006 Diana has made many great strides. Her symmetry has greatly improved. The curvature of her spine is an S-curve (it was a C-curve which is worse) and is decreasing, not increasing! This is unheard of for her age. In 2008 the degree of the curvature lying down was 98 and in 2009 Diana's spine decreased to 90 degrees. This is greatly due to Dr. Masgutova diagnosing that Diana's spine is too big for her core body and that Spinal Galant and Trunk Extension were dysfunctional reflexes. Dr. Masgutova's manipulations and suggesting working daily with these reflexes has greatly helped Diana. If we had not got this information from Dr. Masgutova, she would have been destined for rod surgery. Diana's muscle tone has also improved significantly. We still follow Dr. Masgutova's program today, 6 days a week, and see Dr. Masgutova a couple times a year. At the present time Diana's symmetry is still improving, the scoliosis is still decreasing, and her spine is flexible. Her spasticity is decreasing immensely, the reflexes are integrating. Her muscles, tendons, and ligaments are growing. Her seizures are decreasing and fine motor skills are improving and she is vocalizing a lot more. Diana has started speech development and reading and now her cognition is increasing day by day. We see more and more improvements over the years as we continue to do this program.

Diana during an outing at the Minnesota Arboretum with her mother, JoAnn, and Kaity McKenna, Diana's Personal Care Attendant who does MNRI® along with JoAnn.

Diana is on no medications and has never had any surgeries. MNRI® pays off in the long run versus taking the easy way out (surgeries). Diana's body is currently going through some EXTREME changes (a lot that can be visually seen in the past couple months) since we have been doing Dr. Masgutova's program 6 days a week, almost every week. We are very excited to see such very positive changes coming so quickly now. It seems as though we have been preparing the body for so many years and now it is getting ready to snap back into place. Diana has a left dislocated hip but she is gaining muscle mass around the joint which is then holding the hip in the joint a lot better, and her spine curvature is still decreasing. We continue to gain more and more symmetry throughout the whole body. We are diligently working with her Trunk Extension Reflexes that are lengthening all the ligaments in the legs.

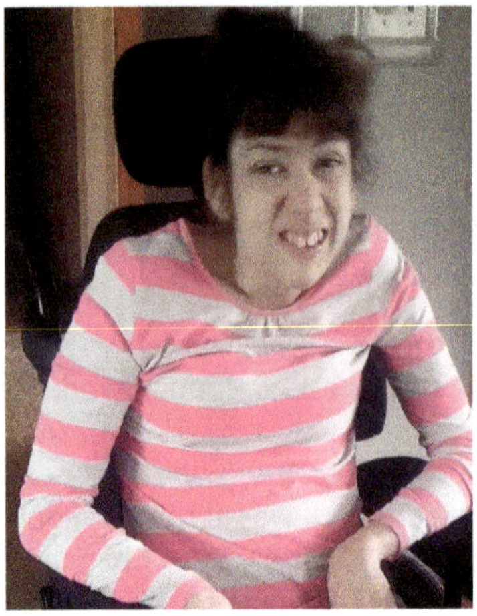
Diana McKenzie today.

One thing that has changed a ton in the past few months is her left side, along with the hip (she was slouching down to the left). We have been doing Breathing Reflex techniques and strengthening her rib cage and other areas so much that she is not slouching over as much anymore. Her bottom left ribs are starting to lay down nicely and aren't as protruded as they were. Everything is really just starting to come together very nicely!

Diana, we congratulate you for your new achievements as one of our first Conference participants. We are so proud of your success! Dear JoAnn, we thank you as a true MNRI® Mom. Your testament that you LOVE to do MNRI® exercises with your daughter is the key for your success! Thank you for this attitude. And thank you to Diana's family for your trust and good work! – Dr. Masgutova & MNRI® Team

Nikola and her Successes

Aneta Michalska, MNRI® Parent, Poland

Introduction

In this story I want to tell about my beloved daughter Nikola, who has cerebral palsy, and her step-by-step success in neurodevelopment with MNRI®.

Nikola was born at 41 weeks (2 days late) with an APGAR score of 9 (1 point less for her skin color). She was discharged from the hospital as a healthy girl.

When she was 4 months old, I noticed that her right-hand was often clamped in a fist. When asleep her hand was relaxed but not completely straight. Kicking with her right foot was also noticeably weaker than with her left.

The doctor (during a vaccination) also pointed out that she tended to bend her body in the shape of a letter S, and I got a referral to a neurologist.

I went to several neurologists but everyone said something different:

"It is normal that she has her hand clamped in a fist."

"The majority of children do the same thing."

"She will grow out of it."

"In the womb she was on her right side, and hence this side is weaker."

None of the doctors thought that this was a serious matter. I, as a mother, believed them, or rather, I just wanted to believe that my child was all right, even though some inner voice told me that it was not so. I also saw that she grabbed her toys with one hand instead of two. She was not using her right hand at all.

But what is this disease? Where did it come from? Why did it happen to Nikola? What should I do? How can we treat her? There were a lot of questions in my head, and I did not know the answers to any of them. However, when my daughter was eight months of age, we began therapy for Nicola with the Bobath method.

Nikola playing.

Diagnosis and Therapies

At the age of nine months, Nikola had an MRI scan, and the diagnosis was cerebral palsy – right side hemiparesis. For me, it sounded almost like a death sentence. My heart stopped beating.

The neurologist told me, "now only therapy, therapy, therapy" was the step to take, and that in a few years the paralysis should not be visible. He also told me that the baby's brain has great plasticity because she is still very young. I knew I had to struggle for the health of my child but internally I was devastated.

Nikola was still receiving Bobath therapy and, when she turned 11 months old, we started using Dr. S. Masgutova's therapy method. Later we used only Masgutova therapy after seeing its effective results for Nikola. At that point she could turn from back to belly, sit herself up by leaning on the healthy left hand, and remain sitting. However, she seemed to forget that she had a right hand.

Results

After three months of rehabilitation using Dr. S. Masgutova's therapy and six months of the Bobath method, Nikola could grab her right hand with her healthy left hand and play with it. She then developed feeling in her right hand and started to use it. She would reach out both hands to be picked up and would support herself easily on both elbows.

Nikola with her sister, above.

After four months, she was able to stretch her hands to the front, pull up her knees slightly, and bring her bottom up. Her right hand continues to get better. With Dr. Masgutova's therapy, her hand is released from the clamped fist, and above all, the thumb is on the outside. She uses her more healthy hand when she eats with a fork and wants to eat by herself. She can take a cup by its handle and drink juice.

In June 2012, she took her pacifier with three fingers of her right hand and raised it to her mouth (although it dropped)! She used to sleep only on her back or side. Since April 2012, she sleeps almost the entire night on her stomach and turns her head from side to side. Through July and August, I was giving Nikola MNRI® therapy treatments myself, and in July 2012, she walked two steps. Since August she has been using her right hand to prop herself up more often.

Nikola enjoying splashing.

Since September 2012 she has been receiving rehabilitation using only Dr. S. Masgutova's method twice a week for two hours (with MNRI® Core Specialist Malgorzata Koczyk from Poland).

After ten months of rehabilitation she began to get up on her feet on her own, and now she can get down from the couch on her stomach. After 11 months of rehabilitation she began to crawl on all fours; the first moves were clumsy and on a clenched fist. Now it's better and you can say that she uses her open right hand to prop herself.

Most importantly, she almost never gets sick, and when she does, it is a slight cold, and she gets over it quickly. It is obvious that MNRI® also helps Nikola's immune system. Each of Nikola's new steps is a happy event for us and we are full of enthusiasm for her future. We plan to continue Nikola's MNRI® treatments and will do everything possible for her recovery and to maximize her neurodevelopment.

We express joy and congratulations to little Nikola for her excellent results! Let positive changes come daily in her life by supporting her learning and strengths! Thank you to her Mom – an amazing and caring Mom! – MNRI® Specialist Team from Poland

Finding your Inner Metamorphosis

Tania Lemus, High School Student, El Salvador, Central America

This is the story of a caterpillar called Tania Lemus; 16 years old, from El Salvador, Central America.

I was born at 28 weeks of gestation; thus my condition of cerebral palsy (CP) and Spastic Diplegia made me have a different life style from the rest of my family.

I spent my first two months in an incubator. Fortunately, the consequences of being so premature merely affected my motor skills, impeding me to walk independently.

Tania Lemus at work with her art.

I started physical therapy (Neuro-developmental treatment) at 4 months old on a daily basis. It was a slow process, but bit by bit, I learned how to lift my head, sit up, crawl, and stand up. My greatest obstacle has always been my gait. I use a wheelchair for long distances and a posterior walker for short ones. My family has been very supportive in providing me with different approaches and taking me abroad to find better ways to improve my condition. When I was 8 years old I underwent surgery (Selective Dorsal Rhizotomy), which helped me release much of my spasticity.

My school life, however, has been pretty much normal. I attend a regular bilingual school in San Salvador, in which I learned to speak English besides my native language, Spanish. Currently I am in the middle of my Junior year in high school (11th grade).

Despite all the efforts made by my therapists, parents and myself, there was a moment where my improvements were not significant anymore. Therefore, my parents started to consider about another surgery as a possibility, but I always refused to the idea. It was then that we found MNRI®.

The first conference I attended was in June 2012, in West Palm Beach, Florida. Everything in this treatment was new for me, but I had the privilege of being accompanied by my mom and therapist. We were impressed to see such amazing changes in my body when practicing the exercises learned in this program. Since then, I have continued doing them at therapy and at home and I can proudly say it helped me get rid of the wheelchair in school and instead, use my walker all the time, even for very long distances.

MNRI® has not only helped me improve my gait, but also has made me more aware of my body. Now, my handwriting is smoother and I do it faster. I also have more symmetry in my face and I have improved my art skills, which has been my passion since a little girl.

I am currently taking the AP Art course in school, which demands a lot of hours of dedication and has time limits. This makes it even harder for me due to the fact that I take longer in completing tasks than everyone else, besides all the rest of homework I have to comply with from other courses. However, I made the decision to take AP Art because, since I receive MNRI®, I have the ability to move and work quicker.

In my AP Art class, I have had the opportunity to work on many different pieces using all kinds of media, such as acrylic paint, color pencils, pen, permanent markers, etc. One of the main themes of my art pieces is comparing the metamorphosis of butterflies to human growth and development, as an inspiration of finding the real secret of the metamorphosis for every single human being. In fact, since I was little I always thought to myself that I was born as a caterpillar which eventually would become a butterfly. Now, I can proudly state that I am on the verge of becoming one! ☺

Examples of Tania's artwork.

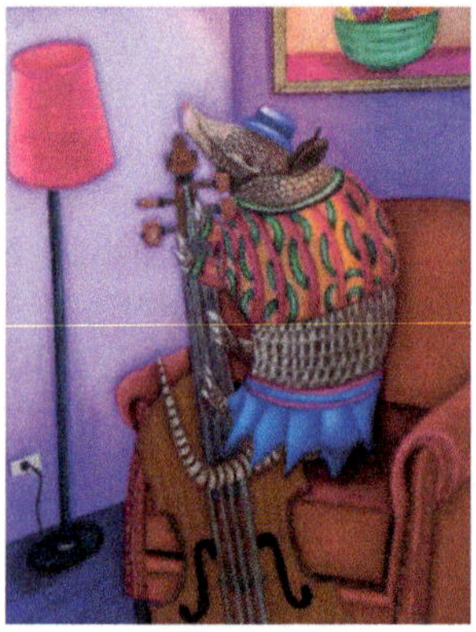

We congratulate Tania for her great results in gross and fine motor coordination and freeing herself for creativity and an amazing journey into art and life! We wish you much success, Tania! And thank you to her dear parents who have supported Tania so well for her future growth and happiness! – MNRI® Team

Running Forward!

Isabelle Renard-Fontaine, Physical Therapist, MNRI® Core Specialist-in-Training, Greenacres, FL, USA

Isabelle Renard-Fontaine

Forty-two kilometers (or 26.2 miles) . . . Wow, I admire these people who somehow manage to run that long! That is the kind of idea that really excites me, and I have always dreamed about it, "Maybe one day, I will become a marathoner, too." At least, that is what I used to think. . .

My name is Isabelle, I am a 43 year old full time working mother of three beautiful children (15, 13 and 9 years old). I have been practicing as a pediatric physical therapist for the past 20 years.

Like many of us, and probably more than ever now, I have been trying as best as I can to find the optimal balance between my family, my professional life, social activities, and opportunities for personal interests or accomplishments. Does that sound familiar to you?

Anyway, one of the key elements working for me in order to keep a modestly successful, and functional 'mental-physical-emotional' equilibrium, has been sports. As we already know, the multiple benefits of physical activity for the well-being of the human race have been proven before. My point here is just to state that I belong to this category of people who became addicted to it; I started competitive triathlon training 9 years ago and without running, swimming, or biking, my body and mind can go downhill quickly. Whether you train and compete at an amateur or professional level, the greatest challenge for any athlete is to stay injury-free. Well, I failed.

About 4 years ago, during a triathlon race I injured my right hip. Wishfully thinking, I tried to convince myself that it was nothing serious. Driven by this well-known addictive physiological and mental need to work out, I very wrongly continued to train for weeks after the race. "It's ok, I have a high pain tolerance," I kept believing, "I'll rest for 2 or 3 weeks and I'll be fine." I trained on and off for months until, out of huge frustrations, I had to accept the seriousness of the problem and decided to take action.

This was the beginning of a series of medical investigations including back/pelvis x-rays, arthrograms, osseous scintigraphy, CT-scans, and MRI. I consulted with a couple of orthopedic specialists (for a first and second opinion), a chiropractor (specializing in sports injuries and who works with the national US Triathlon Team) and finally with a physician specialized in sports medicine. None of the tests mentioned above revealed any clear structural abnormalities, nor soft tissue damages. Stress fracture, bony lesion, labral tear, discs compression, lumbar discal-radicular conflict, spinal stenosis, spondylolysis were all ruled out. However, a few diagnoses were given to me including hip impinchment syndrome, pyriformis syndrome, sciatic chronic pain, degenerative arthritis of the hip joint, and chronic tendinitis of the hip flexors.

As far as treatment, I started with the chiropractor for a few sessions of manual therapy (spinal adjustments) and the use of the Graston method (technique that aims at breaking scar tissues from inflammation located in muscles or around tendons) in addition to a Home Exercise Program (progressive strengthening). This was also combined with the traditional ultra sound and electrical stimulation modalities to reduce pain and inflammation. I felt better but only temporarily. I then tried the Deep Tendon Release (DTR) method for a few visits which I loved very much. However, each time I attempted to resume running, the hip pain (and sometimes lower back pain, too) would reappear immediately. At one point (18 months after the onset of the problem), I agreed to try a cortisone injection into the hip capsule, at least to find out if this was the accurate location of the pain but it did not provide any pain relief. Intermittent complete rest periods were somewhat successful but as soon as I would resume running (or swimming or biking), the pain returned.

As a physical therapist myself, I knew by then that my issues had to be related to misalignment, inadequate posture and/or a possible lack of freedom of movements in my pelvis and lower back. I really had to figure out what was the underlying cause. I went to see a friend of mine, an excellent physical therapist who is also an anatomy professor at St. Augustine University. After assessing my posture and range of motion, one of his conclusions was that I had abnormally limited extension in my lumbar spine (almost no anterior pelvic tilt). He treated me for 8 visits using pretty much a traditional physical therapy approach. I think we got closer to the issue but something was still missing...

It had been 2 years and 4 months since the injury. I was so far away from that time when I used to picture my self running a marathon. Nevertheless, at 41 years old I could not accept that I was going to be forced to quit running, biking, or swimming for good... no way!

Meanwhile, on the professional side, my colleagues (2 pediatric OT's) and I were talking about continuing education. As therapists, we always hope to find 'the' super exciting class from which we come back with novelties, new strategies and 'tricks' allowing us to better treat our young patients.

Well, after hearing about it from a community support group for children with special needs (Bridge to Healing in West Palm Beach, Florida), I attended in June 2011 a course named: *Postural & Dynamic Neurosensorimotor Reflex Integration*. This was my very first introduction to Dr. Masgutova's method!

At the end of this 4 day class, my colleagues and I went through a clear emotional burst. I still clearly remember the excitement of becoming aware of the concept of Reflex Integration but yet this was an incredibly overwhelming feeling. For sure, my expectation of learning new 'tricks' was way more than exceeded. I had no idea at that time, but this course was going to change my life (and I mean all aspects of it) in the most positive way I could have ever hoped for!

After allowing myself several months to digest this newly acquired knowledge, I took a second class, followed by a third, a fourth, and quite a few others (the more you learn about the Masgutova techniques, the more you find out how much more you need to learn). Yes, it is completely addictive, you can't just get enough of it! And I will not even start describing the impact on your limbic system (emotional regulation center) when you start getting involved in the Family Conferences. In summary, after 20 years of being a 'traditional' pediatric physical therapist, a whole new world opened to me and I saw an amazing chance to become a more efficient therapist providing greater positive changes in my patients' lives.

Isabelle, supported by her 3 children, reunited with running.

My mind was so set on a professional focus (it was about me being a therapist) that it took me a little longer to finally wonder if that Reflex Integration approach could also be effective in the treatment of sport injuries? Could this all possibly be good for myself (as the patient)? "What reflexes would be most likely involved in hip/lower back issues?" I asked Dr. Masgutova at the end of a class in September 2012. Led by her unlimited innate desire to help people, before I realized what was happening to me, she had me lying down on the massage table. With Dr. Masgutova on one side and Pamela Curlee on the other, I could feel their four 'golden hands' sensing and addressing my body's struggles (for one hour). I felt like I had won the lottery ticket that night! I also felt changes in my body and most importantly, I received some answers to my questions; I needed to work, at least, on Core Tendon Guard, Spinal Perez, Trunk Extension, and Foot Tendon Guard reflexes.

My next obstacle then was how was I going to do this? These techniques were still new to me and I did not have access to a MNRI® specialist or anyone trained to work on me.

It was not until March 2013, when I attended the MNRI® Archetype Movement Integration course that I realized how much I could actually accomplish independently. I learned that Archetype movements are the basis for inner control and self-awareness, that they are the foundation of primary movements and the blueprint of all movements (and can be done by one's self). I learned that goal oriented, effective, strong, and harmonious movements could only be possible with a proper balance between flexion and extension, stretching and contracting, pushing and pulling. I learned all the techniques, and how to work on the 8 Archetype Motor Patterns (actively) and which pattern serves as the basis for the development of what specific reflexes.

Isabelle, back in the game.

From that day, I rigorously performed my Archetype Movements in addition to a couple other reflex repatterning (i.e. Foot Tendon Guard, Bauer Crawl, Segmental Rolling) on my own and every day. I was able to feel 'pleasant' changes quickly and gradually I resumed swimming, biking, and running.

"Truly, did I just run 3 miles without pain?" I would ask myself. "I should try 4 miles next time." By June 2013, I was enjoying more frequent and longer runs, up to 50 minutes! My body was most likely slowly and properly integrating some important reflex patterns allowing more effective and logical movements during my runs.

On December 7, 2013 one of my long-time dreams became true. After battling my hip injury and being forced to discontinue running for over 2 and a half years, I started it all over again. Without the Masgutova Neurosensorimotor Reflex Integration® method (as well as the strong support of my husband and children), I would not have been able to claim this. I reached a very special finish line 42 kilometers away from the starting point. It took me 4 hours and 30 minutes of enjoyment mixed with deep digging before I realized with the greatest emotion that I had just ran my very first marathon!

Thank you Dr. Masgutova for your profound understanding of brain development and function!

My next dream is to become an MNRI® Core Specialist!

To be continued...

Congratulations on using our amazing program to heal yourself! Thanks for your inspiration! Good health and training to you. – MNRI® Team

MNRI® in Professional Sports

Rok Flander, 2007 Snowboard PGS World Champion, 13 Snowboard World Cup podiums, 6th place Sochi Olympic Winter games

Rok Flander

My experience of the Masgutova Method® felt like an awakening from a deep sleep. I went to work with Dr. Masgutova and her team for two days. I have to say I didn't expect to receive such results! The ability of the MNRI® method to quickly transform long-term physical problems is very extraordinary, not to mention the relief from all the traumas that I was carrying with me on my emotional level. This work brought new life to both.

The main reason that I decided to go and try out the Masgutova program was the fact that I had known how good certain things can feel in life and, for some reason, I was not able to feel that way any more. For quite some time I was asking myself, does it come with the age to begin to feel numb and to lose connection with the fullness of life?

Now I know it has nothing to do with the age. Throughout my sports career and everyday life challenges I have gone through several physical and emotional traumas.

The experience of all these negative emotions and pain have created a confusion in my brain. Suddenly situations that before had always brought me lots of pleasure and joy, now were connected with physical and emotional pain that were caused by injuries. It seemed that this confusion brought me to the point where there was no more joy in my professional sport career as well as in my everyday life!

Rok Flander competing at the Sochi Olympics.

The Masgutova Method® somehow helped my brain and my nervous system to get back on its natural track. Suddenly my body feels 10 kg lighter and the light around me seems to be 20% brighter than before. I have realized there are hundreds of other things that have improved since working with Dr. Masgutova.

The MNRI® program definitely helped my body and soul to rediscover its natural track of growth and development. Walking down the street as well as snowboarding on the Olympic games finally feels full of joy and happiness again! It is important to cherish the gift of life that the universe has given us.

Meeting Dr. Svetlana Masgutova and her team was an amazing experience.

Rok, thank you for your interest in the MNRI® Method and your trust in its techniques. Your inner peace after sessions was radiating and motivated us, also! Our MNRI® team wishes you great health and many new successes.

– S. Masgutova, Ph.D. and MNRI® Team

Creating Great Talent: Maciej Muzaj

Halina Muzaj, Physical Therapist, MNRI® Core Specialist, Mother, Poland

Halina Muzaj

Maciej Muzaj is a 19-year-old young adult who is 6′ 8″ in height. He is a member of the Polish National Volleyball Team and a candidate for the Olympic Team. He's a member Skra Belchatow (for 7-time winners of the Polish Volleyball Championships).

Maciej grew up in an athletic family with a passion for sports. His father is a volleyball player while his mother is a track and field athlete, volleyball player, and archer.

Maciej was born by C-section in the 38th week of gestation. He weighed 3859 g (8.48 lb) and measured 65 cm (25.9 inches). From the very beginning it was known that he will be a very tall man. Maciej was allergic to cow's milk protein and needed a special diet. Even now he only eats goat's milk products that are considered by many doctors to be good for development of the lungs.

Maciej Muzaj in a volleyball uniform.

Being both a physical therapist as well as Maciej's mother provided me with more knowledge and awareness that a child born by C-section requires more stimulation of his development. From the very beginning I was using a lot of tactile stimulation, stroking, squeezing, rubbing, and hugging. I played with him by rolling, spinning, tumbling, creeping, crawling, jumping, and stretching. I used toys with sound and lights to stimulate his vision and auditory systems. All of our games and activities were bringing my son lots of joy and happiness.

Maciej's development was faster than many of the children in his age group. He started teething early and, at the age of 9 months, was walking independently. He had a good memory and, at the age of 2, was able to repeat songs and/or rhymes. He was much taller than other children.

Maciej's growth was fast and needed special attention in order to prevent development of any asymmetries. While in kindergarten we used therapeutic gymnastics and age-appropriate motor games to facilitate proper alignment, strength, and coordination.

My son demonstrated amazing abilities for visual and motor skills. He was able to 'see' a movement and was able to repeat it without the need for much repetition nor practice. (For example at the age of 4 he was able to dribble a basketball without looking at it).

From an early age, these skills were an indication of him becoming a great athlete. Being an active athlete myself, I introduced my son to many sports. Early on he learned how to ride a bicycle, ski, and swim.

At the age of 9, Maciej was diagnosed with dyslexia and dysgraphia (even today he has challenges when reading a traditional clock). At that time I included in his movement routines, Brain Gym® exercises by Paul Dennison Ph.D and the MNRI® Dynamic and Postural Reflex Integration program of S. Masgutova, Ph.D.

These new exercises improved his coordination, attention, and self-confidence.

At the age of 10 Maciej started playing tennis. I chose that sport for him because for the majority of the year practice and games take place outdoors. Additionally, it is a sport that includes many reciprocal, homologous, homolateral movements that develop great motor and visual-motor coordination. Maciek was doing very well and quickly excelled at his skills and led in his group. He enjoyed tennis but it wasn't his favorite sport.

At the age of 13 Maciej independently chose volleyball as a sport that he wanted to practice. He started attending a school with special emphasis on sports and athletic achievements. He participated in practices taking place 6 times a week. His height in 7th grade was 168 cm (5.5 ft) while in 11th grade it was 200cm (6.56 ft). During those years Maciek's growth was approximately 1 cm (0.39 inches) per month.

Because of the rapid growth I decided to continue using Dynamic and Postural Reflex Integration exercises at least once a month. The most crucial reflexes that we worked on were: Asymmetrical Tonic Neck Reflex, Symmetrical Tonic Neck Reflex, Moro, Hand Grasp, Hands Supporting, Flying and Landing, and Trunk Extension.

I strongly believe that thanks to reflex integration his growth was symmetrical, his body stayed healthy and strong, and his athletic talent could continue successfully developing.

At the age of 14 my son was selected for Regional representation in volleyball and the following years continued in Junior representation to finally become selected as one of 8 young athletes among 200 applying to be accepted to the elite sport school Szkoła Mistrzostwa Sportowego in Spala, Poland.

During the first year in Spala, Maciej was selected to represent Poland's Junior Volleyball. He participated in numerous international tournaments. At the age of 18 he became a professional volleyball player for the KPS Skra Belchatow Club (7-time winner of the Polish Volleyball Championships).

That same year Maciej was injured during practice. The strong impact of a ball hitting his face caused asymmetry of the facial bones. It triggered difficulties in balance and problems with his visual-motor coordination. His body's center mass had shifted. During that time I once again used the MNRI® Dynamic and Postural Reflex Integration exercises and Oral-Facial Reflex integration program. After application of the above techniques, his symmetry, coordination, and balance quickly returned to normal.

At the age of 19, despite his diagnosis of dyslexia and the intensity of volleyball practices, competitions and tournaments, Maciej passed the final high school exit exam called Matura and advanced to the Polish Senior National Volleyball Team.

Currently Maciej is a member of the Olympic Volleyball Team to represent Poland during Olympics in Rio de Janeiro.

As Maciej's mother I am extremely proud and happy knowing that he will represent our country while participating in the highest rank of international sporting events. I am deeply grateful to Dr. Svetlana Masgutova, whom I met in 1998, for sharing her knowledge and skills. They allowed me to support the development, talent, and success of my son but also many, many other children.

We congratulate Maciej for his great results in sports and thank his mother, Halina, for being such a great Mother to care so much for his health, development, and talents. – MNRI® Team.

Overcoming the Odds When the Cards Are Stacked Against You

Anna Nameny, MNRI® Parent, Elk River, MN, USA

Anna Nameny

The goal of this article is to demonstrate the effects of MNRI® and the changes that have occurred with my son, George Nameny, since starting this program. His situation is unique because he was introduced late to MNRI® and has a diagnosis of Fragile X.

George was born on February 25, 1996 a healthy 9.5 pound boy. He had difficulty at first latching on, sucking, and swallowing during breast feeding. This resolved on the first day after a lactation nurse intervened. I brought home a beautiful, content, and calm baby who slept through the nights. At two months of age he ran a temperature of 100-103° for three days and was admitted to the hospital with fears he had contracted meningitis. They poked and prodded him and did a spinal tap which was repeated 3 times before the doctor found the right spot. He was sent home with a diagnosis of reflux in his kidneys and was prescribed a maintenance dose of antibiotics. Until the age of five, George was on the maintenance dose of antibiotics until the reflux naturally resolved itself. The doctors also identified (with an ultrasound) that George had a horseshoe kidney. George's temperament changed during and after his hospital stay and he had to be held a certain way, he became extremely fussy, and was resistant to touch. He startled easily to lights, sound, and change in position.

The doctor said, with kidney reflux, George was more susceptible to infections because he was not circumcised. We took George in at 5 months of age for the simple outpatient procedure at our local Children's Hospital. We came home with unstoppable bleeding where the surgery had been performed. Several hours later, we returned for emergency surgery to resolve the bleeding. It was an extremely long day. George wasn't able to eat well, sleep, or be calmed. He cried constantly and it was very distressing to all of us.

George was tested by the school district when he was 25 months to see if he qualified for early intervention services. The school district identified developmental and language delays and began providing in home services. He also qualified for county services which provides a personal care attendant and respite hours in home for support. When George was young, I tried everything the traditional medical model offered but progress was slow and he kept slipping further and further behind his peers. We utilized private occupational, physical, and speech therapy. He was integrated in a regular education preschool and received individual special education services before class. I requested that the school include George in the regular education setting because I knew this was the best place for my son to learn new skills and model appropriate behavior by his peers in a natural setting.

George Nameny in senior high school.

George was quite a handful. My life consisted of waking up every morning and 'chasing' George all day to keep him safe, redirect behaviors, and utilize every teachable moment for him. He and I both 'dropped' from exhaustion at the end of the day and some days this came earlier than others. George had severe separation issues and his activity level resembled someone who was always on the go, full speed with no brakes. He had many temper tantrums daily and only spoke a few words. His fine motor skills were obviously delayed but his gross motor development at the time appeared normal. Looking back, I now recognize there were huge red flags regarding gross motor skills where he developed a lot of compensations.

In 1999, George was given the diagnosis of Pervasive Developmental Disorder and met the criteria for Attention Deficient Hyperactive Disorder. He was taking the highest doses of medication recommended for his age and weight for anxiety, hyperactivity, difficulty focusing and attending, getting stuck, and impulsivity. In 2000, we began further addressing concerns for his anxiety and he was diagnosed with Anxiety Disorder, Not Otherwise Specified. In 2003 after genetic testing, he was diagnosed with Fragile X. In 2006, George was taking five different prescription medications for his symptoms but, by 2007, it was decreased to two prescriptions.

In 2002, George went to kindergarten. He had to be carried into the classroom because he wouldn't walk through the doorway and he needed 1:1 adult support. George could not sit for circle time. His fine motor skills were delayed – he could barely hold a pencil or form letters or shapes, cut with a scissors, or write his name. His only writing ability was to scribble. George's vocabulary was small and communication was limited to a few simple words. (It took until second grade before George could say a complete sentence.) At age 6, according to the Peabody Picture Vocabulary Test-3rd edition, George scored in the first percentile for auditory comprehension, expressive communication, and total language. Some of his common sentences were, "Where do you get?" "I want to play some toys" "I try this on?" His language was characterized as being literal and responding to the concrete. George was preoccupied by his own interests, not the classroom activities. He did best with structured and consistent routines; he did not handle change well because it confused him. Around this age he began to show an interest in his peers. He could play independently; however, he needed an adult for group activities to redirect his attention to the topic of discussion. In elementary, middle, and high school George has been in the regular education setting for a majority of the school day with a 1:1 support.

In February 2006 (at age 10), we began using Brain Gym® activities, especially cross lateral movements. At first I physically moved his body for the movements because he was not able to do the movements independently. Within a short time, he was able to do them on his own and his ability to complete tasks, his tolerance for touch, presence of mind, and verbal and non-verbal interaction increased or improved. George was able to do chores with one prompt and more complex tasks without step-by-step direction. He tolerated light touch even when approached from behind unexpectedly. He had a better understanding of time relating to days of the week and upcoming events, his eye contact improved, and his conversation and social skills increased.

For four years I used individual Pilates sessions for core strengthening with George. I used Rhythmic Movement Training with George in 2007 and noticed his hyperactivity, anxiety, and panic attacks decreased although it was obvious, when I stopped using it daily, the effects soon wore off.

I began using Pattern Learning, a very intense academic program in 2007. George made academic gains but it required extreme repetition and was a slow, time consuming approach. He had improvement in handwriting, telling time, reading, and identifying the value of coins. He learned his full name, birth date, address, and the phone numbers for himself and family members. George has been working with a tutor since 2009 using the Riggs program to help him complete homework and develop more reading skills.

In August 2007, George saw a neuro-developmental specialist who suggested eliminating salt and nitrates from his diet to help decrease fluid retention and support his adrenal function. With these simple changes he appeared less 'puffy'. We continue to avoid salt and nitrates to this day. The specialist also recommended certain movement activities for reflex integration with George. His understanding of verbal information increased along with an ability to sing along with songs on the radio, and his hyperactivity continued to decrease.

I was always searching for something to help George. Then in 2007 one of my professional friends mentioned that she had been invited to sit in on a MNRI® class in Minneapolis, MN. Dr. Svetlana was speaking and demonstrating a technique called Embrace Squeeze. My friend was only there a short time but left amazed. She shared what she had learned with me and I began asking around. Everyone I spoke with respected Svetlana Masgutova's work and held her in the highest regard. My burning desire became to meet Svetlana Masgutva and to learn more about her method and what she was giving to the world. I learned Svetlana was teaching two classes in Minnesota in the summer of 2008 and signed up for both. Our life has never been the same . . .

So in 2009 (at the age of 13), George attended his first MNRI® Family Conference. Immediately we noticed a change is his physical posture. Before the conference, all the muscles in his body were constantly tense and flexed and had been like this since he was 2 months old. After the conference his body appeared relaxed. His eyes were able to track naturally from left to right and back for the first time in his life. After several months of using our MNRI® Home Program, the doctor said he could discontinue the remaining prescription medications. We have used the MNRI® Home Program as recommended – 1½-2 hours a day, 5 days a week of MNRI® – for 1½ years. George has attended four MNRI® Conferences and two MNRI® Clinics since 2009. His social skills have continued to improve – he gives and takes in conversations with others. George has come so far! He played baritone (an instrument) from 6th – 9th grade in band. Academics have come slower but he continues to learn and make progress. He participates in regular school classes with the help of a 1:1 support and modified materials. He goes to his church youth group, school dances, the YMCA, and community events. He has so many wants and dreams including a desire to drive a car someday. I can't get George to quit talking now (and would never try)! He has a cell phone and calls family and friends daily. Sometimes he's on the phone with his Grandpa for hours. George has a positive outlook and is present and happy most of the time. His anxiety has decreased, he demonstrates good problem solving skills, his posture appears more natural, and his fine motor skills have improved drastically. He retains and understands verbal and written information better. There is measurable progress with his physical endurance and speed. Most recently, George stopped shredding his socks – behavior he has exhibited since he was a toddler!

George went through puberty between 2009 and 2010. It was a difficult time for all. He was very interested in the opposite sex and required increased supervision and redirection. The support of the MNRI® team during this time was priceless and I found the Masgutova Home Program very valuable. I was able to demonstrate appropriate touch and provide integrating movements. We survived!

George has had leaps and bounds of growth and development in the last five years. I attribute most of this progress to Masgutova Method®. In George's case, we were able to establish the correct primary reflex patterns and continued integrating them to the place where his body integrates naturally with regular exercise and physical activity. This is ultimately the desired results with reflex integration. Infants with normal circumstances and a typical environment can integrate their reflexes and live a healthy integrated life. I have now witnessed how it is possible for individuals with developmental and reflex challenges to naturally integrate on their own once the correct neurosensorimotor reflex patterns have been established. MNRI® is amazing! It is never too late to integrate with MNRI® but sooner is better. MNRI® empowers parents by supporting us with classes and further educates us by giving special opportunities at Conferences to hear about the method and individual techniques, see how reflex integration works, do the activities and movements with our children, and experience it in our own bodies. Parents, I suggest taking home what

George on the football field.

George with his mother, Anna.

you've learned at MNRI® Conferences, on their website, from Core Specialists, and at MNRI® classes, and use this with your child and family.

Today (2014) George is 17 years old and is on the Elk River High School football team. Recently he was playing defense at practice and intercepted a pass thrown for the offense. He ran over ¾ of the field through the entire offensive line and kept going to the end zone for a touchdown. His team mates, standing on the sideline, were yelling and cheering for George! The further he got, the louder they yelled. Even the coaches joined in! When George reached the end zone, all the players ran out to where he was laying on the ground laughing and smiling, holding the football in the air. George got up and the players surrounded him in a huge circle cheering and giving chest bumps to George and other players! The mood was exhilarating as if they had just won a Super Bowl! Later in the evening George said to me, "Did you see my touchdown, Mom?" I responded, "Of course, I did George! You were awesome!" I now realize how valuable even the practices are for George.

In late 2014 during George's senior year in high school, the *Minneapolis Star Tribune* carried an article about George and the Elk River football team titled, *Elk River Football Plays for Its Emotional Leader.* The article quotes George (who is the kicker) and many of his teammates relating their commitment to each other and their love for the game. The players and coach credit George for helping them keep committed to the game, keeping its relevance to their life in check, and always encouraging them to do their best and win. George has become an inspiration to many.

My family has had many challenges over the last 17 years supporting George through many struggles and teachable moments. I have come to cherish every precious moment past, present, and still to come… I believe George is in a better place today than I ever imagined possible because of MNRI®. I will continue to encourage George to learn and further integrate and give him as many opportunities in life as possible to reach his fullest potential.

George, your growth is amazing, your talents are so great! Thank you for using MNRI® as one of your main methods for reaching all this. Success, good health, and realization of your plans and dreams to you! Congratulations to Anna for the success that came from all your years of dedicated MNRI work with George. We appreciate you being part of the 'team that won the game.' – The USA MNRI® Team Specialists

Incredible Changes in My Son Allen: Our Success at a Family Conference

Kimberly Otero, MNRI® Parent, Oro Valley, AZ, USA

This is a story of my son Allen and how he has responded to MNRI®.

Kimberly Otero

Introduction & History

Allen is 21 years old. He has had challenges both socially and academically his entire life. When he was 18 months old, the doctor determined that his head had stopped growing, and this thrust us into a world of doctors, testing (starting with an MRI), genetic screenings (for Fragile X and other possible conditions), and other tests. Nothing was definitive but there were definitely concerns about delayed development in many areas. We underwent many, many tests for physical and cognitive skills. He qualified for the Special Ed. pre-school program and received OT, PT, and speech therapy. We also had him in summer sessions for sensory integration therapy. Under IDEA, his development fell into several categories from SLD, OHI, and specific learning disability. Yes, we finally agreed to meds so he could function in the classroom. Fast forward and with help and advocates throughout school, he graduated from high school.

Allen at the end of the Conference on the balance board.

My friend, a speech therapist and MNRI® practitioner, recommended we attended a MNRI® Family Conference feeling that Allen could really benefit from the reflex integration therapy, so we tried it in San Diego. I had no idea what to expect or if Allen would be receptive to this therapy, especially when I saw the intense schedule of 6 hours of work a day! Although, I did not understand the entire Assessment, I watched as Dr. Masgutova pinpointed areas of difficulty for Allen. The eye-tracking and the lack of his ability to grip with his hands clearly demonstrated why school work was hard. The gross motor issues, in terms of crossing the midline of his body and balance difficulties, were also evident through her initial Assessment.

The Reflex Assessment Results

At the Conference we got the evaluation of Allen's reflex patterns in scores 'Before MNRI® Program' and 'After', which gave me the opportunity to determine the level of his reflexes and to compare them with changes after the Conference. Many of his reflexes were of a low level of development. The Assessment also showed that his upper and lower limbs reflexes were not corresponding in sensory-motor circuits: he was hyposensitive for stimulation of the upper limbs and hypoactive in motor response, while his lower limbs were hypersensitive and hyperactive. This was one of most important reasons why he couldn't integrate them timely in his childhood and why they couldn't support the development of his motor and learning abilities and skills.

The statistical expression of results of the MNRI® Assessment for Allen is shown below in three groups – reflex patterns serving movements in: A) Sagittal (right-left) plane of the body (Table-1), B) Horizontal (upper and lower parts; Table 2), and C) Dorsal (front-back) (Table 3).

Reflex Patterns Evaluation Results for Allen

Allen	Table 1. Reflex Patterns Serving for Development of Sagittal (Left-Right) Motor Coordination System									
	1	2	3	4	5	6	7	8	9	10
MNRI Program	Core Tendon Guard	Hands Grasp	Hands Pulling	Babkin Palmo mental	Babinski	Leg Cross Flexion Extension	ATNR	Abdominal	Bonding	Foot Grasp
Before	14.25	13.75	12.25	14.75	12.25	14.25	14.25	13.25	14	12.25
After	15	14.75	13.5	15.25	14.25	15.5	15.25	13.75	14.5	13.25

Allen	Table 2. Reflex Patterns Serving for Development of Horizontal (Upper-Lower) Motor Coordination System									
	11	12	13	14	15	16	17	18	19	20
MNRI Program	Automatic Gait	Bauer Crawling	Moro	Fear Paralysis	Hands Support	Segmental Rolling	Landau	Flying/ Landing	Grounding	Head Righting
Before	15	10.25	12.25	12.25	14.25	11.25	13.25	11.75	12.25	12.75
After	15.25	12.25	13.5	13.25	15	13	14.25	14	13.5	13.75

Allen	Table 3. Reflex Patterns Serving for Development of Anterior-Posterior MCS									
	21	22	23	24	25	26	27	28	29	30
MNRI Program	Trunk Extension	STNR	Spinal Galant	Spinal Perez	TLR	Foot Tendon	Spin-ning	Locomoti on	Balancing	Pavlov
Before	14.25	13.75	13.25	12.25	12	12.25	12.25	9.25	12	13.00
After	15.25	14.5	14.25	13.5	14.25	13.75	13.75	9.75	14.25	14.00

Above, the statistical expression of results of the MNRI® Assessment for Allen is shown below in three groups – reflex patterns serving movements in: Table 1, Sagittal (right-left) plane of the body; Table 2, Horizontal (upper and lower parts); and Table 3, Dorsal (front-back).

Below, the graph equivalent of the above Tables.

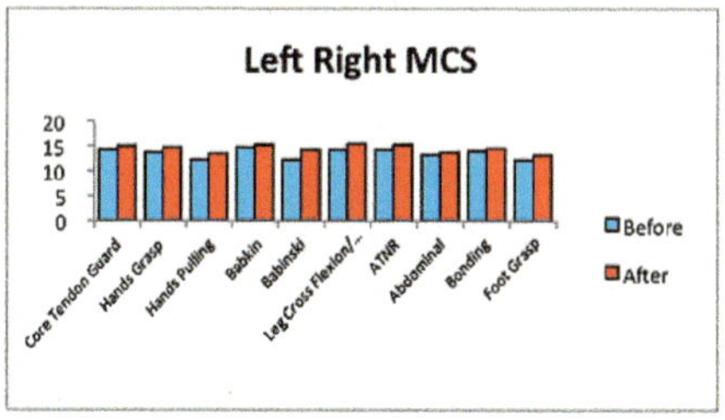

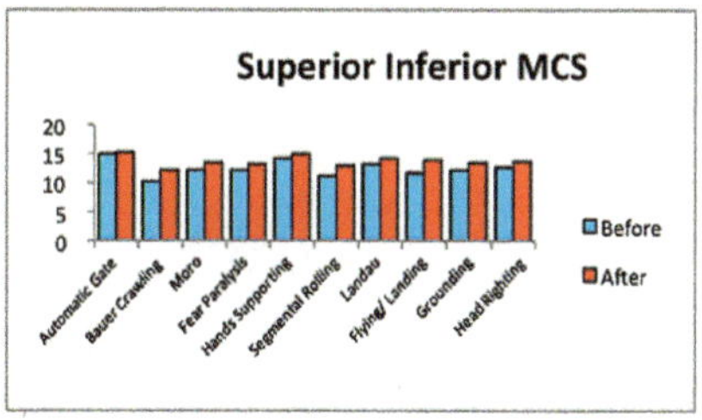

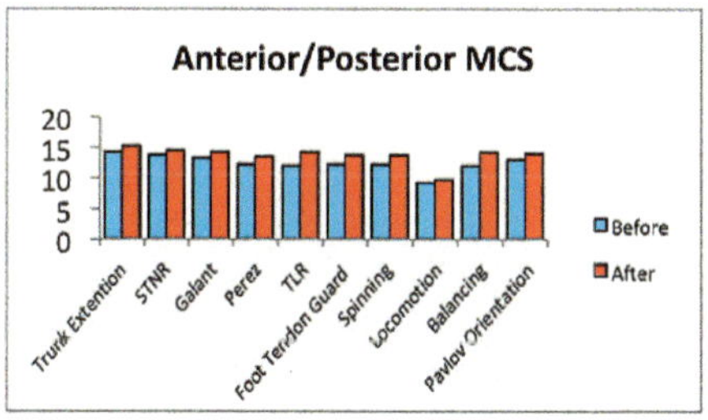

Reflex Patterns Evaluation Results for Allen

Based on the results of the MNRI® Assessment, an individualized program for Allen was established for work at the Family Conference and for a Home Program. The main purpose of the individualized program was to strengthen and support his physical, emotional, and cognitive growth through the work with non-integrated and retained reflex patterns.

Allen's Response to MNRI® Program

We started the 9-day MNRI® program which included MNRI® Reflex Repatterning, Tactile Integration, Neuro-Structural Reflex Integration, and several others. I was so excited at how receptive he was to everything – never a negative word and he willingly went to every session! All the therapists were able to communicate with him and work with him at his level, understanding his inner world, and awakening his curiosity and interest. I could tell many stories about how much he connected during the session of Breathing Reflex Integration program, learning inspiratory process and calming techniques, or doing gross motor skill training during the Archetype Movement Integration sessions.

The first day, he could not stand on the balance board and, by the end, he was standing on it hitting tennis balls on the tennis court. What an amazing change! The Oral-Facial and Visual-Auditory program helped with eye-tracking, convergence, focusing, and his articulation has improved significantly. Allen had missed the important milestone of crawling so working on the Bauer Crawl was also very important. The two best things were that he was so engaged in the every session and that, even after only 9 days, people back home really noticed a difference. They noticed it in how he carried himself, how he spoke more clearly, and how much more confident he seemed – just more present and comfortable in his body.

It has been challenging to carry through with the Home Program due to its novelty and my work schedule but I am starting to take some of the training sessions and we will get back to doing it. He still uses several of the techniques himself, such as the breathing, calming, and other techniques. I am glad we had the opportunity to experience this program and will continue to incorporate aspects of it in our daily lives.

I have to confess that when I saw how much more mobile and verbal Allen was than some of the other participants, I wondered if Allen would really benefit – what a lesson I learned! I am bringing my daughter to an Assessment later this year for her anxiety and panic issues – I believe that anyone would see positive changes from learning more about MNRI®.

Allen, congratulations on your great improvements in learning and courage! We hope you will express your talents even more in your life! We thank your Mom for being so supportive to you studying and applying MNRI®. – Love, The USA MNRI® Team

Change Is Possible – Oskar's Story

Blake Kelly, MNRI® Parent, Edina, MN, USA

My son Oskar has a rare chromosomal deletion on the fifth chromosome. It means he has severe global developmental delay, seizures, hypotonia, and some autistic features. In layman's terms, it means that at 6½ years old, he can't walk or talk yet, and he eats only purees and thickened fluids.

Blake Kelly

I first came across the Masgutova Method® quite by accident about four years ago. After fighting with insurance for years, I had finally gotten Oskar qualified for speech therapy under 'oral motor'. (Apparently NOT SPEAKING does not qualify you for speech therapy.) I took him in for an assessment and first appointment at a well-regarded therapy center, and watched as the therapist fed him a Cheeto and stuck a few vibrating brushes in his mouth. I was crestfallen. This was nothing new, nothing we hadn't already been doing at home and at school for years. I knew this wasn't going to change things for Oskar. I knew he needed something different.

It just so happened that the same week, the center I took him to for occupational therapy was hosting a Mini-Clinic with Dr. Masgutova. They had mentioned that they thought Oskar would benefit from seeing her and so I signed him up. During her Assessment, she told me some exercises she thought would help him, and he had an hour of work from a Core Specialist. I don't think I understood a lot of what happened, but I did feel like at least it was something different than 'Cheeto therapy' and it made sense to me that Oskar needed help that started from the bottom up. He needed those first steps, those primary reflexes he had missed out on somehow.

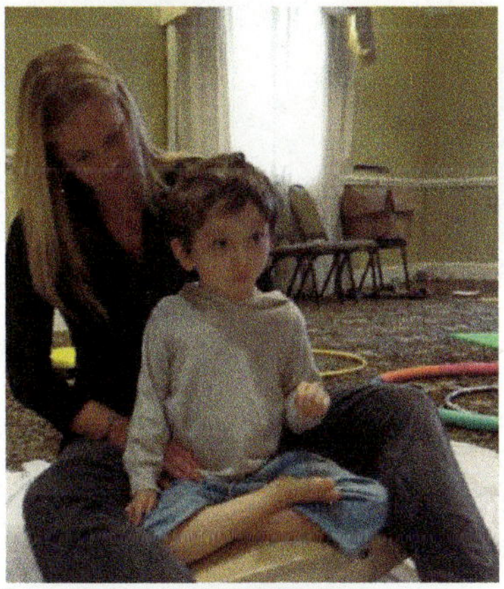

Oskar receiving MNRI® at a Family Conference with Core Specialist Tanya Hallman.

I signed Oskar up for speech with a therapist there who was using the Masgutova Method®. We went every week and I did the five or so exercises that they told me to do with Oskar.

One of the exercises they had me working on was Hands Supporting. Basically, it's the reflex to put your hands out and protect your face when you fall over. Oskar didn't have that one and so when he was sitting he would just tip over. One day, a few months after we started using the method, he started to tip and he put his hands out! "Oh my God!" I thought, "It's working!"

That summer we signed up for an eight-day conference and we just

Left, Oskar showing off his Hands Supporting.

Right, Oskar and Mom goofing off.

attended our second conference in Orlando on January 2013.

The changes I have seen in Oskar are so encouraging. I know the Core Specialists he works with could explain them in terms of reflexes but for me, he is just a different kid. He is generally so much more present. He looks at me when I talk to him. He laughs. He sleeps! For years, we both spent hours awake in the middle of the night. Now if he wakes up he can put himself back to sleep. He is working on crawling; his trunk control and head righting are so much better. He can drink from a straw. He uses his hands now, instead of keeping them balled into fists. His vision and hearing are better. No more glasses to keep his eyes from crossing! He is so much stronger and one of the biggest changes for me, is that now I can carry him on my hip, like a typical toddler. He used to throw himself into extension all the time – arch like he was trying to get away. Now he holds on to me. He was getting so big and strong I almost couldn't carry him any more.

While I don't have a 'big bang' story to relate, all these little changes make a huge difference for Oskar and for me. Some of my friends with special needs kids who have been skeptical of trying 'another new thing' have started Masgutova therapy, too, because of the progress they have seen Oskar make.

A little boy we know with CP told his Mom the other day that when he gets "strong like Oskar, he'll be able to dance." I can't believe that Oskar has gotten to a point where his strength is inspirational for other kids. I can't tell you how far he's come.

I feel like I spent so many years and so much money taking him to therapists who didn't know what to do, who had never "seen a kid like him." The biggest gift the Masgutova Method® has given me is the feeling of hope, of being with therapists who have seen miracles, and seen kids like Oskar and know that change is possible and, if we work hard enough, inevitable.

Oskar with Mom and sister Maddie.

Oskar, you are a real Winner! With all your strength for changes and growth, you can do a lot to move further! We wish you good health and success in your future! Thank you Blake for all your care for Oskar and trust in us as specialists! – The MNRI® Specialists.

Treatment of a Child with Cat's Cry Syndrome

Elzbieta Cieplinska, MA, MNRI® Instructor, Gdansk, Poland

This article describes using the MNRI® program in treating a boy with Cri-du-Chat syndrome (Cat's Cry syndrome). This syndrome is caused by a rare genetic mutation where a portion of chromosome 5 is deleted or otherwise altered. The severity of expression of the syndrome can vary significantly in relation to the amount of genetic material affected. The syndrome is named as it is because the particular sound of an affected baby's crying is very similar to a cat meowing. The syndrome is also characterized by distinct facial features that include widely set eyes as well as an intellectual disability and developmental delays.

Keywords: MNRI® program, Cat's Cry syndrome, cri-du-chat syndrome

Elzbieta Cieplinska

Primary Symptoms of Cat's Cry Syndrome

Cat's Cry syndrome presents itself in the first days of a baby's life in the characteristic sound the baby makes when it cries. A combination of high tones and monotone sounds are similar to a cat meowing. This results from a malformation of the larynx (it tends to be small, narrow, with a rhomboid shape), a small and limp epiglottis, as well as a disorder of the nervous system. This pattern of crying often passes after a couple of months, but may take as long as 10-15 months. A baby with this syndrome will often also have problems with sucking and swallowing, leading to problems with feeding. The baby is often unable to gain weight. An additional set of symptoms of the syndrome include heavy drooling and constipation from the first weeks of life. The syndrome is identified through genetic karyotype testing of a blood sample.

Other symptoms often presented by a baby with Cat's Cry syndrome are cleft lip and palate, cleft spine (spina bifida), heart defects, small retracted jaw, round asymmetric face, flat wide bridge of nose, short bridge of nose, short philtrum (the cleft beneath the nose running down to the lip), increased distance between the eyes, microcephaly, visible forehead tumors, auricles (visible portion of the ears) set low on the head, pale color, improper arrangement of the intestines, kidney defects, inguinal hernia (protrusion of abdominal cavity contents through inguinal canal), lateral flaring of digits of hands and feet, and flat feet.

The Case of Wojtek Drozd

The pregnancy with Wojtek progressed normally until the twentieth week. At that time, a routine ultrasound showed presence of only a single kidney. The doctors assumed that there was no problem, merely that the kidney could not be imaged due to low levels of amniotic fluid. In the ninth month, Wojtek's mother was diagnosed with preeclampsia characterized by edema (swelling) and hypertension (high blood pressure). Based on the risk factors of preeclampsia, it was decided to deliver the baby through C-section without waiting for

Wojtek at a MNRI® Camp.

the due date. Despite the level of discomfort felt by the the mother, 24 hours passed before the procedure was performed. The amniotic fluid was green and there was very little of it. Although Wojtek weighed a healthy 3,090 g. (6 lb. 13 oz.), he received only a score of 5 on the APGAR scale. He was therefore immediately moved to the Neonatal Intensive Care Unit. Wojtek presented with the following problems: significantly retracted lower jaw, difficulty breathing, absence of the Sucking Reflex, propensity for choking, low muscle tone, and microcephaly. His crying was distinctly different from that of other newborns.

A subsequent ultrasound was performed and his 'missing' kidney was found to be partially developed, but probably non-functioning and displaced in the direction of the other, healthy kidney.

After a karyotype genetic analysis was performed, Wojtek was diagnosed with the genetically-based Cat's Cry syndrome.

With occasional visits home, Wojtek spent the next 5 months in the hospital. He frequently caught a variety of infections, primarily those of the pulmonary and urinary tracts. He also had difficulty with swallowing due to the narrowing of his larynx. He developed an inguinal hernia that was operated on when he was 5 months old. Wojtek also had difficulties sleeping. The first time that he slept through the whole night was at the age of 3½ years.

Feeding and eating: Wojtek cried a great deal, had colic, and ate very little. He initially gained almost no weight. It was decided to feed him through a GI tube twice daily for a month. After a period of time, Wojtek's mother worked out a system that allowed him to finally breast feed. The period of breast feeding was very long and, as a consequence, the positioning of his lower jaw improved greatly. In his fourth month, Wojtek was introduced to blended or finely chopped baby food, fed with a spoon. He ate very slowly, as he often gagged or choked on the larger pieces. Incrementally, he was introduced to more coarsely chopped food and even larger pieces that he could bite off and chew. Currently, Wojtek is able to eat independently, using a fork or spoon.

Motor coordination: Wojtek has problems with both fine and gross motor control. One of his challenges is the abduction of the thumb. His mother observed, "Even though he can manipulate a spoon and, to a certain extent, a fork, he's unable to master several other skills. He can't point to objects with a single index finger nor can he coordinate his hands very well. He's unable to perform intricate movements. When he's done eating, the table looks like the aftermath of a bloody battle."

The individual stages leading to walking took a very long time to develop for Wojtek. His creeping, crawling, and sitting up were delayed and stayed on the same level for a long time before reaching the next developmental milestone. He was two years old before he was able to stand up with support and 5 years old before he began to walk.

Intellectual development: Wojtek's intellectual development always proceeded surprisingly well. He reacts to commands, understands the explanations of situations and interpersonal relationships, and he even laughs at jokes.

Sensory stimulation: Wojtek demonstrates a strong preference and need for sensory stimulation especially in the area of his head. His mother said, "My son's need for strong sensory stimulation can be seen in the joy he expresses when, for example, he's pushed hard on a swing, or he is jumping on a trampoline (with help, of course). He likes the sensation when he holds his head against spinning or vibrating toys and speakers. He likes to bite toys and other objects with a rough texture and likes to try foods with a strong, even spicy, flavor. Often, before going to sleep or in a quiet moment, he'll direct my hand to squeeze his head or feet. This calms him down and helps

Wojtek learning to stand up anassisted.

Wojtek on a walk in the forest.

to put him to sleep. As he falls asleep, he often claps his hands loudly."

Results Using MNRI® Therapeutic Program

When Wojtek was three, his parents decided to enroll him exclusively in the MNRI® program, foregoing other forms of therapy. During this time Wojtek exhibited self-aggressive tendencies including: hitting his head against the ground, biting and furiously twiddling his fingers, as well as general aggressiveness, over-excitement, and hyperactivity. He moved around mainly by creeping on all fours. He stimulated his lips with various toys and objects and drooled excessively.

At the very beginning of our work with Wojtek, we used the MNRI® Tactile Integration program. Our sessions took place 4 times a week. By the end of the first session of working with the Tactile Program, Wojtek calmed down and fell asleep for a couple of hours, much to the surprise of his parents. This occurred repeatedly after each therapy session. Incrementally, we began to introduce other MNRI® programs: Dysfunctional and Pathological Reflexes, Oral-Facial, and Visual-Auditory Reflexes. The first time that he slept through the whole night, at the age of 3½ years, was after the MNRI® Tactile Integration program.

Through the application of Dr. Svetlana Masgutova's MNRI® method, we were able to stop a majority of Wojtek's aggressive behavior while also improving his fine and gross motor control, social relationships, and psychological and physical functioning. One of his physical challenges is the abduction of the thumb. After the application of the MNRI® program, that fine motor skill has greatly improved.

Wojtek initially presented with strong auditory sensitivity. He reacted with panic to normal household sounds, such as the dropping of a spoon, by grabbing at his ears, crying, and panicking. Today, he is in panic state much less than before we started the MNRI® program.

Wojtek had difficulties with visual and auditory coordination from the beginning. His ability for eye tracking and convergence was very weak. After application of the MNRI® Visual and Auditory Reflex Integration program, Wojtek's visual and auditory concentration were greatly improved.

Wojtek is now able to walk independently with good stability on an even surface. He eats independently, is able dress and undress with a small amount of help, is able to socialize, and is better able to concentrate. Additionally, even though he still can't speak, he is better able to communicate and react to commands. He has begun to play appropriately with his toys and is better able to demonstrate and execute his own needs. He exhibits almost no aggression nor self-aggression. His head is growing more rapidly. He attends an integrated special needs class at the local school.

Twice yearly he attends intensive rehabilitation camps overseen by Dr. Masgutova. Wojtek's mother, Mrs. Jowita Wallner-Drozd, says, "With full conviction I can say that Dr. Svetlana Masgutova's MNRI® method is the most effective, holistic, and ever-evolving therapeutic method available in Poland that addresses neurological problems in people at various stages of development."

Wojtek attending school, shown here with his mother.

Wojtek, you are such a curious boy, with a determined mind, open for communication and friends. You have great a motivation to learn more about the world. We wish you success in continuing to develop at the same good speed, health, and strength! – Elzbieta Cieplinska & MNRI® the Polish Team

You Are a Winner

Svetlana Masgutova, Ph.D, Creator of MNRI® and Pamela Curlee, MNRI® Instructor,Co-Founder of the Svetlana Masgutova Educational Institute, USA

Svetlana Masgutova

Pamela Curlee

The below text is based on the book by S. Masgutova & P. Curlee – You are A Winner, Trauma Recovery: A New Choice Through Natural Developmental Movement *(2004, 2006). Dr. Masgutova's work in this catastrophe and its resulting analysis became the basis for the MNRI® Program and PTSD Protocol. In the year 1986 (during her work with survivors of the Chernobyl catastrophe) reflex integration techniques were first applied that led to the development of MNRI® as a method and system. Here, in this article, her work became further defined as she helped children survivors of the Ufa catastrophe of Russia in 1989.*

Madness and Despair

The moment I walked into the first room at the children's ward and saw the depth of despair before me, I realized that everything I had ever learned was now being turned upside down.

Madness. That is all there was. I was surrounded by madness, by a situation bigger and more drastic than anything I had ever seen or imagined.

How could I reach these tormented boys and girls? No one should have to endure such suffering. The children desperately needed my help, and I certainly had knowledge, yet I could not think of anything that could reach them in their present state.

Despite all my years of training, teaching, and experience, I had no guidance for a situation such as this. How could this be? I frantically scanned my memory banks, reviewing my logical and systematic training, and found nothing that would help these children. But then, this situation wasn't logical, so why should I expect to find a logical solution? I needed to do something new and different – something I'd never done before. What was it? Where was it? All I wanted to do was reach out with compassion and assist the children in finding ways to be released from their bonds of suffering and pain.

I remember thinking to myself: *Sh-h-h-h! Be quiet, mind. Just listen. I will learn what to do. The children will help teach me. We must teach one another how to move through and beyond this catastrophe. I will teach them, they will teach me, and together we'll find a way for them to move beyond survival.*

So come, children, join me now and help me share this tale, so that others may learn what you have to teach us. Come, children, and help me find the words to let others know how to face a tragedy, how to inspect the shattered pieces of a life and find the hero who was always there.

You, my children, are *winners*. We must tell others how to discover this truth that lies within them, as well. There are natural, simple, proven methods that make all the difference for survival and eventually for the growth and development of who we can become. The secret lies within us, and it involves awakening our natural intelligence. The time has come to tell our story.

The Children on the Train

Many children were invited on a journey on a train because they had excelled in school and had won a vacation as an award for outstanding creativity, talent, and success in their studies. They had put in hours of schoolwork that had finally paid off and were filled with excitement and enthusiasm for this trip to their favorite retreat. With all the necessary preparations made, many had boarded at the train station in Novosibirsk. Their anticipation had built as they eagerly awaited the arrival of the train that would carry them from the forests of Siberia to the holiday camp at Artek on the Crimea, where they would be able to enjoy the warm sun on the shores of the Black Sea.

The train had finally arrived, and the children eagerly boarded, some waving to family members who were seeing them off, others boarded with parents, grandparents, and siblings who had elected to join in this journey of celebration. The train, consisting of twenty cars, also carried other families heading to the city of Adler near Sochi on coast of the Black Sea.

Returning from the Black Sea was another train. Its seventeen cars were filled with families returning from their vacations. These travelers had enjoyed their restful times, and were sharing memories of days at the beach: the food, the sun, the play, and the friendships made. The passengers talked and shared stories as the train gently swayed along the tracks in rhythmic movement, en-route to taking everyone back home.

One train was running behind schedule. Because it had left the station late, the two trains were destined to pass one another as they traveled through the forest 750 miles southeast of Moscow. Both trains traveled swiftly toward a picturesque valley near the town of Asha, nestled in the Ural Mountains of the Bashkir region of Russia. Most of the passengers on both trains had settled in for the night, feeling relaxed and content, and sleeping as best one can on a train.

A Soviet army officer was standing at the window of one train when he noticed an acrid smell. He later reported, "I sensed that something must be wrong, but before I could do anything, there was a glow and then a thunderous explosion."

It would be days before the actual cause of the catastrophe became known.

June 4, 1989

Disasters are always unexpected. I initially heard reports of a railway catastrophe while getting ready to go to work on the morning of June 4, 1989. President Mikhail Gorbachev announced the news of a monumental train accident involving several hundred people. Details were very limited, but he was deeply concerned by the number of lives impacted by such a tragedy. The calamity occurred around 1:10 am., local Bashkir time. President Gorbachev soon followed his announcement with the additional news that the government had suspended plans to have an inauguration of the new Soviet government and legislature that day. The official celebration of the new National Congress of the People's Deputies would be indefinitely postponed. Gorbachev explained that the citizens of the nation must now stop everything to mourn and honor the hundreds of dead and dying people involved in this terrible event.

The entire Soviet press reported the disaster. The political protestors bowed their heads in sorrowful disbelief. As further details came in, the government reported the truth immediately and in an honest way.

It was touching to see such authentic human responses from government leaders. All the people of Russia were feeling a deep grief and sorrow that only tears could speak. There was a feeling of total and unconditional love. Also, a great deal of love and sympathy immediately started pouring in to the Russian people from around the world.

Upon hearing the news of this terrible accident, I knew I must go to the site immediately. Many people would be in need of support for the mind and emotions as well as the body.

Psychology was my deep passion, and my experience as Dean of the Department of Applied Psychology at the University had given me a firm foundation in working with emotions and the mind. As reports started circulating that hundreds of children involved in the accident were being transported to a central location in

Train wreck at Ufa.

the nearby city of Ufa, capital of the Bashkir Republic, I called my university and informed them that I would leave the next day to go to Ufa and volunteer my assistance.

The news would have spread quickly throughout the city. There would be students from the university, many of whom had been in my classes and lectures, who would be able to come to the hospital to help.

Throughout the day, I made personal and professional preparations in order to leave the next morning on the first available flight. Along with the rest of the world, I listened to the news, wanting to learn what had caused such a horrible accident.

As it turned out, June 4 was a monumentally tragic day. News agencies were bombarded with stories of crisis as numerous catastrophic events happened simultaneously, creating worldwide chaos and grief.

In Russia, the announcements about the train tragedy alternated with news of the conflicts in the Uzbek Republic, where thousands of soldiers were sent to the area to uphold curfews imposed in the numerous cities where violence and rioting were causing increased deaths and hardships.

Announcements of the train disaster also continued, stating that hundreds of passengers had perished in the fires and that it would go on record as one of the world's most tragic rail accidents.

The information was interspersed with stories from other countries. News footage from Tehran showed scenes of millions of people who had assembled on June 4 in 90-degree heat to bewail the death of their national leader, the Ayatollah. While these multitudes gathered to pay tribute, eight of the mourners were crushed to death and hundreds of others were injured.

A brief announcement told about dangerous heavy rains falling in Kegalle, Sri Lanka. Wind and storm clouds had dropped more than eight inches of rain in a record period of time. Because of the inability of the land to absorb the moisture, the outcome of this June 4 storm was a raging flood and mud-slide that killed two hundred and fifty people, injuring over one thousand others, and leaving one hundred thousand people homeless.

Continuous pictures and reportage depicting the horrible condition of the train passengers flooded the airways as well as the hearts and minds of the people of Russia. Rumors began to emerge about the possible cause of such a tragedy as authorities continued their search.

The news was again temporarily interrupted to show coverage of military tanks shooting at college students in Beijing's Tiananmen Square. Students had been assembling in this square for two weeks to stage a protest for change. They had been expressing, through song, speeches, and dance, a new vision for the future. Rallying around a collective new thought of political expression, they had camped out together to give voice to their ideas. The military had been outside the Square for the entire two weeks, performing drills and exercises on a daily basis. It was later reported in the Denver Post that "their throaty cadence thundered off the walls of the Great Hall of the People at Mao Tse-tung's mausoleum as they jogged in formation."

At about the same time as the train accident, and also without warning, the military moved in on the collected students in Tiananmen Square. Infantrymen and armored trucks scythed a bloody swathe through hundreds of young men and women. As their voices were brought to silence, the lives of mothers, fathers, sisters, brothers, friends, teachers, and acquaintances were forever changed. The impact of all this simultaneous tragedy was being deeply felt in my country. Had the whole world gone mad?

Finally more news of the trains was broadcast. The cause of the accident was now clear. I will always remember the touching scene on television of Nikolai Ryshkov, chairman of the USSR Soviet Ministry, standing helplessly at the accident site and shedding tears of true compassion. The image of this strong man empathetically weeping served as inspiration to me for weeks.

As the two trains had entered opposite ends of a ravine on their separate tracks, wheel friction had created a spark that ignited liquefied natural gas that had been leaking from thirty-inch pipes nearby. The pipeline carrying liquefied petroleum gas from oil fields in Nizhnevartovsk to refineries in Ufa had ruptured. The leaking fuel contained a mixture of propane, butane, and benzene that partially evaporated, creating a highly combustible formula as it mixed with air.

Rather than investigate a drop in pressure, the operators had turned up the pumps, thus adding to a vast, growing pool of heavy methane vapor that had filled the ravine for three hours prior to the approach of the two trains. The resulting explosion of liquefied gas demolished part of the trans-Siberian railway and killed hundreds of passengers and railroad employees. Many rail cars were crushed and blackened. A shock wave equal to ten thousand tons of TNT threw eleven cars into the air, and twenty-six other cars were burned within ten minutes – seven of them incinerated beyond recognition by the blast of heat that rose to between 1500° and 1800° Fahrenheit. The exploding cloud of gas rose more than 1,600 cubic yards above the earth.

President Gorbachev flew to witness this 'major catastrophe.' More than 250 acres of trees within a three-mile radius were charred to the ground. Windows were shattered in homes and villages up to seven miles away. Approximately 780 of the 1,284 passengers on the two trains died. Many perished immediately, and others eventually died in hospitals. A total of 383 children were known to be on the trains; however, since tickets were not sold to children under the age of eight, the smallest ones were not initially included in the official count.

At the explosion site, Officer Andrej Doncov was the first to start rescuing the surviving passengers.

An engineer on one of the trains, Victor Bezverchij, was thrown from his locomotive and had his skin burned and some of his limbs broken in the explosion, yet he crawled several kilometers to the nearest village to seek help for the passengers. He reported that, before the explosion, even at his train's high speed, he could smell gas hovering like fog at the level of the train windows.

The driver of the second train, Sergej Stoliarov, along with a co-worker, Marat Ganeev, managed to save more than three hundred people by organizing their immediate rescue from the burning train. A nearby villager later wrote, "The air itself caught on fire, and the people of the village breathed the flames while they looked at the fire. Many villagers lost their eyesight. Others died from burns in their breathing passages. If we were two kilometers away from the epicenter and we were literally deafened by the explosion, then what did the unfortunate train passengers endure?"

One boy lying on a stretcher in an ambulance at the Chelyabinsk airport was so severely injured that a news correspondent simply told the viewers, "He's living."

Joining the Team in Ufa

The next day, June 5, I noticed a picture on the news of a young Chinese man standing alone in front of a long row of tanks. He appeared to be demanding that there be a stop to the madness of destruction and chaos. On this day, I was on my way to Ufa to help the children of this disaster face their pain and loss. We too would have to call a halt to the continuing destruction set into motion by our national tragedy.

I knew I had to go. I had to do what I could to help them. It was terrible to hear all the stories on television and the radio. Someone had to go help the victims stop their internal chaos. My heart spontaneously reached out to these children, and I felt sure that my knowledge of psychology (the subject I was teaching in Moscow) would be beneficial.

During the entire flight from Moscow, I reviewed strategies and considered various plans for dealing with this crisis situation. Besides teaching psychology, I had previously organized, trained, and assisted with the psychological recovery teams for Chernobyl and the postwar area of Baku. My ideas were based on my training in psychology and on those prior experiences in dealing with catastrophes.

I found myself thinking over and over again about the various ways to organize people, teams, methods, and procedures to deal with this new challenge. Fluent in the concepts and methods of Carl Rogers, Carl Jung, gestalt therapy, psychoanalysis, and art therapy, to name only a few, I assumed that these psychological tools would be all I would need.

I was willing to face the catastrophe head on, confident that my past experience and knowledge would help me do what was needed. Most importantly, I was coming with a pure intention to just be present for these children as they began their journey back to mental health.

I had also packed up my five-year-old son, Denis. He would stay in Ufa with my parents, who were coming there to give help in this way. None of us realized that this was only the beginning of nearly four months of sixteen-hour days.

Ufa, the capital of Bashkortostan, is an industrial city of almost a million people, built near the western Ural Mountains. It was founded in the 16th century by Ivan the Terrible, as a fort from which to fight off nomads. In the 1940s it became active with industries such as oil refining and the manufacture of petrochemicals, processed foods, synthetic rubber, and electrical and mining equipment.

In the early l980s I had lived in Ufa for several years, and had found it to be a city of excellent universities, offering training in medicine, aviation, mining, and general education. It was here that I had begun my training in pre-medicine, psychology, and foreign languages.

On the day I arrived in Ufa the airport was swarming with people. All flights were being reserved for volunteers coming into the area, while pleasure and business flights had been cancelled. It was a frightening, surreal scene. The sky was filled with helicopters that were to hover like flies for three days solid.

Because of the endless movement of trucks, cars, and helicopters transporting the patients and medical personnel, some of the dazed people who had survived the train accident thought an atomic war had begun. The frenzied presence of air traffic only reinforced this perception.

The majority of the children in the disaster had been brought to a central children's ward for focused care. Those in charge had assigned the children four floors of the building. I came in, explaining that I had a Ph.D. in psychology, and shared my experiences of helping people in psychological trauma. I offered my assistance freely and emphatically. The organizers were extremely happy to see me because they didn't know how to get so many children and parents out of the state of shock.

The organizers immediately assigned me the task of arranging psychological assistance for the children and their parents. Initially our team consisted of twelve psychologists, but every once in a while I would arrive in the morning to discover that a colleague had returned home. The number of professionals on staff continued to diminish, and by the end of our time with the children only three psychologists had been able to withstand the emotional rigor of working in such an environment.

The Thousands Who Came to Help

Help poured in from everywhere. Statistics later documented the scope of the actions of thousands of people who shared their compassion and love in the aftermath of the railway disaster.

Within the first hours following the tragedy:

- All 938 of the initial survivors of the train wreck were evacuated to nearby hospitals, despite the challenges of reaching them in the ravine where the accident had occurred.
- About 800 people from the Ministry of Internal Affairs came to help, along with 500 soldiers. Also, 3,827 people and 545 vehicles arrived from the Civil Defense Service.
- Hundreds of local citizens donated blood, time, and anything else they could give that would help. Within only the first few hours, 425 liters of blood had been donated.
- Britain's Prime Minister, Margaret Thatcher, sent special waterbeds to support the healing of skin transplants for the burn victims.
- Three dozen restaurants supplied food for the workers. In addition, 4,903 individuals and 269 trucks brought in more food.
- For transportation, 550 vehicles and 1,645 buses were brought in.
- A thousand railroad tracks were brought in for the reconstruction, as repair work began on the 350 meters of railway that had been destroyed.
- Repair work began on the 3,000 meters of electric lines and 1,700 meters of telephone and other communication lines that had been destroyed.

The rescue work that began so immediately was highly organized, calling initially on the skills of fire fighters, military people, and medical personnel. Teams came from all over Russia, and support medical personnel came in from the United States, England, Ireland, Australia, Germany, and Israel.

At the site of the accident, a team of 100 doctors and medical personnel set up tent hospitals in order to provide immediate critical care. Thanks to this service, hundreds of people were later stabilized enough to be moved to nearby hospitals. Among them were 196 children. Depending on the severity of their burns, some patients were sent to Moscow, St. Petersburg, Ufa, and other cities.

Doctors, medical students, medical personnel, and psychologists, as well as countless caring townspeople, worked day and night with the children and adults to support their survival, while the transportation of food and medicine continued around the clock.

The true number of volunteers was beyond calculation. In the first two days, 1,600 people from the army and nearly that many civilians came to offer assistance. Fifteen helicopters from the city of Ufa made 276 flights, airlifting 938 victims and transporting 419 doctors to the site of the accident and to hospitals. Overall, well over 7,000 people came to offer their help.

Free telephone service was set up to handle more than 7,000 calls between cities, and 700 telegrams were sent out. Hundreds of relatives of the people on the trains poured into the area to seek their loved ones. For many of these bereaved family members there was no available information, because the remains of 220 of the passengers were so completely incinerated that they could never be found.

An Unprecedented Situation

Nothing could have prepared me for what I saw when I stepped into the children's ward. As I have described, when I first walked down the corridor, I saw hell. It was like walking into Dante's Inferno. Everyone seemed filled with madness. Children ranging from two to nineteen years of age were screaming, crying, shouting in hysteria, or running about frantically because of their pain and difficulty in breathing. Some were just terribly silent as they sat in deep shock. Several children had no limbs, no ears, no fingers, no nose or lips. Many had damaged respiratory systems from the intense heat. All of the children had endured the physical trauma of massive broken bones and burns that covered 40 to 85 percent of their bodies.

The children needed immediate emergency medical assistance as well as psychological assistance. In order to survive, most of these victims had to recover from shock and coma. Due to the extent of their trauma, many of them gave up. During those first few days, one out of every six children died in our arms.

When I first arrived, three of the hospital walls were covered with the names of those who had perished, and each day the list of names grew. Whenever I walked past these walls, my eyes would flood with tears. By the fourth or fifth day, the walls were rapidly being covered with information about children who were in critical states or dying. Passing this area of the hospital became a walk of despair, and I hated the fact that the walls carried only information of death and near-death.

I had no idea how to work with these children who had no perspective on how to survive. I had been determined to come and help others, yet as I walked through the hospital I realized that everything I had learned was now being turned upside-down. My training had not provided me with any tools to help so many children stuck in the depths of physical and emotional trauma. My prior knowledge was based on cognitive skills, yet these children were unable to access the linguistic level of the brain. They were stuck in the brainstem's flight/freeze response.

I felt stunned by my own lack of tools to help me reach into the inner world of these children. When I saw so many children in coma, in shock, or exhibiting symptoms of deep phobia, I began to feel helpless. I was astonished to realize that none of the methods I had learned in my prior psychological and paramedical training could be of assistance in bringing these children to the beginning stages of recovery.

Even though I had extensive academic training and professional experience, my knowledge was primarily for people who had already recovered from the shock stage of Post-Traumatic Stress. My professional approaches relied on words, and here before me were a multitude of children who could not even speak. How could I ever reach this many children? In desperation, I decided to follow my own instincts. I made a decision to use what serves for survival – reflex response regulation and techniques that work with unconscious parts of the brain. Both areas of knowledge – of a reflex and its impact on survival and A. Uznadze's work were familiar to me, as I had used them in my MA paper at the university, "Unconditioned Reflexes, Unconscious Processes and Human Attitude." I also had a little orange book called *Brain Gym®: Simple Activities for Whole-Brain Learning*, by Dr. Paul E. Dennison and Gail E. Dennison, presenting a program on enhancement of brain functions and learning through motor activity.

I became acquainted with Brain Gym® in 1988 through Prof. Irina Dubrovina. Dr. Irina, scientific director of the Psychological Research Institute of the Russian Academy of Education in Moscow, had invited the neurobiologist Dr. Carla Hannaford to give a presentation on the successful use of Brain Gym®. Following her dynamic scientific presentation in Moscow, Dr. Hannaford had thoughtfully left a handful of *Brain Gym®* books as reference for her Russian colleagues.

Dr. Hannaford's research had indeed indicated remarkable results through the use of the Dennisons' movement activities, but it just didn't seem logical that something so simple would have that profound an impact. Due to our lack of understanding of the effectiveness and power of simplicity, most of the books had sat around the institute relatively untouched.

Before I left Moscow to go to the hospital in Ufa, Dr. Natalia Tolstych, director of my psychological research program, suggested that I take along the little orange book, since Dr. Hannaford had reported such remarkable success with the material it contained. I just threw it into my suitcase at the last minute. Upon my eventual return with new respect for the book, Dr. Tolstych fully supported my idea of undertaking scientific research on the use of the Brain Gym® program.

In order to find some way to make a connection with the children in Ufa, I just briefly looked at the pictures in this *Brain Gym®* book. Because these children's injuries and Post-Traumatic stress were severe, I knew that it would be extremely difficult for them to release the patterns of their unbearable stress and emotional pain and corresponding physiological patterns in order to move into the present moment. I decided to begin with three different exercises: Lazy 8s, the Cross Crawl, and Hook-ups.

When I reached the hospital, I approached a child who had been standing and yelling the same sentence for hours, day after day. He cried repeatedly, "We were riding with my grandparents and an explosion happened and my grandmother had fire in her hair."

I looked into the child's frantic eyes, took his hand, and started tracing Lazy 8s. As I did this, with his hand in mine, I repeated his phrases to acknowledge that I had heard his agonized story.

"You were riding with your grandparents and an explosion happened," I said with him, continuing to trace Lazy 8s.

The mutual movement of our hands and arms seemed to stir a sense of surprise and curiosity within the boy. He became distracted from his repetition of hysterical thoughts. I began saying parts of his sentence with him and then broke away from the repeated phrase by adding in other thoughts.

"Yes," I would say, continuing the Lazy 8s, "you were riding with your grandfather and grandmother, and now, look, you are here in this room with me."

He eventually stopped speaking, looked at me, and said, "Yes, now I am here." A nurse standing nearby gave me an understandably grateful look, for she had been hearing the same anguished statement over and over again for hours.

As I turned to see who should receive my attention next, some of the children were running around the ward screaming, "Fire! Fire!" One twelve-year-old boy was lying down, staring off into space, and whispering, "I must go. If I don't, I won't be in time." How I wanted to find a doorway into his mind so I could let him know he was now safe!

I spoke with the children one at a time. With each of them I incorporated the Lazy 8s movement, letting my words honor who they were and the fact that they had experienced the fire. I used movement and words simultaneously, to invite them to step into a new awareness of being present in a reality that was now safe. Time and again I saw how movement was proving to be the means for allowing them to take a step into the present moment, for letting them discover for themselves that they were no longer in danger.

As the children watched or imitated the Brain Gym® movements (www.braingym.org), I continued to explain that I had come to support them and to let them know that they were now in a different time and place that was safe. Even though they were correct about having been in danger in the past, they began to feel that it was safe 'here' because 'here' rested in the present moment of 'now.'

I was always honest with the children. Often people think that if we can forget about a childhood trauma we will 'get over it' as an adult. I do not agree. I learned that it was crucial to accept the children's perception of reality from the past while adding the gift of natural movement. This process invited them to move themselves into the new reality of the present moment. And I found that the Brain Gym® movements I was trying out with them allowed the children to experience the past in their own unique way.

I implemented this process with a whole ward of children. I saw that, if I used Lazy 8s with the children, they stopped repeating their mournful statements. The first day, I only used Lazy 8s as I listened, repeating their words and then changing the last half of the sentence, thus inviting the children to shift their focus from their emotional afflictions and bring themselves into the present moment. I would ask questions, such as:

- "What is your favorite color?"
- "I am here to offer help to you, do you understand that?"
- "What is your favorite food?"
- "You are safe now, do you know that?"
- "Do you like to draw?"
- "What did you eat for breakfast?"

Using Lazy 8s seemed to draw the children safely out of being stuck in the survival and protection functions of the brainstem. They were activating and moving into the higher areas of the brain, areas that allowed them to be logical and to see the present reality of being in another time and space.

The next day I introduced the Cross Crawl to a few boys, but only one could copy the movement. Many of the children were unable to move at all due to their burns, so I would demonstrate the activity and just invite them to imagine or pretend they were doing it with me.

I was reminded of the work of the Russian psychologists and researchers V. V. Lebedinsky and S. Y. Rubinstein in the area of child development, and of the discoveries of these two regarding the use of ideomotoric movement to create changes in a person's experience. These respected Russian researchers originally coined the word 'ideomotoric,' referring to the human ability to create microscopic motor-planning skills, activated in the center of the interbrain (basal ganglia), by watching a motion several times. This creates a mental rehearsal that eventually leads to actual movement. This imitation of movement is based on visualization and imagination and is the same sequence that a child carries out in watching and learning the processes of walking and talking. With time, the mental rehearsal becomes reality. Thus, the initial problem that the children's burns prohibited them from physically participating in the movements was overcome by their mentally experiencing the benefits.

Dr. Nikolaj V. Jermoluk, chief physician at the Ufa clinic, liked the results produced by use of the Brain Gym® movements and reflex 'rerouting' techniques because the children were returning to current reality and releasing their fixation on the accident. Momentary relief would come even to children who were still crying and in shock, even if they were able only to watch or imagine these simple movements and techniques. In addition to working with the children at the hospital, I was soon called to the emergency ward to assist children and staff members there, so that everyone could learn and use these movements. For many of the children, the Brain Gym® movements, along with reflex and tactile integration exercises and words of encouragement, literally made the difference between life and death.

The children were initially stuck in the brainstem response of protection. Their minds had reacted to the horrors of the accident by freezing in shock or frantically running away from any further perceived danger. I understood that the memory and the shock were so intense that the children had become deeply anchored in a state of hypervigilance and overprotection. Yet everyone now noticed that the use of the Brain Gym® movements and tactile and reflex integration techniques were creating a new perspective of safety. As I shared these movements, I spoke clearly to each of the children about being a winner, about finding the inner courage to triumph despite all the odds.

It became clear that, as the children began to watch or do activities we proposed, they began to regulate their time-space perception (between 'tragedy past and safe now'), relax, and manage their pain. This became an important piece of data. Because large amounts of medication had been needed to control the severity of the children's pain, the doctors were concerned about potential side effects. The medical staff members were relieved to see that less medication was necessary when the children were doing reflex activation and Brain Gym® movements.

The doctors started referring to the movements as 'Miracle 8s' and 'the Miracle Cross Crawl.' Within the first three weeks, they asked me to teach the Brain Gym® activities to all the medical students and nursing staff and to the one or two full-time personnel assigned as an aide for each child. Once I had taught the different Brain Gym® movements to the entire staff, I asked them to model these movements each time they approached a child's bed. If the children were able to move, they would follow along doing these developmental movements. If they could not move, we found benefit from just having the child imagine doing the movements while they watched the model. As for the reflex rerouting techniques (the old name of these exercises) I was doing them myself as they were more complicated and demanded more knowledge.

The hospital's medical statistics later showed that, in less than three weeks, the death rate decreased in the wards where the Brain Gym® program was implemented. It was further documented that the rate of healing was much faster than in the wards that did not use these movements.

"Nikolay, You Are a Winner!"

Because shock can be lethal, it was important to release certain children from their immobile state as soon as possible. I found that children who did the Brain Gym® movements were releasing their shock state in two to four days, whereas those who did not do the movements remained in a prolonged state of shock. I also saw that when children who were provided 'reflex re-routing' techniques with use of psychotherapy, were releasing the shock stage of the stress much faster, coming to 'here and now' and normal activities of children, drawling and play.

One twelve-year-old boy, whom I shall call Nikolay, was always hiding his face and body. He would lie in bed and stare at a wall for days, frozen in fear. So great was his need for isolation, that no therapists had been able to get a response from him.

I said to Nikolay, "If you don't want to look at me, that's okay. I'm here to support you and to let you know that I care about you. My heart is open to you."

He started to cry, and told me that it would be better if he were dead.

Thanks to the Brain Gym® exercises Lazy 8s and the Cross Crawl, Nikolay eventually responded to communication. These activities seemed to add considerably to his feeling of safety. In time, I was able to coax Nikolay to draw. He drew a picture of a face covered with lines, with one eye, no nose, big ears, and fangs at the mouth. He gave me the drawing and told me that it was a picture of himself. All mirrors had been removed from the hospital so that Nikolay and the other children couldn't see how they looked. Nikolay had indeed endured many lacerations, including several facial disfigurations. He explained that the drawing was a picture of himself and that the fangs were symbols of anger, calling the picture "the devil with one eye."

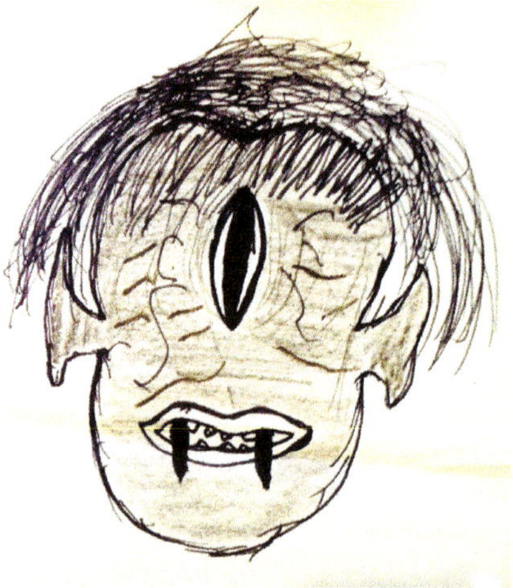

Nikolay's self portrait.

It turned out, however, that Nikolay's depression was caused not only by his injuries, but also by his parents, who when they came to see him were horrified by the thought of their son going back out to face the world. I quickly began to work with the parents. They had put a healthy child on a train, and now that child was dealing with his own shocking physical condition. No wonder the parents required a lot of psychological help and support, as well.

I used art therapy with Nikolay on a daily basis and also had him do the Double Doodle movement from Brain Gym®, which quickly became his favorite. As he worked, I told him that he was young and that his skin would heal. We discussed how it doesn't matter what others think; it only matters what we think of ourselves. "You must know that you are very strong," I said. "You have survived against the odds. As a matter of fact, three times, before today, you have proven to me and the world that you are a *Winner.*"

I explained these ideas to Nikolay in words similar to the following:

"First: In order to come and play on Planet Earth, one must 'catch a ride' on a thought. Now, there are actually trillions and trillions of thoughts, and catching the attention of just one takes a lot of wisdom, speed, and courage. Only one very special thought could become the little boy with your name on it. You beat all the odds and connected with that one thought! You were victorious! Only a *Winner* could do such a thing.

Second: A race was held, and millions of 'racers' were running to cross the finish line in the greatest marathon of all time. Only one racer could beat the odds and win and have his name announced. And look who won. You won! You were triumphant. You were conceived.

Third: Then, you once again proved that you are a *Winner* when you bravely faced and victoriously found your way through the process called birth. No matter how many times barriers were presented, you met and overcame each obstacle and made your entrance into the world. You had won a third time. And now, here you sit, victorious. You have again triumphed over impossible odds, and you are still here. Do you see, you are a *Winner* again and again! You may be afraid that you will not be accepted in this world, but because you're a *Winner*, maybe it means you have come to teach your parents and others how to have understanding and love in the face of all situations. All I know is that you are a *Winner*, and a *Winner* must teach others how to be the same." With this invitation to remember his innate wisdom and courage, and from reflex rerouting, Brain Gym® movements, and art-therapy, this young fellow began to move out of his state of shock and depression and find the courage to create a new future. I knew he would be a courageous person and would stand above his traumatic past.

Touch as a Tool to Inner Peace

The work at this catastrophe allowed us to discover that the role of touch is immense and it can serve as an independent therapy. We discovered several essential key techniques that were highly effective to reduce shock and negative overprotection, thus regulate post-traumatic stress. These particularly were 1) the depth of touch, 2) length and direction of skin stroke, 3) activation of proprioception, and etc. The year 1989 is the year of initiation of the program we now call MNRI® Tactile Integration. Since then this program has been used with great results for thousands of children and adults to release their overprotection and traumatic stress, and to establish their inner peace and strength. Soon, because of the rapid improvements in their ability to move and to be present in the moment, more and more children in the ward started drawing. As their desire to communicate became stronger, they asked for paper, paints, crayons, and anything else that they could use to give them the freedom to express their feelings on paper. The first drawings were full of darkness and pain, reflecting the depths of fear and despair that they were experiencing. The most common picture the children drew was of people running with their hair burning. Within seven to ten days, the pictures changed as the children emerged from their emotional and physical shock.

During the first two weeks, most of the children's drawings had to do with fire, explosions, and train crashes. The Calf Pump movement from Brain Gym® was particularly useful because it helped to release the Tendon Guard Reflex, which tightens in response to fear. This movement looks similar to the warm-up exercise that runners use. Some children were lying in bed and couldn't move, so we needed to work with this reflex and created the specific exercise now called Foot Tendon Guard. We use it widely today as one of the basic reflexes to reduce stress and tension. Because this reflex helps to relax the fight or flight response, it helped to release feelings of anxiety and fears. Through use of the sensory-motor techniques and movements, they were able to swiftly transition from alarm into purposeful, and eventually playful, movements. Although the stages of shock and stress came step by step, I was astonished by the remarkable speed of the progress.

Left, an aerial view of the accident site. Right, Kris' abstract drawing of it.

In the initial stages of his recovery, a young boy whom I shall call Kris drew an abstract picture. He drew the scene of the train accident as if he had viewed it from above the wreckage, showing the center of the explosion and the destruction of the railway.

This particular boy had been driven to the hospital, not flown in a helicopter (which would have given him opportunity to see the sight he drew). A pilot later commented that the boy's picture was a remarkable representation of the view from above.

Kris' second drawing seemed to be more representative of his own physical experience. It was of a burning horse running in the fire. Kris said he was feeling sad not only for the people who had been killed and injured but also for "the thousands of animals that died." As this thought was distressing to him, I used tactile and reflex integration, art-therapy techniques, and also Brain Gym® movements to help Kris let go of these images.

Many of the children's drawings reflected the same physical trauma that they had themselves experienced. For example, one young girl with widespread burns drew animals in dark brown colors with widely opened eyes. She said this was a picture of bears in a fireball. The bears were frightened and wanted to see the daylight instead of the smoke and darkness of the night.

These bears went to the "forest hospital," where other animals tried to help them, but the child initially insisted that they were still afraid of fire.

Naturally the children had a fire phobia. The sight of a worker lighting a cigarette would easily trigger this phobia, and some of the children would run off screaming. Tactile techniques were most sufficient to bring children to inner peace on a non-verbal level. I also had the children sit or stand in the Brain Gym® Hook-ups position or just imagine doing this simple activity.

Bears in the fire.

Over time, the children drew the fire in their pictures smaller and smaller. And as they discussed new ways of looking at the fire, the children would move with reflex rerouting and Brain Gym® movements. When they felt safe to look at fire in a new way, we would draw pictures showing constructive uses of fire. While doing a Brain Gym® movement like Hookups, we imagined sitting around a fire with friends and singing songs about being a winner. At the end we would all do the Cross Crawl. The children's desire for survival was being drawn out of them through the use of art, Brain Gym® movements, psychotherapy, and reflex pattern activation.

Because the explosion had happened at night, other common fears among the children were those of death and darkness. Also, many of the children were afraid of any kind of train or transport.

I offered the children therapy for all of their senses. They had to relearn that it was safe to see, touch, taste, smell, and hear in a normal way again. Because their orientation to the sense of smell was fixated on burning skin, bones, and metal, the slightest scents were often irritating or led to emotional outbursts. The children needed to have their sensory channels experience a re-education to pleasant associations. We explored the contrasts of chocolate, lemon, ice cream, salt, and spices, because nothing was tasty to the children at first.

Our work continued with all of their fears: the fear of fire, of burning people, and of dead people, trains, and treatments. It was obvious that sensory integration work was much more sufficient within reflex rerouting work. I was again astonished at the rapid rate of improvement that occurred daily, and I decided to incorporate Brain Gym® movements into every one of my psychotherapy and reflex rerouting sessions. Later I was able to make a correlation between the Brain Gym® movements and natural developmental reflexes, and to understand why the effects were so profound.

Being a scientist, I promised myself that I would set up indepth studies of the phenomena resulting from these exercises because I knew of nothing else that could take children so quickly from a past tragedy into present reality. Not only did I eventually get Russian universities and scientists to study why reflex integration and Brain Gym® works so well, but I later also had occasion to experience the material first-hand when my personal life was struck by tragedy and loss. Using all of the skills that the children had co-created with me, I later moved beyond the murder of my beloved twenty-year-old sister, Helen, and eventually rediscovered and even deepened my sense of reverence and joy for the gift of life.

Two-Handed Rainbows

As nature has programmed us to survive, then the power of natural survival methods can be used to awaken hope in people who are under stress. In Ufa, I discovered that adding the sensorimotor integration of reflex patterns and developmental movements speeds up and enriches the process of remediation. These techniques and movements to draw out the activity from the brainstem to the cortex, are are the foundation for awareness and conscious control. Only then can we be ready to interpret trauma fully and objectively.

I used the wisdom of natural body movement to give the children a new perspective from which to recall the fires. Homologous movements were exceptionally useful in helping them to move from the memories of the past to the current reality of 'now.'

The children used these movements to release their fear of fire by pretending to stomp on a fire with both feet or pound it out with both hands. The experience of clapping both hands to smother the fire also allowed the children to revisit the strength of homologous movement.

For the same reason, I engaged the children in art therapy, drawing pictures of fire with two hands at the same time, making use of the Brain Gym® Double Doodle activity. For the children, expressing their thoughts and feelings as they drew created a bridge between the past and the present moment. While moving in this way, they told imaginary stories with anger, then with a seriousness, and then with laughter.

The children drew every day with both hands simultaneously, or if they had only one hand, they would draw with that single hand, imagining it was 'made of two hands.' If they drew portraits of favorite people, they were encouraged to stay away from thoughts of loved ones who had died in the crash. Dealing with these memories could only come later, when they could view these truths from a place of inner safety. Creating a goal for the future was a priority.

These young patients loved homologous movements and referred to them as the "rainbow journey." They created rainbows in the air, first starting with their two hands above them and meeting in the center; then they would move their hands in opposite directions, creating an arch. Some children had to imagine this movement, since they could not move due to the severity of their burns. Repeating this activity over and over, they imagined and talked about different colors of the rainbow. I would suggest the color red, which the children associated with fire, last of all, and only if the children seemed ready to handle it. All the other colors were used to bring safety to the color red.

The children often repeated the rainbow movements at different speeds and with various colors and sounds, as well as with stories about strength and wisdom and being a winner. Often, when I would walk into a room to work with a group of children, I would find them telling one another rainbow stories and creating the rainbows in the best form that they could. They seemed to instinctively know that this basic developmental movement was a key to their recovery, because they made repeated, spontaneous use of the movements.

They formed the rainbows with both hands moving outward from the middle; then they would move from the outside and come back to the middle. Instead of being stuck in an extended mode of protection and guard, they would shift their peripheral vision to the experience of relaxed centralization. If one has suffered a trauma, the peripheral vision may become set in hypervigilance as a form of protection. When the visual system is overused in this way by perpetual defense, it becomes exhausted and ineffective. Creating the rainbow with both hands in a playful way allows the eyes to finally release this overprotective posture, creating a more relaxed and centralized visual acuity. In the instance of the children in Ufa, this relaxed visual state would later help them improve their reading skills when they returned to school.

A drawing by a six year old girl: "I survived."

Our work with movement also allowed the children to develop an expanded sense of where they were in time and space. They would move with both hands while they were standing, sitting, and lying down. These movements were physically preparing them to be able to explore different levels of balance and stability with a sense of safety and play. Eventually they would 'walk' across the rainbows and tell stories to one another, which allowed them to move into more advanced developmental movements of a crosslateral nature. Some would cross-crawl over the rainbow, telling funny stories as they moved. I then presented combinations of movements, offering new challenges to some and comic relief to others as I made 'silly mistakes.' Doing the Moro Reflex rerouting and Brain Gym® Cross Crawl with a variety of songs brought both challenge and laughter. Some children actually regained their ability to laugh within five days, while others took a month to rediscover it. Often laughter became a relief valve through which crying and screaming could follow. All this experience allowed us to later create MNRI® 'Visual and Auditory Reflexes Integration', 'Art-Therapy and Reflex Integration', and 'Proprioceptive and Cognitive Intergration' programs for children and adults.

The reflex rerouting and Brain Gym® exercises also created rapid self-education of the mind and body. The children's eyes relearned to feel safe looking in all directions. Later, the children used Alphabet 8s when writing. They started by practicing individual letters in the Lazy 8s on their paper, and then they would write a message like "I love myself" while using this Brain Gym® movement. They became so excited about this achievement that they would follow this exercise by giving themselves a hug. Eventually they used the same process to write each letter of their message sentences, saying, "I am a *winner*" or "I am strong/healing/recovering/developing."

Regaining Self-Worth

I found it important to understand and accept the stages of traumatic stress-grief-loss without making shallow promises, like "time is a doctor," or just "forget this tragedy." The children were teaching themselves, through movement and drawing, that they were winners. Experience must come first, and encoding the experience with words ideally comes second. Otherwise the words are empty and hollow, and carry only the meaning perceived by the person speaking. When the experience naturally draws out the words from within, the meaning is deep and authentic.

I followed the needs of the children by using empathetic listening skills that allowed me to hear, be congruent, and share open and honest feelings from both sides, while doing the reflex exercises and Brain Gym® movements. The important thing was to feel the feelings, follow the children's emotional experience step by step.

Once the sense of survival was experienced and then stabilized by the children, we began preparation for their future, exploring self-image, self-value, and self-worth. The children learned that they must be ready to teach others that they were winners for yet a fourth time, because they had gained strength through their victories. I encouraged them to go out and teach others about this wisdom. The children would have to learn to notice their own behaviors of manipulation and compensation that could put them back in the victim role, thereby causing them to cry out for attention and protection again. They had experienced what it was like to move out of the need for protection and into the place of development. They would have to be ready to explain this to the friends, teachers, parents, and neighbors who might offer them overprotection or go into survival reactions themselves when confronted with these children.

Drawing by a six year old girl, a picture with Lazy 8s (a Brain Gym® technique).

Most of the children had gone through a phase of hating their damaged bodies and faces. I often shared fairy tales, metaphors, and stories of high moral values, in order to help them experience a perspective of self-respect. The following story is one I told to one girl who spoke of her concern about the reactions of others regarding her face:

"A young woman and a young man were falling in love. The girl's face was not very pretty, so she didn't believe her sweetheart when he said he loved her. She felt it was impossible for anyone to truly love her because of the way she looked. The two of them had become very good friends, and yet she continued to feel bad about herself. One day she decided to leave the city and find someone who could change her face. She lived in a remote village in the mountains, but she had heard of a man who had some herbs that could possibly change her looks. She found the man and put on a mask of herbs, waiting to become beautiful. When she returned with the mask on, the boy was frantically running around looking for the girl he loved. As she stood in front of him, he cried because he had lost his best friend and could find her nowhere. The next day she decided to remove the mask and return to him as she was. He saw her and took her into his arms with joy! The face didn't matter. It was the person inside that he loved, and now she understood the true meaning of beauty."

Before leaving the hospital, the girl with the ravaged face came up to me and said, "I am the happiest person in the world, and I am the strongest, as well, because I have survived. I am a *Winner* inside, and I know how to take the *Winner* into my future."

A Resource for Happiness

The strength of the *winner* within is the source of survival and a resource for being happy in any physical, intellectual, or emotional state. The human response to a trauma is characterized by four basic stages: (1) protection for survival (freeze or flight), (2) reaction for survival (fight or flight), (3) growth (stopping to think and learn), and (4) development (implementation and modeling of the new learning).

Accessing the wisdom of primary reflexes and natural movements in the body supports the completion of all four of these stages. Inviting both the physical and the emotional strength of nature to awaken within creates a space where it is safe to let go of any repetitive story of mistakes, failures, and blame.

The children I knew in Ufa met the challenge, gathered their natural internal resources, and moved through the entire experience of survival showing how negative survival and phobias can be transitioned into positive survival, and reflex rerouting, arts, play and Brain Gym® were tools or means for this pathway. The catastrophe taught them that they were heroes. They were brave and strong ones who fought emotional and physical battles against all odds until, eventually, they brought themselves successfully into the new reality and life in front of them. They met fear face to face, released it, and replaced it with the wisdom of the *Winner* inside. From this place of wisdom, the children prepared themselves and others for a new future. Like the boy in Tiananmen Square, they created a picture of courage to show to the world. They also bravely stood in the face of fear and demanded that it stop.

Yet they chose to move into an inner place of simple truth. From the depth of the winner within them, they created a manifestation of the real 'me.' Their self-discovery was based on the experience of facing endless challenges, big and small. Their bravery, wisdom, strength, and inner knowing, which allowed them to overcome countless daily obstacles, remain present to assist themselves through the aftermath of the catastrophe. They showed us all how purposeful movement and touch can simply reconnect us with inner intelligence, which allows us to move forward. We thank them for reminding us of the natural wisdom within the body. Their extraordinary story will long continue to serve as an inspiration to us all.

They did so well! How easy it would have been for these children to choose to continue running away and hide from this fearful event! Now, in 2014, I thank them for the unparalleled lesson they taught me. It is a lesson that went on to help thousands of other children and adults. These children are mature adults now and I hope that they are living with good health and joy in their lives!

The Epilogue

After I returned to Moscow, I initiated extensive research into why work with reflex patterns and the Brain Gym® movements could produce such profound results. The little orange book, *Brain Gym®: Simple Activities for Whole-Brain Learning* by Dr. Paul Dennison and Gail Dennison used in Ufa had shown a deep and lasting impact, just as Dr. Hannaford had reported. The implementation of the Brain Gym® activities had brought remarkable and significant results, and those activities deserved research and investigation so I shared my findings with many of my colleagues, and the research began.

Further investigation of the effect of the sensory-motor patterns had shown that those exercises were sufficient in that they were resembling the reflex patterns in working with post-traumatic protection and survival. This allowed us to create a procedure to work with Post-Traumatic Stress which is presented in the MNRI® manual, *MNRI®: Reflex Integration and PTSD Recovery. Advancement for Children and Adults Experiencing Post-Traumatic Stress* (2013) by S. Masgutova, D. Masgutov, E. Akhmatov.

Dear Children of 1989, Dear Winners of Life in the Asha-Ufa train railway catastrophe. Thank you ALL for your ability to transform survival into a positive state, your courage, and for desire to grow. We are thankful that your message of courage and your ability to find the Winner within will serve as inspiration to many. – Authors, Co-Founders of the SMEI Institute

Post-Traumatic Stress Protocol for Filipino Typhoon/Earthquake Survivors

Anthony S. Grecia, President, Occupational Therapy Association of the Philippines; Kristine Ann Lopez, OTRP; Mara Comagon, OTAP, DRRR; Diane Whiteside, PT, MNRI® Core Specialist and Instructor

Our team arrived at the International airport in Manila, Philippines from the US and Poland.

Anthony Grecia, President of the Occupational Therapy Association of the Philippines, Inc. (OTAP), organized the MNRI® Core Instructors visit to the Philippines. He and his association have consistently committed their organization to serving the Filipino community. This was further enhanced through the formulation of the organization's Disaster Risk Reduction and Response (DRRR) framework and sub-committee. One of OTAP-DRRR subcommittee goals is to enhance the capacity and competency of Filipino Occupational Therapists (OTs) to respond to the survivors of disasters as well as establish a pool of resources (volunteers, financial support, and strategic donations in kind).

Diane Whiteside, Lori Burgess, Anthony Grecia, and Constance Jordan

In mid-October 2013, a 7.2 magnitude earthquake shook the Central Visayas. As a result, many natural and man-made structures crumbled. Basic necessities such as food, water, and electricity were scarce. Help to fill those needs came but survivors are only beginning to get back onto their feet in the affected areas. After the earthquake, the country was then hit by super Typhoon Haiyan (November 2013). Greater needs became apparent which brought about the consideration of rehabilitation services.

The MNRI® team spent their first day preparing a lecture at the Philippine University Medical School. Maja Busma and Teresa Busz reviewed the hands-on techniques; Constance Jordan and Diane Whiteside put together the lecture. We were up late that evening preparing the MNRI® Post-Traumatic Stress Disorder (PTSD) Fast Action Protocol lecture to teach at the Occupational Therapy section of the Medical School. The four of us and Lori Burgess taught to an amazing group of dedicated people.

Anthony said that, through the generosity of Dr. Svetlana Masgutova, they were very grateful to MNRI® for sharing its Post-Traumatic Stress protocols as well as sending their Core Specialists and team leaders (namely Diane, Constance, Lori, Teresa, and Maja). At that time, we felt very grateful to MNRI® for allowing us to assist OTAP-DRRR in meeting their objectives of building a trained group of PTSD specialists as well as sending volunteers to the different areas hit by the earthquake and typoon.

After the day of instruction we were divided into two teams and worked directly with the people in the devastated areas. Maja and Teresa became Team 1 of OTAP-DRRR to respond to the survivors in Roxas City, one of the provinces devastated by Typhoon Haiyan. MNRI® specialists Diane, Constance, Lori, and volunteers of the Occupational Therapy Association of the Philippines became Team 2 of OTAP-DRRR to respond to the orphan survivors of Bohol (specifically the Arms of Love Children's Home Orphanage), the province devastated by the 7.2 magnitude earthquake.

Intervention

Anthony Grecia, President, OTA

Observations on the two day sequence of activities for the house parents/caregivers and orphan survivors:

During these 2 days, Diane, Lori, and Constance worked together. It was amazing to see the MNRI® protocols create a calmer reaction (caused by the release of stress hormones) and become adapted in these people in such short time. Some of the highlights were:

1. After the art activity, sharing, and psychosocial first aid, Diane led our sub group (composed primarily of men and house fathers) in demonstrating the Fear Paralysis protocol. I was able to see how, in general, their postures and breathing patterns slowly loosened and the atmosphere lighten up as the Fear Paralysis protocol was done. I observed this in their cheeks, faces, and conversations. Some of them became more engaging and were willing to share more of their concerns after the protocol was done, considering that they were men and, in general, Filipino men need more time and appropriate context to share feelings.

2. One house father said that he had difficulty sleeping. During the day of the earthquake, the house father was assigned to drive some of the children on a field trip towards a popular falls in Bohol. While driving, he saw rocks fall in front of them and then the road went up. He said that they were fortunate to stop at just the right spot where they did not get hurt. As Diane touched and pressed specific areas of his body, it was amazing to see how the house father comfortably closed his eyes and rested his body on the mat. I noticed how this facilitated rest and sleep using the sleep reflex patterns (Abdominal Sleep Posture). Although it was short, I sensed that the man was really resting and sleeping well. As an occupational therapist, I feel that the sleep reflex can be a major part of Post-Traumatic Stress protocols because sleep is one of the occupational areas affected in times of disasters. I felt that, as Diane facilitated the sleep pattern for the house father, he would then sleep better, have a more adaptive routine the next day, and better fulfill his role as a house father.

3. Another house father shared that he was also in the bus together with his wife, a house mother. When the bus stopped and they saw the rocks falling in front of them and road going up, his wife ran into a nearby banana tree and hugged it. Although this was a very serious and alarming memory, it became a somewhat subtly funny story among the house parents. Some of them joked that the wife hugged the tree rather than her husband. While narrating this story, although the house father was trying to smile a little bit together with the other males in the group, I sensed his tension and suppressed feelings. His posture was quite tight and stooped. His eyes were quite heavy and his smile was forced. I worked on him to process what he really felt and thought while his wife ran and hugged the tree. Diane taught him the Fear Paralysis protocol. Upon meeting him the next day, the house father had a more open and extended posture, with a more pleasant and a more genuine smile. At that point, I saw how the MNRI® protocol can effectively be combined with counseling strategies to promote transitioning from being a victim to becoming a survivor with an easy smile.

4. A highlight was how the tree metaphor was adopted from the MNRI® PTSD protocol and used for the children by Constance. Constance asked that the children pretend they were trees and to draw themselves as a tree with a specific face, showing different emotions. They were tasked to draw two trees – one as a tree during the earthquake and another tree during the present times. Constance emphasized that the use of the tree metaphor makes the survivors temporarily mentally move out of the experience and help them process their emotions. As the children finished their drawing, I was so glad to

Filipino children's drawings of their tree during the tragedy.

Above, more drawings by Filipino children of their experience.

see how the tree metaphor was able to make them express their feelings, even without using words. Some were very particular about drawing the eyes, mouth, and upper branches of the tree to express their feelings. It was also amazing to see how each child's drawings of their two trees were similar or different from each other, reflecting their emotional state in varying times. From a clinical standpoint, I realized the following:

a. The use of trees as a metaphor is very apt during art activities as it is part of nature. It can initiate the processing of the experience from a natural disaster, whether floods, heavy rains, typhoons, or earthquake.

b. The use of two trees (during the disaster and the now) reflect the children's emotional state.

c. The use of two trees (during the disaster and the now) can aid professionals in gauging how children are coping, even if they are not highly verbal.

d. Constance was right in emphasizing that the tree metaphor makes them move the experience out of their body, at least temporarily, to at look at themselves and their experience, and allows them to set new goals for positive survival and returning to a normal emotional state of inner peace.

e. The drawing and coloring of the tree in itself is a process that promotes normalization and routine for children, as this was one of their primary occupations during school days before the disaster.

f. This therapy tool is excellent for the earlier or non-verbal stages of Post-Traumatic stress when the child/adult cannot be reached though verbal instructions

g. MNRI® procedures are built to avoid the trap of re-victimizing people (which happens often with traditional verbal psychological therapy). The goal of MNRI® therapy tools is to also orient people toward a strong future and this is what was so beneficial for our specialists to understand.

My sincerest thank you to MNRI® and Constance for sharing the tree protocol.

5. During day two, we worked on training and capacity building for the house parents. They were taught basic psychosocial counseling techniques and MNRI® Post-Traumatic Stress protocols (e.g. Fear Paralysis, Embrace Squeeze, Moro Embrace, etc.). At this point, the team felt that most house parents had developed a lighter and less stressed disposition, providing a more positive avenue for learning and interaction. It was amazing how MNRI® can hasten the process of coping from victim to survivor. One house mother, who did not attend the psychosocial first aid session or the initial MNRI® protocol joined us later. It was amazing to see how she could easily be spotted among the others

At left, some of the Filipino children's artwork reflecting their feelings.

in terms of her posture, facial expression, breathing, and initiation. Constance also noticed it and requested to work with her first and apply some of the PTSD protocols. We were fortunate again to see how this house mother transformed to become calmer and more open. At the end of the training and capacity building, I also realized that the process of sharing and teaching basic protocols to the survivors makes them feel empowered to a certain point. Some of them were so thankful that they are able to call upon a strategy when they felt overwhelmed or stressed. They also feel empowered that they now have a strategy to use for some of the children. In a sense, learning the protocol itself makes them feel equipped to face challenges and continue coping. I felt that, apart from helping and processing them, we were not leaving them empty handed because they learned a powerful strategy, beyond monetary or material donations. At that moment, I felt so thankful to MNRI® and Dr. Masgutova for sharing her discovery and science with us in the Philippines.

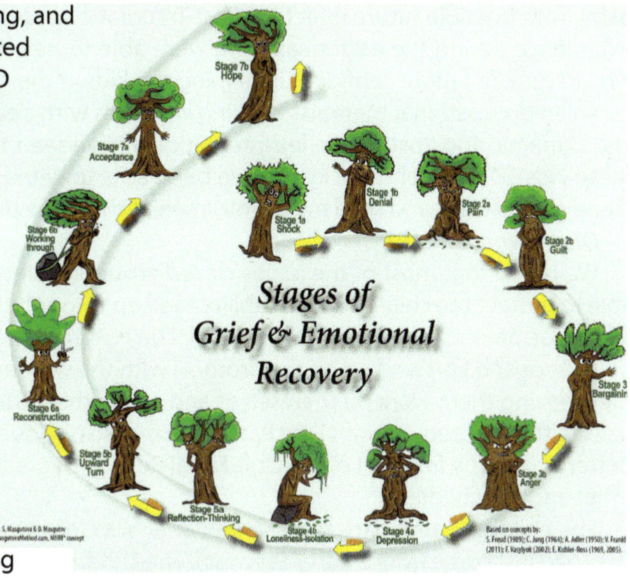

Kristine Ann Lopez, OTRP:

I want to share the earthquake experience from a survivor's perspective and how the reflex integration worked for her:

One social worker from the orphanage told of her experience during the earthquake. She was pregnant at the time and was leading a field trip for a group of kids. As they were on their way to the location, the coaster that they were riding in shook really hard. They saw the roads start to break apart and trees uproot. The children were terrified and she was afraid for the lives of the children. She felt trapped because everywhere they looked, disaster was looming. She said that she felt such a big responsibility and could not let herself crumble at the time of such distress, as she had to take care of the kids. She helped pacify them and led them to a farm lot near this area that she thought was safe. They stayed in this farm lot for several hours, even as aftershocks came, until she heard from a local person that it was finally safe to go home.

It was when they reached their orphanage safely, and all the kids were okay, that she finally thought of her baby, the unborn child in her womb. The fear for her own child's life finally sank in to her. After the earthquake, a series of strong aftershocks came – as if revisiting their past experience. The children tried to stay inside their houses, but they couldn't sleep for fear that the building might collapse on them. They finally decided to pitch tents in the vacant lot in the orphanage, where they slept for a month as the aftershocks subsided.

As our team heard the social worker's story, Constance and Lori led an MNRI® intervention. They did the traumatic experience release on the woman and her unborn child. Lori worked on her breathing pattern, which was notably short and shallow, and performed several Oral-Facial Reflex repatternings to help with adrenocortisol release which could have adverse effects on the unborn baby, if in great amounts. I was lucky to be able to work with them, too, as Constance and I used Embrace Squeeze on her upper extremity, to help her feel safe again through Tactile Integration. Constance also worked on releasing her upper Core Tendon Guard and other breathing reflexes.

Results: After working on her reflexes for 10-15 minutes, her breathing was notably deeper and with better rhythm. Her posture, and overall predisposition changed, as she seemed more at peace. The woman reported that she felt so much better, as if a weight has been lifted. She now felt that her baby was going to be okay after everything that they have experienced. Upon being asked what color she feels and sees, she reported that she sees white – which signifies peace.

We had such a great mission that time. We were able to help the social worker, and possibly even save the baby from possible future effects of Post-Traumatic stress that could have been brought about by the mom's experience during the earthquake. We were able to help many adults and house parents in the orphanage, who in turn, felt more confident in helping the kids in their homes to process their traumatic experiences.

As an occupational therapist working with kids with developmental needs using MNRI® Tactile Integration and Dynamic and Postural Reflex integration, I have seen the Masgutova Method® create positive changes in these kids' lives. I feel so lucky to have been able to witness the method work even with people under Post-Traumatic distress, and help my countrymen at the same time. Masgutova Method® really works!

Diane Whiteside, PT:

We noted that most of the issues circled around anxiety, the feeling of helplessness, the fear of not being able to protect the children, and inability to sleep through the night. Next, it was our duty to teach and work on the house parents with the PTSD protocol. They expressed their worries and that of being unable to sleep well. Constance and Lori worked on the protocol with the women. Each one received a treatment. I worked with the men, having them work on themselves and each other. Anthony asked that we teach them something to take back to the children. I taught Fear Paralysis, Embracing Squeeze, Pyramid Finger Stimulation, Abdominal Reflex patterns, and the first part of the Oral-Facial Reflex Integration Program using easy techniques to support each other and the children.

The children came back from school after we closed with the house parents. They were excited because they had just come from a party at school. The children were gathered into a large circle of about 30. They were given gifts of art and school supplies. Anthony began speaking to the children and encouraged them to start drawing. To work with processing their Post-Traumatic stress, he asked them to draw pictures of their memories of the earthquake. He asked them to talk about how they felt about the disaster, how they felt about their future, about creating courage and inner peace, and setting goals for the future by utilizing the concept of the trees and emotions from Dr. Masgutova's PTSD Manual.

The youngest child, age 5 years, just clung to her housemother and cried whenever she left, not wanting to participate. I walked around the group spending time near the child. I eventually sat next to her. She was accepting. I gave her crayons and she started coloring. Then she began speaking with me. The older girl sitting next to her stated, "You are speaking in English." The child made an effort to communicate. After the group had expressed and released their fears and built a bridge picture to the future, we taught the children how to do the Fear Paralysis technique, to use in stressful situations. Before the program was over, the youngest child and a couple of other girls started singing. This was like a wave and all the children joined in singing. It was very heart warming.

The next day we had the opportunity to reinforce the teaching of the techniques and get feedback from the day before from house parents. The house parents mentioned successes with the techniques. One woman stated, "I put my head on the pillow and closed my eyes. When I opened them it was morning." She and others had slept through the night. We had great success with improvement of their sleep and they said there was a reduction in their anxiety.

I am impressed by the OT association's efforts to create a program that will become a protocol to address the emotional, PTSD aspect of the disaster and take it to their people. I am proud our Polish-American team and honored to have donated our time and skills be a part of the disaster relief effort for the Philippines. I have returned to the USA with a special gift of greater depth and understanding. Anthony and his colleagues were well prepared for this expedition to the islands and created a really good professional team.

Mara Comagon, OTAP-DRRR:

Through setting expectations with the house parents, we learned that, though they are trying to normalize their routine, they have not had much of a chance to process their feelings of the earthquake. They have talked amongst themselves and, at times, resort to humor but they have yet to reach a state of hope. They want to gain knowledge and skills for understanding what they are experiencing, as well as, how to move forward from it. To this end, the house parents committed their attention and cooperation to learn as much as possible. Upon setting expectations between the team and the house parents, they were then split by gender. The team was also split into two groups to implement psychological first aid. The house parents each recounted their most vivid memory of the earthquake and how they felt about the situation then, in contrast to how they feel now. It was apparent that they were still traumatized by the event. The problems identified were anxiety, being easily stressed, difficulty sleeping, having an unclear mind, and difficulty making choices. The house parents stressed that they just do not know what to do during times of disaster or how to recover after it. As they opened up during psychological first aid and the team built rapport with them, they were then open to MNRI® techniques. The techniques demonstrated and used with them were Fear Paralysis, Embracing Squeezing, and Abdominal Sleeping Posture. After the session, the house parents in attendance expressed great appreciation, interest, and enthusiasm over their experience. One even said that, "ikukwento ko sa asawa ko natutunan ko ngayon. Gusto ko siya isama bukas. Gagamitin ko natutunan ko sa kanya para matulungan siya." ("I will tell my wife of what I experienced and learned today. I want to take her with me to attend tomorrow. I will use what I learned to help her.")

Following the session with the house parents was the session with the children. They were asked to draw trees depicting a vivid memory as well as what they felt like during the earthquake and another of the current time with the team. They were then gathered for counseling. The children found it more difficult to express their feelings though it was clear from their drawings that they held negative feelings of post disaster stress. Some were able to articulate their feelings of loss of stability and activation of equilibrium and they said that they were fearful of the aftershocks and hypersensitive to sensations. One told, "Kahit may gumalaw lang ng upuan ko o magulo yung natutulog sa ilalim ng double deck, akala ko aftershock na. (Even just feeling someone shake my chair or when the one I share my bed with is moving about a lot in his sleep makes me panic because I mistake it for an aftershock.)" The children were then asked to imitate the Fear Paralysis exercises to end the session. Some children sang, while others who were hesitant during the counseling still remained quiet and withdrawn.

Later, training in Fear Paralysis, Embracing Squeeze, and Pyramid Finger stretch was demonstrated. The MNRI® team roved in pairs through the house parents and monitored their application of techniques. Then, each house parent was briefed on children at risk and how to further help them. When the house parents were asked how they were throughout the session, they were quite excited to respond. One parent (upon whom the Abdominal Sleeping Posture was demonstrated) said that his mind finally felt clear before sleeping. Another remarked that "Alam mo unang beses ko natulog na napikit ako ng gabi at pagdilat ko umaga na. Diredirestso tulog ko! Ang laking tulong nito sa amin. Gusto ko magamit 'to di lang sa akin, kundi para makatulong din sa iba. (You know what, it's the first time [since the earthquake] that I shut my eyes at nighttime and opened them to find that it was morning. I had uninterrupted sleep! This is a great help to us. I want to use this not only for myself, but to help others as well)." Still another stated, "Ngayon, alam na namin na may magagawa kami pag may nangyaring ganito. Andami naming natutunan. (Now we know that we can do something at times like this. We've learned so much.)" They were all very appreciative of what they learned and experienced. What transpired over just two days was a profound experience where expectations for both the team and house parents were met. To witness people feel empowered as they are given tools, and evolve to be not only survivors but new agents of change is amazing, to say the least. Hearing comments of how the techniques they learned made them feel at ease, lighter, or have their mind become clear was extremely rewarding. With newfound hope, the house parents at Arms of Love Children's Home feel more confident in taking care of, not only the children, but themselves as well.

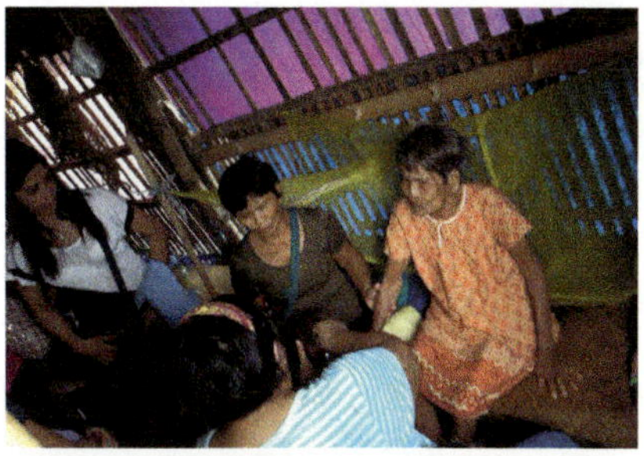

Scenes of Filipinos receiving treatment with Maja Busma to cope with the aftermath of the earthquake and typhoon.

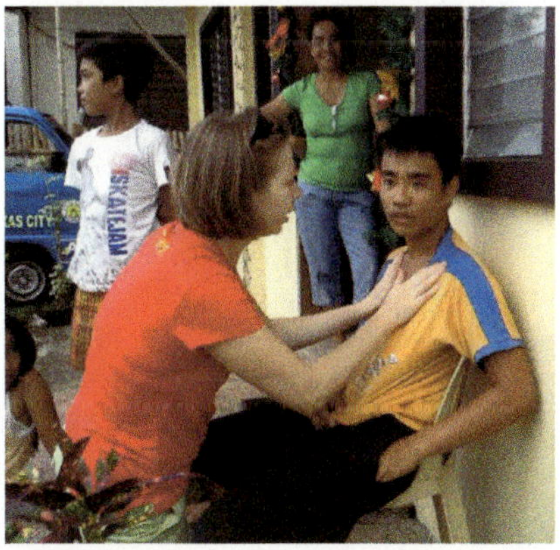

Anthony Grecia, President, OTA:

We are highly grateful and thankful for your recent support to the association, specifically to our OTAP Disaster Risk Reduction and Response Sub-Committee. We have done psychosocial-occupational debriefing and therapeutic activities in the past but it was enlightening to see how the MNRI® Post-Traumatic Stress protocols helped further the most recent disaster survivors to feel better ("gaan ng loob") and feel stronger and hopeful ('lakas ng loob") after our visits. Again, thank you very much Dr. Svetlana for sharing these protocols with us. It is truly one of the wonderful blessings and gifts you have shared with us Filipinos. To all MNRI® Core Specialists who volunteered to join us, thank you very much for sharing your time, expertise, patience, flexibility, and support. Our gratitude goes also to the very kind and generous MNRI® Foundation for sharing their resources for this project. The MNRI® protocols are now incorporated to the sequence of activities OTAP-DRRR initiates and performs while helping the community. Last December 26-30, the protocols were also used while helping Ormoc, Tanauan, and Tacloban, which were all greatly affected by Typhoon Haiyan. It was highly notable that the protocol has now reached the under served far-flung hillside barangay of Ormoc. We, together with David Ford, used the Fear Paralysis protocol and Embrace Squeeze with more than 70 people in a hillside community in a simple front yard. The 'tree' protocol was also utilized to help process around 50 children in a day care center in Tanuan. It was great to see how the Fear Paralysis protocol and other exercises were 'normalizing' the children's worries and leading to a state of inner peace in the wet classroom supported by temporary roofs (made out of tarpaulins). Young lives are continuing to live for the future.

Dear MNRI® Team, it is truly a wonderful blessing and gift you have shared with us Filipinos. To all MNRI® Core Specialists who volunteered to join us, thank you very much for sharing your time, expertise, patience, flexibility, and support. We are very grateful and thankful to the association for the recent support, specifically to the OTAP Disaster Risk Reduction and Response Sub-Committee. Our gratitude goes also to the very kind and generous to Bridge To Healing Foundation and the Svetlana Masgutova Educational Institute for Neuro-Sensory-Motor and Reflex Integration (USA) for sharing their resources for this project. I believe that science, generosity, and collaboration have come a long way for humanity. My sincere gratitude to Dr. Svetlana and MNRI®. – Anthony Grecia, President, OTA, Philippines

Finding His Way Home: The Story of a Soldier Returning from War

Peg Johnson, MNRI® Core Specialist, Madison, WI, USA

Peg Johnson

A wife's account of her husband's PTSD journey:

Before Iraq, Jon was a very humorous, relaxed person. When he returned home, he was in fight or flight mode 24/7. He could not sleep well, produced angry outbursts constantly, and would explode in anger in simple situations such as a glass of water being left out on the counter.

One morning, I came into the kitchen because he was yelling about something, and I found him with his forearm against our 11-year-old son's throat up against a cabinet. Calmly, I asked him, "What was going on?"

He responded with, "I am trying to get his attention."

"We do not use force to get someone's attention," I retorted. Frustrated, he stormed out of the house.

One time a car drove past the house and it backfired. Reacting to the noise, he dove to the ground and covered his head. For the first year and a half, he displayed signs of being stuck in Iraq-military mode. One day, within a month of his arrival home, we decided to drive to Madison, WI. While approaching a bridge, he gunned our Ford Expedition reaching to 90 mph and began to zigzag on the freeway. I began to scream and asked, "What are you doing?"

Jon shook his head and could not figure out what was going on. After a few seconds, I calmly coaxed him into pulling over. He appeared to be lost, as if he did not know where he was.

Jon had issues reintegrating into society. He did not like to talk to people, and fell into a deep depression. His old life seemed like a distant memory that he could not regain. Emotionally, he could not relate to people and did not fit in. Sadly, he could not even attend his daughter's swim meets because of the gun used to indicate false starts. The avoidance of social events and strange behavior continued on for the first year and a half. Oftentimes, I told him that he was scaring the kids and me. Before the war, Jon was never a violent person. With his clown-like personality, he could make anyone laugh. He truly was a wonderful dad and husband.

After living in fear and missing his old life, he decided to begin MNRI® therapy with Peg Johnson. After three sessions with Peg, he returned almost back to normal. Unfortunately, he cannot erase the past, the terrible situations that he faced, and what he saw; but they are mere memories that have been filed on the shelf in his brain. The only residual problems left are his unease with large crowds. Jon oftentimes mutters, "We are a huge target!"

A few years ago, we moved out of state, which stalled Jon's continuance with MNRI® therapy. I strongly believe that continuing on with the MNRI® program would have helped him beat his anxiety towards large social gatherings and events. For now, regrettably, we have learned to avoid events with large crowds.

Peg Johnson's (Jon's MNRI® Core Specialist) account:

I began MNRI® work with Jon after his return from Iraq in 2007. At the beginning, Jon displayed many PTSD symptoms such as depression, isolation, angry outbursts, impaired judgment, and unclear foggy thinking most of the time, that were not a part of his previous history. His wife reported that he would often stand outside in the yard for hours looking up to the sky as if he were lost.

To begin the healing process, integration of the primary sensory-motor reflex patterns – Babinski, Grounding, Foot Tendon Guard, and Core Tendon Guard – were chosen for our MNRI® recovery strategy. Initially, our focus was working with grounding, balance, and stability. Integration of the feet reflexes was essential in helping Jon feel safe after witnessing such horrific sights and sounds that come with war. After integration of these reflexes, Hands Supporting was repatterned, which provided safe postural confidence. His avoidance of new environments and isolation was behavioral evidence that this reflex pattern was retained and needed integration. Additionally, we focused on Fear Paralysis, Moro, Hands Grasp, and both Spinal Galant and Perez Reflexes, as they all were hyperactive.

After just one session, Jon displayed natural awareness of sounds and his surroundings. He attained the ability to return to a safe and healthy 'here and now experience.' His hypersensitive auditory reactions subsided significantly. Jon became more playful and confident in moving forward with life. The aggressive and odd behavior that his wife previously noted ceased and it was as if they had the 'old Jon' back. We continued to meet two more times before he moved out of state.

Dr. Svetlana Masgutova has developed a comprehensive and effective MNRI® PTSD protocol. My experience with Jon and others indicates that MNRI® is the link to positive anchors that significantly impact an individual that has experienced trauma. MNRI® focuses on the restoration and maturation of primary sensory-motor reflex patterns that provide the basis for all neurophysiological development and functioning. In times of severe stress, these reflexes become our allies and are summoned to help us survive. Yet, many chronic health issues can arise as a consequence of these automated survival responses remaining on high alert. The health and well-being of those who suffer with PTSD may continue to be compromised if reflex integration is not addressed.

Using the MNRI® model allows an individual to restore natural self-regenerating programs preventing further stress of natural systems in the body making it an effective tool for treating survivors of traumatic events. In Jon's case, MNRI® truly helped him find his way home.

I first thank Jon for his military service and his courage to be in harms way for the greater good of all. His strength and resolve to restore what had been personally lost from his war experience made all the difference in his recovery from PTSD. It was an honor to serve and support him in this process. – Peg Johnson

The Boston Marathon

Rachel Hansen, Runner, and Tawni K. Lawrence, CLP, LMT, MNRI® Core Specialist and Instructor, Salt Lake City, UT, USA

Rachel Hansen's account of her experience:

I had the opportunity to run the Boston Marathon in April 2013. The race began and I could tell this was a different day! During the entire race I could feel God's love so strongly. The race itself was fabulous. I finished strong and was feeling so overcome that I just kept crying because I felt so happy! But, as soon as I crossed the finish line I had a strong feeling to "get out now". I thought it was because there were so many people around me. I was feeling so claustrophobic and I thought it was just me, but then the message came again so loud, "Get Out"!

Tawni Lawrence

Rachel Hansen

I tried to get out as fast as my legs would take me after running a marathon! I met up with my husband, we celebrated for a quick minute, and then we went up to shower in our friend's hotel on Boylston Street. I went back down with my husband and friends to wait for another friend to finish. I was standing by her and her father while my husband was standing further down the road with another friend.

The first explosion went off and it was far enough away that I couldn't tell what it was or what was happening. It smelled like fireworks and then, seconds later, the second explosion went off across the street 100 yards away from us. Instantly, I ran! I knew something was wrong and so did everyone on that street. As I ran to find shelter, I had no idea if it had been a gunman, an explosion, or just what it was. It was complete mayhem – I was tripping over people as they curled down into a ball for protection. I just kept running to get to protection. I ran to a hotel and hid by the elevators. It wasn't until that moment that I realized I had lost my entire group and, more importantly, my husband!

With shock creeping in, I went back outside to find my him. People were screaming, crying, yelling, and saying things like, "That man lost his leg". Once outside, I saw my friends but not my husband. My friend, who is an ICU nurse, could tell I was slipping into shock. She grabbed my shoulders and said, "Listen, I've seen your husband, he is safe and so are you!" As I went to look for him, I saw a man running down the street with ripped jeans and blood everywhere. I remember thinking and praying, "God, please bless all these people!" I looked again towards the finish line and there saw my husband and he saw me. We reconnected, met up with our party, and went to our hotel that was away from the mayhem.

We turned on the news and pictures of the bombing starting pouring in... people had lost their lives and there were so many other serious injuries – my heart hurt. I kept praying, "Please, God, help them".

As soon as we got back home, I knew I needed to talk to someone. I didn't want to go anywhere – I was so afraid of everything! It was difficult to even let my kids go to school. For 8 months after the bombs my kids slept in our bedroom because I wanted them close to us. I started seeing a psychologist a few days after getting home to help calm the pure anxiety I was feeling. I had huge fears of big crowds and public areas in general. I went to a few different psychologists throughout the year and started doing a little better when being in big crowds and letting my kids be away from me.

A year after the bombing, my husband and I moved back to our hometown. Shortly thereafter, I slipped into a horrible state of fear . . . fear of everything. I was becoming non-functional. I received a prescription for anxiety medication which helped, and I kept going to a counselor to talk through things. My parents felt I should go to an energy/alternative therapist but, for some reason, I resisted. I prayed so hard to know what God would have me do to help myself heal, so I could be able to live a full life of happiness again and be able to function as a wife, a mom, a friend, a sister, and daughter of God.

Rachel Hansen running during the race.

I was then referred to MNRI® Core Specialist Tawni Lawrence – she worked with me using the MNRI® PTSD protocol addressing my reflex patterns. I remember going to her the first time so full of fear that I could barely talk. She knew exactly what was going on. She helped me to understand how talk therapy did help but that my body was also still holding on to trauma from the event as evidenced by my responses and behaviors. She helped me understand how my body was still responding as though it was reliving the Boston Marathon over and over again everyday! The tears rolled down my face as she said this to me. I remember quietly thanking God for placing me into Tawni's care where she could help me to heal and help my body and brain to connect again.

Shortly after the first session with Tawni, I was at an event with my family. I remember thinking, "Wow, there are so many people here – it's so awesome." Then I seriously stopped in the middle of this huge crowd and listened to my own thought for a minute, "There are so many people here – it is awesome!" At that moment I realized my mind and body were connecting again! After just one session I was able to function again, go into huge crowds, and not even think twice about it – and I was not in fear!

Trauma can happen to each person when it is too intense and unbearable, on so many different levels and, as I have learned, there is a way to connect body and mind to be able to heal from these sorts of things and be able to function again! I can't thank a loving God enough for guiding me to healing. Even though this experience was a trauma, I thank God that I was in Boston that day. This experience helped me to become an even stronger person today. I'm so grateful for the ways in which our bodies can heal so we can live a full life again!

Comments from Tawni:

Rachel was delightful to work with, however, she is speaking the truth. She could hardly articulate her thoughts when she came in. Her body was tense, she was breathing erratically, she couldn't con-

Rachel Hansen after finishing the Boston Marathon.

trol her emotions – tears flowed easily, she was nervous about being in my office, and the new space. She felt hopeless and was upset about the direction her life was taking. She was especially sad because she "used to be so full of life and fun." Now, she didn't want to go anywhere and didn't even enjoy running any more. Rachel had hyperactive Fear Paralysis, Moro, and Core Tendon Guard Reflexes – all those that serve for protection and, evidently, their permanent activity led her to a state of emotional instability, fears, and post-traumatic crisis. Also her ATNR was affected as she was sensitive to unexpected sounds and views of crowds which was activating her peripheral vision and thus anchored stress and 'territorial' and self-preservation responses.

I utilized different techniques but the core of the first session was the PTSD protocol. It's so powerful. Rachel left my office feeling relaxed, calm, breathing, and felt hope after the session. I must admit, I was not anticipating such a dramatic change in her after just one session. I knew there would be change but she came back into my office with joy, exuberance, and peace radiating from her as she told me of all the fun things she had been participating in after relieving the traumatic stress. Her comment was, "I'm back." She even showed me a picture of her and her husband out water skiing together – something she didn't believe she could do again.

I am a Core Specialist and Instructor for the Masgutova Method®. While I am trained in many other fields, I have found Dr. Masgutova's PTSD protocol to be invaluable in correcting and minimizing post-traumatic symptoms. I also experienced this to be the case while working with the children from Sandy Hook Elementary in Newtown, CT. Thank you, Svetlana, for healing your own PTSD and turning your pain into a blessing for thousands. You are a gift to the world!

I love the work of Dr. Masgutova – it constantly blesses lives. Thank you, Svetlana!

Thank you, Rachel, for trusting yourself, your body, and me. You're an amazing soul with a passion for life. Your strength and courage bless my life. It's a beautiful reminder that our bodies/spirits want to thrive and have the right to thrive. There is always hope. I'm thrilled beyond words that you have your life back. I, too, truly thank God for sending you my way. Blessings. – Tawni Lawrence

I Want to be Real Again

Josh Orr, Individual and Patricia Wilson-Orr, MNRI® Parent, Shaker Heights, OH, USA

Josh Orr lives in Shaker Heights, a suburb of Cleveland, Ohio. He is the third child of Ron and Pat Orr and has an older sister named Amanda and an older brother named Ron. Josh has autism and has been nonverbal all of his life. He started speech and signing therapy at 7 years old and became proficient in American Sign Language by the time he was 10 years old. As early as 9, it was discovered through letterboard communication at the Soma Halo Clinic in Austin that Josh had an extensive vocabulary and an above average understanding of academic material for his age level. Josh started Lightwriter facilitated communication through educators at Syracuse University at the age of 12.

Dr. Masgutova and Josh Orr

In 2002, when Josh was 10 years old, he experienced PTSD; he was choked at school while being restrained because a teacher thought he knocked over a juice box. We didn't know what happened to him and had neurological testing done. Thereafter, Josh stopped swallowing solid foods and lost 27 pounds. After dropping to 99 lbs. we had a feeding tube inserted and we kept Josh alive through liquids for two years. The cerebral palsy center in Cleveland said Josh had had an 'emotional whip lash'. Later, Josh expressed that he thought that we knew what happened and that is why we took him out of the school program. He said that, because he had food in his throat when he was choked, that feeling terrified him so he stopped swallowing solid foods.

Josh loves biking, social studies, swimming, cooking, and participating in family gatherings for birthdays, holidays, and other celebrations. Presently, Josh is working at the Jewish Community Center, attending classes at Shaker Heights High School, and receives tutoring at home. Josh loves learning and has plans to take courses at Notre Dame College in January, 2015. He will be graduating from high school in May of 2015 and plans to continue his education to become an advocate for nonverbal people with disabilities.

In late 2014 Josh attended the Masgutova Family Conference where he had the opportunity to experience neurosensory reflex integration. Josh had refused to talk about the choking experience for a long time, therefore we where shocked when he started to type about this event during our sessions at the Masgutova Conference through typing on his tablet. He has expressed hope for being a better communicator and participating more fully in his life. Josh's experience has allowed him to move fluidly. With a renewed sense of body awareness and confidence, Josh dreams he can fulfill his goal of going to college after high school like his older siblings. He no longer accepts the self-imposed limits of his disability, and is now practicing verbal communication with an enthusiastic zeal.

At the MNRI® conference on October 25, 2014, Josh wrote:

"i want to feel my body more and i want to bring this [MNRI®] to others in ways to do this. great to be with people who totally realize the real reason we can't show intelligence. i am wanting to do the things that others do feeling this might work to make it possible. have to say want the space that would feel dearly good in saying that i love this method."

And later, Josh wrote:

"Hope I can do whatever I want to do when I am done this week. Either I will be different or ways to change will be obvious. I want to think about really satisfying, all ways--being so darn stuck

I think this really works for ways to participate in my own life. How can Mom learn to do ways [MNRI® techniques] people did today I can want the really good times to continue. My thoughts for sure I feel that it will be pleasing to talk and talking would really satisfy my life and want it badly I believe it is hard because I don't know ways to be the talker that you think I should be Away with your expectations, you think that I feel pleased with the hard things knowing of my body?"

On October 28, 2014, Josh wrote:

"i love all the people here. you perhaps think i am being funny. love is the real time reality that heals. i am so wanting to be healed of the trama that i experienced when i was a younger boy very anxious to tell you that what happened that day was so bad that it from then on everything was different. i was no longer the boy you knew but then became the boy you know today. want to be real again. can it work that way. i want to say that i am hopeful that it can for the first time in my polite way it has been hell. want to start fresh. want a masterful and was as real people say wonderful and satisfying life. want to say that what you have done for me as we came to experience the barely known method easily satisfies the great way to work with my trauma. thinking quite a lot about sad feelings and easily want to be happy again. i am saying that happy is easily the thing that was missing and i want to have that back. trying to want happy is not the same as being happy. i was not being happy and the sadness is all i have known for a long time. table real life real minds want more. i am can love can real saying that the mind needs real casting to be about the satisfying safe life. my life is wanting to greatly change and darn way can be better if sadness exactly sadness dead time love to can because real love really easily deads sadness. saying we are working so hard to make a quiet change and i feel ready for so much more. you got it. its a huge way to change. i want ways to do more. i want to be me again."

Josh, you have an amazing personality, so deep and so special. Your motivation, curiosity, and high spirit are outstanding and now we see such talent – you are dancing! We are so proud of you! You became a very special friend to our Team. Keep your achievements growing! With our congratulations and admiration – Dr. Masgutova and MNRI® Team

Index

C

D

L

M

Q

R

S

T

